Springer Textbooks in Earth Sciences, Geography and Environment

The Springer Textbooks series publishes a broad portfolio of textbooks on Earth Sciences, Geography and Environmental Science. Springer textbooks provide comprehensive introductions as well as in-depth knowledge for advanced studies. A clear, reader-friendly layout and features such as end-of-chapter summaries, work examples, exercises, and glossaries help the reader to access the subject. Springer textbooks are essential for students, researchers and applied scientists.

More information about this series at ▶ http://www.springer.com/series/15201

Alexander Neidhardt

Applied Computer Science for GGOS Observatories

Communication, Coordination and Automation
of Future Geodetic Infrastructures

 Springer

Alexander Neidhardt
Geodaetisches Observatorium Wettzell
Bad Koetzting
Germany

ISSN 2510-1307   ISSN 2510-1315   (electronic)
Springer Textbooks in Earth Sciences, Geography and Environment
ISBN 978-3-319-40137-9   ISBN 978-3-319-40139-3   (eBook)
DOI 10.1007/978-3-319-40139-3

Library of Congress Control Number: 2017941719

Printed on acid-free paper

This Springer imprint is published by Springer Nature
The registered company is Springer International Publishing AG
The registered company address is: Gewerbestrasse 11, 6330 Cham, Switzerland

For the strong women in my life:
my grandma, my mommy, my sister,
and my wife

Foreword

By the President of the International Association of Geodesy (IAG)

Among the modern space geodetic techniques, the Very Long Baseline Interferometry (VLBI) plays a key role as it is the only technique that provides the link between the terrestrial reference system and the celestial reference system and only VLBI allows to determine all five Earth orientation parameters. An increase of the accuracy of VLBI is crucial to meet future requirements as, for instance, defined by GGOS, the Global Geodetic Observing System of the IAG (International Association of Geodesy). Main goals of GGOS are 1 mm accuracy for position and 0.1 mm/year accuracy for velocity of a geodetic observing station on a global scale. Consequently, the International VLBI Service for Geodesy and Astrometry (IVS) released a dedicated design vision of how to fulfill these requirements, including a roadmap with clearly defined measures on how to achieve its goals. Simulations demonstrated that the continuous operation of rather small and compact, fast-slewing radio telescopes, seven days per week, together with observing more than 1,500 different radio sources per day, using a broader bandwidth, and higher data rates will enormously increase the accuracy and general performance of geodetic VLBI.

In September 2005, the IVS Directing Board adopted a final report of an internal working group about the future requirements needed to achieve the above-mentioned ambitious goals as a vision for future geodetic VLBI systems, called VGOS (VLBI Geodetic Observing System). Based on that concept, new antenna types with improved receiving and recording chains, continuous operation, a higher degree of automation, and remote access to the operating systems were developed and implemented in recent years. Meanwhile several national organizations obtained funding for the construction of new VGOS-type radio telescopes or submitted proposals for the implementation of new VLBI equipment. Nowadays, new, fast-slewing radio telescopes with small dish sizes exist already in Australia, Germany, Japan, Portugal, Russia, Spain, and the USA, with Finland, Norway, and Sweden soon to come. While mechanical implementation, high-frequency operation, or electronic upgrades have made good progress, there is still a rather low consensus on a common procedure on how to operate these systems.

Teams in various countries implemented individual software components for operating the telescopes. At sites in extremely remote regions, such as the German Antarctic Research Station O'Higgins on the Antarctic Peninsula, remotely operated observing sessions were successfully run in real-time and showed the huge potential of the new technique to allow for worldwide operation centers. Complete continental networks, such as the AuScope network in Australia, are continuously operated remotely from Hobart, Tasmania. Bandwidth on demand and real-time correlations are performed using software correlators. Most of the participating sites collect their auxiliary data sets with their own monitoring systems for local use.

This book is a first assemblage of possible solutions to implement worldwide VGOS operations. It iteratively and step by step compiles numerous aspects of computer science under an integral view. The book provides the fundamentals for all software developments to design a reliable, stable, and functional common software basis with matching modules. It also provides details on the various techniques needed to implement a general communication infrastructure. Novel features from the automobile industry were exploited and converted to safely design and implement autonomous control systems for space geodetic applications. The integration of existing control systems and their extension by techniques that guarantee secure remote access are explained using the example of VLBI. Finally, the individual concepts are merged to a general, integrative approach to run worldwide observational networks. Each section is complemented with accounts of the underlying theory, many examples, and illustrations. The study greatly benefits from the author's deep insights and many years of experience in operating geodetic systems. Altogether, this book is a genuine compendium, stimulating not only for engineers at the observatories but also for students in the field of geodesy, as well as for interdisciplinary studies.

The IAG as the international body under which the IVS was established more than 15 years ago strongly encourages the further development of VGOS and all its components including the remote control of VLBI radio telescopes.

Prof. Dr.-Ing. Dr. h.c. Harald Schuh
Director Department 1 «Geodesy»
at GFZ, the German Research Center for
Geosciences, Potsdam
President of the IAG (2015–2019)
Past Chair of the IVS (2007–2013)
Past President of IAU Commission 19 «Rotation
of the Earth»
Potsdam, March 2016

Foreword

By the Director of the GGOS Bureau of Networks and Observations

The Global Geodetic Observing System (GGOS) is placing stringent requirements on our space geodesy resources: mm accuracies, 24/7 operations, dramatically increased data flow and storage, real-time data transmission, high reliability, and minimized downtimes. New systems are being planned and built, and current systems are being upgraded to help fill out the core and co-location network to address GGOS needs including the reference frame, precision orbit determination, gravity field, and all of the other Earth, ocean, and atmospheric phenomena that are critical to our understanding of our dynamic planet.

As our observing capabilities continue to evolve, with new instruments on the ground and in space, more and more science will be forthcoming; some predicted will be as predicted and some will be a surprise. Our observatories already have VLBI, SLR, GNSS, DORIS, gravimeters, tide gauges, and other instruments; our space segment includes an array of altimeters, gravimeters, InSAR, magnetometers, and other systems. Expect much more co-location and core site configurations. We are already moving to an era where our instruments will be running continuously; we will be inundated with data, and yet our demand will continue to grow. In many cases, our observations have gone from measuring events to absorbing continuous processes. Our data has grown from megabytes to gigabytes to terabytes and beyond. Real-time and near-real-time data availability and quality control will be required by some of our users. New applications, such as those associated with natural hazard warnings, may not be able to wait. Users are relying on high data quality and comprehensive quality control of data and data products, requiring more extensive information about the data and its environment.

Continuous operation means that our current mode of using operating personnel is becoming financially untenable. We will be relying on autonomous and centralized operations with sufficient imbedded intelligence to keep on-site personnel intervention at a minimum. Real-time communication among systems on site and among sites gives us better coordination and the ability to adjust network operations to maximize efficiency and minimize downtime. As our targets become more numerous and our processes become more rigorous, observing time will become an even more precious resource requiring well-thought-out but flexible strategies.

As complication grows and systems become more complex, we need to be aware of our security vulnerabilities and how they must be addressed.

The backbone of our new systems and the new network is software for operations, communications, automation, remote control, coordination, system and network modeling, security, and real-time data quality control. Software already plays a very important role in the network, but this is really a whole new paradigm. The network is going to look more and more like one organism with the necessary communication and imbedded intelligence to make it work that way.

Some groups have already implemented some form of these features with good success, but we need to start thinking about network-wide capability.

For the first time in publication, *Applied Computer Science for GGOS Observatories* presents the application of computer science to furthering the performance of the GGOS network toward the goals articulated in «GGOS 2020: Meeting the Requirements of a Global Society on a Changing Planet in 2020», edited by H.-P. Plag and M. Pearlman. With very pertinent examples and even relevant code, the book discusses the applications, information flow, strategies for designing for designing information flow networks, coding languages, styles, layouts, and even coding policies; documentation and, for different levels, code testing from the basic programmed unit to the entire software package; and tools and modules that help in the design and help bring improved compatibility among different groups and different levels in the hierarchy. The book examines specific examples of the software applications to SLR and VLBI, GGOS applications that rely on the integrated network, and specific model methods that are critical to proper interpretation on data.

This book is focused on GGOS, but its applications should be of far wider interest in other areas where results depend on network communication, coordination, and automation among different instruments at different locations in different environments.

Dr. Michael Pearlman

Director, GGOS Bureau of Networks and Observations
Director, ILRS Central Bureau
Harvard-Smithsonian Center for Astrophysics
Cambridge, January 2016

Acknowledgment

The successful creation of the practical tasks and the writing of the work would have been not possible without the help, support, and co-work of several people. Therefore, I want to devote this section to these people because I am greatly indebted to them.

First of all, I want to thank my advisors and reviewers of the «Habilitation» who supported me over the long period of working and writing this book. First of all, I want to mention Prof. Dr. Urs Hugentobler, my boss and head of the Research Facility for Satellite Geodesy (Forschungseinrichtung Satellitengeodäsie, FESG) of the Technische Universität München (Technical University of Munich). His gentle character often helped in many discussions and nerve-racking situations. He always supported my daily work and especially the creation of this book. With his tremendous knowledge, he is a very good contact person at any time. This friendly contact was one of the foundations of the work. Next, I want to mention Prof. Dr. Johann Schlichter, who was already my advisor in the field of computer science during my dissertation. He made a multidisciplinary work possible, and I thank him for his agreement to supervise the application of computer science in the field of geodesy. Another thankful word should be spent to Prof. Dr. Dietmar Grünreich, the former president of the Federal Agency for Cartography and Geodesy (Bundesamt für Kartographie und Geodäsie, BKG) who enabled the practical work at the Geodetic Observatory Wettzell, which is mainly funded by the BKG. The inclusion of the tasks into the different software developments at the observatory facilitated the practical orientation of the work and provided possibilities for real software developments. I am really sorry about the fact that Prof. Grünreich had to quit his mentorship because of a personal circumstance. In this situation, I was really pleased that Prof. Dr. Hansjörg Kutterer (current president of the BKG and GGOS chair) agreed to support my request to review the book. It was a long time period with several work phases, so that I want to thank my faculty for the general support and administrative attendance, as well as the BKG administration which enabled many things at the Wettzell observatory. I also have to thank the copyright holders and publishers of the content which was adapted from external sources for their permission and friendly support. Referencing was carried out with the utmost care. Any resemblance to other existing texts or missing acknowledgements are purely coincidental.

Besides all of the official contributors to my «Habilitation», most of the programs, sections, and single parts would not have been possible without the help and co-work of my colleagues at the Geodetic Observatory Wettzell. My colleagues are my combatants to solve most of the daily tasks and developments. I want to thank especially Christian Plötz, Matthias Schönberger, Ulrich Schreiber, Reiner Dassing, Johann Eckl, and Uwe Hessels; the software developers Martin Ettl, Martin Riederer, Katharina Kirschbauer, and Andreas Leidig; and also the secretaries at the observatory, Hannelore Vogl, and at the FESG, Stefanie Daurer. Writing this book was very time consuming and required several months out of my office. But daily shifts, maintenance, administration, and developments went on, and the whole load was taken by the members of my VLBI group at the observatory, which I am allowed to lead since 2008. I want to thank you all for this support. I know that things are not always easy there, but I am proud to chair this group, which provides high-quality VLBI measurements for a very long time, so that the 20 m radio telescope is a well-used core point in the global, geodetic analysis.

Finally, I want to thank for discussions with and ideas of my international colleagues and friends. I was so often privileged to attend international meetings and workshops on such remote and interesting places where I met so many excellent and brilliant scientists and engineers. They quickly accepted me as part of the IVS, where I got the chance to develop and integrate our ideas and implementations. Several observatories now run e-RemoteCtrl as test stations and offer so much constructive feedback. I would like to pick out only a few names, knowing that there are several more, which I surely unjustly forgot to mention. Nevertheless, I thank Ed Himwich, Chris Beaudoin, Jim Lovell, Jonathan Quick, Cristian Herrera Ruztort, Harald Schuh, Axel Nothnagel, Walter Alef, Alessandra Bertarini (the next warm milk is on me), and the whole correlator team at Bonn, the team of the observatory Ny Ålesund, the team at JIVE, and many more. Thank you all!

Last but not least, I want to thank my family. Even if there were some stumbling blocks on our path of life which required our complete attention and our physical and mental power, we stood together and broke down the walls on our way, using our joint forces. You all had so much patience with me during the long time of writing, even if you have had your own sorrows. Therefore, I want to thank you from the bottom of my heart, especially my mommy, whose strong will to survive her diseases gave me a new sense of life; my sister Diana, who was solid as a rock in the midst of all turbulences; and my wife Anita, who always supported me, even if the time together was so reduced. You open new horizons and I am looking forward to our seat bench under our cherry tree, when we will be old.

Finally, my dad should not be forgotten here. He sparked my interest up for technical stuff and engineering with all his technical toys for me. I also want to mention my deceased grandpa whom I dedicated my PhD thesis and who trained me all practical craftsman skills, which are quite useful for applied studies.

But what is a book without the possibility of publication? This print was only possible, because Dr. Johanna Schwarz, Senior Publishing Editor of Springer International Publishing AG, and her team welcomed my manuscript with open arms. Dr. Schwarz always answered my questions and gave useful advises on administrative procedures during the contract negotiations. I also want to mention the attractive and extensive layout process by Mr. Sekar Rajesh and his team of SPi Global - Content Solutions, India. Thank you all for your work and the chance to see my work as a printed book.

Last but not least, all which are not mentioned until now hopefully know which tremendous contribution and impact they had, so that this book was possible in this form. Therefore, I also want to say a general and very heartfelt «Thank you»!

Contents

1	**Introduction**	1
1.1	**Initial Situation**	2
1.2	**Motivation and Purpose of This Work**	6
1.3	**Structure of This Work**	8
1.4	**Who Should Read This Book**	11
1.5	**Used Conventions**	11
1.6	**Software Requirements**	12
1.7	**Web Pages and Contact**	12
2	**Writing Code for Scientific Software**	13
2.1	**Scientific Software as It Is in Most Cases**	14
2.2	**A Suitable Programming Language**	19
2.3	**A Coding Style Guide**	24
2.3.1	An Impression of Coding Layout	27
2.3.2	An Impression of Coding Policies	30
2.4	**Including Legacy Code**	39
2.5	**Documentation**	47
2.5.1	Documentation for the Development Team	48
2.5.2	Documentation for the Users	56
2.5.3	Documentation for Administration and Other Scientists	60
2.5.4	Documentation Landscape	60
2.6	**Code Testing and Code Inspections**	62
2.6.1	Code Quality Metrics	64
2.6.2	Dynamic Code Testing	68
2.6.3	Static Code Inspections	84
2.6.4	Dynamic Code Analysis	86
2.6.5	Testing Landscape	96
2.7	**Version Control**	97
2.7.1	A Suitable Project Directory Tree	98
2.7.2	A Suitable Version Control System	100
2.8	**Continuous Integration**	109
2.9	**Agile Software Development**	117
2.10	**Summary**	128
2.10.1	What Did We Learn?	128
2.10.2	How Did We Use the Contents Learned?	129
2.11	**Questions**	129
3	**Using a Code Toolbox**	131
3.1	**The Idea Behind a Code Toolbox**	132
3.2	**Well-Tested Modules and Components**	144
3.3	**Generative Programming**	148
3.3.1	Classic Solutions for Interprocess Communication (IPC)	149
3.3.2	Remote Procedure Calls	168
3.3.3	Extending Generative Programming for Interprocess Communication	181
3.4	**A Rudimentary Middleware for Controlling of Distributed Systems**	236
3.5	**Summary**	243
3.5.1	What Did We Learn?	243
3.5.2	How Did We Use the Contents Learned?	244
3.6	**Questions**	244

4   Controlling a Laser Ranging System 247
4.1   **Principles of Laser Ranging Systems** 248
4.2   **The Laser Ranging System as an Autonomous Production Cell** 253
4.2.1 Distributed Hardware Control 254
4.2.2 The Construction of the Autonomous Production Cell 277
4.3   **User Interfacing** 283
4.4   **Autonomous Coordination Cell** 300
4.5   **Autonomous Hardware Control Cells** 332
4.6   **Autonomous Data Management Cell** 338
4.7   **Autonomous System Monitoring and Safety Cell** 369
4.8   **Summary** 390
4.8.1 What Did We Learn? 390
4.8.2 How Did We Use the Contents Learned? 391
4.9   **Questions** 391

5   Controlling a VLBI System Remotely 395
5.1   **Autonomous Production Cells as Parts of Multi-agent Systems** 396
5.2   **Principles of Very Long Baseline Interferometry** 400
5.3   **The Used Control System for VLBI: The NASA Field System** 409
5.4   **Extend Existing Control Systems with Multi-agent Abilities** 415
5.5   **Security for the Controlled Systems** 431
5.5.1 Security for Internal Local Access to a Computer 433
5.5.2 Security for External Accesses to a Computer 438
5.5.3 Security for a Complete Distributed System 457
5.5.4 Security for a Complete Observatory 466
5.5.5 On-the-Fly Management of Temporary SSH Tunnels 469
5.6   **Summary** 481
5.6.1 What Did We Learn? 481
5.6.2 How Did We Use the Contents Learned? 482
5.7   **Questions** 482

6   Coordination, Communication, and Automation for the GGOS 483
6.1   **The Global Geodetic Observing System (GGOS) at a Glance** 484
6.2   **Operational Deficits of GGOS** 488
6.3   **Coordination, Control, and Operation of the Terrestrial GGOS Infrastructure** ... 491
6.4   **Communication on GGOS Networks** 494
6.5   **Automation as Key to Deal with 24/7** 504
6.6   **Summary** 507
6.6.1 What Did We Learn? 507
6.6.2 How Did We Use the Contents Learned? 507
6.7   **Questions** 508

7   Outlook 509
7.1   **GGOS Operations as a Revolution** 510

Service Part
Appendix: A Style Guide for Geodetic Software in C and C++ 514
References 524
Index 531

Abbreviations

ACL	Access Control List
ACU	Antenna Control Unit
ADIZ	Air Defense Identification Zone
AES	Advanced Encryption Standard
ANSI	American National Standards Institute
AOP	Aspect-Oriented Programming
APD	Avalanche Photodiode
API	Application Programming Interface
ASCII	American Standard Code for Information Interchange
ADS-B	Automatic Dependent Surveillance-Broadcast
BIOS	Basic Input Output System
BSD	Berkeley Software Distribution
BLOB	Binary Large Object
BNF	Backus –Naur Form
CDDIS	Crustal Dynamics Data Information System
CGI	Common Gateway Interface
CMS	Content Management System
CI	Continuous Integration
CORBA	Common Object Request Broker Architecture
CPF	Consolidated Laser Ranging Prediction Format
CPS	Cyber-Physical System
CPU	Central Processing Unit
CRD	Consolidated Laser Ranging Data Format
CRC	Cyclic Redundancy Check
CSIRO	Commonwealth Scientific and Industrial Research Organisation
CSMA/CD	Carrier Sense Multiple Access with Collision Detection
CVS	Concurrent Versions System
DBBC	Digital Baseband Converter
DBE	Digital Back End
DBMS	Database Management System
DCE	Distributed Computing Environment
DES	Data Encryption Standard
DESCA	Development of a Simplified Consortium Agreement
DGPS	Differential Global Positioning System
DGNSS	Differential Global Navigation Satellite Systems
DHCP	Dynamic Host Configuration Protocol
DNAT	Destination Network Address Translation
DNS	Domain Name Service
DOM	Document Object Model
DORIS	Doppler Orbitography and Radiopositioning Integrated by Satellite

DSA	Digital Signature Algorithm
DSL	Domain Specification Language
DTD	Document Type Definition
DVI	Device Independent file format
EDC	European Data Center
EEPROM	Electrically Erasable Programmable Read-Only Memory
EOP	Earth Orientation Parameters
ERM	Entity Relationship Model
ESA	European Space Agency
EVN	European VLBI Network
FMEA	Failure Mode and Effects Analysis
FPGA	Field-Programmable Array
FTP	File Transfer Protocol
GUI	Graphical User Interface
GDB	The GNU Project Debugger
GCC	GNU Compiler Collection
GGOS	Global Geodetic Observing System
GLONASS	Globalnaja Nawigazionnaja Sputnikowaja Sistema
GMSEC	Goddard Mission Services Evolution Center
GNSS	Global Navigation Satellite Systems
GNU	GNU's Not Unix
GIF	Graphics Interchange Format
GPIB	General Purpose Interface Bus
GPS	Global Positioning System
HEO	High Earth Orbit
HTML	Hypertext Markup Language
HTTP	Hypertext Transfer Protocol
HTTPS	Hypertext Transfer Protocol Secure
HVM	Hardware Virtual Machine
IAG	International Association of Geodesy
ICRF	International Celestial Reference Frame
IDL	Interface Definition Language
IDS	Intrusion Detection System
IEC	International Electrotechnical Commission
IEEE	Institute of Electrical and Electronics Engineers
IF	Intermediate Frequency
IGS	International GNSS Service
ILRS	International Laser Ranging Service
InSAR	Interferometric Synthetic Aperture Radar
IP	Internet Protocol
IPC	Interprocess Communication

IR	Infrared		PPS	Pulse Per Second
ISAM	Indexed Sequential Access Method		PTP	Precision Time Protocol
ISO	International Standards Organization			
ITRF	International Terrestrial Reference Frame		QRFH	Quad Ridge Feed Horn
IVS	International VLBI Service for geodesy and astrometry		QZSS	Quasi-Zenith Satellite System
			RGO	Royal Greenwich Observatory
JIVE	Joint Institute for VLBI in Europe		RAD	Rapid Application Development
JPG	Joint Photographic Experts Group		RADAR	Radio Detection And Ranging
			RAID	Redundant Array of Independent Disks
KVM	Keyboard, Video, Mouse		RFC	Request For Comments
			RFI	Radio-Frequency Interference
LAN	Local Area Network		RGB	Red, Green, and Blue
LASER	Light Amplification by Stimulated Emission of Radiation		RIPE	RACE Integrity Primitives Evaluation
LEO	Low Earth Orbit		RIPEMD	RIPE Message Digest
LGPL	GNU Lesser General Public License		ROC	Remote Operations Center
LIDAR	Light Detection And Ranging		RPC	Remote Procedure Call
LLR	Lunar Laser Ranging		RPCL	Remote Procedure Call Language
			RTAI	Real-Time Application Interface
MAT	Microprocessor ASCII Transceiver		RTCM	Radio Technical Commission for Maritime Services
MCB	Monitor and Control Bus			
MCI	VGOS Monitor and Control Infrastructure		SAP	Service Access Point
MCP	Micro-Channel Plate		SASL	Simple Authentication and Security Layer
MD5	Message Digest nr. 5		SAX	Simple API for XML
MEO	Medium Earth Orbit		SCP	Secure Copy
MTU	Maximum Transmission Unit		SFMEA	Software Failure Mode and Effects Analysis
			SHA	Secure Hash Algorithm
NAT	Network Address Translation		SEFD	System Equivalent Flux Density
NASA	National Aeronautics and Space Administration		SFD	Source Flux Density
NetCDF	Network Common Data Form		SFTP	Secure File Transfer Protocol
NEXPReS	Novel EXplorations Pushing Robust e-VLBI Services		SIP	Session Initiation Protocol
NFS	Network File System		SKA	Square Kilometre Array
NIS	Network Information System		SMTP	Simple Mail Transfer Protocol
NMEA	National Marine Electronics Association		SNAP	Standard Notation for Astronomical Procedures
NP	Normal Points		SNAT	Source Network Address Translation
NREN	National Research and Education Network		SNMP	Simple Network Management Protocol
NTP	Network Time Protocol		SNR	Signal-to-Noise Ratio
NTRIP	Networked Transport of RTCM via Internet Protocol		SLR	Satellite Laser Ranging
			SRAM	Static Random-Access Memory
ONC	Open Network Computing		SSH	Secure Shell
OOP	Object-Oriented Programming		SSL	Secure Sockets Layer
OPC	OPC Unified Architecture		STL	Standard Template Library
OSF	Open Software Foundation		SQL	Structured Query Language
OSI	Open System Interconnection			
			TCP	Transmission Control Protocol
PC	Personal Computer		TIGO	Transportable Integrated Geodetic Observatory
PDF	Portable Document Format			
PDU	Protocol Data Unit		TTY	Teletype
PFB	Polyphase Filter Bank			
PLC	Programmable Logic Controller		UDP	User Datagram Protocol
PNG	Portable Network Graphics		UDT	UDP-based Data Transfer Protocol
POP3	Post Office Protocol Version 3		UML	Unified Modeling Language
			UPS	Uninterruptible Power Supply

URL	Uniform Resource Locator		VoIP	Voice over IP
USB	Universal Serial Bus		VPN	Virtual Private Network
UTC	Universal Time Coordinated		VSI	VLBI Standard Interface
UTF	Unicode Transformation Format			
			WAN	Wide Area Network
VEX	VLBI EXperiment format		WebDAV	Web-based Distributed Authoring and Versioning
VGOS	VLBI Global Observing System		WVR	Water Vapor Radiometer
VLAN	Virtual Local Area Network			
VLBA	Very Long Baseline Array		XDR	External Data Representation
VLBI	Very Long Baseline Interferometry		XML	Extensible Markup Language
VNC	Virtual Network Computing		XP	Extreme Programming

Introduction

1.1   Initial Situation – 2

1.2   Motivation and Purpose of This Work – 6

1.3   Structure of This Work – 8

1.4   Who Should Read This Book – 11

1.5   Used Conventions – 11

1.6   Software Requirements – 12

1.7   Web Pages and Contact – 12

© Springer International Publishing Switzerland 2017
A. Neidhardt, *Applied Computer Science for GGOS Observatories*, Springer Textbooks in Earth Sciences,
Geography and Environment, DOI 10.1007/978-3-319-40139-3_1

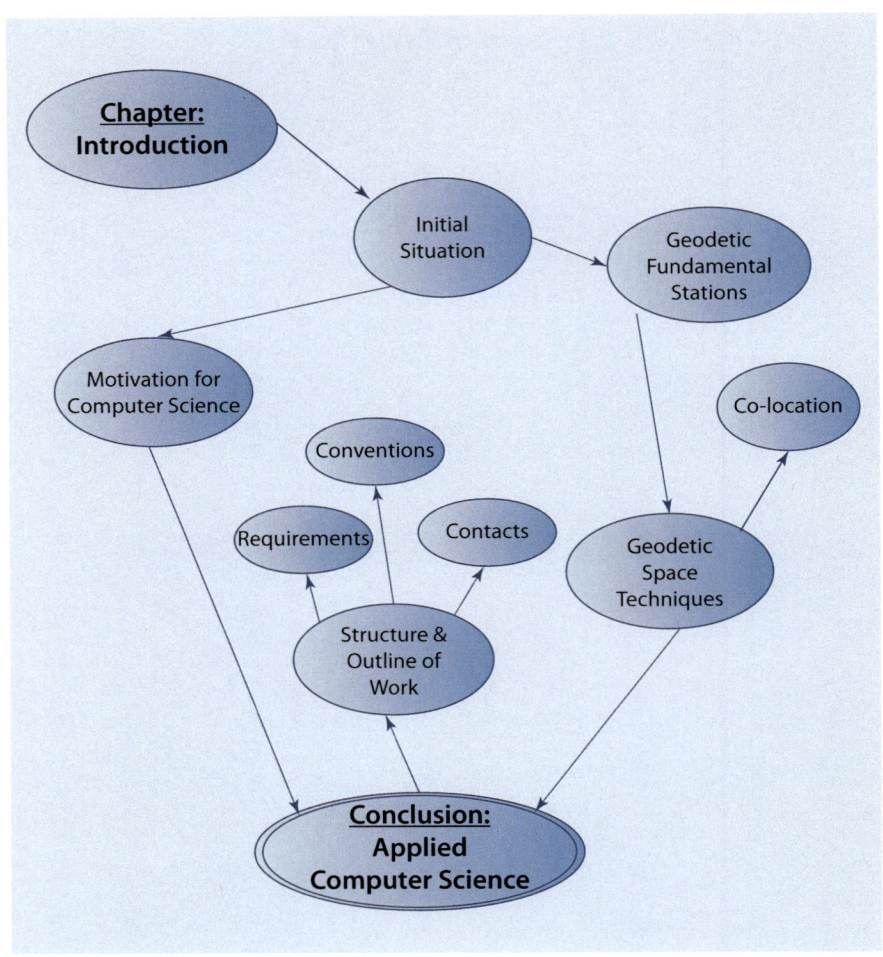

1.1   Initial Situation

Terrestrial Geodesy

The content and concept of this book is generally integrated into the applied field of global geodesy.

Geodesy (The Classic Definition)

> *Definition DEF1.1. of «Geodesy (the classic definition)»* (see Seeber 1988, *l.c. page 1*)
> Geodesy is the scientific discipline that deals with the measurement and representation of the Earth.

Satellite Geodesy

Because the accuracy and precision of measurement instruments increased, new techniques were developed and the new possibilities of space science enabled, to use artificial satellites on orbits around the Earth. As a result, the classic definition of geodesy had to be extended. It became possible to see temporal variations in the position and orientation of the Earth, in its gravity field and in three-dimensional positions on the Earth with a very high degree of precision. Derived from these measurements, new modeling techniques made it possible to describe geodynamic phenomena, such as crustal dynamics, polar motion, and changes in Earth's rotation. Regular observations in different places on the Earth became necessary, and the discipline of satellite geodesy was born (Seeber 1988, *l.c. page 1*). The possibility of using satellites also supported the creation of a global, terrestrial reference frame

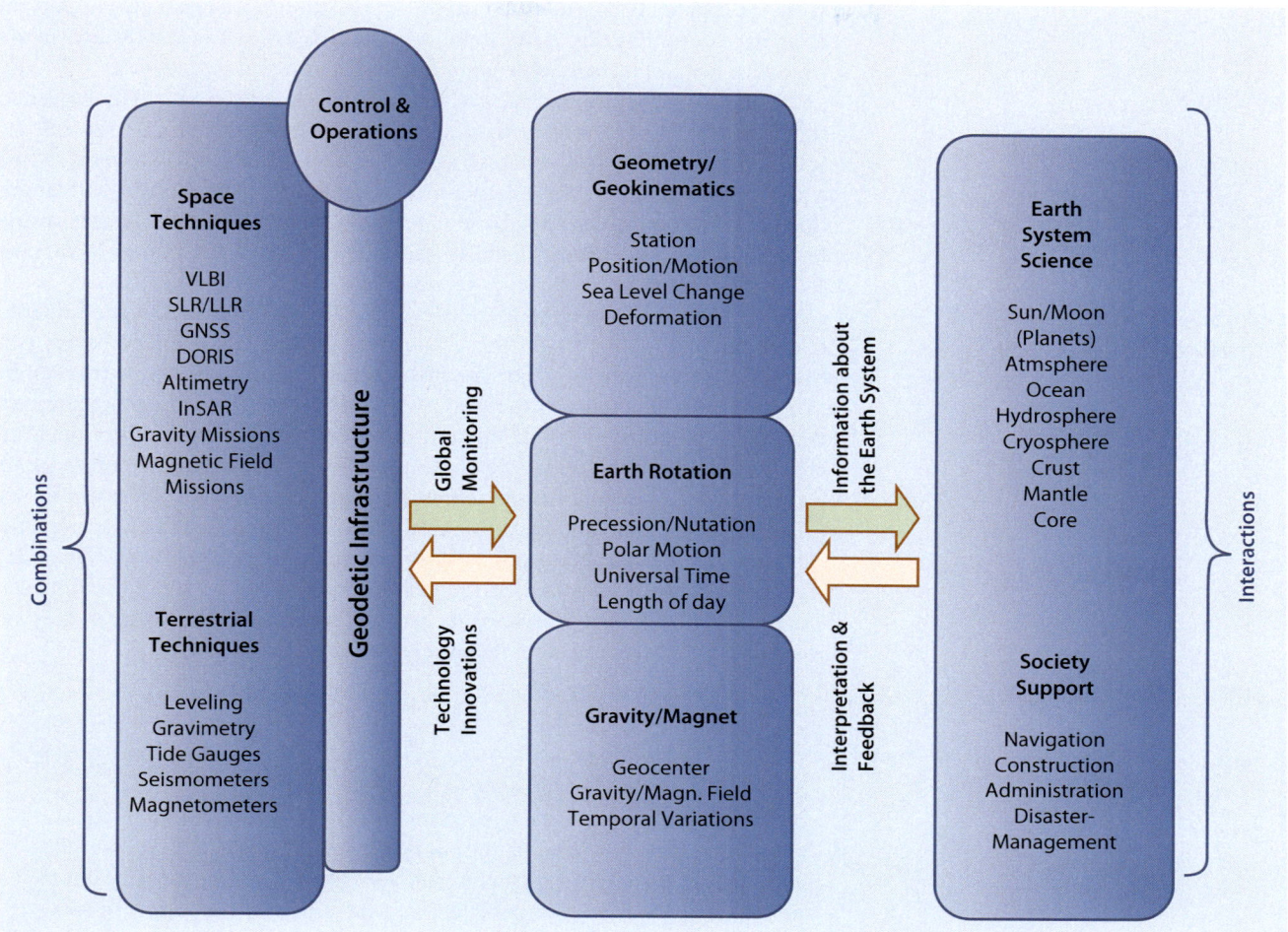

Fig. 1.1 Space geodesy now (as an adaption of (Rothacher 2002, *l.c. page 7*) and (National Academy of Sciences 2010, *l.c. page 22*)), including controlling and operating the geodetic infrastructure; reprinted with permission from ▫ Fig. 1.3, Precise Geodetic Infrastructure. National Requirements for a Shared Resource, 2010, by the National Academy of Science, Courtesy of the National Academic Press, Washington, DC

and enabled scientists to understand the Earth system in more detail (for more details and explanations, see ▶ Sect. 6.1) ▫ Fig. 1.1.

Nowadays, the use of satellites and their observations is far more extensive. Altimetry, Interferometric Synthetic Aperture Radar (InSAR), gravity, and magnetic field missions with special Earth-observing satellites support a global geodesy. Observed targets in outer space many billions of light years away from Earth, such as radio sources like quasars,[1] offer the creation of a stable celestial reference frame and its connection to the terrestrial reference frame on the basis of regular observations with a suitable, globally distributed infrastructure.

Space Geodesy

This infrastructure of space geodetic techniques and (more or less classic) terrestrial techniques allows the global monitoring of Earth parameters to support the modern pillars of geodesy: geometry, Earth rotation, and gravity/magnetic field (see ▫ Fig. 1.1). Separate services offer their observation data to analysis centers, where the data are correlated, analyzed, conditioned, and standardized. These intermediate data are then combined and used for different purposes or use cases in science and society[2] (National Academy of Sciences 2010, *l.c. page 21*). For

Geodetic Infrastructure

---

1   Quasars are cores of galaxies, forcing the emission of huge amounts of energy in different wavelengths.

2   Software engineering defines «use cases» as a list of action items, describing the interaction of a specific user and the used software or program to achieve a specific goal with the software.

society in particular, geodetic output is becoming increasingly important. Managing natural hazards, navigation, or administration of resources is essential for modern life and is often dependent on localization and positioning.

Terrestrial Infrastructure

This general geodetic infrastructure finds its representation on Earth in the form of globally distributed observing systems, using the different space techniques. This terrestrial infrastructure of different observatories represents the Earth as a whole regarding its form and its short-term (daily and shorter) and long-term (years and longer) motions, including deformations. The observatories also deduce the extraterrestrial reference frames and connect them to the terrestrial reference frames ((Schneider 1990, *l.c. page 35 f.*) and (National Academy of Sciences 2010, *l.c. page 21*)).

Co-location of Space Geodetic Techniques

A central role in the terrestrial infrastructure is played by the fundamental stations. These special geodetic observatories run several systems of space geodetic techniques at one location. The instruments at the location of a fundamental station complement each other. Different tests of geodetic models, a redundant determination of observing parameters and the resulting geodetic products, and a comparability of the measuring systems itself and their inertial points (reference points) are possible. Combined with permanent, regular observations, the co-location of the instruments at the fundamental stations is of vital importance in improving the complete geodetic infrastructure (Schneider 1990, *l.c. page 36*), as it is only in these stations that the geometric relations between the different, specific, global reference systems of the different space techniques are well known and locally measurable (Schlüter et al. 2007, *l.c. page 160*).

Fundamental Station

> *Definition DEF1.2. of «Fundamental station»* (see Schneider 1990, *l.c. page 36 ff.*) and (Schlüter et al. 2007, *l.c. page 160*)
> A fundamental station is a geodetic observatory with different permanently, complementary, and redundantly operated geodetic measuring systems, especially of space geodetic techniques, which create a «fundamental point» for global geodesy in a comparable, evaluable, and measurable co-location.

Geodetic Observatory Wettzell

Currently, seven fundamental stations are in operation,[3] which offer at least Very Long Baseline Interferometry (VLBI), Satellite Laser Ranging (SLR) and Global Navigation Satellite Systems (GNSS) systems. They are mostly well distributed in the northern hemisphere. One of them is the Geodetic Observatory Wettzell (see ▢ Fig. 1.2), located in the Bavarian Forest in the southeast of Germany. Its foundation stone was laid in the autumn of 1971, as a first laser ranging system was installed in the former Air Defense Identification Zone (ADIZ) along the German border. Funded by a special research project with the title «Satellite Geodesy» («Sonderforschungsbereich SFB 78»), the site was gradually transformed into a fundamental station. Main investment was the installation of a 20 meter radio telescope for geodetic VLBI in the early 1980s (Schlüter et al. 2007, *l.c. page 160*).

On the Way to a GGOS Core Site

Today the observatory is operated by two institutions: the Federal Agency for Cartography and Geodesy (Bundesamt für Kartographie und Geodäsie) and the Research Facility for Satellite Geodesy (Forschungseinrichtung Satellitengeodäsie) of the Technische Universität München (Technical University of Munich). It co-locates a 20 meter radio telescope, a 70 centimeter laser ranging telescope, several GNSS permanent stations, a large laser gyroscope G (ring laser), as well as required and collocated local techniques, such as time and frequency, a meteorological station, a seismometer, and superconducting gravimeters. These techniques support all the important international services of the International Association of Geodesy (IAG), such as the International VLBI Service for geodesy and astrometry

---

3   The definition, which systems must be available in a fundamental station, differs so that the literature references different numbers of operated stations.

Fig. 1.2 A panoramic view of the Geodetic Observatory Wettzell in 2013, with the new TWIN radio telescopes at the back, the laser ranging telescopes in the center, the old 20 meter radio telescope on the right, the GNSS tower with its antennas on the left, and the rest of the local surveying sensors (Reproduced by permission of Bundesamt für Kartographie und Geodäsie and A. Neidhardt)

(IVS), the International Laser Ranging Service (ILRS), or the International GNSS Service (IGS). The last few years have been marked by the development and installation of new equipment in order to meet the requirements of the Global Geodetic Observing System (GGOS). The GGOS, which was established by the IAG, is an advance to identify future requirements in the field of geodesy. On an abstract level, it describes a design, an implementation, and the administration of a future geodetic Earth observation, which combines all space geodetic techniques, so that the scientific result appears as if it comes from one system. The concept of fundamental stations is continued in the form of GGOS core sites, which must fulfill the special GGOS requirements (see ▶ Sect. 6.1).

For the last ten years, developers at the Geodetic Observatory Wettzell have mainly focused on two projects: the installation of a new laser ranging system in combination with the development of a new control system for all laser ranging systems at the observatory and the installation of a new VLBI Global Observing System (VGOS)-compliant radio telescope pair in combination with the development of new control structures to optimize the operations. In order to run and support those projects at the Geodetic Observatory Wettzell, most of the work of the development teams consisted of writing software combined with some aspects of designing their own suitable hardware. Connecting hardware devices, controlling workflows, scheduling tasks, managing error situations, communicating between distributed equipment, and designing a more automated operation took many years for several project members to develop.

New Systems for SLR and VLBI

The clear and strong impact of applied computer science was apparent, while it also became clear that subjects like coordination, communication, automation, control, and operation of the local geodetic systems were not described in sufficient detail in the official GGOS plans. While the local control systems and automation designs increasingly became reality, situations arose which required a general solution to operate GGOS as one global observing system. A view of state-of-the-art designs from the field of computer science offered ideas and solutions.

Applied Computer Science as Essential Part

This was the starting point of this book and work. The idea was to collect and touch on all essential aspects of computer science which were issued and used during the time period of developments for the new local implementation of GGOS systems at Wettzell. The focus was laid on general programming techniques, generative approaches, ideas for modern control systems, using laser ranging as an

Book About Coordination, Communication, and Automation

example, and solutions for safe and stable remote control for globally distributed VLBI networks. The designs discussed for control and operation might offer enough potential for the future coordination, communication, and automation of a GGOS. Another initial driver was to help other developers of GGOS systems to avoid similar errors, which were made during the software developments at the Geodetic Observatory Wettzell.

1.2   Motivation and Purpose of This Work

Code as Integral Part of Research

Programming is an essential part of modern research. Whether compiling statistics with MathWorks® Matlab® or writing larger analysis programs in other languages, like C and C++, the writing of code is a «hidden ingredient» of many scientific papers. But most researchers use code as a means to an end. Even though the research results are generated by the software, only the scientific results are the integral parts of a scientific paper. Code is just the «secret magic» behind them, about which nobody talks. It is often neglected. Researchers just want their programs to work but do not really quantify or qualify whether they do or not. But the wrong program code can influence the results and might have a dramatic effect on subsequent decisions (or in other words, garbage in - garbage out (Klar 2006, *l.c. page 3*)). Therefore, it is important that programming is also taken as an integral part of research and science. Finding suitable solutions to improve this situation is the main motivation behind this work.

Reviewing Techniques for Code

No scientific paper would be accepted if it did not follow the rules of good scientific work. Papers are proofread and reviewed as a way of ensuring the scientific quality. The same should be applied to the programming code in the back. In a similar way to mathematics, which establishes clear rules, the handling of a source code should follow defined rules. Reviewing processes and a smooth way to reuse the programs are necessary. The problem is that scientific programs are mostly very individual and made a specific purpose, so that others cannot use them directly or cannot understand them for further use. Therefore, many things are developed several times, with new errors and problems. The problem increases, as most scientists and engineers are not trained well enough to write good and functional code, because programming is just touched on in entry-level course and as an add-on to the regular studies. But a system is always only as good as its developer (Klar 2006, *l.c. page 3*). Therefore, another motivation for this work is to make readers aware of these practical issues.

Tree Swing Comic Is Up-To-Date

These experiences hold true for almost every scientific development. In the specific field of this work, this became clear while developing several types of software (and also hardware) for the large measurement systems of the space geodetic techniques at the Geodetic Observatory Wettzell. Training interns and exam students have shown that several essential fields relating to the practicalities of applied computer science were not touched on during their studies. But even the work and discussions with cooperating colleagues have shown that they are often not aware of the basic programming techniques, such as testing their code, using well-known concepts from computer science, or documenting their systems. It was difficult to bring the multidisciplinary team with team members of different ages up to the same level. Even when the developments ended in a final solution which solved the tasks, it was often not the planned and originally designed result. In several cases, it was also accomplished with a reduction of the resulting software quality and/or with far longer development or integration times. But this problem is as old as the history of computer science, stunningly presented in the tree swing comic (see ▣ Fig. 1.3) and in several studies (e.g., see ▶ Sect. 2.1). This is further motivation to try to improve this situation.

Purpose of This Work

Starting with this initial situation, the purpose of this book is to use the field of development tasks at a geodetic observatory to explain techniques from computer science, applied to real challenges. It introduces the terminology used in computer

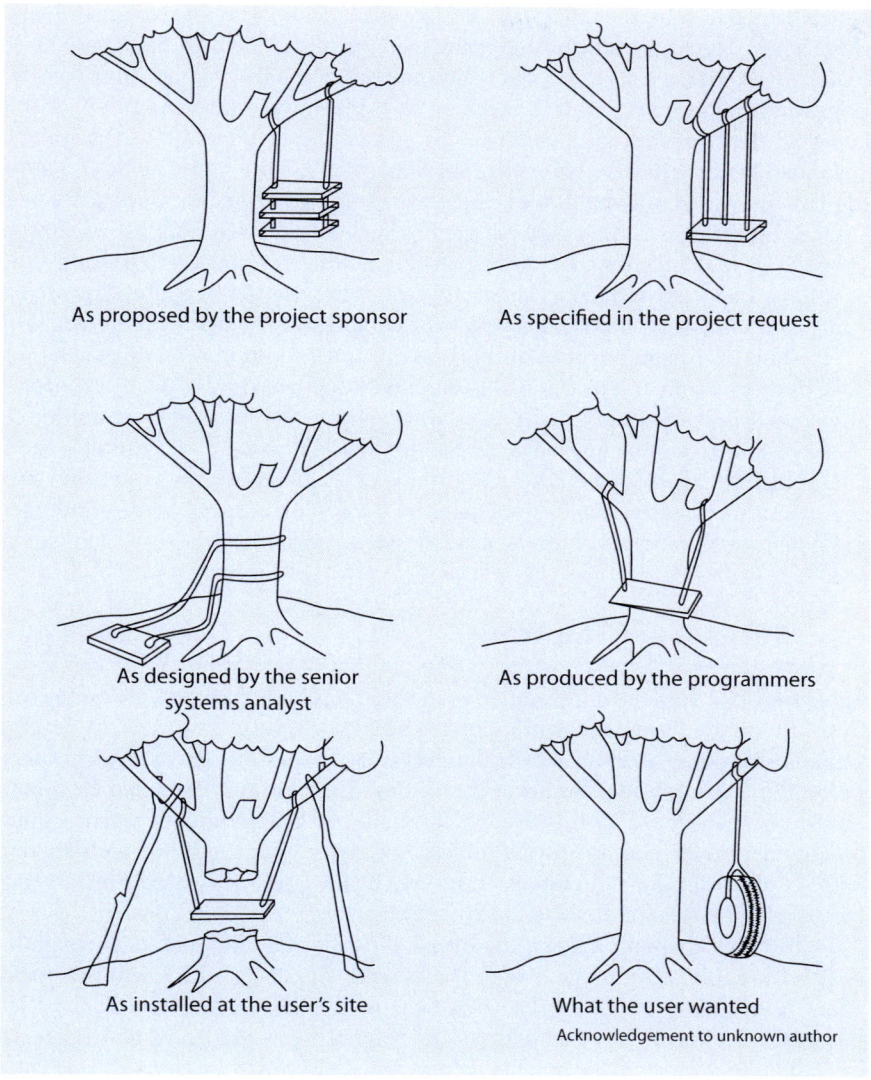

As proposed by the project sponsor

As specified in the project request

As designed by the senior systems analyst

As produced by the programmers

As installed at the user's site

What the user wanted

Acknowledgement to unknown author

Fig. 1.3 The tree swing cartoon from an unknown author (here in the version from the University of London Computer Centre Newsletter No. 53 of 1973 in Meek (2014))

science and explains the necessary theory. While it touches only generally on each subject on the path to its application, related literature is referenced, which can be used to get more detailed information. This introduction should increase the readers appetite to read more about different subjects and to try to implement such things in their own field of work.

The aim is also to give people at observatories a better understanding of the current state-of-the-art solutions and ideas of computer science. But the book also tries to avoid short-term trends and excessively theoretical views. The goal is to support students and observatory staff with the necessary background information, taken from several years of experience in software development and running software projects at a geodetic observatory. Similar to the cover of (Aho et al. 1997) (the «dragon book»), which discusses the principles and realization techniques of compilers in detail and therefore offers the ammunition to fight against the dragon behind so many general computer problems, the idea is to try the same here with related problems and topics in the field of controlling geodetic applications. Even if most of the readers will not develop complete systems for geodetic observatories, most of the ideas and techniques discussed can be adapted to many software developments.

To offer such «ammunition» to fight against the usual problems which appeared during the development tasks at Wettzell, additional code snippets are included,

Better Understanding

Solution Examples in Programmed Form

which helped to solve the tasks. They are like a collection of recipes for the development work. The idea of including larger sections of code, which might not be so interesting for the usual reader, but which help tremendously to find working solutions for programmers with their own tasks, is taken from (Stevens 1990). In this book, all the relevant topics about the programming of Linux networks are discussed and demonstrated with code. The book offered very useful advice for solving problems with different issues which were boring and unclear during my own studies. Even if this coding is not copied in such an excessive way, as in (Stevens 1990), it should include code with internal comments for the explanations and solutions. This means that another motivation is the transfer of coding knowledge and useful solutions, without claiming to be complete or to offer the only valid way.

In short, the main purpose of this book is to try to improve code quality and development work in scientific fields, because good preparation with techniques and knowledge, communicative team work, and a way to combine multidisciplinary subjects are the key skills needed for success. Usually everything is more complicated than anticipated at the beginning of a software development, and having practical recipes can help to reduce the burden of learning by do-it-yourself methods, which sometimes can be painful experiences.

1.3   Structure of This Work

The goal of the work is to introduce given situations and requirements on the way from the classic fundamental stations to the future GGOS core sites, to develop suitable solutions, and to explain the necessary theoretical and technical background in order to understand and deal with those solutions. Each chapter is built on the previous one, and the solutions found in one chapter are the starting point for the ideas and solutions of the following chapter. The result is a construction with six pillars for the main topics, connected by six capstones with the results suggested as a possible solution, as shown in ■ Fig. 1.4.

While this chapter explains the initial situation for this work as a first pillar of the above diagram, it starts with the current situation, which will inevitably affect future geodetic networks. This field is an ideal playground for applied computer science. A lot of daily tasks and requirements are based on the practical implementation of approaches which are well known in computer science. Therefore, the following solutions try to connect the work of a geodetic observatory with practical solutions from applied computer science. They need to introduce the different fields and explain the necessary theoretical background. The rest of this chapter focuses on the structure, requirements, conventions, and contacts for this book.

As the whole book gives a helping hand to those dealing with applied developments for the scientific environment, it is necessary to identify the current situation and field of scientific programming as a second pillar. It is not directly possible to apply the usual developments and processes from commercial projects to the scientific, university-driven environment, as man power, risk management, and project situations are different. Nevertheless, some aspects are also relevant. Version management, documentation, programming style, and testing remain present. It is also crucial to deal with legacy code from former scientific projects that have already closed, where often the main developers, who sat on limited contracts, are not available anymore. Suitable ideas from continuous integration and agile software development fit here in this agile scientific world.

The use of software development structures allows the design of a code base which is modular enough to be useful for almost all developments in the field of a geodetic observatory. This code toolbox is the third pillar in the quest for a general solution and consists of more or less adaptive parts. Adaptations to different problem areas can be reached by modularity but mainly also by generative coding

Six Incremental Pillars

► Chap. 1: Introduction

► Chap. 2: Scientific Software

► Chap. 3: Software Toolbox

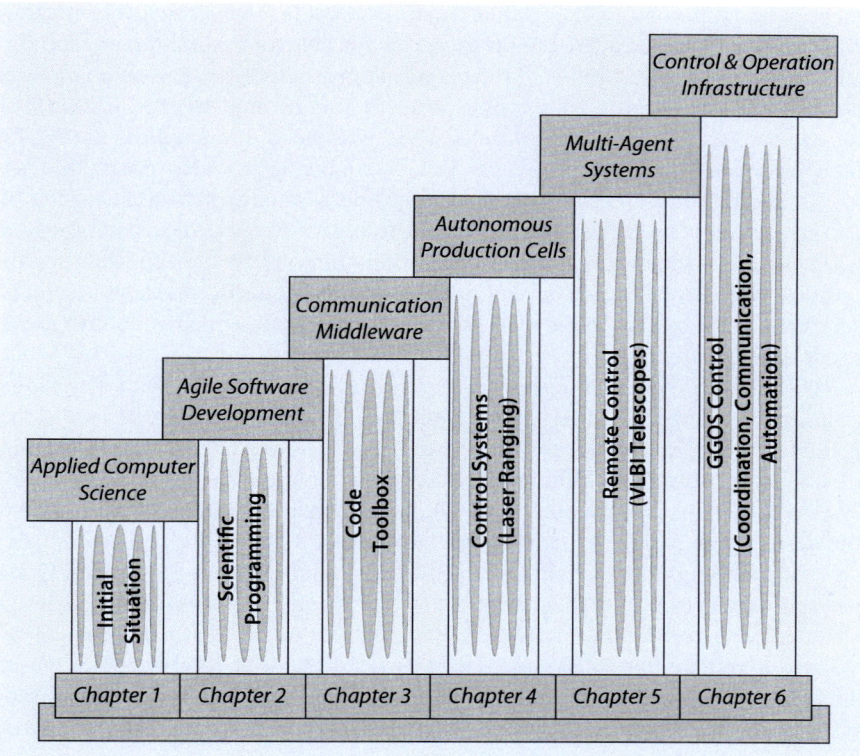

Fig. 1.4 The columns of the work with their connecting capstones to the resulting solution

techniques, where the final source code is produced by a software generator according to a defined interface specification. This is very suitable for the different communication scenarios, which always require the same solution patterns, but which must be adapted to different communication situations. The result is an adaptive interprocess communication using remote procedure calls within a communication middleware as a network backbone.

Having a toolkit with many adaptive tools (a «Swiss Army Knife» for software), makes it possible to start with a real development task: controlling large measuring systems of the space geodetic techniques. Here the focus is put on two systems: SLR/Lunar Laser Ranging (LLR) and VLBI.[4] The first approach further focuses on the creation of suitable control systems and the required theoretical background to deal with distributed hardware. The field of SLR control is a good demonstration here. The resulting concept is a sort of autonomous production cell, known on industrial production lines, for example, in the automobile industry. Adapting this design, four sub-cell structures can be identified: the autonomous coordination for planning and scheduling, the autonomous hardware control cells for direct hardware support, the autonomous system monitoring and safety cell for a parallel acquisition of a system status, and the autonomous data management cell for data handling. Another very important part here is the user's interaction with different (graphical) user interfaces.

▶ Chap. 4: Control Systems (Laser Ranging)

With a structure of autonomous production cells, most of the actions run almost automatically and are only influenced by human interactions. The independence of the systems from direct human control makes it possible to separate the operators from the systems. Then the missing human operator must be represented by a virtual decision maker, which is a specialized software agent for the individual system processes. This agent allows it to send in orders and to retrieve results in a

▶ Chap. 5: Remote Control (VLBI)

---

4   The other techniques do not have so many developments and can mostly be supported by commercial products.

more or less autonomous way. It then becomes possible to safely enable centralized control centers, where a few operators are responsible for several remote and distributed geodetic measurement systems which optimize the necessary manpower during observation shifts. New control strategies can be implemented. This is ideal for the future networks of radio telescopes, like those discussed for the VGOS (originally known as the concept of «VLBI2010»). But large, widely distributed networks, which communicate over the worldwide computer networks, require an increased level of security, as they are prone to hostile attacks. Access management with suitable authentication and authorization, different types of firewalls for the creation of enclaves, and an encryption of data streams and access keys are necessary here. The result is a network of secure and safe agents, which build up a multi-agent system.

▶ Chap. 6: GGOS Control Infrastructure

While each geodetic technique uses specialized agents with individual requirements, they all offer similar interfaces to support a general and global geodetic infrastructure. The abstraction and generalization of such interfaces result in extended control and data flows which combine the different techniques to a system of systems, as proposed for GGOS. Looking at the different hierarchies of the design, it becomes clear that such a system of systems is nothing more than a copy of the design of an agent, enlarged to an overall view on the general global geodetic observing network. Each agent is nothing other than a copy of the design structure of a distributed hardware system and so on, so that each layer of the system-of-systems design is just a copy of itself from a lower-level part. This creates a flexible, self-identical structure, which makes it possible to coordinate common tasks, organize communication between each part and automate most of the working processes to adapt a service-oriented job management.

Introducing Mind Maps

To get a better understanding of the topics in each chapter, the main keywords are shown in the form of a mind map at the beginning of each chapter. These maps start with the topic of the chapter and show paths between subjects and issues found on the way to a specific solution. Both the starting point and the solution are accentuated with bigger bubbles and larger characters. The map also connects the current subjects with those already explained in previous chapters, highlighted with a bubble using text, which is underlined. Therefore, it is strongly suggested that a reader should first have a look at the initial mind maps of the chapters to get oriented to the topic and its subtopics.

Praxis Interwoven with Theory

While each chapter can be read independently, they are linked to each other, as each topping «capstone» of the chapter pillars in ◻ Fig. 1.4 is built up step by step. Also the theoretical background is meshed with the topics of the chapters. The idea is to relinquish separate theoretical sections, which would introduce the required theory of computer science separately. The practical use cases of space geodetic techniques are taken here instead to introduce the theoretical background at the points, where the theory is needed. This links theoretical knowledge from computer science with the practical field of geodesy to interweave it to applied computer science for geodesy.

Code Snippets and Illustrations

As mentioned in the motivations section, the chapters are interspersed with several more or less extended code snippets. They are not explained in detail within the surrounding text blocks, but contain enough comments to explain their structures and workflows. They touch on pitfalls and solutions found in the form of «software cooking recipes» for daily coding work. Additionally, illustrations of the subjects in the text blocks are added to summarize and underscore special parts.

Appendices

One additional appendix is attached to the main chapters, which add further insight to coding style topics touched on in the book. A significant part is the aggregation of general style guide subjects. The appendix offers a simplified summary of the style issues which make life easier for programmers. It would have extended the regular text unnecessarily, but is quite useful in daily work.

1.4   Who Should Read This Book

The book is intended as an interdisciplinary work, which combines the terminology, techniques, and implementations of applied computer sciences and hardware engineering with geodesy. It is something like a compendium for scientists, engineers, and operators mainly at geodetic observatories and especially at future GGOS observatories. For this reason, it uses practical use cases based on the experience of the work at a current fundamental station to show possible solutions from the field of computer science for future GGOS core sites.

Observatory Staff

The work is not only interesting for the staff at observatories, but it is also a general textbook for students mainly in the applied fields of geosciences and applied computer sciences. As the theoretical parts are of a general nature and touch on several disciplines of computer science and informatics, the book might also be a practically oriented guide for other students, engineers, and interested people who want to get a feeling for which practical tasks a dedicated theory from computer science might be helpful. Nevertheless, the book does not claim to be complete.

Students

The book is not a basic introduction. Even though it tries to guide the reader through the chapters, a basic knowledge of computer science and geodesy is helpful. Especially programming knowledge is required to cope with the code sections.

Background

1.5   Used Conventions

The complete work is written in American English. References to literature use the American or Harvard citation directly behind the excerpt or citation with a small modification to reference the page or a range of pages. The scientific content was reviewed by the examination board of Prof. Dr. Urs Hugentobler (head of Research Facility for Satellite Geodesy at the Institute for Astronomical and Physical Geodesy, Technical University of Munich, Germany), Prof. Dr. Johann Schlichter (professor of applied computer science/cooperative systems, Technical University of Munich, Germany), and Prof. Dr. Hansjörg Kutterer (president of the Federal Agency for Cartography and Geodesy, Frankfurt a. Main, Germany). The proofreading was done by Academic Translation and Editing, München/London (contact: Karl Hughes). The proofreading of additional sections was done by Anita Neidhardt.

Language and Citation

The idea is to connect the practical field of a geodetic observatory with the required theory from applied computer science. Unlike in other books, these theories are included in practical examples. Thus, there is no separation between a theoretical section and a practical one. The theory is introduced where it is required and the basics are explained. Further details can be found in the referenced literature. Great importance was placed on a basic explanation of all technical terms. New and popular acronyms are always introduced with their complete form at their first appearance. After that, only the short forms of the acronyms are used. For a quick look to rediscover the meaning of an acronym, an alphabetical list of all acronyms used can be found in the appendix.

Explanation of Theory, Technical Terms, and Acronyms

The code examples are used to underline the facts described with an illustrative example. All code samples have their origin in the development projects at the Geodetic Observatory Wettzell or were specially implemented for this work. They are partly inspired by other codes. But all similarities to other code are just accidental or are referenced.

Use of Source Code

All code samples mainly follow the coding style and policies of the Geodetic Observatory Wettzell. They are focused on C and C++ code (with some excursions to Perl), as these are the main languages used at the observatory. Nevertheless, the

Coding Style and Programming Language

selection of the programming languages is not a general quantitative assessment and only reflects the experiences at Wettzell.

Some brief information about a text block is located in the margin. It should help to get a quick overview of the contents and find text passages about specific subjects. All margin notices are also listed in the index to improve the search for subjects.

The last section of each chapter contains questions about subjects discussed in the chapter. The questions can be used to reflect on learned theory. Most of the questions are taken from real final exams for lessons from the author. The answers can be found on the web pages of the author (see ▶ Sect. 1.7 with contact information).

The work has been completed with the help of several colleagues and also with the output of the different software development teams at the Wettzell observatory. Techniques of software development were tested in small teams (such as pair programming). As representatives of the teams, some of the main contributors are mentioned in footnotes in the different sections to which they mainly contributed.

1.6   Software Requirements

The software described and all selected software packages are focused on the operating system Linux (starting with the kernel 2.6). The tests were performed with different Ubuntu versions and with Debian.

For the compilation, the GNU's Not Unix (GNU) Compiler Collection with «gcc» for the programming language C and «g++» for the programming language C++ was used. Compilation tests were performed starting with version 2.95. The code is principally licensed under the GNU Lesser General Public License (LGPL) if not otherwise stated. Therefore, the code or portions of it can be used under these terms if this book is referenced as its origin.

1.7   Web Pages and Contact

Further information about the content and the author including links to updated software, which is discussed in this book, can be found on the web page of the author as part of the official web site of the Forschungseinrichtung Satellitengeodäsie (Research Facility for Satellite Geodesy) of the Technische Universität München (Technical University of Munich):

▶ http://www.iapg.bv.tum.de/Mitarbeiter/Alexander_Neidhardt/

Information about the Geodetic Observatory Wettzell can be found here:

▶ http://www.fs.wettzell.de

Writing Code for Scientific Software

2.1   Scientific Software as It Is in Most Cases – 14

2.2   A Suitable Programming Language – 19

2.3   A Coding Style Guide – 24
2.3.1   An Impression of Coding Layout – 27
2.3.2   An Impression of Coding Policies – 30

2.4   Including Legacy Code – 39

2.5   Documentation – 47
2.5.1   Documentation for the Development Team – 48
2.5.2   Documentation for the Users – 56
2.5.3   Documentation for Administration and Other Scientists – 60
2.5.4   Documentation Landscape – 60

2.6   Code Testing and Code Inspections – 62
2.6.1   Code Quality Metrics – 64
2.6.2   Dynamic Code Testing – 68
2.6.3   Static Code Inspections – 84
2.6.4   Dynamic Code Analysis – 86
2.6.5   Testing Landscape – 96

2.7   Version Control – 97
2.7.1   A Suitable Project Directory Tree – 98
2.7.2   A Suitable Version Control System – 100

2.8   Continuous Integration – 109

2.9   Agile Software Development – 117

2.10   Summary – 128
2.10.1   What Did We Learn? – 128
2.10.2   How Did We Use the Contents Learned? – 129

2.11   Questions – 129

© Springer International Publishing Switzerland 2017
A. Neidhardt, *Applied Computer Science for GGOS Observatories*, Springer Textbooks in Earth Sciences,
Geography and Environment, DOI 10.1007/978-3-319-40139-3_2

2

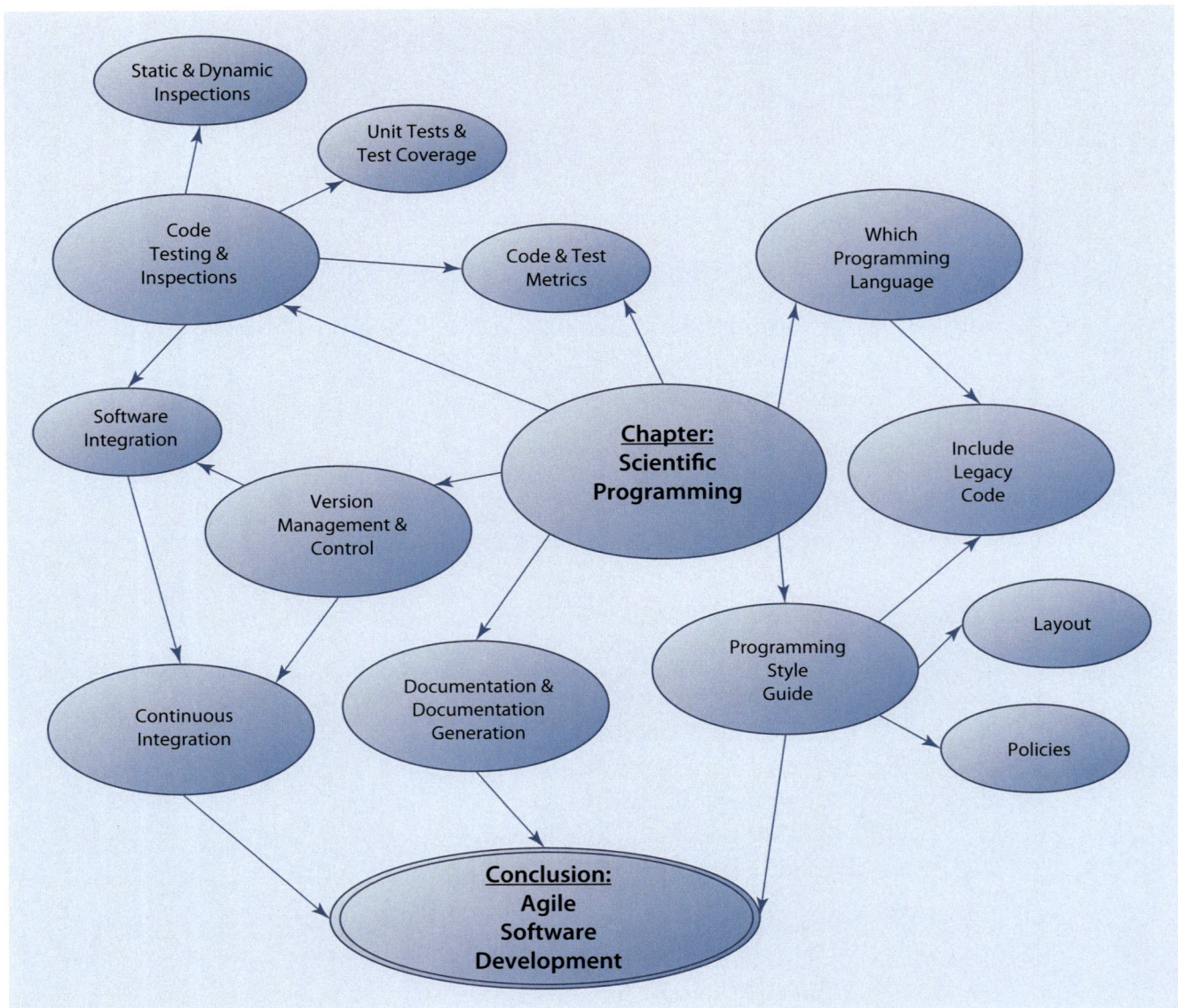

2.1   Scientific Software as It Is in Most Cases

Valuation of Scientific Software

Modern science and research is unimaginable without the use of self-written and in most cases highly specialized software. By analyzing data, simulating model-related items, or simply dealing with different data sets using different formats, computer programs help the researcher to do his daily work. But even though software and its development is an essential part of scientific life, its ranking within the world of science is normally very low.

Software in general is the sum of all executable (computer-)programs on electronic computer hardware[1] (Precht et al. 1994, *l.c. page 51*).

Program

> *Definition DEF. of «program»*
> A (computer-)program is a container (usually defined as a file) with an ordered sequence of instructions, which have the correct structure to be processed by (electronic) computer hardware.

---

1   An explanation of electronic computer hardware, which can be programmed to process data, can be found in ▶ Sect. 4.2.1 about the John von Neumann architecture.

While the user normally just sees the final executable files, software sources consist of several single units. These logical source code parts from which an executable program is compiled normally contain methodical coherent functionalities. These independent, physical software units with their dedicated responsibilities and clearly defined, functional, and external usable interfaces are known as modules (Vigenschow 2005, *l.c. page 24*) or, in another context, components[2] (Fowler and Scott 2000, *l.c. page 127*). They are the basic construction elements represented as single source code files (normally in combination with some header files for declarations, containing the functional interface access points). The code in these files is functionally decomposed into logical task units, combined into functions (or in a wider sense procedures or subroutines[3]). This functional decomposition helps to break down the complexity of an extensive task into manageable, smaller pieces of the task (Fowler and Scott 2000, *l.c. page 98*). The functions contain the instruction sequences in their function body which are needed to fulfill the function task. In some higher-level languages (such as Object Oriented Programming (OOP) languages), the functional units together with the data elements which they commonly use are combined into new class types, where elements of such new types form new programming objects.

Software, Modules, and Functions

While software development in industry has become a significant engineering discipline and a cost-relevant factor (Forbrig and Kerner 2004, *l.c. page 19*), writing software for scientific needs still has the status of a necessary evil. Most students of technical studies learn something about programming. But the courses normally do not teach a structured approach, starting from the need to solve a dedicated task and ending with a high-quality and reusable solution. Programming courses normally just train students in basic skills such as the use of variables, loops, and conditional statements with comfortable, high-level development packages. Therefore, students are able to program, but they have no notion of how to deal with the design, structured development, the implementation workflow, and the maintenance of written code. This means they do not have the skills to deal with a software development process and its life cycle of computer programming projects with their different versions.[4] But modern software is usually written in the form of complex project structures.

Importance of Software

Definition DEF2.2. of «software project» (adapted from (Springer Gabler Verlag (Herausgeber) n.d.))
A software project is a software development task with a clear deadline and designed to create a new innovative software system (for an explanation of software systems, see *DEF2.4.*) or software product of high complexity. Because of the risky character of such a development, software projects require project management, to plan and control the development process and minimize the given risks.

Software Project

Complex systems, as undoubtedly most computer programs are, have a strong tendency to become nontransparent and unstructured. This effect can also be seen in other real-life situations and is normally also independent from programming languages and development tools (see ◘ Fig. 2.1).

Complex System Behavior

In this situation, the following main deficiencies can normally be found in scientific software:

Main Deficits in Scientific Software

— Written from scratch without an identifiable design and expanded on the fly, as needed for the given task.

---

2   For more about modularization with an explanation of modules and components, see ► Sect. 3.1.

3   In a classic view, functions have some input parameters and a return value. Procedures or subroutines only have parameters and separate the code into logic parts.

4   More about the life cycle of computer programs and their development of different versions can be found in ► Sect. 2.9 about agile development and also in ► Sect. 2.7.1 about version control.

2

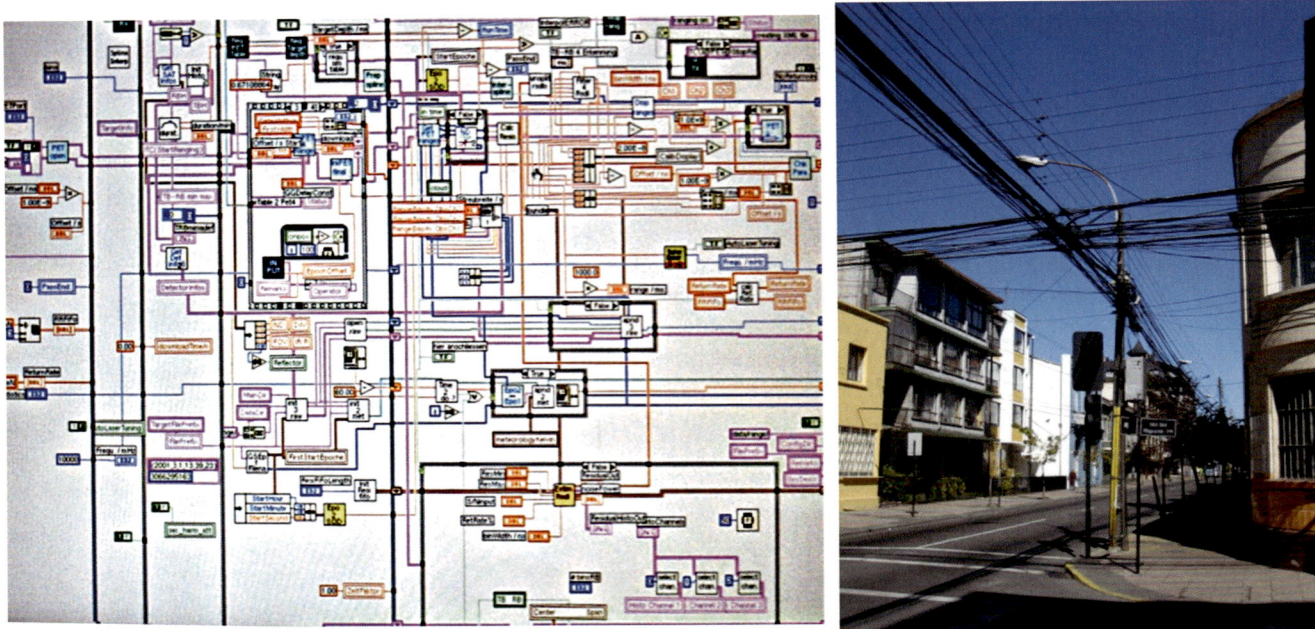

■ **Fig. 2.1**   Complex systems tend to become nontransparent and unstructured: on the left side, a snapshot of the former laser ranging control software at the Wettzell observatory written with the National Instruments™ LabView™ and, on the right side, a snapshot of a power distribution pole taken in South America

— Code which is just used for one particular purpose (e.g., a PhD thesis) and which then disappears over time, as its existence is simply forgotten.
— Specific knowledge is just owned by one single person without cooperation in a team.
— Highly specialized, non-modular code which cannot be used within other developments.
— Successors start from a zero level of knowledge.
— When parts of the code are used further on, additional functionalities are added and similar code parts are written again, because of a lack of knowledge about already existing parts (uncoordinated growth of new features).
— Undocumented, unreadable, unmaintainable code.
— Error-prone, inadequately tested, and unoptimized realizations with local dependencies (e.g., a porting of software to computer platforms or hardware with different conditions entails the risk of changed runtime behavior, so that numerical effects in calculations due to different number representations or racing conditions of threads for processor time due to a slightly changed timing might be the consequence).
— Written in different computer languages with different development environments on different computer platforms, as the individual programmers have their own preferences.
— Missing or not used common terminologies and clearly defined coding conventions.

CHAOS Report About Industrial Software   In general these problems are not too different to the problems encountered in professional, industrial software developments (apart from additional aspects like customer interaction and financial risk). Different studies, such as the often cited but also controversial CHAOS report of the Standish Group (The Standish Group n.d.), try to quantify the criteria responsible for success in software development. According to the results of the study, which has been regularly carried out, not even half of the rated software projects were successful in 2012. Over 40% of the projects were too challenging, delayed, ran over budget, and/or failed to come up

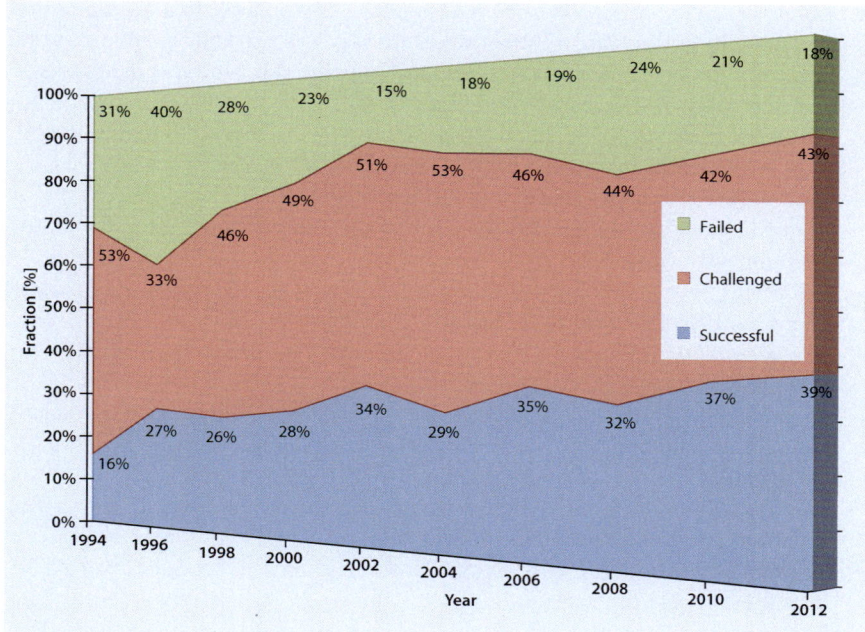

Fig. 2.2 Historic trend of the success of software projects reviewed in the CHAOS report of The Standish Group (data from Dominguez (n.d.) and The Standish Group (n.d.); reproduced by permission of Jorge Dominguez)

with the required features and functionalities. Finally, 18% of the projects directly failed[5] (see ◘ Fig. 2.2, data taken from (Dominguez n.d.) and (The Standish Group n.d.)). Even though the number of successful projects has increased in recent years, taking into consideration the fact that the study has been criticized for not selecting success factors in an ideal way within this evaluation, it nevertheless shows that software engineering is still a critical and difficult task. There is no obvious panacea which leads to high-quality software, which is released on time.

Successful coding within scientific environments is normally not visible as long as the program is unpublished. Software is just an auxiliary technique and the main focus lies in the research result and not in the source code. It is also the case that scientific developers are often their own customers. Therefore, they are able to meet their own needs optimally, as nobody has more knowledge of the subject. But the resulting code mostly does not reflect this. To avoid the deficits, it would be necessary to establish a software coding culture for science. This might follow the well-known quality factors for software (defined in different ways by Boehm, McCall, Dromey, FURPS,[6] and ISO 9126–1 (Rawashdeh and Matalkah 2006) or found in (Rembold 1991, *l.c. page 275 ff.*):

— *Functionality*: Does the software offer the specified functions and do they work?
— *Correctness/accuracy*: Are the results correct and with the specified accuracy?
— *Reliability*: Does the software offer its functionality without constraints over a specific time period?
— *Fault tolerance/robustness*: Is the software robust enough to deal with external influences or error states?
— *Stability/maturity*: How stable is the current version of the software, with respect to error or alert situations?
— *Usability/operability*: How difficult is it to use or operate the software and how much experience is required by a user?

Reflecting the Scientific Needs

Software Quality Factors

---

5   Large, expensive, failing projects are often named «death-march projects» which nobody can stop anymore, but which it is clear will end in disaster.
6   FURPS was proposed by Robert Grady and Hewlett-Packard Co. and stands for **F**unctionality, **U**sability, **R**eliability, **P**erformance, and **S**upportability (Vigenschow 2005, *l.c. page 47 f.*).

- *Understandability/comprehensibility/complexity*: Is it possible to understand the functionality and design of the software and why was a specific result produced?
- *Efficiency/performance*: How fast does the software solve the specified tasks and how many resources are needed for this?
- *Maintainability/supportability*: How much support is needed to maintain and improve the software or keep it functioning?
- *Testability*: How easy is it to detect and locate errors?
- *Portability/configurability*: How easy is it to use and configure the software to different platforms or use case?
- *Reusability/flexibility*: How easy and flexible is it to use the software or parts of it in other projects or for other tasks?
- *Interoperability*: How easy is it to cooperate with external software or in a given environment?
- *Modularity/scalability*: What is the smallest unit which can be used as a separate module and how well can the software be scaled to different classes of complexities?
- *Changeability/modifiability*: How well can the software be adapted to changing situations during its lifetime?
- *Security*: Does the software protect against unauthorized access? How well does the software protect?
- *Integrity*: Does the software keep the data in a complete and unchanged state?

Factors of Success

To fulfill these quality factors and therefore to produce high-quality software, the Standish Group identifies a few factors of success for small projects,[7] which are comparable to scientific, third-party funded developments[8] (see The Standish Group n.d. *l.c. page 3 ff.*):

1. *Executive management support*: skilled, communicative, and engaged executive sponsors or project owners
2. *User involvement*: focusing on real user needs and dealing with user requirements
3. *Optimization of scope*: focusing on the essential parts, which have a high impact
4. *Skilled resources*: having the right, competent, motivated, trained, communicative people doing the right things at the right time
5. *Project management expertise*: project managers with leadership skills, the ability to make connections, and basic process management skills
6. *Agile process*: iterative developments with regular integration stepping-stones and feedback loops in interactive communicating teams
7. *Clear (business) objectives*: using single, common, short-term goals, which are describable as big picture or «elevator statement» in ten words and which can be measured and peer-reviewed
8. *Emotional maturity*: not tolerating overambition, arrogance, ignorance, permanent absence from meeting, or fraudulence, but supporting community, honorable work, and objective awareness
9. *Execution*: rules, standardizations, and formal requirements with small decision pipelines
10. *Tools and infrastructure*: management, software, and quality infrastructure

Quality Metrics

An executive sponsor or project owner needs «[...] an easy and visual way to measure progress» (The Standish Group n.d. *l.c. page 7*). In combination with some metrics, such as the number of code lines per function compared to the average,

---

7   The Standish Group defines all projects with less than one million dollar labor costs as «small projects» (or in the case of the European Union with less than 250,000 Euro labor costs) (The Standish Group n.d. *l.c. page 13 ff.*).
8   The importance of the factors is in descending order.

user satisfaction statistics, or number of version releases in a dedicated time period[9] (Vigenschow 2005, *l.c. page 75*), it is partly possible to quantify and measure progress and also quality factors (see ▶ Sect. 2.6.1). But in general, it is difficult to find suitable metric sets, because they must be adapted individually to each project. Individual test scenarios have to be established to check the defined quality parameters. For this reason, it is quite helpful to reduce the degrees of freedom of these impact factors. This can be done by defining the guidelines each programmer has to follow. These standardized guidelines accompany the whole software development. They are collected in a set of software design rules.

Definition DEF2.3. of «software design rules»
Design rules are a set of guidelines which accompany the whole software development from first analysis steps to the final product. In general they include:
- Guidelines for the programming style
- Guidelines for documentation
- Guidelines for testing
- Guidelines for software integration
- Guidelines for the project workflow
- Guidelines for licensing

Software Design Rules

Of course the design rules are no guarantee of producing high-quality software, but they help to avoid lapses and help to increase the transparency of the design of software systems, which are not only the programs itself but also all the other necessary parts.

Definition DEF2.4. of «software system» (adapted from Sommerville 2007, *l.c. page 5*)
A software system consists of different programs, the necessary configuration files to set up these programs, the documentation for users and developers, and the guidelines during the development and the environment to develop, test, and release the different versions, for example, on web pages with version control techniques.
The following sections describe the most helpful guidelines and methods to write scientific software systems. They are based on the experience of developing software for space geodetic techniques at the Geodetic Observatory Wettzell, and they aim to follow the factors of success. The description starts with rules for the programming and coding style. But prior to this, it is necessary to find a suitable programming language for individual requirements.

Software System

2.2 A Suitable Programming Language

In the same way as it is important to speak the same human language for successful common communication (e.g., English for international communication), it is also important to use a dedicated programming language (or at least a reduced set of languages) to write successful common programs. The use of a common programming language simplifies the work of software development within a team. If the

Criteria to Select a Programming Language

---

9 The number of releases can be visualized with a «heat map» over the source files, where code sequences, which are often changed, are colorized with red and others with less changes get gradually darker colors.

code needs to be reusable in the future, it is especially important to select appropriate programming languages. But this is normally a process which is directly influenced by individual human experience, wishes, and existing knowledge. Everybody has their own favorites when talking about programming languages. But nevertheless, developed code cannot be maintained and reused in an easy way if the different parts are written in many different languages. In general, the decision should optimize the following items and is a compromise for the developers:

- Using a complete formal language
- A varied set of support tools
- Portable to different platforms
- Independence between style and logic
- Programmable with terminal mode and with sophisticated graphical environments
- Support from hardware-level programming to higher-level programming
- Huge community with a varied set of help access
- Optimized and fast in execution time
- Long history with dedicated, backward-compatible standards and tool releases
- Valuable license policies
- A varied set of libraries or existing modules
- Support of different complexity levels from structured programming to object-oriented styles
- Possibility of structuring and designing
- Code strictness, advanced error handling, and a low susceptibility to errors
- Explicit type management to control the numerical precision of mathematical calculations
- Compatibility with other languages

Compiler Versus Interpreter

Besides these items, the first decision to be made concerns compiler and interpreter languages. On the one hand, compilers (as used for C, C++, FORTRAN) use an additional phase to compile (which means to translate) source code into platform-specific, executable machine code, including linking it with external library sources (Brown 2001, *l.c. page 8*), so that the final binary executable can be started immediately for its execution each time. On the other hand, interpreter languages are converted each time the program is called up and during the startup. This enables simple command line interfaces, as program code can be entered directly. But the code of interpreter languages usually requires more runtime during the startup compared to startups of previously compiled executables. Another disadvantage for licensed software is that the source code must be distributed with each release, so that it is easy to breach the copyright. To avoid these disadvantages, most of the modern interpreter languages are also compilable (interpiler techniques).

Compiler

Compilers, such as the GCC with front ends for C, C++, Objective-C, FORTRAN, Java, Ada, etc. (GCC team n.d.), the «clang» compiler with similar front ends (Clang team n.d.), or the Intel® compilers, are powerful tools. Usually the generation of an executable and running of the created program follow specific steps, using these compilation tools. These steps are (see also (Favre-Bulle 2004, *l.c. page 172*) and (Aho et al. 1997, *l.c. page 5*)):

1. Phase 1: Compiling, which also includes the execution of a preprocessor for the macros[10] (of all sources separately), e.g.,
   - `gcc -c source1.c` (compilation of the first C source code file to object code; use the standard include paths to search for header files with function declarations; further include paths can be defined using the «-I» parameter, e.g., `gcc -I../srcext -c source1.c`)
   - `gcc -c source2.c` (compilation of the second C source code file to object code in the same way as described above)

---

10 Macros are replacements in the code, which a preprocessor substitutes with real, final code elements. Therefore, macros are often used for constants or fast, but short code sections, which appear regularly in the source code.

— gcc -c main.c (compilation of the C main source code file (contains the «main»-function) to object code in the same way as described above)
2. Phase 2: Linking and binding
— gcc -o prog source1.o source2.o main.o (linking and binding of the single object files to an executable with the name «prog»; use the standard include paths; special modules can be included by using the «-l» parameter, e.g., «-lm» for the math library; further link paths can be defined using the «-L» parameter similar to the «-I» parameter for the compiling)
3. Phase 3: Loading and executing
— Run the created executable of the compiled program in the command line with «./prog» (load all dynamic and static binary libraries with their pre-compiled code).

Interpreters (e.g., originally Java, Perl, Python) do this during the startup of a pro- **Interpreter** gram call at runtime (Aho et al. 1997, *l.c. page 4*). In the past, interpreters ran the program instruction by instruction. Nowadays Perl, for instance, compiles the source code into a tree of opcodes, which are similar to machine codes, during the program startup, to optimize the runtime behavior (Brown 2001, *l.c. page 9*). Meanwhile both techniques are becoming more similar, as most interpreter languages also make it possible to compile code to an executable, or compiler front ends (e.g., GCC) are able to understand interpreter languages (interpilers). Interpreters are very popular for web programming and as command line languages for scripts. As writing bash or shell scripts with operating system facilities is sometimes a little bulky, because of the reduced language coverage of the pure shell languages, interpreter languages can compensate for such deficiencies while keeping the scripts short. They offer useful utilities, such as the interpretation of regular expressions (see also ▶ Sect. 3.3.3), [11] or an extended and convenient set of system commands. But experience shows that purely compiled code is mostly better optimized, more efficient, and much faster. Therefore compiler languages might be the best choice for binary offered, compact, and independent programs. Interpreter languages are much more convenient for regular shell tasks or startup scripts. Together both offer a powerful environment for all needs. But which compiler language and which interpreter language is valuable for scientific software?

For a long time, FORTRAN was the de facto standard for scientific software. **Former De Facto Standard: FORTRAN** Hence, a lot of existing software packages are written in FORTRAN (e.g., Bernese GPS Software (Bernese n.d.) or some control software for laser ranging and geodetic VLBI). But compared to current programming languages, FORTRAN's structures are a little antiquated. The author's own experience with FORTRAN code shows that most of the programs use common memory blocks and «goto»-statements extensively. The procedure parameters use only call-by-reference[12] so that internal changes of arguments in the functions address the original variables outside of functions, so that they are directly usable without further value returns. The old naming and line length conventions also often result in variable names, which are incomprehensible. It also results in the fact that complex calculations are artificially separated into several lines, so that the line length is not exceeded. New standards such as FORTRAN 2008 (GNU n.d.) may well solve these issues. But a look at the statistics about the most popular languages shows that FORTRAN is no longer very popular in the industry and among open-source programmers (in the

---

11 Regular expressions are strings which define the patterns for a search criterion to be found in other strings by pattern matching (Brown 2001, *l.c. page 231*).
12 Call-by-reference is a method which uses the real addresses of the arguments of functions, so that internal changes are directly visible outside. Call-by-reference is used to return values from functions besides the regular return argument. The opposite to call-by-reference is call-by-value, which creates a local copy of the external argument, so that changes are just local.

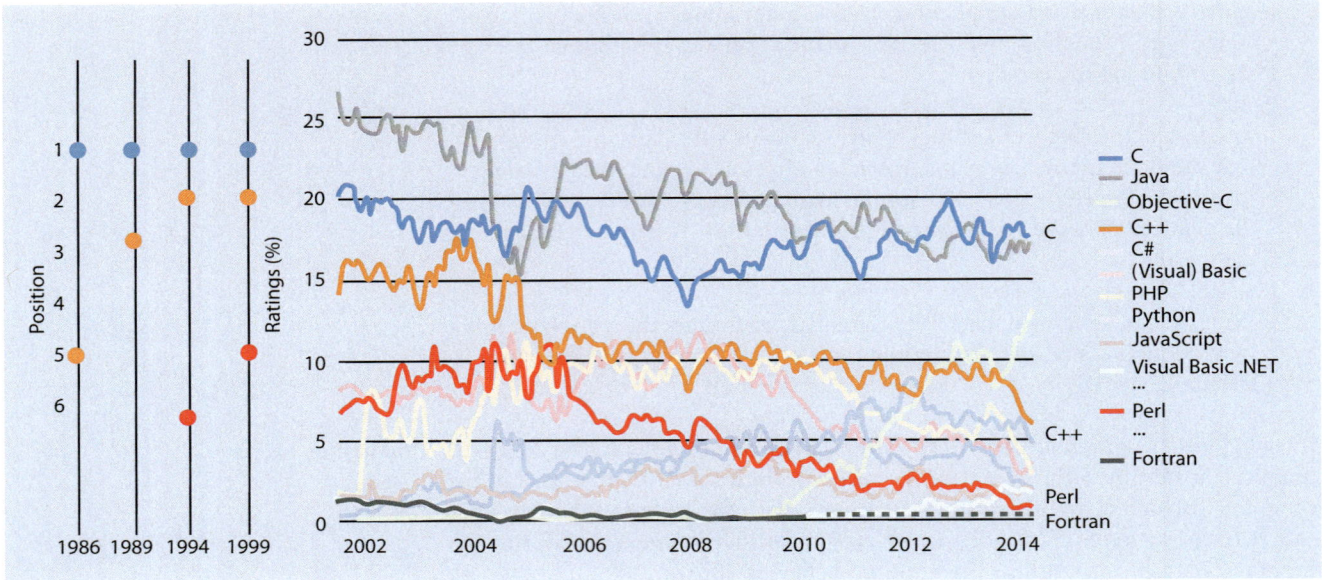

Popularity of Compiler Languages

TIOBE Programming Community Index for April 2014, it was ranked at position 34 (TIOBE Software n.d.); for more about the index, see the next block). Especially this missing acceptance reduces external support according to external code packages or help pages. Whatever the situation, the used language should be able to include FORTRAN code, in order to use existing scientific legacy software.

A look at the most popular languages in the Programming Community Index (TIOBE Software n.d.) reveals a long-term trend, which shows that at least three languages have become very well known in recent years: Java, C, and C++ (see ▣ Fig. 2.3; data taken from (TIOBE Software n.d.); similarities can also be found under LangPop.com (LangPop.com n.d.)). This monthly index rates the number of skilled engineers worldwide, courses, and third-party vendors, using hit rates in search engines, such as Google, Bing, Yahoo!, Wikipedia, YouTube, and Baidu. While some languages may briefly become fashionable and climb up the ratings (e.g., Objective-C in combination with iPhone and iPod programming (TIOBE Software n.d.)), the long-term trend is more interesting for the scientific use of a language. While Java only supports OOP and is not so ideal for low-level, hardware-oriented programming, the combination of C and C++ allows a programming paradigm change between OOP and structured programming.[13] This might be interesting as the design of OOP is difficult for a lot of people who are not familiar with these design patterns. It is much simpler to write structured, modular code in C which can be directly included and encapsulated with C++ to extend it to OOP. C/C++ supports all technical levels of programming from hardware to user interfaces. It is well supported with diverse tools and allows similar possibilities to FORTRAN. A positive point is that the suggested use of C/C++ is stricter in parameter scoping and structuring of the code. Another big advantage for a future trend analysis might be that the Linux kernel itself is programmed in C (Linux Kernel Organization, Inc. n.d.). This offers good prospects for the future existence of the language.

C Details

C provides all the fundamental control structures (conditions, loops, etc.) needed for structured programming. Additional high-level structures, such as parallel operations (threading, forking, etc.) or process synchronizations, are

---

13 Programming paradigms are general classifications of programming languages into specific categories defined by the structure and elements used in the computer programs.

originally not included, but can be added with operating system libraries. Elementary data types can be used to define one's own variables. It is also possible to combine these types with new structure types (Kernighan and Ritchie 1990, *l.c. page 1 ff.*). The supported double precision can be used for most mathematical calculations. The execution speed and memory consumption for this higher precision is quite good in modern computer architectures (Press et al. 2005, *l.c. page 17*). Because of its machine-oriented characteristics, C deals with the elements which most computers offer, like characters, numbers, and addresses. In particular, the powerful but also error-prone addressing with pointers offers an incredibly wide range of possibilities for hardware access, which Java, for example, never had. Functions (usable also in the sense of procedures) with their calling parameters allow the partitioning of code into logical units. Variables within the functions have a local scoping, and the functions themselves can be called recursively. The preprocessor directives allow macros, the inclusion of declarations from other source files (modules) and conditional compilations (useful to adapt the code to different platforms). But this should be used carefully to create an understandable design. The current standard is based on the original language definitions which support the compilation of existing code over decades. A big advantage is that it can also be linked to FORTRAN code modules. Even though C is a very effective and powerful language, there are some disadvantages, for example, the implicit operator associativity, which changes the operator evaluation for different operators in mathematical equations between «from left to right» and «from right to left» (Kernighan and Ritchie 1990, *l.c. page 1 ff.*).

C is a subset of C++ (Stroustrup 1995, *l.c. page 5 ff.*). Therefore, all the powerful constructs of C are also available in C++, including the possibility of low-level hardware access. Several extensions make C++ even more powerful, so that it supports OOP patterns quite well. Multiple inheritance, function overloading, virtual functions, templates, exceptions, and private, protected, and public member variables are suitable programming patterns. For example, templates can be used to formalize the use of macros. The concept of classes and their realization as objects allows the definition of high-level data types, like vectors, strings, lists, maps, queues, or stacks, which are containers for other types. These commonly used and general-purpose algorithms and functions are collected, for example, in the STL. The classes of this library, which are implemented as templates in most cases, allow it to be used in combination with almost all data sets and algorithms, but create a higher degree of complexity, which makes understanding difficult for non-OOP programmers (Schildt 1998, *l.c. page 626*). Modern extensions such as the Boost C++ Libraries are «intended to be widely useful, and usable across a broad spectrum of applications» (Dawes and Abrahams n.d.) and offer reference implementations (Dawes and Abrahams n.d.), addressing disadvantages in STL, like thread safety. For this reason, the combination of C and C++ is ideal for scientific programming, while the differences between both languages are mainly based on the different paradigms (Stroustrup 1995, *l.c. page 5 ff.*).

C++ Details

One short note should also be made about C#: The popularity of the OOP language C# has been rising very fast for the last years. But C# unfolds its full power only with OOP knowledge (similar to Java). The community is quite young, reflected by numerous standards in recent years (Lévénez n.d.). Therefore, it might be too early to propose it as ideal for scientific developments. However, it is worth watching in the future. Another quite strong trend is given in combination with another superset of C, the Objective-C (ObjC) language. Its focus lies in the development of mobile equipment. Apple Inc. is one of its main users (TIOBE Software n.d.), so that it is strongly connected to its product lines and future. Like C#, it is worth taking a look at, but it is too early to make prognoses. Finally, C in combination with C++ with its stable and downwardly compatible standards (Lévénez n.d.) is currently an ideal language combination for scientific software developments.

C# and Objective-C

2

Popularity of Interpreter Languages

Now only the decision about a supporting script language is open. A look at the popularity index (see ▫ Fig. 2.3) shows that there are only Perl and Python as adequate competitors. PHP is too specific for web use. Python is quite popular in scientific environments, and students in particular like to program with it, because of its simplicity and extensive libraries. But one disadvantage makes it difficult for sustainable developments: its indent-dependent scoping. In Python, structuring code blocks are defined by using different indents (white spaces in front of the instructions). At first sight, this is advantageous, as it obliges the programmer to write readable, structured code. But as different editor programs replace the white spaces in different ways with tab stops, it could lead to an implicit change in the code logic after opening and saving an existing program with different editors. The author's own experience showed that this significantly reduces the reusability, maintainability, and robustness of the code. Therefore Perl might still be the choice for a scripting language for productive scientific code.

Perl Details

Perl itself is written in C, so it works quite well with C and Unix/Linux. Similar structures make it easy for C programmers to learn and understand programming. But Perl uses a stricter structure code block definitions, for example, require curly brackets in all cases. Perl offers a very sophisticated use of regular expressions (see also ▶ Sect. 3.3.3). This is quite a powerful way to write file or command inputs as well as one's own interpreters. The powerful data handling with implicit typing allows compact codes, which would be much more extensive in C or C++. With the OOP style of Perl, polymorphism, inheritance, and encapsulation are possible in a similar way to C++ (Brown 2001, *l.c. page 12*). Because of Perl's relationship to C, it is possible to extend Perl code with C sources or to embed a Perl interpreter into a C or C++ source to combine C with Perl or vice versa (Brown 2001, *l.c. page 641 ff.*). With the embedding technique, it is possible to provide an internal scripting system for an existing application by using the extensive text-processing possibilities of Perl. It is also possible to directly call up Perl functions (subroutines) from an external module. But all of these techniques are in some way low-level and raw (Brown 2001, *l.c. page 698 f.*). In general, Perl harmonizes quite well with the selected compiler languages C and C++, which makes life easier. Nevertheless, its use should be limited to the scripting, for example, in combination with the managing of operation system tasks, as Perl programs are too inefficient for higher-level programs and their syntax requires time to get used to.

But each programming language is only as good as its programmer. Therefore, it is necessary to define additional guidelines to ensure the writing of sustainable, understandable, readable, and reusable code. These rules accompany the whole writing process. Attention is focused on C and C++ for most of the programs, and Perl is only used as a general scripting language in combination with startup scripts and operating system tasks. The following guidelines are mainly specified for C/C++. But a lot of the rules can also be adapted to Perl code as well.

2.3   A Coding Style Guide

Coding

Writing code is always an expression of creativity. Nevertheless, it is necessary to guide this creativity into structured and architectural forms. As C and C++ are style independent, they allow very compact and densely packed code. The result could be a very effective but incomprehensible program structure. The following example shows which compactness is possible using the C language. But the question is: which well-known task is solved by these lines of code?

Fig. 2.4 One solution for the «eight-queens puzzle»

■■ **Example 2.1**

Unreadable code (reproduced by permission of International Obfuscated C Code
Contest, ► www.ioccc.org) (Osovlanski and Nissenbaum 2003)

```
int  v,i,j,k,l,s,a[99];                                                    1
main()                                                                     2
{                                                                          3
   for(scanf("%d",&s);*a-s;v=a[j*=v]-a[i],k=i<s,j+=(v=j<s&&(!k&&!!printf(2+"\n\n%c"-(!l<<!j),"   4
         #Q"[l^v?(l^j)&1:2])&&++l||a[i]<s&&v&&v-i+j&&v+i-j))&&!(l%=s),v||(i==j?a[i+=k]=0:++a[i])
         >=s*k&&++a[--i])
     ;                                                                     5
}                                                                          6
```

The answer is an algorithmic solution for the N-queens puzzle. This puzzle is a
very popular problem among computer scientists and a generalization of the eight-
queens puzzle. This was attributed to the chess player, Max Friedrich Wilhelm
Bezzel, in a newspaper about chess and was popularized by Dr. Franz Nauck later
on in the 1850s. Bezzel asked how it was possible to put eight queens on an 8x8
chessboard in such a way that none of them could encounter any other. Only the
standard movements of a queen on horizontal, vertical, and diagonal lines are
allowed. The problem itself is computationally very expensive as 4.4×10^9 arrangements of eight queens on an 8x8 board must be checked to find all 92 solutions (see
■ Fig. 2.4). The puzzle was so popular that the famous mathematician Carl Friedrich
Gauss tried to compete with Nauck, but was beaten (Gierhardt n.d. *l.c. page 3*)
(Gutiérrez-Naranjo et al. n.d. *l.c. page 199*).

N-Queens Puzzle

The above code snippet was the winner of the 7th International Obfuscated C Code Contest, 1990, as the best small program.[14] It finds all possible solutions for the N-queens puzzle for boards of size 4x4 to 99x99. For each solution, the chessboard and the place of the queens are printed. The authors explain that they only used a limited subset of C and that they kept the program as readable as possible. «[I]t contains no C language that might confuse the innocent reader» (Osovlanski and Nissenbaum 2003).[15] The program should demonstrate that any C program can be written with only one artless (but very long) «for» statement and without using a preprocessor-, «if»-, «break»-, «case»-statements, and functions, jumps, or structures (Osovlanski and Nissenbaum 2003).

Coding Style Guide

The example shows that each program is an individual representation of an individual algorithm. It also shows how this can lead to unreadable and barely maintainable code. Even to test the functionality and correctness of the code is difficult, as hidden errors may only appear under specific circumstances in this confusing code. Even errors found by a compiler cannot be located easily in the source lines. Coding style guides are needed to offer an obligatory standard. These coding style guides consist of two parts:
- A layout definition
- Some coding policies

Code Layout

The layout helps with the writing of compliant code, which is more easily ascertainable, as we understand known and familiar structures much better than completely new and different things. This means that different members of the team can get an impression of written code and its functionality more easily, if a common style is used. It helps to structure the code in a better way, which generally supports its readability. The result is not only a well-structured but also a self-documenting code. With clear naming conventions, straightforward scoping, clear visible data, and work flows, the code itself is the best documentation of the internally represented algorithms. It is really powerful in combination with explanatory comments and an automatic generator of developer documentation using the comments (see ► Sect. 2.5.1). In addition to the style, the coding policies offer hints to avoid critical coding techniques and therefore offer a suitable programming framework (Vigenschow 2005, *l.c. page 31 ff.*). For example, they help to avoid possible error cases, such as uninitialized pointers, memory leaks, or out-of-memory scenarios.

Psychology Behind Style Guide Rules

But launching new guidelines should not be an end in itself. Even when style does not express or influence the creativity of a code, launching such style rules is always a psychological process, as the rules touch on individual familiar behaviors of the different developers. It is easier to train new staff to follow the rules than to change the habits of established developers. Nevertheless, strict compliance to the rules is essential, as errors can be reduced and the reuse of the code is better supported. Experience shows that reaching clever compromises helps in some cases to establish such rules. But if someone is unwilling to comply, it is not possible to argue with them by using logical explanations. In such a case, the implementation is only possible if the project and personnel management insist on the rules being used, as the advantages for the whole team, the project, and future enhancements are much greater and more important than satisfying a single person's idiosyncrasy. But in general, programming with the same style can also support team work and a feeling of togetherness.

Existing Style Guidelines

Currently, several guideline sets can be found for C/C++, which build something like de facto standards (e.g., for air vehicle or automotive developments)[16]:
- JOINT STRIKE FIGHTER, Air Vehicle C++ Coding Standards for the System Development and Demonstration Program, December 2005 (Lockheed Martin Corporation 2005)

---

14   With a small change in the present version so that the global variables are from type «int» to also enable the compilation under Microsoft® Windows™, according to the further program author notes for PC users.

15   Reproduced by permission of International Obfuscated C Code Contest, ► www.ioccc.org.

16   Some general hints can also be taken from the Microsoft® *Design Guidelines for Class Library Developers*. NET Framework 1.1 (MSDN Microsoft n.d.).

- MISRA C++:2008 – Guidelines for the use of the C++ language in critical systems (MISRA (The Motor Industry Software Reliability Association) 2008)
- MISRA-C:2004 – Guidelines for the use of the C language in critical systems (MISRA (The Motor Industry Software Reliability Association) 2004)
- Google C++ Style Guide (Weinberger et al. n.d.)
- JPL Institutional Coding Standard for the C Programming Language (Jet Propulsion Laboratory n.d.); as representative for a coding style guide for the scientific field)
  There is also a first set of rules at the Geodetic Observatory Wettzell, which can also be found as revised version in Appendix A:
- Wettzell's Design-Rules für die strukturierte Programmierung unter C und die objektorientierte Programmierung unter C++ (the first version is in German; the translated title is Design-Rules for Structured Programming with C and Object-Oriented Programming with C++) (Dassing et al. 2008)

Most of these standards define hundreds of guideline rules. Normally, if code needs to be compliant to one of these rule sets, it must follow all of them. For most scientific programs which are only developed in a limited sphere, for example, within a masters or PhD thesis, it could be a case of overkill to follow all of the above rules in the style guides, even if all the rules are designed to avoid errors or to produce maintainable code. However, the selection of a small amount of the basic layout styles and programming policies from the long lists of rules can improve the code quite significantly. Therefore, the following subsections should give some idea of what is meant by style. It makes no claim to be complete, but can be used as a starting point for individual styles for C and C++.

Fundamental Layout Styles and Programming Policies

2.3.1 An Impression of Coding Layout

The stylistic instrumentalities can be separated into those that are project specific and those that are source file specific. The project-specific rules define things such as the structure of the project directory tree (e.g., according to the underlying version management system; see ▶ Sect. 2.7) and naming conventions for directories and files (e.g., only lower case characters in combination with underscores «_» and the programming language-dependent extensions, such as «.cpp/.hpp» for C++ sources and «.c/.h» for C sources). The source file-specific parts define the internal structure of a source file (e.g., the order of type definitions, function declarations, and function definitions) or within a class (e.g., the order of declarations as «private», «protected», or «public»). The most important layout techniques for source code are the use of white space characters (white spaces), parenthesis (Vigenschow 2005, *l.c. page 33*), and naming conventions.

Rules for Project and Source File

White spaces are blank characters, tabulator characters, and new or blank lines, which represent a space within the typography of a program code. It is suggested that tabulators should be avoided as a layout mechanism, as different editors interpret them with different spaces. In C and C++, this does not change the syntax of the code (as, e.g., in Python), but it can destroy the clear layout. This can lead to humanly unreadable code. In general, white spaces are an appropriate instrumentality for (Vigenschow 2005, *l.c. page 34*):

Rules for White Spaces Use

- Grouping of logically dependent statements by using empty lines
- Grouping of logically dependent instruction parts in a line by using blanks
- Aligning of instruction parts over several lines with the special case of indents in front of instructions of the same code block or scope by using heading blanks (the programming experience, e.g., shows that four blanks are ideal for indents, offering a valuable structuring mechanism while keeping manageable line lengths)

A clear line structure is particularly suited to supporting readability. Each instruction should be placed on a separate line. Combining instructions as a sequence in one line or within other control elements, like in the heading definition and stop

Rules for Code Instructions

2

criterion of a loop by using a comma separation, is not good style. Complex or long instruction statements should also be separated into several lines. Most of the style guides also define a maximum line length of about 70–80 characters, which is useful for printing the code. But printing is not so relevant anymore, as nowadays most software modules consist of thousands of code lines, so that it is not really practicable to print them. It is more helpful to separate the different parts of an instruction line using single blank characters (e.g., between an identifier name, the assignment character «=» and the value, or between operands and its operators).

Rules for Functions and Methods

While a new line separates instructions, an empty line signalizes the beginning of a new logical context. It is necessary to separate logical, reusable code sequences into functional units to avoid endless and redundant «spaghetti» code. It depends on the designer how the decomposition of the code is handled, but very long function bodies hint at design deficiencies. Function and other control element bodies build a local scope, using parentheses. Therefore, it is useful to use the curly brackets for each block, even if it contains only one single instruction, for example, in an «if»-branch. It is also useful for «case»-branches of a «switch»/«case» condition to clearly mark the according instruction blocks.

Rules for the Indenting

There are several possible layout styles to position these parentheses and the according code block, using indenting. But coding experience has shown that the most intuitive and best readable way is, if parentheses and indents demonstrate clear code block boundaries (Vigenschow 2005, *l.c. page 34*). This means that each code block starts with the instruction (e.g., an «if»-statement or the «while»-condition), which is responsible for the new logical block, followed by the opening curly bracket in the same indent column. Then each following line of the code block uses an additional indent. Indents in this case are a sequence of a fixed number of blank symbols (no tabulator signs, as different editors replace them with different indent margins). Each new, logical block increases the number of used indent sequences, so that all instructions of a special branching depth have the same indentation and are in the same text column. The ending curly bracket is on an empty, new line again on the same indent level as the corresponding beginning one, to show the dependency on the starting bracket and to create a clear scope parenthesis.[17] Experience at the Wettzell observatory has shown that this style is the most intuitive. A sample can be found in Example 2.3.

Rules for Names

Another quite helpful specification deals with the naming of identifiers, like variable, constant, or function names. Even if naming is quite arbitrary, it is useful to define some conventions. A name should directly inform the programmer about the characteristics of a named element. It should be immediately apparent whether it is a locally defined type, a function, or a variable and which type of variable it is or even what the return value of a function is. Also combinations of variables, like arrays or higher sophisticated class types, should be derivable from the name.

Hungarian Notation

An early definition to support such ideas was given by the Hungarian notation of Charles Simonyi (Maguire 1993, *l.c. page xxv–xxviii*). It is an identifier naming convention for variables and functions. It defines name prefixes, which indicate the types or intended use of the named variable or function. As this notation helps to visually identify type mismatches, for example, in the code, it is quite useful. A frequently mentioned disadvantage of this notation is that an action like changing the type of a variable requires altering every aspect of its appearance, where the identifier is used. But exactly this obliges the user to check all locations manually for possible consecutive type mismatch errors which may result from the changed type. Therefore, it also gives hints about implicit type conversions[18] and supports type

---

17   When searching the Internet, the style type can be found under the following denotations: Allman style, «East Coast» style, BSD style, ANSI C style, or also Horstmann style.

18   A very famous type conversion error with catastrophic consequences happened during the launch of the ARIANE 5 rocket of the European Space Agency (ESA), which exploded during its maiden flight. The official report from the inquiry board about the «Flight 501 Failure» on

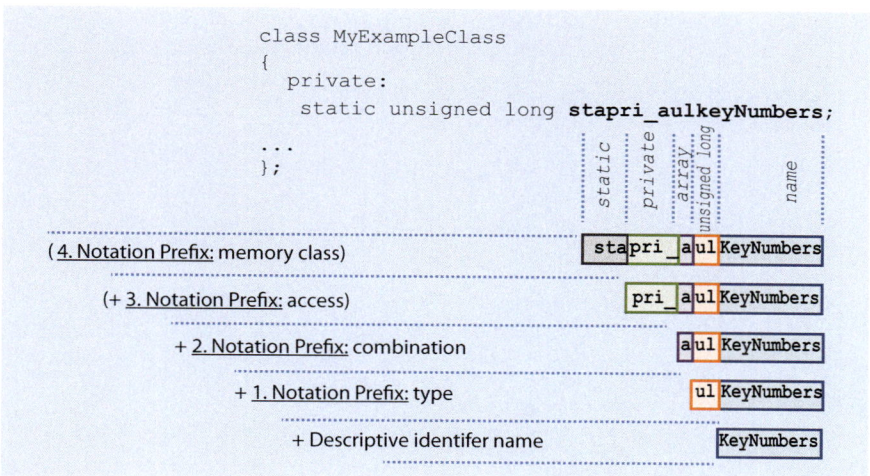

```
class MyExampleClass
{
  private:
    static unsigned long stapri_aulkeyNumbers;

  ...
};
```

(4. Notation Prefix: memory class) — `sta` `pri` `_` `a` `ul` `KeyNumbers`

(+ 3. Notation Prefix: access) — `pri` `_` `a` `ul` `KeyNumbers`

+ 2. Notation Prefix: combination — `a` `ul` `KeyNumbers`

+ 1. Notation Prefix: type — `ul` `KeyNumbers`

+ Descriptive identifer name — `KeyNumbers`

Fig. 2.5 Impression of a possible naming notation

inspections for audits (see ▶ Sect. 2.6.2). But the original notation does not support C++ elements and modern structures so well. Thus, it is reasonable to define a notation set as an extended character combination with adequate acronyms as a prefix to an already descriptive identifier. A possible prefix setup is shown in ▪ Fig. 2.5. The most important extensions are up to the second notation level, which defines the combination information. All higher extensions are more or less optional, even when the access information is quite useful to identify global variables, for example, by using «g_» (for more information about the notation, see ▶ Appendix A.2).

The complete, descriptive identifier follows the «Camel case» convention for the capitalizing of identifiers. This is one of the three most commonly used conventions. «Camel case» states that «the first letter of an identifier is lowercase and the first letter of each subsequent concatenated word is capitalized» (MSDN Microsoft n.d.), e.g., «aulKeyNumber» in the example of ▪ Fig. 2.5.

Camel Case

Another capitalizing convention is the «Pascal case», where all first letters of each subsequent concatenated words are capitalized, even the first one (MSDN Microsoft n.d.). This style is ideal for new type definitions, as, e.g., a class type «MyExampleClass», an enumeration type «SwitchStatesEnum», or a union type «ConverterUnion». In general, variable names and type names are a combination of nouns (e.g., «MinIndex» or «IndexCounter»), while function names should be an imperative combination of nouns and verbs. They describe the activity done within the function body (e.g., «GetFamilyName» which finds the family name or «SumIntegers» which sums up some integers).

Pascal Case

Besides «Camel case» and «Pascal case», the «Uppercase» convention is ideal for constants and include-guard definitions. As often suggested in classic program styles, constants should be completely written with capital letters (e.g., shown in (Kernighan and Ritchie 1990, *l.c. page 37 f.*)). The separation of the concatenated words can be done using the underscore «_». This convention can be used in the same way for «classic» constants, which are defined by preprocessor directives, and for constants, which are defined with the «const»-classifier or within an enumeration definition. Therefore constants may also optionally carry a type extension in

Upper Case

July 19, 1996, describes it like this: «The inertial SRI [= Inertial Reference System] software exception was caused during execution of a data conversion from 64-bit floating point to 16-bit signed integer value. The floating point number which was converted had a value greater than what could be represented by a 16-bit signed integer. This resulted in an Operand Error. The data conversion instructions (in Ada code) were not protected from causing an Operand Error, although other conversions of comparable variables in the same place in the code were protected» (ESA n.d. *l.c. page 4*).

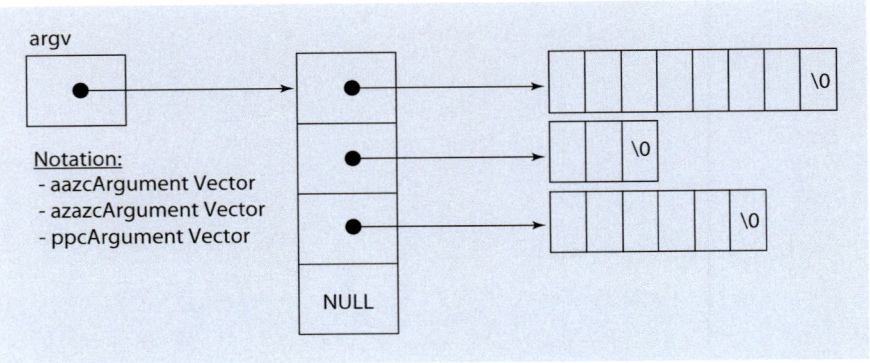

◘ Fig. 2.6   The «argv» vector of a C main function

Disadvantages of Coding Style Guides

the name if necessary. Samples for variable, function, and constant declarations can be found in the Examples 2.3, 2.4, and 2.5.

These naming conventions and style guide rules can give an idea of how style can support and necessitate readable, clear, and self-documenting code (see ► Sect. 2.5.1). But of course there are always pros and cons of such rules. The Hungarian notation and its modified versions in particular are constantly under discussion, as extensive use may reduce the clearness of the identifier naming and as there is always space for interpretations of how to use it. A good example of that controversial discussion can be shown with the argument vector, which is handed over to the main function of a C program (as in the definition «int main (int argc, char *argv[])» (Kernighan and Ritchie 1990, *l.c. page 111*)). It is a pointer to a fixed-length vector of pointers to the argument text strings, which are zero terminated (see ◘ Fig. 2.6). Every programmer normally knows the identifiers «argc» and «argv». If the above-described adaption of the Hungarian notation should be used, e.g., «argv» could be replaced by «aazcArgumentVector» (for an array of zero-terminated character arrays), by «azazcArgumentVector» (for a zero-terminated array of zero-terminated character arrays), or maybe also by «ppcArgumentVector» (for a pointer to pointers to characters). Those replacements are confusing to a greater or lesser extent. It is a matter of taste which version is preferred and if one of them is used. It is preferable to keep the style as simple as possible, in accordance with the general knowledge of programmers, of the team or of the institute. Even though the importance of such notations to avoid implicit type errors decreases when using automated testing and inspection tools in environments of continuous integration (see ► Sect. 2.8), the style guides make it easier for programmers to understand the code as long as they are used with sound common sense. Together with some coding policies which help to avoid classic software errors by following standardized coding rules, coding styles make code much more transparent and error-free, starting from the beginning of the code writing.

2.3.2   An Impression of Coding Policies

Policies for Codes

In general coding policies are rules which help to reduce coding lapses by giving hands-on advice, how specific problems can be or must be solved. Usually coding policies are a collection of numerous guideline rules, as mentioned above, for example, in the MISRA standards. The following section aims to give an impression of what may be defined in such rules in the case of the language C/C++ and how this could improve the code significantly. But these rule sets should be individually adapted to the dedicated coding purpose and software development process, as each rule may not have the same priority for different coding environments. More detailed information can be found in the guideline definitions itself or in appendix A, which may be used as a suitable, low-level, and condensed starting set of coding and design rules (mainly taken from the Wettzell design rules (Dassing et al. 2008)).

In the first instance, coding policies should define behaviors for the whole compiling workflow. They define the project structures in combination with details of the building process (e.g., with make). The use of the code internal preprocessor directives of C/C++ is also related to the compilation workflow. A very important point here is the avoidance of weak or multiple definitions. Usually functions are declared with their calling signature in the header files, which are then included as module interfaces to the source code files (see ▶ Sect. 3.1).

Compiling Workflow

A problem hereby arises if a module is included several times in a final program, so that a function is multiply defined. The compiler cannot assign which one should be used. To prevent such collisions, define or include guards can be used. They are simple macro definitions, which check if a dedicated macro name already exists. If it does not, the include statement is performed and the macro will be defined, so that the next check will run into a false state.

Include Guard

■■ **Example 2.2**
Define/include guard

```
#ifndef  __FILEIO__           1
#define  __FILEIO__           2
...  // Header code            3
#endif  // __FILEIO__         4
```

The coding policies at this level also define if and how the other preprocessing definitions should be used. Usually macros should be avoided as the compiler does not perform the higher code validation checks (like type conformance checks, checks of parentheses balancing, etc.) on them. Similarly, the handling of constants on the basis of preprocessing definitions should also be avoided. The use of the standardized «const» or «enum» declaration allows the compiler to define checkable data types in order to perform a safe initialization and type safety. Even the use of scoping blocks, which keep constants local, is possible. This simplifies the debugging significantly.

Preprocessor Defines vs. Compiler Checkable Defines

Other policies deal with variables and their memory use. For debugging issues, it is ideal to keep constants locally in a scope. Variables in particular should be kept as local as possible to avoid external changes or multiple definitions. Therefore, global variables should not be used. In cases of parallel multi-thread environments (see ▶ Sect. 2.6.4), static variables should also be avoided, as they need special protection, when altered by parallel threads. In the case of arrays like strings, only functions should be used which protect from index overruns. Therefore, functions such as «strncpy» should be chosen instead of «strcpy», as they have the maximum number of characters which can be copied as a function parameter to avoid overruns.

Scoping of Variables

Generally, variables should always be initialized with start values. This is not only convenient for local parameters in the memory stack, for example, in combination with index counters for arrays. It is also a must for pointer variables, pointing to an individual address in the memory heap.[19] As memory pointers can point to each address in the memory, a wrong or initially undefined value can point to anywhere and lead to unpredictable crashes. A consequent initialization of all variables all the time at the definition avoids such problems, even when some compilers cannot find unused variables anymore, as now all variables are touched at least once by the initialization. Some checking tools (see ▶ Sect. 2.6.3) also throw messages, using a consequent initialization, if the variable value is immediately overwritten again after the

Type Casting

---

19 The program memory can simply be separated into stack memory, which is allocated together with the memory for the program code when starting the program and heap memory, which is dynamically allocated during the runtime of the program, each time it is required.

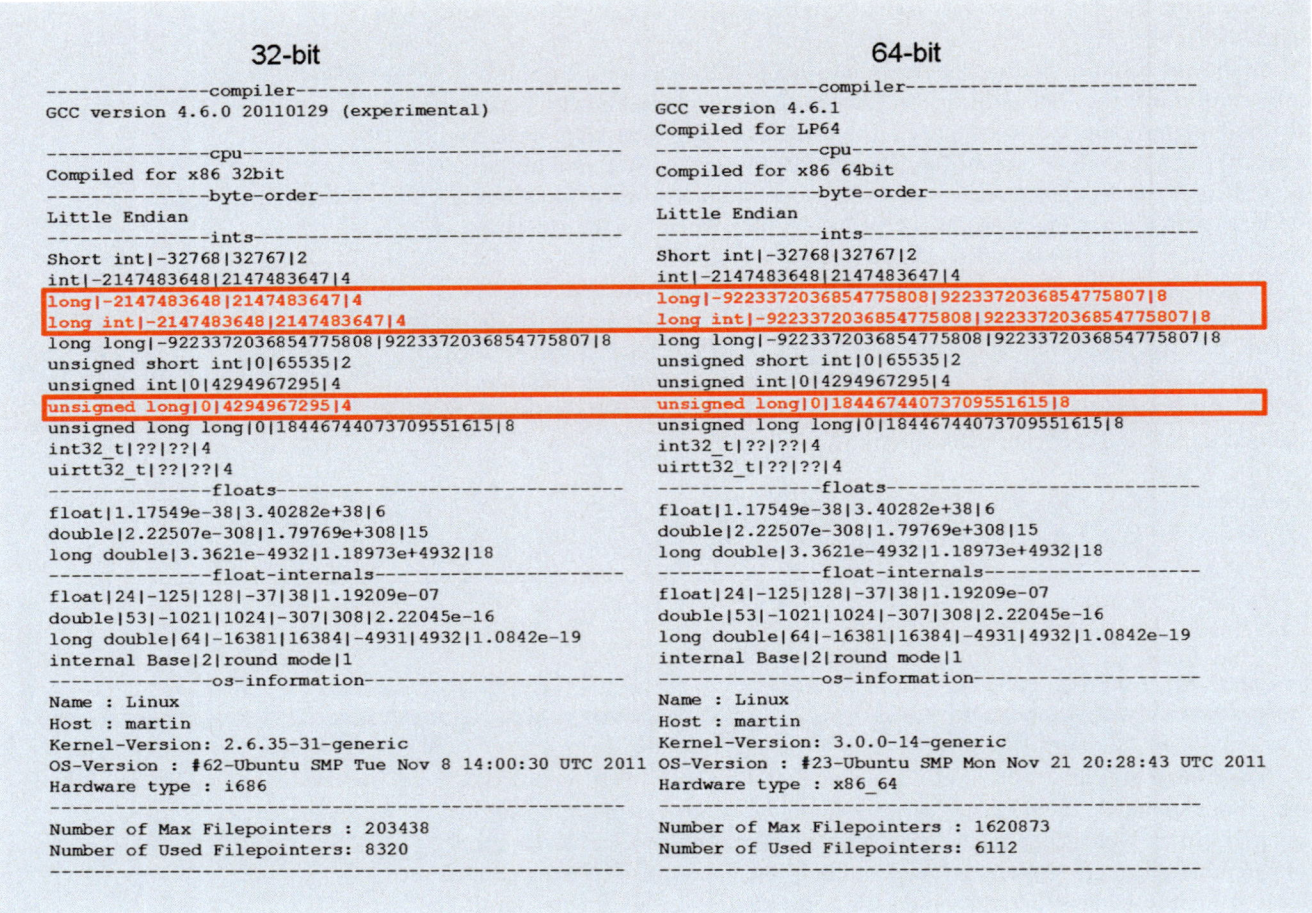

```
                 32-bit                                          64-bit
--------------compiler----------------          --------------compiler---------------
GCC version 4.6.0 20110129 (experimental)       GCC version 4.6.1
                                                Compiled for LP64
--------------cpu---------------------          --------------cpu--------------------
Compiled for x86 32bit                          Compiled for x86 64bit
--------------byte-order--------------          --------------byte-order-------------
Little Endian                                   Little Endian
--------------ints--------------------          --------------ints-------------------
Short int|-32768|32767|2                        Short int|-32768|32767|2
int|-2147483648|2147483647|4                    int|-2147483648|2147483647|4
long|-2147483648|2147483647|4                   long|-9223372036854775808|9223372036854775807|8
long int|-2147483648|2147483647|4               long int|-9223372036854775808|9223372036854775807|8
long long|-9223372036854775808|9223372036854775807|8   long long|-9223372036854775808|9223372036854775807|8
unsigned short int|0|65535|2                     unsigned short int|0|65535|2
unsigned int|0|4294967295|4                      unsigned int|0|4294967295|4
unsigned long|0|4294967295|4                     unsigned long|0|18446744073709551615|8
unsigned long long|0|18446744073709551615|8      unsigned long long|0|18446744073709551615|8
int32_t|??|??|4                                  int32_t|??|??|4
uirtt32_t|??|??|4                                uirtt32_t|??|??|4
--------------floats------------------          --------------floats-----------------
float|1.17549e-38|3.40282e+38|6                  float|1.17549e-38|3.40282e+38|6
double|2.22507e-308|1.79769e+308|15              double|2.22507e-308|1.79769e+308|15
long double|3.3621e-4932|1.18973e+4932|18        long double|3.3621e-4932|1.18973e+4932|18
--------------float-internals---------          --------------float-internals--------
float|24|-125|128|-37|38|1.19209e-07             float|24|-125|128|-37|38|1.19209e-07
double|53|-1021|1024|-307|308|2.22045e-16        double|53|-1021|1024|-307|308|2.22045e-16
long double|64|-16381|16384|-4931|4932|1.0842e-19   long double|64|-16381|16384|-4931|4932|1.0842e-19
internal Base|2|round mode|1                     internal Base|2|round mode|1
--------------os-information-----------          --------------os-information----------
Name : Linux                                    Name : Linux
Host : martin                                   Host : martin
Kernel-Version: 2.6.35-31-generic               Kernel-Version: 3.0.0-14-generic
OS-Version : #62-Ubuntu SMP Tue Nov 8 14:00:30 UTC 2011   OS-Version : #23-Ubuntu SMP Mon Nov 21 20:28:43 UTC 2011
Hardware type : i686                            Hardware type : x86_64
--------------------------------------          -------------------------------------
Number of Max Filepointers : 203438             Number of Max Filepointers : 1620873
Number of Used Filepointers: 8320               Number of Used Filepointers: 6112
--------------------------------------          -------------------------------------
```

Fig. 2.7 Type differences on 32-bit and 64-bit Ubuntu Linux machines show different lower and upper limits and a different byte size for «long» types

initialization. In combination with pointers and dynamically allocated memory, a similar re-initialization should be done after freeing the memory elements. After freeing an occupied memory, the pointer still points to the same address, which is then no longer reserved for the program, which means other programs can allocate it. A consecutive access can then lead to a program crash, which is automatically avoided when the pointer is reassigned with «NULL» after the freeing, as the wrong dereferencing of NULL-pointers can be handled by compilers.

Type Casting

Another important rule concerns type casting, where a variable of one specific data type is converted into a variable of another type. Implicit casts can lead to serious problems, as they may risk a cutoff of significant bits. Therefore, each cast must be an explicit one, done with a static cast or another explicit cast operator.

Type Compatibility of 32 Bits and 64 Bits

Another problem in combination with variables is 32-bit and 64-bit compatibility. As modern hardware architecture mandatorily requires 64-bit operating systems to address the whole memory, the 64-bit compatibility is also becoming increasingly important for scientific software. Another big issue is that programs can calculate with higher accuracy using «long long» or «long double»[20] and that the Unix time in seconds since January 1, 1970, at 00:00:00 o'clock does not end in the year 2038, when the maximum limit of representable numbers with «long» is reached. Figure 2.7 shows the output of byte sizes on a 64-bit and a 32-bit Ubuntu Linux system. A difference can be found for all «long» types,[21] as the byte size and

---

20 Not all 32-bit platforms support «long double».
21 Also the maximum number of file pointers is different, but this is not so relevant for programmers, if they cleanly close opened file pointers after use.

the limits are equal to «int» on 32-bit systems and equal to «long long» on 64-bit machines. As long as binary values are just local on one machine, there should be almost no problem if none of the critical type casts (conversions) are programmed. But if binary values are exchanged either over network or other serial or bus communications, or with binary file transfers, it is recommended not to use «long» at all. If directly «int» or «long long» is used (on systems which support these types), integer mismatches between 32 bits and 64 bits are mostly avoided. Another more professional way is to use official type includes such as «<sys/types.h>», «<inttypes.h>», or «<stdint.h>», where unique types are defined, for example, as «int64_t» for a 64-bit signed integer number or «uint32_t» for a 32-bit unsigned integer number. The advantage of these type definitions compared to the standard C types is that they are the same on all platforms.

Program Workflow

Finally, coding policies define elements of the program workflow. They can define how code should be split into logically dependent functions and how long one function is allowed to be. The standards define several such rules which reduce errors in the workflow, make it more manageable, more portable, or increase performance.

Jumps

A very important rule in almost all guidelines is the prohibition of jump statements such as «goto». Jump statements define code addresses to which the workflow can jump from almost anywhere in the workflow scope. Originally in unstructured languages, this was the only way to create loops to repeat code. But several of these jumps in a large code segment reduce the readability of the code quite significantly, as it is not intuitive to see all jump entry points and from where they originate. Therefore, in higher-level, problem-oriented languages, there are more sophisticated techniques, which do not lead to unreadable and incomprehensive program flows. Conditional «while»-loops or «for»-loops can replace jumps.

Try/Catch Exceptions

In general, this rule can also be extended to exceptions in OOP languages, like C++, as exceptions are nothing more than modern «jump» returns. Therefore «try/catch»-blocks in combination with an internal throwing of an exception should be avoided or «shall only be used for error handling» (MISRA (The Motor Industry Software Reliability Association) 2008, *l.c. page 144*) as defined in the MISRA C++:2008 guidelines. «[B]ecause of its ability to transfer control back up the call tree, it can also lead to code that is difficult to understand. Hence it is required that the mechanism is only used to capture behavior that is in some sense undesirable, and which is not expected to be seen in normal program execution *[Rule 15-0-1]*» (MISRA (The Motor Industry Software Reliability Association) 2008, *l.c. page 144*). The Wettzell design rules therefore only allow jumps and exception handling, if it is kept as local as possible within one scope. Additionally «goto» instructions are only allowed if they jump to a specific error return point at the end of a function (error forward propagation), so that a function just has one entry and one exit point.

Function Arguments

Other coding policies in this sense define the function arguments (function parameters) and the «const»-correctness of the function. For example, the code becomes much faster during runtime if no local copy of an argument value must be assigned after a function is called up. But the usual call-by-value paradigm (e.g., with an argument definition like this: «int iSummand») creates such a function internal copy. In OOP languages such as C++, it is possible to avoid this by using a constant call-by-reference, using an argument definition like this: «const int & iSummand». With this technique, the reference of the variable, which means the address of the original variable, is handed over. Unlike a standard call-by-reference, it is a constant address value, which cannot be changed within the function, without rule breaking.

«Const»-Correctness

The «const»-correctness of a whole function additionally tells the compiler that the function body of a class method does not change member variables (attributes) of the class. OOP languages allow this protection by adding the keyword «const» at the end of the function declaration.

Conditional Statements

Coding policies for the program workflow also deal with errors caused by the evaluation order of expressions. They offer help to avoid general errors, as, for example, those given with undesired assignments in «if»-conditions, where a misused

2

Evaluation Order of Operators

assignment with «=» instead of a comparison with «==» does not produce a compiler error or warning, but changes the program logic dramatically. Therefore, coding policies suggest writing comparisons in «if»-conditions so that the constant is always compared with the variable part, e.g., «if (5 == iIndex)». Using this, the compiler would detect the misused assignment, as nothing can be assigned to a constant.

As the priority and associativity of operators is completely defined in languages like C or C++, but the order of evaluation of expressions is not (except in a few special cases like «&&», «||», «?:»-condition and comma operator (Kernighan and Ritchie 1990, *l.c. page 52*)), a sequence of terms in an expression with different operators can have disagreeable side effects. This means that if the definition of an operator does not guarantee the order of evaluation, it is dependent on the implementation (Kernighan and Ritchie 1990, *l.c. page 193*). In general it is advisable to add dedicated parentheses to prioritize the terms of an expression, as the intuitive code «*piCurrentSolutionNumber++;» does not add one to the value at the dedicated memory address, but changes the pointer and does the dereferencing afterward. The reason for this is that the iteration «++» and the dereferencing «*» are evaluated from right to left. This is different to other operators, which are usually evaluated from left to right (Kernighan and Ritchie 1990, *l.c. page 52*). The combination of an increment operator «++» and similar operators with other operators has further hidden difficulties, as, for example, «++» changes values implicitly without a consideration of possible side effects. The expression «aiArray[iIndex] = iIndex++;» does not define whether the old value of «iIndex» is assigned or the new increased value is used. The order of the evaluation of this expression is not defined by the standard. The same behavior can be expected for function parameters, where «printf («%d %d\n», ++iIndex, aiArray[iIndex]);» can lead to different results with different compilers (Kernighan and Ritchie 1990, *l.c. page 52 f.*).

Policy Dependencies

The circumstances of coding policies are usually also dependent on the development process and its steps of quality evaluations. Other dependencies must be taken into account for the environment or platform for which the program is developed. Two samples illustrate these dependencies. The environment, in which the program should run is particularly important, as it drives the level of coding rules. There is a big difference if code should only run in sequential, linear environments or in parallel or multi-threading systems (see ► Sect. 2.6.4). Parallelism, either in the form of parallel threads within a program process, such as parallel processes, or as parallel programs on different processing cores, usually requires additional knowledge and coding rules if the data and memory sets are shared.

Policies with STL

A big issue here is to avoid the hidden side effects. The use of different STL implementations in particular is prone to such pitfalls. An example is the use of the «std::string»-class, which uses different ways of copy-on-write and reference counting. Because of performance issues, most of the implementations use an internal way to minimize the number of value copies. Copying strings takes a lot of time and is often not necessary if just a read-only copy is created. Therefore, the libraries use a method which copies data only if a write access happens (copy-on-write), so that «two strings share the same data until one of them tries to write to it. At that point, a copy of the data is made, and the two strings go their separate ways» (Oracle Sun Developer Network (SDN) 2000). Internally the references to the shared memory are reference counted. The counter increments and decrements are under special lock protections. But the copying itself is not protected and is usually not performed at the position where the assignment is programmed. This real assignment happens later if the contained string gets manipulated sometime at some time during the processing. This is a critical situation for multi-threaded environments, as changes and read access to the contained string can be unprotected in a parallel way.

Deep Copy

Coding policies must define additional rules for parallel workflows. A possible solution is to forbid the use of such non-thread-safe libraries. Another solution is to define workarounds. A possible workaround for the above string problem could be to only allow deep copy assignments instead of the usual shallow copy.[22] Deep copying always forces the compiler optimization to perform a real copy, even when this requires more performance and time for an assignment. Instead of writing «`strCopyOfName = strName;`» to copy strings, it is thread-safe to write «`strCopyOfName = strName.c_str();`». It reads the string which should be copied as C string, which forces an implicit typecast. This calls up the dedicated assign operator in the STL string class and leads to a real copy without the reference counting problem. Experience at Wettzell has also shown that deleting a string should also not be done, by using «`strCopyOfName = "";`». Instead it is better to define an empty string by using the implicit creation in the standard constructor («`std::string strEmptyString;`») and assigning this one to the string which should be emptied using the deep copy method («`strCopyOfName = strEmptyString.c_str();`»). Similar techniques must be considered if strings are used as function parameters or for the other container classes of the STL library. Experience has shown that the performance on modern machines for the usual string use is acceptable, but there is a considerable advantage in using STL strings on multi-threaded environments.

The paragraphs above show that coding policies can and should always be adapted to specific circumstances. Another adaption can be made, according to the quality evaluations done during the software development. If there are almost no standardized evaluation and test steps during the software development, some rule sets, as suggested in (Vigenschow 2005, *l.c. page 41*), can help to prevent errors while writing and compiling the code. A first example was already mentioned in combination with conditional statements, where the right order of constants and variables in a condition uncover misused assignment operations.

Exploitation of Compiler Abilities

Compared to a more sophisticated way of developing code as defined in strategies like the CI (see ▶ Sect. 2.8), such errors can be detected by automated test environments, which are operated automatically each night or after each release of a new version. These scenarios define several checks with static code analysis, build tests, or automated unit tests, which use internally defined code metrics and error definitions. In such development environments, errors like the one described can be found very easily by the static code checkers. Other errors which change the code behavior can be found with unit tests. Therefore, the coding policies can be reduced, as the metrics are standardized and checked by dedicated utility programs. More about such a setup can be found in ▶ Sect. 2.8 and following.

Quality Profiling

To conclude the section about coding layout and coding policies, it should be taken into account that such rules help programmers to write self-documenting, stable, portable, optimized, and error-reduced code. An important factor in the code development is to detect errors as early as possible. The design and the checks of the design, as defined in the CI (which will be discussed later), help to detect errors and lapses earlier on during the development and not later during the integration test or while the software is already in use. The major goal of such rule sets should be to offer the highest possible level of code quality from the beginning. But the restriction should also be that the rule sets should be slim and sufficiently understandable for the usual developers or scientific code writers to keep them in mind. Methods like the CI can give coders a helping hand to improve their code, by checking metrics automatically and giving helpful descriptions and hints for solution options. But in general, the use of such rules is always a political and psychological question, as a definition is worthless if nobody cares about it in the team and if there are no useful structures to implement it in the development groups. Therefore, only such quality rules should be intended which can be inspected by the analysis tools. All others are basically just weak recommendations.

Early Code Quality

---

22 «A shallow copy means that any reference-type data members in the [... copy] will refer to the same object as the equivalent reference-type data members in the original object. A deep copy means that [...] the entire object graph [... is copied] so that the reference-type data members of the [... copy] refer to physically independent copies (clones) [...]» (Jones and Freeman 2010, l.c. page 627).

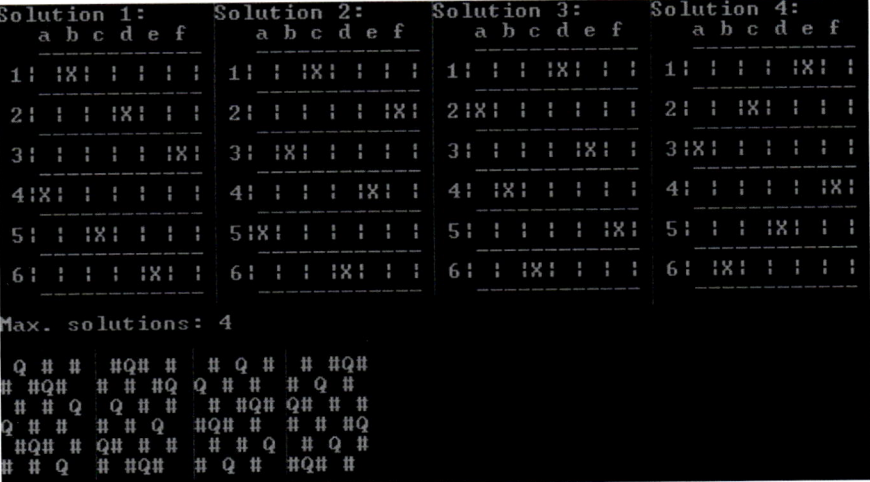

Fig. 2.8 Example of printed solutions from the more readable version of the code (first row) compared to the hack from the 7th International Obfuscated C Code Contest (second row); the original printouts from the programs are vertically arranged

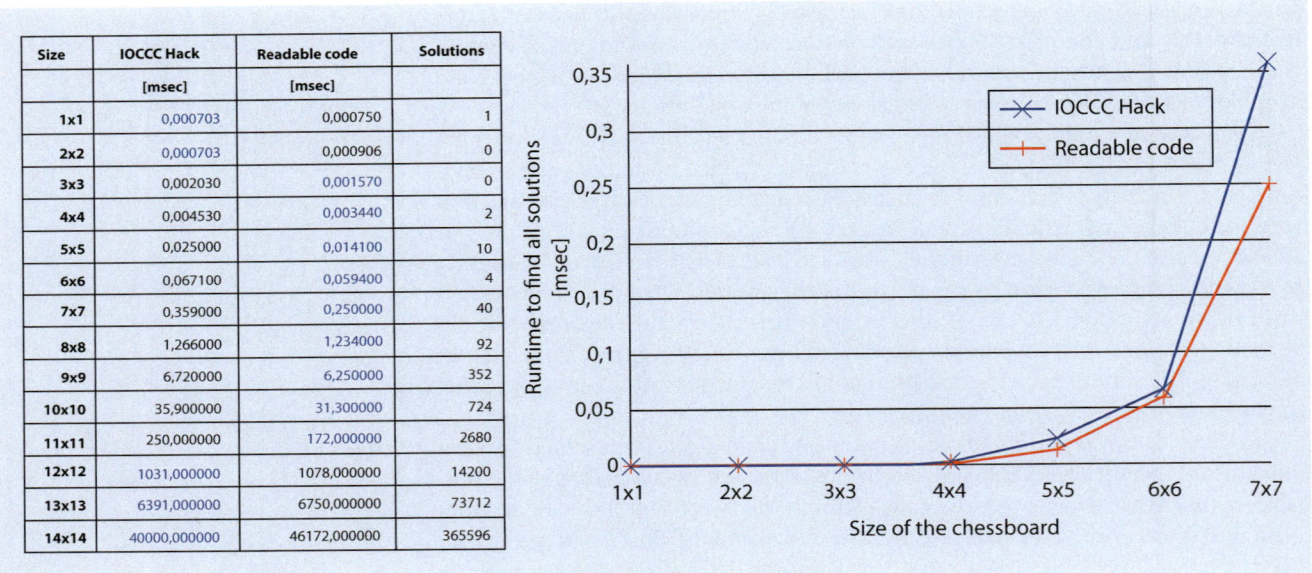

Size	IOCCC Hack	Readable code	Solutions
	[msec]	[msec]	
1x1	0,000703	0,000750	1
2x2	0,000703	0,000906	0
3x3	0,002030	0,001570	0
4x4	0,004530	0,003440	2
5x5	0,025000	0,014100	10
6x6	0,067100	0,059400	4
7x7	0,359000	0,250000	40
8x8	1,266000	1,234000	92
9x9	6,720000	6,250000	352
10x10	35,900000	31,300000	724
11x11	250,000000	172,000000	2680
12x12	1031,000000	1078,000000	14200
13x13	6391,000000	6750,000000	73712
14x14	40000,000000	46172,000000	365596

Fig. 2.9 Simple benchmark between the hack from the 7th International Obfuscated C Code Contest and the more readable alternative to measure the runtime of finding all possible solutions for different chessboard sizes without printing the results

Example Code, Using Styles, and Policies

To give an illustration of a better written code, the following code example also solves the N-queens puzzle, but in a more readable, intuitively comprehensible and almost self-documenting way.[23] An example of the printed solutions of both programs for a 6 × 6 chessboard is shown in ▪ Fig. 2.8. ▪ Figure 2.9 illustrates that performance is not a matter of whether to «hack» short code or not. It is more or less dependent on the algorithm and the compiler optimizations used, which can also be implemented in the form of a readable, self-documenting code, which is much better for code maintenance and code reusability.

---

23 The code is based on a solution, suggested by (Werner n.d. *l.c. page 249 ff.*) but is extended with substantial modifications.

▪▪ Example 2.3

Self-documenting coding layout for the N-queens puzzle recursive backtracking algorithm

```
int iCheckIfQueenPositionIsSave (int iCurrentQueenNumber,                      1
                                 QueenPositionStructType * pSQueenPositionsList) 2
{                                                                              3
    int iListIndex = 0;                                                       4
    for (iListIndex = 0; iListIndex < iCurrentQueenNumber ; iListIndex++)     5
    {                                                                         6
        if ((pSQueenPositionsList[iListIndex].iRow ==                         7
                      pSQueenPositionsList[iCurrentQueenNumber].iRow) ||       8
            (pSQueenPositionsList[iListIndex].iCol ==                         9
                      pSQueenPositionsList[iCurrentQueenNumber].iCol) ||      10
            ((pSQueenPositionsList[iListIndex].iRow +                        11
              pSQueenPositionsList[iListIndex].iCol) ==                      12
                      (pSQueenPositionsList[iCurrentQueenNumber].iRow +      13
                       pSQueenPositionsList[iCurrentQueenNumber].iCol)) ||   14
            ((pSQueenPositionsList[iListIndex].iRow −                        15
              pSQueenPositionsList[iListIndex].iCol) ==                      16
                      (pSQueenPositionsList[iCurrentQueenNumber].iRow −      17
                       pSQueenPositionsList[iCurrentQueenNumber].iCol))      18
            )                                                               19
        {                                                                   20
            return 0;                                                       21
        }                                                                   22
    }                                                                       23
    return 1;                                                               24
}                                                                           25
                                                                            26
void vFindNextQueenPosition (int iChessBoardSize,                           27
                             int iCurrentQueenNumber,                       28
                             int * piCurrentSolutionNumber,                 29
                             QueenPositionStructType * pSQueenPositionsList) 30
{                                                                           31
    QueenPositionStructType SNewQueenPosition = {−1, −1};                   32
                                                                            33
    if (iCurrentQueenNumber == iChessBoardSize)                             34
    {                                                                       35
        (*piCurrentSolutionNumber)++;                                       36
        vPrintChessBoard (*piCurrentSolutionNumber,                         37
                          pSQueenPositionsList,                             38
                          iChessBoardSize);                                 39
    }                                                                       40
    else                                                                   41
    {                                                                       42
        SNewQueenPosition.iRow = iCurrentQueenNumber;                       43
        for (SNewQueenPosition.iCol = 0;                                    44
             SNewQueenPosition.iCol < iChessBoardSize;                      45
             SNewQueenPosition.iCol++)                                      46
        {                                                                  47
            pSQueenPositionsList[iCurrentQueenNumber]= SNewQueenPosition;  48
                                                                            49
            if (iCheckIfQueenPositionIsSave (iCurrentQueenNumber,          50
                               pSQueenPositionsList))                      51
            {                                                              52
                vFindNextQueenPosition (iChessBoardSize,                  53
                               iCurrentQueenNumber+1,                     54
                               piCurrentSolutionNumber,                   55
                               pSQueenPositionsList);                     56
            }                                                             57
        }                                                                58
    }                                                                    59
}                                                                        60
```

■ ■ **Example 2.4**
Self-documenting coding layout for the N-queens puzzle main function

```
#include <stdio.h>                                                              1
#include <stdlib.h>                                                             2
#include <string.h>                                                            3
                                                                                4
enum {LISTSIZE = 99};                                                          5
typedef struct QueenPositionStruct                                             6
{                                                                               7
    int iRow;                                                                   8
    int iCol;                                                                   9
} QueenPositionStructType;                                                     10
                                                                               11
void vPrintChessBoard (int iCurrentSolutionNumber,                             12
                       QueenPositionStructType * pSQueenPositionsList,          13
                       int iChessBoardSize);                                   14
int iCheckIfQueenPositionIsSave (int iCurrentQueenNumber,                      15
                                 QueenPositionStructType * pSQueenPositionsList); 16
void vFindNextQueenPosition (int iChessBoardSize,                              17
                             int iCurrentQueenNumber,                          18
                             int * piCurrentSolutionNumber,                    19
                             QueenPositionStructType * pSQueenPositionsList);   20
                                                                               21
int main (int argc, char * argv[])                                            22
{                                                                              23
    int iChessBoardSize = 8;                                                  24
    int iDigitIndex = 0;                                                      25
    int iListIndex = 0;                                                       26
    int iCurrentSolutionNumber = 0;                                           27
    QueenPositionStructType aSQueenPositionsList[LISTSIZE];                    28
                                                                              29
    for (iListIndex = 0; iListIndex < LISTSIZE; iListIndex++)                 30
    {                                                                         31
        aSQueenPositionsList[iListIndex].iRow = -1;                           32
        aSQueenPositionsList[iListIndex].iCol = -1;                           33
    }                                                                         34
                                                                              35
    if (argc == 2)                                                           36
    {                                                                        37
        for (iDigitIndex = 0; iDigitIndex < strlen (argv[1]); iDigitIndex++) 38
        {                                                                    39
            if (!isdigit(argv[1][iDigitIndex]))                              40
            {                                                                41
                break;                                                       42
            }                                                                43
        }                                                                    44
        if (iDigitIndex == strlen (argv[1]))                                45
        {                                                                    46
            iChessBoardSize = atoi(argv[1]);                                 47
        }                                                                    48
    }                                                                        49
    if (iChessBoardSize < 1 ||                                              50
        iChessBoardSize > LISTSIZE)                                          51
    {                                                                        52
        iChessBoardSize = 8;                                                 53
    }                                                                        54
                                                                            55
    vFindNextQueenPosition (iChessBoardSize,                                56
                            0,                                              57
                            &iCurrentSolutionNumber,                        58
                            aSQueenPositionsList);                          59
                                                                            60
    printf ("Max. solutions: %d\n", iCurrentSolutionNumber);               61
                                                                            62
    return 0;                                                              63
}                                                                          64
```

Self-documenting coding layout for the N-queens puzzle printing function for the chess board

```
void  vPrintChessBoard (int  iCurrentSolutionNumber ,            1
                        QueenPositionStructType * pSQueenPositionsList ,   2
                        int  iChessBoardSize )                    3
{                                                                4
    int  iRowIndex = 0;                                          5
    int  iColIndex = 0;                                          6
                                                                 7
    printf ("Solution %d:\n", iCurrentSolutionNumber );          8
                                                                 9
    printf ("   ");                                             10
    for (iColIndex = 0; iColIndex < iChessBoardSize; iColIndex ++)   11
    {                                                           12
        printf ("%c ", (char) ('a'+iColIndex));                 13
    }                                                           14
    printf ("\n");                                              15
    printf ("   ");                                             16
    for (iColIndex = 0; iColIndex < iChessBoardSize *2; iColIndex ++)   17
    {                                                           18
        printf ("—");                                           19
    }                                                           20
    printf ("\n");                                              21
    for (iRowIndex = 0; iRowIndex < iChessBoardSize; iRowIndex ++)   22
    {                                                           23
        printf ("%2d|", iRowIndex +1);                          24
        for (iColIndex = 0; iColIndex < iChessBoardSize; iColIndex ++)   25
        {                                                       26
            if (iColIndex == pSQueenPositionsList [iRowIndex ].iCol )   27
            {                                                   28
                printf ("X|");                                  29
            }                                                   30
            else                                                31
            {                                                   32
                printf (" |");                                  33
            }                                                   34
        }                                                       35
        printf ("\n   ");                                       36
        for (iColIndex = 0; iColIndex < iChessBoardSize *2; iColIndex ++)   37
        {                                                       38
            printf ("—");                                       39
        }                                                       40
        printf ("\n");                                          41
    }                                                           42
    printf ("\n");                                              43
}                                                               44
```

2.4 Including Legacy Code

Developing software systems does not only mean writing completely new code. Normally there are many existing code lines, modules, or programs available which have been developed and used for years. Many algorithms and tasks are already solved in such software packages. Therefore, it saves time (and costs) to use existing code. But these lines of code include some hidden risks: the developers are often not available anymore, the internal behavior is not well understood, the code normally follows a different style guide (if there is one), and code tests are frequently missing. This code is generally defined as legacy code.

Given Situation and Risks

Legacy Code

> *Definition DEF2.5. of «legacy code»* (see (Feathers 2011, *l.c. page 16*))
> Legacy code is no longer supported («old») code or code from an external source with structures and behaviors which are not entirely clear. This code normally also has a different programming style and reduced or missing tests.

Types

Dealing with such legacy systems (existing program systems) is one of the big challenges for new developments. In scientific environments, many such code modules are available, which solve dedicated tasks, for example, those developed within former research projects. For the further use of such existing code elements, it is necessary to separate them into two parts, which should be handled in different ways:

— Existing, no longer supported, «old» code
— External code from other still active developers and projects

External Code Libraries

External code should be used without any local changes in the external sources to avoid discrepancies between releases and the local copy. Errors found or requests for needed additional functions should always be forwarded to the original developers or should be managed in close cooperation with them. The original code repository will then always be consistent with one's own copy. This goal might be difficult in some cases, if the support from the external developer community is not good. Therefore it is always advisable to only take code from very active and open communities. It is also important to use code libraries with clear developer structures and which are well documented. Another aspect which should be considered is the licensing of the external code, which must be compatible with the licensing of one's own code. Verifying this is not easy, as it is necessary to know the legal rights in detail. Considering all of these side effects, it is advisable to reduce dependency on extrinsic code to a minimum.

Converter Classes for External Code

But, in general, it is not useful to only allow self-written code. Therefore, external sources are quite helpful. If these are already offered in the form of a library, the functional entry points to the library can be directly used as an interface to the external code. But in one's own application architecture, they should be represented as separate and independent components or modules. To avoid a strict dependency (e.g., if the library were replaced someday, the call up of library functions at different program locations would make it difficult), it is useful to encapsulate the external code behind a converter class or module (Oestereich 2005, *l.c. page 164*). Then only one's own interface functions can be used, which hide the external functionality. This technique is also necessary if the extrinsic code is just a loose set of functions or to include code written in different programming languages. It is even possible to mix up languages of different programming paradigms.[24]

Cover and Modify

In the case of «old», no longer supported, but still useful code, the situation is slightly different. It is possible to reuse the code in a rearranged architecture. Therefore, it is not necessary to rewrite the code parts. But it might be quite helpful to just separate the essential functions and maybe to rearrange the internal structure without big changes in the code itself. Normally a very conservative form of risk management is exemplified by the phrase «never touch a running system». But avoiding changes in existing code structures often leads to architectures in which it becomes much more difficult to make changes. This also leads to unclear behaviors and unclear structures. To reduce the fear of change, it is necessary to clearly define which changes are necessary and how to

---

24 The situation is different if a complete package is developed with an external framework, such as a framework for graphical user interfaces. Then the code in question is directly related to that code with all its advantages and disadvantages. But then it makes no sense to write converter.

check whether the changes are correct and will not do any damage to the functionality (Feathers 2011, *l.c. page 32*). A functional way is «Cover & Modify», where a given code structure is covered with suitable tests (Feathers 2011, *l.c. page 33*).

To follow this approach, it is again sensible to encapsulate the given, «old» functionality within a converter class or module. On this basis, it is possible to write appropriate tests for the functional units offered by the new software component, in which the «old» features are hidden. Having these test cases, the «unit tests» (see ▶ Sect. 2.6 about testing), even external code can be reorganized and changed, because running the test will check its correct functionality hidden behind the converter component. This reorganization of software systems is usually called «refactoring».

Unit Tests and Refactoring

> **Definition DEF2.6. of «refactoring»** (see (Osherove 2010, *l.c. page 38*))
> Refactoring means changing an existing code without changing its functionality to improve quality factors such as maintainability, readability, changeability, usability, testability, reusability, interoperability, and modularity (renaming a function or separating a large method into several smaller ones is a typical example of refactoring).

Refactoring

A three-step plan is quite helpful to use with existing legacy code in the form of «Cover & Modify»:

Three Steps of Use

1. Encapsulation
2. Unit testing
3. Refactoring

Using the technique of encapsulation gives several possibilities of combining different sources of compiler languages. Herein it is always simpler to incorporate more easily structured code paradigms into highly sophisticated ones. For example, the use of C code in C++ sources is quite straightforward. In the case of C and C++, it is normally always possible to compile C sources with a C++ compiler, as the C language is completely included in C++. But to use the according language standards for the different languages, it is better to use the specific compilers separately and link the resulting object codes to a final executable afterward. In this case, it is only necessary for the intermediate, internal compiler, and linker lookup tables for the code identifiers to be compatible. Therefore, a dedicated specifier with the keyword «`extern "C"`» must be used to define a block of C code within a C++ environment. It makes sure that the C parts are compiled using the C-specific lookup table naming conventions, even when the compilation is done with a C++ compiler.

Encapsulation of C in C++

It is currently recommended that this classification is used in each C file directly, so that it can be used without any linker errors in mixed environments. But to avoid compiler errors of pure C compilers, the additional «`__cplusplus`» definition should be added. It takes care to just include the specifier for external C code, if a C++ compiler is used. A stable way to include C code in C++ is shown in ◘ Fig. 2.10.

It is much more difficult to use C++ code within a C environment if the according compilers for each language are used and no C++ syntax is allowed to contaminate the C sources. With the converter technique, it is also possible to solve this problem. In a similar manner to the description about how to use «old» code, a converter module covers the C++ code with the OOP constructs. Equivalent functions in the converter translate the C calls into object-oriented calls of methods in

Encapsulation of C++ in C

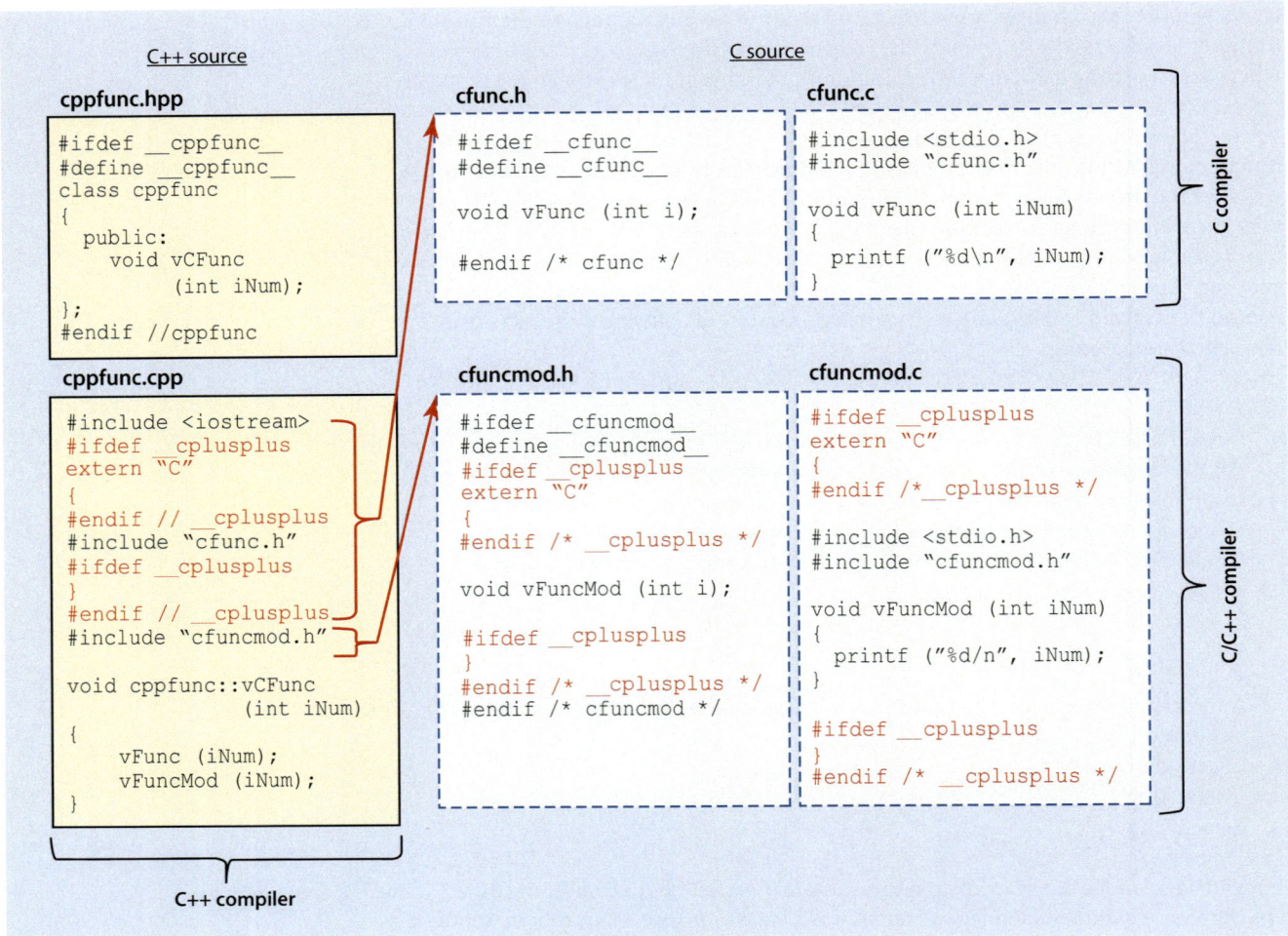

2

Fig. 2.10   A C to C++ conversion in two ways: without changing the sources and restrictions for the compiler use or with a more flexible adaption of the sources, using the keyword «`extern "C"`»

the class (see ◘ Fig. 2.11). The converter module also does all of the memory handling, such as creating and destroying object elements of a class. One possible way of doing this is by using a static, global converter variable of the class type, which can be used as a single realization of the class. But a disadvantage of this approach is that there is only one instance of the class which is shared among the callers. Such a setup is usually implemented with a singleton design pattern[25] and requires specific care in multi-threading environments (see ▶ Sect. 2.6.4), when several parallel tasks can compete for function calls.

Thread-Safe Encapsulation

Therefore, it is better if the converter module also offers a construction (or creation) and a destruction function as pendant to the constructor and destructor of the class, which are encapsulated there. The construction function creates an object of the class in the dynamic heap memory and returns a pointer to it, which is cast into a pointer of the type «`void *`». This pointer is something like a descriptor for the converted functionality, similar to a file descriptor. Each converter function uses this descriptor as function argument to identify the individual object in the

---

25 A singleton pattern is a classic OOP design pattern, which uses an improved global variable to prevent the creation of a secondary object of the same type, so that there is always only one instance of the class available in an application (Alexandrescu 2003, *l.c. page 181*).

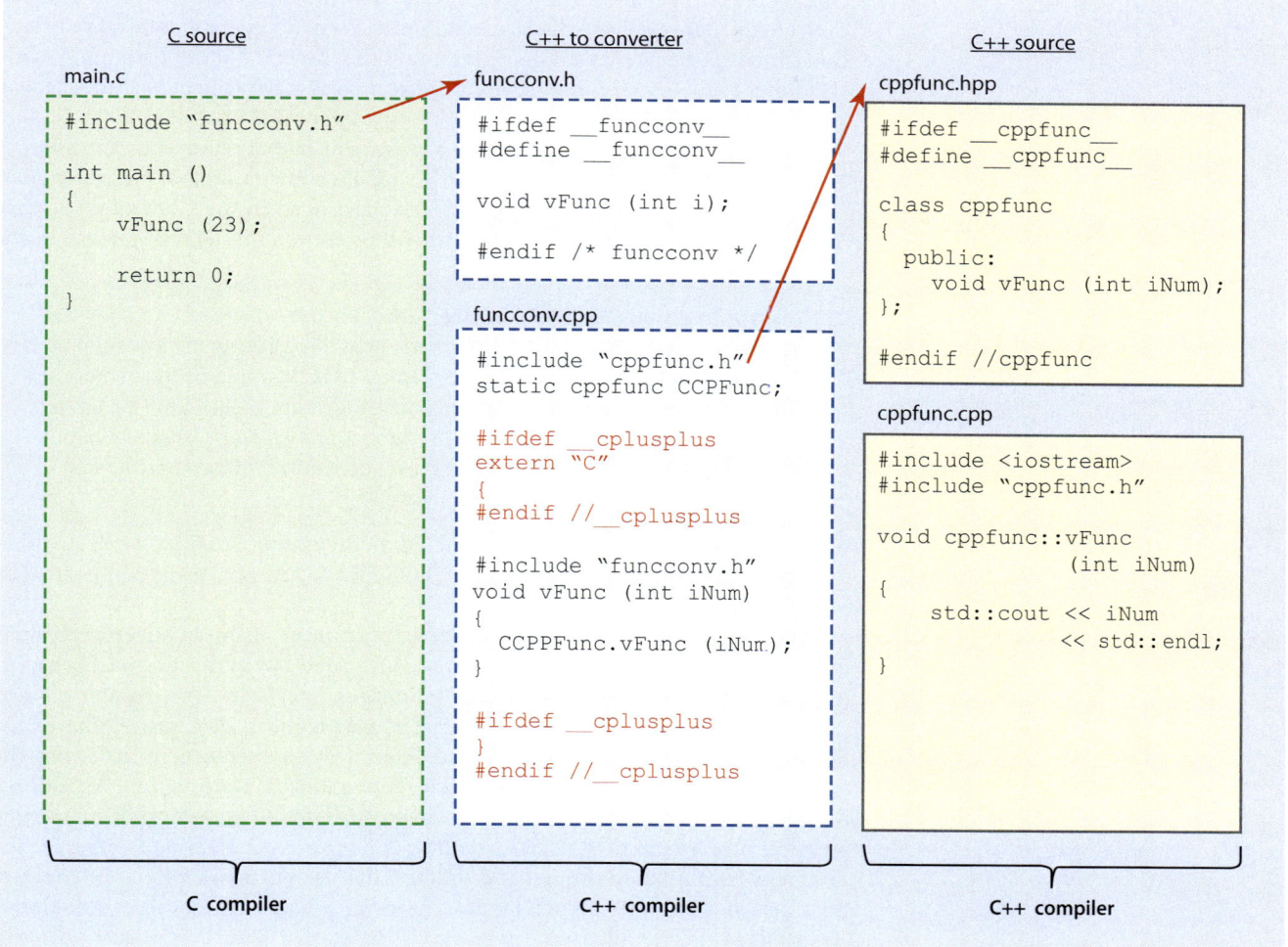

Fig. 2.11 A C++ to C converter which makes it possible to call up C++ code in a C environment using the encapsulation technique

heap memory. This generalized pointer is type-cast back to the original type again in the converter function, so that it can be used to call class methods of the individual object implementation. It is important to explicitly take care of NULL-pointers in combination with such descriptors. But by using this technique, it is possible to represent different constructors, to simulate the copy assignment operator, and to implement individual, separated, and parallel used object variables of the same class (in multi-threading environments). The only restrictions are that the «void *» pointer should only be changed within the converter class (similar to the C file pointer, which is also only used in the print and read functions) and that each source should be kept clean and consistent either as C or as C++ code and not be mixed (e.g., also do not use «//» comments in C sources).

Using the converter technique, it is possible to combine different compiler languages. As most of the legacy sources in scientific environments are written in FORTRAN, the following blocks explain what is necessary to combine C/C++ and FORTRAN. As the new software should use C++ and C, only the inclusion of FORTRAN in C/C++ is considered.

Encapsulation of Other Languages

To do an encapsulation of FORTRAN, several projects use the converter tool «f2c». «f2c» was launched in the early 1990s and is an automatic FORTRAN77 to C converter. It is still very well known as it allows the use of tested FORTRAN code by

FORTRAN Encapsulation with f2c

simply converting it into C. As C offers to express constructs which are not possible in FORTRAN77 (as «storage management, some character operations, arrays of functions, heterogeneous data structures, and calls that depend on the operating system» (Feldman et al. 1995, *l.c. page 1*)), the conversion is a convenient way to «create a portable C program that exploits Fortran source code» (Feldman et al. 1995, *l.c. page 1*). But although the developers «tried to make f 2c's output reasonably readable, [...] [the] goal of strict compatibility with f 77 implies some nasty looking conversions [...]» (Feldman et al. 1995, *l.c. page 2*), also mentioned in the Computing Science Technical Report No. 149 of AT&T Bell Laboratories. The following parts in the report continue with

> » Input/output statements, in particular, generally get expanded into a series of calls on routines in libI77, f 77's I/O library. Thus the C output of f 2c would probably be something of a nightmare to maintain as C; it would be much more sensible to maintain the original Fortran, translating it anew each time it changed. Some commercial vendors [...] perform translations yielding C that one might reasonably maintain directly [...] [but these] generally require some manual intervention. (Feldman et al. 1995, *l.c. page 2*)

Alone this makes it difficult for modern, maintainable architectures. Another aspect is that the new standards of FORTRAN are not really supported by the tool.

FORTRAN Encapsulation with the Converter Module f77.h

It is much better to follow the converter technique, using a converter module, as described earlier. A very useful and sophisticated but in most cases disproportionate realization is the conversion definitions in «f77.h». This header file was created as part of the Starlink Project for astronomical data processing of the Joint Astronomy Centre, Hawaii. The project was shut down in 2005, but the software is still available under the terms of open source (Starlink Joint Astronomy Centre n.d.). The header file contains macros needed to write portable code in a mixed environment of FORTRAN and C. The macros can be used to organize the different naming conventions and linkage, the use of call-by-reference parameters, and the conversion of data types, character strings, logical values, and pointers to dynamically allocated memory (Allan et al. 2008). It is a very powerful converter module. An example of its use in combination with the encapsulation design is shown in ◘ Fig. 2.12 (Allan et al. 2008, *l.c. page 3 f.*). A big advantage of the encapsulation technique besides the ideal possibility for unit tests is that the FORTRAN dependency is not visible in the central C/C++ sources. It is covered by the converter module. The compilation and linking are done by taking the following steps:

1. `f77 -c ffunc.f` (compilation of the FORTRAN source; also «g77» or for newer GNU compilers «gfortran» can be used)
2. `gcc -c funcconv.c` (compilation of the converter source in C)
3. `gcc -c main.c` (compilation of the main source in C)
4. `gcc -o test ffunc.o funcconv.o main.o` (linking of the single object files; if «g77» is used, it is necessary to link the library «libg2c» using «-lg2c» as argument of the linker call to link machine code with FORTRAN capabilities that are not generated by «g77»)

Rudimentary Rules for a FORTRAN Encapsulation

Of course it is possible to combine C/C++ with FORTRAN also in a more rudimentary way as long as the basic differences are considered, which can be found in (Allan et al. 2008) as well as in (Beebe 2001) or (Ippolito 2008). The following lines are a compressed mixture of the solutions to solve the differences between the languages, mentioned in these papers.

FORTRAN Encapsulation of Standard Types

The situation becomes much easier if only the standard types are exchanged and no logical values or pointers are used. The standard types have a direct representation in both languages, as shown in the following table:

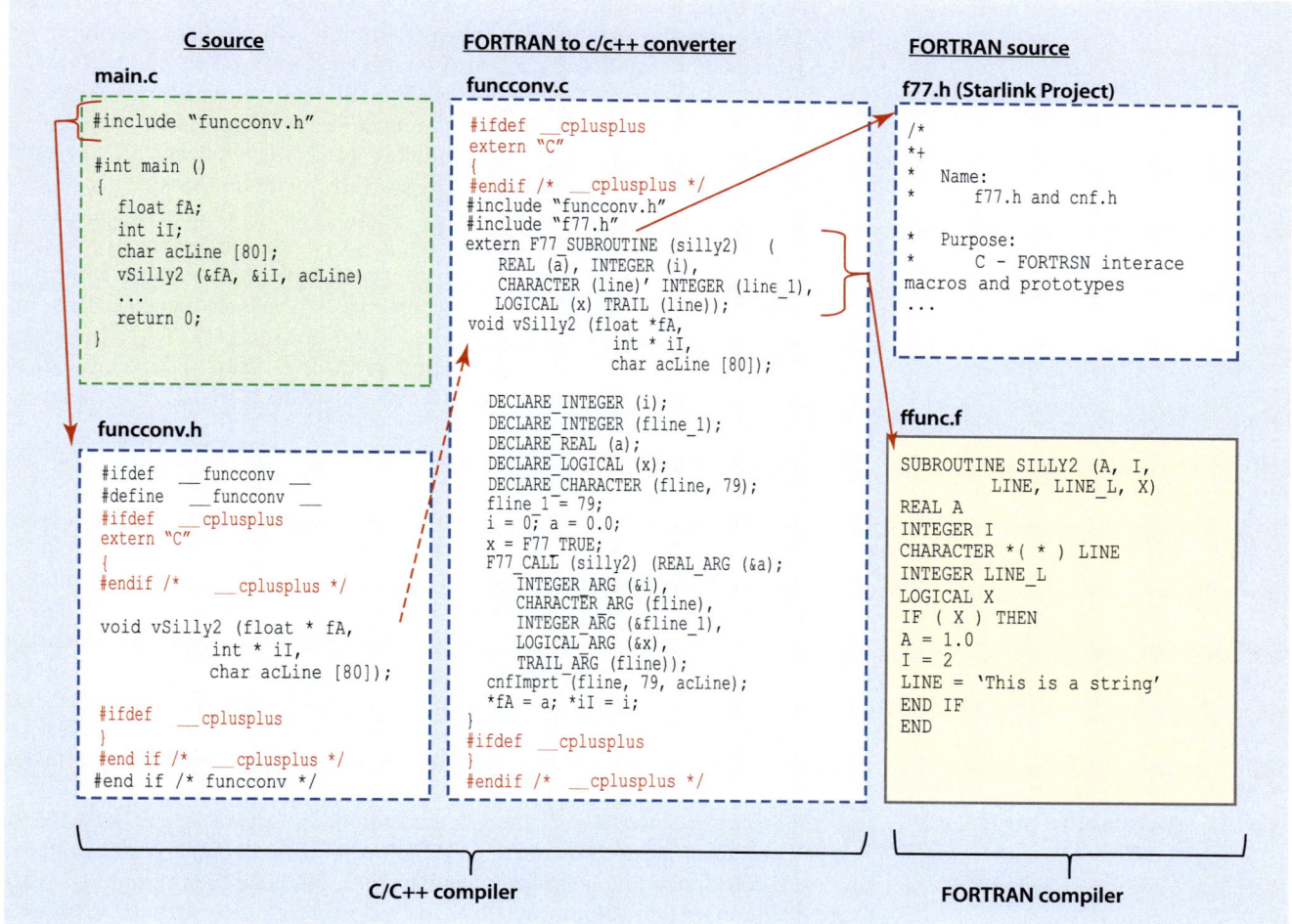

Fig. 2.12   A FORTRAN to C converter using «f77.h» and the encapsulation strategy

FORTRAN	C/C++
integer*1	signed char
integer*2	short, short int
integer, integer*4	int
integer*8	long, long int
logical, logical*1, byte	signed char, bool
real, real*4	float
real*8	double
real*16	long double
complex	struct{float fRealnum; float fImagnum;};
double complex	struct{double dR; double dI;};
integer aiNum(2,3)	int aiNum[3][2];
character*6 acText	char acText[6];
character*6 aacText(4)	char aacText[4][6];

2

FORTRAN Encapsulation of Arrays

Using these standard types, it is also possible to exchange arrays. Usually FORTRAN arrays are used with fixed sizes. But it is also possible to define arrays with runtime dimensions using a dimension declaration (e.g., «REAL Y (MAXYA, MAXYB, 1)» where MAXYA and MAXYB are dimensions, which can be defined as subroutine parameters). The critical point with arrays in mixed environments is that the element order in the memory is different in FORTRAN and C/C++. A 2x3 matrix (two-dimensional array) is defined as an array with the dimensions «[2][3]» in C/C++ (e.g., «int iArray[2][3];») but with the dimensions «(3,2)» in FORTRAN (e.g., «INTEGER ARRAY (3,2)»). Another difference in combination with arrays is the indexing, which starts with 0 in C and with 1 in FORTRAN. The representation of character string arrays is also specific. These are character arrays of a fixed length and filled with blanks in FORTRAN, but arrays with variable length and zero-termination in C/C++. These differences can be equalized within an elementary converter module, which translates between the different representations.

FORTRAN Encapsulation of Function Arguments

Arguments of FORTRAN functions are always addressed, using call-by-reference. This means that all function internal changes are visible outside. If a call-by-value argument should be simulated, which means that a local copy of the variable is used in the function body, the copying must explicitly be written in the converter. FORTRAN functions or (in the terminology of FORTRAN) subroutines do not return types or at least just allow the return of scalars. The FORTRAN naming conventions (for identifiers in general) are not case sensitive, and visible FORTRAN function names end with «_», if used in C programs (e.g., «FOO (...)» in FORTRAN can be called as «foo_ (...)» in C). This is the standard representation in the symbol table, but it may change among FORTRAN compilers, so that at least three different representations can be found in the literature.

FORTRAN Encapsulation Restrictions

These are the most significant differences between C/C++ and FORTRAN. To keep the converter and influences of FORTRAN quite simple, it is also useful to avoid direct access to a common memory block. FORTRAN programmers often use common memory blocks, which are no more than globally defined variables to exchange values over subroutine borders. This is not really a good style in modern designs, as it is hard to read and often incomprehensible (e.g., to find out the type of a variable can be something of an odyssey). Another quite important rule is always to keep the use of resources local. This means that access to a file or to communication equipment should not be exchanged between the different languages. If a file is read in FORTRAN, only the read information should be handed over to the calling C/C++ source.

Encapsulation of Compiler and Interpreter Code

The combination of FORTRAN and C/C++ is a very good example of how to combine compiler languages. In a similar way, it should also be possible to combine other compiler languages with C/C++, as long as the intermediate object files can be compiled and the internal link naming conventions are fulfilled. But if a script or interpreter program needs to be included, there is often only one way to create a parallel process («fork»), where the script is called up inside, using a system call as «system» or «execv».[26] If the input and output are redirected into a communication pipe before the «fork» is processed, it is also possible to interact with the started program (e.g., used in ▶ Sect. 5.5.5 for the «sshbroker»). This is not really an ideal solution, but with the encapsulation technique, using a converter module, it is kept local and is testable. This means that the three steps «encapsulating», «unit testing», and «refactoring» are a general design scheme to work in mixed language environments.

---

26 Here the programming language Perl has a big advantage, as it is possible to directly embed a Perl interpreter into C code (see ▶ Sect. 2.2).

2.5   Documentation

Documentation is often a burden in most projects. It takes a lot of time, which would be better spent on software development. Or in other words, «[...] the best documentation in the world is no excuse if the project [is] supposed to deliver software, but fails to do so» (Rüping 2003, *l.c. page 2*). In scientific environments, in particular, documents which explain software or design papers are usually very rare. For some scientific developments, it might also be a tactic not to provide too much information about the solutions and internal implementations in order to become the only specialist for this implementation. But this is often just an illusion, as stronger teams are often much more successful than single specialists. In addition, local information held in one single brain is also not as accessible for the institute as it would be if the team were able to discuss it together. This situation usually results in stagnation, as people tend to keep their familiar, well-known position and knowledge instead of learning, changing, and validating it with new things.

Documentation as Information Carrier

Therefore, information transfer is one important challenge, especially within scientific processes. Usually it is almost irrelevant which medium is used to transfer the information. But modern, agile software developments (see ▶ Sect. 2.9 and following) tend to devalue written documentation. They proclaim, «[t]he most efficient and effective method of conveying information to and within a development team is face-to-face conversation» (Beck et al. 2001). But to skip written documentation completely is also no solution. Written descriptions are better available over time, even when the original developers are not available anymore. But documentation consists of more than a scientific paper output or a thesis. A lot of scientific programs disappear unused in the obscurity of an institute's archives, without further use, as the implementations and internal algorithms are difficult to understand, even when using the science papers about the software subject. Additional documents are necessary. This is especially the case for software which is written within an exactly determined, contracted, and externally funded project, as documentation often marks dedicated milestones. But endless, confusing papers are not helpful either. Therefore the trick is to hit the spot, which can be defined as «agile documentation» using a light-but-sufficient approach (Rüping 2003, *l.c. page 3*).

Requirement of Documentation

In the case of a scientific software project, it is possible to define three main categories of documents, according to the target group which should be addressed:

- The documents for the development team (design descriptions, coding guides, references, etc.)
- The documents for the users (manuals, tutorials, guides, explanations, etc.)
- The documents for the administration/funding organizations and other scientists (papers, publications, roadmaps, etc.)

Categories

Each of these categories addresses a dedicated target readership which is interested in different aspects of the project (e.g., the developers are mostly interested in design issues and concrete realizations; the administration is mostly interested in the project workflow, timeline, and milestones; and the users are mostly interested in the use of the product and in tutorials about how the product can be used). All of those documents also focus on different levels of detail. Therefore, the documents should be optimized for those groups. The scope and the background needed to read the documents should be kept in mind (Rüping 2003, *l.c. page 24 f.*). The focus should be on the particular and relevant information topics and no more than that. All sections should focus on material which is relevant for the target group and abstracts should offer compact overviews (Rüping 2003, *l.c. page 26 f.*). In general, each project requires an individual set of documents of all categories, which are more or less extensive. It generates the individual document portfolio of the project and the

Documentation Readership

software system. Usually the portfolio can be used as a template for other projects to support the needs of the management, development, and use (Rüping 2003, *l.c. page 28 ff.*). In addition, all documents should be easily available for the individual target reader group. Everything together builds a document landscape which is optimized for the requirements as a light, agile, and informative knowledge base.

2.5.1 Documentation for the Development Team

Written Documentation

Documents for the development team are more or less detailed descriptions about the internal realizations, algorithms, and implementations. They should transport enough information about the project and its design structures to other (maybe also future) developers and programmers. To communicate this information[27] is independent from the transport way or the medium. Therefore, most lightweight and agile developments (see ▶ Sect. 2.9) are based on face-to-face communications (Rüping 2003, *l.c. page 3 f.*). Nevertheless, written documents can improve communication as discussed in the previous section. But the quantity of information does not improve communication. The quality of the information should be on the main focus. Therefore, documentation should be target-oriented to the readers. At this point, the use of the same language with similar or the same meanings is essential to reduce incommensurability in the descriptions.

Long-Term Versus Fast-Changing Documents

Different levels of developer documentations can be identified, dealing with long-term, design-describing strategies or fast-changing, detailed, and realization structure descriptions. As, for example, code details often change more quickly than documents can be updated, while strategy papers are longer lasting, different approaches should be used for these different types of information. The creation and writing of these different document classes must be brought into a suitable workflow. A «light-but-sufficient» approach for developer documentation reduces the additional, detailed, and extensive documents to a minimum. It prevents the developers from expending a lot of effort on documentation and offers accessible, useful, and comprehensive documents within an agile documentation process (Rüping 2003, *l.c. page 3*).

Agile Documentation

> *DefinitionDEF2.7. of «agile documentation»* (see (Rüping 2003))
> Agile documentation is a method which incorporates a number of suggestions to create lightweight, qualitative, sufficient, individual documents, which focus on the individual needs of the different reader target groups. It concentrates solely on the relevant materials and nothing else. Such documents can be created and distributed using a documentation environment, without too much additional time spent, so that the focus is on the code development and not on its documentation.

Documentation: Long-Term, Design-Describing Overview

To form a team and bring all team members onto the same track for the development or to appoint design ideas and restrictions, informative long-term documents build the necessary foundation. As each project starts with a more or less extensive task analysis and its requirements, one of these long-term papers for a project is the specification or requirement analysis. It describes the

---

27 Information is the content and the meaning (semantics) of a message in textual, graphical, or audio-visual form. Ideally, information is free from irrelevant, redundant parts and transmits knowledge about functionality, functioning, principles of operation, methods, techniques, applications, flows and activities (ITWissen n.d.).

common understanding of the task and its environment as the basis for the over-all design (Rüping 2003, *l.c. page 36 ff.*). This specification splits the complex task and its interactions to the users into manageable use cases and offers a more focused view on subtasks. Graphically those use cases can be represented with Unified Modelling Language (UML) diagrams, as they reduce information to a diagram style. But other representations are also suitable, such as written descriptions of the pre- and postconditions, constraints (as real-time requirements), and the task activities. The specification is the framework document of the development.

If the framework requirements and their limiting factors are known, a possible design can be drafted, showing more or less the «big picture» of the main project issues and ideas to solve the tasks. Those design documents are useful throughout the software development process. Documents describing the fundamental ideas, the design decisions, or the architecture for the construction are highly valuable. The documents explain the whole system preferably independent of the platform or language-specific realizations. They abstract over any details which are irrelevant for an overview. The «big picture» describes the entire static system architecture and the basics of dynamic behavior. It explains the design principles, maybe possible alternatives, and also motivations for design decisions. Therefore, for most projects at least 20 pages should be enough, which include links to more detailed documents. It is something like a paper about the software vision or rationale, as a basis for discussions and as a first explanation for general understanding (Rüping 2003, *l.c. page 39 ff.*).

Using this design document, a development plan can result in a roadmap including an approximate time table for the development of the different subtasks. Such additional excerpts can be used as a basis for administrative documents. While more formal development processes define several more documents, the described overview papers should be enough to give a first overall «big picture». But of course they can be extended depending on further needs if necessary. The main driver must be that documentation should not be an end in itself, but should focus on supporting the programming and the comprehensibility of the design. Therefore, working solutions with a self-documenting code are more valuable than extensive documents.

Before an agile method for the creation of code-related, fast-changing documentation is shown, it is worth taking a look at possible documentation realizations for long-term descriptions, as mentioned above. Experience shows that sequentially organized, well-structured text (using chapters and sections) extended with accompanying diagrams offers a very intuitive approach. In particular during discussions, a document history with release notes helps to show differences from previous versions (Rüping 2003, *l.c. page 66 ff.*). Therefore non-binary file formats, as given with LaTeX documents, are very useful in combination with version control systems (see ► Sect. 2.7), as those systems offer versioning with merging and an automatic inclusion of version and release information directly in the text when a new version is committed. Plain text is sometimes preferable to sophisticated Office documents, as those general text formats can be opened on each platform, using the standard editors (e.g., README files). Almost equivalent but with more possibilities for graphical accentuations is the use of Hypertext Markup Language (HTML), as all systems offer browsers to read it. Accompanying, included graphics in standard formats such as Joint Photographic Experts Group (JPG), Graphics Interchange Format (GIF), or Portable Network Graphic (PNG) can be used to give graphical overviews of the architecture, class and call dependencies, or state machines in addition to the written text. Of course it is also possible to use Office packages, but plain text layout systems like HTML or LaTeX offer other advantages: the link references are well organized, the style can be separated from the content, and generative methods (about generative methods, see also ► Sect. 3.3 and

Big Picture

Individual Documents

Documentation Structure and Formats

2

following) can extend the overview documents with faster-changing, code relevant parts automatically, which is difficult in binary and complex Office documents. But neither the tool nor the technique used should be the driver. The qualitative best result should play the key role. Experience has shown that a mixture of HTML, LaTeX, plain text, and Portable Document Format (PDF) caters for all needs and is long-lasting enough to be readable in the future as well.

Documentation: Detailed, Code-Related, and Fast-Changing

The more detailed parts are more or less dependent on the current development or implementation and therefore highly code-related. For such information, an automatic generative process (about generative methods, see also ▶ Sect. 3.3 and following) can be used to generate detailed documents from the self-documenting code (see also the descriptions about code layouts in ▶ Sect. 2.3). With some additional, informative extensions in the code comments, the generative approach can effectively produce detailed, up-to-date, «light-but-sufficient», and less work-intensive documentation in the form of developer references (for the classification of references, see ▶ Sect. 2.5.2). It should also contain at least some links to related long-term descriptions, like «big picture» documents.

Documentation Systems: Doxygen

Documentation systems or documentation utilities offer such a generative approach. They produce resulting documents in different multiple output and target formats (like Microsoft® Word, HTML, or LaTeX) from the source code. Online documentation with HTML, in particular, is a very useful, flexible, and suitable possibility for fast-changing documents. A very well-known, open-source tool is «Doxygen». Doxygen was developed by Dimitri van Heesch and can be used to automatically create documentation for source code in different programming languages (Helmke et al. 2007, *l.c. page 111*), like C++, C, Java, Objective-C, Python, IDL (Corba and Microsoft flavors), FORTRAN, VHDL, PHP, C# (van Heesch n.d.), and with some filter extensions also for some other languages such as Perl. Doxygen can be downloaded for different platforms (Windows™, Linux, etc.) and as a source code package. In addition «Graphviz» should be installed as a graphic package (also available for different platforms), to use for drawing call graphs. Graphviz is an open-source graph visualization software and takes descriptions of graphs in textual form to produce images in commonly used formats (Graphviz n.d.). Doxygen uses it to produce the informative call graphs between functions.[28]

Doxygen on Plain Code

Doxygen can be simply used directly on the self-documenting code without any additions. It uses a predefined configuration file, which can be created using the Doxywizard. This is the Doxygen Graphical User Interface (GUI) end offering a «Wizard» style with predefined settings and an «Expert» style for more sophisticated definitions. While Doxygen is usually called as a command line tool (by just calling «`doxygen [ConfigFile]`») or in the Makefile (see ▶ Sect. 2.6.2), it can also be started by using the GUI. Then it interprets the syntax of the code files and walks recursively through the code folder hierarchies. This basic use can be effectively used to get a quick overview about large, external source code distributions, as Doxygen can visualize relations between various elements graphically (van Heesch n.d.).[29] But the full volume of functionality is supported if the code itself is extended with dedicated Doxygen comments.

Doxygen on «Doxygen-Commented» Code

Comments are always recommended in source code as they support the intelligibility of the code and as they are tightly connected to the individual elements of the code. Therefore, they should not only explain what can be easily retrieved from

---

28 Call-graphs are graphical presentations which show the structure how functions call each other in a program. These calls are usually not ordered according to the calling time of each function. They just show the interactions between the functions of a program. Ordered graphs are the sequence charts of UML.

29 For the following, see the web pages of (van Heesch n.d.).

the code itself (e.g., if a comment describes a following loop like this «// A for loop to count the index»). They should offer additional meta-information.[30] This meta-information can be directly used to generate the short-term developer documents about the code. The only requirement is that the specific comments must be classified as Doxygen comments. They are characterized with the comment classifiers «///...comment...» instead of «//...comment...» in C++ or «/*! ...comment... */» instead of «/* ...comment...*/» in C.[31]

As Doxygen generates sophisticated follow-up documents, it also allows highlighting with style markups in the comments. The classifiers for those accentuations are the classic HTML tags, such as «<b>» for bold or «
» for a new line. In addition, Doxygen offers its own identifiers in the text, which can be included with the usual comments and which are specially interpreted during the generation, to add or specify particular information. This Doxygen classifier set is very extensive. But usually a few are enough to generate suitable, graphically well-designed, and manageable documentation. It is advisable to standardize the most commonly used elements for one's own projects.

Text-Highlighting

■ ■ Example 2.6

File header with the suggested Doxygen comments

```
/****************************************************************/
/*! \file queenspuzzle.c
    \brief Solving the N-queens puzzle <br>
    \authors A. Neidhardt
    \date 2011-08-05

    The program finds all possible solutions for the
    N-queens puzzle (position N queens on a NxN chess board,
    so that no queen is in danger to be attacked by the others)
    for boards of size 1x1 to 99x99. For each solution the chess
    board and the place of the queens are printed. <br>

    \attention The code bases on a solution, suggested by
               Werner, Matthias (Algorithmen und Programmierung.
               Skript zur Vorlesung.
               http://osg.informatik.tu-chemnitz.de/lehre/old/ws1011/aup/AuP-script.pdf
               (Download 2011-08-05) l.c. page 249 ff.) but is extended
               with extensive modifications.
*****************************************************************
    <b>External available precompiler definitions / constants:</b><br>
       n.a. <br>
    <b>External available type definitions / classes:</b><br>
       n.a. <br>
    <b>External available namespaces:</b><br>
       n.a. <br>
    <b>External available exception handles:</b><br>
       n.a. <br>
*****************************************************************/
```

1
2
3
4
5
6
7
8
9
10
11
12
13
14
15
16
17
18
19
20
21
22
23
24
25
26
27
28

To further support the consistent appearance of the code, not only should the structure and arrangement of the source code sections be similar, it is also reasonable to define a structure for the comment header of each file and the comment headers for each function. Example 2.6 shows such a file header, including dedicated Doxygen tags. It starts with the filename (behind the tag «\file») followed

Doxygen Source File Header

---

30 Meta-information is additional information about given data to improve its comprehensibility and shareability for users (Kretzschmar and Dreyer 2004, *l.c. page 83*).

31 By the way, all asterisks, comment delimiters, etc., do not appear in the final documentation.

by a brief description (behind «\brief») and the authors info (sequence of authors behind «\authors») and the startup date of the development of the code file (behind «\date»). The following text is completely copied into the documentation and is used as detailed description. If there are specific notifications, which should be disclosed, the «\attention» tag can be used. The following text is completely copied into the documentation and is handled as detailed description. The parts following the header text describe externally available definitions or compiler settings like pre-compiler definitions, type/class definitions, namespaces, or exception handles. They can be used in the style of an itemization.

Doxygen Function Header

In a similar way, the module functions or class methods can be described. Example 2.7 shows a suggestion for a function comment header and the Doxygen comments, included with comments in the function body. The suggested header consists of three parts, where the first one is just the function name in «non-Doxygen» style, as the name is derived automatically by Doxygen. The second part is the Doxygen-related section, describing all the arguments which are handed over to or returned from the function, using «\param». The description shown uses a simple convention for each parameter, consisting of the name, a brief description, and the handover direction flag in between («->» for an input parameter value, «<-» for an output parameter value, and «<->», if the value is input and returns an output). Alternatively, the Doxygen format for the definition of the direction of the parameter can also be used («[in]» for an input parameter value, «[out]» for an output parameter value, and «[inout]», if the value is input and returns an output), which must then directly follow the «\param». The Doxygen section also contains the «\return» value description of the function in the same notation as the handover parameters. Special notifications can also be added, which are used in a similar way to the file header («\attention»). The third part is just informative and not included in the Doxygen documentation. Experience has shown that the function history, function authors, and so on just expand the documentation, while such information is not directly used by other developers. But it is necessary if other developers who work with the same code want to get in contact with the original programmers. Therefore, it should still be available from the original source.

Doxygen Inline Comments

While the rest of the function signature and its dependencies are derived automatically, Doxygen comments should be added to the function body to offer additional background info. It can be separated into variable definitions and operations. Both parts are headlined with corresponding highlighted text titles (e.g., «/*!<b> Variables</b>*/» in the example shown). Those titles can be followed by explanatory text. Variables are described using the list classification «\li». The operation workflow is just described with a more or less extensive description always directly in the line before code instructions which is associated with.

Include Documenting Images and Graphics

A very important fact for developer documentation is that usually plain, written text on its own is not able to explain algorithms properly. Difficult scenarios can often be better explained by using images, sketches, or just camera snapshots of whiteboard drawings, created during developer discussions. For this reason, it is always ideal to combine plain text with a supporting graphical description besides the automatically generated call or dependency graphs. These images should ideally be directly inline within the documentation text. This is also possible with Doxygen. After defining a source directory for images of the common types (e.g., JPG), the images can be directly included using the «\image» tag.[32] It defines the output type, the name, and the caption of the used image (see Example 2.7 after the operation description text).

---

32   But the included images should already have the correct output size.

■■ **Example 2.7**

Function with the suggested Doxygen comments

```
/*************************************************************        1
 * function vFindNextQueenPosition                                   2
 ************************************************************/        3
/*! Find position for the next queen using a recursvie algorithm     4
    \param iChessBoardSize -> Size of the chess board                5
    \param iCurrentQueenNumber -> Current number of the queen which was    6
           positioned and should be checked                          7
    \param piCurrentSolutionNumber <-> Current number in the ascending numbers    8
                                 of solutions                        9
    \param pSQueenPositionsList <-> Row/column list of queens positions    10
    \return —                                                        11
    \attention Recursive backtracking algorithm                      12
 ************************************************************/        13
/* author A. Neidhardt                                               14
   date 2011-08-05                                                   15
   revisions                                                         16
   2011-08-05 Original (A. Neidhardt)                                17
 ************************************************************/        18
void vFindNextQueenPosition (int iChessBoardSize ,                   19
                             int iCurrentQueenNumber ,               20
                             int * piCurrentSolutionNumber ,         21
                             QueenPositionStructType * pSQueenPositionsList)    22
{                                                                    23
    /*!<b> Variables </b> */                                        24
    QueenPositionStructType SNewQueenPosition = {-1, -1};           25
           /*! \li SNewQueenPosition = New queen position structure with    26
                             row and column fields */               27
                                                                    28
    /*!<b> Operations </b> */                                       29
    /*! The algorithm is recursive. The function places a queen step by step    30
        on the columns of the current row, which was handed over. The starting    31
        first row (= 0) is handed over by the main program. The recursive    32
        function checks if the positioned queen is safe, using the dedicated    33
        function "iCheckIfQueenPositionIsSave". If it is safe, it calls itself    34
        again with the next line. If all queens are set and the maximum number    35
        of possible rows is reached, it prints the found solution. Then it    36
        returns to the previous calling state on the previous row and continues    37
        setting those queens on this row. A possible step-by-step behavior for    38
        a 4x4 board is shown in the following image:                39
        \image html 4QueensProblemExample.jpg "A possible step-by-step behavior for a 4x4    40
           board."                                                   41
    */                                                              42
                                                                    43
    /*! Check if the last queen for the current solution was placed and the    44
        chess board configuration is complete. */                   45
    if (iCurrentQueenNumber == iChessBoardSize)                     46
    {                                                               47
        /*! If chess board configuration is complete print solution */    48
        (*piCurrentSolutionNumber)++;                               49
        vPrintChessBoard (*piCurrentSolutionNumber ,                50
                    pSQueenPositionsList ,                          51
                    iChessBoardSize );                              52
    }                                                               53
    else
```

2

```
{
    /*! If chess board configuration is not complete find new position */      54
    /*! Start at the next empty row */                                          55
    SNewQueenPosition.iRow = iCurrentQueenNumber;                               56
    /*! Position the new queen on each empty column at the current row */       57
    for (SNewQueenPosition.iCol = 0;                                            58
             SNewQueenPosition.iCol < iChessBoardSize;                          59
             SNewQueenPosition.iCol++)                                          60
    {                                                                           61
                                                                                62
        pSQueenPositionsList[iCurrentQueenNumber]= SNewQueenPosition;           63
                                                                                64
        /*! Check if queen is safe at this position */                          65
        if (iCheckIfQueenPositionIsSave (iCurrentQueenNumber,                   66
                                          pSQueenPositionsList))                67
        {                                                                       68
            /*! If queen is safe, let it there and find safe position          69
                for next queen by calling the same function                     70
                "vFindNextQueenPosition" recursively again with the            71
                following number of the next queen */                          72
            vFindNextQueenPosition (iChessBoardSize,                            73
                                     iCurrentQueenNumber+1,                     74
                                     piCurrentSolutionNumber,                   75
                                     pSQueenPositionsList);                     76
                                                                                77
        }                                                                       78
        /*! If queen is not safe, continue with searching columns */           79
    }                                                                           80
}                                                                               81
/*! If solution is complete or no safe position was found, return to          82
    previous run in the recursive workflow tree (backtracking) */             83
}
```

Documenting Mathematic Formulas

It is also possible to include mathematical formulae that appear in the documentation text, using the LaTeX notation. These formulae must be put in between a pair of «\f[... \f]» and Doxygen needs to know the installation directories of a LaTeX compiler of «dvips» to convert Device Independent file format (DVI) files to PostScript files and of the GhostScript interpreter.[33] It is even possible to set links to other online documents, for example, about the design and «big picture» by simply using a Uniform Resource Locator (URL),[34] which is automatically replaced by an HTML link in the documentation (van Heesch n.d.). The resulting output of Doxygen in HTML style for the function «vFindNextQueenPosition» from the N-queens puzzle is shown in ◻ Fig. 2.13.

Avoidance of Documentation Redundancies

It is also possible to avoid redundant information in the Doxygen function header and the function declaration by combining both. A possibility is shown in Example 2.8. It produces similar documentation where only the parameter section is at the end of the function documentation. Which style is favored depends on the personal preferences of the developers and of the project specifications. In the context of this work, the first style is preferred.

---

33 A LaTeX compiler reads the LaTeX sources and usually generates a DVI file, which then must be converted into a PostScript file. PostScript is a description language for the style of a document, including the content, which can be interpreted, for example, by the GhostScript interpreter. Most printers also directly understand PostScript.

34 URL is a standardized nomenclature to reference and locate Web and network resources.

```
void vFindNextQueenPosition ( int          iChessBoardSize,
                              int          iCurrentQueenNumber,
                              int*         piCurrentSolutionNumber,
                              QueenPositionStructType*  psQueenPositionsList
                            )
```

Find postion for the next queen using a recursvie algorithm

Parameters:

iChessBoardsize	- > Size of the chess board
iCurrentQueenNumber	- > Current number of the queen which was positioned and should be checked
piCurrentSolutionNumber	<- > Current number in the ascending numbers of solutions
psQueenPositionsList	<- > Row/column list of queens positions

Returns:
-

Attention:
Recursive backtracking algorithm

Variables

SNewQueenPosition = New queen position structure with row and column fields

Operations

The algorithm is recursive. The function places a queen step by step on the columns of the current row, which was handed over. The starting first row (=0) is handed over by the main program. The recursive function checks if the positioned queen is safe, using the dedicated function "iCheckIfQueenPositionIsSave". If it is safe, itself again with the next line. If all queens are set and the maximum number of possible rows is reached, it prints the found solution. Then it returns to the previous calling state on the previous row and continues setting those queens on this row. A possible step-by-step behavior for a 4x4 board is shown in the following image:

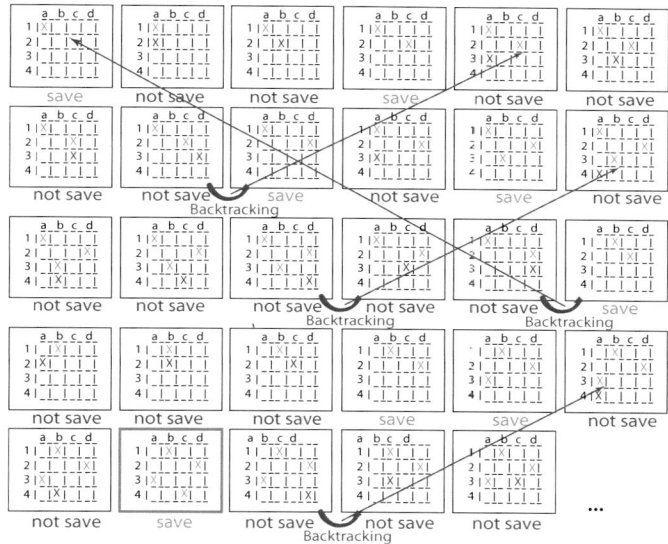

A possible step-by-step behavior for a 4×4 board.

Check if the last queen for the current solution was placed and the chess board configuration is complete.

If chess board configuration is complete print solution

If chess board configuration is not complete find new position

Start at the next empty row

Position the new queen on each empty column at the current row

Check if queen is safe at this position

If queen is safe, let it there and find safe position for next queen by calling the same function "vFindNextQueenPosition" recursively again with the following number of the next queen

If queen is not safe, continue with searching columns

If solution is complete or no safe position was found, return to previous run in the recursive workflow tree (backtracking)

Defintion at line 206 of file queenspuzzle.c.

Here is the call graph for this function:

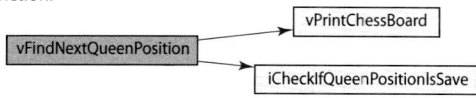

■ **Fig. 2.13** The Doxygen-generated HTML documentation from the function in Example 2.7

■ ■ **Example 2.8**

Function with a suggested Doxygen header without redundancies

```
/********************************************************
********************************************************/
void /*! \return No return value */
vFindNextQueenPosition
(
    int iChessBoardSize              /*! [in] Size of the chess board */,
    int iCurrentQueenNumber          /*! [in] Current number of the queen which was
                                            positioned and should be checked */,
    int * piCurrentSolutionNumber    /*! [inout] Current number in the ascending numbers
                                            of solutions */,
    QueenPositionStructType * pSQueenPositionsList /*! [inout] Row/column list of queens
            positions */
)
/********************************************************/
/*! \brief Find position for the next queen using a recursvie algorithm
    \attention Recursive backtracking algorithm
********************************************************/
{
    ... // Function body
}
```

1
2
3
4
5
6
7
8
9
10
11
12
13
14
15
16
17
18
19

Doxygen in the Development Process

In the way described, Doxygen has several advantages for agile documentation. It reduces the effort required by developers to write documents in parallel with the programming. The generation can be automated and all relevant files can be added to the project, so that the documentation can also be generated during the building process of the software if Doxygen is installed on the machine. In such a way, only the general documents need to be distributed. The rest can be generated with the latest version of the code. The HTML documentation is particularly useful, as the links offer a very intuitive and flexible navigation through the code and the related documents. In combination with some additional tools from the continuous integration workflow (see ► Sect. 2.8), spell checkers, for example, the documentation relevant parts in the comments can directly be checked for lexical language correctness. Doxygen fits perfectly into the development landscape of scientific programs. But it only shows its full potential if the dedicated Doxygen comment styles are used.

2.5.2 Documentation for the Users

User Documentation in Scientific Projects

While the previous section described the developer requirements to work together in the project and to reuse modules or source code in one's own implementations, this section focuses on the user's need to simply use a complete, existing, and implemented software system, for instance, control software for a telescope. Even though scientific software is usually used by scientists and even though they are used to dealing with complex systems, lower-level documentation is also helpful, especially if the software is also used by junior scientists, as is often the case for control software for telescopes, where students work as operators. Adequate documents about procedures and schedules are necessary. This aspect is often neglected during scientific software development, and therefore such documents which give a helping hand to operators are often missing.

Documents for Users

Reading user-dedicated documentation, the descriptions just touch on the higher-level interaction between user and the user interface in the sense of use cases. At this level, there is generally no need to explain all-embracing, internal workflows or implementations. These user documents describe the functionalities and features, which are offered by the software or offer solutions for frequently asked questions, such as what to do if a specific error appears. A general description of the tasks solved, the required resources and environment, some help scenarios, and error descriptions to solve problems are suitable (parts of the activity context in (Barker 2003, *l.c. page 67*)). But the agile idea can be kept in mind, which means that the amount of information the user needs should be determined (Barker 2003, *l.c. page 66*) and also how the transfer of that information can be integrated in a task-oriented way into the users' working environment (Barker 2003, *l.c. page 10*).

User documentation can be separated into three main levels of support: guidance, reference, and teaching. Ideally, these levels should correspond to the user's requirements (Barker 2003, *l.c. page 92*). Classification of User Documentation

Guidance (see (Barker 2003, *l.c. page 63 ff.*)) accompanies the user step-by-step through a concrete task. It consists of step-by-step and how-to instructions, on which it lays its main focus point. Examples in the scientific environment are (quick) installation, getting started guides, or operational guides, which describe step-by-step what must be done during a specific activity. Such documents are very convenient for junior scientists and students. Enumerations are usually a means to an end (e.g., to start an observation, first do this, second do that, etc.). Including screenshots is usually very helpful, especially if they are used in a parallel format. In this format type, prose texts, enumerations, and screenshots are cross-referenced, for example, to explain entry mask or dialog boxes of a GUI. Usually, highlighted notes, tips, cautions, and warnings are also integrated to illustrate points, where the user should be really careful or where useful alternatives are reasonable. These documents can simply be just text files (such as `INSTALL.txt` files) or more sophisticated office, LaTeX, or HTML documents. Other representatives of guides within the programs are embedded or interactive help outputs. Part of this category are, for example, simple legends, explaining colors in graphs, etc., or hover and roll-over texts, which appear while the user is lingering with the mouse pointer over a graphical element. An example of such a guidance document can be found in ◘ Fig. 2.14 (Lovell et al. n.d.), showing the documentation Wiki[35] of the new geodetic VLBI network AuScope in Australia (Lovell et al. n.d.). Guidance

References (see (Barker 2003, *l.c. page 92 ff.*)) are support documents and focus on lookup sections. They consist of itemizations of command descriptions, menu explanations, definition and function lists, error messages, or frequently asked questions. The target reader group consists of experienced users who want to look up specific requests. Indexes or search capabilities are therefore quite important. Examples are simple `README.txt` files with installation details, history and changes lists, new feature lists, errata, compatibility requirements, and so on. The Linux man pages (see ◘ Fig. 2.15) or «-h» program call parameters, which print a help output also belong to this category of reference. They point out keystrokes, parameter lists, definitions, brief descriptions, and command summaries. They serve as job aid references. They are the most important elements needed to work efficiently with software. They usually follow the same style pattern and are organized alphabetically or are searchable. In this style, they are self-explanatory in most cases. References

Tutorials (see (Barker 2003, *l.c. page 30 ff.*)) are learning lessons or learning modules. They consist of graphically and didactically well-structured information. The target reader group consists of novice users or beginners. In scientific developments, tutorials are not common except in university study courses. In agile environments, it is better to do such teaching in regular face-to-face-meetings about concrete subjects. Exception to this are demonstrations. Demonstrations illustrate specific parts and limited versions of the program, showing a real-world Tutorials

---

35 A Wiki is a hypertext web system to offer, find, change, and discuss different topics in a fast way. The idea is to offer a simple, browser-based opportunity to read and edit text.

```
Datei  Bearbeiten  Ansicht  Suchen  Terminal  Hilfe
progstarter(1)                    User Commands                    progstarter(1)

NAME
       progstarter - Creates  a  project  sample folder.

SYNOPSIS
       progstarter [PROJECT-NAME]... [OPTION]...

DESCRIPTION
       Creates  a  project  sample folder in consideration of the Design-Rules
       of the Geodetic Observatory Wettzell.

       -without option,
              creates a folder called [PROJECT-NAME], with:
                   bin/
                   conf/
                   conf/[PROJECT-NAME].conf
                   make/
                   make/Makefile
                   obj/
                   rpc/
                   src/
                   src/C[PROJECT-NAME].cpp
                   src/C[PROJECT-NAME].hpp
                   src/main.cpp
                   srcext/

              Use this option, if you want to start a simple program / driver,
              which will be extended next time.  It's also possible to connect
              the idl2rpc-procedure later.

       -s, --server
              creates a folder called [PROJEKT_NAME], with:
                   bin/
                   conf/
Manual page progstarter(1) line 1/127 34%
```

◻ **Fig. 2.15** A real-life example extract from a Linux man page as a reference document for a tool from a student's project to set up the project directories

application or an important use case of the program. Usually a brief printed tutorial paper is useful. Demonstrations are also quite useful to train future developers in a specific given task in the predefined environment.

In general, several of the aspects discussed above are combined in the wide field of user's manuals (Barker 2003, *l.c. page 209*). In printed or online form, they offer mainly a reference, but in combination with guides and sometimes with tutorials. Such sophisticated documents are often not available for scientific code as long as the programs are not popular in a wider user community. Then other users request such manuals. Manuals should focus on the user's needs and should be performance-oriented (Barker 2003, *l.c. page 383*). They should be manageable, well structured, and comprehensive in a very general, descriptive, linguistically simple, and technically user-level way. These manuals should be intelligible for all. | **User's Manuals**

Manuals should always follow the same style. Templates in LaTeX or HTML are ideal. Additionally, they should also use language which is suitable for the knowledge level of the user (see (Barker 2003, *l.c. page 383 ff.*)). A first necessity is to write about actions instead of defining functions. The main idea behind this is to show the user what he can do with a specific function. The sentences should be in the active voice following the sentence style: subject, verb, and object. This also keeps the sentences simple and comprehensible. Short sentences without similar-looking words and without technical terms support the readability for users. Imperative voices with parallel sentence structures in enumerations offer direct contact to the user. Manuals are very suitable in combination with prose passages about concepts and suitable graphics which are cross-referenced in the texts. Nevertheless, it takes a lot of work to produce good manuals. It is better to focus on the main topics of the work being described and make these descriptions high quality, instead of producing endless, useless explanations. | **The «Right» Language**

2.5.3 Documentation for Administration and Other Scientists

Documents for Administration and Other Scientists

The documents for administrative or scientific needs are not the main focus of this work. But two specific types should be briefly mentioned: scientific project papers and scientific result papers.

Scientific Project Papers

Administrative project documents are often documents for the funding organizations. They are usually defined in cooperation agreements, the funding contracts and guidelines, or in the institute guidelines. An example is the document templates for the framework programs of the European Union, such as the Development of a Simplified Consortium Agreement (DESCA), which «is a comprehensive Model Consortium Agreement for Horizon 2020. Initiated by key FP7 stakeholders and updated for Horizon 2020 in consultation with the FP community, it offers a reliable frame of reference for project consortia» (DESCA n.d.). Those documents are usually the first deliverables of a project, which need to be handed over to the funding agencies or to the administration. Style and coverage can be individual. In most cases, they are similar to the «big picture», overview, and planning documents from the developer documentation. They contain more or less the same information about project development, roadmaps, work packages, retrieved results, project timelines, and specifications. They are often excerpts from the developer documents or vice versa. A difficulty is that those administrative documents are normally also combined with the financial planning of man power or material investments. They are judicially binding, assess foreground and background knowledge of the developments included, and depict legal rights, such as software licensing. Thus, it is important to include the appropriate specialists in this process.

Scientific Result Papers

On the other hand, there are scientific publications which usually present the scientific results of the work. These are scientific innovations as results of the use of the developed software. The focus is not on the software description itself, but more on the topic in which it is integrated. Other more technically oriented, scientific papers describe technical innovations. They describe new process technologies or algorithms and their realizations. The target journals and therefore the scientific reader groups are different. But for both counts: «Since the early modern forms of institutionalization of science in the 17th century, scientific findings are only recognized when they have been published and laid open to criticism and scrutiny» (Deutsche Forschungsgemeinschaft Geschäftsstelle (DFG) 1998, *l.c. page 73*). Therefore, all of those published documents should follow the rules and recommendations for good scientific practice such as the «Proposals for Safeguarding Good Scientific Practice» (Deutsche Forschungsgemeinschaft Geschäftsstelle (DFG) 1998). Scientific papers, proposal articles, journal contributions, and so on are the scientists' focused outputs. Compared to the other forms of documentation, they are essentially summaries with a specific focus on the work results for which the code itself plays a subordinated role.

2.5.4 Documentation Landscape

Offering Documentation

While all the previous sections discuss which documents are necessary in an agile project style and how they should look or how they could be created easily and generatively, it is also important to offer the documents in an organized way to the groups of people who are involved. Poorly organized, not easily accessible, but perfectly written document archives that cannot be found are useless for offering expertise to other team members. Therefore, it is also vitally important to create a document landscape besides the documentation culture (for the following, see (Rüping 2003, *l.c. page 120–149 ff.*)).

> *Definition DEF2.8. of «documentation landscape» (see (Rüping 2003, l.c. page 120 f.))*
> A documentation landscape represents a suitable environment to offer documents to involved participants. It should offer a structure which works, such as mental maps, mind maps, or organizing trees to organize the documents for the target groups, which navigate through such intuitively over-linked pathways to retrieve or add project or code relevant information.

Documentation Landscape

In such a documentation landscape, it should be easy to retrieve or add information for all participants. The generating of documents for the developers should be done automatically in regular cycles. It makes almost no difference whether the documents are organized in well-structured, hierarchical file system directories or online as web pages with hyperlinks as pathways through the documents. The organization of the avoidance of redundancies in such structures is of far greater importance. But in some cases, online documentation has some advantages: it can be accessible for a wider range of target readers all over the world, it can offer easy search mechanisms, and it offers an intuitive way to «travel» through the document landscape and combine simple HTML text documents with sophisticated graphical browser outputs. Nevertheless, web pages also require well-managed, manual structuring (e.g., a Wiki is just as good as its ability to find things, except for the search function over all entries).

Organizing Documentation

As the documentation grows and changes in versions during the project, it is important to archive such revisions. Therefore it is not only important and necessary to have the source code version controlled. The documentation archive should also be managed in a version control system (see also ▶ Sect. 2.7). This is especially useful for plain text documents as the version control system merges new document versions automatically with existing ones. In this way, it deals with changes from different editors, while such systems also tag the revisions and releases.[36] In this archive, redundant documents should be avoided. The version control system usually offers the option of including external parts as links to the originals to avoid duplicates. The version control archives should also be saved to backup archives.

Organizing Document Versions

A combined documentation Wiki additionally offers better exchangeability of information over the web and supports communication methods, for example, message blocks about dedicated topics. There is a possibility of collaborative, interactive, but yet asynchronous communication between distributed participants, which is ideal in agile software projects (see ▶ Sect. 2.9). Wikis also offer sophisticated search mechanisms. But there must be a clear rule set for how new information should be structured and where it should be included, to keep clear documentation organization. The web pages should contain searchable guides, tutorials, or manuals but can also contain complete user manuals in platform-independent formats like PDF, which can be printed as a compendium. This combines the aspects of online search and reader-friendly printouts. It is also useful in web environments with HTML or a Content Management System (CMS) to

Docu-Wiki

---

36 A revision is a change in the document. A release is a finished version in a frozen state, which is named and delivered to the target group.

2

separate the layout from the content. This allows a very fast adaption to new layouts and supports reuse of the documents.

A final but very important aspect in the documentation landscape is also that developers, users, and other participants get notifications after add-ons, updates, or changes. Each significant change should trigger such a notification with short explanations, what is changed, but without the change itself. Version control systems can be administrated to trigger such email notifications automatically, triggered by document commits (see also ▶ Sect. 2.7). But even manually written emails distributed with email exploders have the same effect. All changes should not alter or reorganize known structures too often, as human users like well-known structures, and as it is always expensive to check if everything is still correct after such a reorganization.

Such modern documentation archives help to share information between team members and users as well as supporting agile techniques. Even though documentation is important, it should not be an end in itself. The main focus should be on functional, working, and qualitative code. While on the one hand, documentation supports the understanding of the code to reduce design errors, a qualitative code can only be reached with sophisticated software testing mechanisms.

2.6   Code Testing and Code Inspections

Dealing with Software Errors

Fulfilling the software quality factors (see ▶ Sect. 2.1) should be one of the main drivers for software developers and the process in which the software is implemented. But quality is often more an intuitive feeling than a measurable parameter. User satisfaction is especially dependent on psychological factors. In the case of errors and malfunctions, it is also quite difficult to give quantitative statements about whether a software product is correct or not. Writing error-free software is the idealistic dream of programmers, while reality reveals something completely different. In general, there is no software without any errors. That's a fact! Therefore, it is important to deal with this situation by performing tests.

Disadvantages of the Usual Testing

Each program developer tests his software. But these tests are usually executed within a very specialized and limited environment. Tests are always just snapshots influenced by the set of checks of the tested object. During the writing of code, several tests are normally used to check functionality. After the writing, these tests are forgotten or partly deleted, so that they cannot be used anymore if the source code is changed or extended. Another aspect is that developers of code normally have their specialized view on their own code. The tests usually just verify things which are in the mind of the developer. Therefore, the tests will be passed, as the software is only fed with expected values. But tests from another person or the users might fail. Another aspect is that developers only have limited ways to test hardware and that they are also limited by time, as they have to finalize their developments. But software on a 32-bit operating system with one dedicated version of a compiler might behave differently on another 64-bit operating system with a different compiler.

Safety Critical Approach with FMEA

Safety critical components and situations which represent a danger for people or complete hardware systems require a specific approach. It is gaining importance if hazardous equipment is controlled, such as lasers which are not eye-safe or if the degree of automation is increased, so that machines have to make decisions and must provide all necessary safety mechanisms. A specific risk analysis for each component is essential at the beginning of each iteration (see ▶ Sect. 2.9 about iterations in software development processes) which develops such safety critical components. A suitable method is the Software Failure Mode and Effects Analysis (SFMEA) which is a special adaption of the general Failure Mode and Effects Analysis (FMEA) which was originally established in military and space projects. Nowadays, FMEA is a standard procedure for reliability and risk analyses. The FMEA approach identifies critical functions and components, their mode of functioning, their potential effects on the system safety, and related consequences for humans and the hardware. The use of FMEA techniques in the early development phases reduces later costs significantly (Spillner et al. 2011, *l.c. page 270 ff.*).

Geodetic Observatory Wettzell	SFMEA TEST Protocol	Date: Date Iteration: 1 Page: 1/ xx
Component/Module		

Test object:

Part of the component (e.g., function)

Function:

Which functionality is intended to be done

Risk analysis

Failure mode:	Potential effects:	Serverity (S):	Probability (P):	Detection (D):	Risk Priority Number (RPN):
Failure situation	Failure consequences	1 – 10 (++) (––)	1 – 10 (++) (––)	1 – 10 (++) (––)	S x P x D

Test cases:

Failure description / test technique:	Acceptance criteria / metric parameter:
Failure situation and used test methods to detect the failure	Specified: Actual: Parameter 1 _____ _____ ok ☐ not ok☐ Parameter 2 _____ _____ ok ☐ not ok☐

Test result: Accepted ☐   Not accepted ☐

Comments / suggestions :

Notes and recommendations

Authorization: Name: _____   Function: _____   Date: _____   Signature: _____

_____   _____   _____   _____

Fig. 2.16 Example test protocol with FMEA aspects for a risk analysis

Test Protocol

It is not possible to give a complete description of FMEA here. Nevertheless, Fig. 2.16 shows an intuitive example protocol for code tests with FMEA aspects. A similar analysis and test documentation are used to record the status of the system safety of SLR systems (see ▶ Sect. 4.1) at the Wettzell observatory. Each safety-relevant component is characterized with such a description. Critical parts of the component are identified (test objects) and their function is described. Failure descriptions and suitable test scenarios are specified for each individual test case. Acceptance criteria and metric parameters (see next section) make it possible to classify the acceptability of the specific test case. Suggestions recommend improved techniques or different methods if a test is not accepted. Finally, the responsible people sign the document.

Risk Analysis

The protocol in Fig. 2.16 is extended with a risk analysis which has to be made at the beginning of an iteration. It is used to optimize test cases. Such a risk analysis describes potential failure modes or failure situations and their consequences. To improve the classification, an evaluation of the risk with occurrence probabilities and a classification of the severity are used. These probabilities are combined with the feasibility of detecting or preventing of error situation. Each parameter is classified with a number between 1 (low risk) and 10 (high risk). The numbers are determined by the team members which are specialists for the different topics relevant for the test object. The multiplication of all probabilities defines the risk priority number (Spillner et al. 2011, *l.c. page 274 ff.*). Test objects with high-risk priority numbers have to be considered first and require extended test cases or might be showstoppers for the iteration or the project.

Right Method of Testing

Therefore, testing is not always testing. There is a quality difference in testing itself and there is a need for the right interpretation of test results. This requires quantitative criteria which can be measured. Those criteria are defined as code test metrics. These metrics are not only a must for industrial products but also scientific code should check such controlling parameters. This is easily done with automated checks and inspections, using existing tools. But it also requires a suitable team structure and interaction, with the necessary awareness of testing. To establish such a testing landscape, it is first of all necessary to define the meaningful metrics with the focus on the main advantage of expressiveness in scientific environments.

2.6.1 **Code Quality Metrics**

Test Metrics

> *Definition DEF2.9. of «test metrics» (see (Spillner et al. 2011, l.c. page 303))*
> Test metrics or software metrics are parameter sets for software tests, checks, and inspections, which allow a quantitative statement about the quality factors (see ▶ Sect. 2.1) of software and projects, using measuring methods and theories. The ascertainment of the parameters must be performed regularly during the development, and the interpretation must be clear and transparent for the team.

Three Main Code Metric Dimensions

Metrics are mainly aimed at the three principal aspects: the dimensions, the complexity, and the quality (Sneed et al. 2010, *l.c. page 9*). The dimension metrics measure the sizes or volume of code or inherited parts, such as the number of files, code lines, instructions, functions, logic paths, data accesses, and data elements (Sneed et al. 2010, *l.c. page 23 f.*). The complexity metrics measure structural, conceptional, and algorithmic relations and dependencies. But complexity is very subjective. A solution is a comparison between the parts needed to reach the specification goal and the parts which are not (code entropy). But in general, programmers need clear structures and guidance, as they inevitably complicate matters (Sneed et al. 2010, *l.c. page 50 ff.*). The quality metrics measure and try to interpret the measured numbers to quantify the quality factors. They define the degree of fulfillment of software properties compared to the specified requirements, user satisfaction, or purposeful realization (Sneed et al. 2010, *l.c. page 67 ff.*).

Metric Categories for Scientific Developments

There are different categorizations for software metrics in the literature. But at least two categories seem to be very manageable for scientific developments (as a simplification of (Kan 2003, *l.c. page 85*)):

- *Code or product metrics*: Quantitative statements about the software quality and the resulting product are released like code size, complexity, features, performance, errors per 1000 lines of code, number of warnings, tested compilers, tested operating systems, and so on.
- *Project or development process metrics*: Quantitative statements about the project progress, the team structures, the team member behaviors, and the development steps, like bug removal rates, response times, development times, estimated times versus real-times, costs, number of developers, number of add-ons compared to updates, number of contributions, rates of contributions, contributed lines of code in a specific time interval, cumulative bug statistics, and so on.

Purpose of Metrics

The first category offers feedback to the developers to improve the code and to clean up errors.[37] It also supports the comprehensibility of the software, as it allows a classification of relationships between the modules of the software and the internal realizations themselves. Therefore, they can support interpersonal communication, because elusive, abstract notations gain by reliable numbers. Those metric parameters can also be used to compare software parts or complete programs (Sneed et al. 2010, *l.c. page 7 f.*). The second category allows a very suitable evaluation of the development and the behavior of the development team. It is a suitable controlling instrument for projects, as the knowledge of the actual project parameters is a compulsory condition for the supervision. These measurements are also usable for prognoses about the project itself (costs, time, delays, risks, etc.) or for

---

37 Errors are program states which cause a fault resulting in a failure of the program logic within a dedicated runtime environment (Spillner et al. 2011, *l.c. page 4*). They can be caused by internally wrong code sequences or by unexpected interactions from external sources.

future projects (Sneed et al. 2010, *l.c. page 8*). These measurements should be operated regularly to identify changes in the project and the development (e.g., health states of a software or project) (Sneed et al. 2010, *l.c. page 15*). The first category should be observed in each project of different sizes, as its parameters quantify the software quality. The second one becomes necessary if the team increases, so that several contributors must reach a development goal together. In flat organization structures with small hierarchies, it is particularly important to have this feedback for a self-regulation with personal responsibility, which treats all contributors democratically in the same way.

A metric parameter, measure, or «count» (in the measure theory) is a set of symbols, labeling, or numbers, which describe reference numbers and attributes of the measured object. They are obtained using an observation strategy and are usually supported by measurement tools. The measurement can show direct parameters (as lines of code or the byte size of the executable to describe the code or program size) or can give indirect information (such as mathematical/programming operator density within the instructions as a measure for the readability of the code). There are several, sophisticated standards for software quality metrics, such as the well-known ISO 9000 or the IEEE 1061 (Spillner et al. 2011, *l.c. page 304 f.*). But if the developed product of the scientific program is not specified for industry norms, those standards are too oversized for small scientific developments. It is also suitable for small projects to define the goals of the development, which can be detected by answering questions about the degree of fulfillment. Small metrics with reference parameters and actual measured counts can then be used to analyze these questions (Sneed et al. 2010, *l.c. page 15 f.*). This also helps small projects to remain on track to the desired goal.

Counts, Metric Parameters, and Measures

Some examples of useful measures for software are collected in the following list (inspired by (Ristanovic CASE n.d.)). While the first measures are quantitative statements about the code, the final parameters offer some facts about the project and its dynamic.

Characteristic Factors for Scientific Developments

- Total number of lines of code as a parameter for the size of the developed software
- Maximum and average number of lines of code of the modules as a parameter for module sizes and package weights
- Number of code lines in modules in proportion to the total number of code lines as a parameter for code distribution
- Number of code lines per function in relation to the complete number of code lines
- Number of local and global variables and its proportion or the number of external elements as a parameter for scoping
- Number of comment lines as a parameter for the size of documentation
- Number of comment lines in proportion to the total number of lines (for the whole project and for modules) as a parameter for the code explanation
- Number of jump statements as a parameter for code workflow complexity
- Logical depth of instructions as a parameter for readability and complexity
- Maximum and average length of lines as a parameter for readability
- Number of logical and mathematical operators per instruction as a parameter for compactness
- Runtime until error or crash as a parameter for the dynamic behavior of functionality in real use
- Number of detected errors and warnings from the static analysis as a parameter of correctness
- Number of detected errors from the dynamic runtime tests as a parameter of correctness
- Number of errors and warnings relative to 1000 lines of code as a parameter for error or defect density
- Number of traced bug-fix requests from the users as a parameter for correctness of a release

- Number of covered lines (or workflow paths) in the unit tests as a parameter showing the test coverage
- Benchmark results, such as speed and runtime numbers on reference machines as a parameter of runtime performance
- Weekly calculated, cumulative number of errors as a parameter of bug-fix processes and project states
- Number of new code line contributions per week as a parameter for a suitable project progress
- Number of updated code line contributions per week as a parameter of adaptions
- Length of the to-do-list as a parameter for the project size
- Length of the to-do-list over time as a parameter for the project development
- Factorized, rated items on the to-do-list categorized as mandatory, important, and nice-to-have, as a more detailed parameter of the project development
- Number of fixed bugs per week (bug-fix rate) as a parameter of maintenance behavior
- Number of active developers per time as a parameter for project costs
- Number of developer contributions relative to the number of contributed lines of code per commitment as a parameter for the review steps and a suitable progress
- Number of processed, passed and failed unit tests as a number for functional quality
- Time intervals between releases, updates, and upgrades as a number for the integration process

Measuring Process and Its Classification

Such counts or measures are observed within a measurement process. While most metrics are used according to the individual development steps, it is very useful to automatically run the evaluation as often as possible in agile environments. This can help to find failure situations as early as possible by doing acceptance tests and unit tests. Testing has an accentuated importance in the development, as the tests control the implementation of the specifications. Dedicated frameworks with predefined test sets are used. It is not necessary to measure everything. Therefore prioritization is necessary to separate the tests into the test class criteria of critical, important, or desirable checks. The test observations run automated, using suitable tools or self-written scripts. To offer a comparability of the code-based results, it is necessary to standardize and beautify the code according to the layout definitions and coding policies. After this, the real measurement data acquisition can be executed, running the unit tests, code checks, build tests, and so on. The results are saved and archived for interpretation by the developers (Spillner et al. 2011, *l.c. page 41 f.*) and (Crowther n.d.)). This requires a suitable graphical presentation of the results in the form of diagrams, graphs, or other plots, because the measuring process just produces raw data sets. There are several possible diagram styles, such as X/Y-plots, pie charts, multidimensional Kiviat graphs,[38] or cockpit dashboards with a snapshot collection of graphical outputs at a glance as if in an airplane cockpit (Sneed et al. 2010, *l.c. page 323 f.*).

Code Beautifier

To beautify the code before testing, tools can be used such as the Artistic Style 2.02 («astyle»). Artistic Style is a source code indenter, formatter, and beautifier for the C, C++, C#, and Java. The beautifier program takes existing source code and reformats it according to the defined options to standardize the style automatically for all used sources[39] in order to prepare them for the (automated) code

---

38 A Kiviat graph is a multi-vector line graph which shows how multiple variables interrelate.
39 The source code at the Geodetic Observatory Wettzell is beautified using the following options: `astyle –style=ansi -S -w -Y -p -C -c -N <source-code-file>`, where «–style=ansi» sets the scope parenthesis style to Allman/ANSI style, «-S» defines the «switch» indent style, «-w» defines the preprocessor indent style, «-Y» defines the scope closing style, «-p» defines the operators space padding style, «-C» defines the «class»/»struct» indent style, «-c» defines the conversion of tabs into blanks, and «-N» defines changes in the original file (for more information, see (Davidson and Pattee n.d.)).

quantification with a code metric (Davidson and Pattee n.d.). Two style settings are especially important here: the indent options and the bracket style options. The indent option replaces mixtures of different white spaces (tabulators, blanks) with blanks only. The bracket options define the indent style for the scopes parenthesis, such as the Allman/ANSI style (see ▶ Sect. 2.3.1).

Before the interpretation can be started, it is necessary to test the intelligibility of such results (Sneed et al. 2010, *l.c. page 323 f.*) and to classify them into intuitive risk categories. The risk categories (severities) of the results can be high, medium, or low. The cause of result with a high severity must be fixed immediately. The cause of medium-severity risk parameters must be fixed during the next steps of the project. Low-severity parameters are not critical, but their origin should be fixed if there is time. A final release must fulfill a predefined level of fixed problem results and must fit into a predefined level of well-measured results. This is especially important for errors and warnings found (Crowther n.d. *l.c. page 5*).

Metric Parameter Severities

A good example to explain the different interpretations according to the categories is the number of errors detected. The total number of errors and warnings detected provides information about the current state of the software quality as a snapshot. If the errors are calculated in proportion to the total number of lines of code or to 1000 lines of code, this parameter becomes comparable to other projects. With a focus on modules, it is also possible to identify critical modules either because of complex tasks or because of bad coding. The cumulative number of errors, however, shows that if the code becomes more stable, the errors are being fixed by the team and that the state of the project is progressing. It can also partly give hints about team behavior and if the administrative structures work. Nevertheless, only errors that have been detected can be evaluated, which does not provide comprehensive information about the general error situation of the complete software.

Parameter Interpretation

Another example is the number of lines of code. While the total number just gives the information about the development size, a view on contributed lines per time and the contribution frequency offers information about the use of copy and paste for external code libraries (huge jumps in short periods can be hints) and also about the commit behavior of the contributors (agile work thrives on daily updates and progress). The distribution of code lines over the modules or over functions and the average size of modules or functions offer additional information about the design and its implementation. Large modules or functions with huge bodies suggest insufficiently structured code. The number of nested instructions (e.g., the logical depth as a number of hierarchy levels of conditional statements) or a missing proportional balance between comments and instructions exposes code which is difficult to maintain.

Interpretation of Test Results

But the quantitative results are not absolute markers. There is always a range of interpretation. For example, an escalating increase of errors from one test or inspection to the next does not necessarily mean there is a decrease in the quality of the software. The number needs a verification. In relation to an also escalating increase of the total lines of code, it could give another interpretation, as new code can have more errors than an already stabilized one. It could also have a simpler explanation. If the code inspections are carried out with automated tools, an upgrade of a tool can also result in more detections, as usually the tool is also enhanced and might detect more error classes than before. Therefore, besides measurement errors, where parameters are collected in the wrong way, the interpretation is also critical and not always easy. The interpretations must also be made transparent for the team members to prevent unfair blame being cast or negative control structures. It is also very important to have a close look at correlations as a statistical method. There is not always a linear correlation, so that correlation coefficients might be weak. It might be necessary to use other

Land Mines in Test Results

correlation calculations, such as rank-ordered approaches. But a correlation alone does not lead automatically to causality. It is necessary to have a logical or timed change in the data with a significant, real correlation in the observational data (Kan 2003, *l.c. page 77 ff.*).

Therefore, code quality analysis is not so simple. It must start with a change in the behavior of the team members and a willingness to accept and deal with errors and the feedback about errors in one's own source code, and it should result in adequate interpretations, controlling techniques, supervision, and appropriate decisions. While in companies, independent teams or specialized team members do such tests and inspections, the methods are underestimated within scientific projects. But already a very reduced set of techniques can improve software in general.

2.6.2 Dynamic Code Testing

Everybody has an intuitive understanding of tests. It follows more or less the following definition:

Software Tests

> *Definition DEF2.10. of «software tests»* (see (Vigenschow 2005, *l.c. page 20*))
> A software test is a run of a software or of parts of it as a test object to verify and validate properties by means of a comparison between actual values/behaviors and reference values/behaviors as a dedicated control sample under predefined constraints and conditions.

Verification and Validation

A suitable verification demonstrates the functionalities of the developed software. But to improve the quality and find errors, a more rigorous approach is necessary which shows errors and validates the correctness or possible error sources.[40] Therefore, it is better to run the software to the limits during testing. Tests for bad cases are much more interesting than tests for the good ones, so there are usually more tests for the worst scenarios. This is difficult to learn, as usually the developer knows too much about his written code and tends to demonstrate the good runs which function efficiently and well (Vigenschow 2005, *l.c. page 20 ff.*). All indications of errors might be seen as attacks against the abilities of the developer.

Gray Box Tests and Test Data

It is better if the developers are part of the testing team or write the tests themselves. In this situation, most tests are gray box tests as a combination of black box tests and white box tests. Black box tests verify the interface to a written module without knowing anything about the internal realization by checking limits and critical values as input (Forbrig and Kerner 2004, *l.c. page 163 ff.*). White box tests validate internal workflows, conditions, and instruction chains, so that each instruction is run at least once (Forbrig and Kerner 2004, *l.c. page 167 ff.*). The combination to gray box tests is often not completely objective. They are designed as a chain of different test cases, using specific, predefined preconditions, test input data, expected output reference data, and final postconditions (Spillner et al. 2011, *l.c. page 15*). This data can offer more objectivity. It consists of input sets and expected or unexpected reference values or runs (Vigenschow 2005, *l.c. page 20 ff.*). The actual data or runs are then compared to the expected references after the run of the test

---

40 Verification checks if the specified functionality is reached, while validation checks if the results are correct.

chain. Extreme values are ideal here like inputs of zeros, integer limits, empty strings, large strings, or floating point numbers with rounding problems (as 0.1 for single precision float numbers). Nevertheless, complete tests are more or less illusory. But it is possible to combine input situations to equivalence classes. These are groups of input data which are equivalent in the sense that they produce similar error cases (Vigenschow 2005, *l.c. page 60 ff.*).

After the detection of errors, it is necessary to respond to the test report and analyze the errors. The goal is to find the exact cause and origin of the error. This is not always easy in combination with the use of dynamic memory referencing on the heap (segmentation faults), because the error effect can be greatly delayed during the test run in a totally different context when the infected memory is used again. After finding the position, it is necessary to identify the real cause of the error. A simple debugging is easily done by entering tracing text outputs, by entering combinations of test data only to the identified and localized code sequence, or by using more sophisticated development and debugging tools to follow the variables, registers, and memory changes step-by-step (Vigenschow 2005, *l.c. page 25*) (see also ▶ Sect. 2.8 about Eclipse).

Debugging

In the practice of scientific, small- to medium-sized projects, directly written unit tests and system/integration tests are very suitable as test techniques. The work has shown that a combination of both supports agile techniques best.

Unit Testing

The most powerful but also most manageable technique is given by unit tests, even though it might be time-consuming during the development. Every software developer tests his own code dynamically during the development by running the program on a real platform (Forbrig and Kerner 2004, *l.c. page 162 ff.*), because he has to check if the code does what he expects. But then the usual way is to delete the code lines of the test again after the functionality is validated. Some developers keep the tests in their own subroutines or in comment lines. This is much better, but even here the test can only be performed after preparation and only with the knowledge of the developer. But these tests can be the key to a code life cycle if it should be ported to other platforms, extended with new functionality in existing functions (e.g., a function should have additional conditional decisions) or changed in functionality (a function should do something different, but influences other functions). So why not keep the test for further validations and prepare it directly in the form of a test suite for easy further use? It is precisely here that unit tests become necessary.

Usual Test Culture

> *Definition DEF2.11. of «unit tests» (see (Osherove 2010, l.c. page 26 and 31))*
> Unit tests are automatable, additional test codes to validate assumptions about the results and logical behaviors of code functions, which should be tested in the form of a comparison between assumed reference values and actual results. Such a test should be repeatable, readable, simple, trustworthy, and usable for each developer.

Unit Tests

Unit tests are specified for small pieces of code with any logical structure. Small instruction sets are usually functions (or in OOP nomenclature, methods). Therefore, unit tests only validate functions and their behavior and no more. They mask the rest of the software and take a look at functions without any relation to the rest and without any assumptions about. In general, unit tests act as black box tests. Because of this, it is sometimes advantageous if the tests are written by people without knowledge of the internal code. But also the knowledge of a developer can help to test dedicated logical paths in the function. These are conditional statements (if-else, switch-case), loops, or another code which changes internal states (Osherove 2010, *l.c. page 26 ff.*).

Logical Paths

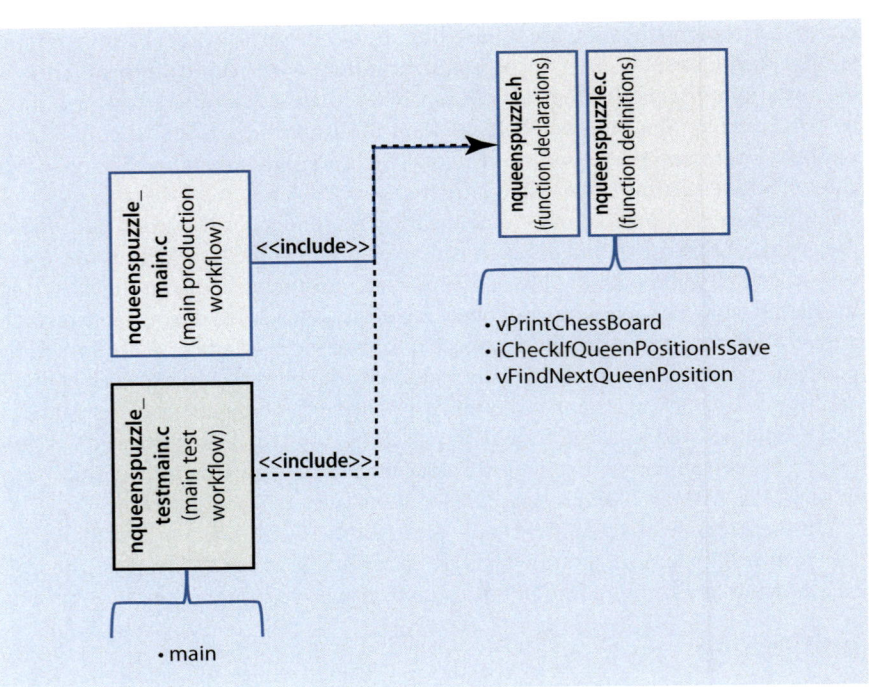

◘ Fig. 2.17   Test-friendly structure with re-usable code modules (using the N-queens puzzle example)

Test-Driven Development

Particularly in the context of test driven development, unit tests are a means to an end. This technique proposes that tests must be written before the actual code is produced. Then it can be directly checked against the unit tests to validate and partly verify the functionality (Osherove 2010, *l.c. page 36*). The problem is that most experience comes during the code generation, especially in developments in new fields of science or in student projects. Regular refactoring changes tests and results in time-consuming updates of the tests. Therefore, it is only suitable in projects in which the complete structure is almost clear or defined earlier and where well-experienced personnel are involved. In all other projects, it is advisable to write the tests as early as possible but at latest during the development of the critical functions. These tests should be kept as references and should be integrated into a test suite or framework.

Test Friendliness

A simple example relating to the previous N-queens puzzle should illustrate what is meant by the use of unit tests. If the error-free run of the function «iCheckIfQueenPositionIsSafe» (see Example 2.3) should be checked, it is necessary to add additional test routines. It is not advisable to include the test code in the final production code, but the situation of the example program where the complete code is located in one source code file does not support a separation. Therefore, it is necessary to structure the code in a «test-friendly» way, which means making the code reusable in the form of modules (see ► Sect. 3.1). This entails separating the N-queens functions from the module with the main function which uses the functionality, by dividing it into two modules «nqueenspuzzle.h/.c» (the «.c»-file contains the function bodies as definitions and the «.h»-file contains the declarations of the functions and used constants) with the calculation functionality and the «nqueenspuzzlemain.c» with the main workflow control as shown in ◘ Fig. 2.17. A simple unit test would be to use another main function which runs several test cases. Each of these test cases prepares the preconditions before a function call, calls the function with the test input data, and checks the resulting, actual output data against the reference data.

Before the test code can be created, it is necessary to think about which tests should be processed. There are different perceptions: as the developer knows a great deal about the internal behavior of a function, he will test the behavior in conjunction with the idea which he had in mind during the writing of the function. Therefore, a first test view is to check at least one test case for each test class. Test classes are characteristic situations for a function which demonstrate the different logical paths through the function logic. Each test case must be passed successfully. For «iCheckIfQueenPositionIsSafe» in the example, it is possible to find 17 classes which are representative enough. Assume a queen is positioned in row 4 and column 4, it is necessary to check the vertical, horizontal, and both diagonal danger zones directly around that queen (test cases 01 to 09). Also the queen's position itself should be tested, as this position is not allowed. There are nine test cases which should return 0 (for «false») if the queen is not safe (see 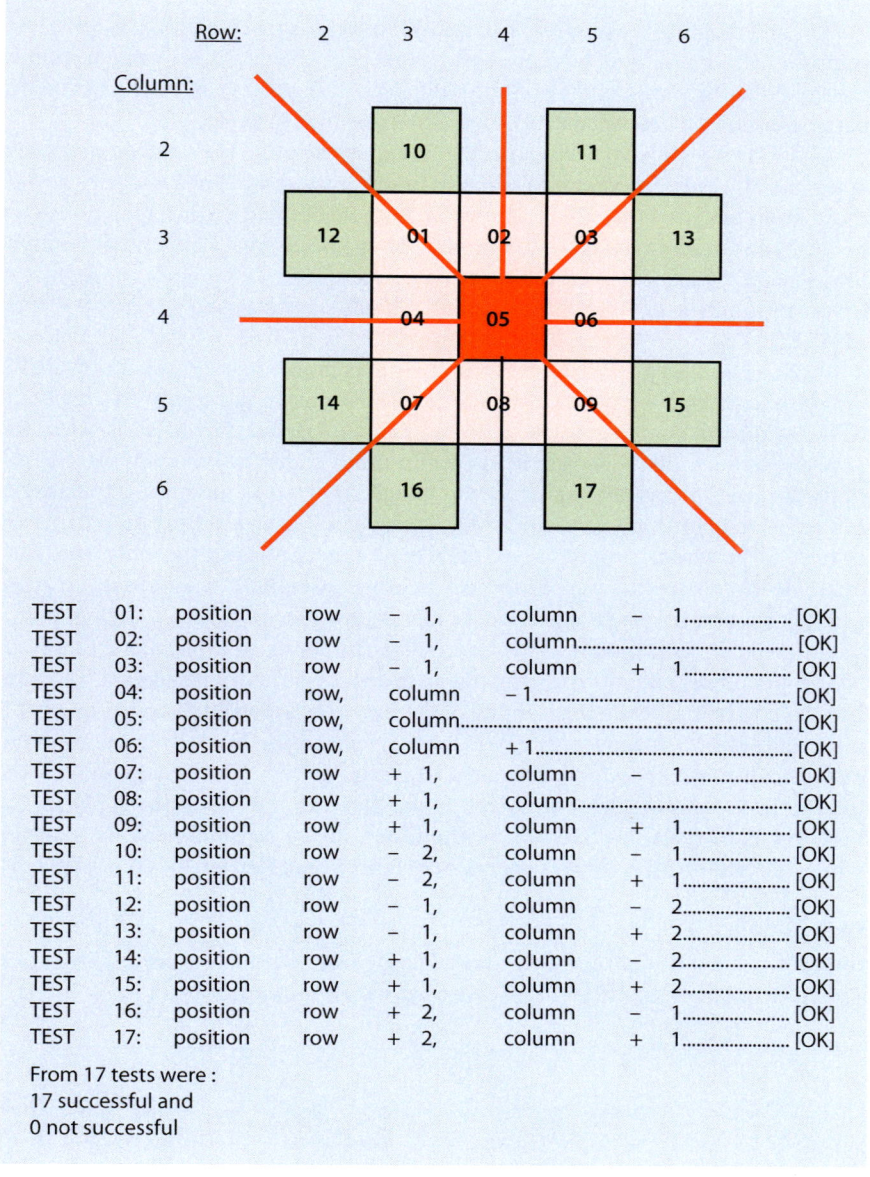 Fig. 2.18).

«False Cases»

```
TEST  01:   position   row    - 1,    column    - 1.....................[OK]
TEST  02:   position   row    - 1,    column...........................[OK]
TEST  03:   position   row    - 1,    column    + 1.....................[OK]
TEST  04:   position   row,   column  - 1..............................[OK]
TEST  05:   position   row,   column.................................[OK]
TEST  06:   position   row,   column  + 1..............................[OK]
TEST  07:   position   row    + 1,    column    - 1.....................[OK]
TEST  08:   position   row    + 1,    column...........................[OK]
TEST  09:   position   row    + 1,    column    + 1.....................[OK]
TEST  10:   position   row    - 2,    column    - 1.....................[OK]
TEST  11:   position   row    - 2,    column    + 1.....................[OK]
TEST  12:   position   row    - 1,    column    - 2.....................[OK]
TEST  13:   position   row    - 1,    column    + 2.....................[OK]
TEST  14:   position   row    + 1,    column    - 2.....................[OK]
TEST  15:   position   row    + 1,    column    + 2.....................[OK]
TEST  16:   position   row    + 2,    column    - 1.....................[OK]
TEST  17:   position   row    + 2,    column    + 1.....................[OK]

From 17 tests were :
17 successful and
0 not successful
```

◘ Fig. 2.18 Basic test cases for the N-queens puzzle function «iCheckIfQueenPositionIsSafe» and the output of the tests

«True Cases»

In addition, it is necessary to validate characteristic positions which are safe. Therefore, it is possible to split the chessboard into eight zones along the danger lines. Each of these zones is a characteristic test class, where at least one test case should be processed (here test cases 10 to 17). These are the representative test classes to validate and verify the specified functionality (see ◘ Fig. 2.18), as it is not possible to check all possibilities on a N x N board, where N is any optional number.[41] All of these test cases must return 1 (for «true»), as a queen is safe there.

Test Case Structure

Each test case realization has the same style in the test main function. Example 2.9 shows one possibility for such a realization. Each test starts with the preparation for how to reach defined preconditions (here a correctly filled array). Then the test is processed, by setting the test input data (here the position of the queen which attacks and relative to this the position of a queen which is attacked) and running the test. The resulting, actual values are saved by collecting the return values. After that, the results are compared to the reference values (here «true» or «false» answer states). This evaluation of the test should return the information that a test failed and which one failed without interrupting the program run until all tests are processed. Additionally, final statistics with the number of processed, passed, and failed tests are quite helpful to classify the quality state of the software (the derived numbers can be used as metric parameters for the quality control). The output is shown in ◘ Fig. 2.18.

Safety Tests

The tests which have been explained are quite helpful. They can be processed automatically even on different platforms in the given setup. But a wider view with more distance shows that not all important tests are defined according to necessary behaviors of module interface functions. For example, what happens if the input list is empty (e.g., «iCheckIfQueenPositionIsSafe (1, NULL)») or if the index of the queen which should be checked is greater than LISTSIZE (e.g., «iCheckIfQueenPositionIsSafe (LISTSIZE+1, aSQueenPositionsList)») or if that index points to a list position which is not yet filled with valid position data? These situations do not appear in the assembled, integrated, complete software from Example 2.3, but may happen, when the new N-queens module is used by others in other contexts. Precisely these situations can lead to crashes (segmentation faults), as external memory is touched. It can result in inaccurate, unpredictable return values. Such segmentation faults are critical for external programs, as they may appear delayed or only sporadic. Therefore, to enable the use of modules in other contexts, it is important to avoid situations where a misuse is provoked or even possible.

Limits and Extreme Situations

In general, tests should always check for the limits of a function. Empty or NULL-parameter situations also belong to these categories for input data. The limit tests help to demonstrate the stability of the function and validate its behavior. It is important even if it appears to work correctly according to its own environment and specification. Therefore, tests are also design drivers. This again confirms the importance of writing the tests at least in parallel with the code writing phase. The best way is to trust the intuition of the tester, in the sense of «error guessing», which is based on a developer's experience (Forbrig and Kerner 2004, *l.c. page 171*).

■■ **Example 2.9**

Simple realization of one test case for the N-queens function «iCheckIfQueenPositionIsSafe» (here test case 01)

```
int  main ( int  argc,  char  *  argv [])           1
{                                                   2
    /*!<b> Variables </b> */                        3
    int  iListIndex = 0;        /*! \li  iListIndex = Index  for  the  list  of    4
                                       solutions  */                              5
```

---

41 For the simple Example 2.3, it would be possible, as the chessboard is limited to 99x99 positions, but it does not make sense, as there will not be new information about the quality of the functionality by doing this.

```
QueenPositionStructType  aSQueenPositionsList [LISTSIZE];              6
                             /*! \li  aSQueenPositionsList = Row/column list of    7
                                    queens positions */                8
    int  iResult = 0;          /*! \li  iResult = Result of a function used for     9
                                    the test assert */                10
    int  iTestNumber = 1;      /*! \li  iTestNumber = Test counter */    11
    int  iNumberOfFailedTests = 0;   /*! \li  iNumberOfFailedTests = Failed    12
                                    test counter */                13
    int  iNumberOfSuccessfulTests = 0;   /*! \li  iNumberOfSuccessfulTests =    14
                                    Succesful test counter */          15
                                                                       16
    /*!<b> Operations </b> */                                          17
    /*! Preparation: Initialize all positions in the position list with −1 as   18
                start value */                                         19
    for (iListIndex = 0; iListIndex < LISTSIZE; iListIndex++)           20
    {                                                                  21
        aSQueenPositionsList [iListIndex].iRow = −1;                    22
        aSQueenPositionsList [iListIndex].iCol = −1;                    23
    }                                                                  24
                                                                       25
    /*! Test case run: process the test with the test input data */     26
    printf ("TEST %02d: position row − 1, column − 1 .... ", iTestNumber++);  27
    aSQueenPositionsList [0].iCol = 4;                                  28
    aSQueenPositionsList [0].iRow = 4;                                  29
    aSQueenPositionsList [1].iCol = 3;                                  30
    aSQueenPositionsList [1].iRow = 3;                                  31
    iResult = iCheckIfQueenPositionIsSave (1, aSQueenPositionsList);    32
                                                                       33
    /*! Validation: Test the asserts */                                34
    if (iResult == 0)                                                  35
    {                                                                  36
        printf ("[OK]\n");                                             37
        ++iNumberOfSuccessfulTests;                                    38
    }                                                                  39
    else                                                               40
    {                                                                  41
        printf ("[NOK]\n");                                            42
        ++iNumberOfFailedTests;                                        43
    }                                                                  44
                                                                       45
    /* ... */                                                          46
}                                                                      47
```

The additional test classes described also show limits in this reduced and simple test environment. Crashes are not handled by the above testing system. They stop any further processing of the responsible test case, but cannot be interpreted. Additionally, the code in Example 2.9 can become unreadable and unmaintainable if a large number of tests is implemented. OOP structures can help a little here, because of the inheritance and the exception handling with try/catch-blocks.[42] But sophisticated test environments need further techniques to automate the runs, standardize the tests, simulate relationships between modules and hardware, and control program runs from external.

Limits of Simple Solutions

A first step toward such environments is automated builds. There are several possibilities with different tools to enable them (e.g., also Eclipse). The «Makefiles» are still very popular in scientific developments (maybe in combination with a

Automated Builds with «Make»

---

42 In OOP environments, it is necessary to classify local parameters and methods with «protected» access rights, so that the derived test class can access them. In the structured programming environments, as given with C, it might be necessary to work with «extern» declarations in the test main module if functions are not accessible via the header files.

configuration tool such as «./configure» from the GNU Build System to derive the dependencies automatically). The GNU Make (GNU n.d.) with the tool «make» (or «gmake» on some systems) interprets «Makefiles», in which a developer defines dependencies between the source files. It simplifies the steps of compilation as described in ► Sect. 2.2, reduces the burden of dependencies, and takes care to detect, compile, and link the latest available source versions. Each time a source file is changed, «make» interprets the dependencies in the «Makefile» and recompiles the necessary parts. It uses the last modification times of the source, object, and executable files stored to decide which file needs to be updated during the build process (Stallman et al. 2010, *l.c. page 1*). Example 2.10 and Example 2.11 show a suitable Makefile with a basically automated dependency creation (see target «`depend:`»), which can be adapted to other projects.

■■ **Example 2.10**
Makefile to compile and link the N-queens puzzle test

```
# ================================================================
# DEFINITIONS
# ================================================================
# Needed programs:
# make, echo, cp, mv, awk

BASISDIR = ../
SRCDIR = $(BASISDIR)src/
OBJDIR = $(BASISDIR)obj/
BINDIR = $(BASISDIR)bin/
USR_LOCAL_DIR = /usr/local/include

# Detect the number of cores of the machine, on which the make is executed.
# This helps to set the -j parameter of the make call, to speedup the build
NPROCS=1
OS:=$(shell uname -s)
ifeq ($(OS),Linux)
    NPROCS:=$(shell grep -c ^processor /proc/cpuinfo)
endif

# Compiler command
# ───────────────
# if no c++-compiler is specified, choose g++ by default
ifndef $(CPPCOMPILER)
        CPPCOMPILER:= g++
endif
COMPILER_FLAGS = -I$(SRCDIR)\
    -I$(USR_LOCAL_DIR)\
    -Wall -pedantic -Wshadow -Wconversion
COMPILER_CALL = $(CPPCOMPILER) -c $(COMPILER_FLAGS) $< -o $@

# Linker command
# ──────────────
LINKER      = $(CPPCOMPILER)
LINKER_FLAGS  = -L$(OBJDIR) #-lm -lpthread -lcrypt
LINKER_CALL   = $(LINKER) -Wl -o $@ $^ $(LINKER_FLAGS)

# Targets
# ───────
all: nqueens
nqueens: $(BINDIR)nqueens

# ================================================================
# LINKER-RULES
```

```
# ==========================================================       47
$(BINDIR)nqueens  :  $(OBJDIR)queenspuzzle.o\                      48
     $(OBJDIR)queenspuzzlemain.o                                  49
  $(LINKER_CALL)                                                  50
                                                                  51
# ==========================================================       52
# SOURCES                                                         53
# ==========================================================       54
SRCS  =   $(SRCDIR)queenspuzzle.c\                                55
   $(SRCDIR)queenspuzzlemain.c                                    56
                                                                  57
                                                                  58
                                                                  59
# ==========================================================       60
# SUPPORTING—RULES                                                61
# ==========================================================       62
                                                                  63
clean:                                                           63
  rm —fR  $(OBJDIR)*.o                                           64
  rm —fR  $(OBJDIR)*.a                                           65
  rm —fR  $(BINDIR)nqueens                                       66
                                                                  67
touch:                                                           68
  @find  $(SRCDIR) —name "*.c*" —type f —exec touch {} \;        69
                                                                  70
rpc:                                                             71
  # Nothing  to  do  with  Remote  Procedure  Calls  generation  in  this  project    72
                                                                  73
install:                                                         74
  # Nothing  to  install                                         75
                                                                  76
prepare:                                                         77
  # Nothing  to  prepare                                         78
                                                                  79
build: clean  prepare  rpc  depend                               80
  # after  generating  the  dependencies  it  is  necessary  to  call  make  again    81
  make —j$(NPROCS)  all                                          82
                                                                  83
depend:                                                         84
# Create  dependencies                                          85
  @cat < /dev/null > makedep                                     86
  @for  i  in  ${SRCS}; do \                                     87
    (echo $$i; $(CPPCOMPILER) $(COMPILER_FLAGS) —MM $$i >> makedep); done   88
# Write  full  path  information  before  each  *.o—dependency   89
  @echo 's/.*\.o:/$$(OBJDIR)&/g' > eddep                         90
  @sed —f eddep makedep > makedep1                               91
  @mv makedep1 makedep                                           92
  @rm eddep                                                      93
# Add  the  rule  to  each  dependency                          94
  @echo 'a' > eddep                                              95
  @echo ' $$(COMPILER_CALL)' >> eddep                            96
  @echo '.' >> eddep                                             97
  @echo 'g/.*\.o:/i\' >> eddep                                   98
  @echo ' $$(COMPILER_CALL)\' >> eddep                           99
  @echo '.' >> eddep                                            100
  @echo '1,1d' >> eddep                                         101
  @echo 'w' >> eddep                                            102
  @ed — makedep < eddep                                         103
  @rm eddep                                                      104
  @echo '# DO NOT EDIT THIS FILE HERE.' > .Makefile.dep         105
  @cat < makedep >> .Makefile.dep                               106
  @echo '# DO NOT EDIT THIS FILE HERE.' >> .Makefile.dep        107
  @rm makedep                                                   108
                                                                109
.Makefile.dep:                                                 110
  @cat < /dev/null > .Makefile.dep                              111
                                                                112
include .Makefile.dep                                          113
```

2

Automatically created Makefile dependencies for the previous Makefile as separate file Makefile.dep

```
# DO NOT EDIT THIS FILE HERE.                                              1
$(OBJDIR)queenspuzzle.o:  ../src/queenspuzzle.c ../src/queenspuzzle.h      2
  $(COMPILER_CALL)                                                         3
$(OBJDIR)queenspuzzlemain.o:  ../src/queenspuzzlemain.c ../src/queenspuzzle.h  4
  $(COMPILER_CALL)                                                         5
# DO NOT EDIT THIS FILE HERE.                                              6
```

Definitions in a Makefile

In general, a Makefile consists of two main parts: the definitions and the target rules. Variables, which should be used within the make process, are defined in the definition section. In the example shown, there are directory settings, which define the locations of the sources, the object files, and the binary (executable) files. All of these settings are defined in the form of variables. Another part detects the number of processor cores to speed up the compilation. Finally, the compiling and linking calls are assembled as variables, which define the settings for the compiler flags such as the include path settings with «-I» and the additional compiler flags to control the optimization, warning outputs, and so on. The linker flags, such as library paths (option «-L») and needed libraries (option «-l»), are also defined here.

Targets in a Makefile

The following target rules always have the same style: «target : prerequisites ...» followed by the «recipe», which must have a tabulator as indent. «target» appoints the result file, which is produced by this rule. «Prerequisites ...» are the input files, on which the target is dependent. Usually the recipe is one or more program calls as a conversion rule which uses the prerequisites to create the target file if any of the prerequisites change (see Example 2.11). Rules do not necessarily need recipes if only dependencies in the build process are relinked (Stallman et al. 2010, *l.c. page 3 f.*). The target description for «all» is one of these re-links, which just indicates that a complete make should build «nqueens». This activates the whole compilation chain, as «nqueens» is again dependent on the object files. These files are dependent on the source files. Therefore, «make» processes all rule recipes to create the object files first, which are then linked to the executable (Stallman et al. 2010, *l.c. page 5 f.*).

Support for Automated, Complete Builds

Recipes can also be independent from prerequisites. This can be used to write rules which support the compilation. Usually they describe cleanups, installation sequences, preparation sequences, or they do an automated dependency generation (see Example 2.10 in the second part of the file). The Makefile example shown creates its dependencies between source files and object files itself. It uses the list of source files in the variable «SRCS» and runs several commands to create an additional file, «Makefile.dep» (see Example 2.11), which is then included to define new target rules to convert source code files to object files. A complete compilation of all sources in the executable can be started by using the target «build:», which can be activated with the command «make build» (the output is shown in ▢ Fig. 2.19). It runs all the steps needed to build the software system with all its intermediate and final files. This build call can be used to automate the builds and run tests in an automated form. The test script first uses the «make build» to build the test and then starts the executable in its environment (different to the suggestion in (Duvall et al. 2011, *l.c. page 67 ff.*), where the test start is integrated in the build script).

Build Output as Test Feedback

With the above Makefile, it is possible to build the whole system with a single automated command. It can also be used for local and private working builds during the development, because after the complete creation, the «make» call without parameters just compiles changed files according to their last modification times. Additionally, it can be used to create the complete, deployable software integration or regular, automated builds for a rapid feedback. But to implement automated builds, it is not only necessary to have build scripts. Hard-coded, relative, or absolute

```
rtwadm@ubuntu:~/Software/tests/nqueenspuzzle/make$ make build
rm -fR ../obj/*.o
rm -fR ../obj/*.a
rm -fR ../bin/nqueens
# Nothing to prepare
# Nothing to do with Remote Procedure Calls generation in this project
../src/queenspuzzle.c
../src/queenspuzzlemain.c
# after generating the dependencies it is necessary to call make again
make -j2 all
make[1]: Betrete Verzeichnis '/home/rtwadm/Software/tests/nqueenspuzzle/make'
g++ -c -I../src/ -I/usr/local/include -Wall -pedantic -Wshadow -Wconversion ../src/queenspuzzle.c -o
../obj/queenspuzzle.o
g++ -c -I../src/ -I/usr/local/include -Wall -pedantic -Wshadow -Wconversion ../src/queenspuzzlemain.c
-o ../obj/queenspuzzlemain.o
g++ -Wl -o ../bin/nqueens ../obj/queenspuzzle.o ../obj/queenspuzzlemain.o -L../obj/
make[1]: Verlasse Verzeichnis '/home/rtwadm/Software/tests/nqueenspuzzle/make'
```

Fig. 2.19 Output of the compilation run, using the described Makefile

include paths and references should be avoided (instead «-I» and «-L» parameters can be used) (Duvall et al. 2011, *l.c. page 67 ff.*). Then the output of the build process is the first test result (see the ▶ Sect. 2.6.3 about static analysis). All compilation warnings should be analyzed and treated as errors. Compilation errors show integration problems, especially if the builds are processed on different platforms.

Software development becomes more structured with the described build and unit test techniques. Then the following steps are taken to implement a new source (see also (Osherove 2010, *l.c. page 155 ff.*)):

Software Creation Workflow

1. Change or add source code and adapt the Makefile.
2. Use private, local builds to compile the changes.
3. Write some unit tests, mainly to test extreme situations and limits.
4. Run the integration build (of the program with the test main).
5. Run the unit tests.
6. Evaluate the feedback of the tests in relation to other projects.
7. Use the automated unit tests to create nightly builds and test feedbacks over all projects to generate project statistics (see ▶ Sect. 2.8 about continuous integration; test feedbacks can be easily converted to HTML pages using text to HTML converter or Perl scripts, to offer them via web servers to the developer community).

For the nightly builds and test feedbacks, it is necessary to create an environment for the executable, which allows it to compile all sources and to run all tests (see ▶ Fig. 2.20). If the code is dependent on external hardware, on other external software, or on other programs, such as databases which are not available on the test system, it is necessary to use stubs. Stubs are controllable proxies or collaborators for existing dependencies in the system, which act like dependent elements

Stubs for External Dependencies

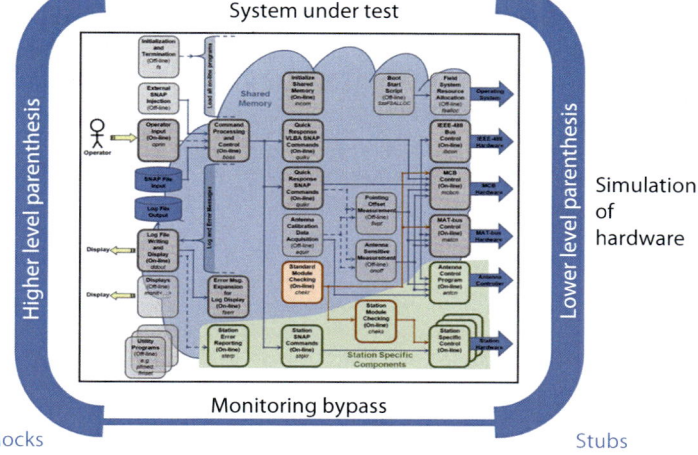

Fig. 2.20 The test setup with mock objects (higher-level parenthesis) and stubs (lower-level parenthesis)

2

Mocks for System Imitations

Wettzell Unit Test Suite

Use Example of the Wettzell Simple Unit Test Suite

Automated Unit Tests

(Osherove 2010, *l.c. page 67 f.*). Stubs can become very complex and take additional time to develop. They also just simulate behavior which might not always be the same as it is in the real system. In combination with real hardware, the simulation of the timing of real systems is difficult with stubs. But they are the only possible way to verify and validate software if the dependency cannot be dissolved.

On the other hand, unit testing also needs to be able to control software and treat it with external inputs to force activities. Therefore, the unit tests must run within a test code: the test environment. This special external dependency is called «mock». A mock object is an imitation of a system part which runs the test to decide if it failed or not. It checks if the tested object interacts with the imitation as expected (Osherove 2010, *l.c. page 101 f.*). The difference to the stubs is that stubs realize lower-level simulations of dependencies, such as the inclusion of external libraries, while mocks realize imitations of dependencies on higher levels, such as user interfaces. The code in Example 2.9 can be seen as something like a mock module (even if the literature would specify it as a «test spy» (Osherove 2010, *l.c. page 104*)).

Sooner or later, the reduced, simple tests as shown in Example 2.9 are too inconvenient to deal with project needs. Sophisticated unit test environments are important especially for larger projects (Forbrig and Kerner 2004, *l.c. page 156*). There are several open-source environments, such as the CppUnit (n.d.), which help to set up their own sophisticated tests. Another one is the lightweight environment of the Wettzell Simple Unit Test Suite.[43] It consists of the following modules, which must be linked to a user-written test module:

- «`simple_testsuite.cpp/.hpp`», which contains the necessary test suite methods, like running of the test, processing of assertions, and printing of reports
- «`testrunner.cpp`», which contains the main function of processing all registered tests or just a selection of them, defined by the calling parameters
- «`simple_shell_macros.h`», which contains macros for the highlighting of the shell outputs in the unit reports
- «`simple_tokenize.hpp`», which contains a basic pattern-matching tool, to split up text strings according to predefined text patterns (tokens)
- one or more user-written, individual test source files (test modules)

Each test module contains a test class, which is derived from the test suite class «TestFixture» (see Example 2.12, which now uses the same test as described in Example 2.9 but with the Wettzell test suite). In this example, the test class is something like a converter class, whose methods call the basic C functions which should be tested. A similar style can also be used to test legacy code, as described in ▶ Sect. 2.4. The test class additionally contains a function «`void run()`», which is started by the test runner of the Wettzell Simple Unit Test Suite. «`run()`» consists of a sequence of test cases, defined as macro calls with the according, individual test case function (in the example «`InDanger01`» for the test case 01; see ▶ Sect. 2.6.2). This function processes the test preparations, the test situation itself, and defines assertions, in this case with the «`ASSERT_EQUALS`» macro, which compares the committed values to start the assertion handling. It prints colorized report outputs and final statistics (see ◘ Fig. 2.21).

Each module is extended with such a test class and the test runner is started automatically via «cron jobs»[44] during the night. The output is then converted to HTML to be transmitted via the intranet.

---

43 The importance of such automated quality tests in fields found in the Geodetic Observatory Wettzell was regularly shown by a colleague, Martin Ettl. Untiringly he developed his own compact unit testing framework, which improved the quality of all the projects tremendously. The resulting Wettzell Unit Test Suite adapted structures and ideas from the test suite used by «Cppcheck» (see ▶ Sect. 2.6.3) and extended them. Therefore special thanks should be given to him for that work at this point here!

44 Cron jobs are timed processes, started by the operating system and defined in a «crontab»-file.

Fig. 2.21 Output of the «simple_testsuite» run for the N-queens test example: **a** with a successful test (first output) and **b** with failure assertions (second output)

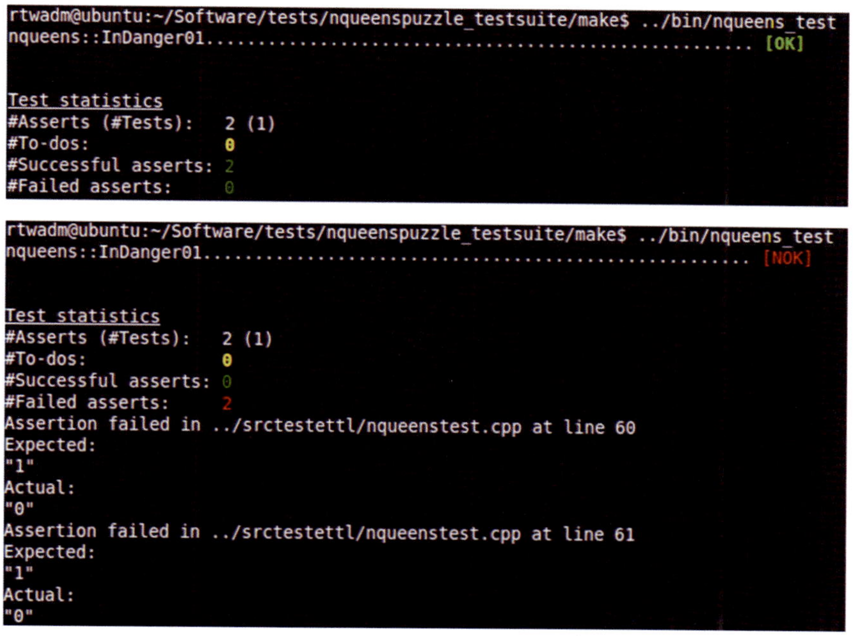

▪▪ Example 2.12

Test class, using the Wettzell «`simple_testsuite`» environment

```
#include <string>                                                       1
#include "simple_testsuite.hpp"                                         2
#include "queenspuzzle.h"                                               3
                                                                        4
/// Test class ("mock" class)                                           5
/// Usually inherits from the unit test suite and                       6
/// from the class which should be tested                               7
/// e.g. class <test> : public TestFixture, public <classtotest>        8
class TestNQueens : public TestFixture                                  9
{                                                                       10
    public:                                                             11
        // Constructor                                                  12
        TestNQueens()                                                   13
            : TestFixture("nqueens")                                    14
        {                                                               15
        }                                                               16
                                                                        17
    private:                                                            18
        // Converter function to call C function uiCheckIfQueenPositionIsSave  19
        // from nqueenspzzle.h/.c                                       20
        unsigned int priv_uiCheckIfQueenPositionIsSave(int iCurrentQueenNumber,  21
                        QueenPositionStructType * pSQueenPositionsList,  22
                        unsigned short & usQueenIsSaveAnswer)            23
        {                                                               24
            return uiCheckIfQueenPositionIsSave(iCurrentQueenNumber,    25
                        pSQueenPositionsList,                           26
                        &usQueenIsSaveAnswer);                          27
        }                                                               28
                                                                        29
        // All operated tests are called in the function "run"          30
        void run()                                                      31
        {                                                               32
            TEST_CASE(InDanger01)                                       33
            /// ... all the other test calls                            34
```

```
    }

    /// Test 01: position row − 1, column − 1
    void InDanger01()
    {
        QueenPositionStructType aSQueenPositionsList[LISTSIZE];
                    /*! \li aSQueenPositionsList = Row/column list of
                                    queens positions */
        unsigned short usAnswer = 0; /*! \li usAnswer = Function answer used for
                                    the test assert*/

        /// Preparations
        for (int iListIndex = 0; iListIndex < LISTSIZE; iListIndex++)
        {
            aSQueenPositionsList[iListIndex].iRow = −1;
            aSQueenPositionsList[iListIndex].iCol = −1;
        }
        aSQueenPositionsList[0].iCol = 4;
        aSQueenPositionsList[0].iRow = 4;
        aSQueenPositionsList[1].iCol = 3;
        aSQueenPositionsList[1].iRow = 3;
        /// Unit test and asserts
        ASSERT_EQUALS(0, this −>priv_uiCheckIfQueenPositionIsSave(1,
                                    aSQueenPositionsList,
                                    usAnswer) );

        ASSERT_EQUALS(0, usAnswer );
    }

    /// ...
    /// All the other tests
};

/// Register the new test class at the test suite
REGISTER_TEST(TestNQueens)
```

Line numbers: 35 36 37 38 39 40 41 42 43 44 45 46 47 48 49 50 51 52 53 54 55 56 57 58 59 60 61 62 63 64 65 66 67

Test Coverage as Metric Parameter

As the test coverage can be used as a quality parameter in the metrics, it is important to identify which lines of code are tested by the unit tests in relation to those which are not (Forbrig and Kerner 2004, *l.c. page 168*). A complete test is almost impossible, but a code or test coverage of 95% is quite good (Forbrig and Kerner 2004, *l.c. page 155*). Test coverage which is greater than 75% is also quite acceptable compared to untested or sporadically tested code. Tools such as «bcov» help to produce coverage information without recompiling a program. It uses breakpoints and therefore it effectively debugs the program automatically, so each breakpoint is removed after it is hit. This produces detailed coverage information with minimal runtime overheads. It can be used to produce human readable text outputs but also more attractive HTML feedback (Neumann n.d.). An example output for the unit tests at the Geodetic Observatory Wettzell is shown in ◘ Fig. 2.22.

Test Feedback History

Ideally, all of the feedback numbers are stored during the lifetime of a project, to be able to iteratively verify the given situations over the whole project period. It would also be necessary to back up the test input data for that purpose (Forbrig and Kerner 2004, *l.c. page 157 ff.*). If the unit test and code coverage feedbacks are used as snapshots of current states, they improve the code quality and reduce interface misuse tremendously. It was only from thinking about tests for the function «iCheckIfQueenPositionIsSafe» of Example 2.3 that hidden possibilities for a misuse were detected which would possibly have caused crashes. Any issues found can be intercepted in an improved version of the code by additional tests of function input arguments with the according exceptions (the updated code is shown in Example 2.13). This is an unbeatable demonstration that tests should always be developed in parallel with the writing of the functional source code (Forbrig and Kerner 2004, *l.c. page 160*). Time-consuming problems and redesigns happen only if the software is not well planned before, as then each change of interfaces will cause several changes in the tests which use them. Therefore, it is necessary to have a clear project design with all

Coverage Report

Current view:	directory
Command:	../bin/testrunner --output=ascii
Date:	Fri Jul 6 02:18:05 2012	Instrumented lines: **61856**	Instrumented statements: **94671**
Code covered: **55.0 %**		Executed lines: **34004**	Executed statements: **51014**

Directory	Coverage		
/build/buildd-glibc_2.7-18lenny7-i386-lqGp8g /glibc-2.7/build-tree/glibc-2.7/sysdeps/generic/		100.0 %	7 / 7 lines
/build/buildd-glibc_2.7-18lenny7-i386-lqGp8g /glibc-2.7/build-tree/i386-libc/csu/		100.0 %	11 / 11 lines
/home/subversion/codecheck/trunk/scripts/coverage /temp/simple_testsuite/rpc/test/		42.6 %	11474 / 26945 lines
/home/subversion/codecheck/trunk/scripts/coverage /temp/simple_testsuite/rpc/test_auth/		24.2 %	656 / 2710 lines
/home/subversion/codecheck/trunk/scripts/coverage /temp/simple_testsuite/rpc/test_errors/		14.5 %	393 / 2710 lines
/home/subversion/codecheck/trunk/scripts/coverage /temp/simple_testsuite/src-test/		79.1 %	12705 / 16072 lines
/home/subversion/codecheck/trunk/scripts/coverage /temp/simple_testsuite/src/		43.7 %	155 / 355 lines
/home/subversion/codecheck/trunk/scripts/coverage /temp/simple_testsuite/srcext/Decimal/		86.8 %	1482 / 1708 lines
/home/subversion/codecheck/trunk/scripts/coverage /temp/simple_testsuite/srcext/Xdouble/		77.1 %	1157 / 1500 lines
/home/subversion/codecheck/trunk/scripts/coverage /temp/simple_testsuite/srcext/constant/		100.0 %	87 / 87 lines
/home/subversion/codecheck/trunk/scripts/coverage /temp/simple_testsuite/srcext/interpol/		86.4 %	57 / 66 lines
/home/subversion/codecheck/trunk/scripts/coverage /temp/simple_testsuite/srcext/arbitrals/		29.4 %	117 / 398 lines

◻ **Fig. 2.22** A coverage report, as typically created by «bcov» for the unit tests at the Geodetic Observatory Wettzell (it is possible to zoom into the files where untested parts are highlighted)

single interfaces first before tests are implemented (see the agile software development in ▶ Sect. 2.9) in order to avoid long-lasting iterations.

▪▪ Example 2.13

Updated, more stable realization of the «`iCheckIfQueenPositionIsSafe`» function from the N-queens puzzle project

```
/**********************************************************
 * function iCheckIfQueenPositionIsSafe
 **********************************************************/
/*! Check if a queens position is safe, so that the the queen can't
    be attacked by the others and return 1 if it is safe
      \param iCurrentQueenNumber -> Current number of the queen which was
             positioned and should be checked
      \param pSQueenPositionsList -> Row/column list of queens positions
      \param pusQueenIsSafeAnswer <- Answer of test (0=not safe, 1=safe)
      \return unsigned int <- Error code (0=ok, 1=error)
      \attention -
 **********************************************************/
/* author A. Neidhardt
   date 2011-08-05
   revisions
   2011-08-05 Original (A. Neidhardt)
 **********************************************************/
unsigned int uiCheckIfQueenPositionIsSafe (int iCurrentQueenNumber,
                        QueenPositionStructType * pSQueenPositionsList,
                        unsigned short * pusQueenIsSafeAnswer)
{
    /*!<b> Variables </b> */
    int iListIndex = 0; /*! \li iListIndex = Index to run through the
                           position list */

    /*!<b> Operations </b> */

    /*! Init return value */
    *pusQueenIsSafeAnswer = 0;

    /*! Check input values, if valid for:<br>*/
    /*! - size lower then LISTSIZE<br>*/
    if (iCurrentQueenNumber >= LISTSIZE)
        return 1;
    /*! - list not empty<br>*/
    if (pSQueenPositionsList == NULL)
```

```
1
2
3
4
5
6
7
8
9
10
11
12
13
14
15
16
17
18
19
20
21
22
23
24
25
26
27
28
29
30
31
32
33
34
35
36
```

```
return 1;                                                                              37
/*! — list continuously filled <br>*/                                                  38
for (iListIndex = 0; iListIndex < iCurrentQueenNumber ; iListIndex++)                  39
{                                                                                      40
    if (pSQueenPositionsList[iListIndex].iRow < 0 ||                                   41
        pSQueenPositionsList[iListIndex].iCol < 0)                                      42
    return 1;                                                                          43
}                                                                                      44
/*! — not yet filled elements are initialized with −1<br>*/                            45
if (iCurrentQueenNumber < LISTSIZE −1)                                                 46
{                                                                                      47
    if (pSQueenPositionsList[iCurrentQueenNumber+1].iRow != −1 ||                      48
        pSQueenPositionsList[iCurrentQueenNumber+1].iCol != −1)                        49
    return 1;                                                                          50
}                                                                                      51
                                                                                       52
/*! Check if queen is in danger of an attack from one of the existing queens          53
    in the list. <br>                                                                  54
    Four conditional decisions are made concerning horizontal, vertical and           55
    the both diagonals. The condition criteria are calculated, using the              56
    indexes.<br>                                                                       57
    The following image shows the decision condition using the                        58
    row and column indexes as criterion (quc is represented by index                  59
    iCurrentQueenNumber and i is represented by index iListIndex)                      60
    \image html 8queenspuzzlehitconditions.jpg "The conditions for the decision."     61
*/                                                                                     62
for (iListIndex = 0; iListIndex < iCurrentQueenNumber ; iListIndex++)                  63
{                                                                                      64
    if ((pSQueenPositionsList[iListIndex].iRow ==                                      65
                    pSQueenPositionsList[iCurrentQueenNumber].iRow) ||                  66
        (pSQueenPositionsList[iListIndex].iCol ==                                      67
                    pSQueenPositionsList[iCurrentQueenNumber].iCol) ||                  68
        ((pSQueenPositionsList[iListIndex].iRow +                                       69
          pSQueenPositionsList[iListIndex].iCol) ==                                     70
                    (pSQueenPositionsList[iCurrentQueenNumber].iRow +                   71
                     pSQueenPositionsList[iCurrentQueenNumber].iCol)) ||                72
        ((pSQueenPositionsList[iListIndex].iRow −                                       73
          pSQueenPositionsList[iListIndex].iCol) ==                                     74
                    (pSQueenPositionsList[iCurrentQueenNumber].iRow −                   75
                     pSQueenPositionsList[iCurrentQueenNumber].iCol))                   76
        )                                                                              77
    {                                                                                  78
        /*! Return 0 if the queen is in danger */                                      79
        *pusQueenIsSafeAnswer = 0;                                                     80
        return 0;                                                                      81
    }                                                                                  82
}                                                                                      83
                                                                                       84
/*! Return 1 if the queen is out of danger => queen is safe */                         85
*pusQueenIsSafeAnswer = 1;                                                             86
return 0;                                                                             87
}                                                                                      88
```

Test Case Generators

Nevertheless, one should bear in mind that tests are only as good as their developers. Other unit testing frameworks follow another system: they use an automatic unit test case generator (e.g., the «API Sanity Checker» (Linux Foundation, The n.d.); see also ► Sect. 3.3 about the use and advantages of generators). These generators create reasonable, randomized input data for the «fuzzing» of library functions to compose different test cases. The tests are designed to find segmentation faults, aborts, program hangings, and so on, caused by different, unexpected input arguments. This random input generation is quite helpful, as usually developers of programs tend to just test positive and already known behaviors. Therefore, they just feed the tests with inputs which usually will not crash the program, as the bugs have already been cleaned. The randomized inputs or argument values which test the representation limits help to generalize the tests. But they also have disadvantages. The test generators usually do not know anything about semantically correct combinations of parameters. Therefore, they do not always check logical, real scenarios. In some cases, fixing the bugs found might solve problems which would never appear in real configurations. Therefore, it is again up to the developer to verify whether an input bug which has been found is relevant and critical or not.

System and Integration Testing

While unit tests are ideal as development accompanying quality-control mechanisms, it is also necessary to test and at least verify the functionality of the complete system over all. The most positive unit test feedbacks do not reveal anything about whether the software system fulfills the specified functionality in the correct way. In this situation, it is necessary to run integration tests.

> *Definition DEF2.12. of «integration tests»* (see (Osherove 2010, *l.c. page 27*) and (Forbrig and Kerner 2004, *l.c. page 161*))
> Software integration testing is a systematic test progress, where several software modules or components are verified and validated as a combined, integrated group together with the necessary hardware to check the complete, specified functionality and its interfaces.

Complete System Tests

Integration Tests

System tests are special cases of final integration tests, which take place at the end of a project before or during the transition phase in order to hand over the software to the user according to a requirement specification (Forbrig and Kerner 2004, *l.c. page 162*). In this case, they are at least acceptance tests. In combination with graphical user interfaces, such tests are quite important. They test a complete logical behavior, from an external input to a final, expected result or behavior. But as final integration tests are only performed at the end of the project, such system tests are necessary during the project to continuously verify the interfaces between different modules from different developers.

System Tests

In real agile projects, it is ideal to run such integration tests regularly, preferably every two weeks. During such a test, the individual developers of the different, currently integrated modules meet and run their software together. Any errors detected should be noted in a short protocol (what was tested and which error or success was seen) to determine the next steps and the time for the next integration test. These more complex tests are independent from the more individual, simpler, and more compact unit tests which can be processed automatically or by each developer alone. The integration tests are then also (technical or informal) reviews (see the following ▶ Sect. 2.6.3), which can be combined with audits,[45] where the developers analyze the common code together for potential errors or discrepancies. Bug fixing or fixes in the program workflow can directly be integrated and checked within the unit tests later on.

Integration Tests as Dynamic Reviews

In this scope, it is worth being aware of a few other review techniques which help to integrate code. «Pair programming» (developing code together in groups of two people) and «buddy testing» (two team members test their code together) are here quite promising for small scientific developer groups (see also ▶ Sect. 2.9 about agile teams) (Spillner et al. 2011, *l.c. page 164 f.*). Informal, common learning and information exchange has a high acceptance especially in the agile

Buddy Testing and Pair Programming

---

45 An audit is an interactive, quality checking discourse to verify the implementation and conformance of a development to guidelines and specified functionalities.

development style, which is founded on informal meeting techniques. Here it is advisable that such meetings should be held regularly.

In general, such integration tests are systematic and selective human code inspections, normally in combination with a program run. But code inspections can also be automated with static code checkers.

2.6.3 Static Code Inspections

Reviews

Not all code inspections require a mandatorily dynamic program run. Human reviews, technical or just informal,[46] can take the form of meetings to discuss the strengths and weaknesses of a design, a software implementation or another project-related part. The object to test must exist at least as a written, preferably formal document. Possible, structured group reviews are informal or technical for the management or the project team, where the code can also be analyzed manually (just on a sheet of paper). These code «walkthroughs» are manual code inspections and support an exchange of experience and knowledge (Forbrig and Kerner 2004, *l.c. page 150 ff.*). But they are «painful» and expensive (The MathWorks, Inc. 2010, *l.c. page 2–5*). Automated static code inspections with analyzer tools can reduce this burden.

Static Code Inspections

> *Definition DEF2.13. of «static code inspections» (see (Forbrig and Kerner 2004, l.c. page 153))*
> Validation of a program or component by using analyzer tools which search for commonly error-prone sections without running the program.

Compiler Outputs

Compilers are also such analyzer tools, as they process further analysis besides the syntax checks. By using the cross-reference list, undeclared variables, inaccessible code, implicit type conversions, and uninitialized variables can be found. Compilers are the first and always accessible level of static code analyzers.

Analyzer Tools

Additional static analyzer tools inspect similar things but also check further common errors, which compilers do not find. Such tools also validate common conventions and standards (Forbrig and Kerner 2004, *l.c. page 153*). Sophisticated tools can detect missing states, state transitions, or other inconsistencies (Spillner et al. 2011, *l.c. page 347 f.*). The techniques used to find such issues are pattern matching, regular expressions (see also ► Sect. 3.3.3),[47] syntactical and formal methods, or abstract interpretation as used in the tool «PolySpace®» (The MathWorks, Inc. 2010, *l.c. page 2–5 and 2–8*). Abstract interpretation is a general theory for the approximation of the semantics of discrete dynamic systems such as computations of programs. Safety, correctness, or soundness is established by proving that a soundness relation is satisfied (Cousot 1996, *l.c. page 324*). Other tools like the open-source tool «Cppcheck» (Cppcheck n.d.) use regular expressions to define search patterns, which are used in a rule file to provoke matching error outputs (Marjamäki n.d. *l.c. page 1 ff.*).

---

46 A technical review verifies the compliance to specifications and guidelines according to a defined protocol together with experts with long experience in the field of work. Technical problems should be found and solved. Discussion is very welcome. Informal reviews are similar but reduced in complexity and time or personnel costs. They are usually just a discussion between developers (Spillner et al. 2011, *l.c. page 163 f.*).

47 Regular expressions are strings which define the patterns for a search criterion to be found in other strings by pattern matching (Brown 2001, *l.c. page 231*).

Static analyzing tools can usually find the following issues (compare (Cppcheck n.d.), (Cppcheck n.d. *l.c. page 3*), and (The MathWorks, Inc. 2010, *l.c. page 8–4 and A-9*)):

1. Errors – bugs found in the code
   - Out of bounds checking
   - Division by zero checks
   - Check the exception handling safety
   - Check for uninitialized variables and unused functions
   - Memory leaks (e.g., undeleted heap memory)
   - Resource leaks (e.g., unclosed file pointer)
   - Illegal references or dereferences of a pointer
   - Overflows
   - Code correctness
   - Check for invalid use of STL
   - Nonterminating calls or loops
   - Unreachable code
   - Inheritance errors
   - Correctness of return values
2. Warnings – suggestions for a defensive programming to prevent bugs
   - «scanf» use without width limits or overflow protection
   - Missing initializations
   - Operator self-assignments
   - Redundant assignments to itself
3. Style – stylistic issues, e.g.:
   - Unused functions, variables or value assignments
   - Redundant code
   - «const»-correctness
   - Order of variable use and limits checks
   - Exception catches as references
   - Consecutive return, break, goto or throw
   - Duplicate branches in conditional statements
   - Missing constructor, copy constructor, destructor or copy assignment operator
4. Performance – suggestions for a faster code, e.g.:
   - Unchanged function parameters as constant call-by-reference instead of call-by-value input
   - Operator use (e.g., the unary increment/decrement operator used as a prefix is faster than when used as a suffix)
   - Use of constructor initialization lists
   - Inefficient variable checks (e.g., for emptiness)
5. Portability – suggestions to improve the code usage on different platforms and with different compilers, e.g.:
   - 32-bit/64-bit portability
   - Obsolete functions of an older standard
   - Non-reentrant functions[48]
6. Information – informational messages that might be interesting for the developer

The numbers of issues encountered over time can be used as a metric parameter to quantify the software state and its development progress. Cumulative plots are particularly good at offering information about the bug-fixing behavior and culture in a team. It is best if the output of the checks, the statistics, and the plots are presented as web pages to the developer team. The generation of these checks should be processed automatically each night. An example plot of a real-life project at the Geodetic Observatory Wettzell is shown in ▪ Fig. 2.23.

---

48 In a multi-processing environment, a reentrant function can be interrupted by a hardware interrupt while processing and safely restarted again without any change in its behavior.

2

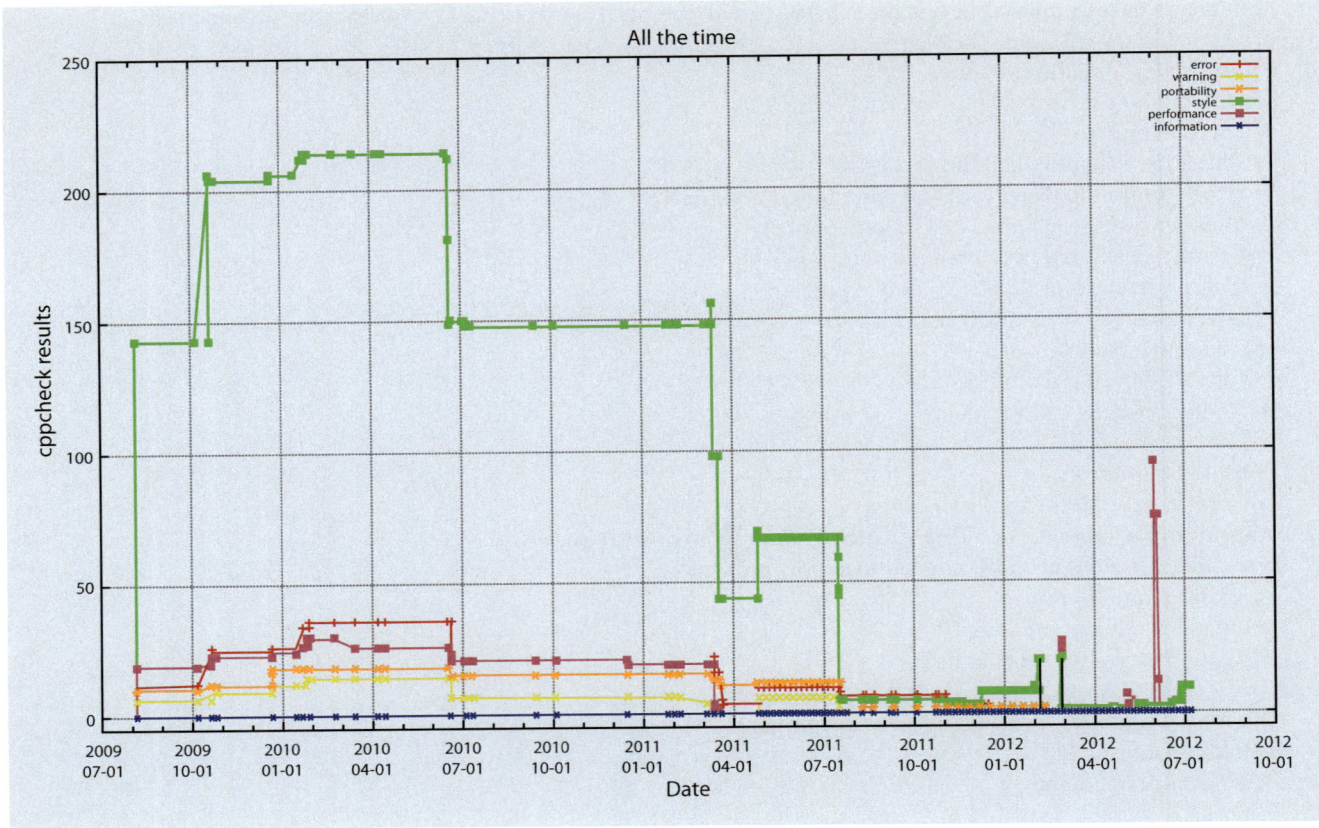

False Positives

A disadvantage of automated, static code inspections is the danger of «false positives». False positives are messages about flaws from the analyzer tools, which are not real error situations. As the checker tools just use logical mechanisms and do not analyze the complete semantics of the code, correct code sequences, which accidentally fit a specific flaw pattern, can give rise to such unintentional messages. Sudden increases in the number of errors are also not always a sign of rapidly decreasing software quality, as they might be the result of newly installed checker tools with new test cases or of one error in a module which is replicated in other parts. This means that the output of such analyzer tools must be analyzed and evaluated in a time-consuming, human post-processing (Spillner et al. 2011, *l.c. page 347 f.*). But nevertheless, it is quite helpful that they offer stable and high-quality software in combination with the runtime checks. Therefore, there are also further methods besides the unit tests, to check, profile, and debug the software dynamically especially under parallel workflow conditions.

2.6.4   **Dynamic Code Analysis**

Operating Systems and Multitasking

Special techniques and analysis are necessary under systems with concurrent use of resources and parallel tasks. Current, modern operating systems[49] support

---

49   Operating systems are a software abstraction level above the hardware to administrate resources, such as the memory, file systems, or input/output devices. They offer interfaces and services via system calls (functions from an operating system library (Tanenbaum 1995, *l.c. page 22 f.*)) to operate and share devices, manage users, do networking, and handle parallel tasks (Tanenbaum 1995, *l.c. page 5 ff.*).

multitasking functionalities even on single processor core machines instead of the former batch processing with its sequential run of programs. The key concept for this (virtual) parallelism is given by processes and threads.

> *Definition DEF 2.14. of «processes»* (see (Tanenbaum 1995, *l.c. page 17*))
> Processes are programs in execution in a single address space, together with all information necessary for the execution in the operating system (as, e.g., swap/core images, program data, instruction counter, hardware register content, etc.) to solve one dedicated task.

Processes

> *Definition DEF 2.15. of «threads»* (see (Singhal and Shivaratri 1994, *l.c. page 15*) and (Tanenbaum 1995, *l.c. page 617 ff.*))
> A thread or «lightweight process» executes a portion of a program (with all the necessary information) within the same process. Therefore, a process can be separated into several, concurrent threads in the same address space, which reduces the overheads for data sharing and switching between the threads significantly.

Threads

Processes themselves can create one or more other child processes. Those processes run parallel, which means that they share the same resources as the processor (Tanenbaum 1995, *l.c. page 17 f.*). This sharing is administrated by the operating system according to fairness in processor time, efficiency in processor use, response times for waiting users, and throughput of task instructions in a specific time interval. To handle this, the operating system takes control at regular time intervals and decides which process gets access to the resources. The process which is currently running gets deactivated. Its running information is saved as a core image in the memory. The runtime information of the newly selected process is copied into the active space, and the reactivated process gets processor time and access to the processor. Such a method, where ready processes are suspended, is an example of preemptive scheduling.[50] The most well-known, fairest, and simplest method in this category is round-robin scheduling (see ◘ Fig. 2.24 and also ▶ Sect. 4.4). Each process gets a specific time interval, the quantum. After such a time period, the active process is suspended and the next in the cyclic list gets the allowance to run. Therefore, each process gets processing time, but no process is prioritized. But sometimes it is helpful if a process can be privileged over others, for example, if essential or interactive tasks compete against long-running batch jobs for the printing queue. Users might not be happy if there is no reaction to button clicks for a while, because there are several other processes in the background which share the processor to an equal degree with them. Other priority-scheduling methods solve this, but have to deal with preventing discrimination. Each process must get the resources after a specific time period, without taking priority, so that it does not «starve» in getting the resources (starvation) (Tanenbaum 1995, *l.c. page 79 ff.*). In principle the same is valid for threads which offer a simpler handling of parallel tasks, as everything is managed within one program and one memory space.

Preemptive Scheduling

According to the definitions, data exchange between processes is always a transfer between different address spaces in such a parallel system (see Interprocess Communication (IPC) in ▶ Sect. 3.3.1). This can be done using signals, messages, or a shared memory. The last one is a memory area to which both processes have

Shared Memory

---

50 Non-preemptive scheduling, run-to-completion, or cooperative methods allow the process to run until it ends or until it returns the resources by choice.

2

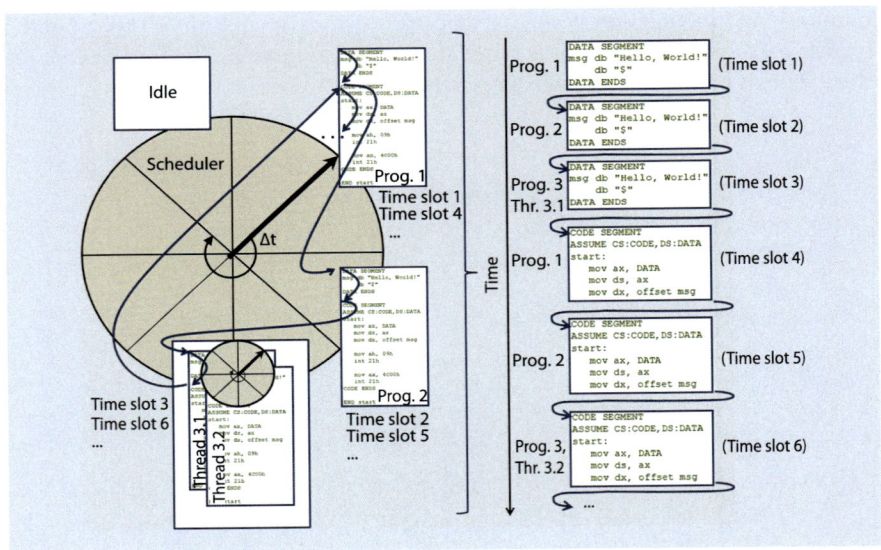

◻ **Fig. 2.24**   Schematic diagram of the round robin mechanism with three processes, where process three has two parallel «lightweight processes» or threads in the same address space (a process is a program in execution; the idle process is an operating system process, which is always executed if no other is ready to run)

write and read rights. The situation with a shared memory is always given for threads as they already share the same address space. But this leads to a problem in parallel processing: critical sections.

Critical Section

> *Definition DEF2.16. of «critical section»* (see (Singhal and Shivaratri 1994, *l.c. page 15 ff.*))
> A critical section is a code segment in a thread or process code in which a shared resource (shared variable, shared device, etc.) is accessed also by other parallel running threads or processes.

If this access happens from different threads or processes at the same time, the integrity of the resource might be violated (see ◻ Fig. 2.25) (Singhal and Shivaratri 1994, *l.c. page 15 ff.*).

Race Conditions

The result is race conditions. These are time critical, parallel workflows, in which the access time of each process to the shared resource decides which result is produced. The behavior is not deterministic anymore, as it depends on several external conditions, manipulating the time. There are two main types of such conditions: data races, where several parallel processes compete to modify a shared memory data, and message races, where messages on a network or in the system compete against each other to reach the receiver. The results are either inconsistent variables or communication and synchronization conflicts (Vigenschow 2005, *l.c. page 88 f.*). These conflicts impose missing, infinite, or unnecessary wait times (Vigenschow 2005, *l.c. page 86*), which force resynchronizations, and exception handling, etc., and are critical errors in real-time systems (see also ▶ Sect. 4.2.1).[51]

---

51 Real-time systems have hard time constraints with hardware restrictions especially in the controller segment, so that a specific task must be processed within a specified delay time or must start precisely at a specified start time after an interrupt. Violations are unacceptable errors (Vigenschow 2005, *l.c. page 221 ff.*). Several operating system

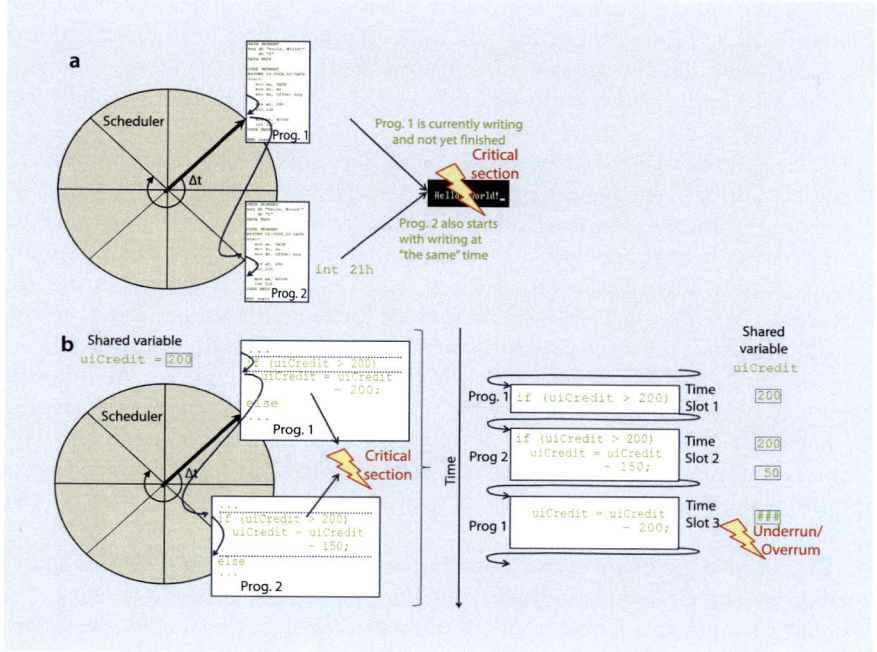

■ **Fig. 2.25** Two examples for critical sections: **a** where two hardware level printings of two processes occupy the device at the same time and **b** where a parallel checking of a condition and changing an unsigned integer value leads to a value overrun

To solve this problem, synchronization mechanisms are necessary, so that only one process can access the shared resource at any one time (mutual exclusion). A well-known mechanism to deal with such mutual exclusion scenarios is semaphores, which provide a simple, sufficient, and general scheme for all kinds of synchronization (Singhal and Shivaratri 1994, *l.c. page 15 ff.*).

Mutual Exclusion

> *Definition DEF2.17. of «semaphore»* (see (Singhal and Shivaratri 1994, *l.c. page 16 f.*) and (Stevens 1990, *l.c. page 148 f.*))
> A (binary) semaphore is an integer variable with two atomic operations to synchronize concurrent threads or processes: «test&set» (decrement) the variable to block and «release» (increment) it to unblock a critical section. Only the task, which successfully sets and blocks the semaphore is allowed to operate the critical section. Each semaphore also has a queue, in which the blocked requesters are waiting for access.

Semaphore

The start count of the integer variable of a semaphore defines the number of possessors allowed to a critical section. Binary (1 and 0) semaphores allow only one process to run the critical section. They are like traffic lights for shared variables with the colors «red» (value 0: requester must stop and wait) and «green» (value 1: requester can continue, but has to switch the state to red). The atomic character of the two semaphore operations is quite important here. The «test&set» operation consists of two subcodes: a conditional statement to

Semaphores as «Traffic Lights»

---

extensions for different operating systems are available, which add a real-time scheduling to a standard operating system (Langmann 2010, *l.c. page 284 ff.*). One representative of such extensions for Linux operating systems is Real-Time Application Interface (RTAI), which enables the writing of applications with strict timing constraints (see (RTAI Team 2010)).

2

Mutex

check if the semaphore can be decremented and the actual decrement. An interrupt after the test condition and a preemptive switch to a process doing the same would lead to an inconsistent state. Both would wrongly get the permission to acquire the semaphore and enter the critical section (as shown in ◘ Fig. 2.25).

Therefore, operating systems offer libraries to handle this mechanism to create, lock, and unlock a semaphore (as described for UNIX systems in (Stevens 1990, *l.c. page 148 ff.*)). There are also additional libraries, for example, for the programming of graphical user interfaces, which offer similar techniques, such as named «mutex» classes (from mutual exclusion). Such a mutex is usually a binary semaphore, defined as OOP class, which is specialized for threads. In a similar way to binary semaphores, only one thread can lock a semaphore and access the critical section at one time. From this point on, only the semaphore owner can unlock it. Additionally there are non-blocking tests implemented (Smart et al. 2006, *l.c. page 451 f.*) or even sophisticated mechanisms to avoid self-blockage. But semaphore operations must be carefully implemented. Missing release operations can result in inconsistent states of the resource, in deadlocks or starvation[52] (Singhal and Shivaratri 1994, *l.c. page 17*).

Dynamic Heap Elements in Parallel Systems

In general, it is difficult to check parallel threads for correctness if they use shared memory or shared resources (Singhal and Shivaratri 1994, *l.c. page 17*). Missing or wrongly used semaphores can result in unpredictable behaviors under specific circumstances. The current processing situation and resulting errors are strongly dependent on the timing situation between concurring tasks on the system. The same program can have a totally different behavior at another time or on another machine with other determining factors. Therefore, it is even more complex to debug the errors found, as the situation might not be reconstructed again. It becomes even more complex if heap memory elements are shared. In this case, it is not sufficient to protect the pointer change itself in the form of a critical section. The complete access to the referenced data structures must also be protected. This is a general difficulty in the OOP design, as copying operations are processed implicitly in copy constructors and copy assignment operator. When these methods are not designed as critical sections with semaphores, they are not thread-safe. This is the case in the STL.

Pitfalls with STL Strings

The string class in particular has some hidden pitfalls. It uses the technique of reference counting. If a string is assigned to another string, no real copy is performed until the content of the copy is changed. The copy just contains a reference to the original memory (shallow copy, see ▶ Sect. 2.3.2). It counts the references to this content, so that it is not deleted while it is referenced by others. This works quite well in sequential programs and improves performance, as real copies are not always necessary. But it results in crashes in parallel systems with threads, because the real copies are processed later when a change requires them. This normally happens outside a critical section, even if semaphores provide protection during the assignment. Therefore, it is necessary to force deep copies (real copies of all elements at the time of the assignment, see ▶ Sect. 2.3.2) within a semaphore-protected critical section instead of shallow copies, forced by the reference counting technique. The constraint is that the performance is reduced, because of the actual copy operations. Example 2.14 shows a solution for STL strings, which was found and is used in projects with threads at the Geodetic Observatory Wettzell.

---

52   Starvation is an infinite waiting of one thread or process to get access to a critical section. Deadlocks block the whole system, for example, if two processes wait for each other, so that the blockage cannot be changed by the system itself anymore (Vigenschow 2005, *l.c. page 85 f.*).

■ ■ **Example 2.14**

Deep copy solution for STL strings to avoid reference counting errors in preemptive systems

```
// ...                                                                      1
                                                                           2
enum {OK=0, NOK};                       /// Enumeration to define error codes (OK=0, NOK=1)   3
                                                                           4
unsigned short usBlockHandlerFlag = 0;  /// \li  usBlockHandlerFlag = Flag to identify the   5
    semaphore handler
semvar<char> cProtectVariablesSemaphore; /// \li  cProtectPrivateVariablesSemaphore = Wettzell  6
        simple semaphore class object
std::string strString1;                 /// \li  strString1 = Empty example string   7
std::string strString2 = "Hello, world!"; /// \li  strString2 = Filled example string   8
std::string strEmptyString;             /// \li  strEmptyString = Empty string for string   9
    deletes
unsigned short usErrorValue = OK;       /// \li  usErrorValue = Error variable for error   10
    codes
                                                                           11
// ...                                                                     12
                                                                           13
/// Acquire binary semaphore and block other requesters                    14
try                                                                        15
{                                                                          16
    cProtectVariablesSemaphore.vBlock (ulBlockHandle);                     17
}                                                                          18
catch (...)                                                                19
{                                                                          20
    return NOK;                                                            21
}                                                                          22
                                                                           23
/// Critical section                                                       24
try                                                                        25
{                                                                          26
    /// Force a deep copy of STL-strings                                   27
    strString1 = strString.c_str();                                        28
                                                                           29
    /// Force a deep copy delete                                           30
    strString2 = strEmptyString.c_str();                                   31
}                                                                          32
catch (...)                                                                33
{                                                                          34
    usErrorValue = NOK;                                                    35
    goto release_semaphore;                                                36
}                                                                          37
                                                                           38
/// Release binary semaphore and unblock other requesters                  39
release_semaphore:                                                         40
try                                                                        41
{                                                                          42
    cProtectVariablesSemaphore.vUnblock (ulBlockHandle);                   43
}                                                                          44
catch (...)                                                                45
{                                                                          46
    return NOK;                                                            47
}                                                                          48
                                                                           49
// ...                                                                     50
```

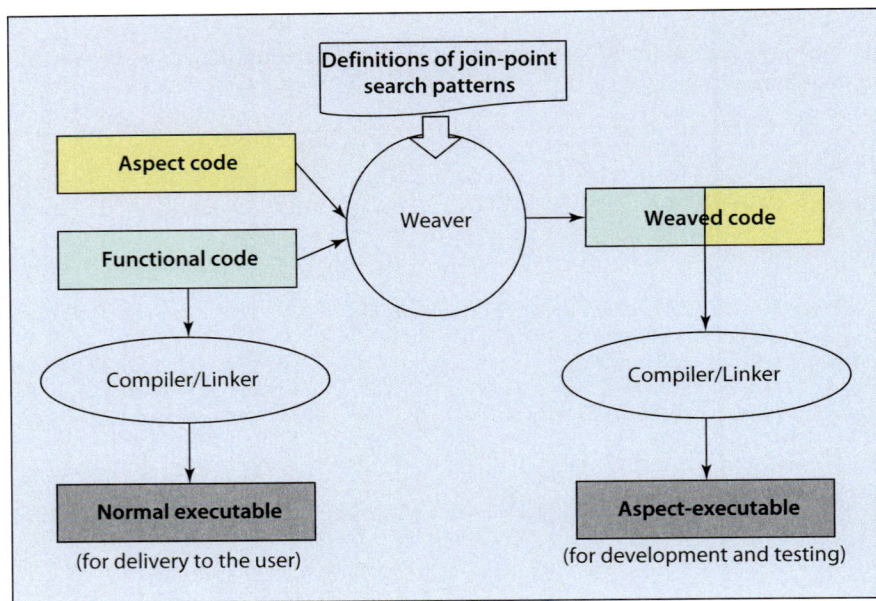

Fig. 2.26 Mixing functional code with aspect «advice» code by a weaver (a generator program as preprocessor before the actual build) to create an executable of the original project code or of the generated weaved code

Such problems are especially difficult to find without suitable testing methods. Thus, it is necessary to run integrated programs under special circumstances and with adequate debugging output in preemptive systems to analyze the dynamic behavior.

Dynamic Code Analysis

> *Definition DEF2.18. of «dynamic code analysis»*
> A dynamic code analysis is a system (integration) test to validate the dynamic behavior in multitasking systems. Usually it is necessary to use additional techniques or tools to get runtime protocols of the current parallel processing.

Aspect-Oriented Programming (AOP)

A simple way to get a first impression of the runtime situation is to insert print statements into the existing code. With this technique, it is often possible to get a first location of a synchronization error (e.g., infinite waiting). A problem is that the code is weakened by additional statements which have nothing to do with the original task of the source code (defined as cross-cutting concerns) and are only used in the tests and not in the final, delivered release of the software. A better solution for that user-defined tracing, logging, and performance checking is given by the approach of Aspect Oriented Programming (AOP). Aspects in the AOP deal with specific concerns, such as how to extend the logic of an existing source code with tracing information. Therefore, AOP defines point-cuts, where such a possible activity can be integrated into the code. Usually these points are defined with regular expressions (see also ► Sect. 3.3.3). A generator program (see more about generative programming in ► Sect 3.3), the «weaver», uses the defined patterns and a separate «advice» source code of the aspects (a code part which is offered in addition to the project code and which should be processed at the position of a point-cut) to merge the original project code with the additional aspect code. The additional code sections appear in the newly produced files at each position of a match (join-points) after the weaver run. This embedded, new code is not relevant for the original logic of the project task, but helps to implement the defined concern task such as the tracing (Wunderlich 2005, *l.c. page 20 ff.*). The developer can now decide if he will build the original released code or the generated weaved code (see ▣ Fig. 2.26).

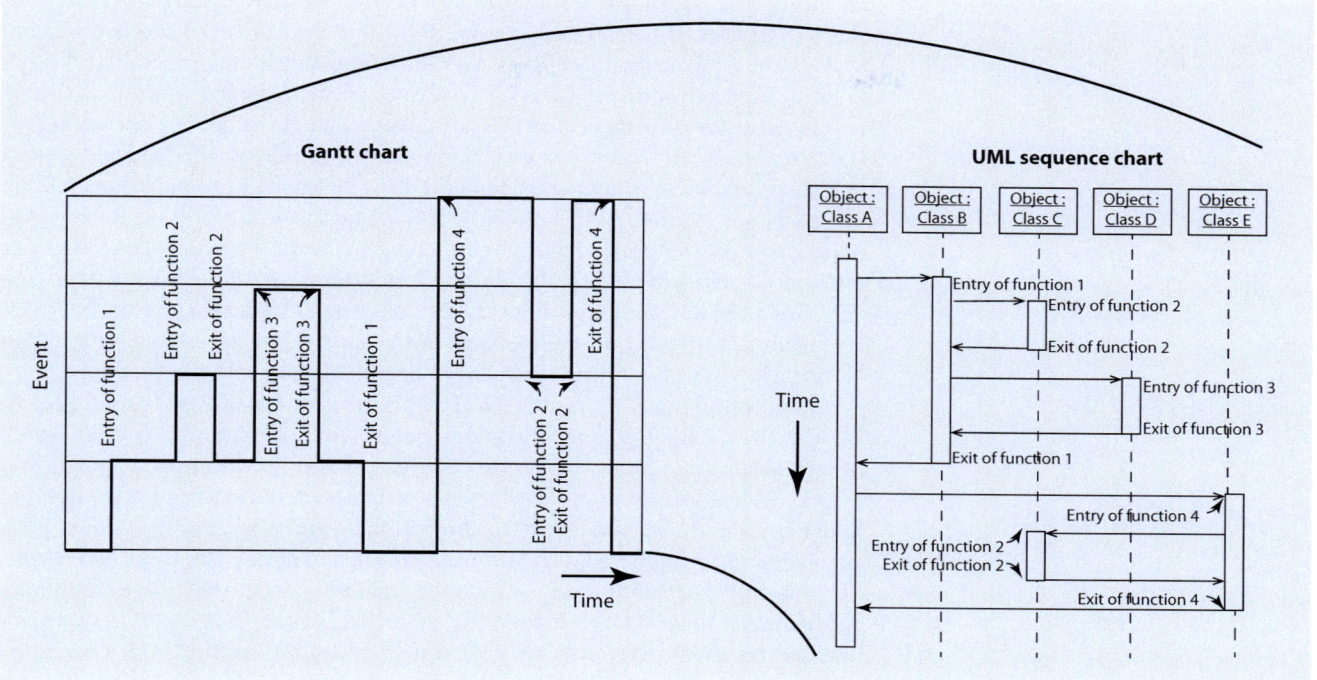

▪ **Fig. 2.27** A Gantt chart can visualize the chronological runtime behavior of a program and therefore communication or synchronization problems in preemptive systems (they can also easily be converted to UML sequence charts)

Several frameworks are available to generate aspect-oriented code, especially for Java applications tracing (Wunderlich 2005, *l.c. page 87 ff.*), even when the popularity of AOP has already reached its zenith. In the field of tracing, logging, and performance checking in particular, the ideas of AOP can be taken to write one's own, rudimentary solutions, using Perl and its ability to search with regular expressions. A Perl-written generator can search for point-cuts in functions at the entries, the exits (return statements and also exception throws) and if necessary at the jump or even conditional statements. Such an approach is used within the communication generator in ▸ Sect. 3.3.3. Each of the join-points found is extended with print statements to write a timestamp (in milliseconds or better) and a function identifier to either a file or to the screen. The results are tracing protocols about the chronological runtime behavior of programs, processes, or threads on a particular machine. They can be used for interpretations in combination with communication and synchronization errors on one machine. The results can also be used to produce Gantt charts,[53] which visualize production or process states over time under real circumstances and are runtime representations of UML sequence charts (see ▪ Fig. 2.27) (Neidhardt 2005, *l.c. page 110 f.*). It is not really possible to make comparisons between processes which are distributed over different machines, as the clocks on the different machines do not run synchronously. Additionally, the created runtime information from one machine must be interpreted carefully, as the timestamps are always adulterated by the time needed for the logging itself. These influences affect all spying techniques of the dynamic code analysis.

Similar information can also be retrieved by controlling a debugging tool (see also page ?? about debugging) with breakpoints (similar to the coverage generation with «bcov» in ▸ Sect. 2.6.2). Debugging tools (debuggers) or profiling tools are also

AOP Solutions for Tracing and Its Interpretation

Debugging Tools

---

53 Originally such charts were designed by H. L. Gantt in 1919 to evaluate the performance of industrial plants (Klar 1995, *l.c. page 135 f.*).

an appropriate way to search for the origins of crashes caused by race conditions[54] on shared variables, which are much more difficult to find. A very famous debugging tool is «The GNU Project Debugger (GDB)». It allows the developer to start the program in the environment of the GDB, define specific stopping conditions (breakpoints), analyze the state of variables, and manipulate them directly during the run (GDB team n.d.). It is only necessary to activate the production of debugging output in the executable by switching on compiler flags for the build (e.g., for the GCC the option «-g», (Stallman n.d. *l.c. page 72*)). In some cases, it is also advisable to use optimized code for the debugging, by switching on the optimization flags of the compiler for the debug build (e.g., for the GCC the option «-O», (Stallman n.d. *l.c. page 93*)). The program can then be operated step-by-step to trace the evolution of variables during runtime with the feedback of the appropriate debugging information. Nevertheless, it is still difficult to find race condition flaws with this method.

Profiling Tools   A better and much more successful way is the use of specialized profiling tools. Besides the debugging possibilities described, profiling frameworks offer several other profiling programs, which can also be used for an automated detection of errors. «Valgrind», for example, is an instrumentation framework for the building of dynamic analysis tools. The tools can automatically detect many memory management and threading bugs and profile a program in detail. The Valgrind distribution currently includes six production-quality tools and three additional experimental tools (Valgrind developers n.d.) which can be started via command line options in the style «valgrind--tool=<tool-name> [valgrind-options]» «test-program» «[test-program-options» (Seward 2011, *l.c. page 9*):

— A *memory error detector* («--tool=memcheck» (Seward 2011, *l.c. page 47 ff.*)), which can detect C/C++ memory flaws, such as illegal memory accesses like over- or underrunning of heap blocks or overrunning of the stack; undefined values; accessing of freed memory; incorrect freeing like double freeing or mismatches between malloc/free, new/delete, and new[]/delete[]; and overlapping memory copying («memcheck» serializes thread instructions to simplify the internal structures, which might change the behavior of multi-threading programs (Seward 2011, *l.c. page 20 ff.*))

— Two *thread error detectors* («--tool=helgrind» (Seward 2011, *l.c. page 101 ff.*) and «--tool=drd» (Seward 2011, *l.c. page 116 ff.*)) for the detection of synchronization errors in C, C++, and FORTRAN programs that use the POSIX threading primitives, «pthreads», to find misuses of the library, potential deadlocks arising from lock ordering problems, data races, lock contentions between threads, or false sharing of cache between threads on different processor cores («drd» is more sophisticated and supports all multi-threading paradigms such as locking, message passing,[55] automatic parallelization,[56] and software transaction memory[57])

— A *cache and branch-prediction profiler* («--tool=cachegrind» (Seward 2011, *l.c. page 72 ff.*)) to find misuses or performance intensive uses of the different memory cache levels[58]

— A *call graph generating cache and branch-prediction profiler* («--tool=callgrind» (Seward 2011, *l.c. page 88 ff.*)) to record the runtime

---

54 In a race condition, the timing and therefore the time order of single instructions defines the result of an algorithm.

55 Message passing is a communication between threads, using messages over a communication network.

56 Automatic parallelization splits the program into parallel processable parts during compilation.

57 Transactions are communications which are processed completely or not at all if there is an error.

58 Cache buffers are copies of recently used memory blocks in the memory and administrated by the operating system to optimize the access performance, such as reducing the number of hard drive accesses (Tanenbaum 1995, *l.c. page 382*).

call history among functions in a program's run such as a call graph (similar to the call graphs described for the automatic generated documentation; see ▶ Sect. 2.5.1)

- A *heap profiler* («--tool=massif» (Seward 2011, *l.c. page 132 ff.*)) to measure the used heap (and optional stack) memory size of a program including the bookkeeping and alignment bytes to optimize the speed and swapping behavior[59]
- A *heap/stack/global array overrun detector* (experimental «--tool=exp-sgcheck» (Seward 2011, *l.c. page 151 ff.*)) to find overruns of stack and global arrays by using an heuristic approach, derived from an observation about the likely forms of accesses, because, for example, it is highly likely that the same memory will be accessed again
- A *second heap profiler* that examines how heap blocks are used (experimental «--tool=exp-dhat» (Seward 2011, *l.c. page 145 ff.*)) to track the allocated heap blocks and inspect every memory access to find, for example, excessive memory allocation, useless or underused allocations, or never used areas and create statistics about them (usable as metric parameters)
- A *SimPoint basic block vector generator* (experimental «--tool=exp-bbv» (Seward 2011, *l.c. page 154 ff.*)) to find basic blocks of a program as linear sections of code with only one entry point and one exit point, which can be used by the SimPoint method to order blocks with similar behavior for a simulating extrapolation of the runtime behavior for a total program without running a complete benchmark

The dynamic code analyzing tools, like Valgrind, are designed for a non-intrusive check of existing executables. Independent from a selected tool, they run the program in a synthetic Central Processing Unit (CPU) (see ▶ Sect. 4.2.1), provided, for example, by the Valgrind core. This takes control over the program and reads debugging information from the executable and associated libraries (if these are compiled with such information, for example, with the «-g» option of the GCC). This debugging information is used for error outputs to identify the source code locations. Within the synthetic CPU, the selected profiling tool extends the existing code with additional instrumentation code, so that, for example, the Valgrind core can then continue the execution of the instrumented code. The analyzer simulates and checks every single instruction in the test program. Thus, the profiling also includes all dynamically linked libraries (e.g., the GNU C or X11 libraries; ▶ see Sect. 4.3), which might produce additional error outputs. These prints, which are not interesting for a specially tailored program, can be suppressed by using the analyzer flags offered (Seward 2011, *l.c. page 3 ff.*).

Non-intrusive Checks

For daily use, the memory and heap error detectors (like «memcheck» and «helgrind» in the Valgrind framework) are the most important. Memory problems and data races (such as the reference counting problem in STL strings) can be detected and located along with possible deadlocks of the whole system. These are the usual flaws in multi-threading environments. A problem is that the profilers slow down the processing. In the case of Valgrind, programs run 10–50 times slower than natively, which is dependent on the type of tool and the amount of additionally included instrumentation code (Seward 2011, *l.c. page 3 ff.*). Additionally, the results must be investigated carefully in a human and time-consuming post-processing, because the dynamic code checkers can produce «false positives» similar to the static code analyzing tools. False positives can be the result of higher compiler optimization if the code does not fit well for that (Seward 2011, *l.c. page 4 ff.*).

False Positives

Nevertheless, testing is essential for high-quality software. A functional and sophisticated testing landscape can be established for the whole team by using the testing tools, a structured approach, and the necessary human and psychological

---

59 Swapping is a technique used by the operating system to swap memory blocks to the background memory (e.g., hard drive space) if not enough main memory is available. If swapping starts, the processing times increase, as hard drive accesses are much slower than main memory accesses (Tanenbaum 1995, *l.c. page 372 f.*).

comprehension. The establishment of such an environment takes time initially, but once established supports developers in all projects over the whole period with the quality feedback necessary.

2.6.5 Testing Landscape

The techniques, methods, and tools described are an ideal environment for programmers to improve the quality of their programs. Nevertheless, such techniques are rarely established in scientific development teams. While they are a must in the business projects of companies, they are not required so often in university institutes, so are commonly not used or even not known. But it would be so easy to offer a suitable code testing landscape.

Testing Landscape

> *Definition DEF2.19. of «testing landscape»*
> A testing landscape represents a suitable environment to offer metrics, techniques, methods, and tools to developers for dynamic code tests, static code inspections, and dynamic code analysis. It should be organized as uninvasively and unobtrusively as possible, should be easily accessible and available, and should regularly give psychologically neutral and constructive feedback for team meetings and agile discussions.

Costs and Benefits in the Risk Management

The problem in scientific fields is that it is difficult for scientific code writers to rate the advantages and costs of such large and more structured tests. Even test managers in business projects are also often compelled to balance budget and resource requirements for tests with their potential benefits. Metrics and their associated qualitative and quantitative statements are necessary to demonstrate the benefits for a reduction of production risks, because of the tests used. The more errors that can be found in the early production phases, the more reduced the costs are for the later development phases, for costs in the later use of the software or for follow-on costs. But increasing the requirements also raises testing costs. Economically the optimum for that situation is the intersection of the adverse trends of follow-on costs (caused by errors) and of costs for quality management. But this is a complex field of work (Spillner et al. 2011, *l.c. page 82 f.*). It is even more complex for scientific needs, as such quality mechanisms are less than secondary to the primary development and research goals, for which the programming is nothing more than a useful appliance.

Prioritization of Tests

Nevertheless, it is worth having such a testing landscape as part of a scientific development project. The techniques allow the developer to ascertain that a software really does what it is designed for (e.g., if satellite passages are calculated, it is important to know if the functions used always calculate in the correct way or if there are situations where the results are wrong, e.g., because of precision errors in iterative floating point arithmetic). This is quite useful for deliverables in third-party funded research projects. This verification and validation is much more important if the software is used to control huge hardware systems, like those found at observatories, where telescopes must be controlled in a safe way for humans and the systems themselves. However, the funds are normally only available for the primary scientific goal. Therefore, it is important to determine which tests or aspects of the tests are most useful and what their benefits are. This classification and consideration should include testing the most important parts of the final system first in order to get a general evaluation of the complexity, constraints, and testability for further more detailed tests. It is also advisable to use existing and often scalable testing methods and tools as often as possible (Spillner et al. 2011, *l.c. page 80 ff.*).

In general, the use of a testing landscape is a sign of a constructive testing and error-handling culture. Errors are normal, especially in complex systems. To produce no errors is impossible! Therefore statements like «my software is without errors» or «we make no errors» are signs of the mishandling of error situations in the development. Developers and teams should learn to deal with errors constructively. This does not mean laying the blame on the guilty person or covering up errors and hoping to fix them stealthily. It means using error situations to learn from them. The team must lose the fear of making errors or of being unable to find the cause of them. Additionally, it must learn how to handle them to find a solution. Hasty reactions can often turn a minor problem into a disaster at a later date. To ensure a fruitful team environment, it is necessary that on the one hand, the team leadership offers enough freedom to allow people to develop their own ideas, role models to reduce operative chaos, identification with interesting tasks, and a constructive management of mistakes (Vigenschow 2005, *l.c. page 213 ff.*). On the other hand, the team members must have the ability to deal with and to accept criticism, a feeling for error sources (error guessing) and the ability to find them; to distinguish between important and unimportant tasks; to have enough discipline, accuracy, endurance, a high frustration tolerance, and the necessary knowledge (e.g., for multi-threading problems and operating system mechanisms); and to be team-minded and communicative. Organizational blindness with statements like «we have always done it this way» and an obsession with the technical aspects of optimizing tests or extending them are bad signs for a constructive, purposive testing culture (Spillner et al. 2011, *l.c. page 288 ff.*).

Testing and dealing with the results are always related to the psychology of the team members. It is always a difficult thing to find the ideal way to motivate people sufficiently to manage errors without suggesting that it does not matter at all that there are errors. Experience has shown that each team is different in this point. But it has also shown that a suitable testing landscape can improve project quality tremendously. Automated tests and inspections play the key roles here. These automated machines also help the team to accept errors more easily, as a criticism from a machine is less intimidating than a remark from another team member.

Culture for the Constructive Handling of Errors

2.7 Version Control

One problem in software development is keeping the different, historical development states with their different source files manageable. This becomes even more complex when more than one developer writes code for a project. Local, manual solutions with understandable filenames and an included time tagging quickly exceed the limits of a manual procedure. Thus, it is necessary to use a suitable version control system, a program suite which takes the burdens of such complex problems from the shoulders of the programmer.

> *Definition DEF2.20. of «version control system» (see (Wassermann 2006, l.c. page 15 ff.))*
> A version control system is a file server which manages files and directory structures and adds a versioning to the administrated documents, while it traces their historic evolution. It allows safe and easy multi-user access and has mechanisms to control revisions, tag versions with identifiers, and split side branches for parallel development stages.

Version Control System

2

There are several version control systems available for software versioning, dealing in similar ways with the version and revision management. But before one of the most popular is explained and a description given of how it can support scientific developments in particular, it is necessary to define a suitable and easily controllable structure for project code.

2.7.1 A Suitable Project Directory Tree

Hierarchical Document Tree

For a common use of sources for centralized builds (see ▶ Sect. 2.6.2 (Duvall et al. 2011, *l.c. page 74 ff.*)), it is generally useful to organize directories for projects of software system always in the same way.[60] Therefore, each developer can navigate through the source code structures even if they come from others. But such structures are also the basis for the version control, as they create a logical structure for change logs. Each project is thereby a directory tree with logical folders, containing the according files and documents in a hierarchical way. The directory names and the filenames follow the coding layout (see ▶ Sect. 2.3.1).

Repository Project Directories

The first level of this hierarchy represents the repository view of the complete software project[61] (the explanation for «repository» follows in the next section) (Wassermann 2006, *l.c. page 47*).
— *Trunk:* the main branch of the project software
— *Branches:* the derived development sidelines of the project software
— *Tags:* the named release snapshots of the project software, which represents a published and tagged version

Program Directories

The second level represents the program systems. It is a combination of logically dependent programs, scripts, individual basis modules, and the necessary global parts (such as documentation documents, global data, or global configurations), which are widely used in the program system. A memorable name for a program system should be used to identify the appropriate directory. It may also be helpful to use several hierarchical layers at this level to arrange the logically dependent parts (e.g., the first system level is named «system_monitoring», which contains the subsystems «main» with the programs of the central runtime environment of the monitoring system, the subsystem «sensors» with programs for the individual monitoring sensor, and so on). The main part of the program system consists of different, self-contained programs or scripts (independent executables of a software system, such as programs to control the hardware) or individual basis modules (generally independent file containers with a collection of internally dependent code, which is required in several programs, such as a file with definitions of constants or a file with routines to convert different representations of time formats[62]). Memorable names for the programs, or respectively modules, are also convenient. The program and module folders again contain further subdirectories, which separate the different elements into manageable units (e.g., intermediate files are separated from source code or from the finally resulting executables; see the following list). It is advantageous to avoid symbolic links[63] in the version-controlled trees to be compatible with all platforms and other systems, even if some of the version control systems can deal with them on some platforms. It is advisable not to change a once defined structure too often, as it always has some influence on the building process, for

---

60 See ▶ Sect. 2.1 about the definition of a software project and ▶ Sect. 2.1 about the description of software systems.
61 A complete software project in this field is a complete software system with all the necessary parts, like program files, configuration files, documentations, and so on, plus structural mechanisms to represent and organize the project life cycles.
62 For more about modularization with an explanation of modules, see ▶ Sect. 3.1.
63 Symbolic links are references to other files in the file system.

example, on the content of the Makefiles. A possible complete program structure, which is used very successfully at the Geodetic Observatory Wettzell, is shown in the following itemization. But not all appointed directories are mandatory. Empty ones can be left out.

- **<program system name>:** specific folders of logically dependent programs or modules (it might be helpful to use several, hierarchical sub-folders here)
  - **<program/module name>:** all source code parts of a dedicated program, which must be compiled together get executables (C/C++ or legacy code in FORTRAN etc.); it contains the descriptively named sub-folders with an internal structure of
    - **bin:** contains the executable binaries
  - **conf:** contains all configuration files
  - **data:** contains needed data for processing or testing
  - **doc:** contains all program-specific documentations (eventually in different, logical sub-folders; partly generated with a documentation generator, see ► Sect. 2.5.1)
  - **make:** contains the Makefiles or Microsoft®Visual C++®project files for the program
  - **obj:** contains the intermediate object files
  - **rpc:** contains generated remote procedure code (see ► Sect. 3.3.3)
  - **src:** contains all program-specific source files (including the relevant header files; the source files can be organized with subdirectories)
    - **src<identifier>:** sub-folder with specific source code files which logically belong together (for instance, to control dedicated hardware which is specific to the program and is not worth developing as a separate, program-external module)
  - **srcext:** contains subdirectories with source code of program-external modules, e.g., from a software toolbox (see ► Chap. 3; each toolbox module has its own Makefile and consists of source code files with the relevant header files; it can also be a complete external project directory tree)
  - **<individual directories>:** specific parts which cannot be assigned to the previous folders
- **<script name>:** all source code parts which can be interpreted as scripts (Perl or Shell scripts); it contains the descriptively named sub-folders with an internal structure of
  - **conf:** contains all configuration files
  - **data:** contains needed data for processing or testing
  - **doc:** contains all program-specific documentations (eventually in different, logical sub-folders; partly generated with a documentation generator, see ► Sect. 2.5.1)
  - **run:** possible scripts with a complete start setup to run the program for tests
  - **src:** contains all program-specific source files (the source files can be organized with subdirectories)
    - **src<identifier>:** sub-folder with specific source code files which logically belong together (e.g., to control dedicated hardware which is specific to the program and is not worth developing as a separate, program-external module)
  - **srcext:** contains subdirectories with source code of external modules, for example, from a software toolbox (see ► Chap. 3; each toolbox module has its own Makefile and consists of a single source file with the according header file or again of a project directory structure)
  - **<individual directories>:** specific parts which cannot be assigned to the previous folders
- **sql:** contains Structured Query Language (SQL) files with database commands (the files can be organized with subdirectories)
- **make:** all global Makefiles, which can be used to build the whole program and to install scripts and executables in the operating system-specific

directories (another possibility is to hold the Makefile directly in the program root folder)
- **doc:** all global program documentations (see ▶ Sect. 2.5) separated in
  - **developer_manual:** contains all global developer relevant information (partly generated with a documentation generator, see ▶ Sect. 2.5.1)
  - **user_manual:** contains the manual for users, who want to run and use the program without changes to the code
- **conf:** all global configuration files which are used for more than one subset of program tasks
  - **data:** all global data sets which are used for more than one subset of program tasks
  - **<individual directories>:** specific parts which cannot be assigned to the previous, global folders

Each version-controlled project should fol such a structure, which fits into each available version control system. It should not usually be necessary anymore to change it extensively. Therefore, it is quite important to design the directory structure in the best way, as redesigns (e.g., splitting of code repositories) usually destroy the log history or can cause inconsistencies in the directory structure in the version control system. The structure described is the result of the development organization at the Geodetic Observatory Wettzell, is used for all projects there, and has not been changed for several years now.

2.7.2 A Suitable Version Control System

Code Version Management Tools

Usually version control systems use similar techniques. But they are user friendly, complex, and distributable to varying degrees and have different unique features. The Eclipse Developer Report of the year 2012 (Eclipse Foundation n.d.) shows that there are three main version control systems in use: Subversion®(SVN), Git, and Concurrent Version System (CVS). While CVS is still only a historic relict in the world of version control, Subversion® from Apache™ is still the market leader (according to the report with about 45%). But Git is increasingly gaining in popularity in the distributed open-source community.[64] There are pros and cons for both systems. Therefore, it is necessary to take a look at the general principles of version control systems to understand the techniques of versioning better. This is done in the following sections in the example of Subversion®(SVN).

Repository

What is common to all is the use of a code repository. In the case of Subversion®, the repository is a centralized memory in the form of a file system tree. The repository is administrated by a file sharing server, to which several client programs establish a connection to read, change, or add content (for more about client-server-architectures, see ▶ Sect. 3.3.1). Writing files to the central repository provides files and directory structures which are local working copies on a specific computer to other developers. Then they can read them, perform an update, or check out new files to get their own, local working copy of these files and directories. Besides this simple file server functionality, the repository server tracks all changes over time to be able to restore each historical version besides the latest one (head version). This meta-information enables queries about versions, such as ascertaining the identity of the developer who changed a file on a dedicated day (Pilato et al. 2009, *l.c. page 1 f.*).

Versioning Model «Lock-Modify-Unlock»

The problem with sharing files among different users is to allow parallel use and even the option to change the different files without the conflict of mutual

---

64   Git was originally designed by Linus Torvalds to support the development of the Linux operating system kernel (Loeliger 2010, *l.c. page 1*).

overwriting. There are two versioning models to solve that problem. The first one is the «lock-modify-unlock» solution. This allows changes only from one user at a time. He needs to lock a file which should be changed before it can be modified. After modification, the file is committed to the repository and unlocked. From this time on, it can be updated in other different working copies. The problem with this solution is that locks can be forgotten by the developers, change actions might be serialized unnecessarily because of the mutual wait, and relations between files can result in conflicts in multi-user systems, as content relations are not considered.

Therefore, the second versioning strategy deals with this situation better by using a «copy-modify-merge» solution. Most version control systems use this strategy. They use the local working copies, which are checked out from the repository to the local file system. These copies can be changed in parallel by the developers on their systems. Each one makes his changes in his own copy of the project tree and local files. After finishing this task, the first developer commits his changes to the repository. If the second one wants to do the same and has changes in a file which was changed by somebody else and is already committed by this developer, the client system detects that the file is no longer up-to-date. An update forces the local client to merge the content automatically if there are no conflicting regions in the file in which both developers made changes in overlapping areas of the code. A new, merged version of the file is created, which can then be committed.

If there are conflicts, for example, changes of the same lines of code, human interaction is necessary to resolve them (Pilato et al. 2009, *l.c. page 2 ff.*). In this case the version of the file, which was of local origin before the change, and the new externally changed version of the file are downloaded and added to the local copy. This local version is also extended by conflict blocks, which mark the conflicting regions and show the differences between both conflicting contents. The developer can now decide whether to fix all conflicting regions or to adopt one of the files completely. In any case, a manual, human merge is necessary here to resolve the conflict (Pilato et al. 2009, *l.c. page 36 ff.*). This works quite well, as conflicts are usually very rare. This strategy also works with binary files, for which Subversion® also registers the changes (deltas), which is quite efficient for the file system volume. Additionally, it detects binary files automatically (Wassermann 2006, *l.c. page 323 f.*) and (Pilato et al. 2009, *l.c. page 416*)).

The complete communication to get a working copy or to commit changes is operated over the standard network and the Internet. The communication can use different protocols (for more about networking and protocols, see ▶ Sect. 3.3.1) or protocol combinations, e.g., (Pilato et al. 2009, *l.c. page 211 ff.*) and (Wassermann 2006, *l.c. page 141 f.*)):

— *svn:* This is the raw Subversion®protocol, which is usually on port 3690. It is offered by a specific Subversion®server «svnserve» on the repository machine, which can only talk to Subversion®clients. The protocol itself is state-based and quite fast. The authentication is simply done by using the Simple Authentication and Security Layer (SASL), which maps the user credentials to the operating system or private user files of different authentication servers. The read and write access can be granted over the whole repository or specified paths, and the communication can be optionally encrypted with available SASL techniques. There is no logging implemented and no direct web viewing available. Replication servers can be just read-only, when using this protocol. But in general the whole setup is quite simple and an ideal starting point for small teams and beginners.

— *svn+ssh:* This protocol combination is similar to the raw «svn» with similar restrictions and features, but tunnels «svn» over Secure Shell (SSH) , which is usually on port 22. This is offered by the standard SSH server in combination with the specific Subversion®server on the repository machine. A big

Versioning Model «Copy-Modify-Merge»

Merge Conflicts

Communication, Protocols, and User Credentials

difference is the authentication. First of all SSH is used to encrypt the whole communication. It is also used to exclusively authenticate users, using the real system accounts of registered users. As the SSH server always spawns a private «svnserve» per user, the access can only be granted to the whole repository. The access is a little bit slower than the raw «svn» protocol, as all transfers must be tunneled and therefore encrypted and decrypted again. But it is a safe way also over the whole Internet, as long as a routing via SSH is possible.

- *http:* This is the standard Hypertext Transfer Protocol (HTTP), which is also used for the transfer of web pages to web browsers. It usually uses port 80 and is offered by a web server, such as the Apache web server, in combination with the plug-in package «mod_dav_svn», which extends HTTP with the Web-based Distributed Authoring and Versioning (WebDAV)/DeltaV protocol. This is used to coordinate distributed writing, changing, and administrating of data files and document structures. It includes mainly a lock mechanism, namespaces, the use of metadata, and the version control «DeltaV» (Neidhardt 2005, *l.c. page 159 f.*). The authentication is organized by the Apache server, for example, with private user credential files. Access rights similar to «svn» can be granted to the whole repository or specified paths. The web server itself logs all HTTP requests, which can be sent from different WebDAV clients. The repository content can be scrolled via a web browser, and it is possible to implement transparent write-proxy replications. The disadvantages are that the setup is a little more complex and that the performance is not as good as with «svn» alone, as there are additional components involved. A big advantage is that HTTP is usually opened and available in firewalls, so that it is quite simple to cross such security boundaries without big configuration changes there.

- *https*: This is the standard Hypertext Transfer Protocol Secure (HTTPS) for secure network communications and quite similar to HTTP, but with a secure layer and encryption (Secure Socket Layer (SSL)). The rest of the features are the same as for HTTP. It is usually offered on port 443 by a standard web server if the HTTPS packages are enabled.

URLs

For the user of such a repository system, it does not matter which protocol is used. It just appears in the URL as a prefixed classifier. The rest is known from the daily use of web pages. A typical checkout command for the Subversion®system, for example, looks like an HTTP request in the following form:
«svn co http://svnserver.svn/svn/tools ./tools»
A request with the SSH-tunneled «svn» looks quite similar and has the following form:
«svn co svn+ssh://[username@]svnserver.svn.wtz/home/svn/tools ./tools»

The explanation of the command structure can be found in Example 2.15 . The two commands show the different use of protocols, authentication methods, and details for directory trees. To simplify handling, most of the clients save the user credentials for reuse if the server requests them again. Only as a result of such a request will the client send the authentication information to the server. The client checks if there are calling parameters, defining the credential information («--username»/«--password»), or if there are cached credentials. If neither is valid, it requests the authentication data again from the user (Pilato et al. 2009, *l.c. page 106 ff.*).

Working Copies and Revisions: Checkouts

The usual work with Subversion® is based on the local working copies. Such a copy is created with a checkout which fetches the directories from the repository and brings them into the local file system. It is possible to check out whole repositories or even just their single directories (Wassermann 2006, *l.c. page 93 f.*). Each directory of the copy also contains an additional, administrative subdirectory «.svn». This directory contains all the information which the Subversion® client needs. The most important entries there are the URL for the working copy, revision numbers (see later), user-defined properties, and original copies of the files. These are used to speed up the work without permanent requests to the server (Wassermann 2006, *l.c. page 99 ff.*) and (Pilato et al. 2009, *l.c. page 23*)). The checkout does not yet change the data on the repository and forces no locks there. Each parallel user can check out the same files or can update them, which means fetching all alterations from the repository.

If the changes on the local copies are finished, they can be committed to the repository. It is again possible to commit complete repository contents, just subdirectories of them, or single files. But new files or directories must be added explicitly to the administration of Subversion®. The commit supports attaching a log message («-m» parameter of the client as shortcut for «--message»), which is added as meta-information to the committed data. These messages build the change log information together with a set of lexical differences between the previous and the new version («diff») (Wassermann 2006, *l.c. page 94 ff.*).

Commits

Each of these successfully, newly committed versions are a new revision. The creation of such a revision is atomic, so that either all changes are processed or none. It gets a revision number, which is automatically incremented by 1 compared to the previous revision. Therefore, it is possible to check out older revisions, by using the according revision number (after the client parameter «-r», which is a shortcut of «--revision»). It is also possible to use revision keywords, e.g., «HEAD» for the latest revision or «PREV» for the previous one. Another possibility is the selection of a specific date (Wassermann 2006, *l.c. page 27 ff.*).

Revisions

The use of dynamically changing revisions in the directory tree «trunk» is the usual, daily contact with the version control system. But often it is also necessary to freeze a revision to a final release with a dedicated name (tag) for a final delivery to the users of the software or to keep a specific state in the software development process. Such a tagged version is nothing more than an SVN-copy «svn cp»[65] of a revision into the «tags»-directory, which is part of the first level of the previously defined project directory tree (see ▶ Sect. 2.7.1); only the use of externals needs special treatment (see the following block about external properties). Subversion® does not support tag meta-information and does not know anything about the new tagged release. It is just a directory tree like each other in the «trunk» directory. Therefore, it is in the developers' hands to avoid changes of these copied, tagged versions (Wassermann 2006, *l.c. page 117*).

Tags

The use of branches is similar to this. They are created in the same way, but their conception is more dynamical and they are not statical for the rest of the time. They are used to separately develop a parallel alternative of the «trunk» data stock, for example, to test a special feature which should not influence the main trunk, or to diversify different versions of one source code stock for different purposes. It is an independent derivation of the main development branch at a specific time. The interesting thing is that those parallel branches can be joined or merged again into one version or to include updates from one branch to the other. This can be done by «svn merge», which merges the changes from one branch into another. Conflicts during this process must be resolved manually in a similar way to the technique which is used during updates (Wassermann 2006, *l.c. page 108 ff.*). The different parallel development lines are illustrated in ▪ Fig. 2.28.

Branches

Daily work with Subversion® has shown that some commands are regularly used. The following list in Example 2.15 is a collection of such commands, assembled as a guideline for projects at the Geodetic Observatory Wettzell.

Commands

---

65 Copying with the Subversion® command does not destroy the historical information, e.g., about the revision changes in the past.

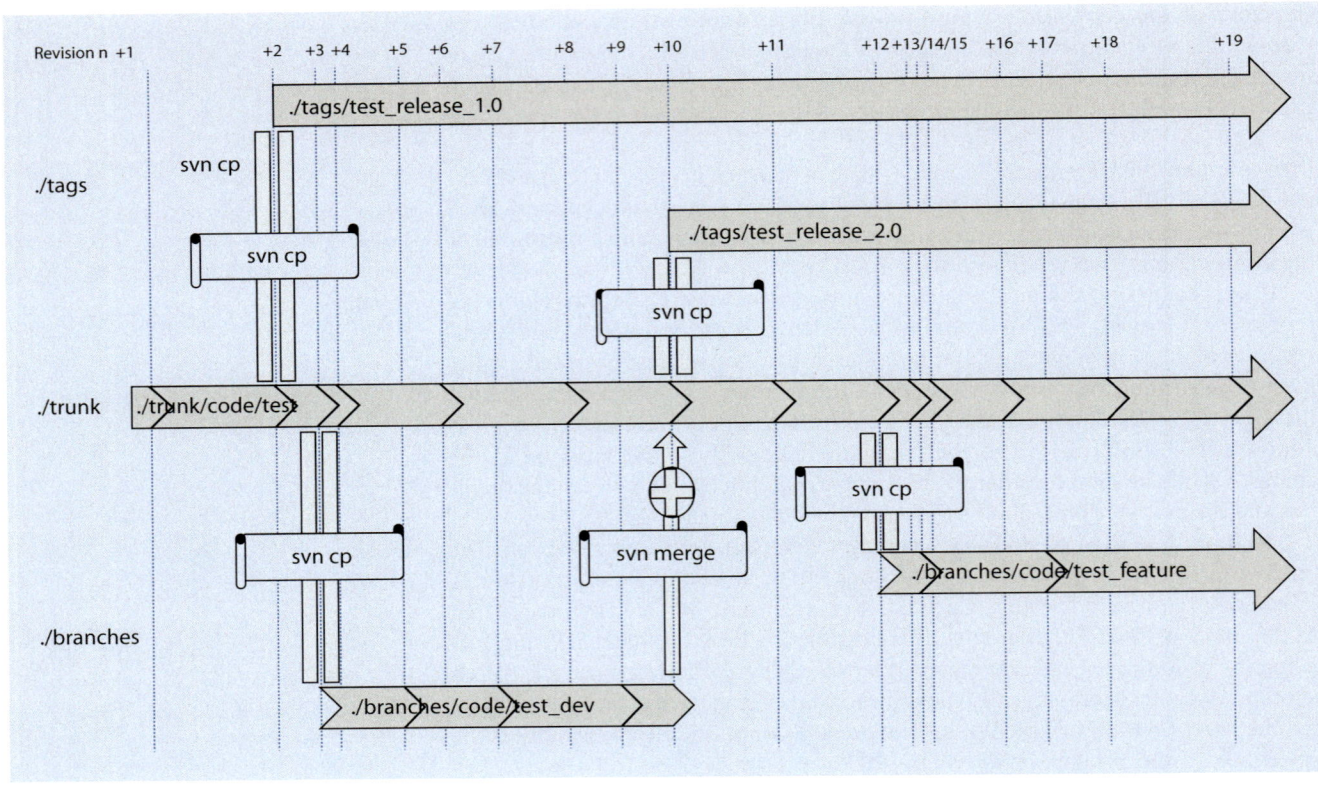

Fig. 2.28 Illustration of the trunk as a main development branch and its tags and parallel branches of a version controlled document tree

■ ■ **Example 2.15**
Collection of the most used SVN commands

```
##################
# SVN User Manual #
##################
# Notice: <repository_url> must be replaced by the real URL of the
#         SVN server

# ==============================================================
# The first import of a new project (here with the name "tools")
# (root rights are necessary on the SVN server system)
# ==============================================================
# It is necessary to log in as root
# Create the new repository
svnadmin create ——fs-type fsfs <repository_directory>/tools
# Change the access rights of the new repository,
# e.g. for access by the user "apache" and the group "httpd"
# to read, write and execute
chown —R apache:httpd <repository_directory>/tools
chmod —R 770 /home/src/svn/tools
# Locally create first directory level
mkdir ./tools
mkdir ./tools/trunk
mkdir ./tools/branches
mkdir ./tools/tags
# Import (first commit) of the new project structure
svn import —m "basic structure" http://<repository_url>/tools

# ==============================================================
```
```
 1
 2
 3
 4
 5
 6
 7
 8
 9
10
11
12
13
14
15
16
17
18
19
20
21
22
23
24
25
26
27
```

```
# Checkout of the new project from the repository as common user      28
# (Only complete directories can be fetched)                          29
# ================================================================     30
svn co http://<repository_url>/tools ./tools                          31
                                                                      32
                                                                      33
# ================================================================     34
# Commit changed parts to the repository                              35
# (<commit_log_text> should be replaced by an individual message)     36
# ================================================================     37
svn commit -m "<commit_log_text>" ./tools                             38
# Also single files can be committed                                  39
svn commit -m "<commit_log_text>" ./tools/trunk/doc/<file.ext>        40
                                                                      41
                                                                      42
# ================================================================     43
# Update local parts with changed parts from the repository           44
# ================================================================     45
svn update                                                            46
# Also single parts can be updated                                    47
svn update ./tools/<directory>                                        48
                                                                      49
                                                                      50
# ================================================================     51
# Solve local conflicts manually                                      52
# ================================================================     53
# The conflicting regions in the conflicting files must be resolved   54
# manually.                                                           55
svn resolved ./tools/trunk/doc/<file.ext>                             56
                                                                      57
                                                                      58
# ================================================================     59
# Changes of file system structures                                   60
# ================================================================     61
# Delete a file, which is version-controlled                          62
svn rm ./tools/trunk/doc/<file.ext>                                   63
# Copy a file or directory tree (used for branches and tags)          64
svn cp ./tools/trunk/doc/<file.ext> ./tools/trunk/doc/<newfile.ext>   65
                                                                      66
                                                                      67
# ================================================================     68
# Read the change log                                                 69
# ================================================================     70
# Change log of a whole project                                       71
svn log http://<repository_url>/tools
# Change log of a project starting at revision 100
svn log -r 100:HEAD http://<repository_url>/tools
# Summary information about the project
svn info
```

Another supporting feature of the version-control system helps to increase the maintainability tremendously. In an ideal project tree, external modules are not hosted as internal copies. If the same source code is multiply used in a project, it is difficult to track changes to all separate code files. This is quite important if bugs which have been fixed are propagated. Therefore, Subversion® offers the possibility of setting a specific property: svn:externals (Pilato et al. 2009, *l.c. page 92*). Generally, properties are meta-information for the revisions. They are not version-controlled, so that the historic states cannot be reconstructed again. But they add additional attributes to a revision or a directory tree, which can be used in scripts («hooks») for the repository administration (Wassermann 2006, *l.c. page 297 f.*). Such additional information is the composition information which external modules have to include in the project source. The SVN client makes it possible to define a list of directories in combination with the URL reference from where they should be fetched (see Example 2.16). After such a definition, it is possible to update the local working copy with these external repositories

Properties to Define Externals

(«externals»), where the external code is stored. After that, the updating of projects in which the externals are used directly fetches the bug-fixed or updated parts as local copy, so that changes can be easily propagated. To also register the external properties in the central repository, they must be committed (Pilato et al. 2009, *l.c. page 91 ff.*).

Centralized Module Base

Therefore, externals help to use centrally administrated modules, which are general enough to be useful for different projects (see ▶ Chap. 3 about a software toolbox). This is possible without using real copies, which would complicate the maintenance, as changes would have to be retraced manually. With the external properties, it is done automatically during each checkout with all the advantages of the version control. A single disadvantage is that externals are something like links. This means that copies of them to create, for example, a tag might change the tagged version if the module referenced by an external changes. Therefore, it is necessary to change externals in tags or releases to hard copies of the referenced modules.

■■ **Example 2.16**
Creation of external properties

```
###################                                                     1
# SVN User Manual #                                                     2
###################                                                     3
# Notice: <repository_url> must be replaced by the real URL of the      4
#         SVN server                                                     5
                                                                        6
# ===============================================================       7
# Creation of external properties svn:externals                         8
# ===============================================================       9
# Create the directory, which holds all externals in the working copy  10
# (according to the directory definitions "srcext")                    11
mkdir srcext                                                           12
# change into the new directory                                        13
cd srcext                                                              14
# Edit the external properties (maybe it is necessary to set the       15
# used editor before, e.g. with "export EDITOR=vi")                    16
svn propedit svn:externals .                                           17
# It opens the editor with the external properties                     18
# Enter the externals list in the style local directory, white space and 19
# URL reference to the remote repository, e.g.                         20
timecalc http://<repository_url>/tools/trunk/modules/timecalc         21
constant http://<repository_url>/tools/trunk/modules/constant         22
orbitcalc http://<repository_url>/tools/trunk/modules/orbitcalc       23
# Save the changes and close the editor                                24
# An update shows if the properties are correct                        25
svn up                                                                 26
# The commit publishes the new properties to the other developers      27
svn commit —m "Add externals properties"                               28
```

Properties to Replace Keywords

Another useful property setting for each file is the replacements of keywords in source code files: svn:keywords. Different keyword settings can be switched on, which forces the SVN client to run a pattern matching for these keywords in the source files each time a commit is processed. The keywords in the files, which are enabled for pattern matching, using additional properties, are then replaced by meta-information about the revision, as (Wassermann 2006, *l.c. page 324 f.*):

- *$Date$*: Date and time of the latest change
- *Rev*: Revision number of the latest change

- *$Author$*: Name of the developer who made the commit
- *URL*: Complete URL of the file in the repository (only interesting if the repository is also accessible over the Internet and not only locally in the institute's network behind a firewall)
- *Id*: Summary as combination of filename, revision, date, time, and author (date and time are usually in Universal Time Coordinated (UTC)[66]) The keyword properties must be set for each file and for each keyword individually (see (Pilato et al. 2009, *l.c. page 344 f.*)), e.g., for Id with
- «`svn propset svn:keywords «Id» <file.ext>`»,[67] or can be deleted again individually, e.g., with
- «`svn propdel svn:keywords «Id» <file.ext>`».
  The inserted information is quite helpful to identify the revision, the age of the file, and the responsibilities for the file if the file appears somewhere separately or if it is used as an external for another project.

To support this useful information, it is appropriate to define standardized comment headers and trailers for the source code files, which are added before and after the real source code, for example, for the documentation generators (see Example 2.17). Together with the already mentioned stylistic extensions of the source code, the SVN header and trailer complete the source file style.

SVN File Header with Keywords

■■ **Example 2.17**
SVN file header and trailer with keywords

```
/*********************************************************************
 * SVN:  Version  Control  with  Subversion
 * _____
 * $Id:  idl2rpc.pl  3628  2012-06-06  12:41:42Z  neidh  $
 *********************************************************************/

// ...
// Source  code
// ...

/*********************************************************************
 * END OF FILE (SVN:  Version  Control  with  Subversion )
 * _____
 * $Id:  idl2rpc.pl  3628  2012-06-06  12:41:42Z  neidh  $
 *********************************************************************/
```

| 1 |
| 2 |
| 3 |
| 4 |
| 5 |
| 6 |
| 7 |
| 8 |
| 9 |
| 10 |
| 11 |
| 12 |
| 13 |
| 14 |
| 15 |

Until now the focus has been on Subversion®. But there are several other version control systems available. Generally, the ideas and in most cases also the terminologies are similar. Knowing one system and being able to work with it makes it easy to work with another. Each of these available systems has some advantages and also some disadvantages. Subversion®, for example, is easy to understand and to install and supports a centralized architecture, which makes it ideal for scientific developments in a university institute with a small group of involved developers. Merging however can be an inconvenient mess, which is often only possible after a coordinating communication between the different developers (but this should not be a problem in agile developments, see ▶ Sect. 2.9). What makes Subversion® unbeatable are the detailed property possibilities, especially the inclusion of

Pros and Cons of SVN

---

66 UTC is the coordinated universal time based on atomic standards and corrected with leap seconds according to the Earth's rotation parameters.
67 <file.ext> must be replaced by the real filename for which the property should be enabled.

2

Git and its Differences

externals. Other version control systems have many more restrictions or fussy details. But Git in particular is becoming increasingly popular, especially for distributed open-source projects. Therefore, a following final overview should characterize it generally.

Git is a very young development, started in the year 2005. The design goals were to allow large numbers of distributed developers to offer good performance, to force change logs, and to demonstrate responsibilities clearly. Git is a completely distributed system[68] which preserves the integrity and uniqueness of its data identifiers by using a secure hash function[69] (Loeliger 2010, *l.c. page 2 f.*). Its memory is organized as a content addressable storage system, using such hash values as indexes, so that branches can easily be tracked. Git generally takes files as binary (Binary Large Object (BLOB)). It uses commits similar to SVN. But these are snapshots of staged files. These are files or file states from a local copy, which are labeled as ready to be committed (Loeliger 2010, *l.c. page 71 f.*). Because of its distributed character, Git supports the easy handling of several remote repositories («remotes»), besides the local one (Loeliger 2010, *l.c. page 196 f.*), to which data can be pushed or from which data can be pulled. Each duplicated repository contains the complete data and is therefore a backup. But it is difficult just to work with repository parts alone. As Git is a content tracking system, it can identify the same content in different files to reduce the used memory. This means that it has no problems with varying appellations of files with upper and lower case characters (Loeliger 2010, *l.c. page 34 ff.*).

Human Skills

Version control systems are quite useful. They offer the possibility of reconstructing each state of a project again at any time. Therefore, confusing amounts of backup archives are not necessary anymore. Alterations and changes can be easily propagated to new versions and different developers. They can work in parallel on the different versions. Therefore, version control systems make the software project life easier. But nevertheless, they cannot replace communication and coordinated work. Social skills are still the key features for successful projects. Human inspections, discussions, and human communication in team meetings to exchange information about the project content need to happen anyway. The team members should trust each other in the team and the versioning system. The developers involved should therefore try to follow a regular routine. They should always run an update in the morning when they start their work. They should try to commit small, manageable changes regularly, using their own user credentials and should not check in files in the name of anybody else. This is relevant, as generally the files rights correspond to the credentials of the user who made the first checkout. Developers should also always make sure to enter change log messages. They should commit their daily work in the evening after finishing as long as it is compilable and not destructive for others. Following these few basic rules makes revisions traceable. Generally, these are also the qualifications which make the use of all the tools already described successful for a project. Then they can be combined to get a CI of software.

---

68 See ▶ Sect. 3.3.2 about the definition of distributed systems.
69 Hash functions are methods, for example, to calculate an addresses information from the content of a string to get a direct reference and therefore a direct access to the position, where the string data are stored. The problem is, that mapping is not unique. The memory where data can be stored is limited so that an infinite number of possible contents must be mapped to a finite set of addresses. Thus, this first hash function can lead to conflicts as reference calculations of several contents result in the same address. It needs an additional exception function to solve the conflict by finding another reference with a reversible and reproducible method. A good and secure first hash function reduces the number of conflicts to a minimum.

2.8   **Continuous Integration**

Having all of the previously described techniques and the suitable, partly automated environment supports a further step toward a continuous development procedure for high-quality software even with small, scientific teams, namely, continuous integration CI.

> *Definition DEF2.21. of «continuous integration»* (see (Duvall et al. 2011, *l.c. page 27*))
> Continuous integration is a software development practice, where software is frequently (at least once per day) and continuously integrated into a centralized environment to reduce integration problems. Each integration is automatically compiled, verified, and validated with automated builds, tests, and inspections to develop cohesive software more rapidly.

Continuous Integration

While the methods and tools described are already very powerful in the hands of single, structured working developers, they are much more powerful as a general environment on a centralized server. This server can be used for all projects and each developer can join the development practice quite easily. Everything starts with a wider definition for the build. Building software in the sense of CI is more than compiling the single parts, as done in simple software or code builds all the time during the development. It should also include things like testing, inspections, documentation, and deployment preparation and forms a real CI build. Each change forces a build of this category. Usually this is partly done by the developer privately, including local resources (Duvall et al. 2011, *l.c. page 4 ff.*). But it can also be run on a centralized server or a server farm under different conditions (e.g., operating system releases, compilers, 32-/64-bit platforms, and so on) using centralized software assets. These include one's own and third-party source components and libraries, configuration files, data files, build scripts with according environment settings and installation scripts, and therefore the complete software system. The centralized and more generalized approach avoids software which is only used on one dedicated developer machine and also helps to test older environments (such as historic but still used compiler versions) (Duvall et al. 2011, *l.c. page 74 f.*). To reduce the complexity of the different environment sets, the reduction of programming languages and external dependencies is quite important (for the discussion about programming languages, see ► Sect. 2.2).

Centralized Software Assets

The starting point for that is to have a consistent, logical directory structure (see suggestion in ► Sect. 2.7.1) with clearly defined contents according to a coding style guide (see ► Sect. 2.3). These can be retrieved selectively from a centralized software asset, so that only compiling-relevant parts can be fetched, avoiding, for example, documentation files, which are unnecessary for the compilation. The centralized software assets are mainly supported by the version control system (see in ► Sect. 2.7.2). This is the heart of CI. It should have all files under version control, which a developer or user needs for a first build and installation without the necessity to separately fetch them from other external sources. In some cases, it is also useful to add files or environment parts for particular developing situations (images of a given, historic situation) (Wiest 2011, *l.c. page 31 f.*).

Version Control System

The next step is to use centralized CI builds. They should be able to be started using a single build command, so that just by calling this command, all necessary actions are operated. They should also run without human interactions, and the

Centralized Continuous Integration Build

separate project parts should have their own scripts, to avoid unmaintainable coupling (Duvall et al. 2011, *l.c. page 67 ff.*). At the Geodetic Observatory Wettzell, a shell script does all of the essential things. The main part is the code build, which is the compilation of the software under a dedicated environment using a single «make build» call (for the basics, see ► Sect. 2.6.2). The used Makefile prepares and cleans the environment to reduce dependencies on previously existing code. It supports the different build types. The private code build is used by single developers or a pair of developers (pair programming, see ► Sect. 2.6.2) on their local machines. The integration build is used by the CI server to build the head version of the software mainline («trunk»), checked out from the repository. A release build runs necessary CI builds and prepares a deployable release as a «tagged» version if the processed tests and inspections are successful. The focus of CI lies in the continual integration builds. For this a dedicated integration build machine is used. It has the recommended system resources for all of the software parts which should be regularly integrated (e.g., database integration, legacy code assets, used external code assets, libraries, etc.). It has access to and at least checkout rights at the version control system (Duvall et al. 2011, *l.c. page 78 ff.*).

Continuous Self-Testing and Self-Inspecting

The goal is to always have easily deployable software. For this reason, the integration builds are run continuously. But it is also necessary to evaluate the build status to decide if the software is in a state which allows a new release or not. This requires an automated code testing and inspection mechanism. The first result is the compiler output, which is the first test feedback (see ► Sect. 2.6.2). The compiler parameters should be set to detect the issues which are relevant for the software (for GCC things like «-Wall -pedantic -Wshadow -Wconversion»). Warnings should be handled like errors. A successful code build is entirely free of errors and warnings. Another important piece of information is the result of the automatically processed unit tests as a form of dynamic code testing (see ► Sect. 2.6.2). It also offers the derived information about the code test coverage (see ► Sect. 2.6.2). It is possible to test more than the function units in additional component tests to see the cooperative interaction between them. Important counts can also be retrieved from the automatic run of static code inspections (see ► Sect. 2.6.3). Here it is suitable to check only the important things which need a response and avoid a flood of warnings about secondary issues. Then the thresholds, values which define whether the integration has failed, can be set more restrictively. A good inspection and therefore a good integration should have no flaws at all. Finally, it is also quite healthy for the software to check its behavior under long-running conditions in the form of system tests, using automated dynamic code analysis tools (see ► Sect. 2.6.4). In any case, it is necessary to alter the configuration files and the configurations at all (Duvall et al. 2011, *l.c. page 77*). But it is quite difficult to automate complete system tests (Wiest 2011, *l.c. page 35 ff.*), for example, because of interactive GUI interactions.

Virtualization of Environments

In practice, it is necessary to test different environments (hardware, operating systems, compilers, user rights, etc.). For this, it is quite helpful to use the technique of virtualization.[70] Different operating system images with individual environments can be run in parallel on one computer or server, using virtual machines. This is ideal for the CI under different environments and configurations. Two such virtualization frameworks should be mentioned: the market leader «VMware®» (VMware n.d.) with its «VMware® Player» and the open-source industry standard for virtualization «Xen® hypervisor»[71] (Xen n.d.). Xen® is a powerful, efficient, and secure feature set for virtualization for different architectures and supports a wide range of guest operating systems. CI tests on Xen® can be arranged as

---

70 Virtualization is a technique to emulate a physical hardware virtually. It is used to run several «virtual», emulated machines with their own operating systems on real hardware with a specially prepared, virtualization operating system or in a special player. Each virtual machine looks like a real machine for the users of the virtual machine.

71 A «hypervisor» is a virtual machine monitor, which runs on a guest system and offers everything for a virtual machine.

paravirtualization, where the test systems are modified for the virtualization and run on the «hypervisor». Another possibility is Hardware Virtual Machine (HVM), where the unchanged virtual machine images run on a real, specific hardware hypervisor in a similar way to the final hardware (Xen n.d.). While Xen® is open source, «VwWare®» is licensed. Only the player can be downloaded for free. The handling itself is quite easy and for CI both offer ideal ways to reset virtual machines to a defined starting condition each time before a new test is started (Wiest 2011, *l.c. page 40*).

But even if there is no chance of a test virtualization, CI can be installed on a single piece of hardware. Then it is advisable to reduce the environment to the real-life scenario, to compile and test as close to the real application area as possible (Wiest 2011, *l.c. page 39 f.*). Generally, it is then important that the build mechanisms provide scheduled runs and runs on-demand. Scheduled processes are driven by time. This means the CI builds with their tests run once per hour or once per day, perhaps scheduled by the «cron job» (see ► Sect. 2.6.2). Large tests can be scheduled during the night in the off-hours of the offices. Run on demands can be used for private builds as well as for complete and time-consuming software tests or system tests. Both mechanisms can be realized easily. Faster, regular feedback is offered if the build mechanism polls for changes at the version control system or if this triggers the build run via an event (event-driven). Each change directly starts a new integration and the necessary tests. Therefore, it directly offers interesting feedback about the complete system in the «trunk» path for the developer after each run (Duvall et al. 2011, *l.c. page 80 f.*). The primary goal is to have many, frequent integrations, ideally after each change with rapid feedback, which should be as fast as possible.

To enable progress in bug fixing according to the CI, suitable, accessible, and well-arranged feedback is a means to an end. In general, this feedback should be done in the right way (browser, email, mobile text message, etc.), with the right information (status, reports, and results) for the right people (developers, managers, etc.) at the right time (at least daily) (Duvall et al. 2011, *l.c. page 206 f.*). A three-step approach is suggested, consisting of an active notification, a public status display, and detailed reporting information on demand. The active notification can use emails. These can be sent automatically or manually by a responsible CI manager. They should be informative and detailed enough for the receiver to be able to fix the problem directly. A problem here is that those emails can quickly become annoying spam if too many of them arrive or if duplicate emails with the same information arrive. For this reason, the team at the Geodetic Observatory decided to use selective emails, sent manually by a responsible team member.

Public status displays are much more informative and helpful. Such displays can be offered on a web page which is updated at least once a day, to which each participating developer has access rights and a simple access at that. For this type of status information, a display concept with symbols comparable to traffic lights or light-emitting diodes in green, yellow, and red has prevailed. The complete CI feedback about all projects can be arranged as a status matrix. In this matrix, each red light symbolizes a broken integration, each yellow light symbolizes a successful integration with warnings, and each green light symbolizes a completely successful integration (see a solution such as smiley matrix in ◘ Fig. 2.29). Such a matrix gives a quick overview of the states of all projects as a snapshot. To get more detailed information, it is necessary to request conditioned information about the status and project details or flaws found. By using the matrix concept on a web page, links to further details can be offered. Therefore each smiley is a link to the relevant information about the CI build section and follow-on web pages. These can be detailed protocols with information about and locations of errors found or statistics according to metrics or project states (Wiest 2011, *l.c. page 40 f.*). For example, for some third-party funded projects, it is necessary to release time sheets for participating developers, showing the percentage of personal development in man hours. This information can be easily taken from the statistics about the version control system in the CI technique (see ◘ Fig. 2.30).

Mechanisms for Continuous Integration Builds

Continuous Feedback

Public Project Status Displays

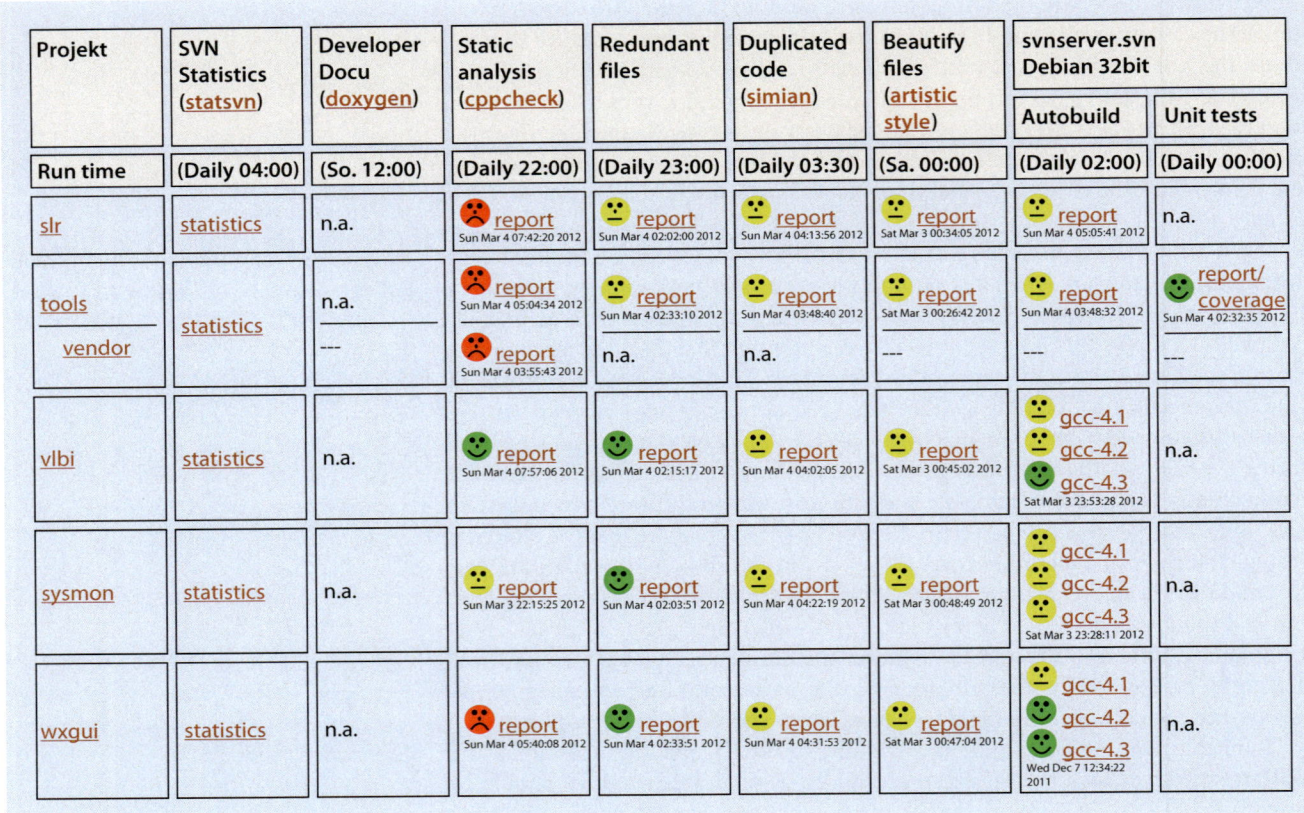

Fig. 2.29 Example for a public status display of the CI build of all projects at the Geodetic Observatory Wettzell using a matrix with smileys («*red*» symbolizes a failed integration with errors, «*yellow*» symbolizes a successful integration but with warnings, and «*green*» symbolizes a completely successful integration)

Continuous Documentation

As CI build runs autonomously without human intervention, it is also an ideal environment to add other automatic generations to that workflow, such as documentation generation (see ▶ Sect. 2.5.1) (Duvall et al. 2011, *l.c. page 249*). The build scripts can also start the documentation generation with Doxygen. Therefore, the documentation tools must be part of the build environment on the CI machine. As it makes no sense to include the complete generated documentation directly in the version-controlled source structure, because of the amount of fast-changing, code-redundant information in the documentation, the generated descriptions should directly be published to a download server. Every developer can then download the latest version of the documents from there.

Additional CI Jobs

Other tasks such as code beautifying and so on, can be handled similarly and should also be run during the CI build. But for that, it is necessary that the CI server has also commit rights to the version control system, so that it can directly check in the beautified code files. Beautifier changes are also always hints on code violating the specifications.

Continuous Deployment

Software integration tests are a preliminary stage for system tests and therefore also for a release test, which is necessary for a deployment of the software.[72] In this way, the CI can be used to prepare releases on demand, and this is precisely the main goal of CI: releasing working software at any time for any environment (Duvall et al. 2011, *l.c. page 191*). Hereby it becomes necessary to also automate the release procedure. The external definitions of the version control system Subversion® are helpful here. Release contents can be directly assembled using the external references to directory trees in the repository. The CI build can directly compile and test these external copies during the automated integration tests. If a release is created, it is

---

72 Deployment of the software means handing the release out to the customer or user.

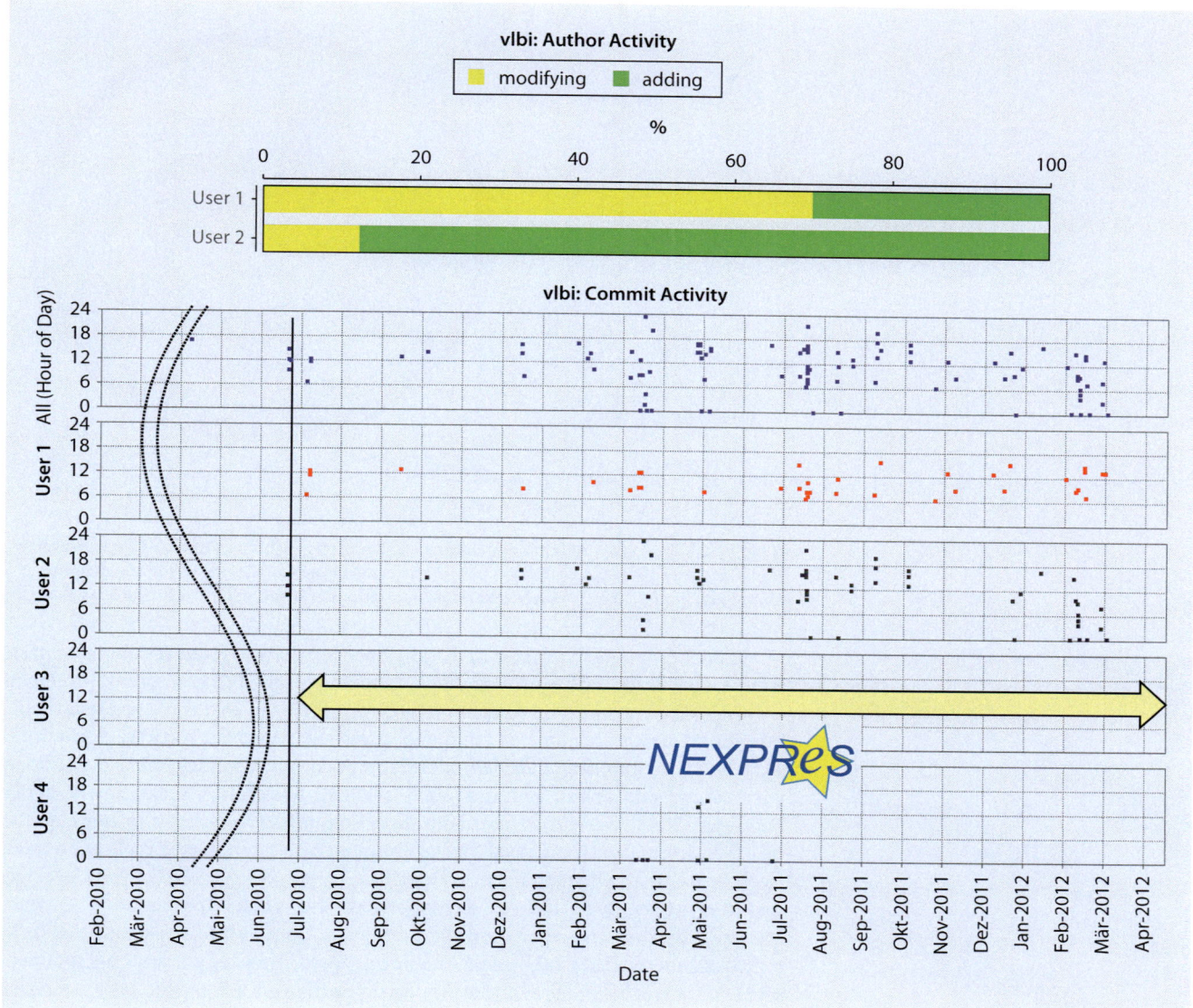

Fig. 2.30 Example of project participation statistics (here for the task 3 in work package 5 of the NEXPReS project) derived by CI from a version control system. A scatter plot is used with the project days on the x-axis and the day time on the user-dedicated y-axes. Each dot is a commit activity in the version control system. The other bar chart shows the percentage of code that has been added or changed by each developer (Note: real usernames are replaced, because of data privacy)

necessary to check out the content of the release directory with the externals in a tagged sub-folder. This tagged version can be used as a separate static branch in the tag directory tree on the version control repository. This can be done using the specific commands of the version control client (e.g., for SVN a copy in the tag tree must be created using «svn cp»). All files are then tagged as a static release. But creating such a separate tree is usually not enough. It is often also necessary to create a distribution or archive file, which can be published on a web page. This step is handled in a similar way to the documentation generation: the new tagged version can automatically be archived, compressed, and finally uploaded to a download server in the web. With this strategy, it is possible to publish releases every day (compare also (Clark 2006, *l.c. page 98 ff.*)). Releasing can then be separated into main releases and intermediate releases, which can be separated by the tag classifications.[73]

---

73 A suitable classification consists of a major release number and a sub-release number, for example, in the form of «Ver1.56», where 1 is the major release number and 56 is the sub-release identifier. In this sense, major releases realize completely new features, while sub-releases are more or less bug-fixed versions.

2

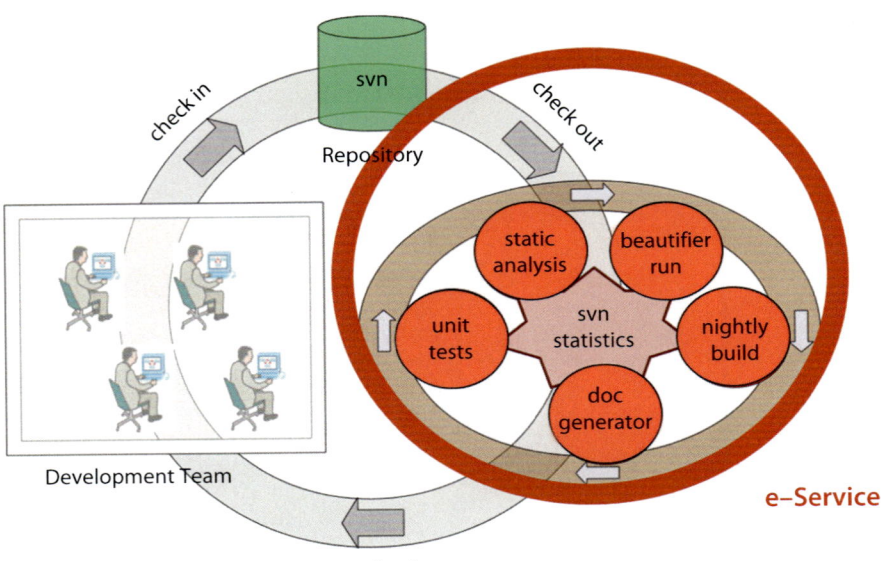

◘ **Fig. 2.31**   The continuous integration cycle as used at the Geodetic Observatory Wettzell and offered by a private initiative via web page to dedicated open-source projects of the geodetic community (figure taken from a poster about continuous integration by M. Ettl and adapted)

Build and Release Management Software

Besides proprietary, script-based, individual solutions (e.g., the development team at the Geodetic Observatory Wettzell uses a hierarchical set of Perl scripts, producing web pages as reports on an Apache web server), there are also dedicated continuous integration server solutions or build and release management software frameworks available. They implement the centralized server core component and also additional features and add-ons or plug-ins, such as email notification services, distributed build-slave knot realizations, user administration, ticketing systems for bugs, and so on. The servers can be accessed via web interfaces, programmable application interfaces, or command line interfaces (Wiest 2011, *l.c. page 45 ff.*). The Eclipse Developer Report of the year 2012 (Eclipse Foundation n.d.) shows that «Apache Ant™», «Apache Maven», «Jenkins», «Oracle's Hudson» (originally from Sun Microsystems), and the previously mentioned GNU «make» are mainly used. They offer everything for open-source CI server solutions to create process supporting environments. They are ideal if the main focus lies in software production, so that the necessary effort to learn about and maintain the external tools is in proportion to the amount of developed software products. Otherwise simple script solutions, which are easily understandable and completely in the hands of small scientific teams do a better job (for scientific institutes).

Useful Tools

The development teams at the Geodetic Observatory Wettzell use several separate open-source tools, which are combined into their own proprietary CI system (see ◘ Fig. 2.31) consisting of a set of hierarchically arranged Perl scripts for the CI build. Currently it is a sequential processing triggered once a day as a cron job. It presents the results via generated web pages, using an Apache web server. This CI system is also offered to a dedicated group of external developers in the geodetic community on an external web server, so that they can build and check their own code assets with the selective tools.[74] It is almost entirely automated and can also deal with archive files. The used open-source tools are[75]:

— *Version control statistics*:
  — StatSVN: Create statistics about the version control system status
    ▶ http://statsvn.org/

---

74 Currently the service exists as a free «e-Service» of the «e-Control Software» environment on the web page ▶ http://www.econtrol-software.de. Each project has its own user rights and credentials.
75 All web pages were available on December 16, 2015.

- *Coding style*:
  - Artistic Style 2.02: Beautifies the code according to the defined coding style
    - ▶ http://astyle.sourceforge.net/astyle.html
- *Code build*:
  - GNU make: Automatic code builds with different compilers
    - ▶ http://www.gnu.org/software/make/
- *Code testing*:
  - Wettzell Unit Test Suite: proprietary unit test suite for C++/C to process unit tests
  - bcov: Detecting the test coverage of the unit tests
    - ▶ http://bcov.sourceforge.net/
- *Static code inspection*:
  - Cppcheck: Static code analysis
    - ▶ http://sourceforge.net/projects/cppcheck
  - codespell: Spell check of program and text files
    - ▶ https://github.com/lucasdemarchi/codespell
  - nsiqcppstyle: Find non-reentrant functions in code
    - ▶ http://code.google.com/p/nsiqcppstyle/
  - Flawfinder: Find security problems
    - ▶ http://www.dwheeler.com/flawfinder/
  - PScan: Detect common printf/scanf format errors
    - ▶ http://deployingradius.com/pscan/
  - Simian: Detect duplicated code
    - ▶ http://www.harukizaemon.com/simian/
  - Proprietary shell development at the Wettzell observatory: Detect redundant files in the repository
  - Proprietary Perl development at the Wettzell observatory: Detect project style flaws
- *Dynamic code analysis*:
  - Valgrind: Local dynamic code analysis on demand
    - ▶ http://valgrind.org/
- *Documentation generation*:
  - Doxygen: Generate developer documentation
    - ▶ http://www.stack.nl/~dimitri/doxygen

But all of the best tools are useless if the developers do not use them correctly. Seven practices should be considered which work well for projects with CI (Duvall et al. 2011, *l.c. page 39 ff.*):

Developer Practices

1. *Commit code frequently*: Developers must commit code early and frequently to obtain the benefits. Therefore, it is necessary to make small changes, instead of changing huge parts, which may also affect several, different components. These small pieces can be tested well with unit tests for correctness, and the commit logs give a clear, step-by-step image of the evolution of the software. In any case, it is necessary to commit after finishing a task. Therefore, a task should not take longer than one day. One commit per day by each single developer is the least that should be done. But developers should avoid commits at the same time (experience at Wettzell has shown that commits at the same time are very rare, as scientific teams have very flexible working times). One working practice is to update code in the morning and commit new features at least once in the afternoon or preferably after each small change.

2. *Run private builds*: Developers should always run complete, private builds after finishing a change or a task. For this, they should update their local working copy from the repository first. The existing code together with the new code must pass all existing and all new unit tests before a commit. Commenting and committing not working code sections is strictly forbidden.

3. *Do not commit broken code*: Every developer should take care not to commit code which does not work correctly. A successful private build, including at least some standard tests, is a must.

4. *Fix broken code immediately*: Broken builds on the server should be fixed immediately by the developer who caused the failure. Warnings should be interpreted like errors. There should be an implicitness and automatism, and developers should not be blamed for broken builds (experience at Wettzell has shown that some developers become afraid to do new commits, as these might break the existing software; here it is necessary that they become used to private builds and unit tests and that the team does not focus on laying the blame on someone). Working on software causes effects and errors. That's usual! But these problems are there to be fixed immediately.

5. *Write automated tests*: Each part of the software should be covered with tests. These tests should be automated, using environments, like the Wettzell Unit Test Suite. The tests should be arranged in the program or module directory tree to clearly state the dependencies. These exact tests should also be used on the automated CI server.

6. *All tests and inspections must pass:* An essential rule is that all tests and inspections must pass. Test and inspection errors must be handled in the same way as compiling errors. They are hints at possible quality risks (apart from possible «false positives»).

7. *Avoid getting broken code*: Broken trunks should not be checked out, even when workarounds are possible. The developer who is responsible should fix the problem. Therefore, it is necessary that developers become sensitive to CI feedback and that they take the results seriously. Human communication is here, as in most cases, a very important factor, which should be supported in agile developments (see the following section).

Supporting Environment: Eclipse

To reduce the efforts of CI, suitable development environments might be helpful. One of them is «Eclipse». Eclipse is an open-source development environment and framework from IBM for software projects in different programming languages, such as Java, C, and C++. It offers a graphical user interface in the form of a sophisticated «editor» with several additional components and optional features. Eclipse simplifies the programming itself, for example, with an extended helping system, code highlighting for a better reading, or with a content assistant with popups, showing possible and defined variables and functions with their signatures according to entered identifier parts (Bauer 2011, *l.c. page 116 f.*). Even though there might be an initial barrier to overcome in order to learn and deal with all suitable features, Eclipse and its C/C++ Development Tools can help programmers to write better code.

Eclipse and Continuous Integration

In addition to supporting the ideas of the CI approach, Eclipse supports and implements additional features for version control, static and dynamic code analysis, debugging, and documentation generation. These features are mostly based on previously mentioned tools in the form of plugins. For example, Eclipse allows the use of Subversion®, using the plugin «Subclipse». It allows all known actions from Subversion within the framework of Eclipse (Bauer 2011, *l.c. page 320 f.*). For a static code analysis, Eclipse offers the feature «Codan». After its activation, it is a rudimentary, graphical static code checker for issues like assignments in conditions, ineffective statements, recommended catching by reference, and so on (Bauer 2011, *l.c. page 113 ff.*). A Valgrind plugin includes the dynamic code analysis of Valgrind transparently in Eclipse and offers graphical outputs, for example, about the memory consumption of the program (Bauer 2011, *l.c. page 377 ff.*). It is very helpful to have the possibility of starting programs with the help of a launcher in a graphical debugging mode with a special debug perspective on the code. The debug mode makes it possible to set line, data, address, and event breakpoints to stop the program at specific points or under specific conditions. Then variable contents can be monitored or expressions can be controlled variably. Other further views make it possible to check special registers, signals, or memory contents (Bauer 2011, *l.c. page 259 ff.*). To support the documentation generation with «Doxygen», a feature can be switched on which supports the integration and editing of Doxygen comments with special highlighting in the editor and the support of the content assistant (Bauer 2011, *l.c. page 259 ff.*).

Even though Eclipse is a very powerful tool, which makes the programmers' life much easier and directly helps them to deal with aspects of CI, a general and centralized solution is indispensable. In any case, the use of CI can help to reduce assumptions about the code (such as sentences like «in principal it should work») (Duvall et al. 2011, *l.c. page 24*). Functionality and to some extent correctness can be validated directly after each change by running a build. But CI is not self-organizing. It stands and falls with the acceptance of the developers. Information and its flow is the key. The administrative structure is also another factor. Therefore, the organization of projects and the planning of project phases are very important drivers for a project, especially in scientific developments, where often prototypes are the goal.

2.9   Agile Software Development

Software development in the field of science is usually applied science. New methods, solutions, implementations, and so on are created in the form of process engineering to support another scientific goal. Therefore, prototypes are often the goal and result of such work, which demonstrate the functionality of an algorithm or of a workflow. The industry puts its main focus on saleable products with short development times. Scientific developments have to deal with more risks, as it is not always clear if it is possible to convert the scientific idea into programming code. Continuous learning accompanies the implementing process. Single developers or small individual teams with flat hierarchies work cooperatively and often with a personal interest in finding new solutions. Their personal interest is often driven by masters or PhD theses. A lot of discussions are necessary to find an optimal, new way for which the specifications are not completely clear at the beginning. Therefore, different technical prototypes must be evaluated and tested. User manuals and documentation are secondary and only important if they are necessary to provide information for new team members. The developers in research institutes usually accompany a development from half a year (e.g., students) to a few years (PhD students or employees on contracts for work and labor). The length of time depends on funding.

Requirements of Scientific Software

A project must adapt to these parameters. But this is not simple! Classic development processes (see, e.g., in (Spillner et al. 2011, *l.c. page 31 ff.*) or (Leffingwell and Widrig 2003, *l.c. page 23 ff.*)), such as the sequential[76], V-[77], W-[78], or spiral[79]

Classic Models: Sequential, V, W, or Spiral

---

76   Sequential, top-down- or waterfall-models organize the project as a strict sequence of the activities: requirement analysis, design, coding/implementation, test, system integration, deployment, and operation with maintenance (adapted from (Leffingwell and Widrig 2003, *l.c. page 24 f.*)). Steps to backward activities are not allowed or restricted to avoid development delaying cycles.

77   The V-model is another sequential development process which aligns verification and validation activities to a similar development chain as in a classic waterfall model. If the tests fail, restricted steps backwards are allowed for fixing bugs. The graphical representation of the activities looks like a big «V» (adapted from (Spillner et al. 2011, *l.c. page 33 f.*)).

78   The W-model is a more detailed V-model which accentuates the affiliation of the different development phase to adequate test cases, each with a test preparation, test execution, and debugging. Therefore the graphical representation looks like a big «W». Cycles are included, because of the debugging and test case specifications. They are planned early and parallel to the development documents. But the model is in principle just a sequential process with some cycles.

79   The projects in a spiral model start with a series of risk-driven prototypes, followed by a waterfall-like process for the actual product development. Therefore the spiral model starts

2

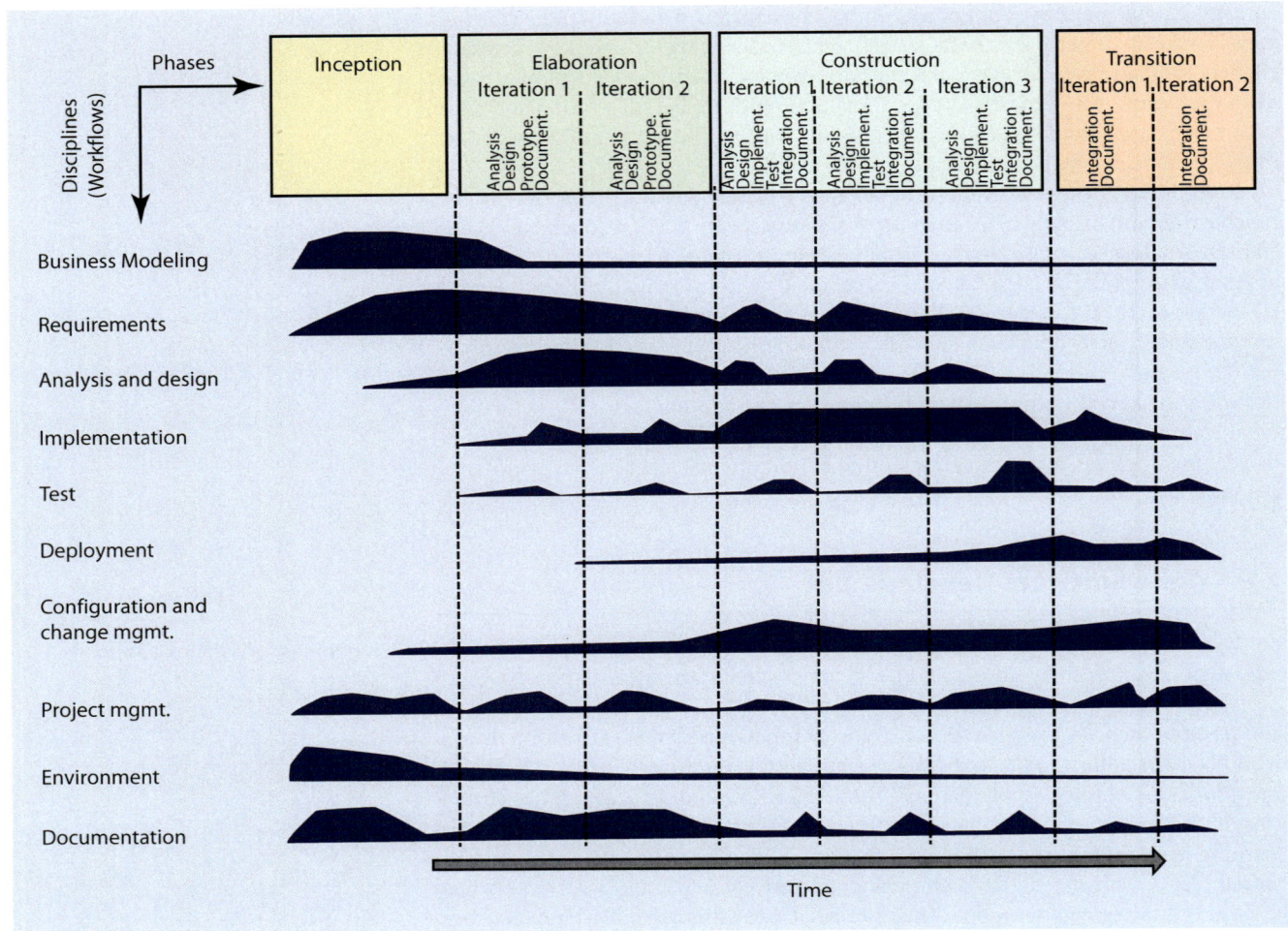

◘ Fig. 2.32 The phases and process workflows or disciplines of the Unified Software Development Process (adapted from Spillner et al. (2011 *l.c. page 37*) and Vogel et al. (2011 *l.c. page 317*); reproduced by permission of dpunkt.verlag Heidelberg)

models, are too static, even though some offer cyclic steps backward in the case of problems found. They define clear structures on a clearly defined specification with all acceptance details at the beginning of the project. They are usually plan- and time-driven and have some problems with short-term changes of specification parameters and do not deal with newly learned things during the development. But it is inevitable that programmers will increase their knowledge through being engaged in their work. It is precisely this new knowledge that can improve the development and change decisions or solution designs. Therefore, the classic models are too antiquated for modern scientific needs.

Iterative Software Development Processes

Processes which can adapt to project-accompanying changes over all phases of the project would be ideal. This is only possible with iterative methods, like those defined in the Unified Software Development Process of the company rational (see ◘ Fig. 2.32, adapted from (Spillner et al. 2011, *l.c. page 37*) and (Vogel et al. 2011, *l.c. page 317*)[80], where the final point of a previous activity is the starting point of a new analysis and

---

with a better requirements planning and concept validation at the beginning. The problem is the consumption of time at the beginning and before the final development (Leffingwell and Widrig 2003, *l.c. page 26 ff.*). But after finalization of the pretests, a usual top-down model starts.

80   The graphs over time for each discipline demonstrate the expenditure of time for that individual workflow according to the iterations in the project phases. Documentation is officially not part of the Unified Software Development Process and was extended to draw attention to continuously developing software documentation.

design for the following activity. This following activity is then an increment of the previous one, as it takes the existing parts and adds new features to them. But it is also an iteration of the existing code, as it changes and adapts these sources according to feedback or new knowledge. The iterative steps stretch over the whole project, from the first planning phases to the final deployment. They split the project into its phases over time. The different project workflows or disciplines, such as analysis, design, implementation, test, and deployment, pass through these phases. They are more or less present over time and in the different phases. For example, on the one hand, more time is spent on analysis and design at the planning phase of the project, while expenditure of time decreases during the construction. On the other hand, it is necessary to have the planning documents to start the implementation, so that the time spent on implementing things increases during the construction phase of the software.

The first steps of such iterative processes are concerned with planning a general outline and an architectural design at the beginning of the project (big picture). Therefore, the process starts with an «inception» phase, which defines the framework for the project on the basis of meetings, first feasibility analyses, and a general cost estimation. It is usually very short and succinct. It has a smooth transition into the «elaboration» phase. This period is used to specify the requirements, using «user story/feature cards»[81] and «use cases»[82] to plan user interactions. With the user interactions (user interface), it is possible to design the upper layers of the software architecture on the basis of required features for the dedicated user interactions. This can be used for a construction plan which defines the workflows with its milestones and various constraints. Having all of these is the basis for a risk analysis.[83] The result of the elaboration phase is then a detailed project frame with all specifications and the used terminology. It arranges the workflows and defines software release times for the following phase, the «construction». The construction deals with the implementation of the specified features using all of the techniques, described in the previous sections. As most of the iteration steps are taken at this stage, it also has to deal with frequent refactoring of the code (see also ▶ Sect. 2.4). Each iteration consists of dedicated activity sequences for the current iteration, such as analysis, design, implementation, test, integration, and documentation. (Beta-) releases are produced and the project ends with the final deployable version, the final release. This phase between the last beta-release and the final release is the «transition». No new features are added during this phase. It is only used to fix bugs and to train the users (Fowler and Scott 2000, *l.c. page 11–29*).

Such iterative processes adapt better to requirement changes. The changes are active throughout the life cycle of a project. The requirements are understood early at a certain level of detail but refined and extended iteratively over time. Another advantage is that multiple releases are developed over time. Even if the project may be badly scoped and the software may have a lack of functionality, the delivered product can be used and has value to the user. It is also a good basis for further developments (Leffingwell and Widrig 2003, *l.c. page 30 f.*).

Each iteration should follow the **SMART** criteria to define its goals. **SMART** stands for (Steffens 2015, *l.c. page 7 ff.*):

- **S**pecific: The goal and tasks of an iteration have to be specific and clear. Weak phrases must be avoided.
- **M**easurable: The result of an iteration reached can be measured using the defined metric parameters of the quality factors.

Development Process Phases

Advantages of the Iterative Process

SMART Criteria for Iteration Goals

---

81 User stories are very short, textual descriptions of requirements, user interactions, and software features. They are organized with cards, where each card holds one story (Wolf and Bleek 2011, *l.c. page 39 f.*).

82 Software engineering defines «use cases» as a list of action items, describing the interaction of a specific user and the used software or program as a software scenario to achieve a specific goal with the software (Fowler and Scott 2000, *l.c. page 36*).

83 Risk in scientific projects is different to industrial. While industrial projects must result in a specific, salable product, the gain in knowledge and insight is the prior goal in science.

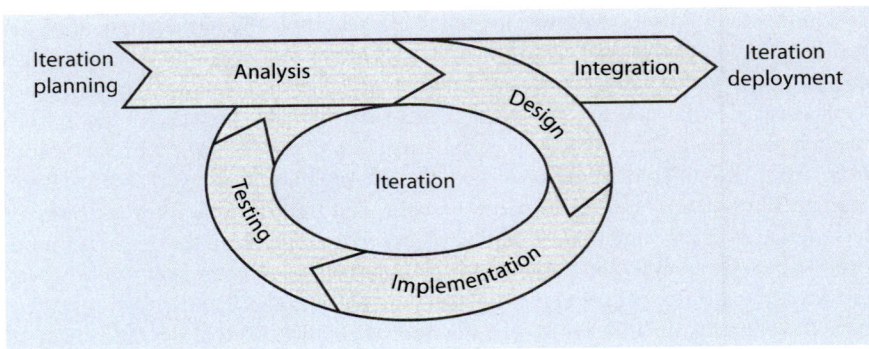

Fig. 2.33 Iterative processes of agile developments without predefined intermediate phases as in the example of Scrum sprints, which are short iterations with the steps: planning and assessment, analysis, design, implementation, acceptance testing, integration, and deployment

- **A**chievable/**A**ssignable: The goal of an iteration is realistic and achievable by clearly identified people within a specific time interval.
- **R**elevant: The goal of an iteration is relevant for the whole project and increases the value of the software.
- **T**ime-bound: The completion of an iteration has a fixed deadline.

Agile Philosophy

In relation to iterative development processes, another idea is in everyone's mind: agile development. Representatives are Extreme Programming (XP) (Wells 2009), Scrum, Lean, Feature Driven Programming, or Pragmatic Programming. A first look directly shows that agile development is not another iterative development process. The original idea does not define any phases or workflows at all. The initial driver was that most of the development processes were too oversized and static to adapt well to the fast changing requirements of today's world. They put processes and tools over the abilities of individuals and human communication. They do not deal with the fact that it is almost impossible to plan complete costs, risks, schedules, and changes at the beginning of a project. Developers also learn during the project work and can use this new knowledge to improve and correct their subsequent work. Even the iterative processes described with the parallel disciplines adapt too slowly here. Agile developments have short iterations and thus small incremental steps (usually in time periods of one day). They start with a first assessment and planning of the coming work. Then a suitable method and solution is analyzed and selected. The advised implementation is designed and gets developed. Finally the implemented software is reviewed and tested. The accepted software from the finished incremental step is integrated and committed to the repository to deploy it. The review and status of the integration regulate the planning of the following incremental step (see ◘ Fig. 2.33). The idea of agility is more or less a mindset which describes a philosophy about software development. It defines agile values, which are described with agile principles (Smith and Sidky 2009, *l.c. page 5 ff.*). The values and the principles are part of the Agile Manifesto (Beck et al. 2001):

Agile Values

Agile values behind the Agile Manifesto:

> **Individuals and interactions** over processes and tools
> **Working software** over comprehensive documentation
> **Customer collaboration** over contract negotiation
> **Responding to change** over following a plan (Beck et al. 2001)

Agile Principles

The 12 agile principles behind the Agile Manifesto:
1. «*Our highest priority is to satisfy the customer through early and continuous delivery of valuable software*».
2. «*Welcome changing requirements, even late in development. Agile processes harness change for the customer's competitive advantage*».
3. «*Deliver working software frequently, from a couple of weeks to a couple of months, with a preference to the shorter timescale*».

4. *«Business people and developers must work together daily throughout the project».*
5. *«Build projects around motivated individuals. Give them the environment and support they need, and trust them to get the job done».*
6. *«The most efficient and effective method of conveying information to and within a development team is face-to-face conversation».*
7. *«Working software is the primary measure of progress».*
8. *«Agile processes promote sustainable development. The sponsors, developers, and users should be able to maintain a constant pace indefinitely».*
9. *«Continuous attention to technical excellence and good design enhances agility».*
10. *«Simplicity – the art of maximizing the amount of work not done – is essential».*
11. *«The best architectures, requirements, and designs emerge from self-organizing teams».*
12. *«At regular intervals, the team reflects on how to become more effective, then tunes and adjusts its behavior accordingly».* (Beck et al. 2001)

At first sight, this philosophy does not really simplify a developer's life or help to enhance software development. It is also not totally new or totally different to the earlier iterative methods. But because of the focus on short iterations and on communicative skills, it is possible to derive several agile practices (for project management and for technical realizations), which can really be used during scientific developments. In principle, agile practices can be adapted to each (iterative) software development process. But it is better to adapt the software process to agile practices, to change from a clear plan-driven approach to a more feature-driven one (Smith and Sidky 2009, *l.c. page 9 f.*). It embraces changes, plans and delivers software frequently, uses human-centric methods with technical excellence, and collaborates with participants (Smith and Sidky 2009, *l.c. page 14*). This is the paradigm change in agile development. It focuses on working software features, developed in a social community, and not by strictly adhering to plans. Having working software without all the features is better than finishing the whole development on time, but having error-prone programs. But what does this mean for software development? While there are several official representatives of agile developments, the following describes an agile process which is currently being used successfully in scientific projects in the Geodetic Observatory Wettzell (bachelor/masters or PhD theses, third-party funded project work packages and local developments).

Agility does not come by itself. The administration and the team itself must do a lot to promote it. This was also a long learning process at the observatory, and in some cases, it still is. But an efficient agile development does not end in itself. It genuinely improves software and the way it is created. The techniques are quite simple even if it is difficult to find the right way to use them productively.
◘ Figure 2.34 shows the resulting iterative development, which is the result of several process improvements over the last few years.

The iterative process used defines three virtual phases: idea, big picture, and construction. It is not a strict structure. Actually there is no need for a sequential flow, as the phases and their activities can be performed at any point in the agile project. It means that the «big picture» documents are living documents and experience changes even during the life of the project. This is the reason it is known as «virtual» (Smith and Sidky 2009, *l.c. page 116 f.*). According to agile documentation, the «idea» phase is nothing more than a general headline for the project and should be kept very compact. It is the first envisioning of the project and nothing more than an «elevator statement». The idea behind it is to describe a project in a concise and interesting way, short enough for all the relevant keywords to be said in the time needed for an elevator trip (Smith and Sidky 2009, *l.c. page 140 f.*). Experience has shown that the idea can be arranged as the title page of a presentation, with one headline sentence and one subtitle.

Need for Agile Practices

Agile, Iterative Process

Virtual Phases

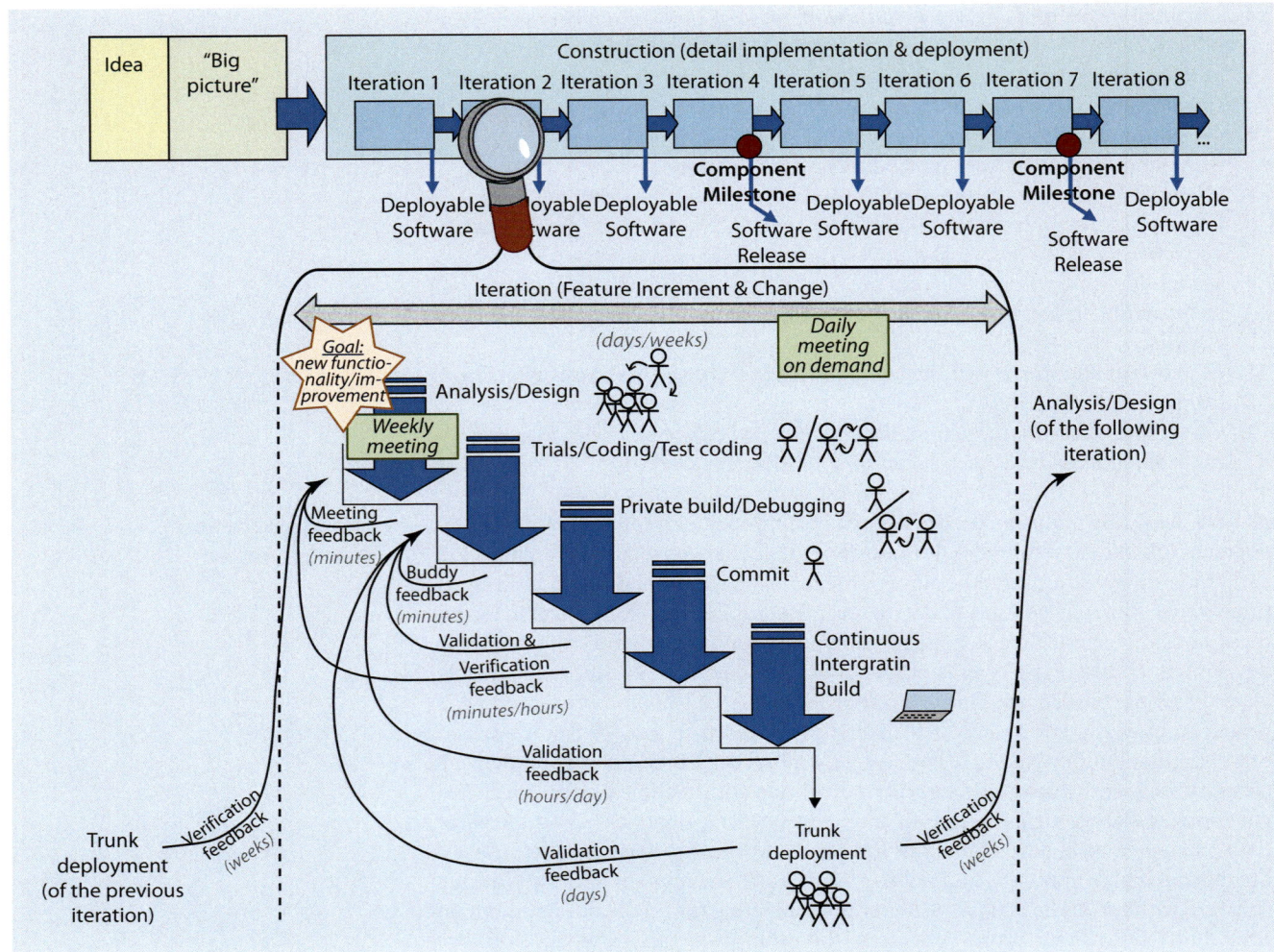

◘ Fig. 2.34 The phases and feedback iterations of an agile development, used at the Geodetic Observatory Wettzell for scientific software developments

This idea headline is then described in more detail to give an overview as a «big picture» (see ▶ Sect. 2.5.1). It can again be arranged as slides of a presentation. 20 to 30 pages should be enough. Each page should describe a feature and its requirements and therefore form a feature (or story) card. A feature in this context is one required functionality of the software, which is realized with one or several programming functions (which can be seen as story points). The main driver should be simplicity. Each feature card slide has:

— A headline (the feature name)
— A short description (including a first idea of story points/functions)
— A priority (according to importance, value, and risk; features with «quick wins» or in other words «low-hanging-fruits», with the most impact are prioritized (Wolf and Bleek 2011, *l.c. page 47*); functions or decisions, which are not yet necessary, are low priority and not produced on stock, in the hope that they might be useful someday (Wolf and Bleek 2011, *l.c. page 94*))
— Dependencies on other cards (sequentializing and grouping)
— Constraints (requirements, other involved parties, etc.)
— Possibly some quick hand drawings as graphical descriptions

Everything which needs more than one page to be described is too complex for an iteration. This description can be used for the science funding acquisition needed for the following, iterative «construction» phase (ideas in this block are adapted from (Smith and Sidky 2009, *l.c. page 154 f.*), (Smith and Sidky 2009, *l.c. page 170 f.*), and (Wolf and Bleek 2011, *l.c. page 37 ff.*)).

Creating the «big picture» is usually done by a single scientist, the project initiator, and leader. He is comparable to a product manager in industry. The initiator usually has the idea and already a first big picture in mind. How he gets further information is up to him. Ideally, it is a result of discussions and an iterative process, to which specialists have already been invited. This can happen in workshops or symposia or simply during coffee breaks. It is ideal to include knowledge from various functional areas and different points of view. This avoids strictly centralized decisions, which might fail, because of a missing «panoramic view» of the scientific requirements. The project leader also becomes the agile team leader (see (Smith and Sidky 2009, *l.c. page 66 ff.*)) in the following process. Even when self-organizing team members force design changes and discover their own solution paths, the agile leader has the final decision. He has at least a veto right, can «shepherd» the team members (help, guide, encourage, mentor, absorb interruptions (Smith and Sidky 2009, *l.c. page 77*)) to keep them on track, and must have the option to change the team composition (Wolf and Bleek 2011, *l.c. page 37 ff.*). In an institute or observatory with several parallel developments, another way of coming to decisions is also possible: the constitution of an agile core team. At the Wettzell observatory, experience has shown that it is ideal to include different leaders from different disciplines. Developments are shared anyway between the teams, so it is pragmatic to also discuss and exchange information about the projects (adapted from (Smith and Sidky 2009, *l.c. page 66 ff.*). The core team must have understood what agility means and which techniques are necessary (e.g., continuous integration, frequent meetings, coding styles, etc.). Without this acceptance, agile attempts are doomed to failure.

After getting the funding, two things must be done first: the acquisition of suitable agile project employees and the planning of the construction and releases. For agile developments, small teams are ideal. They are flexible and need less administration in order to come to decisions. The project employees should be 100% part of the project, without any additional commitments (Wolf and Bleek 2011, *l.c. page 78 f.*). This can be a problem in scientific institutes, as staff are often engaged in a number of other projects as well as being involved in university teaching if it is not a direct third-party funded position. But for funded projects, agile structures are ideal, as team members are only employed for the research project and can be selected according to requirements. The team members should have good social skills, like communicability, high technical background, being a generalist, flexibility, ability for teamwork, motivation, and so on. They should also appreciate the value of the product. Those who are «just doing a job» do not have the ideal attitude, as this stance leads to undermining, passive-aggressive behavior (Smith and Sidky 2009, *l.c. page 140 ff.*), which must be gently corrected by the core team. The fortunate thing in scientific projects is that usually young engineers who have a personal interest in the selected research apply for them. Experience at the observatory has shown that the social skills of colleagues are more important for projects than time plans, tools, and equipment. Team sizes of two to three people are ideal (see ◘ Fig. 2.35). Additionally, having a team who are clearly focused on the project is extremely valuable. The location of the team members should be as close as possible to support direct contact and frequent meetings and discussions in parallel with the official team meetings.

After gathering the team, the planning of the iterative construction can start. The principles are to implement something like «time boxing». This means that iterations should be manageable and completed within a given maximum time period. The process described has iteration times of three to seven working days. It is important that not all iterations are planned in detail at the beginning. The most important iteration is the first one, as it has to implement the project environment, development machines, and administration workflows (Smith and Sidky 2009, *l.c. page 196*). As a whole, the created iteration plan then identifies sequential requirements and parallel possibilities, orders the iterations according to their priorities, and gives the first indication of when each step will be taken and what must and should be part of it. This helps to adapt to changes during the development. Delays mean that iterations and therefore features are not implemented and shifted to a later time if they are less important. The main focus is on high-quality,

Agile Team Leader and Core Team

Agile Team

Planning of the Iterative Construction

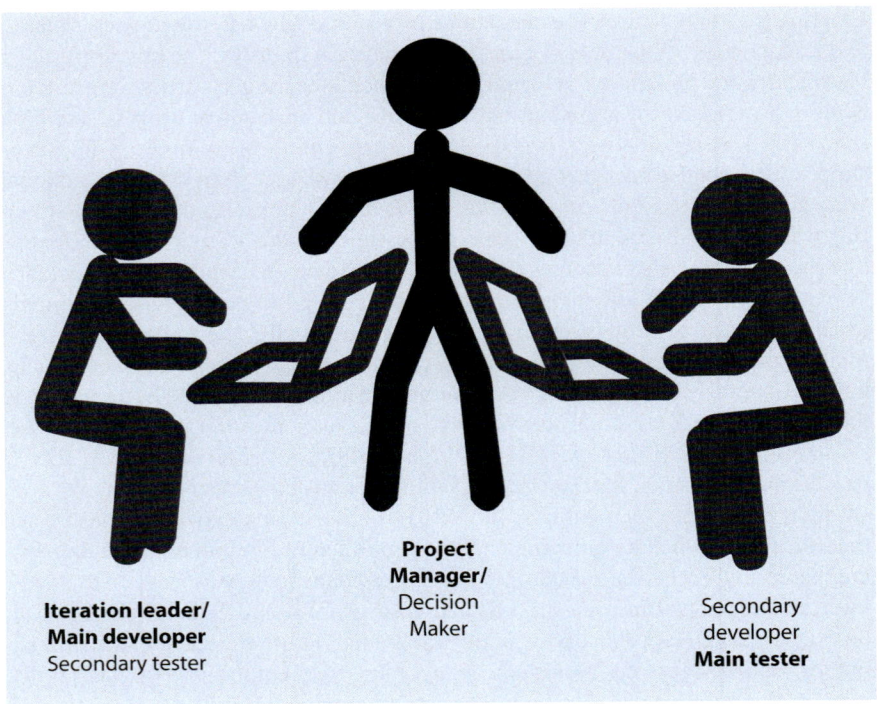

Fig. 2.35 An ideal agile team: the iteration leader is the main developer for the iteration, a secondary developer is the main tester of code produced in the iteration, and the project manager has the competence to make final decisions if the developers cannot agree

high-priority features instead of on the total amount of functionality (Wolf and Bleek 2011, *l.c. page 52 f.*). With this iteration gross planning, it is possible to plot the release dates on the time line over the usual fixed total research project time. These points mark the deliverables. At these points, the software is published as a new version, while at least at each iteration, the software is published as a new trunk, «beta» version. The result is an agile roadmap for the project.

Virtual Activities of One Construction Iteration

As described, the construction in this agile development is split into iterative steps, where each one implements a new feature (increment) or changes/improves an existing one (the direct goal of each iteration). One feature is directly translated into one compact iteration. Each iteration itself consists of six virtual activities: analysis&design, trials&coding&test coding, private build&debugging, commit, continuous integration build, and trunk deployment. The activities seem to follow something like a waterfall model, but use feedback structures, which are iterative cycles within the literal iteration. Because of its short time periods, the construction is very flexible. The feedback not only determines current iterations but subsequent ones and can even change parts of the «big picture» plan.

Iterative Analysis and Design, Agile Meetings

Analysis and design concentrates mainly on the features which should be implemented in the current iteration. Ideally, it is done in a group of developers in the weekly meetings. The purpose of those meetings is to give the group the chance to discuss the project and give mutual feedback and concentrate on what was done during the last week, which problems occurred, and what should be done during the following week. The agile leader moderates such meetings and «shepherds» the participant's discussions. The feedback is direct, should be constructive with a feedback structure, and can be done within a few minutes (at most one hour). Similar meetings can even be held daily on demand. The meetings are a retrospective review of the process. They allow the team to identify what went well and what did not. Clear questions and discussions may also help to find the causes of any problems (Smith and Sidky 2009, *l.c. page 298 f.*). The meetings should not be used to cast blame on the work of the developers. Another advantage of regular meetings is that they can be used to communicate decisions from the agile leader as well as from the individual iteration development team (Wolf and Bleek 2011, *l.c. page 94*). Everybody can learn from errors, even when they were made by others. Experience at the Geodetic Observatory Wettzell has shown that such meetings can be ideally combined with manual, functional integration tests of the software. Parallel meetings are used to train the team in agility, technical issues, and the workflows at a frequency of 2–3 weeks. The training sessions should be organized as real knowledge transfers and should be held by the team members themselves (compare also (Wolf and Bleek 2011, *l.c. page 79*)). The time spent for all of the meetings and the preparation of each team member is not an end in itself. It pushes the common knowledge of the whole team.

The coding itself is split into writing of trial code (prototyping of features), the programming of the code with real functional features, and the creation of according tests (e.g., unit tests). This is done by individuals or preferably in the form of pair programming (see also ▶ Sect. 2.6.2, (Wolf and Bleek 2011, *l.c. page 70*)). The best results with the shortest development times can be reached in such teams of two people. Discussions about interfaces, problems, or solutions can be done face-to-face. This «buddy-feedback» is therefore very valuable. Another advantage is that more than one person gets knowledge about a dedicated feature and its realization. This is quite important to avoid project failure risks (known as «truck factor», which means: how great is the likelihood that the project will fail if one team member is «killed in a truck accident»?), because of staff quitting. It avoids specialist knowledge, which is often a problem with specialized scientists or employees, speculating on contract continuations on the assumption that they are irreplaceable. The basis for this is that there is no personal ownership of the code. The complete code is owned by all the members of the agile teams. Therefore, everybody can change anything in the existing code assets logged in the change log of the version control system. It benefits from the use of coding style guides and self-documenting code. In a similar way, errors can be fixed. If possible, the developer who finds a bug can fix it directly. If he cannot, he should take care that the developer responsible for it is informed. A suitable way is also to prepare a patch and send it to the specific developer (see Example 2.18) (Pilato et al. 2009, *l.c. page 123*). Then he must fix it with the highest priority. This behavior has caused and in some circumstances is still causing the most problems in the projects, as most developers have a fear of destroying code of someone else inadvertently. The solution for this dilemma is the unit tests. Each code requires such test codes and each change must pass them as a first validation. This private build and debugging is an essential process to avoid commits of not functioning code. The private builds also offer a verification and validation feedback in time cycles of minutes to hours. Finally, if something totally unexpected happens, the version control system allows the developer to turn the time back to a correct version (compare also (Wolf and Bleek 2011, *l.c. page 82 f.*)).

Iterative Coding and Private Testing

■ ■ **Example 2.18**

Use of «svn diff» (differential file content comparison) and the «patch»-command (Pilato et al. 2009, l.c. page 123)

```
###################                                                            1
# Create a patch #                                                             2
###################                                                            3
# Create a file with all differences between the working copy and the base version (version    4
        before changes)
# (see SVN description and also man page with "man diff")                       5
svn diff <working_copy_path> > patchfile                                       6
                                                                               7
############################                                                    8
# Patch an existing version #                                                   9
############################                                                    10
# Apply a created patch to a local working copy                                11
# (see man page with "man patch")                                              12
patch -p0 < patchfile                                                          13
```

Iterative Committing, Continuous Integration, and Deployment

After all of these private activities, the code can be committed and checked with a public continuous integration build. Only working and completely buildable code is allowed to be committed. But it is not only the final code of an iteration that should be committed. Ideally, small, manageable code updates should also be committed, as long as they are usable by others and support the features. The automated checks of the continuous integration run regularly or on demand and process all of the inspections, tests, and statistics reports. They offer a validation feedback from various other machines in periods of days. The plots derived allow an evaluation and measurement of the project state. Errors which are detected must be fixed immediately with the highest priority by the team, as the code assets on the repository are accessible by all the other agile teams. Further graphical information about progress might also be a feature burn down graph, which marks all of the planned features in relation to the already implemented ones over time (split in percentage counts for required and nice-to-have features) (Wolf and Bleek 2011, *l.c. page 110*). Each of these commits results in deployable software, so that releases can be deployed with already stable software versions within a few days. The releases always implement the most important features at the current project state (Wolf and Bleek 2011, *l.c. page 122 f.*). Additionally, the common code repository and mutual use of source «cross-fertilize» all the agile projects of the different agile teams and have a huge impact on the progress and quality of projects. This impact is one key feature to improve coding all over the given team structures at an observatory. It ensures mutual influences between projects because of ideas knowledge and source code transfers (see also ◘ Fig. 2.36) but also requires discipline and team spirit.

Example: Students Project and Agile Training

Finally, a working example can be shown, using projects in student internship or bachelor and masters theses. The following example is the description of a software development during a student internship, to develop a new version for a data acquisition system as part of a system monitoring, collecting height and temperature data about the Wettzell radio telescope. This previous sentence is already the

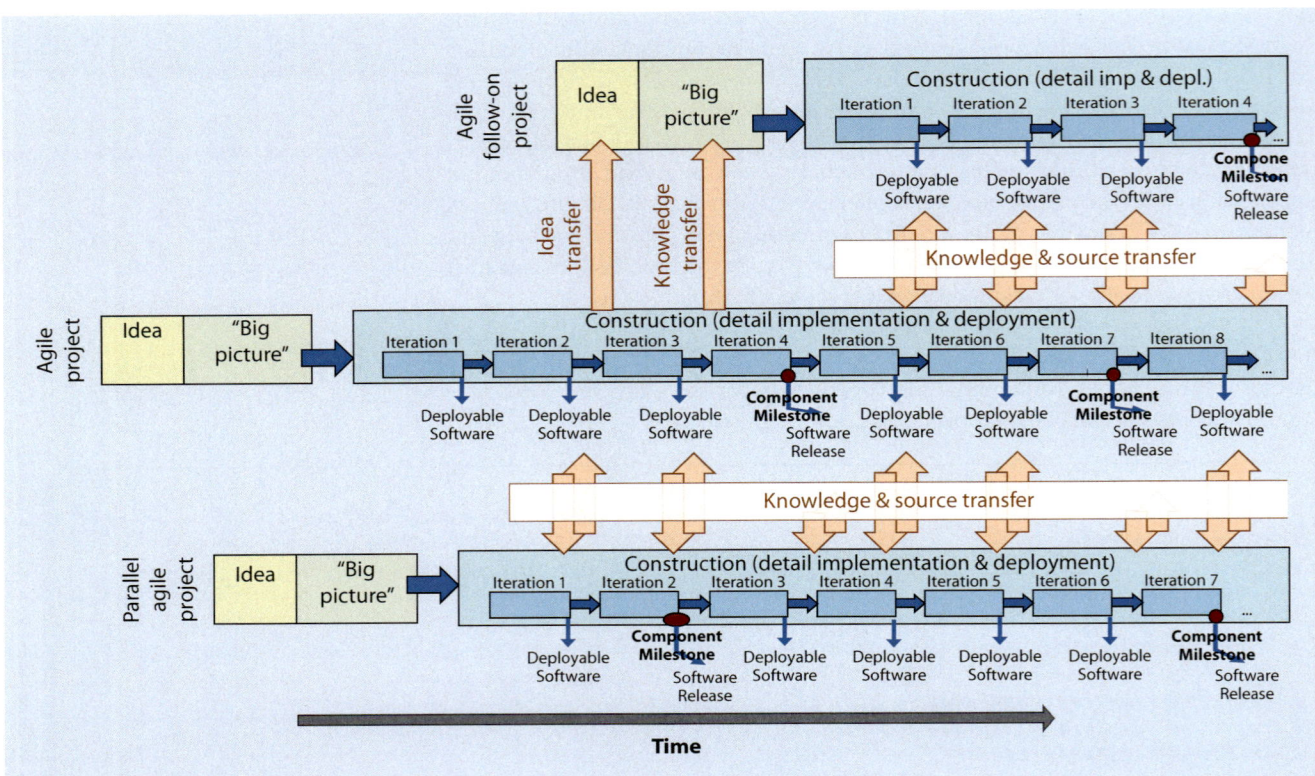

◘ **Fig. 2.36**   Mutual influence of agile developments with ideas, knowledge, and source code transfer

«elevator statement». The following planning resulted in the big picture document (a handwritten sheet, created during meetings with the student) and an iteration plan. The new software had to use existing hardware, accessible via serial communication. All the documents for the controllers and the sensors were available. The software had to implement a new and stable client-server communication to request the data from remote. Therefore the following features and iterations were identified:

1. Repeat, read, and learn the basics about the C/C++ programming, including: find personally suitable books and learn a few basics.

2. Write a module to control the common serial communication hardware in a computer (a module was already available, so that it had only to be proofed for suitability to the project with partial modifications).

3. Write a module which implements the proprietary communication protocol of the sensor controller with the subtasks (story points): sending a request, resynchronization, and interpreting the answer (with additional features to set controller properties).

4. Write a server, using the communication platform of the observatory to offer prepared and human-readable data to the observatory network with the parts (story points): initialize the server, use a configuration file for the properties, and answer a request for a complete data set (with additional features to set the server and the controller properties and to request additional, individual data separately).

5. Write a command line client to test the server with the parts (story points): reading a command (prior commands first), processing the request, printing the result.

6. Write a client which could be included in the radio telescope control system to log the values into a log file (similar to the command line client, only without command line interpretation).

7. Write a graphical user interface which frequently plots the results.
   The project was limited in time to about 18 weeks. The above list already represents the found prioritization. Each item represents a feature which could be translated directly into an iteration. Most iterations follow a sequence. Only the final three can be performed in parallel. Even if the client for the radio telescope control system has a higher priority than a command line client, the command line client is quite important for verification and validation issues, so that it was favored. The iteration plan was mainly created by the internship leader, but in communication with the student. For him, it was the first software project of all and he only had a basic knowledge of programming. Therefore, the first step was a direct learning phase, using a tutorial book. The first real project iteration was the implementation of the module for the proprietary communication protocol. The analysis and design included the reading of the manual and the planning of a general structure for the module and its functions. In parallel, the student tried first simple code snippets, using the serial communication module. Iteratively he created the module by doing continuous refactoring guided by the internship leader. The student had the opportunity to have regular meetings on demand (which he took more often at the first iterations), but had to join a frequent, weekly (or each second week) meeting. There he had to present the work of the past week, had to explain problems and solutions he had found, and had to predict the work of the coming week. The meetings were attended by different engineers, who could give feedback on the presentation and the work. The weekly slides offered an ideal documentation for the later internship report. In parallel, he got individual training units from the leader, and he participated collaboratively at international teleconferences about system monitoring standards to understand the key tasks.

   Iteration-by-iteration and guided by the leader, the student developed the most important features with his own ideas. The codes created were frequently built and tested locally and afterward committed to the repository, but had almost no influence on other projects, as the system monitoring task was very specialized. During the internship time, almost all (except for the last) features could be realized, so that the transition to the new monitoring system was possible well in time. During an additional voluntary iteration, the student optimized the performance of the server and made a further refactoring, while keeping the functionality. At almost any stage of the development, a deployable, working code was available, even if it was not always perfect.

Similar project styles are currently being used for almost all coached projects at the Wettzell observatory. It was a long learning process with different pitfalls (such as insufficiently working communication, social problems between team members, fears, tools which were not completely functional, and so on). The learning process also led to projects which might have been able to come to a better result as they developed. But continuous learning is one of the main principles and very welcome in agile behaviors, and therefore recent projects have already shown some success, such as the realization of a work package in the project Novel EXploration Pushing Robust e-VLBI Services (NEXPReS) or an attempt to train a programming beginner with the agile training method, so that he was able to implement a complete meteorological data acquisition system for a new meteorological station within nine weeks. Nevertheless, this was only possible by setting up existing modules and components, which are available as central code repository. By using such a toolbox, agile training and agile methods unfold their whole potential.

2.10   Summary

2.10.1   What Did We Learn?

This first chapter offers the background which is useful for every software project. Writing software is a complex task if we want to follow the well-known quality factors. In the first section, we learned about the differences between compilers and interpreters to select suitable languages for programming control tasks and scripting. Having a suitable programming language does not automatically produce high-quality software. The style which is used for writing the source code is essential to keep the programs readable and comprehensive. The coding layout defines the indenting, naming, structure, and so on, of the program code written. Code policies add an enhancement in stability, functionality, and correctness. Rules which define how specific code should be written automatically reduce the possibility of errors. A big block is also the inclusion of existing legacy code. Suitable converter classes with clear interfaces keep the changes local and hide external or legacy code to make it more manageable. Documentation for developers, users, and administration is essential if the use of software should be continued after a project ends. Automatic document generators help to implement an agile documentation landscape where especially developer documents are created daily, including the newest release. Wikis or online manual pages support users to do their daily jobs, using the software.

Style and suitable documentation is one side of the coin. The other side of the coin is defined by high-quality testing mechanisms and a suitable process to develop software. Test metrics provide quantitative classifiers to characterize the software quality. Even if they should be discussed individually, they offer reliable numbers. Unit tests and a form of test-driven development is the foundation for dynamic code testing to get a suitable test coverage over the entire software. Tools for static code inspections can additionally find critical issues and flaws. The dynamic behavior of the software with all its processes, threads, and memory which is allocated dynamically during runtime is only visible if tools for dynamic code analysis are used. We discussed scheduling mechanisms, where critical sections must be protected with semaphores. All tests are integrated into a testing landscape which supports continuous integration. The method of continuous integration provides daily releases with tested, up-to-date software from a central repository. This repository is implemented with a version management system which tracks all changes and updates of source code files and directory structures. Using continuous integration means that each developer gets a daily overview of the current status of the software developed and the project status. This ideally supports agile development processes which put their focus on the development team and not on tools. Agile processes assign different workflows of a project to iterative phases.

2.10.2 How Did We Use the Contents Learned?

The techniques learned are essential for all software projects. They are the foundation for software development. Therefore, the chapter offers example implementations. Coding style, coding policies, documentation, and unit testing were shown with example code to solve the N-queens puzzle. Starting with unreadable code solving the N-queens puzzle, each section improves the quality of the code by adding increasingly better style. Including Doxygen tags provides a possibility to create documentation of the software using HTML. Unit tests using the Wettzell test suite demonstrate how the functional quality can be improved with testing. The knowledge of scheduling mechanisms helped to improve the dynamic code analysis.

Experience from daily work at the Wettzell observatory and implementations used at the observatory demonstrate the use of continuous integration, version control systems, and a suitable agile development. The daily output from tools related to continuous integration are used to create web pages which give a daily overview of the project status and the software quality. The methods provided by version control systems are used to support the different branches during a software life cycle. We also used the experience of the work at Wettzell about psychological factors to improve the iterative development.

2.11 Questions

1. What is the difference between a compiler and an interpreter? What is about runtime efficiency of compiler and interpreter code?
2. What are «quality factors» for software?
3. The selection of a programming language is always a compromise for the developers. What do you keep in mind when choosing?
4. What happens during the linking and binding in the compilation?
5. Why are all forms of naming notations (such as the Hungarian notation) still under discussion? What are the pros and cons?
6. What is the difference between code layout and code policies? Why do policies help to get a stable code?
7. What is the risk with pointer operations?
8. What is «const»-correctness? Where is it used?
9. Which function call would you expect is faster: the call with the argument «*double dInput*» or the call with the parameter «*const double & dInput*»?
10. Why should you write «if»-conditions with constants in the following way: «*if (5 == uiState)...*»?
11. What is «shallow copying»? What should be forced instead in multi-threading environments?
12. Why do we avoid Camel case or Pascal case for filenames?
13. What is legacy code? Which technique improves the use of legacy code?
14. Which documents do you know are related to software documentation?
15. What does the Doxygen tag «*[inout]*» mean for function arguments?
16. Which advantage does a documentation landscape offer to users?

17. Which categories for test metrics do you know in the field of scientific software and what is the purpose for them?
18. Why is focusing on one single count (such as the produced number of lines per day or the increase of found errors per day) not really ideal to evaluate the quality of a software?
19. What does a code beautifier do?
20. Compare static code inspections with dynamic code analyzing?
21. What are «false positives»?
22. What is the difference between «stubs» and «mocks»?
23. Which inputs would you prefer for unit tests: limits and extreme values or inputs which are expected for the functional task? Why?
24. Explain the principle of «round robin» and preemptive scheduling?
25. An operating system is an additional software layer to do what?
26. What is starvation and what is a deadlock?
27. What does «non-reentrant» mean?
28. What is a «semaphore»? What is «mutual exclusion»?
29. What are the phases of a sequential model, a top-down model, or a waterfall model?
30. What is the difference between «disciplines» and «phases» of a software development process?
31. Which main phases are used in iterative, agile development processes?
32. Which disciplines or workflows do you know while software is developed? Please characterize them according to their chronological assignment to the development phases.
33. What is continuous integration?
34. What is a «weaver»?
35. What is UML and for what is it used related to documentation?
36. What are hash functions? What would make them more effective?

Using a Code Toolbox

3.1   The Idea Behind a Code Toolbox – 132

3.2   Well-Tested Modules and Components – 144

3.3   Generative Programming – 148
3.3.1   Classic Solutions for Interprocess Communication (IPC) – 149
3.3.2   Remote Procedure Calls – 168
3.3.3   Extending Generative Programming for Interprocess
        Communication – 181

3.4   A Rudimentary Middleware for Controlling of Distributed
        Systems – 236

3.5   Summary – 243
3.5.1   What Did We Learn? – 243
3.5.2   How Did We Use the Contents Learned? – 244

3.6   Questions – 244

© Springer International Publishing Switzerland 2017
A. Neidhardt, *Applied Computer Science for GGOS Observatories*, Springer Textbooks in Earth Sciences,
Geography and Environment, DOI 10.1007/978-3-319-40139-3_3

3

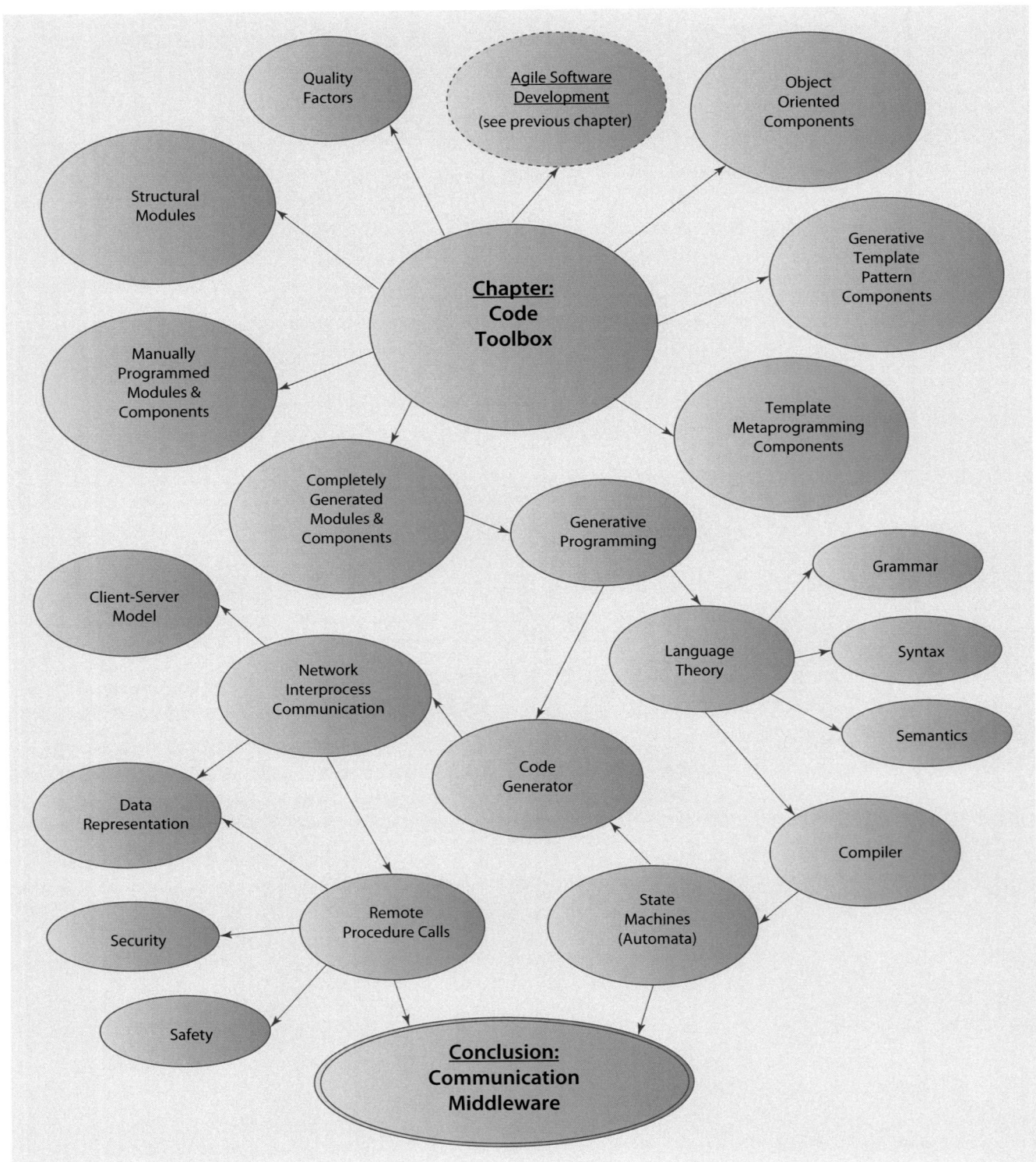

3.1   The Idea Behind a Code Toolbox

Sharing Prefabricated «Off-the-Shelf» Building Blocks

Through observing the software development process in agile environments (see ► Chap. 2), it has become clear that the different projects influence each other. But they do not only benefit from information and new knowledge transfer. Already existing source code can also be shared, iteratively improved, and extended over project borders (see ◘ Fig. 2.36). This is especially supported by the commonly shared code assets in the version control system. But it only works if the code there is designed as prefabricated «off-the-shelf» building blocks which are reusable, comprehensive, and compact enough for the construction of other larger software systems (Brookshear 2012, *l.c. page 314*). Therefore, it is important to design large software systems as a composition of (ideally) reusable, separately implementable, and testable units: the modules or components. This is the principle of modularity.

Definition DEF3.1. of «modularity» (see (Rembold et al. 1991, *l.c. page 277 f.*)) Modularity is the decomposition of a complex software system into manageable, compact modules and components.	Modularity

Definition DEF3.2. of «module» (see (Rembold et al. 1991, *l.c. page 277 f.*)) A module is a sequence of program code which can be designed, implemented, and tested separately. A module consists of a description or declaration of its interface (of externally callable functions), a declaration of its internal data variables and types, and a definition of its functions (the according function bodies of the interface functions).	Module

Definition DEF3.3. of «component» A component is a module in OOP designs. Because of the design criteria in OOP, components use a more sophisticated and abstract interfacing on the basis of classes. Member functions (methods) are defined there, which can be classified as private (only internally available), protected (available internally and for derived components), and public (externally accessible as interface functions). The class concept of components also allows all OOP possibilities, such as inheritance, overloading, templates, and so on.	Component

In general, modules are nothing more than source code files with a more or less loose collection of functions. Usually one module is separated into a header file with the interface function declarations and a source code file with the function bodies and internal, module-global data variables. But assembled functions in a module should have something in common. The best modules maximize internal binding, the cohesion. Cohesion is very important for the maintainability of a module. Functionality which belongs together is easier to understand. But there are different forms of cohesion. The weakest is the logical or administrative cohesion. It groups functions with activities which process logically similar things, such as a function set which can convert different time formats. Most of the modules are like this. The functional cohesion, where all parts of a module are focused exclusively on one activity, is much stronger. But a high functional cohesion can split software activities into numerous single files, which are then again difficult to maintain. Therefore, another cohesion definition focuses on data orientation. All functions dealing with the same data are assembled. Internal cohesion interacts with the goal to maximize the independence of the modules from one another, so that changes in a module are kept as local as possible.

Cohesion influences the (control) coupling, which is the interrelationship between the different modules or components. Control coupling means that functions from one module call functions from another module and pass all the necessary values during this call as parameters. This coupling of modules should always be concentrated on the control logic. Data coupling may be allowed in some specific cases, such as in systems with shared memory. But usually, it is necessary to maximize independence among modules (compare (Brookshear 2012, *l.c. page 311 ff.*) and (Rembold et al. 1991, *l.c. page 278*)).

(margin notes: Cohesion, Coupling)

Parametrization

But not only cohesion and coupling define the quality and the classification of modules and components. They also make it possible to parametrize specific problem cases and solutions. The level of parametrization defines the flexibility and adaptability of the modules and components to specific application requirements. In detail, parametrization means that not everything is completely defined and static according to the structure and functionality of the software unit. A simple example can be given in OOP environments by comparing standard classes and template classes. While standard classes have fixed type bindings, templates allow the use of the same code in combination with changing types. Therefore, templates can be better parametrized than classes. In this case, the design of the final product is more compact, and the amount of code can be reduced. But definitions and handling become more complex inside of the template classes. This means that on the one hand, better parametrization offers higher flexibility for change and a more compact design of the final product. On the other hand, the components or modules become increasingly complex.

High Parametrization

The adaptability with the usual programming mechanisms has its limits, as it is defined and limited in the programming languages. But if the parametrization should be further increased, only the use of generative methods is possible. Generative programming means that complete code blocks, modules, and components are generated according to a highly parametrizable generation description, which follows a specific formal logic, and can be interpreted and understood by a generator tool. It is possible to implement a high parametrization by using this generative programming.

Information Hiding

Generally, each software system can be decomposed using such modularity. Often it is a more or less loose assembly of functions. The OOP paradigm with its classes also allows a combination of logical and data-oriented cohesion, as methods of a class usually work on the same data which is also a private or protected part of the class. This increases cohesion and offers a better way of hiding information. «Information hiding» is the restriction of information to a very limited section of code. It is important, because it avoids confusing effects of data changes in other modules. Data corruption can be one consequence, without information hiding. The protection of information as internal data, but also program structures and internal data types, can be ensured by designing cohesion and coupling in an ideal way. But it is already well supported if using the OOP class protections. It is one of the main techniques for functionality abstraction. Users of the functionality do not have to know too much about the internal implementations. They can just use the existing functionality or features of the declared interface and trust in the correctness of the implementations (verified by unit tests and the CI, see ▶ Sect. 2.8). The whole component acts like a «black box».

The extension of modularity, suitable parametrization, and a way of supporting information hiding make it possible to create sets of (more or less abstract) software tools for different needs, which can be collected in a (preferably centralized) code toolbox.

Code Toolbox

> *Definition DEF3.4. of «code toolbox»*
> A code toolbox is a collection of useful, prefabricated, «off-the-shelf», specified, and highly specialized source code modules or components as tools, where each tool solves a specific type of task in a well-tested and verifiable way without redundant code. The tools in the toolbox are available for everybody in the development, can be combined like construction blocks, and can be extended along new requirements.

Categorization of Toolbox Elements

Such a code toolbox tries to implement software units which build an ideal combination of cohesion, coupling, parametrization, information hiding, and complexity. The modules and components of such a toolbox can be classified according to these categorizations (see ▫ Fig. 3.1). Modules are static implementations of more or less loosely coupled standard solutions. Components add the object-oriented design mechanisms of information protection, abstraction, and inheritance. Templates allow a first level of a parametrization of data types. Template metaprogramming

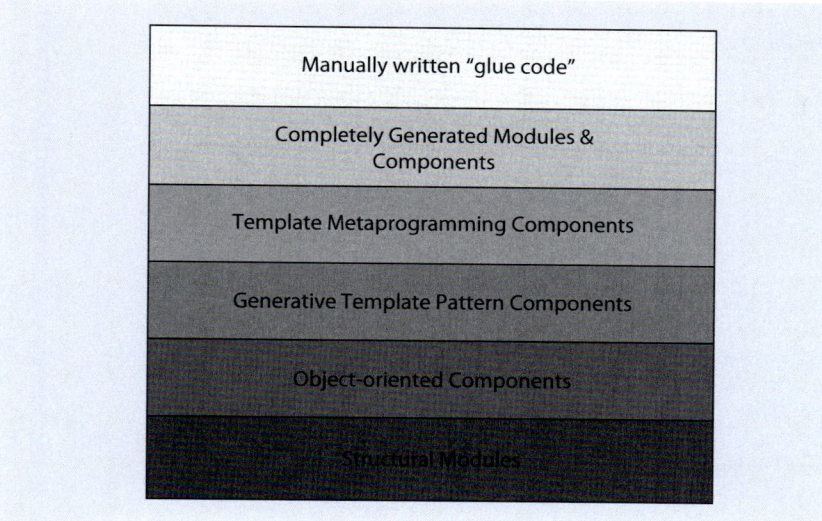

Fig. 3.1 The hierarchical elements for an adaptable modularity in a code toolbox

offers a parametrization of internal functionality, for example, for a performance increase. Generators enable the parametrization of model solutions (in sets of modules and components) according to a specified metaprogram and use case. Finally, the manually written code allows all variances and is used as «glue code»[1]

A collection of more or less loosely coupled functions following the structured programming paradigm (e.g. in C) falls under the category of structural modules (e.g., see Example 3.1 with function declarations to realize a serial communication). It is the simplest combination of code parts to build logical units, even though the optimization of coupling and cohesion requires the right degree of sensitivity and experience. Nevertheless, splitting a complex task into subtasks and the subsequent required functionalities, which are then grouped, should be the minimum of each development.

Structural Modules

■ ■ Example 3.1

C module «`simple_serial.h`» with functions to realize a serial communication (source code extract taken from the «`simple`»-toolbox of the Geodetic Observatory Wettzell)

```
/* ... */                                                                      1
                                                                               2
#ifndef __SIMPLE_SERIAL__                                                      3
#define __SIMPLE_SERIAL__                                                      4
                                                                               5
#ifdef __cplusplus                                                             6
extern "C" {                                                                   7
#endif                                                                         8
                                                                               9
    /*******************************************************************       10
     * Defines:                                                       *        11
     * Used Errorcodes                                                *        12
     *  ───────────────────────────────────────────────────────────  *        13
     * Additional Info:                                               *        14
     *                                                                *        15
     *******************************************************************/      16
#define NO_ERROR            0                                                  17
#define ERROR_NOTAERRNO     1    /* Errornumber is not of type errno */        18
#define ERROR_SAVEATTR      2    /* Saving the attributes for a later restore doesn't work */   19
#define ERROR_ALLOC         3    /* Can't allocate more memory */              20
#define ERROR_NOATTRSAVED   4    /* There were no attributes saved for a retsore */   21
```

---

1   The following explanations mainly focus on the interface declaration to show the main differences.

```
#define  ERROR_RESTOREATTR   5   /* Can't restore saved attributes to device */
#define  ERROR_GETATTR       6   /* Can't get attributes from device */
#define  ERROR_SETATTR       7   /* Can't set attributes of device */
#define  ERROR_OPEN          8   /* Can't open device */
#define  ERROR_CLOSE         9   /* Can't close device */
#define  ERROR_NOTOPEN      10   /* Device is not yet opened */
#define  ERROR_NOATTR       11   /* Can't find attributes in saved attribute list */
#define  ERROR_NOTTY        12   /* Filedescriptor doesn't point to a serial device */
#define  ERROR_NOTASPEED    13   /* The speed is not allowed */
#define  ERROR_FLUSH        14   /* Can't flush device */
#define  ERROR_DATABITS     15   /* Wrong number of databits */
#define  ERROR_PARITY       16   /* Wrong parity */
#define  ERROR_STOPBITS     17   /* Wrong stopbits */
#define  ERROR_FLOWCTRL     18   /* Wrong flowcontrol */
#define  ERROR_WRITE        19   /* Flushed writing didn't work */
#define  ERROR_READ         20   /* Read didn't work */

    /*******************************************************************
     * Functionprototypes:                                            *
     * Interface functionality                                        *
     *_____    *
     * Additional Info:                                               *
     *                                                                *
     *******************************************************************/
    /* Convert error code into a error text message */
    unsigned int uiErrorToText (char * pcErrText,
                                unsigned int uiMaxLength,
                                int iErrNo);
    /* Serial connection attribute administration: read attributes of the serial
       hardware device and save them*/
    unsigned int uiSaveAttributes (int iFDtty);
    /* Serial connection attribute administration: use saved attributes and
       restore them at the serial hardware device */
    unsigned int uiRestoreAttributes (int iFDtty);
    /* Open a serial connection */
    unsigned int uiOpenTTY (char * pcTTYName, int * piFDtty);
    /* Reset a serial connection */
    unsigned int uiResetTTY (int iFDtty);
    /* Close a serial connection */
    unsigned int uiCloseTTY (int iFDtty);
    /* Check if a serial connection is opened */
    unsigned int uiIsTTYOpened (int iFDtty);
    /* Change the attributes of the serial hardware device */
    unsigned int uiSetStandardAttributes (int iFDtty,
                                          unsigned int uiSpeed,
                                          unsigned int uiCharacterSize,
                                          unsigned int uiParity,
                                          unsigned int uiStopBits,
                                          unsigned int uiFlowControl,
                                          int iInit);
    /* Write characters to serial hardware device and flush hardware buffer */
    unsigned int uiWriteWithFlush (int iFDtty, char * pcText,
                                   unsigned int uiLength);
    /* Read characters from serial hardware device and clean buffer before */
    unsigned int uiReadWithClear (int iFDtty, char * pcText,
                                  unsigned int uiLength);
    /* Local unit test main function */
    int iSimpleSerialTestMain ();

#ifdef __cplusplus
}
#endif

#endif /* __SIMPLE_SERIAL__ */

/* ... */
```

Related to coupling and cohesion, object-oriented components also follow a similar concept, while the higher OOP design mechanisms of the languages promote better optimization. They additionally add the mechanism of information hiding mentioned and increase reusability in good OOP designs. The goal in creating such a toolbox is to achieve more functionality with less effort expended on code size and development (Vandevoorde and Josuttis 2003, *l.c. page 301*). Supporting techniques of the OOP languages are overloading, inheritance, virtualization, polymorphism, or class interface abstraction.[2] They increase the design possibilities. Nevertheless, a clear, reusable, and optimal design is quite difficult and requires much more experience than is necessary for the simpler structural modules. A typical sample for OOP components can be found in Example 3.2 with a class declaration to implement the basics of an online chat.

■■ **Example 3.2**

C++component «`simple_chat.h`» with methods to realize the basic functionality for an online chat (source code extract taken from the «`simple`»-toolbox of the Geodetic Observatory Wettzell)

```cpp
// ...

#ifndef __SIMPLE__CHAT__
#define __SIMPLE__CHAT__

#include <list>
#include <string>
#include <map>

class simple_chat
{
    protected:
        // map to save messages according to arrival order (= line number)
        std::map<unsigned long, std::string> priv_CMsgList; // Line number (=key), message (=
            value)
        // map to save user aliases according to derived user identifications
        std::map<unsigned long, std::string> priv_CAliases; // ulUserId (=key), alias (=value)
        // map to save user specific read pointers according to user identifications
        std::map<unsigned long, unsigned long> priv_CCurrentReadLine; // ulUserId (=key), Line
            (=value)
        // current write pointer
        unsigned long priv_ulCurrentWriteLine;
        // size of internal message stack
        unsigned int priv_uiStackSize;
        // random seed
        unsigned int priv_uiRandBuffer;
    protected:
        // clean-up stack to defined stack size
        unsigned short usCleanUpStack ();
```

```
1
2
3
4
5
6
7
8
9
10
11
12
13
14

15
16
17
18

19
20
21
22
23
24
25
26
27
```

2　The idea behind these OOP techniques is to create individual behavior using a set of classes of a specific object-oriented design. Overloading allows methods of the same name to have different parameter sets. Inheritance brings two classes into a relationship which makes it possible to inherit methods or attributes from another class. Those inherited elements can then be used as if they were local ones. Virtualization uses the classifier «virtual» to overwrite methods dynamically. This offers a form of polymorphism, which means that a pointer to an object of a polymorphic class can be cast in either the one or the other relative of the classes family and adopts then either the one or the other occurrence of the virtual methods inside. Finally, this polymorphism also allows abstract base classes in which none of the methods are implemented. They must implement the method bodies individually only if other classes are derived from them. This can be used to standardize class interfaces of complete class families.

```cpp
public:
    // simple_chat => constructor                                           28
    simple_chat ();                                                         29
    // simple_chat(CIn) => copy constructor                                 30
    simple_chat (const simple_chat & CIn);                                  31
    // ~simple_chat => destructor                                           32
    ~simple_chat ();                                                        33
    // operator = (CIn) => assign operator                                  34
    simple_chat & operator = (const simple_chat & CIn);                     35
    // usRegisterAlias (strUser, ulUserId) => registration of an user alias 36
    unsigned short usRegisterAlias (std::string strUser,                    37
                                    unsigned long & ulUserId);              38
    // usDeleteAlias (ulUserId) => delete a registered user alias           39
    unsigned short usDeleteAlias (unsigned long & ulUserId);                40
    // usAddMsg (ulUserId, strMsg) => insert a new message with registered user  41
    unsigned short usAddMsg (unsigned long & ulUserId,                      42
                            std::string strMsg);                            43
    // usReadNewMsgs (ulUserId, strMsgs) => read all new messages since last read  44
    unsigned short usReadNewMsgs (unsigned long & ulUserId,                 45
                                 std::string & strMsgs);                    46
    // usReadAllMsgs (ulUserId, strMsgs) => read all saved messages         47
    unsigned short usReadAllMsgs (unsigned long & ulUserId,                 48
                                 std::string & strMsgs);                    49
    // usClearAllMsg () => clear all saved messages                         50
    unsigned short usClearAllMsg ();                                        51
    // usPrintAllMsgStdout (ulStartLineNumber) => print messages to stdout  52
    unsigned short usPrintAllMsgStdout (unsigned long ulStartLineNumber = 0);  53
    // usCheckUserId (ulUserId) => check if user identification is registered  54
    unsigned short usCheckUserId (const unsigned long &ulUserId);           55
    // usCheckAlias (strUser) => check if user alias is registered          56
    unsigned short usCheckAlias (std::string strUser);                      57
    // usGetCurrentUserId (strUser, ulUserId) => convert user alias into user  58
    //    identification
    unsigned short usGetCurrentUserId (std::string strUser,                 59
                                      unsigned long & ulUserId);            60
    // usGetCurrentAlias (ulUserId, strUser) => convert user identification into user  61
    //    alias
    unsigned short usGetCurrentAlias (unsigned long ulUserId,               62
                                     std::string & strUser);               63
    // define size of message stack                                        64
    unsigned short usSetSizeOfMsgStack (unsigned int uiSize);               65
    // return size of message stack                                        66
    unsigned short usGetSizeOfMsgStack (unsigned int & uiSize) const;       67
};                                                                          68
                                                                            69
#endif // __SIMPLE__CHAT__                                                  70
                                                                            71
// ...                                                                      72
                                                                            73
```

Generative Template Components

But even with these OOP techniques, the collection of methods in a software component is still very static in data type conventions. For example, a function which calculates a mathematical equation is always fixed in data types for its parameters and return values. If another type is used, it is necessary to duplicate the function for the additional data type or to use abstract pointers. This increases redundancy and reduces maintainability, so that it will be necessary to insert changes or bug fixes into all the redundant code parts. It also increases code size and development time unnecessarily. While structured modules, such as in the programming language C, have no solution for this design dilemma, OOP languages, such as C++, offer a suitable construct: the templates. Templates are functions or classes written for data types which are not specified in the class declarations (e.g., see Example 3.3 with a template class declaration for semaphore-protected variables of different types). The specification of the data type happens later in the declaration of the individual variable with an additional argument. Template classes support scoping, naming, and type checking like the usual data types. A template defines the structure of a family of related classes in a type-independent way (Vandevoorde and Josuttis 2003, *l.c. page 8*). This is only possible because a generative compiler replaces all placeholders for types with the correct data types during the compilation, checks their correctness, and generates the final machine code.

▪▪ Example 3.3

C++template component «simple_sem.h»with methods to realize the basic functionality for semaphore protected variables (source code extract taken from the «simple»-toolbox of the Geodetic Observatory Wettzell)

```
// ...                                                         1
                                                               2
#ifndef __SIMPLE_SEMVAR_H__                                    3
#define __SIMPLE_SEMVAR_H__                                    4
                                                               5
// Define needed in older compilers to set GNU mode           6
#ifndef _GNU_SOURCE                                            7
#define _GNU_SOURCE                                            8
#endif // _GNU_SOURCE                                          9
                                                              10
#include <string>                                             11
#include <time.h>                                             12
                                                              13
class semvar_throw                                            14
{                                                             15
    private:                                                  16
        unsigned short _operation_state;                      17
        std::string strErrorMsg;                              18
                                                              19
    public:                                                   20
        semvar_throw ()                                       21
            : _operation_state(1)                             22
        {                                                     23
        }                                                     24
                                                              25
        semvar_throw (const semvar_throw & CIn)               26
            : _operation_state (CIn._operation_state)         27
            , strErrorMsg(CIn.strErrorMsg.c_str())            28
        {                                                     29
        }                                                     30
                                                              31
        ~semvar_throw ()                                      32
        {                                                     33
        }                                                     34
                                                              35
        semvar_throw & operator= (const semvar_throw & CIn)   36
        {                                                     37
            _operation_state = CIn._operation_state;          38
            strErrorMsg = CIn.strErrorMsg.c_str();            39
            return *this;                                     40
        }                                                     41
                                                              42
        unsigned short _usGetOperationState () const          43
        {                                                     44
            return _operation_state;                          45
        }                                                     46
                                                              47
        std::string _strGetErrorMsg () const                  48
        {                                                     49
            return strErrorMsg.c_str();                       50
        }                                                     51
```

```
        void vSetThrowState (const std::string & strErrorMsgIn, unsigned short
            usOperationStateIn)
        {
            strErrorMsg = strErrorMsgIn.c_str();
            _operation_state = usOperationStateIn;
        }
};
template <class VarType> class semvar
{
    private:
        /// internal variable memory
        VarType m_CVar;
        /// the pointer to CSemaphore class
        CSemaphore m_CSem;
        /// timeout settings
        int m_iSemTimeoutMSec;
        /// Handle set when blocking is active
        unsigned long m_ulBlockHandle;

        void vSetThrowState (const std::string & strErrorMsgIn, unsigned short
            usOperationStateIn)
        {
            strErrorMsg = strErrorMsgIn.c_str();
            _operation_state = usOperationStateIn;
        }
};
template <class VarType> class semvar
{
    private:
        /// internal variable memory
        VarType m_CVar;
        /// the pointer to CSemaphore class
        CSemaphore m_CSem;
        /// timeout settings
        int m_iSemTimeoutMSec;
        /// Handle set when blocking is active
        unsigned long m_ulBlockHandle;
    private:
        // Try to get semaphore
        unsigned short usEnterSemaphore ();
        // Give semaphore back
        unsigned short usExitSemaphore ();
        // Forbidden copy-constructor
        semvar (semvar<VarType> & CSemVar);
        // Forbidden asignment of semvar to semvar
        semvar & operator= (const semvar<VarType> & CSemVar);
    public:
        // Standard constructor
        semvar () throw (semvar_throw);
        // Standard constructor with parameter for timeout
        explicit semvar (int iSemTimeoutMSec) throw (semvar_throw);
        // Destructor
        ~semvar () throw (semvar_throw);
        // Asignment of VarType to semvar
        semvar & operator= (const VarType & TVar) throw (semvar_throw);
        // Insert value into semaphore variable
        void vInsertSemVal (VarType TVar) throw (semvar_throw);
        void vSetSemVal (VarType TVar) throw (semvar_throw);
        // Get value from semaphore variable
        VarType CGetSemVal () throw (semvar_throw);
        // Explicit blocking methods and the additional access methods
        void vBlock (unsigned long & ulBlockHandle) throw (semvar_throw);
        void vUnblock (unsigned long & ulBlockHandle) throw (semvar_throw);
        void vSetSemVal (unsigned long & ulBlockHandle, VarType TVar) throw (semvar_throw);
        VarType CGetSemVal (unsigned long & ulBlockHandle) const throw (semvar_throw);
};
```

```
// ================================================================================    99
// template definitions of semvar                                                     100
// ================================================================================   101
                                                                                      102
                                                                                      103
template <class VarType> unsigned short semvar<VarType>::usEnterSemaphore ()           104
{                                                                                     105
                                                                                      106
// ...                                                                                107
                                                                                      108
#endif                                                                                109
```

A closer look shows that this concept can be used very flexibly and in many ways, for example, for the type definition of vectors, matrices, lists, queues, and so on, which all benefit from type-independent algorithms. An implementation of these container classes is standardized in the STL. Thus, the STL is one first choice for a suitable toolbox, available for a suitable modularity in programming with pre-fabricated «off-the-shelf» building components.[3] As the templates are nothing more than design pattern realizations of code stereotypes, they can also be used to solve recurring problems in the software design. There is also a growing collection of such design patterns in complete frameworks, like «.NET» of Microsoft˙. These design patterns are one or more template classes, which can be combined as objects of an OOP design to avoid coding errors by standardization in the form of centralized design realization. Examples are the singleton pattern or smart pointers. The singleton pattern improves global variables by implementing singleton objects which have exactly one instance in an application and no more. The design pattern ensures and enforces that such objects exist only once in the memory, using language constructs (Alexandrescu 2003, *l.c. page 182 ff.*). Smart pointers, however, combine syntactical and semantical aspects of pointers, but avoid common errors with regular pointers as they take control of memory access, reference counting, and the lifetime administration of dynamic memory. Externally they behave like the standard pointers of different types, while internally they implement the design pattern with difficult, but powerful templates (Alexandrescu 2003, *l.c. page 217 ff.*). They are powerful, prefabricated black boxes, combinable to solve design issues.

STL and Design Patterns

Templates are extremely powerful as they use the generative aspects in the compiler. It is not necessary to implement the complete functionality during the source code writing, as already seen for the type binding. The generative parts can be generated later and without human interaction during the compilation, according to predefined mapping rules. This generative process of a compiler can be fully utilized with metaprogramming. It is a technique of writing code, which is executed by the compiler to generate the actual, further code, that implements the desired functionality. It can be used to write a program which changes itself before it is executed. The original meta-code is part of the source code, for which it generates a bit of further code. This means that the developer-defined computation happens at compile time instead of at runtime, so that the processing at runtime becomes much faster. The motivation to do that is a performance benefit for the runtime of the program and also a source code reduction (see (Vandevoorde and Josuttis 2003, *l.c. page 301 ff.*) and (Ettl 2012, *l.c. page 105 ff.*)). But the speedup (e.g., for a benchmark calculation, which is programmed by (Ettl 2012, *l.c. page 112 ff.*) up to about 14 times the speed of conventional programming) is usually accompanied by difficult development and debugging, because of the incomprehensible, recursively generated code. Additionally, the result generated is compiler dependent (Ettl 2012, *l.c. page 119*). But it is precisely these replacement

Template Metaprogramming

---

3   Another would be the Boost C++ Libraries (see ▶ Sect. 2.2).

characteristics of templates which can be used to compute functionality at compile time, as their recursive instantiation mechanism implements a primitive recursive language (Vandevoorde and Josuttis 2003, *l.c. page 301*). It can be shown that the language is Turing complete,[4] so that it can compute all (also nontrivial) computations computable by a computer (Vandevoorde and Josuttis 2003, *l.c. page 312*). The resulting, nontrivial technique is «template metaprogramming» (specific to C++).[5] An illustrative sample of a for-loop unrolling is given in Example 3.4 (adapted from (Ettl 2012, *l.c. page 111 ff.*)).

■ ■ **Example 3.4**

Use of template metaprogramming in the case of loop-unrolling (Ettl 2012, *l.c. page 111 ff.*)

```cpp
#include <iostream>

// ************************************************
// Template metaprogramming replacement for
// for (int iIndex = 0; iIndex <= 10; iIndex++)
// {
//        std::cout << iIndex << std::endl;
// }
// ************************************************
// First template to implement the general recursive rule
// for an incrementing for-loop
template <int iIndex> void ForIncrement ( )
{
    ForIncrement <iIndex-1>();
    std::cout << iIndex << std::endl;
}
// Second template to specialize the end of the recursion
template <> void ForIncrement <0>() {}

// ************************************************
// Template metaprogramming replacement for
// for (int iIndex = 10; iIndex >= 0; iIndex--)
// {
//        std::cout << iIndex << std::endl;
// }
// ************************************************
// First template to implement the general recursive rule
// for an decrementing for-loop
template <int iIndex> void ForDecrement ( )
{
    std::cout << iIndex << std::endl;
    ForDecrement <iIndex-1>();
}
// Second template to specialize the end of the recursion
template <> void ForDecrement <0>() {}
```

---

4 Turing completeness defines computational completeness, which means that an algorithm can be simulated by a simple state machine which manipulates symbols on a memory tape according to predefined replacement rules. The concept was described by Alan Turing (compare also (Schöning 1997) *l.c. page 94 f.*).

5 For more information about this generative technique using static metaprogramming in C++, see also (Czarnecki and Eisenecker 2000, *l.c. page 397 ff.*).

```
// ************************************************
// Main
// ************************************************
int main()
{
    // The pre-compiler generates the loop-unroling here,
    // so that there is no more loop here during runtime.
    ForIncrement<10>();
    ForDecrement<10>();
    return 0;
}
```

37
38
39
40
41
42
43
44
45
46
47

But besides it being difficult to understand, to debug, and to maintain programs with template metaprograms, there is another disadvantage: metaprogramming generation in the compiler does not make it possible to extend the syntax. It is limited to the specific notation defined by the programming language (Czarnecki and Eisenecker 2000, *l.c. page 340*). But a generalized code generation technique would commonly be very powerful. It makes it possible to keep the library scaling linear for whole families of similar code. This is because code can be parametrized according to the requirements, starting from a general model for a problem case. Therefore, not everything must be programmed in a library, as it can be generated for a specific case (Czarnecki and Eisenecker 2000, *l.c. page 333*). The result is an active library, which can be ideally parametrized to application programmers' requirements (Czarnecki and Eisenecker 2000, *l.c. page 142*). As metaprogramming is not completely ideal for these needs, it is necessary to use additional, external tools before compilation: the generators. They work in a very similar way to the generative process in a compiler. The generators discussed here take code models or code templates and a metaprogram to generate code variations, which are parametrized to the given situation for which they should produce code (see ◘ Fig. 3.2). Usually the metaprograms are not part of the code models anymore. They are separated and use a different syntax (Klar and Klar 2006, *l.c. page 15 f.*). It is defined in a separate metaprogramming language, the Domain Specification Language (DSL). This way of developing software separates software development into domain engineering and application engineering. The domain engineering designs and develops the generator for a specific field of work, for which it should be used. This family of similar solutions is the domain. An application engineer can then take the generator without changes to implement a specific realization of the domain model in his application, doing a parametrization for a specific type of application. This type of parametrization is done with the DSL (Klar and Klar 2006, *l.c. page 3 ff.*).

Generative Programming

Most things can be decomposed into the above building blocks, which can be more or less parametrized for the final application. The main principle behind this strategy of modular systems is: «Don't repeat yourself».[6] Redundancy is in most cases a critical parameter in large software projects. But even with all of these modular blocks, it is necessary to stick the single parts together. This «glue code» forms the final application, as manually written code is the most flexibly adaptable part. But this code is also the most expensive one as it must be written manually, individually debugged, and also individually maintained (Klar and Klar 2006, *l.c. page 2*). Therefore, it is ideal for a fast development to reduce these individual parts

Manually Written «Glue Code»

---

6   This is also known as the DRY principle, as a combination of the first characters of the words.

3

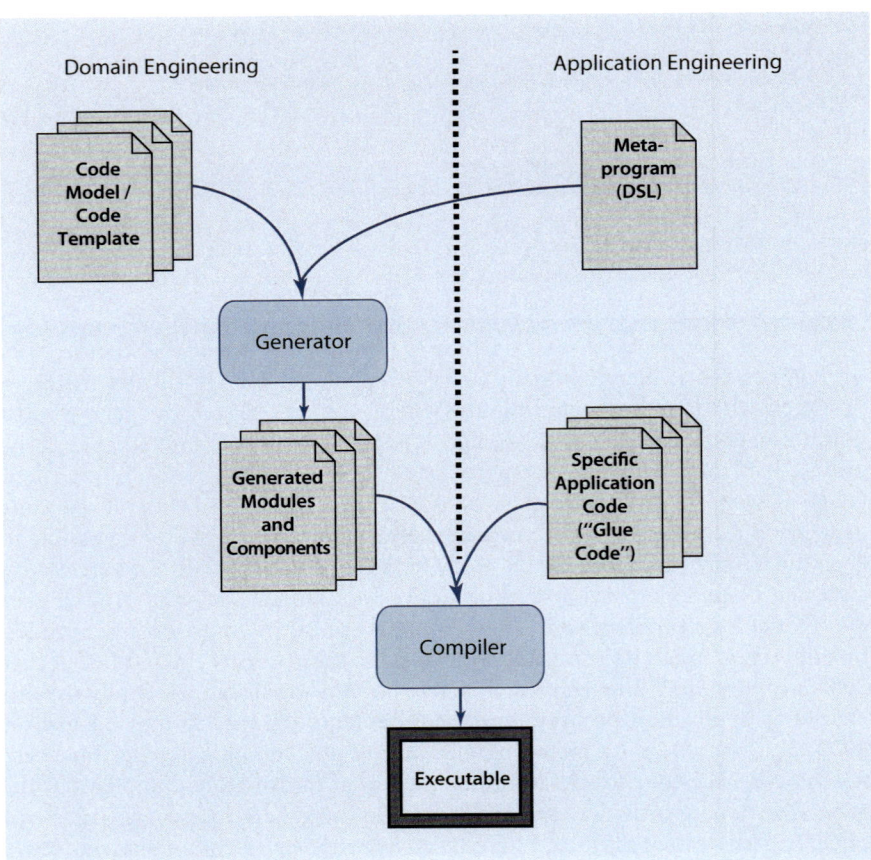

☐ **Fig. 3.2** The principle of code generation to get parametrized applications for a dedicated specification using a code model and a metaprogram for a generator

to a minimum by extending the above modular blocks in a suitable, «living», permanently growing, and shared code toolbox.

Elements of a Code Toolbox

In summary, this code toolbox consists of tools of all the aspects described from more structural and static modules to highly generative components and generators. Together with the manually and individually written code parts, they result in the code toolkits available for the development of new software. A powerful code toolbox is the «backbone» of a successful and purposeful development. It is also clear that agile techniques with their shared code assets which can be adapted and improved by each member of the team can become the ideal catalyzers for powerful, fast, new scientific developments.

3.2   Well-Tested Modules and Components

Centralized Code Assets and Continuous Integration

As described, prefabricated «off-the-shelf» building blocks are a big benefit for the construction of software. A further important task is to offer these units to all the projects and the different teams involved. This can be done using the version control system and its code assets. In combination with the CI and its automated checking tools (see ▶ Sect. 2.8), the code tools can be offered as well-tested stereotypes for the code implementations that are usually necessary. If unit tests offer a high coverage and significant metrics counts are available, every developer using the tools from the toolbox can trust in their functionality and correctness. They must also be sure that they can add or update these existing code assets if new functionality is required or an existing functionality fails under particular circumstances. For this reason, the developers need quick feedback after the changes, which is part of agile development techniques (e.g., see ▣ Fig. 2.29).

Therefore, each development group of a scientific institute should have the possibility of accessing a centralized collection of existing, well-tested modules and components of the institute. These version-controlled sources are not only an archive of the coding work of the institute, but also if these modules and components follow the coding guidelines and are self-documenting, they are the foundation for new implementations. Searching for and using suitable existing code, which can directly be included, speed up research projects. But this code is not an example solution in the sense of «look, there is a similar solution for your problem», which can only be used as an inspiration for an individual solution. The centralized modules should directly be used in the sense of modularity. Local duplicates, besides the external working copies from the version control system, are not necessary anymore. The benefits of such centralized construction kits are fast and trustworthy implementations (e.g., the hardware control of the new antennas of the Wettzell twin radio telescope was successfully written within two weeks, using existing modules for the communication tasks and adding the individual «glue code» for the implementation of the individual communication protocol).

Institute's Construction Kit

Almost every problem already has a solution in the worldwide community. But ideally, code snippets are accessible in the archive in the form of sample code. For example, each geodetic system implements its own algorithms according to the generally described specifications. This means that similar implementations are developed redundantly. An example is the transformation matrices to convert between the different coordinate systems. These transformations are required everywhere, so that several implementations can be found in the geodetic community. This leads to profoundly heterogeneous implementations and therefore to time-consuming, parallel, and unnecessary developments. An ideal solution would be to offer code archives not only centralized for separate institutions but also as global, community-centralized, scientific construction kits.[7] It would improve geodetic science significantly, as the scientists could then concentrate on the really important things, instead of repeatedly implementing and debugging the same things. It is a false conclusion that only by keeping the individuality of system code can new and independent scientific outputs be produced. The tools only support the basic tasks which are required in each higher-level research development.

Scientific Construction Kit

The idea behind the establishment of such a centralized, version-controlled toolbox archive is a main goal at the Geodetic Observatory Wettzell. It started with the basic structural modules, OOP components, and generative template pattern components. It was important to implement the parts which were needed in a specific phase of development, and not what might be needed or interesting in further steps. The written code parts were grouped into suitable construction blocks, which have been continuously expanded and upgraded since then. The result is called the «simple tools» of Wettzell. The directory structure of this toolbox is quite important, as the contained tools are included in each project as an external property link in the version control system (see ▶ Sect. 2.7.2). Therefore, changes in the main directory structure or the file and directory names will influence all the projects in which the tools are included. But if the basic structure is clearly defined, updates in the tools can be handled quite simply, and all related projects are automatically updated as well. Nevertheless, these tools are the most

Wettzell's Simple Code Toolbox

---

7   Similar things can already be found for astronomical tasks in different open-source libraries from different developer teams.

critical elements in the whole repository, as bugs inside the tools are propagated to many other development assets. This disadvantage is managed with the CI techniques and can be neglected because of the huge advantages of fast and stable coding. A subset of the Wettzell tools is summarized in the following itemization[8]:

1. *General code tools*: Can be useful for each software project
   - *Component «simple_crc»*: A component for the calculation of Cyclic redundancy check (CRC) checksums (see ▶ Sect. 5.5.5 about Message Digest Nr. 5 (MD5))
   - *Component «simple_aes»*: A component for the encrypt and decryption with the Advanced Encryption Standard (AES) and key lengths of 128, 192, or 256 bits (see ▶ Sect. 5.5.1 about symmetric block ciphers)
   - *Component «simple_md5»*: A component for the calculation of MD5 checksums with cryptographic hash functions of different numbers of bits (see ▶ Sect. 5.5.2 about MD5)
   - *Component «simple_run_length_encode»*: A component with compression functions, using the run-length encoding
   - *Component «simple_debug»*: A module with macros for debug outputs with different verbosity levels and a trace log in a program
   - *Component «simple_stl_util»*: A component with utility methods, which solve known problems in the STL for common requirements
   - *Component «simple_string_util»*: A component with helper methods for standard strings, such as trimming; removing empty lines, line feeds, or whitespaces; or replacing line feeds with carriage returns; etc.
   - *Module «simple_tail»*: A module with functions to continuously read lines from the tail of a file if new lines are added
   - *Component «simple_tokenize»*: A component with methods to split standard strings at the position of specific character patterns («tokens»)

2. *Operation and control code tools*: Are interesting for all operational and controlling activities on a dedicated operating system
   - *Component «simple_sem»*: A component to implement semaphores and semaphore-protected variables
   - *Component «simple_serial»*: A module with functions necessary for a serial communication, such as opening and closing serial devices, sending and receiving character messages, and so on
   - *Component «simple_socket»*: A module with functions necessary for an Internet socket communication over the network, such as opening and closing communication sockets, sending and receiving character messages, and so on, for the Transmission Control Protocol (TCP) and the User Datagram Protocol (UDP)
   - *Component «simple_psqlquery»*: A component with the SQL interface to the PostgreSQL database management system
   - *Component «simple_structured_conf»*: A component which implements a hierarchical configuration file parser and administration with access mechanisms to the parameters, using tree paths
   - *Component «simple_chat»*: A component which implements all functions for an Internet chat administration
   - *Component «simple_role_manager»*: A component with methods for user authentication and authorization by offering different user roles with different access rights

---

8   It represents the current iteration state. A refactoring is scheduled according to the project requirements and in the form of agile developments from time to time.

- *Component «simple_network_util»*: A component to extract network information from the operating system, such as network addresses, hardware interfaces, network statistics like lost packages or sent packages
- *Component «simple_filesystem_util»*: A component class with methods to check specific attributes of files, such as block devices, named pipes, symbolic links, file existence, modification times, etc.
- *Component «simple_system_info»*: A component to read out system parameters, such as number of cores, number representation limits, operating system kernel version, host name, used memory size, etc.
- *Component «simple_program_monitor»*: A component to start, administrate, and monitor programs of an operating system
- *Module «simple_shell_macros»*: A module with a collection of Unix shell macros to highlight, underline, and colorize the text, printed to the shell (of type «bash»)
- *Component «simple_stopwatch»*: A component which implements the functionality of a common stopwatch to measure time
- *Component «simple_compiler_info»*: A component to search installed compilers and to collect information

3. *General scientific code tools*: Can be useful for all scientific software projects
   - *Module «constants»*: A module in C containing all necessary physical constants and units accessible with standardized access functions
   - *Module «timecalc»*: A module in C with all necessary conversion functions to convert the different time representations (which currently usually use UTC as basis[9]), such as Gregorian Date (GRD), Julian Date (JD), Modified Julian Date (MJD), Global Positioning System (GPS) Date, or Unix Epoch (continuous seconds since January 1, 1970 00:00:00), and to extract specific parts, such as the day of the year, the fraction of seconds of the day, or Julian days since January 2000. Also conversions to readable strings in the usual calendar formats are included.
   - *Component «decimal»*: A component which implements a decimal representation of numbers instead of a binary to have an accurate representation of decimal numbers
   - *Module «interpol»*: A module with functions for a Hermite and Lagrange interpolation
   - *Component «simple_floatingpoint_util»*: A template component for a scientific floating point mathematics with error terms, mantissa extraction, maximum round error detection, etc.
   - *Component «simple_string_mathlib»*: A component with functionality to interpret mathematical expressions in strings
   - *Component «Xdouble»*: A component which implements an «Xdouble» type with higher floating point precision (quad precision)

4. *Specialized scientific code tools*: Are interesting for a specific scientific task or observing system
   - *Module «defines»*: A module in C with definitions of standardized structures for site coordinates in X/Y/Z or longitude/latitude/height, topocentric target angles, Earth-centered body-fixed target vectors, and Earth Orientation Parameters
   - *Module «cpf_file_common»*: A module to read in files in the Consolidated Laser Ranging Prediction Format (CPF) with the orbit predictions of the satellites for laser ranging
   - *Module «orbitcalc»*: A module with functions to convert between different coordinates in different reference frames (X/Y/Z to X/Y/Z with velocities

---

9 Currently the leap seconds are not completely treated while changing to continuous time scales.

XV/YV/ZV and to topocentric angles in azimuth/elevation/range or star positions in the celestial reference frame in right ascension/declination to topocentric angles in azimuth/elevation, etc.), to calculate satellite orbits, to calculate the approximated sun position for a sun avoidance mechanism and for the calculation of satellite illumination by the sun, and to calculate lunar reflectors with inbound and outbound corrections of the two distances, passed by the laser to and from the moon

— *Component «eurolas»*: A component to send the EUROLAS status of a laser ranging system to a central logging server of the laser ranging community

The code toolbox at Wettzell currently also contains complete program systems, which can be used for different projects:

— *Program «sshbroker»*: A program to control standard SSH clients for a safe but automated connection management (see also ▶ Sect. 5.5.5 about the function of this program)

— *Program «simple_copy_and_paste_detector»*: A program to detect copy and paste sections in files

— *Program and components «simple_testsuite»*: The Wettzell Unit Test Suite to perform unit tests (see ▶ Sect. 2.6.2)

— *Program «progstarter»*: A program, which creates a complete directory and building environment for a new project, using the project standards of the Geodetic Observatory Wettzell

Script Language Tools and Vendor Tools

The code assets are extended with the toolbox of scripting language files. The script files are arranged in a similar style to the code tools. Besides these observatory specific code and script assets, code packages from different vendors are located in the code toolbox and can be extracted using the external properties of the version control system. These vendor-specific code assets, such as the telescope control unit for the laser ranging telescope from Zeiss, the source code examples for laser ranging from the National Aeronautics and Space Administration (NASA) Web page of the ILRS or also the CI tools, are in a specific «vendor» directory. This directory is separately checked and inspected in a similar way to the observatory code with the programs of the CI. Issues detected are directly reported to the responsible external vendors. This can mostly be done by sending bug-fix patches, using the diff/patch technique (see ▶ Sect. 2.9).

Generated Modules and Components

As described, generators are also part of the code toolbox. One of these is Doxygen, which is located in the vendor directory. It is used for automated documentation generation using in-line code attributes (see ▶ Sect. 2.5.1). Another is the observatory specific generator for individual network communication middleware: «idl2rpc.pl». This generator is at the heart of all operations and of all projects for control tasks at the Wettzell observatory. It is a complete self-made generator system, which helps to improve and standardize the development of distributed control systems for the requirements of an observatory. Because this method of generative programming is so central and powerful, it is necessary to take a closer look at the mechanisms behind it.

3.3 Generative Programming

Generating code is very efficient for tasks with similar characteristics. These tasks usually solve a specific class of programming problem. If similar code lines need to be written several times with only small changes, the generative idea can improve the development and stability of software tremendously. A typical example is the IPC.

3.3.1 **Classic Solutions for Interprocess Communication (IPC)**

Processes on the same machine, or on machines which are distributed over a communication network,[10] often have to communicate with each other to coordinate or synchronize their activities or to process common tasks (Brookshear 2012, *l.c. page 146)*. This Interprocess Communication, or IPC, is the essential data exchange mechanism between processes. There are different IPC techniques, which are mostly useful on the same machine, such as shared memory or communication pipes.[11] Another form of IPC is the communication sockets. These can not only be used to connect processes on the same machine, they are also able to realize a data transfer all over a communication network (Stevens 1990, *l.c. page 183)*.

Different Mechanisms

> *Definition DEF3.5. of «communication socket»* (see (Stevens 1990, *l.c. page 280 ff.*))
> Communication sockets are application interfaces for the message exchange, which provide all required system calls to connect an application to a physical transport medium (e.g., the Ethernet), using a transport protocol stack (see next definition) for sending and receiving data.
> The transport protocol stack implements a specific communication protocol between the different communication partners.

Communication Socket

> *Definition DEF3.6. of «communication protocol»* (see (Halsall 1996, *l.c. page 14)* and (Halsall 1996, *l.c. page 168 ff.*))
> A communication protocol is a set of rules for the exchange of character-oriented or bit-oriented data on a communication network link. The protocol defines the message format (syntax) and its meaning (semantics) together with a flow and error control (synchronization), saying which order of the messages is allowed and how errors in this flow should be treated. The syntax and synchronization can be described with finite state machines (see ► Sect. 3.3.1).

Communication Protocol

To connect a user-written application to a physical network like the Ethernet, a layered interface is used (for the following, see (Halsall 1996, *l.c. page 11 ff.*)). There are two very important open system standards which describe the implementation of such a stack of communication layers: the International Standards Organization (ISO) Open System Interconnection (OSI) reference model and the TCP/Internet Protocol (IP) protocol suite. While the OSI with its physical, link, network, transport, session, presentation, and application layers is more or less a theoretical reference construct, TCP/IP is widely used for the Internet. Its protocol specifications are in the public domain without any fees or licenses, so that it is extensively used by commercial and public authorities. Therefore, the following is confined to the TCP/IP protocol suite (even when its layers find some representational layers in the OSI model). The TCP/IP protocol stack only uses four layers (see also ◘ Fig. 3.3).

Open System Standards

---

10 A communication or computer network is a physical connection between different, autonomous computers to implement communications between each other. Today usually Ethernet is used, which is a set of standardizations for a wired or wireless communication network. Classifications for such networks are, e.g., Local Area Network (LAN) for local, institutes network or Wide Area Network (WAN) for larger, worldwide networks.

11 A mechanism of the operating system to stream data between processes using an unidirectional first-in, first-out memory buffer.

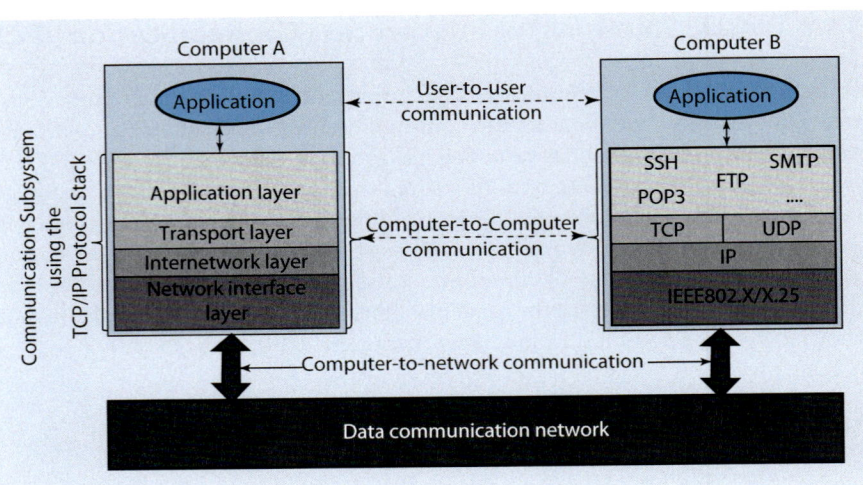

Fig. 3.3 The TCP/IP protocol stack with the four layers and related examples for protocol standards: network interface (IEEE802.X/X.25), internetwork (IP), transport (TCP and UDP), and application (such as SSH for security, SMTP for the sending of emails, POP3 for receiving emails, FTP for file transfers)

Network Interface Layer

1. *Network interface layer*: This layer is nearly comparable to layers 1 to 2 of the ISO OSI reference model. It defines the mechanical and electrical interface and the data link control (with framing of messages into blocks of bytes (Halsall 1996, *l.c. page 112*), keeping the data transparency with bit synchronization (Halsall 1996, *l.c. page 112 f.*), and performing an error control).

Internetwork Layer

2. *Internetwork layer*: This layer is nearly comparable to layer 3 of the ISO OSI reference model. It is responsible for the addressing with IP addresses (Halsall 1996, *l.c. page 486 ff.*) and (Halsall 1996, *l.c. page 496 f.*)), the routing to find a correct and suitable way to the target machine (Halsall 1996, l.c. page 504 f.), establishing the call and the fragmentation,[12] and reassembly of packages along the way to adapt to different data package sizes (Halsall 1996, l.c. page 501 f.).

Transport Layer

3. Transport layer: This layer is nearly comparable to layer 4 of the ISO OSI reference model. It implements an end-to-end message transfer with addressing of communication ports for each application and the management of the connection either as connection-oriented with data flow control and retransmissions (using TCP) or as connectionless without a flow control and a best-try transfer approach (using UDP) ((Halsall 1996, *l.c. page 442 ff.*) and (Halsall 1996, *l.c. page 18*)).

Application Layer

4. *Application layer*: This layer is nearly comparable to layers 5 to 7 of the ISO OSI reference model. It connects the synchronization with the application, organizes the message interchange, controls the access, transforms data representations (Halsall 1996, *l.c. page 16 ff.*), and realizes the end-user application as session-oriented (with an assignment of a user to a specific communication flow) or session-less (without an assignment of a user to a communication flow) communication.

IP Addresses and Port Numbers

If a message is sent, it flows down through the protocol stack. Each layer takes the message from the previous one and packs it with a layer-specific header (and partly with a trailer) and, in this way, forms the Protocol Data Unit (PDU) of the

---

12 Fragmentation is a mechanism to split large packages into several smaller ones to adapt to smaller, maximum package sizes.

dedicated layer. In a similar way to the Russian «babushka» or «matryoshka» dolls, from layer to layer, the original message becomes densely packed and increases. Each package of individual information added by the layers, for example, in the headers, supports the individual tasks of the layer (Halsall 1996, *l.c. page 430*). The Internet work layer adds specific address information to allow a routing through the network. Usually it is still the IP Version 4 address based on a tuple of four bytes (32-bit integer), which specifies the address class, the network identifier, and the host identifier.[13] Usually the four bytes are separated with dots and written as decimal numbers. Currently Version 4 is being replaced by Version 6 of the IP addressing with a 128-bit integer number to allow a larger number of individual, registered networks and also hosts. But both versions of the IP addresses generally implement the addressing of a machine (Halsall 1996, *l.c. page 496 ff.*). In addition to the addressing of computers in a network, the addressing of a specific application access point is necessary to connect two applications with each other. This is implemented by using port numbers in the transportation layer. A port is a 2-byte integer number and therefore a number from 1 to 65,535, which is assigned to a specific communication socket of a specific application (Halsall 1996, *l.c. page 642 ff.*). Each application can locate a desired communication partner process with the IP address and port number. Generally known application protocols are officially assigned to «well-known» ports, such as the HTTP to port 80. This means that both pieces of information can be used to protect local networks securely (see ▶ Sect. 5.5.3). After passing all the stack layers, the extended PDU is sent as framed bit stream over the physical network cable.

The IPC is usually not symmetrical, as there is a requesting and a replying part. The requester processes a specific task, and if it needs remote support, it requests it from a waiting service provider process. This processes the requested activity and returns the result to the requesting process. This popular setup is the client-server model, where the requesting process is the client and the serving process is the server (Brookshear 2012, *l.c. page 146*).

> *Definition DEF3.7. of «client-server model»* (see (Singhal and Shivaratri 1994, *l.c. page 80*))
> In a client-server-model, a client requests a service from a server by message exchange. The server waits for requests and processes them. It replies with the result in the form of an answer message to the requester. The initiation of this communication is always performed by the active client.

Client-Server Model

The use of the socket interface is very similar to file operations (for the following blocks, see the detailed descriptions of the Berkley sockets at (Stevens 1990, *l.c. page 280 ff.*)). They are also controlled with system calls (here declared in the include file «<sys/socket.h>»). The system calls use file descriptors to further identify an already established connection within a program. But unlike file operations, sockets need more options and operations. The operations must set up an asymmetric relationship between server and client. The workflow for a connection-oriented communication with TCP is shown in ▣ Fig. 3.4 (adapted from (Stevens 1990, *l.c. page 283*)). There the server creates a socket with the call of «socket». It assigns the administration of the connection and returns a file descriptor. After that, this descriptor can be used to bind the new socket to an address and a port with the call

Use of Connection-Oriented Sockets: Create Socket

---

13 The address class specifies the size of a network. The network identifier is the address of a registered network at the Network Information Center, and the host identifier is the address of the machine in the network.

3

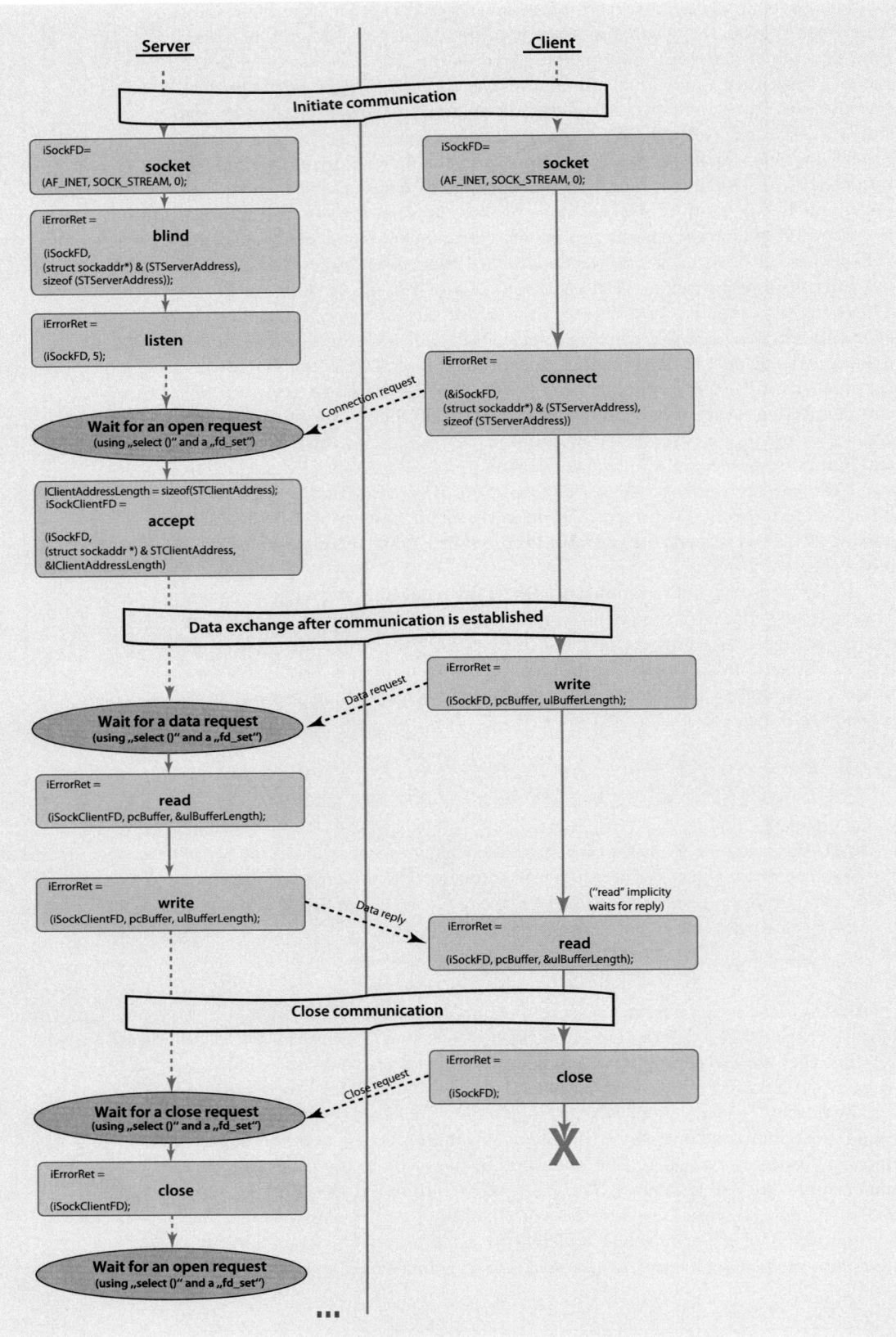

Fig. 3.4 The workflow of socket system calls for a connection-oriented socket communication with TCP (Adapted from Stevens (1990) *l.c. page 283*)

of «bind», using the structure «struct sockaddr». A TCP socket interface of the server has to be marked as passive with the activation of «listen», so that it waits on incoming connection requests. The client on the other side of the communication can itself create a socket with «socket». But unlike the server, the client does not need to bind to an address and port, as the client always establishes the connection-oriented communication, which implicitly defines the connection endpoint. The address and port of the client are sent implicitly to the server during the connection establishment.

The initiation of such a connection-oriented communication is processed by the client using «connect» together with the server address and port, coded in the structure «struct sockaddr». Usually each system call which communicates with the communication partner uses a sequence of protocol messages over the network and a chain of activities in the stack as a consequence. The «connect» call, for example, tries to establish a connection, which forces the protocol stack of TCP/IP to find the address in the network, to search a route, and to perform the three-way handshake.[14] The server on the other side sleeps until a connection request arrives. As a server can usually handle more than one client in parallel, the waiting is implemented with «select». «select» continuously monitors the listen port for activities and returns if a connection request or another activity is detected. Then the complete set of available and active file descriptors in «fd_set» can be checked to see if there is a connection, some data, or a close request. Each case forces a different activity on the newly active file descriptor derived from the set of file descriptors. If a connection request appears, the server accepts the communication with «accept», which establishes the communication channel between the server (with its specific IP address and port) and a client (with its IP address and port). «select» can also be used to define timeouts with the structure «struct timeval». This forces the waiting routine to return after the specified time has expired. Such behavior can be used if the server needs to do frequent other tasks apart from waiting for client requests.

After the establishment of the connection, the data exchange can be performed. For a connection-oriented communication, this is done with the usual «write» and «read» function calls, using the derived file descriptor for the client connection from the file descriptor set. The activity detection on the connected ports is performed with the already described «select» call. The data itself can be binary or text character sequences which are packed as messages. The complete communication stability, such as package ordering, error detection, retransmission, and so on, is guaranteed by the lower levels of the stack. Usually, each package is sent at least three times with a timeout of 20 seconds if errors appear. If all retransmissions fail, the sending returns with an error. The delays and other attributes of the sockets can be configured or checked using, for example, system functions for the input/output control with «ioctl», for the manipulation of open file descriptors with «fnctl», or directly for the manipulation of socket options, such as «setsockopt». For example, «ioctl» is a very useful way to check whether a data message has arrived or if the connection should be closed.

Use of Connection-Oriented Sockets: Initiate Communication

Use of Connection-Oriented Sockets: Data Exchange

---

14 A three-way-handshake is a special procedure to establish a connection-oriented communication using a sequence of three communication messages. The first one from the client requests the connection using a SEQ flag and an initial sequence number for the communication packages in the transport header. The second one from the server accepts the request also using the SEQ flag and its own sequence number in combination with an acknowledgment flag ACK and an additional acknowledgment sequence number, which is the sequence number from the client plus one. The third message again from the client confirms the acknowledgment of the server by setting the ACK flag, and returning the sequence number of the server plus one as the acknowledgment sequence number (Halsall 1996, *l.c. page 652 ff.*).

Use of Connection-Oriented Sockets: Close Communication

If the client wants to end a communication, for example, after the data exchange, it can just «close» the file descriptor. This forces the close sequence of the protocol in the lower layers of the stack. If the close request appears at the server side, it can again be detected using the combination of «select» and the set of opened file descriptors. If the specific descriptor is found, the server also processes the close function and the communication is ended. It is quite important for the complete operating system that opened file descriptors are also closed again, as the number of them is limited. Unclosed ones are potential resource leaks, which might block the machine after a while. Another item here is that most operating systems keep the file descriptors as used for a while (usually a few minutes), which should protect others from reusing it too soon. This could lead to problems if large communications are still transmitting, while one socket is closed and another is reopened on the same port. Applications which do not want this behavior need to change it by using the «SO_LINGER» flag in the structure «struct linger» during the socket creation between «socket» and «bind» (for more, see (Stevens 1990, *l.c. page 341 ff.*) and Example 3.5). This change of behavior is quite useful if the connection is established again quite fast, for example, after a crash.

■■ ■ **Example 3.5**

Sample function from generated code for Sun RPC with the settings for «SO_LINGER» between «socket» and «bind» to avoid, that a server delays the closing of a listen socket

```
/*********************************************************************       1
    function  uiInitInetServer                                             2
 *******************************************************************/      3
/*!               Initialize and open server socket                        4
    \param        unsigned int uiPort -> Port of the server socket          5
    \param        int * piSocket -> Pointer to the newly created socket      6
    \param        unsigned int uiSocketProtocol -> Identifier for TCP (=1) or UDP (=0)   7
    \return       unsigned int <- Errorcode (0 = ok, 1 = error)             8
 *******************************************************************/      9
/*  author        Alexander Neidhardt                                      10
    date          2010                                                     11
    revision      —                                                        12
    info          —                                                        13
 *******************************************************************/      14
unsigned int uiInitInetServer (unsigned int uiPort,                        15
                               int * piSocket,                             16
                               unsigned int uiSocketProtocol)              17
{                                                                          18
    struct sockaddr_in STServerAddress; /*! STServerAddress = Server address */   19
    int iReuseAddr = 1;        /*! iReuseAddr = Flag to set the socket option SO_REUSEADDR */  20
    struct linger SLinger;   /*! SLinger = Structure to set the socket option SO_LINGER */  21
                                                                           22
    /*! Check port correctness */                                         23
    if (uiPort < 1 || uiPort > 65535)                                     24
    {                                                                      25
        return 1;                                                          26
    }                                                                      27
                                                                           28
    /*! Create socket */                                                  29
    if (uiSocketProtocol == 0)                                            30
    {                                                                      31
        /*! UDP = SOCK_DGRAM */                                           32
        if ((* iSocket = socket (AF_INET, SOCK_DGRAM, 0)) == -1)          33
        {                                                                  34
            return 1;                                                      35
        }                                                                  36
    }                                                                      37
```

```
    else
    {
        /*! TCP = SOCK_STREAM */
        if ((*iSocket = socket (AF_INET, SOCK_STREAM, 0)) == -1)
        {
            return 1;
        }
    }


    /*! Set option to restart the server on same port immediately after a close */
    /*! Reuse address allows to bind to a dedicated address after a crash */
    iReuseAddr = 1;
    if (setsockopt(*iSocket, SOL_SOCKET, SO_REUSEADDR, &iReuseAddr, sizeof(iReuseAddr)) < 0)
    {
        iOSremotectrlCloseSocket (*iSocket);
        *iSocket = -1;
        return 1;
    }
    /*! Set linger time to 0 forces the socket close immediately after close */
    SLinger.l_onoff = 1;
    SLinger.l_linger = 0;
    if (setsockopt(*iSocket, SOL_SOCKET, SO_LINGER, &SLinger, sizeof(SLinger)) < 0)
    {
        iOSremotectrlCloseSocket (*iSocket);
        *iSocket = -1;
        return 1;
    }


    /*! Bind socket to the local address */
    STServerAddress.sin_family = AF_INET;
    STServerAddress.sin_addr.s_addr = htonl (INADDR_ANY);
    STServerAddress.sin_port = htons ((unsigned short)uiPort);
    if (bind (*iSocket, (struct sockaddr *) & STServerAddress, sizeof (STServerAddress)) ==
        -1)
    {
        iOSremotectrlCloseSocket (*iSocket);
        *iSocket = -1;
        return 1;
    }


    /*! To activate "listen" on the socket is not necessary for Sun RPC */
    /*
    if (listen (*iSocket, 5) == -1)
    {
        iOSremotectrlCloseSocket (*iSocket);
        *iSocket = -1;
        return 1;
    }
    */
    return 0;
}
```

Lines 38–87

For connectionless communications with UDP, the sequence of system calls differs slightly from connection-oriented sockets. The main difference is that the client does not establish a connection which means that the client does not «connect» and the server needs not to «accept». In terms of socket communication, both partners are more or less equal. Both create a socket with «socket» and both bind it to an IP address and port with «bind». Then both can send data with «sendto» or receive data with «recvfrom». Both communication functions use directly coded addresses and ports in a structure «struct sockaddr» to directly address the communication partner. A call for sending directly leads to a single PDU flowing through the protocol stack as a user

Use of Connectionless sockets

3

Module for Socket-Based IPC

datagram (Halsall 1996, *l.c. page 643 ff.*). As there is no established connection, the error handling is also reduced significantly. Lost messages only lead to several send retries, where the number of attempts can be defined earlier. The reduced safety overhead improves the throughput of data for a specific time period, especially over long-haul connections. This is the reason UDP is often used for fast data streams which are controlled with a parallel lightweight TCP connection, such as those implemented in the «Tsunami UDP Protocol», which is used for fast transfers of large observing data from radio telescopes (see ▶ Sect. 6.4) (Wagner n.d.).

All of the previously described system calls have to be used correctly in the workflow to establish and process a successful IPC communication. Both partners need to handle errors and define and react to timeouts. They need to use various methods to avoid endless delays (e.g., by using a version of the «connect» call which uses a timeout as shown in Example 3.6) and problems with rebinding after a crash (as shown for a close of sockets without a delay, such as in Example 3.5) or to allow parallel clients at the server (by selecting the right activity after a request arrival, such as in Example 3.7). Therefore, the whole workflow can be structured within a module or component which groups the different functions for the different phases of the communication in a suitable way and deals with all exceptions and parameters. At the Geodetic Observatory Wettzell, this is implemented with the code toolbox module «simple_socket». It is a black box with interface functions to handle either client or server communication on a functional level which combines low-level system calls to a higher communication level, so that all the really important additional parameters can be managed implicitly.

■ ■ **Example 3.6**

Sample function from the «simple_socket» module to open a socket and connect a client to a server, but realizing a predefined timeout for the connect

```
#ifdef  __cplusplus                                                      1
extern  "C"  {                                                           2
#endif                                                                   3
                                                                         4
    /**********************************************************          5
        function   uiClientOpenSocketWithTimeout                         6
    **********************************************************/          7
    /*!          Open  socket  at  client  side                          8
        \param   char * pcIP -> String with IP-address                   9
        \param   unsigned int uiPort -> Port for connection (server must listen to that)  10
        \param   unsigned int uiSocketProtocol -> Define used protocol (0=UDP, 1=TCP)   11
        \param   unsigned int uiTimeoutSec -> Timeout in seconds        12
        \param   SimpleSocketType * SSocket <- If open succeeded the resulting socket   13
                 descriptor
        \return  unsigned int <- Errorcode (0 = ok, >0 = error)         14
    **********************************************************/          15
    /*  author   Alexander Neidhardt                                     16
        date      14.06.2012                                            17
        revision  —                                                     18
        info      —                                                     19
    **********************************************************/          20
    unsigned int uiClientOpenSocketWithTimeout (char * pcIP,            21
            unsigned int uiPort,                                        22
            unsigned int uiSocketProtocol,                              23
            unsigned int uiTimeoutSec,                                  24
            SimpleSocketType * SSocket)                                 25
```

```
{
    /*!<b> Variables </b> */
    struct sockaddr_in STBindAddress;    /*! \li STBindAddress = Structure with IP address
        and port */
    struct hostent * hostent_ptr;        /*! \li hostent_ptr = Structure with host
        identification */
    struct timeval ConnectTimeout;       /*! \li ConnectTimeout = Structure with timeout
        settings for "select" */
    int iSocketOptions = 0;              /*! \li iSocketOptions = Bit mask for all flags,
        defining the socket options */

    /*!<b> Operations </b> */
    /*! Set socket protocol */
    SSocket->uiSocketProtocol = uiSocketProtocol;

    /*! Create socket */
    if (uiSocketProtocol == SIMPLE_SOCKET_UDP)
    {
        /*! UDP */
        if ((SSocket->iSockFD = socket (AF_INET, SOCK_DGRAM, 0)) == -1)
        {
            return 1;
        }

        STBindAddress.sin_family = AF_INET;
        STBindAddress.sin_addr.s_addr = htonl (INADDR_ANY);
        STBindAddress.sin_port = 0;
        if (bind (SSocket->iSockFD, (struct sockaddr *) & STBindAddress, sizeof (
            STBindAddress)) == -1)
        {
            close (SSocket->iSockFD);
            return 2;
        }

        /*! Define address of server */
        SSocket->STServerAddress.sin_family = AF_INET;
        SSocket->STServerAddress.sin_port = htons((uint16_t)uiPort);
        /*! Call reentrant helper function to get host by name */
        if ((hostent_ptr = pSGetHostByName(pcIP)) == NULL)
        {
            close (SSocket->iSockFD);
            return 2;
        }
        memmove(&(SSocket->STServerAddress.sin_addr.s_addr), hostent_ptr->h_addr,
            hostent_ptr->h_length);
    }
    else
    {
        /*! TCP */
        if ((SSocket->iSockFD = socket (AF_INET, SOCK_STREAM, 0)) == -1)
        {
            return 1;
        }

        /*! Derive address information and port */
        SSocket->STServerAddress.sin_family = AF_INET;
        SSocket->STServerAddress.sin_port = htons((uint16_t)uiPort);
        /*! Call reentrant helper function to get host by name */
        if ((hostent_ptr = pSGetHostByName(pcIP)) == NULL)
        {
            close (SSocket->iSockFD);
            return 2;
        }
        memmove ( &(SSocket->STServerAddress.sin_addr.s_addr), hostent_ptr->h_addr,
            hostent_ptr->h_length);
```

26
27
28
29
30
31
32
33
34
35
36
37
38
39
40
41
42
43
44
45
46
47
48
49
50
51
52
53
54
55
56
57
58
59
60
61
62
63
64
65
66
67
68
69
70
71
72
73
74
75
76
77
78
79
80
81
82
83

```
    /*! Connect address with socket but dealing with timeout */
    if (uiTimeoutSec == 0)
    {
        ConnectTimeout.tv_sec = 20;
        ConnectTimeout.tv_usec = 0;
    }
    else
    {
        ConnectTimeout.tv_sec = uiTimeoutSec;
        ConnectTimeout.tv_usec = 0;
    }

    if ((iSocketOptions = fcntl(SSocket->iSockFD, F_GETFL, NULL)) < 0)
    {
        return 3;
    }
    if (fcntl(SSocket->iSockFD, F_SETFL, iSocketOptions | O_NONBLOCK) < 0)
    {
        return 3;
    }
    /* Idea for connect with timeout from
       http://exodeveloper.ch/tutorials/C/Connection-Timeout, Oct. 2011 */
    if ((connect(SSocket->iSockFD, (struct sockaddr *) & (SSocket->STServerAddress),
        sizeof(SSocket->STServerAddress))) == -1)
    {
        if (errno == EINPROGRESS)
        {
            fd_set wait_set;
            FD_ZERO(&wait_set);
            FD_SET(SSocket->iSockFD, &wait_set);
            if (select((SSocket->iSockFD) + 1, NULL, &wait_set, NULL, &ConnectTimeout
                ) == 0)
            {
                close(SSocket->iSockFD);
                return 3;
            }
        }
        else
        {
            close(SSocket->iSockFD);
            return 3;
        }
        if (fcntl(SSocket->iSockFD, F_SETFL, iSocketOptions) < 0)
        {
            close(SSocket->iSockFD);
            return 3;
        }
    }

    return 0;
    }

#ifdef __cplusplus
}
#endif
```

84
85
86
87
88
89
90
91
92
93
94
95
96
97
98
99
100
101
102
103
104
105
106
107
108
109
110
111
112
113
114
115
116
117
118
119
120
121
122
123
124
125
126
127
128
129
130
131
132
133
134
135
136
137
138

▪▪ Example 3.7

Sample function from the «`simple_socket`» module to select an active socket
at the server and to decide what to do: accepting a connect wish, process a communicate or close the connection

```c
#ifdef __cplusplus                                                          1
extern "C" {                                                                2
#endif                                                                      3
                                                                            4
   /*******************************************************************     5
         function    uiServerSelectWait                                     6
   *******************************************************************/     7
   /*!            Receive message                                           8
      \param    SimpleSocketType * SSocket -> Socket to listen to           9
      \param    unsigned long ulTimeoutSec -> Possible timeout in sec in combination with  10
         ulTimeoutUSec (both equal 0 means no timeout)
      \param    unsigned long ulTimeoutUSec -> Possible timeout in usec in combination with 11
         ulTimeoutSec (both equal 0 means no timeout)
      \param    void (*fp)(int) -> Function pointer to a callback function which is called  12
         when a message is detected
      \return   unsigned int <- Errorcode (0 = ok, 1 = timeout, >1 = error) 13
   *******************************************************************/     14
   /*   author    Alexander Neidhardt                                       15
        date       11.09.2008                                               16
        revision   —                                                        17
        info       —                                                        18
   *******************************************************************/     19
unsigned int uiServerSelectWait (SimpleSocketType * SSocket, unsigned long ulTimeoutSec,  20
      unsigned long ulTimeoutUSec, void (*fp)(SimpleSocketType))
{                                                                           21
   /*! <b> Variables </b> */                                               22
   int iSelectRetVal;                    /*! iSelectRetVal = Return value of select */   23
   int iFD;                              /*! iFD = Variable file descriptors */          24
   int iRead;                            /*! iRead = Flag to check if a client wants a   25
         connection close */
   socklen_t ClientAddrLength;           /*! ClientAddrLength = Size of client address   26
         info length */
   int iFDNewClient;                     /*! iFDNewClient = File descriptor for a new     27
         client */
   struct timeval STTimeout;             /*! STTimeout = Timeout settings for select */   28
   static struct sockaddr_in STClientAddress;  /*! STClientAddress = The client address  29
         info of a new client */
   SimpleSocketType SNewPortSocket;      /*! SNewPortSocket = Socket of requesting client  30
         */
   int iCountDataReadyChecks;            /*! iCountDataReadyChecks = Used for data ready   31
         checks to repeat delayed checks */
                                                                            32
                                                                            33
   /*! <b> Operations </b> */                                              34
   SNewPortSocket = *SSocket;                                              35
                                                                            36
   if (ulTimeoutSec == 0 && ulTimeoutUSec == 0)                           37
   {                                                                       38
      /*! No time out in select */                                        
      iSelectRetVal = select (FD_SETSIZE, &g_STFDSet, (fd_set *) 0, (fd_set *) 0, NULL);  39
                                                                            40
      /*! Error while select */                                          41
      if (iSelectRetVal < 0)                                             42
      {                                                                   43
         return 2;                                                        44
      }                                                                   45
```

```
        /*! Timeout */
        if (iSelectRetVal == 0)
        {
            return 2;
        }
    }
    else
    {
        /*! Select with timeout */
        STTimeout.tv_sec = ulTimeoutSec;
        STTimeout.tv_usec = ulTimeoutUSec;
        iSelectRetVal = select (FD_SETSIZE, &g_STFDSet, (fd_set *) 0, (fd_set *) 0, &
            STTimeout);

        /*! Error while select */
        if (iSelectRetVal < 0)
        {
            return 2;
        }

        /*! Timeout */
        if (iSelectRetVal == 0)
        {
            return 1;
        }
    }
    /*! Check filedesctriptors */
    for (iFD = 0; iFD < FD_SETSIZE; iFD++)
    {
        if (FD_ISSET (iFD, &g_STFDSet))
        {
            if (SSocket->uiSocketProtocol == SIMPLE_SOCKET_UDP)
            {
                /*! UDP directly calls callback function */
                (*fp)(*SSocket);
            }
            else
            {
                if (iFD == SSocket->iSockFD)
                {
                    /*! A new client wants to get in contact => accept and insert into
                        filedescriptor set*/
                    ClientAddrLength = sizeof (STClientAddress);

                    iFDNewClient = accept (iFD, (struct sockaddr *) & STClientAddress, &
                        ClientAddrLength);
                    FD_SET (iFDNewClient, &g_STFDSet);

                    /*! Set in the filedescriptor array */
                    if (iFDNewClient >= FD_SETSIZE)
                    {
                        return 2;
                    }
                    g_uiOpenFD[iFD] = 1;
                }
                else
                {
                    /*! Check if a client wants to stop communication */
                    /*! Some operating systems has a delayed ioctl response => try it 4
                        times an nap a little bit between */
                    for (iCountDataReadyChecks = 0; iCountDataReadyChecks < 4;
                        iCountDataReadyChecks++)
```

```
                {
                    struct timespec t, t1;
                    if (ioctl (iFD, FIONREAD, &iRead, sizeof(iRead)) < 0)
                        return 1;
                    if (iRead)
                        break;

                    /* struct timespec
                     {
                         time_t tv_sec;    // s
                         long   tv_nsec;   // ns
                     }; */
                    t.tv_sec  = 0;
                    t.tv_nsec = 10000000;
                    nanosleep(&t, &t1);
                }

                if (iRead == 0)
                {
                    /*! Close communication */
                    close (iFD);
                    FD_CLR (iFD, &g_STFDSet);
                    g_uiOpenFD[iFD] = 0;
                }
                else
                {
                    /*! Client wants to communicate => call function */
                    SNewPortSocket.iSockFD = iFD;
                    (*fp)(SNewPortSocket);
                    return 0;
                }
            }
        }
    }
}

    return 0;
}

#ifdef __cplusplus
}
#endif
```

	104
	105
	106
	107
	108
	109
	110
	111
	112
	113
	114
	115
	116
	117
	118
	119
	120
	121
	122
	123
	124
	125
	126
	127
	128
	129
	130
	131
	132
	133
	134
	135
	136
	137
	138
	139
	140
	141
	142
	143
	144
	145

But this extended socket module or component simply implements the hardware access and simplifies only the communication for IPC itself. It neither defines any higher-level messages nor the specific, correct sequence of these request and response messages for a specific communication protocol. This means that the developer of a proprietary IPC has to define an individual data representation and a specific higher-level communication protocol himself (Bloomer 1991, *l.c. page 10*), using the improved socket module. Essentially, the developer has to define his own «language» for his communication. The correct sequence of messages between the communication partners and adequate exceptions requires syntax checks and suitable error management. Even the messages themselves must be interpreted correctly. This means that they need to be checked lexically and syntactically before they can be semantically interpreted to trigger an activity.

Proprietary Communication Language

Message and Dataflow Correctness

While the general communication flow can be implemented more or less simply, if unique message identifiers are used, the scanning for correctness of the internal message structure is not always so simple. All those checks belong to the type-3 languages in the language theory, which are the regular languages in the Chomsky hierarchy of languages (see ☐ Fig. 3.15). This is the simplest type of language in the classification of formal computer languages ((Schöning 1997, *l.c. page 17 ff.*) and (Schöning 1997, *l.c. page 27 ff.*); see ▶ Sect. 3.3.3). Therefore, the communication sequences can be described with regular expressions,[15] using also directed state graphs. A direct translation into code is implemented with conditional if/else statements in combination with character or string comparisons. But often a clear, complete, formal description is missing for most proprietary socket communications, so that the error control is not complete. This can lead to unpredictable behavior if error-prone messages arrive or the higher protocol sequence is corrupted. Operators of telescopes still encounter such problems if a hardware device needs to be rebooted, because another computer cannot get in contact anymore. This problem is often a result of a «grubby» socket communication, which is independent from the coding of the messages (binary or text based).

Binary Messages

But the errors in the messages are not usually a result of the communication channel itself, since the protocol stack (especially the IP layer) takes care that a message arrives completely and correctly, for example, according to the byte order and protected from interferences on the physical line (unlike a serial communication where only low-level send and receive functions are used directly on the serial device and the medium is not always stable). It is not really necessary to add additional CRC checksums at the end of the messages. But binary communications in particular are burdened with specific problems. One of these is that usually additional information (metadata) is missing in the message itself. Binary messages are normally just a sequence of assembled binary numbers and character representations. The byte position decides how real numbers can be reassembled.

Binary Messages Assembly

For example, if four integer numbers are sent, the first four bytes build the first integer number, while the next four belong to the second integer number, and so on. If the binary message is misinterpreted as a sequence of short numbers, eight short numbers with completely wrong values are read. This means that a binary message is interpreted without any knowledge of whether the read data is really integer numbers or not[16] and whether the sender has spoilt the communication with wrong byte combinations such as character symbols. Another big problem nowadays is that some numbers differ in byte sizes in 32-bit and 64-bit operating systems (e.g., see ▶ Sect. 2.3.2). It is quite delicate to communicate with binary streams between such machines. But even if the bit sizes are similar, the problem of endianness can crash a communication. Not every operating system uses the same byte order for binary values which is defined as «little endian» or «big endian». Even the bit order cannot be the same on different machines (see ☐ Fig. 3.5).[17] Even though nowadays most computers use the same formats, binary messages are bulky and unsuitable for IPC.

Text Messages

But text messages are also not so suitable for generalized socket communications as they seem at first glance. As they commonly use American Standard Code for Information Interchange (ASCII) or American National Standards Institute (ANSI) code[18] for the character representation, the first problem appears when the

---

15 Regular expressions are character strings with mathematical descriptions for syntax rules of a type-3 language, using pattern matching (Brown 2001, *l.c. page 231*).

16 A simple example can be given in combination with editors. If an editor program reads a binary file, it interprets the single bytes without knowledge of the binary structure, using the ASCII table, so that the single, resulting characters make no sense.

17 This is often a problem on serial lines.

18 These are character codes where printable signs are assigned to bit sequences of 7 or 8 bits. Coded numbers in these character codes are not equal to binary representations of the

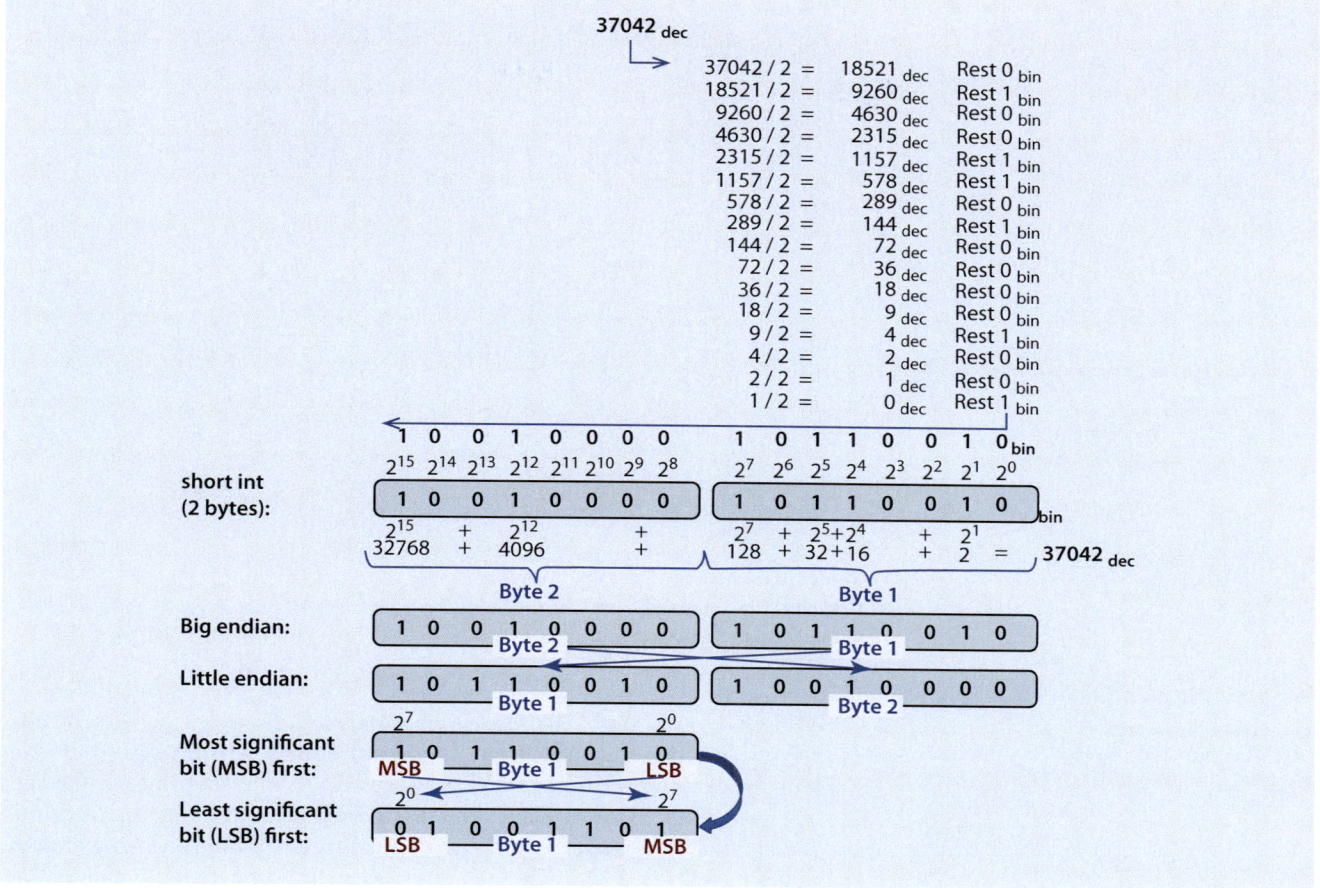

◘ Fig. 3.5 The binary representation of decimal integer numbers and the endianness (byte order) and bit order problem

characters are interpreted with country-specific representations of 8-bit characters (such as Ä, ä, Ö, ö, Ü, ü in the German alphabet). Messages with such characters should be avoided by reducing to just 7-bit ASCII. Another possibility would be to use a Unicode encoding such as Unicode Transformation Format (UTF), which combines an optional number of bytes to a character code sign using internal header and piggyback bits. But then, the interpretation must be extended to such new character combinations using additional libraries.

In general, text messages use individual text patterns, which are combined into a specific message. The rules for checking messages usually belong to the regular languages. The individual text message is a sequence of optional characters with more or less additional meta-information to interpret character blocks. A typical example can be shown using a real-life sample from the operation of Mark5 data recording systems at radio telescopes. If the client of the control system sends the request message:

«dir_info?»

the server on the Mark5 recording system answers with the hard drive status as a response, for example,

«!dir_info? 0 : 0 : 0 : 2007665700864 ;».[19]

Proprietary Text Messages

---

numbers. There are also coded control characters in the code, which are originally assumed from the tele-typewriter systems.

19 The hard drive status gives detailed information about the position of read and write pointers and the total capacity of the drive stack (here of 2 terabytes).

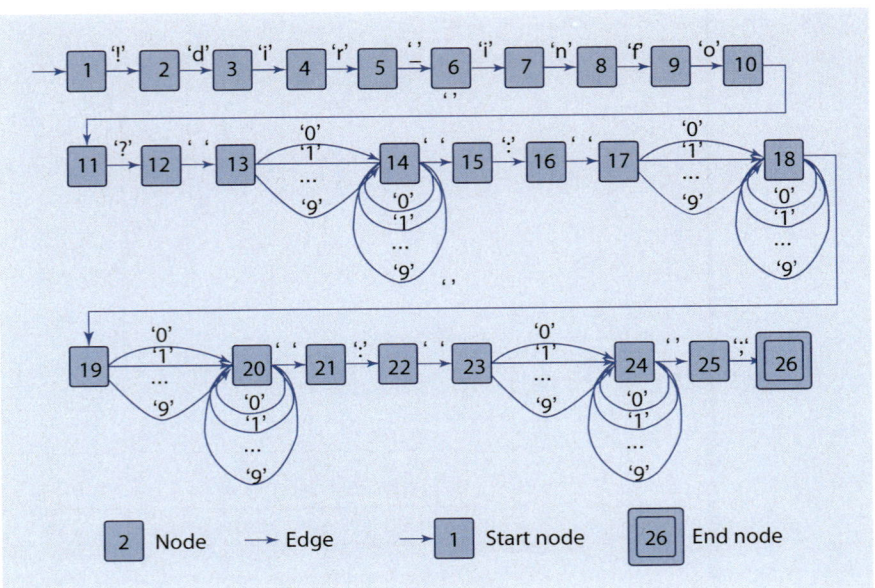

◘ Fig. 3.6 Directed state graph, representing the regular expression to interpret and check the replied answer of the style «`!dir_info? 0:0:0:2007665700864;`» from the Mark5 data recording system

Directed State Graphs

While later in ▶ Sect. 3.3.3 formal grammars are discussed in more detail, a short explanation should be given at this point to show how difficult a correct and stable use and interpretation of text messages is. To interpret the message and check it for correctness, a regular expression should be used which can be printed as a state graph (state chart). A state graph is a directed graph consisting of nodes and edges (see ◘ Fig. 3.6). Each node is represented by a cycle with an identifier and represents a particular state. Each edge is represented by an arrow from one state to another. The edge defines what state changes are allowed for which the label of the edge defines the condition. Only the defined changes (edges) are possible. Undefined changes end up in an error state. The graph also has two prominent states: one start state with an arrow without a label, where the walk through the graph starts, and at least one or more final states. Only if a final state is reached is the graph successfully passed (Brookshear 2012, *l.c. page 474*). Therefore, these directed graphs can ideally be used to check the syntax of the messages. The example in ◘ Fig. 3.6 shows the state graph for the above reply message from the Mark5. For a correct validation of the message, the checking has to follow the graph and stop at one of the end nodes (in ◘ Fig. 3.6, this is node number 26). All other patterns end in an error state (not drawn in the figure). Messages are only valid if they pass the graph correctly, so that a reliable socket protocol must perform similar regular checks for each message to guarantee correct functioning.

Finite State Machine and Regular Expressions

The program which can understand and process such state graphs is a finite state machine (see ▶ Sect. 3.3.3). In principle, such a machine (automaton) is nothing more than the sequence of if/else conditions described with text pattern matching. Each edge is one condition, which forks into a logic program path to the following state or to an error state. To avoid unmaintainable, nested, conditional statements, the processing can be split into a loop, in which a «`switch`» statement is used to check integer state numbers. Each case of the «`switch`» statement realizes the different conditional decisions for the individual state to transit to the following allowed state and changes the integer state number appropriately before the next loop is started. If an end state or an error state is reached, the loop ends. For each message, it would be necessary to create such a state graph and an according finite state machine to offer correctness and stability for the communication. To simplify this, some programming languages (such as

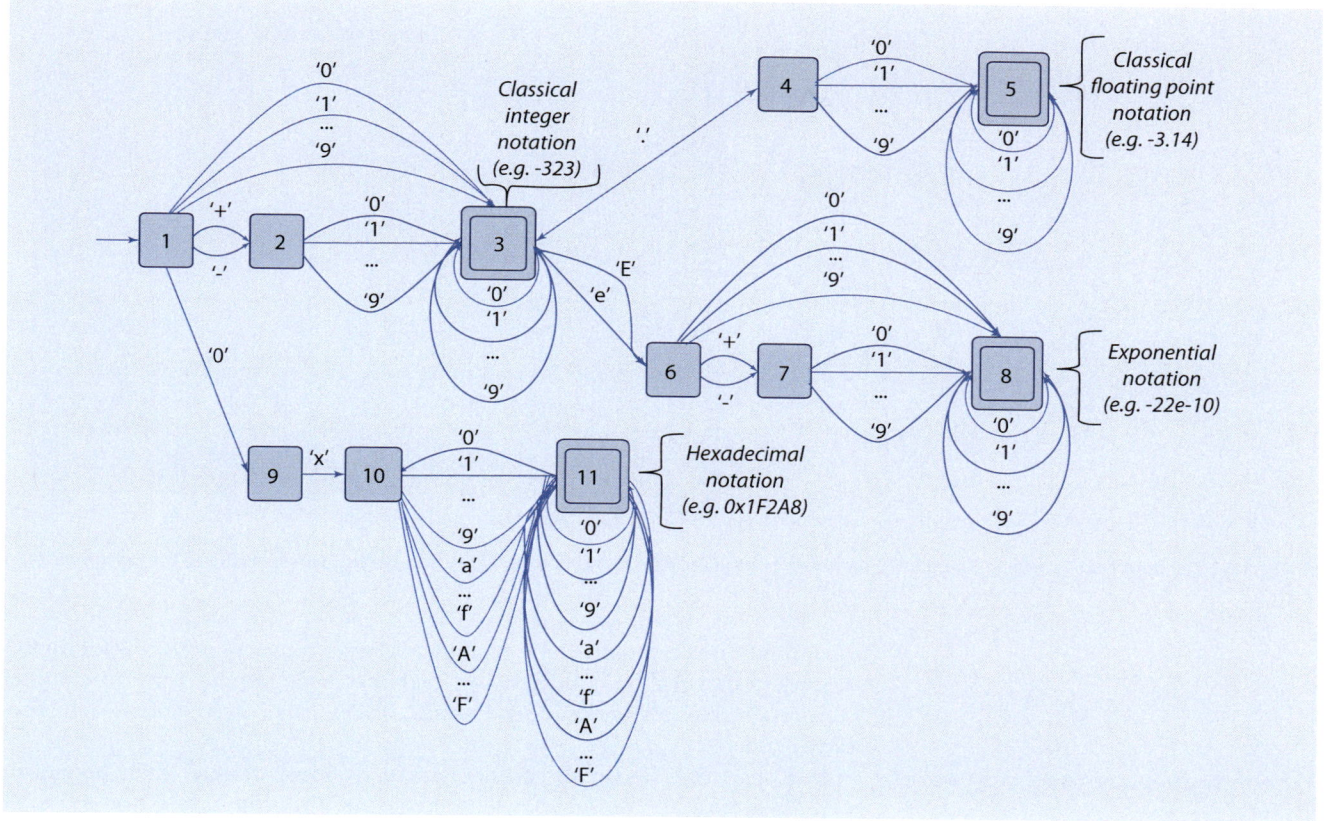

◘ **Fig. 3.7**   Possible directed state graph, representing the regular expression to interpret and check floating point numbers (without a control of the binary number limits)

Perl) offer a regular expression interpreter, which provides the possibility of defining the syntax as a sequence of pattern definitions. The pattern definition is then checked against the string which is under inspection. The comparison returns «true» or «false», which can be used for a conditional decision. The above state graph to interpret the Mark5 message can be described in Perl using pattern matching with the following regular expression: «$strMessage=~/^!dir_ info\?\s\d+\s:\s\d+\s:\s\d+\s:\s\d+\s;$/»[20]. Nevertheless, it is not easy to find all possible states and to define message checkers in a correct and stable way. This can especially be seen in ◘ Fig. 3.7, where a possible state graph to analyze floating point numbers in text messages is drawn.

Another problem with messages of ASCII or ANSI character codes is that they need much more memory space for the same information than binary representations. For example, if a complete floating point number, such as the «circle constant» PI with 3.1415926535897932384, with its complete presentable precision is transmitted, it would require 21 single character bytes, instead of just 8 bytes for the same information in the binary double presentation. This is at least three times more. As long as the maximum size of the communication message does not overrun the maximum size of a PDU, so that a higher fragmentation can be avoided, which would cost additional processing time, a larger message does not contribute too much to the communication costs. But often

Memory and Processing Time Consumption for Text Messages

---

20 Where the character string in $strMessage is checked with the pattern description between «/^» and «$/». The pattern itself is a combination of ASCII characters and reserved signs, where, e.g., «\?» represents the «?», as the same sign is used to mark optional parts, «\s» is a single whitespace, «\d» is a single digit, and «+» marks a repetition of the previous sign at least once (Brown 2001, l.c. page 242 ff.).

3

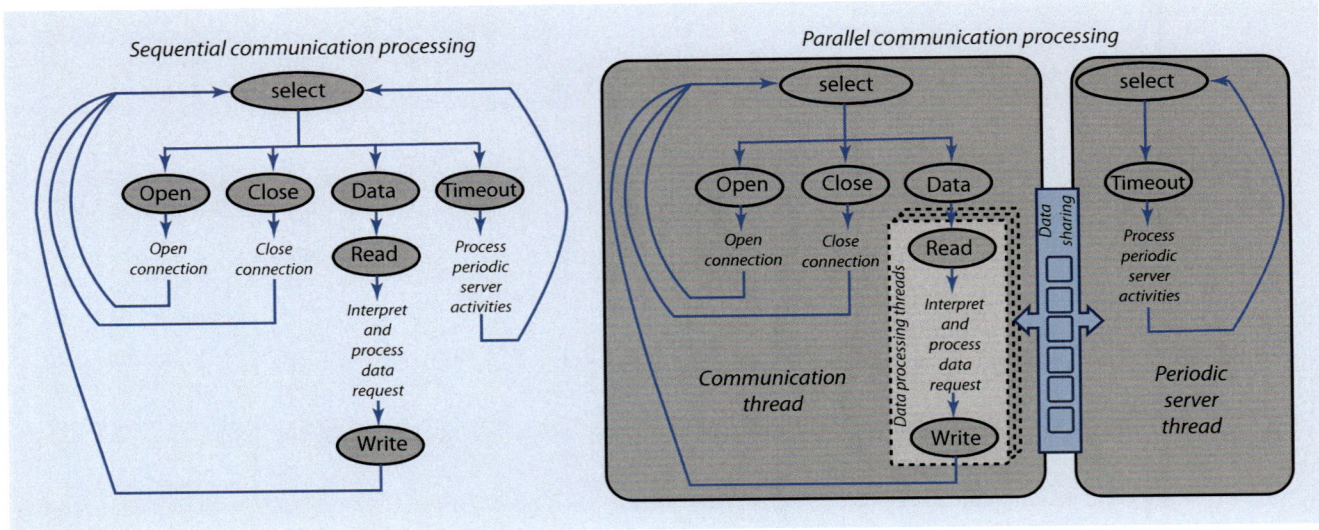

Fig. 3.8 The difference between a sequential processing of the socket communication (*left*) and a parallel which improves the performance

this limitation is a problem, for example, if larger matrices with several double values are transmitted. Additionally, text messages require a more complex interpretation to retrieve the single values, which usually also increases the processing time.

Performance Improvements with Parallel Threads

It is clear that IPC with sockets is actually much trickier than previously posted if almost all eventualities for a stable and safe communication are to be caught. But it is not only the interpretation of the messages that can present some challenges. The performance-oriented realization of the communication partners has also to be considered individually for each communication. The trick is to parallelize sequential requests. The usual implementation of a server-side communication with «select» shown in ■ Fig. 3.4 and demonstrated in Example 3.7 is sequential. But it can also be parallelized using threads. As explained in ■ Fig. 3.8, the periodic server activities can be sourced out to another thread of the same server process. In this way, the periodic tasks do not block incoming communication requests and its resource competition is managed automatically by the operating system. This is a good way to separate the placing of an order by a client from the real processing of the according task in a periodic processing loop. A difficulty then is that information transfer can only be implemented with shared memory, which needs to be protected carefully as a critical section to avoid race conditions, starvation, and deadlocks (see ▶ Sect. 2.6.4). Another very sophisticated implementation can also parallelize the processing of data requests, so that the mutual blocking of parallel requesting clients is reduced to a minimum.

Availability of the Server

Another aspect which should be considered is the availability of the communication service of the server. As errors can always happen, it is a question of how the system and especially the service-offering server react to them. If the server is no longer available after a crash, all of the connected clients are affected and may also be blocked and disturbed in their processing. Manual restarts are not a suitable solution, as most services run as daemon processes in the background.[21] Therefore, safe IPC servers must have adequate mechanisms which restart automatically after a crash. This is usually implemented with a parallel process which controls the service offering process. This «watchdog» process

---

21 A daemon is a server process which is started by the operating system during the startup and which is available automatically in the background, so that the applications and users do not directly see it.

runs independently from memory and processing of the controlled process, whose existence is permanently checked by this parallel routine. It can be implemented by forking an additional process for each communication server. At this juncture, a parent process creates a child process with the system call «fork». This duplicates the current process, and each one continues processing after the fork instruction. The return value of «fork» then decides which code part is processed by which process. The return value is the process identifier number, which is a unique number for a process in the operating system. If it is «0», the processing continues with the code of the child. Otherwise, it continues with the parent code (except if it is «-1», which symbolizes an error of «fork»). After forking, both processes are independent from each other, but as they are relatives, the parent process can use this relationship to monitor the child process. The parent process can wait for the end of the forked child processes and can operate consecutive activities, such as restarting the child process with a new «fork» system call (Stevens 1990, *l.c. page 60 ff.*). The code snippet in Example 3.8 demonstrates the principle of this solution. It is taken from a solution for Remote Procedure Call (RPC) at the Geodetic Observatory Wettzell (see ▶ Sect. 3.3.3).

■■ **Example 3.8**

Realization for a «watchdog» process using the process forking mechanism to create a monitoring parent and a controlled child process.

```
/* ... */                                                                1
                                                                         2
    while (1)                                                            3
    {                                                                    4
        /*! Create new process which runs the service server */         5
        g_iPIDDad = getpid ();                                           6
        g_iPIDSon = fork ();                                             7
        switch (g_iPIDSon)                                               8
        {                                                                9
            case -1:                                                    10
                fprintf (stderr, "%s", "unable to fork process.");      11
                exit (1);                                               12
            case 0: /*! Child */                                        13
                /*! Create all signal handler */                        14
                signal (SIGKILL, vTermHandlerSon);                      15
                signal (SIGQUIT, vTermHandlerSon);                      16
                signal (SIGTERM, vTermHandlerSon);                      17
                signal (SIGINT, vTermHandlerSon);                       18
                atexit (vExit);                                         19
                /*! Init server management: create all objects for the socket */  20
                /*! (here it is also necessary to define a flag and all functions,  21
                    which allow the server to end, if the user wants to stop it,    22
                    so that the parent process does not restart it again, e.g.      23
                    realized with a kill signal to the parent process) */           24
                                                                        25
                /* ... */                                               26
                break;                                                  27
            default: /*! Parent (= watchdog process) */                 28
                /*! Create all signal handler */                        29
                signal (SIGKILL, vTermHandlerDad);                      30
                signal (SIGQUIT, vTermHandlerDad);                      31
                signal (SIGTERM, vTermHandlerDad);                      32
                signal (SIGINT, vTermHandlerDad);                       33
                /*! Wait on exit of son and force fork again at the beginning of while */  34
                wait (& iStateVal);                                     35
```

```
        SCurrentTime = time(NULL);                                              36
        pSCurrentTimeStruct = gmtime_r (&SCurrentTime, &SCurrentTimeStruct);    37
        fprintf (stderr, "server terminated undefined and restarts on %04d-%02d 38
            %02d:%02d:%02d\n",
                        pSCurrentTimeStruct->tm_year+1900,                      39
                        pSCurrentTimeStruct->tm_mon+1,                          40
                        pSCurrentTimeStruct->tm_mday,                           41
                        pSCurrentTimeStruct->tm_hour,                          42
                        pSCurrentTimeStruct->tm_min,                            43
                        pSCurrentTimeStruct->tm_sec);                          44
                                                                                45
        continue;                                                               46
    }                                                                           47
  }                                                                             48
                                                                                49
  /*! Should never be reached */                                               50
                                                                                51
/* ... */
```

All in all, the implementation of IPC with sockets is a huge task, which has to be individually adapted to the specific requirements of the current situation for which it should be used. Each additional communication needs similar developments and again additional development time even if prefabricated modules or components are available for the socket handling. An individual pairing of request and reply messages needs to be implemented within a higher-level communication protocol, an individual data representation needs to be designed, the addressing of remote machines has to be organized, and all of these should be done with sufficient awareness of system and communication failures (Singhal and Shivaratri 1994, *l.c. page 87 f.*). All of these parts are usually implemented with individual «glue code» (see ▶ Sect. 3.2). But all of the aspects described just form occurrences or classes of a similar task. Therefore, wouldn't it be good if there were a way to hide the communication and its system specifics? Wouldn't it be helpful if there were no need to deal with the communication protocol and the according error handling each time, so that a local function call looks like a remote function call and belongs to a specific class of methods for communication, automatically including all the required subfunctions and protocol messages? This would certainly make IPC less painful.

3.3.2 Remote Procedure Calls

Local Function Calls

One of the oldest and still simplest ways to make IPC less painful is RPCs. Remote Procedure Calls, as the name already indicates, are procedure or function calls on another remote computer. The appearance of an RPC is very similar to a local use of a function. A local function must be defined somewhere (e.g., in a module) with a function name as identifier, a return value, a list of function parameters or arguments, and a function body, which contains the instructions which the function operates on call. External function declarations are known from header files, which can be included in source code files, so that the function's calling structure is known. If such a local function is called, the workflow changes into the function body and processes the instructions which are located there. After finishing, the processing returns from the function body with a return value back to the calling instance.

Remote Function Calls

In a similar way to this well-known and well-understood local call of a function, RPCs extend this mechanism for a transfer of control and data to another remote machine (see also the comparison of local and remote calls in ▪ Fig. 3.9). The «local-seeming» Remote Procedure Call is automatically translated into a communication over a communication medium to hide all of the challenges described earlier for IPC (Singhal and Shivaratri 1994, *l.c. page 88*). The communication in the background of an RPC starts with a «request» message to the remote system, which is automatically interpreted and converted to a function call on the remote machine. There the processing of the function body is performed in a

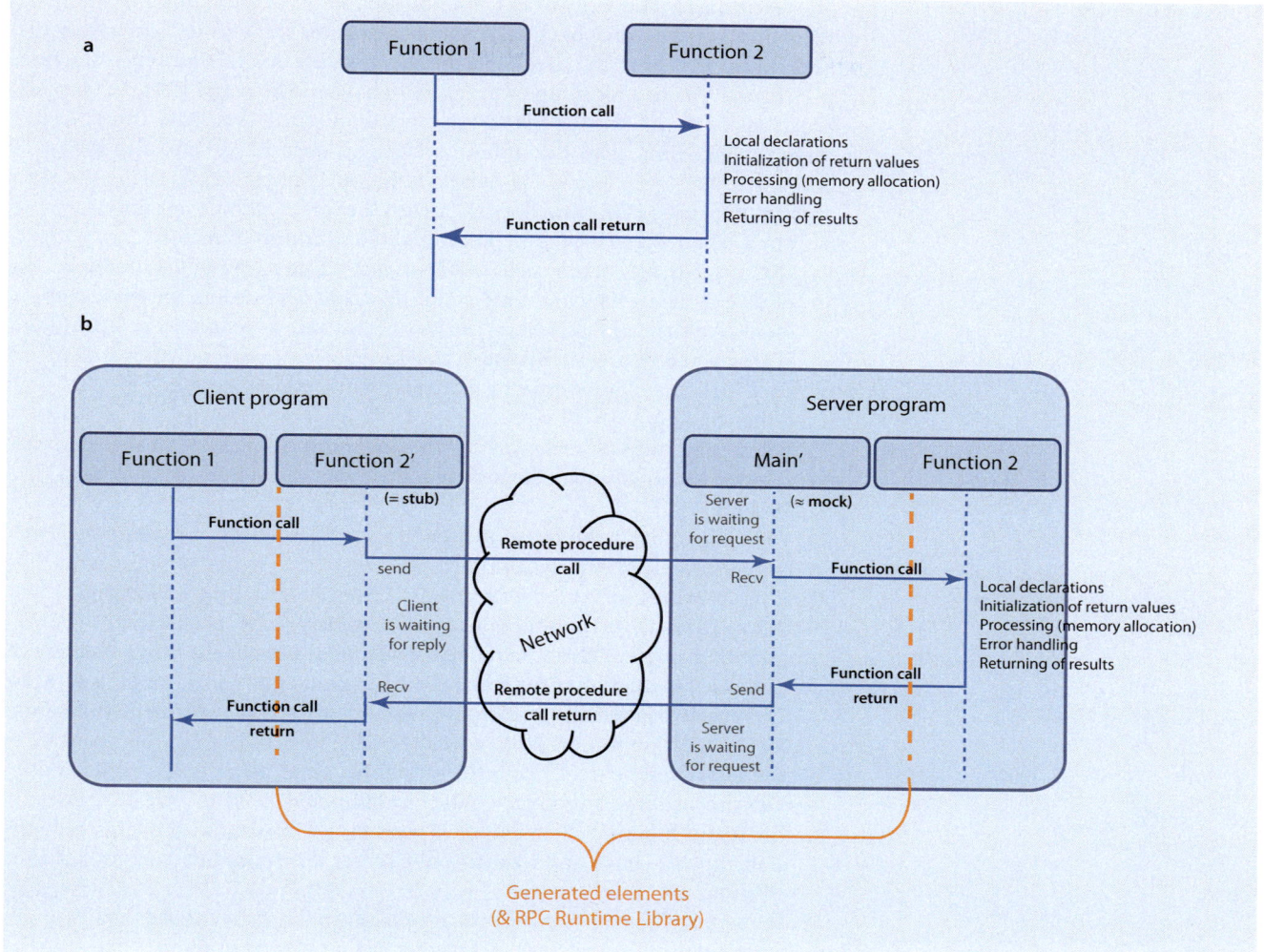

◘ Fig. 3.9 The difference between **a** a local function call and **b** a Remote Procedure Call. For the calling function «Function1» and the executing function «Function2», everything looks similar. The RPC mechanism just generates a complete network communication between both functions using a client stub «Function2'», which locally represents «Function2», and a server with the server stub «Main'» which handles the incoming requests (and simulates the environment like a mock object)

similar way to the local call. After finishing, it comes back with the return value, which is then again translated into a «reply» (or «response») message, which returns to the calling system. It is then converted into a local return, so that the complete instruction flow looks just like a local call, while all of the communication activities are locally not visible by the function caller (Stevens 1990, *l.c. page 723*). Because of the structure of the application flow with «request» and «reply» messages, and a requesting and a replying process on the two machines involved, it is again a client-server model (see ▶ Sect. 3.3.1). This means that RPCs are sophisticated realizations of the client-server architecture with the improvement to automatically transfer control and data between two interacting machines.

Remote Procedure Call (RPC)

> *Definition Definition DEF3.8. of «Remote Procedure Call (RPC)»* (see (Singhal and Shivaratri 1994, *l.c. page 88*))
>
> RPCs are procedural control flows and dataflows, similar to the standard local procedure or function calls in a program, but which extend this mechanism with a transfer of control and data to another remote machine. The machines involved have no shared memory, and the communication between them follows the request-reply pattern of a client-server model, where the client needing a service invokes a procedure at the server offering this service.

Stubs and Their Use

To ensure the functionality of an RPC, both endpoints of the communication work with stubs[22] (see also ▶ Sect. 2.6.2 about the use of stubs in test environments). This means that both the client and the server interact with «dummy» functions, playing the role of a local function call (Singhal and Shivaratri 1994, *l.c. page 88*). This is quite a similar situation to the one found in test scenarios. Thus, an RPC takes the following steps (see (Stevens 1990, *l.c. page 723 f.*) and also ◘ Fig. 3.9, where «Function1» calls the remote function «Function2» using the client stub for «Function2» and the server with the server stub in «Main'»):

1. The client makes a local function call to the «dummy» function «Function2'» of the client stub, corresponding to the actual function «Function2», which is remote. For the client application, the client stub looks like the actual function. The task of this stub function is to convert the function and its arguments into a standardized data representation and to pack it all together into one or more communication messages («marshalling»).

2. The communication messages can then be sent as requests via a socket to the remote machine using system calls to the communication stack (see also ◘ Fig. 3.3).

3. The PDUs are transmitted to the server using connection-oriented (TCP) or connectionless protocols (UDP).

4. A server with a server stub «Main» on the remote machine waits for the incoming requests from the client (as described for the usual socket communication in ◘ Fig. 3.3). The server stub processes the unmarshalling to separate the function identifier and the function arguments from the network message. It also reconverts it from the standardized data representation of the communication protocol into a locally usable format for local function call.

5. The server stub «Main'» calls the local function «Function2» and hands over the retrieved arguments to the local function.

6. The server function «Function2'» processes the data with the instructions in its function body and returns to the server stub with the return values after finishing.

7. The server stub converts the return values into the standardized data representation and packs it all together again into one or more communication messages («marshalling»).

8. The messages are sent back in reply to the client stub via socket communication.

9. The client stub «Function2'» reads the PDU using the socket system calls. It performs the unmarshalling and converts the return values from the communication message into a locally usable return format. After this, it returns like each common local function to the client.

RPC Realizations

There are several implementations of RPC libraries available and used for complete communication environments, such as the Distributed Computing Environment (DCE) from the Open Software Foundation (OSF) (Neidhardt 2005, *l.c. page 215*). But one of the most widely used and easily available RPCs on each Linux operating system are from the Open Network Computing (ONC), offered by the company Sun Microsystems (Stevens 1990, *l.c. page 725 f.*) (which is now part of the company Oracle˚). This original ONC RPC is nowadays also known as Sun RPC and since 2009 has been available in the «Glibc» library under the open-source license of the Berkeley Software Distribution (BSD). As it is part of all Linux and Unix operating systems, it is easily usable without installations (Diedrich 2009), so that

---

22 Compared to testing environments, the client-side proxy works like a stub, while the server-side proxy is more like a mock object.

the following discussions concentrate on this specific RPC implementation, which was originally described as ONC standard in the papers Request for Comments (RFC) 1057 (for the Sun RPC itself) and RFC 1014 (for the used data representation)[23] (Stevens 1990, *l.c. page 732*).

The aim of RPC s is to hide the complete code for the network communication behind the stubs. Therefore, the application does not need to take care of communication details, such as the socket system calls and the translation of data to and from a communicable representation. Nevertheless, the communication takes place with great many disadvantages, problems, and errors. Therefore, it is important that the communication library for RPC considers the following design issues (see (Stevens 1990, *l.c. page 725–732*), (Stevens 1990, *l.c. page 737–740*), and (Singhal and Shivaratri 1994, *l.c. page 88–92*)):

- *Binding*: Binding describes the address and port of a remote service according to the description for sockets (see ▶ Sect. 3.3.1) so that it can be contacted by a client. But RPCs use an additional method to locate servers, via a service identification for the type of service which is offered by a remote server. The identification is a combination of a program, a version, and a procedure number in combination with the IP address of the server machine (Stevens 1990, *l.c. page 733*). The server registers itself at a centralized binding server and transmits the service identification. It can use an implicit port binding, where the port is defined by the binding server or the port can be defined explicitly, so that a developer specifies the port binding. The central register is then something like the «yellow pages» for RPC services on a system or in a network. In the case of Sun RPC, the binding server is the «portmapper» process «portmap». It is a well-known daemon[24] on port 111[25] on the server machine. The «portmapper» is responsible for mapping between essential services. It allows direct access to both the server and the client. The server uses this connection to register a specific server dispatch routine, which is accessible via a specified transport protocol (using the library function of Sun RPC «svc_register» (Bloomer 1991, *l.c. page 377*)) or to unregister a registered routine (using «pmap_unset» to remove registration and «svc_unregister» to erase the own entry (Bloomer 1991, *l.c. page 357 and 379*)). The client uses it to find a service with the function «clnt_create», using the IP address, the program number, the version number, and the transport protocol. It first contacts the «portmapper» to retrieve the binding information of the server, so that it can then contact the server directly (clients cache this port mapping information, so that they do not have to contact the «portmapper» for each request). A further version of such a mapper for more abstract addressing and arbitrary transport protocols is the more general «rpcbind» daemon (Bloomer 1991, *l.c. page 11*), which is usually still available on modern Linux systems.
- *Data representation*: The big difference between local and remote activities is that data types and representations (e.g., of arrays, pointers, etc.) might be different on different machines. Therefore, it is quite important to define a suitable data representation to avoid the endianness difficulty (see ▶ Sect. 3.3.1), which can be used to convert other platform data representations. Sun RPC uses the External Data Representation (XDR) for this conversion. XDR

Design Issues

Binding, e.g., with «Portmapper»

Data Representation, e.g., with XDR

---

23 Documents in the series of the «Request for Comments» (RFC) sections of «The Internet Engineering Task Force» are papers which are on the way to becoming a standard or are already standards.
24 A daemon is a server process which is started by the operating system during startup and is available automatically in the background.
25 An alternative port on Sun Solaris systems is port 32,771 (McClure et al. 2006, *l.c. page 171*).

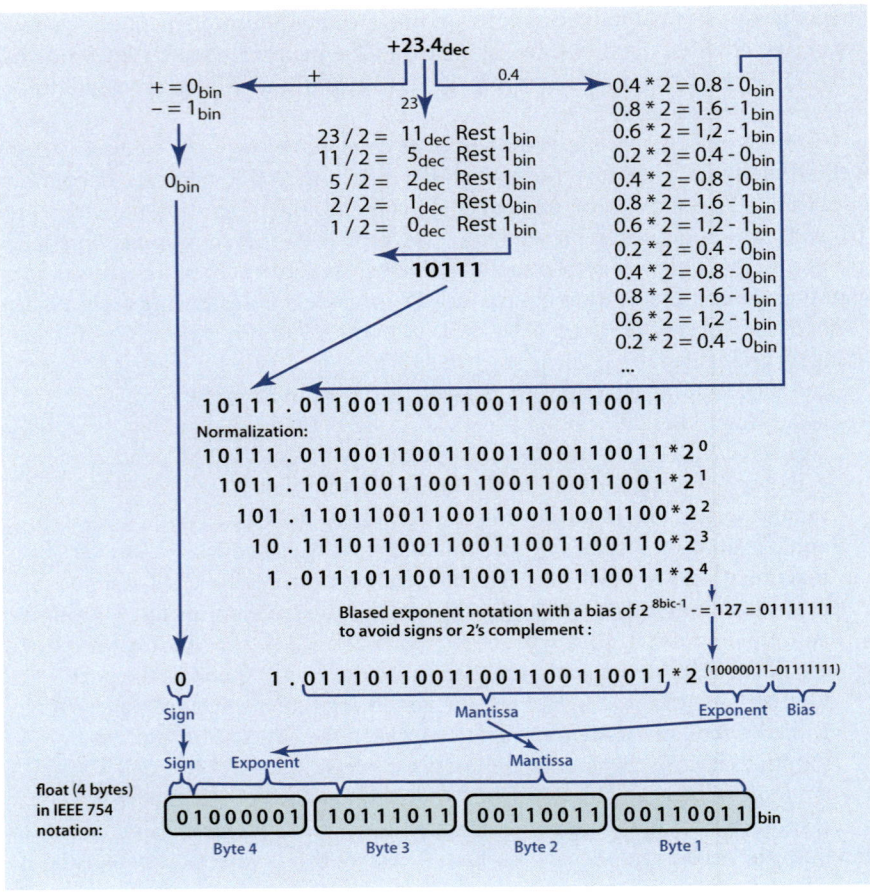

□ **Fig. 3.10**   A floating point representation in IEEE notation, as used in the XDR format of Sun RPC

uses implicit typing (only the value is transferred as binary over the network and not the type of the value) in a single canonical data representation. It only defines big-endian (or most significant byte first) as the only byte order allowed, uses the two's complement for signed integer,[26] and uses the single floating point representation standard from the Institute of Electrical and Electronics Engineers (IEEE)[27] (IEEE 754, see □ Fig. 3.10; see (Rembold et al., 1991, *l.c. page 172 ff.*) and (Brookshear 2012, *l.c. page 53 ff.*)). The data representation allows any complying machine to share data independently from local differences. The Sun library offers the according XDR functionality

---

26 The two's complement is a way to represent negative integer numbers in the binary system by negating the positive binary number bit-wise and adding the binary one to it, to avoid a «minus» zero if only the first bit is used as a sign bit.

27 All floating point representations deal with some problems in combination with limited numbers of bits in a computer. They not only define a maximum and minimum representable number. They also define a minimum representable difference between two floating point numbers, the floating point accuracy («eps»). In combination with rounding methods defined, even arithmetic operations can be influenced. For example, it can happen that a subtraction of a very small floating point number from a very big one does not change the big one (absorption). It could even happen that mathematically unsolvable equation systems can be solved in a wrong way doing floating point arithmetic in a computer. Another problem is the deterministic chaos in iterative calculation processes, so that starting at one specific iteration, the correctness of the iterative results is chaotic (see (Peitgen et al. 1998, *l.c. page 57*)).

(usable with the header file «`<rpc/xdr.h>`») to convert data in both ways. Both the client and the server need to use these functions to convert their data into this standard format (Bloomer 1991, *l.c. page 341 ff.*). This supports binary representations for ASCII character signs, integer, long integer, float, double, and also Boolean values or value combinations up to one-dimensional arrays with fixed or variable lengths (heap elements are defined by a structure with the pointer to the dynamic memory element and the length of the memory element).

— *Parameter and result passing*: While passing parameters by value is quite simple, passing heap elements by reference does not work as addresses would not be valid on the remote machine. The solution is to copy the heap elements and to create a copy of the client heap structure at the server. This is not a simple job. Therefore, Sun RPC does not emulate pointers for a call-by-reference. It is necessary to convert them into the previously described structure for arrays with variable lengths. Multidimensional arrays must be converted into a suitable one-dimensional array of variable length (Bloomer 1991, *l.c. page 46*). In the classic definition of Sun RPC, each function has only one argument and one return value, so that more than one argument or return value must be packed into a structure. Newer Sun RPC implementations allow C-like function definitions. But the use of classic structures is better to sustain the compatibility to ONC libraries. — Parameter and Result Passing

— *Transport*: Sun RPC supports connectionless communications with UDP as well as connection-oriented communications with TCP. Experience has shown that message sizes with character strings of variable lengths need to be lower than the PDU size of UDP if a connectionless communication is used. To separate the single data structures in the communication byte stream, Sun RPC uses 4-byte integer values at the beginning of the structure to define the total size of the following structure. — Transport Protocols

— *Error handling and safety*: The error handling on remote server machines is a difficult task, as program crashes after computing failures are difficult to handle and to debug. Other error sources are caused by communication failures. If messages are lost, repeated calls can invoke procedures several times, which have to be solved using suitable call semantics (see the next item). Also client crashing creates a problem if they sent a request shortly before the crash. In this case, there is no client left to receive the reply. Those abandoned server invocations are request orphans. Sun RPC implements automated repeats of the requests at the client side for both protocol types. After a predefined number of failed repetitions, direct error returns and timeouts allow the client to perform decisions in his workflow to manage the error. The server has almost no error handling to check error states apart from an RPC invocation of a local server function. — Error Handling and Safety

— *Call semantics*: In view of the possible errors, the call or execution semantics is important for remote calls. The client cannot be sure if a call has really been processed after an error has been returned. Therefore, detected failures can force the client to repeat the request, and remote functions can be invoked several times. But RPCs themselves are stateless. This means that they do not remember a call. Therefore, it is part of the server function development to make a server stateless or «stateful» (Bloomer 1991, *l.c. page 12 f.*), by using, for example, global memory variables. But then the call semantics plays a key role for a correct result of «stateful» functions. If several calls do not affect the result, the function is «idempotent». Typical of this type are requests for calculations or snapshot values such as the server time. Those data-returning requests do not change the state of the server at — Call Semantics

3

all. File operations, however, are critical in this case, such as several, uncontrollable calls of a function, which, for example, adds text to a file or changes the state of the remote file in an unintended way. The following general call semantics can be defined for all RPCs:

- «*Exactly once*»: This is the most difficult semantics for an implementation for remote calls, as an invocation of a function after an RPC is forced to occur only once (in the same way as local calls).
- «*At most once*» or «*zero or none*»: In this case, a correct run creates exactly one call, while an error means that one or no run was performed (for «at most once», it can also mean that the call was only partly processed). Sun RPC, like most of the RPC implementations, implements this «at most once» semantics, using 4-byte-long unique transaction numbers «xid» in the message frame for each request from a client. The server uses this number to assign replies to the correct requests. In the case of UDP, an option allows the caching of requests to identify retransmissions (Sun Microsystems, Inc. 1988, *l.c. page 8*), which then can be answered with an already created result of a previous invocation. In general, the «at most once» semantics is best supported by using the connection-oriented TCP for highest safety.
- «*At least once*»: This can be implemented very easily. If the call succeeds, one invocation is performed. A failed call means that no, one, a partial run, or at least a few runs are activated.

Call Correctness

- *Call correctness*: The call correctness for local or remote functions is given if an order of the function calls forces a similar order of data modifications performed by the functions. The RPC can implement this with sequentialization of incoming requests at the server side without considering the different sending and transfer times. This implements a local call correctness and not a distributed one over a whole network, where all events are chronologically ordered, like for network-wide time distribution and clock mechanisms.

Performance

- *Performance*: Generally, RPCs are worse than local calls as the whole communication takes time. The performance is very dependent on the communication network and its traffic load. But even locally used RPCs have reduced performance as the conversion of formats on both sides of the communication partners is wasteful. Therefore, it would be better to use other IPC techniques (pipes, shared memory, etc.) for communications on the same machine. Nevertheless, from the perspective of a manageable software design, it is preferable to keep the same interfaces. But as RPC implementations do not usually differentiate between local and remote use, applications have to accept reduced performance if they want the simplification of the network programming using one communication interface. Literature such as (Stevens 1990) *l.c. page 731*, numeralizes the RPC performance to be at least ten times worse than local calls.

Security

- *Security*: As the RPC technique needs access to a third-party remote system to invoke tasks, it is also a security issue to allow such calls. But usually these risks are not handled by the basic RPC implementation, even when they offer some rudimentary authentication mechanisms. It is usually part of the application development, for example, to encrypt messages to protect them against spying. It is part of the server machine itself to handle access rights and to protect against unauthorized accesses and other destructive attacks. Nevertheless, RPC mechanisms are often a security leak, especially in combination with implementations which enable buffer overflows (see ▶ Sect. 5.5.1). Sun RPC offers three general authentication mechanisms (Sun Microsystems, Inc. 1988, *l.c. page 11*) (besides these, proprietary application authentications can also be implemented individually):

- *Null authentication*: If the client does not know its identity or the server does not care about it, no authentication is necessary («AUTH_NULL»).
- *UNIX authentication*: If an authentication identifier is used which is similar to a UNIX or a Linux system with an arbitrary identifier number, a machine name, the user identifier, and the group identifiers of the client, the UNIX authentication («AUTH_UNIX») can be used.
- *Data Encryption Standard (DES) authentication*: The DES authentication («AUTH_DES») implements a UNIX-independent universal username, user identifier, and group identifier with verifications, so that credentials cannot be faked easily. The server uses the client's credentials with the encrypted timestamp from the client as verification criteria (see also ▶ Sect. 5.5.1).

The major advantage of Sun RPC is that the technique is well known, included in each Linux operating system in the «Glibc» library (or can be obtained as sources), and is therefore well tested and proven. But the library alone would not improve the IPC programming completely. It offers the above design issues but still requires individual programming for the utilization of the interface functions, as each communication is parametrized to the individual application. This is the starting point for the use of a generator, which is here a protocol compiler. Such a compiler is offered for the Sun RPC: the «rpcgen». — **RPC Application Parametrization**

To create new individual RPC applications, the first step is to parametrize the individual communication protocol (for the following blocks, also see (Stevens 1990, *l.c. page 732–738*) and (Bloomer 1991, *l.c. page 3–10*). This is done with a specification file (usually with the file extension «.x») using the Remote Procedure Call Language (RPCL) for «rpcgen». A sample specification file in the classical notation, where remote procedures only have one argument and one return value, so that structures are defined for both, is shown in Example 3.9. It defines the constants and enumerations used by the client and the server (in the example, it is used for the error codes «OK» and «NOK») and specifies the structures, unions, and type definitions («typedef») which are used by the functions of the RPCs (in the example, these are the structures «struct OUTSAPtestiGetUnix TimeInSecondsStruct», «struct INSAPtestiSetUnixTimeIn SecondsStruct», and «struct OUTSAPtestiSetUnixTimeIn SecondsStruct» with the derived type definitions «OUTSAPtestiGetUnix TimeInSeconds Type»,«INSAPtestiSetUnixTimeInSecondsType», and «OUTSAPtestiSetUnixTimeInSecondsType»). The definition style is very C oriented. After these type declarations, the description of the interface «SAP_test» with the interface identification number «0x29bb1d72» and in the version «SAP_VERS_test» of number «1» is added. It contains the declarations of the remote procedures «sap_test_igetunixtimeinseconds» (procedure number 1) and «sap_test_isetunixtimeinseconds» (procedure number 2).[28] The program numbers are 4-byte integer values, defined with the hexadecimal notation of C, starting with «0x». But Sun regularizes the assignment for private users to numbers between «0x20000000» (decimal number 536,870,912) and «0x3fffffff» (decimal number 1,073,741,823). — **Defining the Communication Protocol**

---

28 In the design definitions of the Wettzell observatory, each remote function is defined as SAP, which is just a proprietary agreement.

■ ■ **Example 3.9**

Sun RPC specification file to generate two remote functions to get and to set the UNIX time of the remote machine (the specification uses the classical notation, where remote procedures only have one argument and one return value)

```
/*********************************************************
 * CONSTANTS
 **********************************************************/
/*! Constants for the error return
    => are converted to "#define" preprocessor statements
    (another possibility would be to define the statement as "%#define",
    because '%' lines are passed through) */
const OK = 0;  /*! No processing error in the remote function */
const NOK = 1; /*! Processing error in the remote function */

/*********************************************************
 * STRUCTURES
 **********************************************************/
/*! The structure with the individual name "OUTSAPtestiGetUnixTimeInSecondsStruct",
    which contains the UNIX time in seconds and the return value for the reply of the
    function "sap_test_igetunixtimeinseconds" */
struct OUTSAPtestiGetUnixTimeInSecondsStruct
{
    int _Ret;
    long lUnixtimeInSeconds;
};
typedef struct OUTSAPtestiGetUnixTimeInSecondsStruct OUTSAPtestiGetUnixTimeInSecondsType;

/*! The structure with the individual name "INSAPtestiSetUnixTimeInSecondsStruct",
    which contains the UNIX time in seconds and the return value for the request arguments
    of the function "sap_test_isetunixtimeinseconds" */
struct INSAPtestiSetUnixTimeInSecondsStruct
{
    long lUnixtimeInSeconds;
};
typedef struct INSAPtestiSetUnixTimeInSecondsStruct INSAPtestiSetUnixTimeInSecondsType;

/*! The structure with the individual name "OUTSAPtestiSetUnixTimeInSecondsStruct",
    which contains the return value for the reply of the function
    "sap_test_isetunixtimeinseconds" */
struct OUTSAPtestiSetUnixTimeInSecondsStruct
{
    int _Ret;
};
typedef struct OUTSAPtestiSetUnixTimeInSecondsStruct OUTSAPtestiSetUnixTimeInSecondsType;

/*********************************************************
 * INTERFACE DECLARATION (with all of the remote procedures)
 **********************************************************/
/*! Definition in the classic Sun RPC style, where remote procedures only have one argument
    and one return values */
/*! Define the individual program identifier "SAP_test" and the program number "0x29bb1d72" */
program SAP_test {
    /*! Define the individual version identifier "SAP_VERS_test" and the version number "1" */
    version SAP_VERS_test {
        /*! Define a remote procedure with the individual name
            "sap_test_igetunixtimeinseconds"
            and the procedure number 1; the function should return the UNIX time in seconds
            since Jan. 1st., 1970 00:00:00 and an error code with the above definitions from
            the remote machine */
```

1
2
3
4
5
6
7
8
9
10
11
12
13
14
15
16
17
18
19
20
21
22
23
24
25
26
27
28
29
30
31
32
33
34
35
36
37
38
39
40
41
42
43
44
45
46
47
48
49
50
51
52
53
54
55

```
OUTSAPtestiGetUnixTimeInSecondsType sap_test_igetunixtimeinseconds (void) = 1;    56
/*! Define a remote procedure with the individual name                            57
    "sap_test_isetunixtimeinseconds"                                              58
    and the procedure number 2; the function should set the UNIX time in seconds  59
    on the remote machine and should return an error code with the above          60
    definitions */                                                                61
OUTSAPtestiSetUnixTimeInSecondsType sap_test_isetunixtimeinseconds (              62
    INSAPtestiSetUnixTimeInSecondsType) = 2;
} = 1;                                                                            63
} = 0x29bb1d72;                                                                   64
```

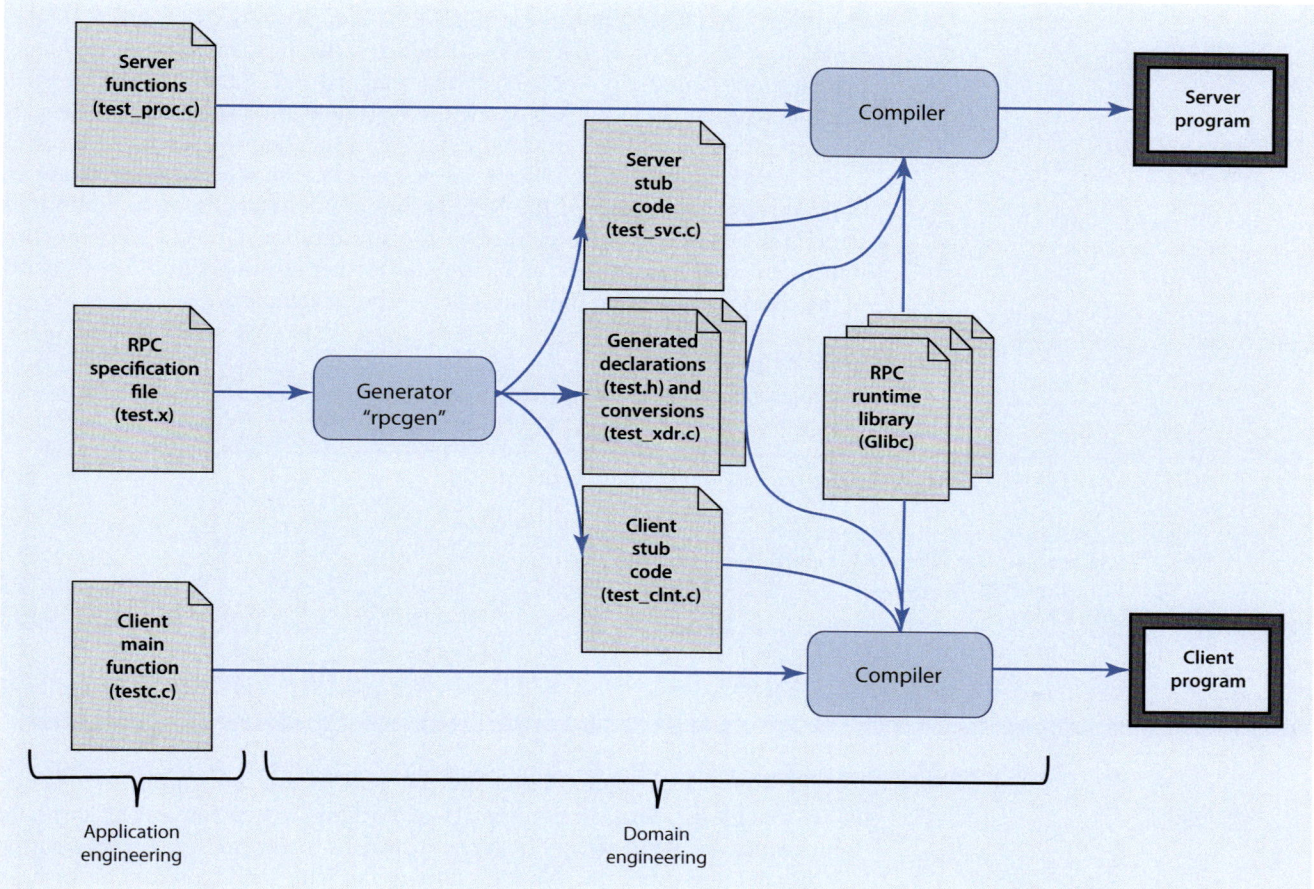

Fig. 3.11 Generative programming with Sun RPC and the included interface generator «rpcgen» (adapted from (Stevens 1990, *l.c. page 733*))

After specifying the interface, it is used to generate the communication specific modules. The complete process of generating and compiling Sun RPC communication-relevant files is shown in ◘ Fig. 3.11 (adapted from (Stevens 1990, *l.c. page 733*)). The first step is to start the RPC protocol compiler «rpcgen» with the command line sequence, shown in the following itemization for the above protocol specification file «test.x».

Generating the Protocol Modules

— «rpcgen -M test.x», to create all communication-relevant module files from the specification file «test.x» as a multi-threading safe version («-M»)
— «rm -fR test_svc.c», to delete the server stub module «test_svc.c», generated during the previous step

- «rpcgen -M -s tcp -s udp -o test_svc.c test.x», to create the server stub file again from the specification file for a parallel use of the transport protocols TCP and UDP

It creates the source files for the client stub module «test_clnt.c», the server stub module «test_svc.c», and a general header description file «test.h», with the remote procedure declarations, which are used on both sides in the client, and the server stub. «rpcgen» also generates an additional procedure with procedure number 0 («null procedure»), which is used for pinging a server[29] and measuring the round-trip delay.[30]

Writing and Compiling the Client

The client stub functions can be used directly within a client main program (e.g., in the file «testc.c», as in ◘ Fig. 3.11). After including the general header file «test.h», the client can create an RPC connection using the RPC library function «clnt_create» with the IP address, the program and version identifier, and the transport protocol type. The returned RPC handle must be used as additional argument to the call of one of the generated remote functions (e.g., in the case of a function with procedure number 2 by using the generated function «sap_test_isetunixtimeinseconds_1»). The other arguments for the multi-thread safe version of the RPC call are pointers to the input and output structures, whose types are defined in the specification file. The RPC function returns an error code, which can be used for further error handling. The local call of the client stub function then looks as shown in Example 3.10.

■ ■ **Example 3.10**
The local call of the client stub function

```
if (((clnt_stat) SRetVal_1 = sap_test_isetunixtimeinseconds_1    1
    (                                                             2
        (INSAPtestiSetUnixTimeInSecondsType *) &SInput,           3
        (OUTSAPtestiSetUnixTimeInSecondsType *) pSOutput,         4
        (CLIENT *) pSRPCHandle                                    5
    )) != RPC_SUCCESS)                                            6
{                                                                 7
    /* ... ERROR */                                              8
}                                                                 9
```

To feed the function, it is necessary to create the input structure and fill it with the values which should be sent to the server within the request messages. This is simple C programming. After the call, the values of the return structure can be used for further processing. Such calls can be repeated as needed depending to the application workflow. At the end of the client program, the RPC handle needs to be destroyed again, using the library function «clnt_destroy». The manually written client source code file together with the generated client stub, the header definition file, and the RPC runtime library «Glibc», which is hosted in the operating system libraries, must then be compiled and linked together to create the client executable.

Writing and Compiling the Server

The server main function is implemented and automatically generated in the server stub code as a module file with the name «test_svc.c». There are also the dispatcher functionalities, which process incoming requests and call the relevant server functions, which are usually located in a source file with «_proc» in

---

29 The name «pinging» is derived from the submarine technique, where sonar signals, used to detect foreign targets, sound like a «ping».

30 Round-trip delay is the time needed for sending of a request to a server processing the request with a remote procedure invocation at the server, returning the reply to the client, and receiving it at the client.

its name, such as «`test_proc.c`» for the previous sample. In order to complete the module design, there is also a suitable header file. The bodies of these server functions must be written manually. Their function head declaration must fit the style of the Remote Procedure Calls in the dispatcher routine. For multi-threading environments, it follows the declaration in Example 3.11 for the previously specified function with procedure number two:

■■ Example 3.11
The server-side structure of a RPC function, called by the dispatcher

```
extern  bool_t  sap_test_isetunixtimeinseconds_1_svc     1
(                                                         2
    INSAPtestiSetUnixTimeInSecondsType * pSInput,        3
    OUTSAPtestiSetUnixTimeInSecondsType * pSOutput,      4
    struct svc_req * pSSvc                                5
)                                                        6
{                                                        7
    /* ... function body, which processes the RPC functionality */   8
}                                                        9
```

In a similar way to the equivalent client function, a server function defines the pointers to the input («`pSInput`») and output («`pSOutput`») structures as arguments. An additional structure with the calling information is another argument («`pSSvc`»). The function returns a Boolean error value as a return value. The input pointer can be directly dereferenced to get the transferred data from the client in the structure. The output pointer must be created as a local, static variable. The static declaration is necessary, as it is returned to the server stub, so that the values must be available also after returning from the function, which would destroy locally scoped, non-static variables. The defined RPC functionality is then processed in the function body. It returns after this processing, which requires a suitable return value. The processing of a server function blocks all incoming requests, which are queued. Therefore, it is advisable to split remote procedures into small, less time-consuming, manageable functions. All of the specified server functions need to be manually written in this style. The server stub source in «`test_svc.c`», the procedure realization code in «`test_proc.c`», the header definition file, and the RPC runtime library «Glibc» then need to be compiled and linked together to create the server executable.

The created executables can then be started on different machines and should be able to communicate as long as the IP address is correct. The rest is handled automatically in the internal stubs. The method described uses the «portmapper» to bind the server to an RPC service. After starting the server, the registration at the «portmapper» can be checked on the command line by using the command «`pmap_dump`» (see the «man pages» of Linux for more information). It shows a list of all registered RPC servers with the program number, program version number, transport protocol, port, and an optional alias for the service (the first server in the list is usually the «portmapper» itself with two entries, one of which is for TCP and the other for UDP). The RPC server aliases are statically defined in the Linux operating system file «`/etc/rpc`», where alias names are assigned to program numbers. The information from «`pmap_dump`» can also be used to restart all servers, when the output is dumped to a file, which is then piped as input to the command «`pmap_set`» (see the «main pages» of Linux for more information).

The method described so far always uses the «portmapper» to request the binding information of a server. This is not ideal in certain circumstances, as the port 111[31] of the «portmapper» needs to be accessible on the network. But as described, Sun RPC does not offer any security. This means there is a potential risk of network attacks (McClure et al. 2006, *l.c. page 171 ff.*). As the port of the «portmapper» is only expected

Running Server and Client

Direct Access to the Server Without a «Portmapper»

---

31 Or also the alternative port 32,771 on Sun Solaris systems (McClure et al. 2006, *l.c. page 171*).

at this known reserved port, it is even difficult to tunnel it over a secure SSH connection (for further information, see ▶ Sect. 5.5.2), because well-known ports cannot be forwarded without administrator rights. Experience has shown that it is often quite useful to skirt the «portmapper» and contact the server directly. This can be arranged using an independent, classically created socket using the system call «socket» (see ▶ Sect. 3.3.1) for UDP or TCP, the Sun RPC library functions «clntudp_create» for the opening of a UDP transport, or «clnttcp_create» for the TCP transport. Both functions adopt the respective sockets, define the program and version number of the server service, and set additional socket attributes. Further socket attributes can be set like described for common communication socket[32] (see (Bloomer 1991, *l.c. page 371*) for more information). Therefore, no additional request is necessary any longer to connect directly to a server. This is quite useful, as the individual server ports can be tunneled without problems in order to offer a direct, secure connection. The disadvantage is that port changes can no longer be derived via a centralized binding database.

Distributed System

In summary, RPC offers a common infrastructure for IPC over different systems. It is a basic infrastructure which makes it possible to create distributed applications using prefabricated elements which can be parametrized. The support of a communication platform for distributed tasks is in the age of grid computing[33] and cloud computing,[34] an essential part of a sophisticated implementation (in allusion to (Brookshear 2012, *l.c. page 148 f.*)). Almost all modern computing solutions, for example, for controlling of complete production halls or single systems with several, parallel tasks, are based on such distributed systems.

Distributed System

> *Definition DEF3.9. of «distributed system»* (see (Singhal and Shivaratri 1994, *l.c. page 71*))
> A distributed system consists of several computers with their own operating systems which are connected over a communication network to solve a specific task together. The individual computers share no memory and have no common clock. They communicate with each other using message exchange over the communication network, preferably with a standardized communication platform.

Simplify the Development of Distribute Systems

The point of using RPC is to simplify the development of distributed systems (Stevens 1990, *l.c. page 724*). Even though the implementation of the Sun RPCs already supports a lot of the simplification mechanisms to define individual, parametrized communication data representations and protocols, it is still a little bulky in the handling. A look at the protocol specification file of Example 3.9 gives a feeling about the extensive way to define remote procedures using large structures. This does not really fit into OOP environments and offers less safety and protection from security attacks. Implicit changes of interface versions cannot be managed. Only a reduced mechanism to handle heap elements is available (multidimensional arrays, such as those often used for database tables, are not supported and have to be converted into one-dimensional arrays). The RPC communication platform also does not take care of parallel workflows in the server task and does not deal with server crashes, for example, by using a watchdog to detect them. Nevertheless, the basic ideas to abstract the communication implementation are very good. This means that Sun RPC is a good starting point for specific extensions which offer various solutions for the problems described. These extensions can be added as additional modules and components. It completes the RPC concept with implementations which are quite helpful for distributed systems. But as there is a further necessity to parametrize the prefabricated construction blocks to

---

32 But here the sockets have not to be bound or connected as in common socket programs.
33 Grids are more loosely coupled networks of distributed computers which solve a specific task such as large calculations together.
34 Clouds are shared computers of a network, where tasks and data are individually distributed to the single, participating computers.

individual communication issues, the extension of the generative method seems to be much more promising than just adding additional modules.

3.3.3 Extending Generative Programming for Interprocess Communication

The previously described use of «rpcgen» and the Sun RPCs are part of application engineering. It uses predefined realizations and models in the generator and the RPC runtime library. These parts were developed as domain elements by Sun Microsystems. The generator or, in the present case, the Sun RPC protocol compiler «rpcgen», together with the protocol specification and some command line parameters, are used to parametrize the created source to a specific use case. These parameters are the characteristics that distinguish between the different application fields. In this field, the application types are focused on IPC. As discussed, Sun RPC offers all the basic techniques for the generative programming of IPC. But the bulky handling, lack of safety and security solutions, and the integration into OOP environments require a more intuitive approach to map the application requirements to the implementations of remote procedures. This means, for the following discussion, that it is not only necessary to take a look at the field of application engineering with an existing generator. It is necessary to change parts of the domain itself and to work in domain engineering which changes or develops new generators. This also requires an extension or change of the existing parametrization possibilities as a consequence. As mainly the interface specification is used for such parametrization, an improved DSL and additional generator parameters must be designed. Since the basics of «rpcgen» are quite promising, it can be used as the basis of a generative process, which needs to be extended with an additional generative layer. For this, it is essential to also take a look behind the generative process and its theory (following the notations from (Klar and Klar 2006, *l.c. page 3 ff.*).

Domain Versus Application Engineering

An Origin Domain Specification Language

As already mentioned, it is necessary to extend the possibilities of the DSL for new generator abilities. As the Sun RPC protocol language is not changeable without changes in the Sun «rpcgen», it is better to add an additional layer around it, using a new interface language for the generation of new, sophisticated source code elements. Well-known and already existing elements from the Sun RPC protocol language can be reused and need only be converted within this process. Because of this, the well-tested and widely distributed Sun RPC with all of its functionalities can be used further on, while new layers offer the required additions. The result is a generator system, which means a set of different generators working together to create the final desired result. But this is not possible without having a knowledge of formal languages and the theory behind them, because it is necessary to interpret languages and configurations for a higher parametrization.

DSL for a Generator System

Defining a new language means (for the following theoretical blocks, see (Schöning 1997, *l.c. page 11–17*), (Savage 1998, *l.c. page 153–208*), and (Brookshear 2012, *l.c. page 268–274*)) to define a finite set of syntax rules (usually described as «syntax»), which describe what combinations of symbols from a suitable language alphabet are allowed, in order to build the «sentences» of the new language. These «sentences» are expressions entered as character strings, which must be interpreted. The interpretation must check for lexical and syntactical correctness and is used to create the semantic relationship (or semantics, which is the meaning of the interpreted expression elements) for the following code generation. The language alphabet here is a set of lexical symbols, such as the definition of a remote function declaration. Each of these expressions (or «sentences») of the language must follow the defined syntax rules of allowed combinations exactly. Each expression is described as a sequence of alphabet symbols. Undefined expressions are not allowed, and therefore not understood by the language interpreter. ◘ Figure 3.12 demonstrates this general style for a typical function declaration (a sample

Alphabet, Syntax, Semantics

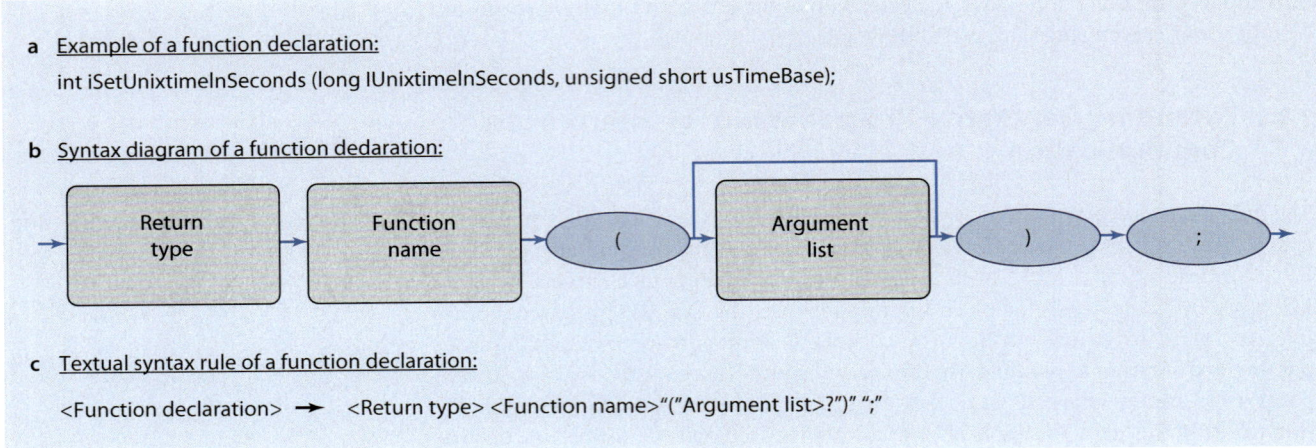

a Example of a function declaration:

int iSetUnixtimeInSeconds (long lUnixtimeInSeconds, unsigned short usTimeBase);

b Syntax diagram of a function dedaration:

c Textual syntax rule of a function declaration:

<Function declaration> ➔ <Return type> <Function name>"("Argument list>?")" ";"

Fig. 3.12   A possible syntax diagram for the declaration of a function head (e.g., in **a**) in the form of a syntax diagram or syntax graph in **b** and as a textual syntax rule in **c**

function, following this setup, can be found in Example 3.9). The function declaration is split into alphabet symbols or reserved words of the language (enclosed in ellipses) and terms that require a further description with another syntax rule (enclosed in rectangles). The alphabet symbols and reserved words are taken from the language alphabet and therefore form the terminals (keywords), as they no longer need to be replaced. The terms which are described with further rules are appointed as nonterminals. The sequence or concatenation of such symbols can be drawn as a syntax diagram (see the upper part of ▪ Fig. 3.12). But it is not only a real sequence: it can also define parallel alternatives of sequences, for example, for the «Argument list» of the function, which can be a real list of arguments (main sequence line) or which can be left empty (a parallel arrow from the opening bracket to the closing bracket). Besides the graphical syntax diagram, the syntax rules can also use a compacter description with a textual notation (see the lower part of ▪ Fig. 3.12).

Textual Syntax Notation

The textual description uses a notation, which is similar to the Backus – Naur – Form (BNF) (see (Aho et al. 1997, *l.c. page 32 ff.*)), but with some extensions, inspired by the Perl regular expressions (see (Brown 2001, *l.c. page 242 ff.*)). The extensions offer more compactness and comprehensibility of the notation, but can be directly replaced by syntax rules just using concatenations and alternatives. The notation uses the following elements:

```
<...>          := Nonterminal symbol
→              := Symbol  before  this  arrow  is  defined  as
                  sequence after the arrow
«..».          := Terminal symbol
blank          := Concatenation between symbols (sequence)
|              := Alternative (equal to OR)
(...)          := Group of symbols
(...){n:m}     := Repetitive sequence with a minimum of n and
                  a maximum of m repetitions
[...]          := Set  or  range  of  single  symbol  characters
                  (e.g., [a-zA-Z] for all letters)
?              := Previous symbol is optional (e.g., <SYMBOL>?
                  or also (<SYMBOL>)? and equal to (<SYMBOL>)
                  {0:1})
+              := Previous symbol repeats at least once (e.g.,
                  <SYMBOL>+  or  also  (<SYMBOL>)+,  which  is
                  equal to <S> → <SYMBOL> | <SYMBOL> <S>)
*              := Previous symbol repeats zero, one or more times
                  (e.g., <SYMBOL>* or also ((<SYMBOL>)+)?, which
                  is equal to <S> → (<SYMBOL> | <SYMBOL> <S>)?)
```

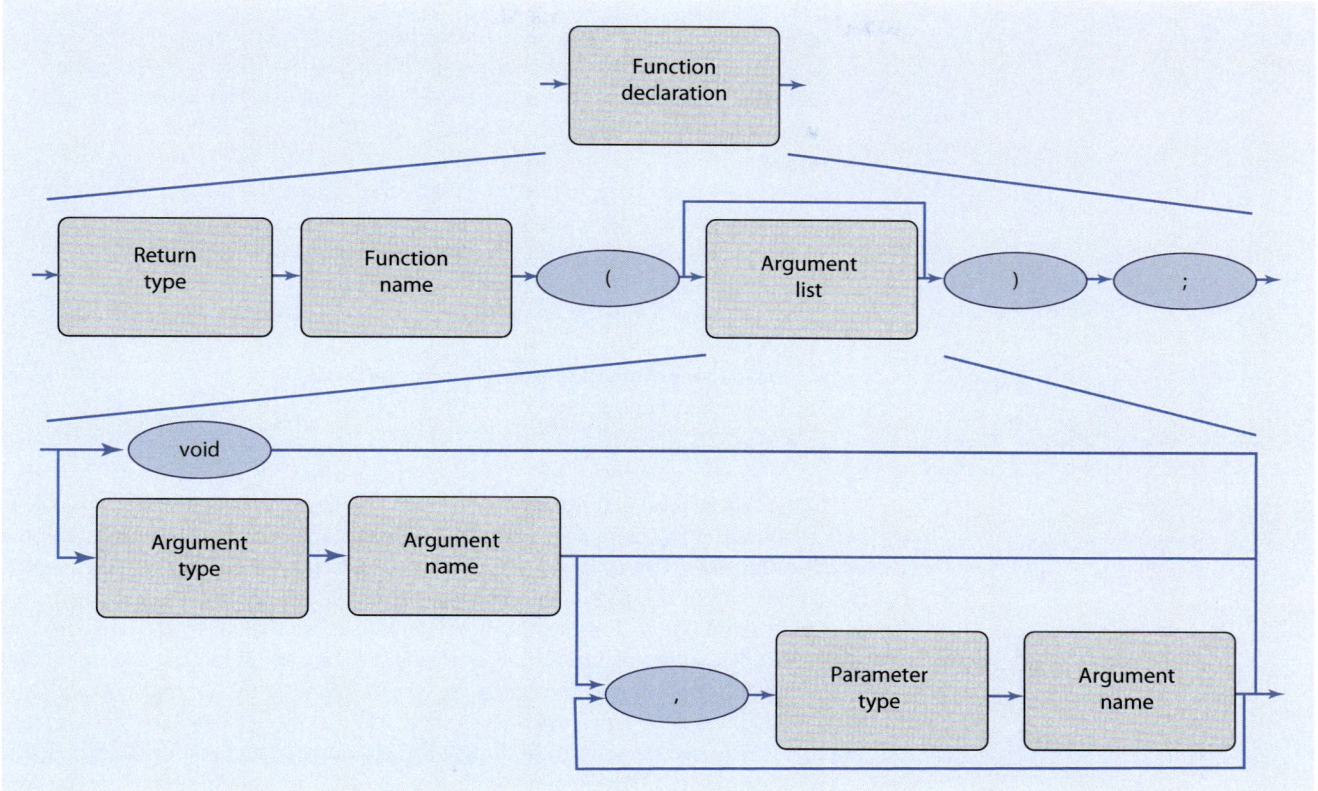

□ **Fig. 3.13**   A part of the syntax tree replacement chain for the description of function declaration

Having such a syntax description, a language interpreter reads input symbols and follows the syntax rules to check the validity of the combination entered. Following the replacements and replacing all nonterminals then unfold a syntax tree (see □ Fig. 3.13). The syntax tree describes all decomposition steps (or interpretation paths) to each final expression, starting with a first nonterminal (which is «Function declaration» in □ Fig. 3.13). Each new layer shows how the nonterminals of the previous level are decomposed, while the sequence with alphabet symbols of an entered expression is obtained step by step. The syntax tree represents the generator's interpretation of the input language. For IPC generation, the input language is the DSL. For an unambiguous interpretation of DSL, two different paths through the syntax tree for one input expression of the language are not allowed. To reach such a distinct tree of syntax rules, it is necessary to use a particular scoping of sequences with parentheses, which combines symbols and expressions into one interpretation block.[35]

The complete description of a language based on an alphabet of symbols and defined with a set of syntax rules, replacing nonterminal symbols with a sequence of terminal and nonterminal symbols, is a formal language grammar. The grammar defines the complete language. Each assumed statement in the language can be checked for syntactical correctness using the defined grammar. Starting with one prime syntax rule, each expression of the possibly infinite set of expressions can be combined by the finite set of syntax rules.

Syntax Tree

---

35 An example can be given for if/then/else instructions and is known as the «dangling else» problem, where it must be clearly defined by the syntax rules which branch of a nested conditional sequence belongs to which condition decision. Usually this problem is solved with the agreement to interpret expressions always from left to right and to combine expressions into a block, using parentheses.

3

Grammar

> *Definition DEF3.10. of «grammar»* (see (Schöning 1997, *l.c. page 13 f.*))
> A formal (mathematically describable) grammar (see also (Savage 1998, *l.c. page 181*)) describes a language mathematically using a tuple of five elements which are used for the structured decomposition of expressions of the defined language:
> 1. A finite set of nonterminals (or variables, which are the placeholders for a further syntax rule to process a decomposition)
> 2. A finite set of terminals (or symbols of the language alphabet)
> 3. A finite set of syntax rules (or production rules) which describe how a nonterminal symbol is replaced by a sequence of other nonterminals and terminals
> 4. A start symbol out of the set of the nonterminals
> 5. A finite set of end states

Chomsky Hierarchy

Using this general definition of a (formal) grammar, it is additionally possible to classify the resulting languages according to their complexity. Still the most famous classification is the Chomsky hierarchy, defined by Noam Chomsky. He took a look at the complexity of the production rules of formal languages and divided them into four types from 0 to 3. The classification is organized so that all type-3 languages are automatically a subset of type 2, all type-2 languages are automatically a subset of type 1, and so on, which forms the hierarchy (see ◘ Fig. 3.15, (Schöning 1997, *l.c. page 17 ff.*) and (Savage 1998, *l.c. page 182 ff.*)). The complexity also defines which processing tool is required for syntax checking, so that a given expression can be verified to see if it belongs to a language or not. Therefore, each grammar system can be converted into an individual processing machine. The differences between the language classes are:

Regular Languages

1. *Type-3 languages or «regular languages»*: All languages which can be produced with syntax rules, where a nonterminal is only replaced by a terminal or sequence of one terminal and a nonterminal, are regular languages (e.g., «A→b» or «A→bA»[36]). A typical representative is the regular expressions (see the description in ► Sect. 3.3.1). To process expressions of these languages, a finite state machine is necessary (see ◘ Fig. 3.14 and also ► Sect. 3.3.1, where it has already been explained for the interpretation of communication messages) ((Schöning 1997, *l.c. page 27 ff.*) and (Savage 1998, *l.c. page 158 ff.*)). A finite state machine reads input signs (e.g., from an «input tape») and codes the possible combinations with defined states. Each read word or sign which is allowed enables a transition from one state to another, using a defined edge between both states. The state machine has a starting state, several paths through the chart of all states, and the processing ends at clearly defined final states. Each time the processing ends at such a final state, the read input combination is accepted. All other situations reject the read combination. This forms a complete state graph with all paths (also compare ◘ Fig. 3.12).

Context-Free Languages

2. *Type-2 languages or «context-free languages»*: All languages which can be produced with syntax rules where a nonterminal is replaced by a sequence of terminals and nonterminals in different orders are context-free languages (e.g., «A→BacDd»). Context-free means that the left nonterminal of a syntax rule can be replaced directly by the right part of the rule without attention being paid to the context in which it is embedded. This type of language makes it possible to count an arbitrary number of symbols, such as opening parentheses to add an adequate number of closing parentheses. A typical representative is the

---

36 Uppercase characters are nonterminals here. Lowercase characters are terminals.

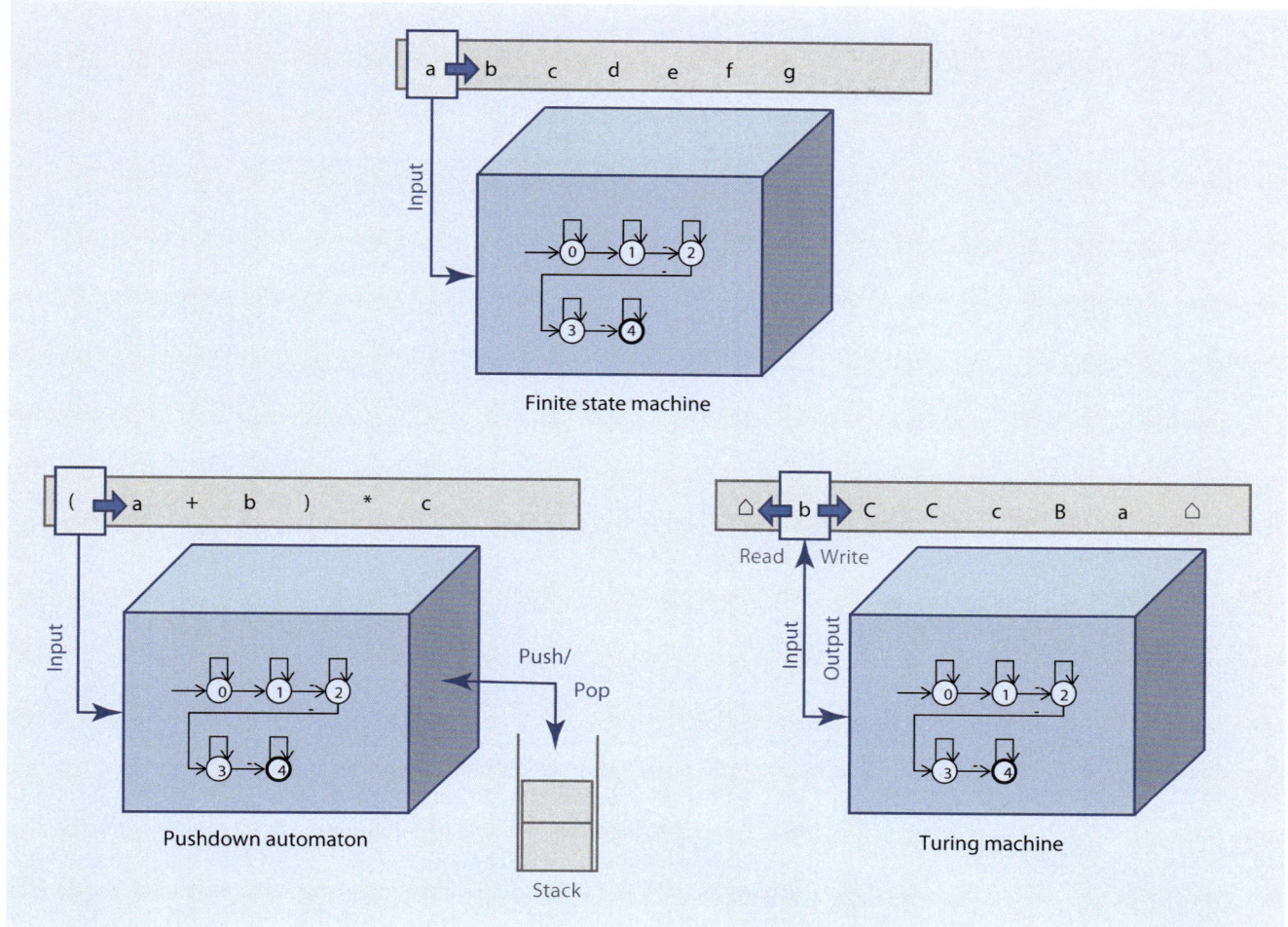

◘ Fig. 3.14 The different types of machines (automata) to interpret the different languages

mathematical expressions (as in the (E)xpression-(T)erm-(F)actor grammar for algebraic expressions (Brookshear 2012, *l.c. page 270 f.*)). The same type is necessary for nested constructs of modern programming languages. To process expressions of these languages, a nondeterministic pushdown automaton (stack machine, see ◘ Fig. 3.14) is necessary ((Schöning 1997, *l.c. page 51 ff.*) and (Savage 1998, *l.c. page 177 ff.*)). A pushdown automaton (stack machine) is a state machine, which additionally uses a primitive stack memory of the type last-in, first-out. The memory is similar to a stack of paper, where it is only permissible to put new sheets on the top and to take them again from the top. Each time an input sign is read which should be counted, an element is pushed to the stack. Each time the counterpart of the previous sign is read, the element is taken out again from the stack. Therefore, an input combination read is accepted if a defined final state is reached and the stack is empty again. All other situations reject the read combination.

3. *Type-1 languages or «context-sensitive languages»*: All languages which can be produced with syntax rules where a sequence of nonterminals and terminals is replaced by a sequence of nonterminals and terminals with the same number of or more elements on the right side of the rule than on the left are context-sensitive languages (e.g., «aCb→acb» or «aCb→aCBb»). It means that now the context around the nonterminals on the left side of a syntax rule decides if the rule can be used for a decomposition. To process expressions of

Context-Sensitive Languages

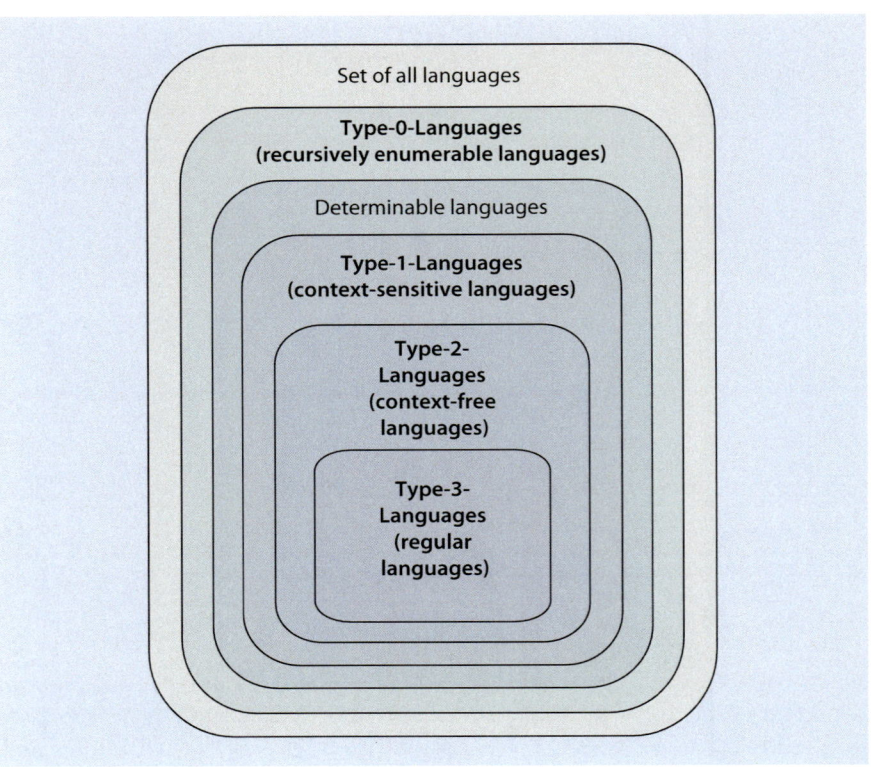

Fig. 3.15 The Chomsky hierarchy of formal languages

these languages, a Turing machine is required (see ▣ Fig. 3.14), which has a linear bounded input tape ((Schöning 1997, *l.c. page 79 ff.*) and (Savage 1998, *l.c. page 183 ff.*)). A Turing machine is a mathematical construct of a state machine with a memory. The construct was defined by Alan Turing. The memory is implemented as a tape used for input and output. The Turing machine uses an input alphabet and an internal working alphabet. A read/write pointer can be shifted on the tape by one element to read symbols from and write symbols on the tape. As Turing machines are general computing concepts in theoretical computer science, they can be used to show the computational completeness of a language. This means it is able to represent all instruction combinations like loops and conditions which are required by a computer with a Turing machine (see also information about the Turing completeness in ▶ Sect. 3.1). A Turing machine produces output from a read input, and the run is completed if a defined final state is reached and the converted output is completely written on the tape.

Recursively Enumerable Languages

4. *Type-0 languages or «recursively enumerable languages»*: All languages which can be produced with syntax rules where a sequence of nonterminals and terminals is replaced by a sequence of nonterminals and terminals of an arbitrary number and order of (also different) elements are recursively enumerable languages (e.g., «aCb→aB»). The verification of whether an expression belongs to the language is not determinable anymore. But the fact that the processing does not end within a deterministic time gives no information about whether the expression is wrong. Only unbounded Turing machines can be used for those languages ((Schöning 1997, *l.c. page 79 ff.*) and (Savage 1998, *l.c. page 220 ff.*)). As this type of language is not relevant for the following discussion, it is not necessary to take a closer look at it.

Interface Definition Language

With this theoretical knowledge, it is now possible to define an individual, new language: a specific DSL. The extended RPC generator can then use this new DSL in the form of a specific Interface Definition Language (IDL), which is a mixture of higher programming language elements (comparable to C++ and other languages) and the basic «.x» definitions, used by «rpcgen». The language does not use nested constructs to avoid too much complexity. Therefore, the main structure can be decomposed with an elementary regular grammar. In addition to the standard variable types, dynamic, multidimensional heap arrays should be possible by using pointer arguments in the argument list of functions (to reduce the complexity, no static arrays are allowed, so that every array must be converted to reside in the dynamic heap memory). An array argument then consists of a pointer to the array (e.g., «pArray_val») and one length or size argument per dimension of the array. The automatic generation of unique argument names for the dimensions can be reached by counting the dimension, which is then concatenated to the identifier of the array, e.g., «pArray_len0» for the first dimension, «pArray_len1» for the second, and so on. To enable this counting, the state machine must consequently count the number of dimensions and must also create internal copying loops as nested program elements. This requires a context-free language. The syntax checks and conversions of such a language can be realized as a program, which is written with a high-level language (e.g., like Perl), already supporting regular expressions or counter variables. As all usual programming languages are Turing complete, the algorithm for the syntax checks in such a program can be translated into a state machine, which «eo ipso» manipulates symbols and a memory according to predefined replacement rules (see also ▶ Sect. 3.1). This means by using the syntax rules of a high-level language, it becomes possible to implement the checks of syntax rules of another, different language like the new IDL.

The separation of the IDL symbols can be done with an optional number of whitespaces. But at least one whitespace should be used. The symbols themselves can be arranged individually, so that no fixed format is required. Therefore, the new IDL is a format-free language. The application developer can organize the written RPC definitions in the IDL in a way that enhances readability by humans with individual indentation (according to (Brookshear 2012, *l.c. page 268 f.*)). The readability and comprehensibility are also supported by an ability to write comments directly into the IDL definitions. Comments in this context begin with «//» and end at the end of the line, in which they begin (similar to C++).

Free-Format Language

A new IDL for the extended RPC functionalities can be defined by the following grammar syntax rules where the rule which decomposes the nonterminal symbol «IDL» is the start point (nonterminals are completely written in upper case, while terminals are in double quotes; the rest follows the notation of regular expressions under Perl, but defines a context-free language (see (Brown 2001, *l.c. page 242 ff.*)), and should be very intuitive):

IDL Grammar

```
<IDL> → <CONSTANTDEFINITION>*<TYPEDEFINITION>*<INTERFACEDEFINITION>

<CONSTANTDEFINITION> → "const" <CONSTANTIDENTIFIER> "=" <CONSTANT>
     ";"
<CONSTANT> → <CONSTANTIDENTIFIER> | <STRING> | <NUMBER>
<CONSTANTIDENTIFIER> → <IDENTIFIER>

<TYPEDEFINITION> → "struct {"<TYPEVARIABLE>+"}" <TYPEIDENTIFIER>
     ";"
<TYPEVARIABLE> → (<TYPE> | <TYPEIDENTIFIER>) <VARIABLEIDENTIFIER>
     ";"
<TYPEIDENTIFIER> → <IDENTIFIER>
<VARIABLEIDENTIFIER> → <IDENTIFIER>

<INTERFACEDEFINITION> → "interface" <INTERFACEIDENTIFIER>
     (<VERSIONSTRING>)? "{"<INTERFACEMETHOD>+"};"
<INTERFACEMETHOD> → ("void" | <TYPE> | <TYPEIDENTIFIER>)
     <METHODIDENTIFIER> ("!ASD")? "("<PARAMETERLIST>");"
<PARAMETERLIST> → <PARAMETER> (","<PARAMETER>)*
<PARAMETER> → ("in" | "out" | "inout") ("string" | <TYPE> |
     <TYPEIDENTIFIER>) <PARAMETERIDENTIFIER> "<>"*
```

3

```
<INTERFACEIDENTIFIER> → <IDENTIFIER>
<PARAMETERIDENTIFIER> → <IDENTIFIER>
<METHODIDENTIFIER> → <IDENTIFIER>
<IDENTIFIER> → [a-zA-Z_][a-zA-Z0-9_]*
<STRING> → \".*\"
<VERSIONSTRING> → #[a-zA-Z0-9_]+
<NUMBER> → "0x"[0-9]* | [0-9]* | [1-9][0-9]*³⁷
<TYPE> → "short" | "unsigned short" |
        "int" | "unsigned int" |
        "long" | "unsigned long" |
        "longlong" |
        "float" |
        "double" |
        "char"
```

IDL File

Generally, the programming of interfaces using the new IDL is similar to the already familiar use of the DSL in the «.x» files of Sun RPC. The interface in the form of the new IDL must be manually written to an IDL file just as with the DSL. It consists of two optional parts and one mandatory part, which must be arranged sequentially, as defined in the grammar. Thus, an IDL file usually has the following three parts:

1. Constants definitions
2. Type definitions
3. An interface definition

IDL Constant Definitions

Constants are in this context optionally selectable values, which are accessible in the server modules as well as in the client modules. The generator uses these definitions to produce constants definitions in C style with «#define» statements, which are available in the client as well as in the server. These definitions can be used directly within the application code.

IDL Type Definitions

Type definitions allow combined structure types as in C. But only the standard, rudimentary data types can be used here. It means that no dynamic pointer elements or arrays are allowed. This constraint is necessary to avoid nested, hierarchical constructions which would require a recursive managing of all related value elements. Without such hierarchical elements, IDL types defined can be directly mapped into the resulting code.

IDL Interface Definition

Last but not least, all of the interface communication methods which are used for remote access are defined in the *interface definition*. It uses a very compact design. The interface is comparable to a class in C++ which contains all declarations of the interface methods (function heads of the remote procedures). Only one interface definition is allowed per IDL file, so that no inheritance strategy is realized in the IDL. An interface method is like a conventional method, familiar from C++. The difference is that each argument must define its transport direction, using a prefix symbol to the type («in» means a transport from the client to the server as part of a request message, «out» means a transport from the server to the client to get a remote value as part of the reply message, and «inout» means to send an argument value to the server and receive the changed value from the server). The dimension of a dynamic array of any type is defined by adding «<>» at the end of the argument definition (like in «rpcgen» specifications). For each dimension, an additional «<>» must be added. The generator then generates additional, individual length arguments for each dimension. If the multidimensional array uses strings, the string sizes can vary in length at each position of the array. No fixed length is compulsory. A sample file is shown in Example 3.12.

---

37   rpcgen does not allow any floating points here.

■■ **Example 3.12**

IDL - file, using the new language for the RPC generation

```
const MYCONST1 = 0;
const MYCONST2 = 0x3333;
const MYCONST3 = 12345;
const MYCONST4 = "Hallo";
const MYCONST6 = MYCONST1;

typedef struct {
    char cVar1;
    char cVar2;
    char cVar3;
    char cVar4;
} MYTYPE;

typedef struct {
    unsigned int iVar1;
    int iVar2;
    MYTYPE SVar3;
} MYTYPE_COMBINATION;

interface test #20081023_v2
{
    // This is a test comment
    void vFunc ();

    int iGet ();
    unsigned int uiGet ();

    void vSet (in int iVal);
    void vSetUnsigned (in unsigned int iVal);
    void vGet (out int iVal);
    void vSetAndGet (inout int iVal);

    void vSetString (in string strText);
    void vGetString (out string strText);
    void vSetAndGetString (inout string strText);

    void vSetMulti (in int iVal, in string strText);
    void vGetMulti (out int iVal, out string strText);
    void vSetAndGetMulti (inout int iVal, inout string strText);

    void vSetArray (in unsigned int Array <>);
    void vGetArray (out unsigned int Array <>);
    void vSetAndGetArray (inout unsigned int Array <>);

    void vSetMultiArray (in unsigned int Array <><><>);
    void vGetMultiArray (out unsigned int Array <><><>);
    void vSetAndGetMultiArray (inout unsigned int Array <><><>);

    void vSetStringArray (in string Array <>);
    void vGetStringArray (out string Array <>);
    void vSetAndGetStringArray (inout string Array <>);

    void vSetMultiStringArray (in string Array <><><>);
    void vGetMultiStringArray (out string Array <><><>);
    void vSetAndGetMultiStringArray (inout string Array <><><>);

    MYTYPE SGetType ();
    void vSetType (in MYTYPE Val);
    void vGetType (out MYTYPE Val);
    void vSetAndGetType (inout MYTYPE Val);
};
```

```
1
2
3
4
5
6
7
8
9
10
11
12
13
14
15
16
17
18
19
20
21
22
23
24
25
26
27
28
29
30
31
32
33
34
35
36
37
38
39
40
41
42
43
44
45
46
47
48
49
50
51
52
53
54
55
56
57
58
59
60
61
```

A comparison between Example 3.12 with the new DSL and the Sun RPC specification file in Example 3.9 directly demonstrates the more intuitive and object-like structure of the new IDL. It is still textual, but simplifies the design of new remote applications and hides the classic structures of Sun RPC. Besides the new, more sophisticated style, the new generator can also change and add special features to implement safety and security (see ▶ Sect. 3.3.3.4). The new IDL offers a fixed, separate DSL which must be implemented by a separate language translator (see also (Czarnecki and Eisenecker 2000, *l.c. page 138*)). It is a costly development, but as all parts are adaptable if specific generator knowledge is applied, it is also a valuable way to reduce costs later for all other projects in which it is included. Therefore, with this grammar definition, it is now necessary to start the domain development, writing the new generator which is able to understand, interpret, and act on the new language.

Domain Development of «idl2rpc.pl» as a Perl-Based IPC Generator

Generative Domain Model

The goal of this section about how to write one's own generator is to create a generative domain model for an active library which supports IPC. One part of the generative domain model is the configuration possibilities (parametrization), which is achieved primarily by the previously defined DSL, and secondarily also with parameters to control the generative process. Together with the implementation of required components and their combinations (solution space) and the application-oriented concepts and features (problem space), they offer a complete generative domain model (Czarnecki and Eisenecker 2000, *l.c. page 132*). It is implemented with a specifically written generator and new components which implement the interpretation and conversion of IDL definitions to usable IPC source code in the form of a code converter.[38]

Compiler Phases

Such a converter is something like a compiler. Therefore, it is appropriate to take a look at the technical principles of compiler engineering (for a general overview of the following blocks, see also (Aho et al. 1997, *l.c. page 1–29*). Each compiler usually has six phases, which can be grouped into a front end and a back end part (see ▯ Fig. 3.16). The front end combines the lexical analysis, the syntax analysis, the semantical analysis, and the generation of intermediate code. The back end processes the code optimization and the final, real code generation. In parallel with these phases, administration and storage for language symbols are necessary. Errors during the processing of input source code are handled with a global error processing. As the compilation requires pattern matching (see the following blocks), a suitable environment is important. The following development uses Perl and its sophisticated use of regular expressions for a suitable pattern matching.

Lexical Analysis

The first step of the compilation is lexical analysis. This analysis is done by the «scanner». It reads input signs from the source code file and creates a sequence of symbols which are then used by the syntax analysis. Both interact very closely with each other. As the scanner has to prepare a useful symbol stream, it also cleans the input character stream from whitespaces (blanks, tab stops, etc.) and from comments (e.g., comments can be started with «//» and continue to the end of the line), because these signs are not important for the syntax analysis. The scanner also saves the current line number, so that errors can be assigned to the correct position in the source text. The identification of symbols and its type is usually done with pattern matching, while the patterns are described with regular expressions. The lexical analysis uses the mechanisms

---

38 Besides the following approach to writing a generator, public code generators such as «yacc» or «bison» can be used, which interpret an inline form of a DSL in the code and are also supported by programming frameworks, such as Eclipse (Bauer 2011, *l.c. page 223 ff.*).

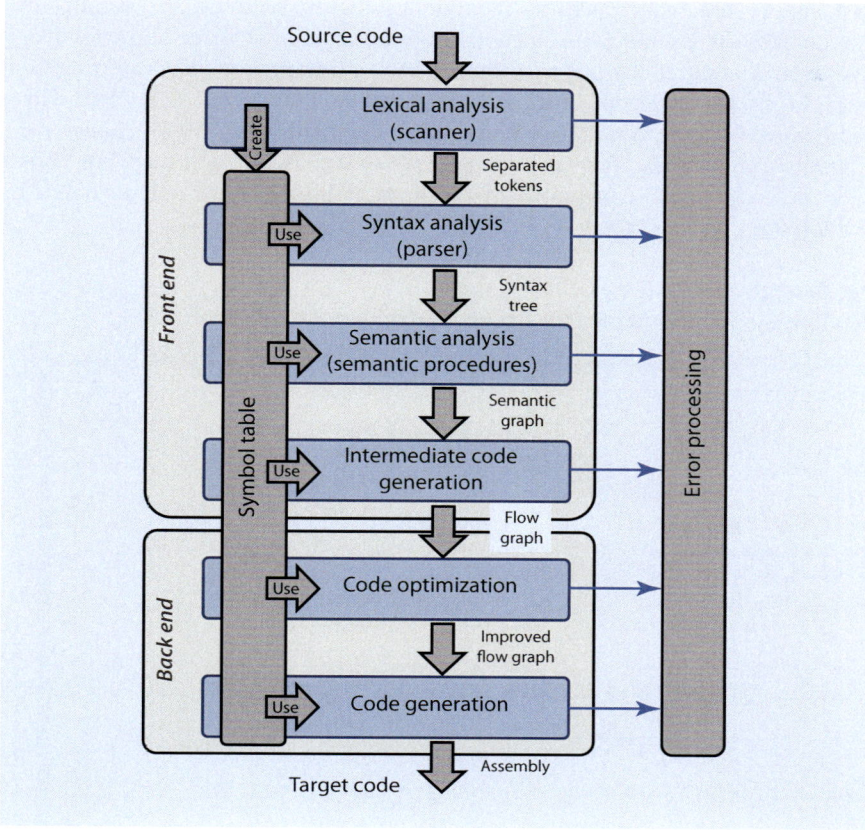

○ **Fig. 3.16** The phases of a compiler (Adapted from Güting and Erwig (1999 *l.c. page 4*) and Toal (n.d.); reproduced by permission of Ray Toal)

of regular languages (see ▶ Sect. 3.3.3) to find representations of symbol patterns, which are named «lexemes». Lexemes can be handled as one unit (also called «tokens») during the complete analysis and are the terminals of a language syntax (e.g., keywords, operators, constants, literals, or punctuations). Therefore, the scanner must, for example, decide which type of symbol was found. This additional information is collected in the symbol attributes on which the parser can run further decisions.

At the scanning phase, it is not always possible to detect errors, as the scanner has no overview of the complete syntax. Therefore, a scanner cannot decide if a text pattern is an identifier or maybe a misspelled reserved word, as this is only derivable in the context of the syntax rules. But the scanner can detect situations where no symbol pattern matches. In this case, it throws a lexical error (Aho et al. 1997, *l.c. page 101 ff.*).

Lexical Error Handling

A possible implementation of a scanner (written in Perl) for the symbols of the previously defined IDL is shown in Example 3.13.[39] As Perl directly offers regular expressions as language constructs which realize the textual rules of the regular syntax (see ○ Fig. 3.12), they can be used for pattern matching in the scanner.[40] Otherwise, it would be necessary to map the syntax rules of the state graph to conditional statements, using «if/then/else» or «switch/case» blocks. As the new

Scanner Example

39　Usually the program part for the syntax analysis starts the scanner each time a new symbol is required for processing. The example solution solves it in a different way, as the scanner reads the file and passes the tokens found to the parser subroutine. This is a more straightforward handling for the simple IDL.

40　The «if/then/else» constructs in Example 3.13 are part of the higher-level syntax checks of the context-free grammar in syntax analysis.

IDL has to use whitespaces as separators between symbols, the scanner can directly use the lexeme found without any additional knowledge of the following symbols. Therefore, it is not necessary to look ahead for a correct interpretation using a «lookahead» operator, as is required in FORTRAN compilers where whitespaces do not always have any significance (Aho et al. 1997, *l.c. page 135*). The language is a free-format language (see ▶ Sect. 3.3.3), as it does not matter how the symbols are formatted in the lines as long as they are separated by whitespaces.

■■ **Example 3.13**

Realization of a scanner for the defined IDL

```
# >————————————————————————————————————————————————————<     1
# > Subroutine:  Scanner                                 <     2
# >              Prepares input: delete white spaces, delete '\n' etc. and   <     3
# >              hand tokens over to parser              <     4
# > Parameter:   $IDLFilepath —> Filepath to IDL—file    <     5
# > Return:      <— Errorinfo: Error (0=ok, 1=error)     <     6
# >                            Error message             <     7
# >                            Line number               <     8
# >                            Line content              <     9
# > Author:      A. Neidhardt                            <    10
# > Date:        02.02.2007                              <    11
# > Revision:    —                                       <    12
# > Info:        —                                       <    13
# >————————————————————————————————————————————————————<    14
sub Scanner {                                                 15
    my ($IDLFilepath, $ProgramNumberPrefix, $MultiThreading, $Authenticate)=@_;   16
    my $Line = "";          # Read line                       17
    my $FileContent = "";   # Complete file content           18
    my $OrigLine = "";      # Originally read line             19
    my $Error = 0;          # Errorcode                        20
    my $ErrorMsg = "OK";    # Errorcode                        21
    my $Count = 0;          # Linecounter                      22
    my %ParserState;        # State                            23
    my $Token;              # Token content                    24
    my $TokenType;          # Identified token type            25
    my $LineRest;           # Rest of the scanned line         26
    my @NewTypes;           # With typedef defined new types   27
    my $NewType;            # Iterator for types               28
    my @NewConsts;          # With const defined new constants 29
    my $NewConst;           # Iterator for constants           30
    my @DefinedIdentifiers; # Identifiers of current namespace 31
    my $CurrentNamespace;   # Current namespace of identifiers 32
    my $NumberOfNamespaceFunctions; # Current number of functions at namespace   33
    my @SplitLine;          # Helping element to split line into command and comment   34
                                                              35
    # Check if IDL—file has extension ".idl"                  36
    if (!($IDLFilepath =~ /.*\.idl$/)) {                      37
        return (1,"File ".$IDLFilepath." has wrong extension!",0,"");   38
    }                                                          39
                                                              40
    # Calculate MD5—hash for version ID                       41
    open (FileHandle,"< $IDLFilepath") or return (1,"Can't open file ".$IDLFilepath."!",0,"");   42
    while (defined ($Line = <FileHandle>)) {                  43
      $FileContent = $FileContent.$Line;                      44
    }                                                          45
    close (FileHandle);                                       46
    $g_IDLFileMD5 = md5_hex ($FileContent);                   47
                                                              48
```

```perl
# Open file (old style open, so that older perl versions can handle it)
open (FileHandle ,"< $IDLFilepath") or return (1,"Can't open file ".$IDLFilepath."!",0,"");

# Read lines until needed information is completed
while (defined ($Line = <FileHandle>)) {
    $Count = $Count + 1;
    $OrigLine = $Line;
    # Replace white spaces
    $Line =~ s/\s+/ /g;
    $Line =~ s/^\s+//g;
    $Line =~ s/\s+$//g;
    # Delete comments starting with //
    if ($Line =~ /^.*\/+\/.*$/) {
        @SplitLine = split(/\//, $Line);
        $Line = $SplitLine[0];
        if (!defined $Line) {
            $Line = "";
        }
    }

    while (length($Line) > 0) {
        # Qualifier
        if ($Line =~ /^interface(\W.*)?$/) {
            $Token = 'interface';
            $TokenType = 'INTERFACE';
            ($LineRest) = ($Line =~ /^interface(.*)/);
        }
        elsif ($Line =~ /^const(\W.*)?$/) {
            $Token = 'const';
            $TokenType = 'NEWCONST';
            ($LineRest) = ($Line =~ /^const(.*)/);
        }
        elsif ($Line =~ /^typedef(\W.*)?$/) {
            $Token = 'typedef';
            $TokenType = 'NEWTYPE';
            ($LineRest) = ($Line =~ /^typedef(.*)/);
        }
        elsif ($Line =~ /^struct(\W.*)?$/) {
            $Token = 'struct';
            $TokenType = 'STRUCT';
            ($LineRest) = ($Line =~ /^struct(.*)/);
        }
        elsif ($Line =~ /^in(\W.*)?$/) {
            $Token = 'in';
            $TokenType = 'IO';
            ($LineRest) = ($Line =~ /^in(.*)/);
        }
        elsif ($Line =~ /^out(\W.*)?$/) {
            $Token = 'out';
            $TokenType = 'IO';
            ($LineRest) = ($Line =~ /^out(.*)/);
        }
        elsif ($Line =~ /^inout(\W.*)?$/) {
            $Token = 'inout';
            $TokenType = 'IO';
            ($LineRest) = ($Line =~ /^inout(.*)/);
        }
        elsif ($Line =~ /^short(\W.*)?$/) {
            $Token = 'short';
            $TokenType = 'TYPE';
            ($LineRest) = ($Line =~ /^short(.*)/);
        }
```

49
50
51
52
53
54
55
56
57
58
59
60
61
62
63
64
65
66
67
68
69
70
71
72
73
74
75
76
77
78
79
80
81
82
83
84
85
86
87
88
89
90
91
92
93
94
95
96
97
98
99
100
101
102
103
104
105
106
107
108
109
110

```perl
      elsif ($Line =~ /^int(\W.*)?$/) {                   111
          $Token = 'int';                                 112
          $TokenType = 'TYPE';                            113
          ($LineRest) = ($Line =~ /^int(.*)/);            114
      }                                                   115
      elsif ($Line =~ /^long(\W.*)?$/) {                  116
          $Token = 'IDL2RPCLongType';                     117
          $TokenType = 'TYPELONG';                        118
          ($LineRest) = ($Line =~ /^long(.*)/);           119
      }                                                   120
      elsif ($Line =~ /^longlong(\W.*)?$/) {              121
          $Token = 'long long';                           122
          $TokenType = 'TYPE';                            123
          ($LineRest) = ($Line =~ /^longlong(.*)/);       124
      }                                                   125
      elsif ($Line =~ /^float(\W.*)?$/) {                 126
          $Token = 'float';                               127
          $TokenType = 'TYPE';                            128
          ($LineRest) = ($Line =~ /^float(.*)/);          129
      }                                                   130
      elsif ($Line =~ /^double(\W.*)?$/) {                131
          $Token = 'double';                              132
          $TokenType = 'TYPE';                            133
          ($LineRest) = ($Line =~ /^double(.*)/);         134
      }                                                   135
      elsif ($Line =~ /^char(\W.*)?$/) {                  136
          $Token = 'char';                                137
          $TokenType = 'TYPE';                            138
          ($LineRest) = ($Line =~ /^char(.*)/);           139
      }                                                   140
      elsif ($Line =~ /^string(\W.*)?$/) {                141
          $Token = 'string';                              142
          $TokenType = 'TYPESTRING';                      143
          ($LineRest) = ($Line =~ /^string(.*)/);         144
      }                                                   145
      elsif ($Line =~ /^void(\W.*)?$/) {                  146
          $Token = 'void';                                147
          $TokenType = 'VOID';                            148
          ($LineRest) = ($Line =~ /^void(.*)/);           149
      }                                                   150
      elsif ($Line =~ /^unsigned(\W.*)?$/) {              151
          $Token = 'unsigned';                            152
          $TokenType = 'UNSIGNED';                        153
          ($LineRest) = ($Line =~ /^unsigned(.*)/);       154
      }                                                   155
  # Qualifying character                                  156
      elsif ($Line =~ /^{/) {                             157
          $Token = '{';                                   158
          $TokenType = 'ACCOLADEOPEN';                    159
          ($LineRest) = ($Line =~ /^{(.*)/);              160
      }                                                   161
      elsif ($Line =~ /^}/) {                             162
          $Token = '}';                                   163
          $TokenType = 'ACCOLADECLOSE';                   164
          ($LineRest) = ($Line =~ /^}(.*)/);              165
      }                                                   166
      elsif ($Line =~ /^\(/) {                            167
          $Token = '(';                                   168
          $TokenType = 'BRACKETOPEN';                     169
          ($LineRest) = ($Line =~ /^\((.*)/);             170
      }                                                   171
      elsif ($Line =~ /^\)/) {                            172
          $Token = ')';                                   173
          $TokenType = 'BRACKETCLOSE';                    174
          ($LineRest) = ($Line =~ /^\)(.*)/);             175
      }                                                   176
```

```
    elsif ($Line =~ /^,/) {
        $Token = ',';
        $TokenType = 'COMMA';
        ($LineRest) = ($Line =~ /^,(.*)/);
    }
    elsif ($Line =~ /^;/) {
        $Token = ';';
        $TokenType = 'SEMICOLON';
        ($LineRest) = ($Line =~ /^;(.*)/);
    }
    elsif ($Line =~ /^<>/) {
        $Token = '<>';
        $TokenType = 'VARLENGTH';
        ($LineRest) = ($Line =~ /^<>(.*)/);
    }
    elsif ($Line =~ /^=/) {
        $Token = '=';
        $TokenType = 'EQUAL';
        ($LineRest) = ($Line =~ /^=(.*)/);
    }
    # Variable identifiers
    elsif ($Line =~ /^[a-zA-Z_]/) {
        ($Token) = ($Line =~ /^([a-zA-Z_][a-zA-Z0-9_]*).*/);
        $TokenType = 'IDENTIFIER';
        ($LineRest) = ($Line =~ /^[a-zA-Z_][a-zA-Z0-9_]*(.*)/);
        if (($Token =~ /^_.*/) ||
            ($Token =~ /^basicping$/) ||
            ($Token =~ /^basicresetserver$/) ||
            ($Token =~ /^basiccheckidlversion$/) ||
            ($Token =~ /^basicauthenticate$/) ||
            ($Token =~ /^basiclogout$/)) {
            $Error = 1;
            $ErrorMsg = "Lexical error (reserved identifier name) at line";
            goto Scanner_Return;
        }
        # Check if identifier is a known type
        foreach $NewType (@NewTypes) {
            if ($Line =~ /^$NewType(\W.*)?$/) {
                $TokenType = 'TYPE';
            }
        }
        # Check if identifier is a known constant
        foreach $NewConst (@NewConsts) {
            if ($Line =~ /^$NewConst(\W.*)?$/) {
                $TokenType = 'CONST';
            }
        }
    }
    # Version identifiers
    elsif ($Line =~ /^[#]/) {
        ($Token) = ($Line =~ /^([#][a-zA-Z0-9_]*).*/);
        $TokenType = 'VERSION';
        ($LineRest) = ($Line =~ /^[#][a-zA-Z0-9_]*(.*)/);
    }
    elsif ($Line =~ /^\!ASD(\W.*)?$/) {
        $Token = '!ASD';
        $TokenType = 'NOASD';
        ($LineRest) = ($Line =~ /^\!ASD(.*)/);
    }
    # Variable content
    elsif ($Line =~ /^\d\D/) {
        if ($Line =~ /^0x\d+/) {
            ($Token) = ($Line =~ /^(0x\d+).*/);
            $TokenType = 'NUMBER';
            ($LineRest) = ($Line =~ /^0x\d+(.*)/);
        }
```

```perl
            else {
                # Single number
                ($Token) = ($Line =~ /^(\d)\D.*/);
                $TokenType = 'NUMBER';
                ($LineRest) = ($Line =~ /^\d(\D.*)/);
            }
        }
        elsif ($Line =~ /^-?[1-9]\d*/) {
            ($Token) = ($Line =~ /^(-?[1-9]\d*).*/);
            $TokenType = 'NUMBER';
            ($LineRest) = ($Line =~ /^-?[1-9]\d*(.*)/);
        }
        elsif ($Line =~ /^".*"/) {
            ($Token) = ($Line =~ /^(".*").*/);
            $TokenType = 'STRING';
            ($LineRest) = ($Line =~ /^".*"(.*)/);
        }
        # Error at line
        else {
            $Error = 1;
            $ErrorMsg = "Lexical error at line";
            goto Scanner_Return;
        }
        ($Error, $ErrorMsg) = ParseToken ($TokenType, $Token, $IDLFilepath,
                            \%ParserState, \@NewTypes, \@NewConsts,
                            \$CurrentNamespace, \@DefinedIdentifiers,
                            \$NumberOfNamespaceFunctions,
                            $ProgramNumberPrefix,
                            $MultiThreading, $Authenticate);

        if ($Error) {
            goto Scanner_Return;
        }
        $Line = $LineRest;
        $Line =~ s/^\s+//g;
    }
}

# Finish state check
if (!($ParserState {'Part'} =~ /^IDL_FINISH$/)) {
    # Finish state is not reached
    $Error = 1;
    $ErrorMsg = "IDL-File is not completed at line";
    close (FileHandle);
    return ($Error, $ErrorMsg, $Count, $OrigLine);
}

Scanner_Return:
close (FileHandle);
if ($Error) {
    return ($Error, $ErrorMsg, $Count, $OrigLine);
}
else {
    return ($Error, $ErrorMsg, 0, "");
}
}
```

243
244
245
246
247
248
249
250
251
252
253
254
255
256
257
258
259
260
261
262
263
264
265
266
267
268
269
270
271
272
273
274
275
276
277
278
279
280
281
282
283
284
285
286
287
288
289
290
291
292
293
294
295
296
297

Besides checking lexical correctness, the scanner creates a symbol table. This is a data structure to save and administrate information and attributes about the constructs of the source code. The information can be modified or added to by following phases of the compilation (Aho et al. 1997, *l.c. page 73*). In the given case, just a reduced version of a symbol table in the form of the hash arrays «@NewTypes» and «@NewConsts» (see explanation of hashes in ▶ Sect. 2.7.2) in combination with the explicit parameters «$Token» and «$TokenType» for the currently evaluated lexeme is used. The scanner uses the hashes to identify multiple definitions of identifiers and offers the symbol table to the following syntax analysis. As the currently evaluated lexeme and its symbol type are not saved in the symbol table, it is directly handed to the syntax analysis subroutine. This has additional memory variables to save the program (interface) name and other relevant identifiers frequently during the compilation.

The second step is the syntax or syntactical analysis («parsing»). It is done by the «parser». The parser creates a parse tree (starting with a root node and «growing down» to leaf nodes), which is a representation of the syntax tree, defining the syntax of a grammar. The parser checks the grammar. There are usually two implementation classes useful for this check: bottom-up and top-down methods. As the names suggests, they create the parse tree from different starting points. The bottom-up method starts with the leaves of the tree and climbs up through the nonterminals to the root. The top-down method starts at the nonterminal root element and tries to build up the whole tree with all its leaves using the input symbols. This is usually a recursive process and forms a predicative parser. The top-down method can be implemented very efficiently and for specific grammars as a sequential processing of an input stream of symbols from the scanner (Aho et al. 1997, *l.c. page 50 ff.*). For the defined IDL, it is possible to construct a nonrecursive predicative parser which unrolls to standard loops.

An implementation for the new language is based on the syntax graph in ▢ Figs. 3.17 and 3.18. Because the grammar for the IDL does not use nested or paired constructs, a very straightforward finite state machine can be implemented using elementary conditional statements. The IDL definition implicitly avoids the possibility of ambiguities, as is usually possible in combination with variable «else» branches where it is not clear to which condition the «else» belongs to in a nested code (see, e.g., (Aho et al. 1997, *l.c. page 245*) or also the discussion about the «dangling else» problem, see ▶ Sect. 3.3.3). Only the use of multidimensional arrays with an optional number of dimensions complicates it a little. As there is no limit for the dimension, it is necessary to count the number of dimensions, so that each dimension is represented by an individual length argument attached to the value in the argument list of a remote function. This means that the argument definition:

> «inout string strArray <><><>»

in the source IDL is converted to:

```
«std::string *** & strArray_val,
unsigned int & _strArray_len,
unsigned int & _strArray_len1,
unsigned int & _strArray_len2»41
```

in the method definition of the target code. To enable this, it is necessary to infinitely count the number of dimensions. This is implemented with a simple pushdown automaton to simplify the processing. This means that for the generation of the multidimensional arrays, the principles of context-free languages are

Symbol Table

Syntax Analysis

Syntax Graph

---

41 For the «idl2rpc.pl» generator, it is defined that generated identifiers from the generator domain always start with an «_» to separate them from the variable names of the user application.

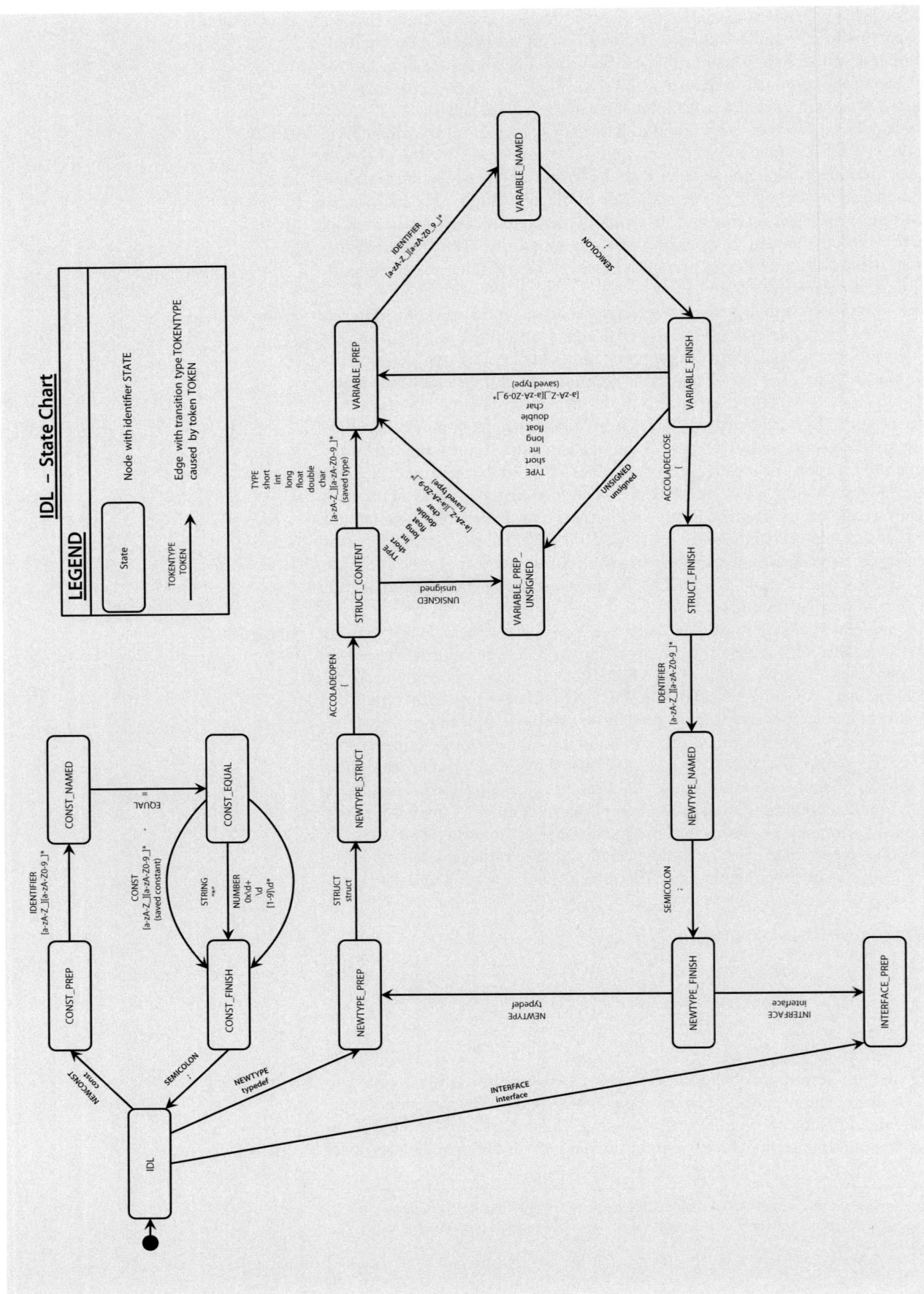

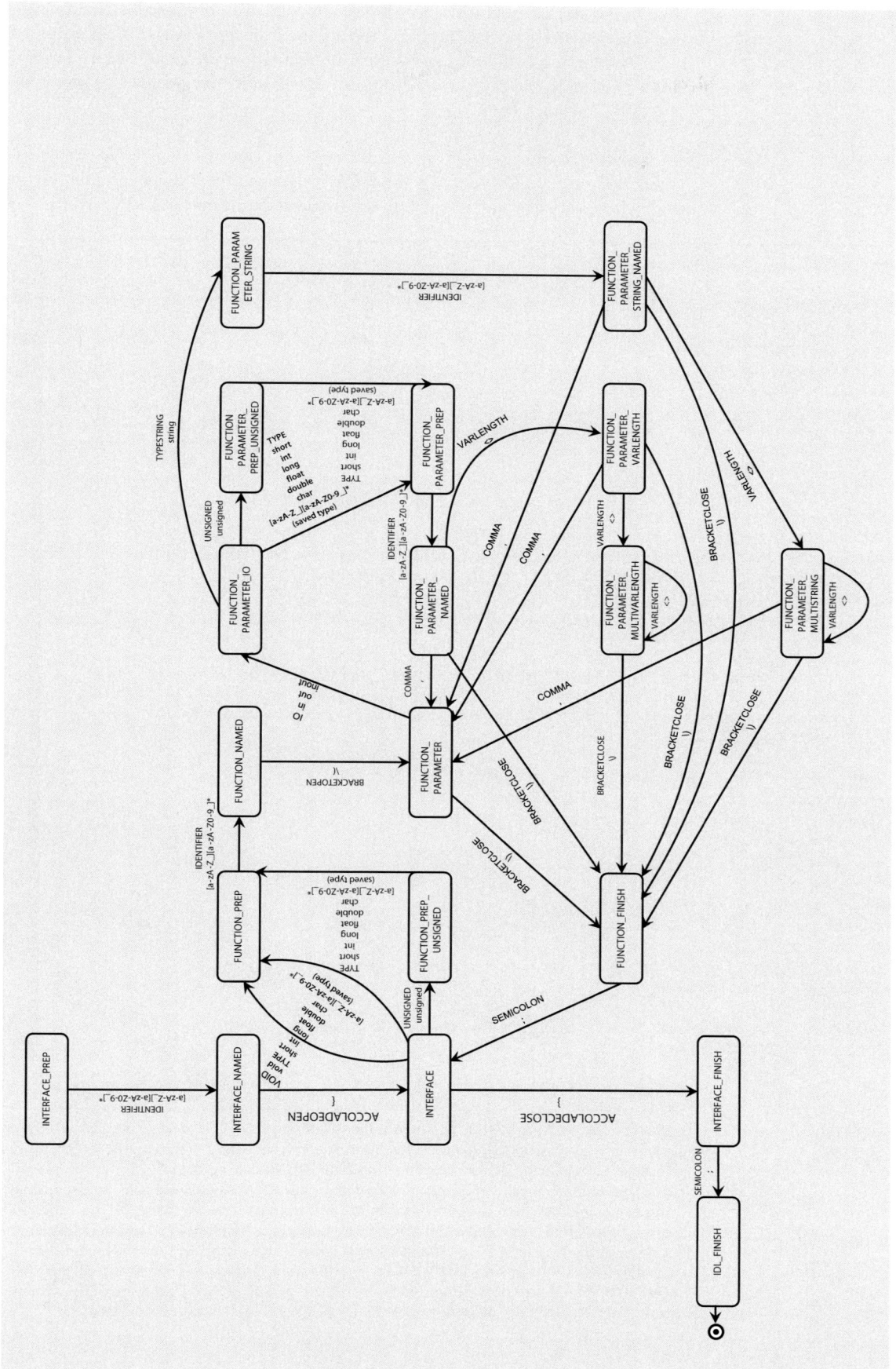

Fig. 3.18 The complete state graph of the top-down parser for the defined IDL (part 2)

3

used. It is also clear that it is necessary to copy the multidimensional arrays into a one-dimensional array for Sun RPC during the transfer. To do this, an unlimited number of nested FOR loops with different indentations are required (see Example 3.14).[42] Indenting also ensures the use of context-free languages (see also (Aho et al. 1997, *l.c. page 193 ff.*)).

■ ■ Example 3.14

Generated nested FOR-loops in the target language of the client module to copy multi-dimensional string arrays into the single-dimensional Sun RPC array

```
if (strArray_val == NULL)                                              1
{                                                                      2
    SInput._strArray_len = 0;                                         3
    SInput._strArray_len1 = 0;                                        4
    SInput._strArray_len2 = 0;                                        5
    SInput.strArray.strArray_len = 0;                                 6
    SInput.strArray.strArray_val = NULL;                              7
}                                                                      8
else                                                                  9
{                                                                     10
    SInput._strArray_len = _strArray_len;                            11
    SInput._strArray_len1 = _strArray_len1;                          12
    SInput._strArray_len2 = _strArray_len2;                          13
    SInput.strArray.strArray_len = 0;                                14
    for (Index_strArray_len = 0; Index_strArray_len < _strArray_len; Index_strArray_len++)  15
        for (Index_strArray_len1 = 0; Index_strArray_len1 < _strArray_len1;                 16
            Index_strArray_len1++)
            for (Index_strArray_len2 = 0; Index_strArray_len2 < _strArray_len2;             17
                Index_strArray_len2++)
                SInput.strArray.strArray_len += ((unsigned int) strArray_val [              18
                    Index_strArray_len][Index_strArray_len1][Index_strArray_len2].length())
                    +1;
    if ((SInput.strArray.strArray_val = (char *) malloc (sizeof(char)*SInput.strArray.      19
        strArray_len)) == NULL)
    {                                                                 20
        prot_strError = "Can't allocate enough memory";              21
        CInterfaceExcept.strErrorMsg = prot_strError.c_str();        22
        CInterfaceExcept._operation_state = 1;                       23
        __usInternalError = 1;                                       24
        goto sap_test_vtest_1_Return;                                25
    }                                                                26
    pcCopyRunner = SInput.strArray.strArray_val;                     27
    for (Index_strArray_len = 0; Index_strArray_len < _strArray_len; Index_strArray_len++)  28
    {                                                                29
        for (Index_strArray_len1 = 0; Index_strArray_len1 < _strArray_len1;                 30
            Index_strArray_len1++)
        {                                                            31
            for (Index_strArray_len2 = 0; Index_strArray_len2 < _strArray_len2;             32
                Index_strArray_len2++)
```

---

42 A specific problem with string arrays is that the strings can have variable lengths, which are at least an additional, varying dimension. Therefore, it is necessary to sum up the lengths of all elements to determine the length of the single-dimensional arrays for the transfer with a Sun RPC. This is not necessary for number arrays with multidimensions, where only the dimension sizes can be multiplied to get the complete size of the single-dimensional array. This additional, varying dimension also complicates the copying process, as for number arrays, the specific position of an element in the multidimensional array, such as «aiArray[5][8][3]», can be calculated as «aiTransferArray[5*8*3]». To copy the varying strings, it is necessary to use a position pointer (here «pcCopyRunner»), which must then be shifted by the size of the copied string. Therefore, it is not possible to directly reference an element by a simple address.

```
              {
                  strncpy (pcCopyRunner, strArray_val [Index_strArray_len ][Index_strArray_len1 ][
                      Index_strArray_len2 ].c_str(), strArray_val [Index_strArray_len ][
                      Index_strArray_len1 ][Index_strArray_len2 ].length());                      33
                                                                                                  34
                  pcCopyRunner += strArray_val [Index_strArray_len ][Index_strArray_len1 ][
                      Index_strArray_len2 ].length();                                              35
                  *pcCopyRunner = '\0';                                                            36
                  pcCopyRunner++;                                                                  37
              }                                                                                    38
              delete [] strArray_val [Index_strArray_len ][Index_strArray_len1 ];                 39
          }                                                                                        40
          delete [] strArray_val [Index_strArray_len ];                                           41
      }                                                                                            42
      delete [] strArray_val;                                                                      43
      strArray_val = NULL;                                                                         44
  }                                                                                                45
  prot_ulRoundTripByteSum += (SInput.strArray.strArray_len*sizeof(char));                         46
```

But as Perl is a full programming language and therefore Turing complete (see
► Sect. 3.1), it is even able to work with grammars for higher language classes than
type 2. The stack is herein explicitly replaced by the usual memory variables for counting and intermediate code saving. This offers the possibility of nonrecursive, predicative parser realization, where the memory must be explicitly administrated, instead of
using the implicit handling of recursive parsing algorithms.

To optimize the compilation, the number of internal runs must be reduced. More   **Semantic Analysis**
phases are usually combined into one run, so that the intermediate information can
be kept in the memory. In this way, scanning, parsing, semantical analysis, and the
intermediate code generation are meshed. At a specific time, still unknown parts can
be labeled with code attribute tags, which are replaced later when the information is
available (Aho et al. 1997, *l.c. page 25 f.*). This means for the generation that the parse
states in the parse tree are interspersed with semantical actions, which then activate
code generation procedures (Aho et al. 1997, *l.c. page 46 f.*). The semantical analysis
checks meaning errors and forces semantic decisions. In the case of the defined IDL,
it splits between «in», «out», and «inout» arguments and forces the individual code
generation of the different communication directions. For example, for the input
arguments with the prefix «in», copying of elements can only be processed before the
remote function is called, while for output arguments with the prefix «out», the copying happens afterward (for «inout», respectively, before and afterward). Therefore,
the semantic analysis collects and evaluates additional type information. For complete programming language compilers, this evaluation and correctness of type
information are obligatory and an essential part of the compilation (Aho et al. 1997,
l.c. page 9 ff.).

The final step of the front end part is intermediate code generation. Usually   **Intermediate Code Generation**
compilers write a simply producible and convertible intermediate code as a pro-
gram for an abstract machine. It simplifies the code generation for different target
languages and offers a higher abstraction for the code optimization (Aho et al.
1997, *l.c. page 569*). But this part is limited for a code converter like «idl2rpc.pl».
Because internally most parts of the code are stored as templates in the form of Perl
strings, they can be seen as master models for the final code. They use an abstract
definition, which can be the basis for different target languages (Klar and Klar
2006, *l.c. page 156 ff.*),[43] while only specified placeholders in the templates are
replaced during the conversion. These replacements are the best way to

---

43   Nevertheless, the generator is currently optimized for C/C++, so that an adaption to other
     target languages would require additional changes to the converter code.

parametrize the code for specific applications. This is similar to the comment tags for document generation with Doxygen (see ▶ Sect. 2.5.1) (Klar and Klar 2006, *l.c. page 162 ff.*). The converter only searches for the placeholder tags and replaces them with real code, or uses them for other processing steps (see Example 3.15, which shows the interspersed semantic and generative instructions for the different edges «COMMA», «BRACKETCLOSE», or «VARLENGTH», leaving from parser state «FUNCTION_PARAMETER_VARLENGTH»; see also the state graph for this state in ◘ Fig. 3.18).

Oblique Transformation

The converters use compositional and transformational refinement steps to replace code tags and to generate new modules. Compositional refinements (also defined as vertical or forward refinements) are improvements of the code which keep the structural behavior of the definition. For example, the interface definition produces an equivalent class definition for the client and the server. But it also adds additional modules and splits the server and the client code into several layers from the higher communication classes to the lower RPC communication modules. This changes the structure taken from the source language using transformational (or horizontal) refinements. The combination of both is defined as oblique transformation, which decomposes the higher interface definition into RPC modules, selects an ideal representation of an algorithm (e.g., for the copying of multidimensional string and number arrays), and specializes and concretizes the realization (Czarnecki and Eisenecker 2000, *l.c. page 341 ff.*). In the case of the «idl2rpc.pl» generator, this is a partly iterative process which generates different, chronological, improved implementations as copies of the templates for the intermediate code (transformation engine with iterative rewriting rules (Czarnecki and Eisenecker 2000, *l.c. page 352*)). Another intermediate code is the Sun RPC specification file with the extension «.x», which is also generated through this process.

■■ Example 3.15

Interspersed semantic decisions for input and output arguments and calls of code generation and replacement functions for the different edges leaving from parser state FUNCTION_PARAMETER_VARLENGTH

```
# ...                                                                         1
    if ($rState -> {'Part'} =~ /^FUNCTION_PARAMETER_VARLENGTH$/) {            2
        if ($TokenType =~ /^COMMA$/) {                                        3
            $rState -> {'Part'} = 'FUNCTION_PARAMETER';                       4
            ($$rCurrentNamespace) = ($$rCurrentNamespace =~ /(.*)\.\w+$/);    5
            if (ParameterIsInput ()) {                                        6
                AddToRPCX_FUNCTION_INPUTSTRUCT (";\n");                       7
            }                                                                 8
            if (ParameterIsOutput ()) {                                       9
                AddToRPCX_FUNCTION_OUTPUTSTRUCT (";\n");                     10
            }                                                                11
            AddToRPCABSTRH_DEFINITION_BLOCK (",");                          12
            ReplaceAtRPCABSTRH_DEFINITION_BLOCK ("ğ","");                   13
            AddToRPCCLIENTCPP_DYNAMICARRAY ($MultiThreading);              14
            AddToRPCSERVERPROCCPP_DYNAMICARRAY ();                          15
            AddToRPCSERVERCPP_DYNAMICARRAY ();                              16
            ReplaceTemplateAtClientClassRPCMethodDefinition ("ğğDeRefğğ","");  17
            ReplaceTemplateAtServerClassRPCMethodDefinition ("ğğDeRefğğ","");  18
            ReplaceTemplateAtClientClassRPCMethodDefinition ("ğParametersğ",",");  19
            ReplaceTemplateAtServerClassRPCMethodDefinition ("ğParametersğ",",");  20
            ReplaceTemplateAtServerProcRPCMethodDefinition ("ğParametersğ",", ");  21
            ResetRPCCLIENTCPP_TemporaryTemplates ();                       22
            ResetRPCProcRPC_TemporaryTemplates ();                         23
        }                                                                 24
```

```
        elsif ($TokenType =~ /^BRACKETCLOSE$/) {                                 25
            $rState -> {'Part'} = 'FUNCTION_FINISH';                              26
            ($$rCurrentNamespace) = ($$rCurrentNamespace =~ /(.*)\.\w+$/);        27
            if (ParameterIsInput ()) {                                           28
                AddToRPCX_FUNCTION_INPUTSTRUCT (";\n");                          29
            }                                                                    30
            if (ParameterIsOutput ()) {                                         31
                AddToRPCX_FUNCTION_OUTPUTSTRUCT (";\n");                         32
            }                                                                    33
            AddToRPCABSTRH_DEFINITION_BLOCK (")");                              34
            ReplaceAtRPCABSTRH_DEFINITION_BLOCK ("ğ","");                       35
            AddToRPCCLIENTCPP_DYNAMICARRAY ($MultiThreading);                   36
            AddToRPCSERVERPROCCPP_DYNAMICARRAY ();                              37
            AddToRPCSERVERCPP_DYNAMICARRAY ();                                  38
            ReplaceTemplateAtClientClassRPCMethodDefinition ("ğğDeRefğğ","");   39
            ReplaceTemplateAtServerClassRPCMethodDefinition ("ğğDeRefğğ","");   40
            ReplaceTemplateAtClientClassRPCMethodDefinition ("ğINVariableDefinitionğ","void *  41
                vInput = NULL;");
            ReplaceTemplateAtClientClassRPCMethodDefinition ("ğINVariableğ","vInput");        42
            ResetRPCCLIENTCPP_TemporaryTemplates ();                            43
            ResetRPCProcRPC_TemporaryTemplates ();                             44
        }                                                                       45
        elsif ($TokenType =~ /^VARLENGTH$/) {                                    46
            $rState -> {'Part'} = 'FUNCTION_PARAMETER_MULTIVARLENGTH';           47
            if (ParameterIsInput ()) {                                           48
                AddLENVARIABLEToRPCX_FUNCTION_INPUTSTRUCT ();                    49
            }                                                                    50
            if (ParameterIsOutput ()) {                                         51
                AddLENVARIABLEToRPCX_FUNCTION_OUTPUTSTRUCT ();                   52
            }                                                                    53
            NextRPCX_LENVARIABLE_FOR_PARAMETER ();                              54
            ReplaceAtRPCABSTRH_DEFINITION_BLOCK ("ğ","*ğ");                     55
            AddToRPCABSTRH_LENVARIABLE ();                                      56
            ReplaceTemplateAtClientClassRPCMethodDefinition ("ğğDeRefğğ","*ğğDeRefğğ");      57
            ReplaceTemplateAtServerClassRPCMethodDefinition ("ğğDeRefğğ","*ğğDeRefğğ");      58
            NextRPCCPP_LENVARIABLE_FOR_PARAMETER ();                            59
        }                                                                       60
        else {                                                                  61
            $Error = 2;                                                         62
            $ErrorMsg = "Syntactical error at line";                           63
        }                                                                       64
        goto ParseToken_Return;                                                 65
    }                                                                           66
# ...                                                                           67
```

During the compilation of target code, there are several starting points for an optimization of the resulting code (see also (Aho et al. 1997, *l.c. page 718*). In the case of the «idl2rpc.pl», a static optimization is done in the domain development. All templates are manually optimized to produce fast and easily readable code. Additionally, all code modules are inspected with the methods of the CI (see ► Sect. 2.8). Flaws found, such as resource leaks, performance leaks, or style issues, are directly improved in the template code.

Static Code Optimization

A further, more dynamic optimization concerns the handling of Sun RPC elements. Input to and generated code from «rpcgen» gets revised to avoid warnings, misuses, or performance issues. An example is the use of a sequence of variable definitions of the same data type in structures of the RPCL file as input to «rpcgen». The code produced is not always optimal in combination with «idl2rpc.pl» generated code. Therefore, dummy variables of the size of one byte are interspersed between these sequences of variables to optimize the resulting code. Another dynamic optimization cleans all unused variables (such as «register int32_t *buf;»), which are sometimes generated by «rpcgen». To avoid unnecessary output and

Dynamic Code Optimization

multiply appearing outputs of warnings during the final compilation, the unused variables are deleted in the generated code or an use is simulated (e.g., «(void) buf;») is implemented. Also the replacement of the general server loop function «svc_run()» by an individual, domain-specific function «_vPrivate_svc_run ()», which is started from a domain-specific main function, can be seen as dynamic code optimization. It replaces the original, «rpcgen» created main function in the «_svc.c» file and allows the creation of individual sockets with a timeout-controlled «connect» call and an active «SO_LINGER» flag (see Example 3.6 and Example 3.5). Additionally, the handling of crashes with a parent process as a watchdog (see Example 3.8) and several other improvements is possible, such as signal handlers for process shutdowns. All of these optimizations are also parametrized with calling arguments of «idl2rpc.pl», which are selected by the application developer according to the application environment in which the generated code is used.

Code Generation

Using «idl2rpc.pl», the generation first starts with the writing of the RPCL file (with the file extension «.x»). This file is then used as input to the external «rpcgen» of the Sun RPC suite, which runs several times with different call arguments to produce all the required RPC files. The files produced are just intermediate states, because the dynamic optimization reads and adapts them again. All templates which are Perl strings in the code generator are then taken to produce all additional «idl2rpc.pl»-specific code files. All placeholders are replaced by the parametrized, real values from the IDL file or from internal generative processes, according to the IDL definitions and the arguments of the «idl2rpc.pl» call. The code generation can assume that the intermediate code is correct (Aho et al. 1997, *l.c. page 627*). Each finished and optimized template is then written to an individual target file. All of this is the real generative process.

Error Handling

During all the phases of the generation, errors must be detected and communicated to the application developer. Therefore, the standard compilers use recovery strategies, such as the panic mode recovery. The scanner deletes characters from the input language stream until a usable character for a useful symbol is found. This allows the compiler to finish the processing and to report all possible errors at once (Aho et al. 1997, *l.c. page 107*). «idl2rpc.pl» follows a much simpler error handling strategy. As the IDL is usually very compact, the generator detects errors and directly stops after a detection with an error message, indicating the line number of the error detected, the currently analyzed token, and the line content. If errors in the parameter list of the command line call appear, then a help page with all the alternative parameters is shown.

«idl2rpc.pl» Phases

All the steps described form the phases of the code generation with «idl2rpc.pl». They have several similarities to the standard phases of a compiler, but have a number of small adaptations, which are shown in ◘ Fig. 3.19. The reason for these additional generative overheads is the simplification of an interface definition, the integration of an OOP style (see the next section), the use of multidimensional arrays in the heap memory (e.g., for matrices), improvements in error handling with improved socket modules and an implicit watchdog process, the possibility of a simple parallelization with a possible handling of critical sections, individual authentication mechanisms, safety and security improvements with better version checks in the RPC functions, and so on. It opens up the possibility for improved application engineering on the basis of this improved domain engineering.

In addition to the domain engineering tasks, knowledge of compiler designs helps with the evaluation of input files, which follows a specific format, because it can usually be described with a specific grammar for the format (see ► Sect. 4.6). Thus, the language theory and the compiler theory are essential for several aspects of applied computer science.

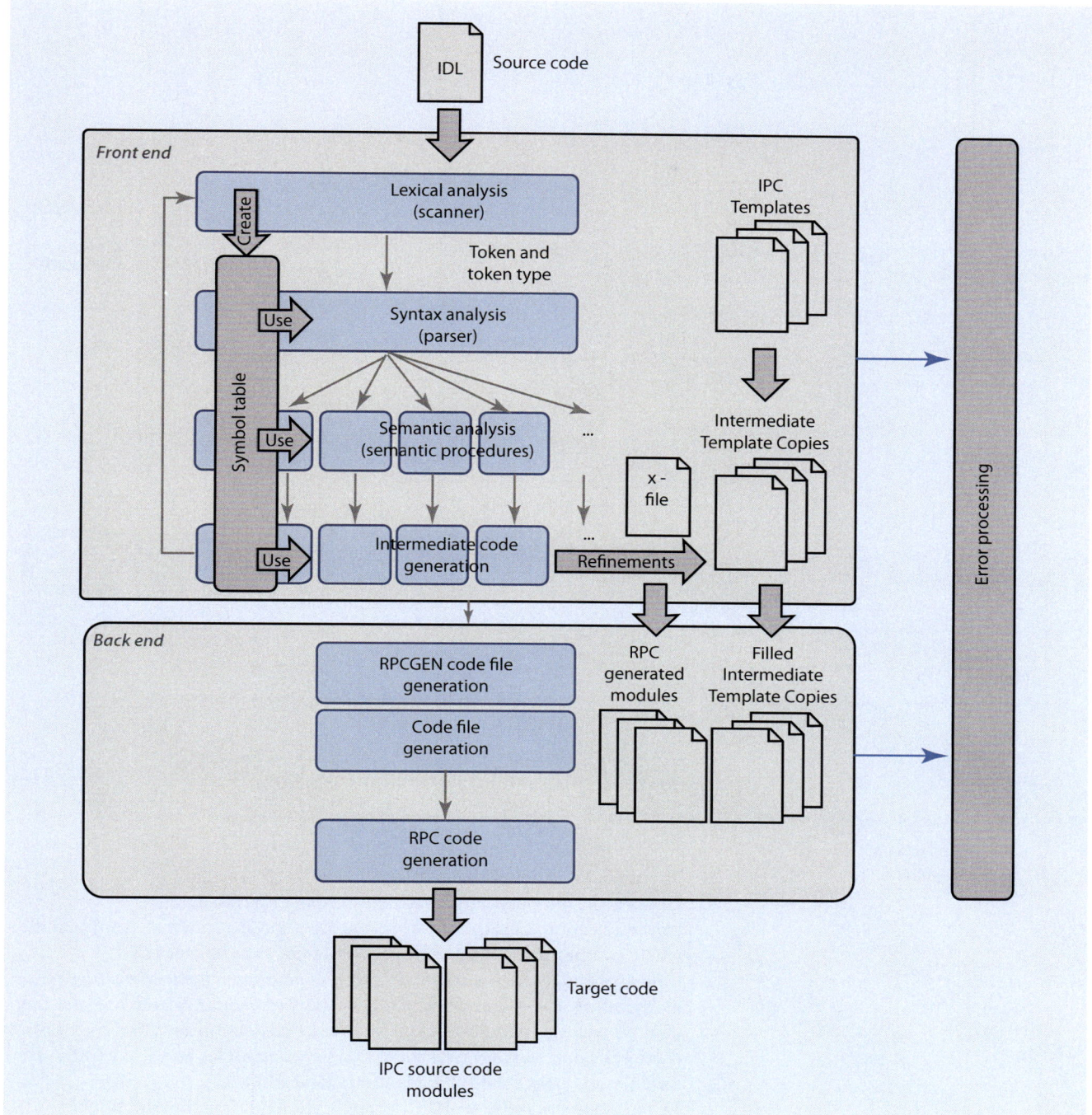

◘ Fig. 3.19 The phases of the idl2rpc.pl generator

Application Development with the New «idl2rpc.pl» Generator

Application engineering with the generator «idl2rpc.pl» is very similar to the generation with the Sun RPC generator «rpcgen». After specifying the interface with the new IDL, it is used to generate communication-specific modules. The complete process of generating and compiling the communication-relevant files is shown in ◘ Fig. 3.20 (compare also ◘ Fig. 3.11 about the original «rpcgen» process).

Generating the Modules

3

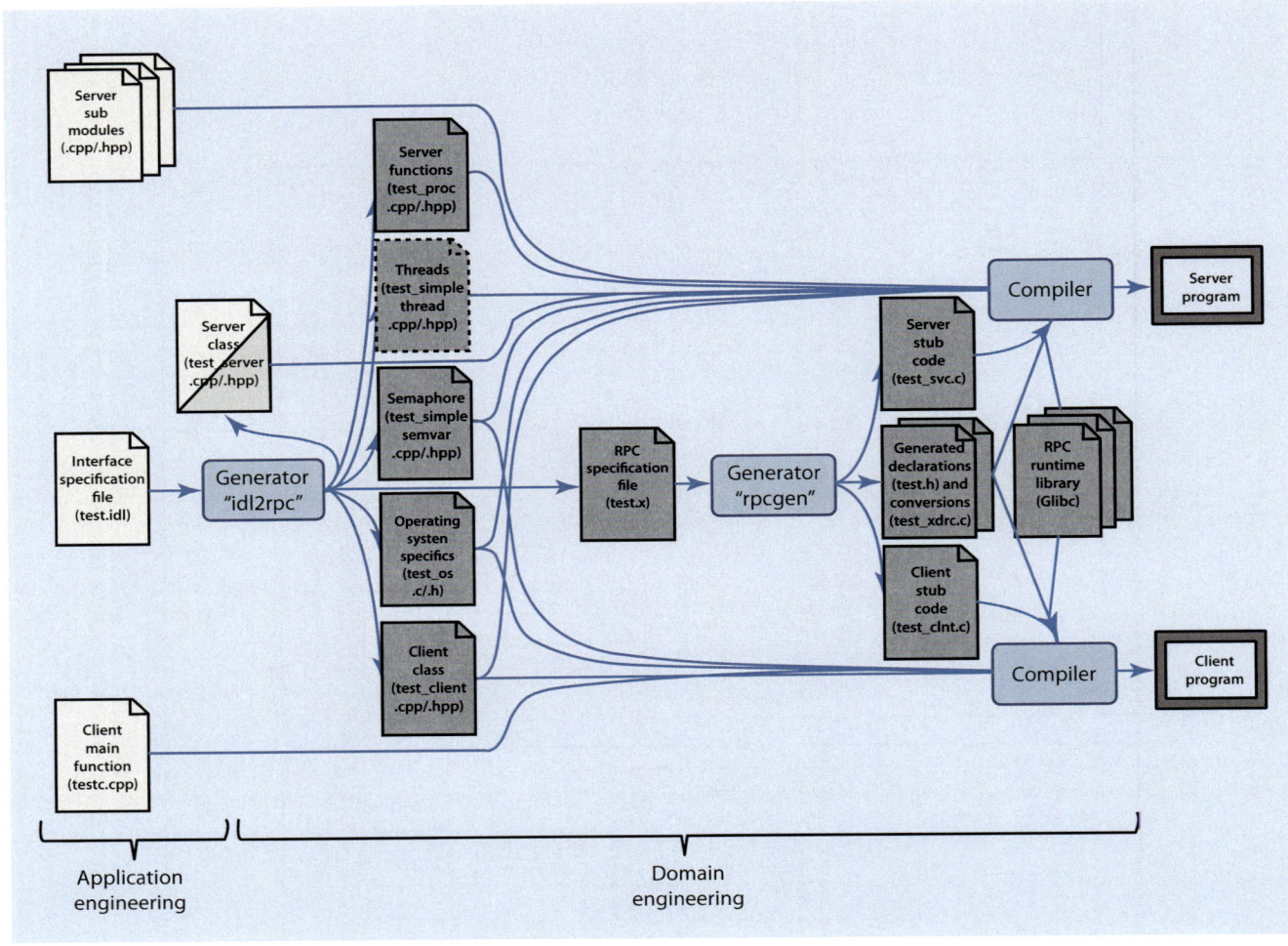

Development Steps

The application development has to follow the following steps:

1. *Defining the interface*: First of all, the developer has to define the communication. This is done by setting up a specific interface definition file where communication functions are declared with the new IDL.

2. *Calling of «idl2rpc.pl» with the IDL file and additional parametrization values as argument*: The call of the generator «idl2rpc.pl» automatically creates the communication modules needed for the defined communication functions. It creates the Sun RPC-specific client and server modules and also additional modules to operate with threads and critical sections.

3. *Filling the server skeleton class*: The application developer has to fill the generated skeleton function bodies of the server with the actual server service functionalities. Here he can use external library modules, which are called from the server functions.

4. *Writing of the client main*: On the client side, the application developer has to write his own client main, which creates an object of the generated client class and calls the appropriate interface functions.

5. *Compiling of server and client*: The final step is to compile the server and the client code including all the generated and manually written modules and components necessary (see also ▶ Sect. 3.1 about the differences between modules and components).

6. *Starting of server and client*: The complete communication over the network is then automatically available and processed in the background of the remote function calls, after the start of server and client.

Command Line Call

As the generator «idl2rpc.pl» is a Perl script, it needs a Perl environment on the machine where the application will be built. The Sun RPC environment with the runtime libraries is also required. The general call of the generator is «perl./idl2rpc.pl <IDL-file>» (or also with an implicit Perl activation with «./idl2rpc.pl <IDL-file>» if Perl exists in the directory «/usr/bin/perl»[44]). It uses the standard settings and creates all the code files needed for a standard communication without further parametrization. Already existing files are saved as backup files, adding a «~» sign (tilde) at the end of the file extension. Developer written and therefore already existing parts between the specific tags of the server skeleton are also kept (e.g., method bodies are not changed during the generation process as long as the method name is not changed). The generation mixes C and C++ files. To identify them, the extensions are different: «.c/.h» for C and «.cpp/.hpp» for C++, as suggested about coding style guides. ◘ Figure 3.20 also shows which files must be linked together for server and client. For the development of an application, only the light gray parts are important, as they must be developed by the application engineer.

Further Parameterizations

In addition to the standard run, an additional parametrization is possible using call parameters. It controls details of the generated modules. The following help output is an overview of possible parameters of «idl2rpc.pl»:

```
idl2rpc.pl ([-SLT <sec>/<usec>] |   --> Server    loop    timeout    for
                                        periodic activities (sec > 0
                                        and/or usec > 0)
            [-TLT [<sec>/<usec>]])--> Server loop timeout for peri-
                                        odic activities using threads
                                        (sec > 0 and/or usec > 0,
                                        without  time  setting  peri-
                                        odic activity is called only
                                        once => user written timing)
            [-CT <sec>/<usec>]     --> Client request timeout for
                                        each request (sec > 0 and/
                                        or usec > 0)
            [-CRT <sec>/<usec>]    --> Client retry timeout after a
                                        request   retry   is   started
                                        (sec > 0 and/or usec > 0)
            [-PCL <number>]        --> Program classification code
                                        >= 600 and <= 1070, stan-
                                        dard is 700
            [-TCP|-UDP]            --> Mainly used transport
                                        protocol
            [-PTCP <port>]         --> Fixed    port    for    UDP
                                        connections
            [-PUDP <port>]         --> Fixed    port    for    TCP
                                        connections
            [-CL]                  --> Create only client
            [-NMT]                 --> Reduced  safety:  no  multi-
                                        threading safety
            [-NOW]                 --> Reduced safety: no watchdog
                                        for the server
            [-NOM]                 --> Reduced safety: no watchdog
                                        and  no  main  function  for
                                        the server
            [-AUTH]                --> Activate transfer of authen-
                                        tication    credentials    and
                                        activate authorization checks
            [-ASD <sec>]           --> Automatic    safety    device
                                        («deadman» safety control)
                                        (sec  >=  0  specifies  the
                                        reaction time after client
                                        contact is lost)
            [-V]                   --> Print only idl2rpc version
                                        information        without
                                        processing
            <IDL-filepath>.idl     --> IDL-filepath
```

---

44   The Perl script of the generator defines this in the first line with «#!/usr/bin/perl –w».

3

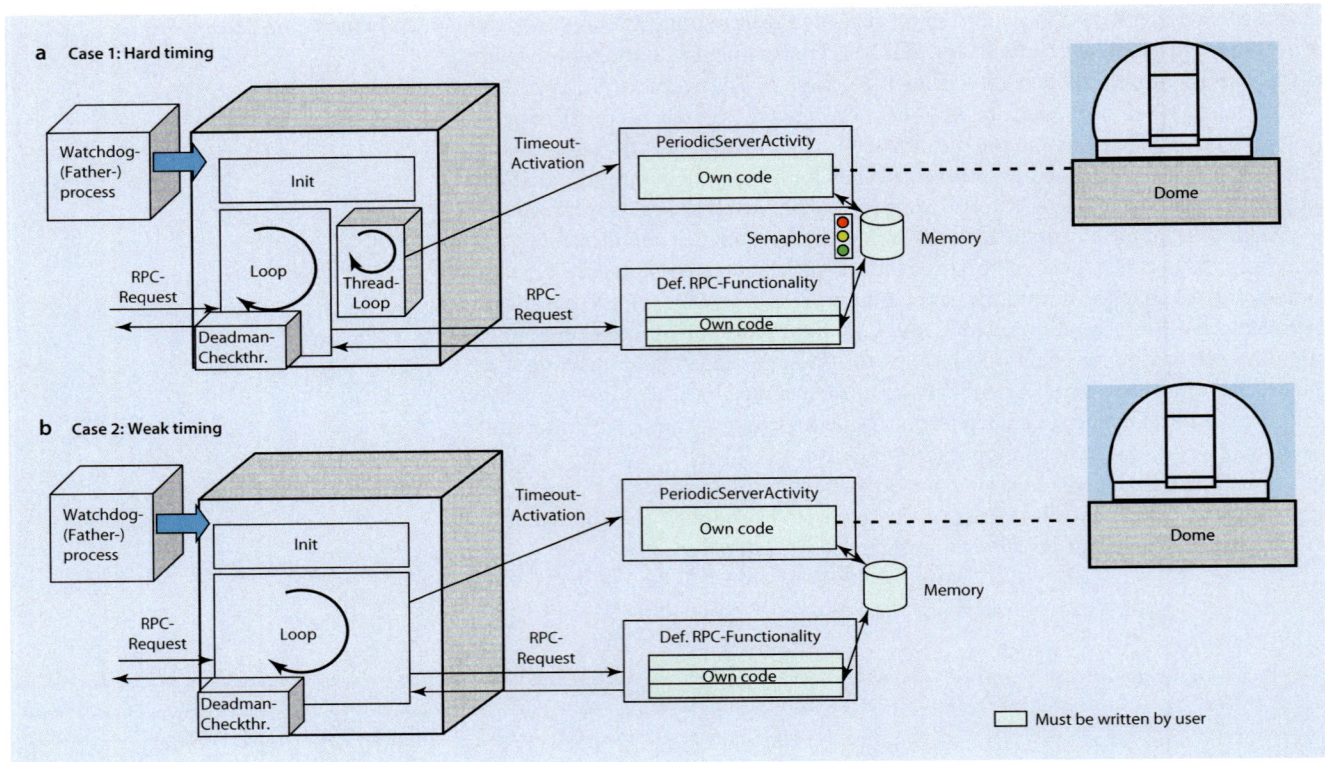

Fig. 3.21 The difference between weak and hard timing for periodic activities of the server which are not triggered by a client

— *SLT or TLT*: If the server is used to process periodic activities which are not triggered by a client, it is possible to activate a periodic server loop, the server loop timeout («SLT» parameter). The defined seconds and microseconds (values greater than or equal to 0) are used for a sleep between two activations of the periodic method «_vPeriodicServerActivity». This method is then started frequently, but without the use of threads. Therefore, it is a concurrent run to the asynchronously arriving client requests. All server method calls are sequentialized, so that it just offers a weak timing with a small jitter of the start time even though the sequencing of all methods is internally optimized. If the asynchronous functions are very time-consuming, it can lead to the effects of starvation, so that the periodic method does not get processing time anymore. This effect can be completely avoided by using threads for the incoming RPCs and the periodic method. The thread loop timeout («TLT» parameter) is similar to the previous «SLT» for starting the method «_vPeriodicServerActivity», but uses an additional, parallel thread in the server for incoming RPCs. This means that asynchronous requests from the clients do not interrupt the timing of the periodic task (see also **Fig. 3.21**). Because of the parallel threads, the use of semaphores (from the module «..._simple_semvar.cpp/.hpp», which is also generated) is mandatory to protect the concurrent access of variables (see also ▶ Sect. 2.6.4). If the TLT parameter is used without second and microsecond settings, the periodic server method is called only once and no implicit timing is started. This can be used to write an application-specific timing loop in the method «_vPeriodicServerActivity».

— *CT*: The «client timeout» («CT» argument) or client request timeout defines in seconds and microseconds (values have to be greater than 0) after which time period a request must reply. CT is valid for UDP as well as for TCP.

— *CRT*: The «client retry timeout» («CRT» argument) is only valid for UDP transfers. It defines the time period in seconds and microseconds (values greater than 0) after which a reply is sent again. Big timeout values (e.g., several years) make sure that a UDP resending is never done.

- *PCL*: Usually the program number of Sun RPC connections is generated implicitly under «idl2rpc.pl», using the interface name and an «idl2rpc.pl» identifier as prefix (the value «700» is defined). But the calculation is similar to a simple hash function. Because of that, it is possible for more than one interface name to map to the same program number. For this case, and to allow the parallel use of more than one running server of the same type (e.g., one working production system server and one development system server, where specific care for concurrent access on hardware must be taken by the application developer) at the same time on the same machine, the «program classification code» («PCL» argument) can be used to define a different prefix. Values from decimal «600» to «1070» are allowed for that. PCL is only relevant for «portmapper» registrations.
- *TCP or UDP*: The server always listens to UDP and TCP, so that the programmer of the client is free to choose which transport protocol he would like to use. Without a specific setting, UDP is the preferred protocol of the client. But with the TCP/UDP parameter, it is possible to explicitly set the initially used protocol type.[45] If the communication is tunneled over SSH (see ► Sect. 5.5.2), TCP is the protocol of choice.
- *PTCP/PUDP*: Normally the server uses the «portmapper» (which listens on port 111) to offer a first contact point to its services. To set the port of the server to a fixed one (e.g., to easily pass a firewall with no limitations), a port for TCP or UDP can be defined with PTCP or PUDP. If such a fixed port is used, the client directly contacts the server using the defined port without the use of the «portmapper».
- *CL*: If only a client is built, for example, on a remote machine, the «client only» (CL) parameter forces the generation of client modules only.
- *NMT, NOW*, and *NOM*: These parameters are used to reduce the automatically generated safety if the server is integrated into an external safety system. «NMT» («no multi-threading») switches the multi-threading stability off in the Sun RPC Remote Procedure Calls. «NOW» («no watchdog») deactivates the watchdog, and «NOM» («no main») changes the main function of the server to a conventional function, which can be called from outside for an external control.
- *AUTH*: Activates the authentication and authorization transport mechanism. «idl2rpc.pl» does not include the authentication checking itself, but offers an additional way to transfer authentication credentials. This is an additional mechanism different to the Sun RPC techniques.
- *ASD*: Normally the server does not respond on client communication losses. With the «automatic safety device» («ASD» argument), the server monitors the incoming activations from the clients. After the defined time period in seconds without any contact from a responsible client, the server calls the special method «_vAutomaticSafetyDeviceAlert» once directly after the loss and the method «_vAutomaticSafetyDevice AlertRepeated» each following second until contact with the lost client is again established. These methods can be used to stop special, critical actions when no responsible client is available, for example, after a connection blackout. It is similar to the «deadman control» in trains, where the driver has to push a button regularly to prevent a safety stop of the train. In principle, it offers a heartbeat monitoring, which requires regular requests (see also ► Sect. 3.3.3 about safety). In the IDL file, the automatic safety can also be deactivated for specific methods by adding «!ASD» to the function name.
- *V*: Prints only the current version of the «idl2rpc.pl» generator.
- *<IDL-filepath>.idl*: The last parameter of the generator call is the IDL file with the user-defined interface methods. It must always have the extension «.idl».

---

45 If dynamic arrays are used, TCP is mandatory. Otherwise the remote call will end up in an exception at the client.

Client Application: Opening

On the client side, the application programmer has to create an additional file with the main program of the client. It includes the interface methods of the client class, generated in the file «*<INTERFACEFILE>*_client.cpp/.hpp».[46] The client main program must contain an instance of this interface class. This object processes all defined remote operations. In addition to the user-defined remote procedures, the generator creates administration methods, offering possibilities of handling the interface connection, for example, «_usOpenInterface(...)» to establish a connection. The connection can be established using the following possibilities:

- The standard constructor and a later call of «_usOpenInterface(...)»
- The constructor with IP address or IP alias as argument for an implicit open
- The copy constructor

Client Application: Closing and Copying

The connection can explicitly be closed by using «_vCloseInterface()». Another implicit possibility is the destructor, which is automatically called before the object is deleted. Interface instances can also be assigned directly by using the «=», because the assignment operator is overloaded. Then it is a real duplication of the interface object, so that after an assignment, two separated client access points exist.

Client Application: Administration Methods

The additionally and automatically generated administration methods of the client are here in detail:

- *Methods for the class administration*: Constructor, constructor with integrated opening of the communication, copy constructor, destructor, and assign operator.
- *_usOpenInterface (...)* opens a new RPC connection to a server with the specific IP address and starts its internal administration.
- *_vCloseInterface ()* closes the connection to a server.
- *_usIsClientConnected ()* checks if a connection has been established and processes a ping to the server. It returns «0» if the connection is not yet opened and «1» if it is.
- *_strGetErrorMsg ()* returns the last thrown error message and resets the internal message memory.
- *_ulPing ()* offers a possibility of simply pinging the server. It returns the round-trip delay of the remote call in milliseconds.
- *_vResetServer ()* resets the server by activating an «exit ()» call at the server, which leads to a restart of the server, done by the watchdog.
- *_usCheckIDLVersion () and alternatives* compare the IDL version of the specification file and of the generator between client and server. The alternative methods are able to return the server and client values of the versions and split the result of the check into one for the specification file and one for the generator version.
- *_ulGetRoundTripDelay () and alternatives* return the round-trip delay for the last method call in milliseconds (and optionally more precise in microseconds).
- *_usGetRequestTimeout (...) and _usSetRequestTimeout (...)* return or set the current value for the request timeout in seconds and microseconds to define the time period after which a call must return.
- *_usGetRetryTimeout (...) and _usSetRetryTimeout (...)* return or set the retry timeout for UDP connections in seconds and microseconds, which defines the time period after which a UDP call must be repeated.
- *_usGetTransportProtocol (...) and _usSetTransportProtocol (...)* return and set the currently used transport protocol (0 = UDP, 1 = TCP).
- *_usGetTCPPort (...) and _usSetTCPPort (...)* return and set the current TCP port value for connections to a server.
- *_usGetUDPPort (...) and _usSetUDPPort (...)* return and set the current UDP port value for connections to a server.

---

46 *<INTERFACEFILE>* is replaced by the filename of the interface specification file, e.g., if the interface specification is in the file «*test*.idl», the resulting client class is in the file «*test_*client.cpp» and «*test_*client.hpp».

A sample code for a main program of a client is shown in Example 3.16   Client Application Sample

■■ Example 3.16

Sample main program for a client, which uses the generated code from the interface definition in Example 3.12

```cpp
#include <iostream>
#include <string>
#include "test_client.hpp"

int main ()
{
    test_client CClient;

    // Open interface
    std::cout << "Open interface" << std::endl;
    try
    {
        if (CClient._usOpenInterface ("soswdb.sosw"))
        {
            std::cout << "Can't open interface" << std::endl;
            return 1;
        }
        if (!CClient._usCheckIDLVersion ())
        {
            CClient._vCloseInterface ();
            std::cout << "Wrong IDL versions" << std::endl;
            return 1;
        }
    }
    catch (_interface_throw CError)
    {
        std::cout << "[ERROR] " << CError._strGetErrorMsg() << " (" << CError.
            _usGetOperationState() << ")" << std::endl;
        return 1;
    }
    catch (...)
    {
        std::cout << "[ERROR] Unknown error" << std::endl;
        return 1;
    }

    // Do a ping
    std::cout << "Do a ping: ";
    try
    {
        unsigned long ulRoundtripTime;
        ulRoundtripTime = CClient._ulPing ();
        std::cout << "Roundtrip time = " << ulRoundtripTime << " msec" << std::endl;
    }
    catch (_interface_throw CError)
    {
        CClient._vCloseInterface ();
        std::cout << "[ERROR] " << CError._strGetErrorMsg() << " (" << CError.
            _usGetOperationState() << ")" << std::endl;
        return 1;
    }
    catch (...)
    {
        CClient._vCloseInterface ();
        std::cout << "[ERROR] Unknown error" << std::endl;
        return 1;
    }
```

```
                                                                        57
// Close interface                                                      58
std::cout << "Close interface" << std::endl;                            59
try                                                                     60
{                                                                       61
    CClient._vCloseInterface ();                                        62
}                                                                       63
catch (_interface_throw CError)                                         64
{                                                                       65
    std::cout << "[ERROR] " << CError._strGetErrorMsg() << " (" << CError.
        _usGetOperationState() << ")" << std::endl;                     66
    return 1;                                                           67
}                                                                       68
catch (...)                                                             69
{                                                                       70
    std::cout << "[ERROR] Unknown error" << std::endl;                  71
    return 1;                                                           72
}                                                                       73
                                                                        74
return 0;                                                               75
}
```

Server Application Skeleton

On the other hand, the application programmer must write the server functionality. Therefore, the server skeleton in the file «<INTERFACEFILE>_server.cpp/.hpp»[47] is provided. The skeleton contains generated code sections as placeholder for application-specific code. These areas are marked with tags in specific comment lines. The content between those tags is retained during further generation processes. The tags themselves are C++ comment lines and mark the beginning and the end with a comment statement.

The code attributions used in the server header file are:

— *Prioritize user includes between*
«USERDEFPRIORINCLUDEBEG» and «USERDEFPRIORINCLUDEEND»:
Here is space for «include» statements to include external modules which are used by the application server and must be included before all the others.

— *Normal user includes between*
«USERDEFINCLUDEBEG» and «USERDEFINCLUDEEND»:
Here is space for usual «include» statements to include external modules which are used by the application server.

— *User definitions of member variables / attributes between*
«USERDEFATTRIBBEG» and «USERDEFATTRIBEND»:
Here is space for member variables as attributes of the server class which are used by the server application.

— *User declarations of member methods between*
«USERDEFATTMETHODBEG» and «USERDEFATTMETHODEND»:
Here is space for the declaration of member variables as attributes of the server class which are used by the server application.

The code attributions in the server source file are:

---

47 *<INTERFACEFILE>* is replaced by the filename of the interface specification file, e.g., if the interface specification is in the file «*test*.idl», the resulting server class is in the file «*test_*server.cpp» and «*test_*server.hpp».

— *Normal user includes between*
 «USERDEFINCLUDEBEG» and «USERDEFINCLUDEEND»:
 Here is space for the usual «include» statements to include modules which are
 used by the application server and should not be included via the server
 header file.
— *User declarations of member methods between*
 «USERDEFATTMETHODBEG» and «USERDEFATTMETHODEND»:
 Here is space for the declaration of member variables as attributes of the
 server class which are used by the server application.
— *User initializations of member variables/attributes between*
 «USERDEFATTRIBBEG» and «USERDEFATTRIBEND»
 in the constructors, destructor, and the assign operator:
 Here is space for the initialization of member variables as attributes of the
 server class which are used by the server application.
— *User method bodies between*
 «USERDEFMETHODBEG» and «USERDEFMETHODEND»:
 Here is space for the user-defined method bodies for administration methods
 (e.g., «`_vInitServerActivity()`» for instructions during the startup,
 `_vPeriodicServerActivity` for the periodic tasks,
 `_vOnExitServerActivity` for instructions during the shutdown, etc.)
 as well as for generated functions from the IDL file.

A sample code for a filled server skeleton is shown in Examples 3.17 and 3.18.　　Server Application Sample

■■ Example 3.17

Filled sample header file of a skeleton for a server, which uses the generated code
from the interface definition in Example 3.12

```
/***************************************************************/      1
/*! \file                                                              2
 * \brief Interface code file defined at test.idl and generated by idl2rpc.pl<br>   3
 ***************************************************************/       4
                                                                       5
                                                                       6
#ifndef __test_server__                                                7
#define __test_server__                                                8
                                                                       9
#include <rpc/rpc.h>                                                  10
#include "test_interface.hpp"                                         11
#include "test_simple_semvar.hpp"                                     12
                                                                      13
// USERDEFINCLUDEBEG: Userdefined includes                           14
// USERDEFINCLUDEEND                                                 15
                                                                      16
class test_server : test                                             17
{                                                                     18
    std::string priv_strError;                                       19
    unsigned short usStopServerLoop;                                 20
    // USERDEFATTRIBBEG: Userdefined attributes                      21
    test_semvar<int> CValue; // CValue = Semaphore protected variable  22
    // USERDEFATTRIBEND                                              23
    // USERDEFATTMETHODEBEG: Userdefined attribute methodes          24
    // USERDEFATTMETHODEEND                                          25
                                                                      26
// ...                                                               27
                                                                      28
};                                                                    29
                                                                      30
// ...                                                               31
```

3

■ ■ **Example 3.18**

Filled sample source file of a skeleton for a server, which uses the generated code from the interface definition in Example 3.12

```
/*****************************************************************/
/*! \file
 * \brief Interface code file defined at test.idl and generated by idl2rpc.pl<br>
 *****************************************************************/
#include <stdio.h>
#include <rpc/rpc.h>
#include <string>
#include "test_interface.hpp"
#include "test_server.hpp"
// USERDEFINCLUDEBEG: Userdefined includes
#include <iostream>
#include <time.h>
#include <sys/time.h>

// ...

/*****************************************************************
 *  class      test_server
 *  function   test_server
 *****************************************************************/
/*!             Constructor
 *  \param   —
 *  \return  —
 *****************************************************************/
/*  author     Alexander Neidhardt
 *  date       14.05.2007
 *  revision   —
 *  info       Part of the idl2rpc.pl — generator!
 *****************************************************************/
test_server::test_server () throw (_interface_throw) : test()
{
    try
    {
        priv_strError = "Everything ok";
        usStopServerLoop = 0;
        usServerInitAlert = 0;
        uiUDPPort = 0;
        uiTCPPort = 0;
        usPortSetActivated = 0;
    }
    catch (...)
    {
        _interface_throw CInterfaceExcept;
        usServerInitAlert = 1;
        priv_strError = "Can't init interface";
        CInterfaceExcept.strErrorMsg = priv_strError.c_str();
        CInterfaceExcept._operation_state = 1;
        throw (CInterfaceExcept);
    }
    try
    {
        // USERDEFATTRIBBEG: Userdefined attributes
        CValue.vSetSemVal(0);
        // USERDEFATTRIBEND
    }
```

```
1
2
3
4
5
6
7
8
9
10
11
12
13
14
15
16
17
18
19
20
21
22
23
24
25
26
27
28
29
30
31
32
33
34
35
36
37
38
39
40
41
42
43
44
45
46
47
48
49
50
51
52
53
54
55
```

```
    catch  (...)
    {
        _interface_throw  CInterfaceExcept;
        usServerInitAlert = 1;
        priv_strError = "Can't init interface";
        CInterfaceExcept.strErrorMsg = priv_strError.c_str();
        CInterfaceExcept._operation_state = 1;
        throw  (CInterfaceExcept);
    }
}

// ...

/****************************************************************
 * class test_server
 * function _vPeriodicServerActivity
 ****************************************************************/
/*! Periodic tasks done at the loop time (defined by user)
 * \param —
 * \return —
 ****************************************************************/
/* author Alexander Neidhardt
 * date 14.05.2007
 * revision —
 * info Part of the idl2rpc.pl — generator!
 ****************************************************************/
void  test_server::_vPeriodicServerActivity ()
{
    // USERDEFMETHODEBEG: Userdefined methode body
    try
    {
        int iVal = CValue.CGetSemVal();
        iVal++;
        CValue.vSetSemVal(iVal);
    }
    catch (test_semvar_throw CError)
    {
        // Sepcial error handling
        // ...
    }
    catch (...)
    {
        // Unknown error handling
        // ...
    }
    // USERDEFMETHODEEND
}

// ...

/****************************************************************
 * class test_server
 * function iGet
 ****************************************************************/
/*! Generated interface methode. See interface definition file! (defined by user)
 ****************************************************************/
/* author Alexander Neidhardt
 * date 14.05.2007
 * revision —
 * info Part of the idl2rpc.pl — generator!
 ****************************************************************/
int  test_server::iGet () throw (test_interface_throw)
```

56
57
58
59
60
61
62
63
64
65
66
67
68
69
70
71
72
73
74
75
76
77
78
79
80
81
82
83
84
85
86
87
88
89
90
91
92
93
94
95
96
97
98
99
100
101
102
103
104
105
106
107
108
109
110
111
112
113
114
115
116
117

3

```
{
    // USERDEFMETHODEBEG: Userdefined methode body      118
    int iRet = 0;                                        119
    try                                                  120
    {                                                    121
        iRet = CValue.CGetSemVal();                      122
    }                                                    123
    catch (test_semvar_throw CError)                     124
    {                                                    125
        // Sepcial error handling                        126
        return 1;                                        127
    }                                                    128
    catch (...)                                          129
    {                                                    130
        // Unknown error handling                        131
        return 1;                                        132
    }                                                    133
    return iRet;                                         134
    // USERDEFMETHODEEND                                 135
}                                                        136
                                                         137
// ...                                                   138
                                                         139
```

Besides the new OOP style of generated code and a better handling, additional modules also allow more safety. One part is the explicit use of threads in combination with semaphores, as shown in the previous examples (see also ▶ Sect. 2.6.4). But there are additional techniques included to increase the safety of the client, the RPC communication, and the server.

Safety

Generally, safety describes intuitive definitions about reliability, availability, stability, and protection of the designed system, its components, and its users. A general definition of a basic safety might be the following[48]:

Safety

> *Definition DEF3.11. of «safety»:* (see (Langmann 2010, *l.c. page 512*))
> Safety is the lack of and the protection from unexpected fluctuations of the system behavior, even under unpredicted, internal, or external influences. Safety is the foundation for the high availability of a properly working and reliable system which behaves as specified.
> Therefore, safety ensures that the functionality works as expected. There are some overlaps to the definition of system security (see more about security in ▶ Sect. 5.5) which protects the system from inexpedient use especially from unauthorized access from outside the system or from changes in behavior because of unexpected inputs. A stable system deals with such influences from the external world according to a predefined scheme. In a well-designed system, all layers offer a high level of safety which can be trusted to and used from higher layers, for example, system monitoring and safety structures (see ▶ Sect. 4.7). The following blocks describe the lower levels of safety, offered within the communication layer of «idl2rpc.pl».

Client Safety

All of the internal accesses to the real RPC communication in a client are defined within a critical section, which is implemented with an internal semaphore in the client class (see also ▶ Sect. 2.6.4). Parallel calls from different parallel client threads

---

48 See ▶ Sect. 4.7 for more about the safety of a complete system.

automatically protect the communication from reentrant uses. This increases the stability of the client in multi-thread environments, usually in combination with graphical user interfaces. Additionally, most of the activities are defined with error codes to identify the cause of an error. Errors in the RPC communication are always reported with C++ exceptions, which must be handled as shown in Example 3.16. Two functions are available for that: «`_strGetErrorMsg()`» returns a human-readable error message and «`_usGetOperationState()`» returns a state information about the current call. It offers information about the processing progress by using an integer number of the following values:

- 0: Operation completely processed.
- 1: Operation started but not finished.
- 2: Operation caused errors.
- 3: Operation produced a memory access error.
- 4: Operation had too many dynamic heap parameters (max. 256 dynamic parameters are allowed).
- 5: Operation cannot be processed with the current transport protocol settings of UDP because of its large data volume.
- 112: Client function and server function are incompatible (hash check failure).
- 120: Authentication with username and password is incorrect (authentication failure).
- 150: Client versions (of IDL specification and of generator) and server versions are different (version failure).
- 230: Authorization failure.

The communication uses the safety techniques of the lower protocol layers. For example, TCP offers a connection-oriented safety. Sun RPC adds safety to a connection by offering the port mapping with the information about a program and version number plus the procedure number. The data representation with XDR provides data safety on different machines. This is very reliable as long as a client for a dedicated RPC communication does not try to connect directly to a server with another specification of an RPC communication. Direct connections do not check the program and version number and have only the procedure number as a criterion to identify remote procedures. At best, such connection attempts result in server crashes. In the worst case scenario, they can run a remote procedure with illegal arguments. This is the point where safety issues meet security issues (see ▶ Sect. 5.5.2).

Communication Safety

To prevent this, «idl2rpc.pl» adds a version check with a version string to each open activity of an interface connection. This version check includes the comparison of MD5 checksums (see ▶ Sect. 5.5.2 about MD5) of the server with those from the client. The MD5 checksums are derived from the generated code from the interface definitions. After each successful opening of a socket connection, the client sends its checksums from the IDL interface file and from the «idl2rpc.pl» generator program to the server, where they are validated with the checksums of the server (see ▶ Sect. «Check versions» in the code of Example 3.19). If the comparison fails, the connection is closed again. Additionally «idl2rpc.pl» adds MD5 checksums in front of the data of each RPC request. The function checksums are calculated during the generation of the code from the client function code of the individual remote function. The checksums are validated by the server at each request and before all other activities in the server code. If the server sum is different to the received client sum, the request is rejected and the remote function is not activated (see the ▶ Sect. «Initial checks» in the code of Example 3.19). The only possibility for an attacker is to send data combinations which are a mismatch for the lower Sun RPC layer, for example, during the processing of dynamic arrays. If this data incompatibility leads to a memory mismatch in the Sun RPC part, the server crashes, but is restarted by the watchdog again. As the Sun RPC code is an external package, this cannot be avoided. But nevertheless unauthorized functions cannot be started from remote using the checksum validation, and this offers the safety required. It protects the system from critical external activities.

Communication Checksums

■ ■ Example 3.19

The use of MD5 checksums for the check of program and function versions.

```
/********************************************************************
 *   function   sap_eremotectrl_basiccheckidlversion_1_svc
 ********************************************************************/
/*!          Always usable, given RPC-method to check idl and interface version the server
 *   \param    INSAP_eremotectrl_CheckIDLVersionType_ * pSInput -> Input structure with function
 *          parameter
 *   \param    struct svc_req * pSSvc -> Input structure with server/client info
 *   \return   OUTSAP_eremotectrl_CheckIDLVersionType_ * <- Answer
 ********************************************************************/
/*   author    Alexander Neidhardt
 *   date      30.10.2008
 *   revision  -
 *   info      Part of the idl2rpc.pl - generator!
 ********************************************************************/
extern   bool_t sap_eremotectrl_basiccheckidlversion_1_svc (
     INSAP_eremotectrl_CheckIDLVersionType_ * pSInput, OUTSAP_eremotectrl_CheckIDLVersionType_
     * pSOutput, struct svc_req * pSSvc)
{
    unsigned short usMethodID = g_SServerMethods->MID_usCheckIDLVersion;
    /// RPC-in/out-structures
    unsigned short __usInternalError = 0;
    std::string strIDL2RPCVersion;
    (void)pSSvc;    // Reduce compiler warnings because of unused variables
    strIDL2RPCVersion = pSInput->strIDL2RPCVersion.strIDL2RPCVersion_val;
    std::string strInterfaceVersion;
    strInterfaceVersion = pSInput->strInterfaceVersion.strInterfaceVersion_val;
    static OUTSAP_eremotectrl_CheckIDLVersionType_ SOutput;
    // ============================================================
    /// Variable init
    SOutput.strIDL2RPCVersion.strIDL2RPCVersion_val = NULL;
    SOutput.strIDL2RPCVersion.strIDL2RPCVersion_len = 0;
    SOutput.strInterfaceVersion.strInterfaceVersion_val = NULL;
    SOutput.strInterfaceVersion.strInterfaceVersion_len = 0;
    SOutput._Ret = 0; // NOK
    SOutput.iInterfaceVersionOK = 0; // NOK
    SOutput.iIDL2RPCVersionOK = 0; // NOK
    SOutput._operation_state = 0; // Operation processed
    // ============================================================
    /// Fill in-structure
    _vResetDynVarMemory();
    // ============================================================
    /// Initial checks
    (void)usMethodID; // Reduce compiler warnings because of unused variables
    if (g_SServerMethods->_usAuthorize (pSInput->_uiUserID, usMethodID))
    {
        SOutput._uiUserID = pSInput->_uiUserID;
        SOutput._operation_state = 230; // Wrong user identification
        *pSOutput = SOutput;
        return (bool_t)1;
    }
    SOutput._uiUserID = pSInput->_uiUserID;
    if (strncmp ("abaec8a24557094e507aba76213373f0", pSInput->_acFunctionMD5, strlen ("
        abaec8a24557094e507aba76213373f0")))
    {
        SOutput._operation_state = 112; // Wrong MD5 checksum
        goto sap_eremotectrl_basiccheckidlversion_1_svc_Return;
    }
    // ============================================================
    /// Set return parameters
    if ((SOutput.strIDL2RPCVersion.strIDL2RPCVersion_val = (char *) malloc (sizeof(char)*(
        strlen ("V2.0/4537-2013-02-27-170306(2de772ecf5a1d691b43a0e6fa8e5cbfd)")+1))) == NULL)
```

```
{                                                                              57
    __usInternalError = 1;                                                     58
    SOutput._operation_state = 3; // Memory error                             59
    goto sap_eremotectrl_basiccheckidlversion_1_svc_Return;                   60
}                                                                              61
strcpy (SOutput.strIDL2RPCVersion.strIDL2RPCVersion_val, "V2.0/4537-2013-02-27-170306(2  62
    de772ecf5a1d691b43a0e6fa8e5cbfd)");
SOutput.strIDL2RPCVersion.strIDL2RPCVersion_len = sizeof(char)*(strlen("V2   63
    .0/4537-2013-02-27-170306(2de772ecf5a1d691b43a0e6fa8e5cbfd)")+1);
if ((SOutput.strInterfaceVersion.strInterfaceVersion_val = (char *) malloc (sizeof(char)*(  64
    strlen("#20121017002(8be51eec43ac928c508da95bea67c214)")+1))) == NULL)
{                                                                              65
    __usInternalError = 1;                                                     66
    SOutput._operation_state = 3; // Memory error                             67
    goto sap_eremotectrl_basiccheckidlversion_1_svc_Return;                   68
}                                                                              69
strcpy (SOutput.strInterfaceVersion.strInterfaceVersion_val, "#20121017002(8  70
    be51eec43ac928c508da95bea67c214)");
SOutput.strInterfaceVersion.strInterfaceVersion_len = sizeof(char)*(strlen("#20121017002(8  71
    be51eec43ac928c508da95bea67c214)")+1);
// =========================================================                   72
/// Check versions                                                             73
if (strIDL2RPCVersion == "V2.0/4537-2013-02-27-170306(2de772ecf5a1d691b43a0e6fa8e5cbfd)")  74
{                                                                              75
    SOutput.iIDL2RPCVersionOK = 1; // OK                                      76
}                                                                              77
if (strInterfaceVersion == "#20121017002(8be51eec43ac928c508da95bea67c214)")  78
{                                                                              79
    SOutput.iInterfaceVersionOK = 1; // OK                                    80
}                                                                              81
if (SOutput.iInterfaceVersionOK && SOutput.iIDL2RPCVersionOK)                 82
{                                                                              83
    SOutput._Ret = 1; // OK                                                    84
}                                                                              85
free (pSInput->strIDL2RPCVersion.strIDL2RPCVersion_val);                      86
pSInput->strIDL2RPCVersion.strIDL2RPCVersion_val = NULL;                      87
pSInput->strIDL2RPCVersion.strIDL2RPCVersion_len = 0;                         88
free (pSInput->strInterfaceVersion.strInterfaceVersion_val);                  89
pSInput->strInterfaceVersion.strInterfaceVersion_val = NULL;                  90
pSInput->strInterfaceVersion.strInterfaceVersion_len = 0;                     91
// =========================================================                   92
/// Return jump point                                                          93
sap_eremotectrl_basiccheckidlversion_1_svc_Return:                            94
// =========================================================                   95
/// Clean up                                                                   96
if (__usInternalError)                                                        97
{                                                                              98
                                                                               99
        free (pSInput->strIDL2RPCVersion.strIDL2RPCVersion_val);              100
        pSInput->strIDL2RPCVersion.strIDL2RPCVersion_val = NULL;              101
        pSInput->strIDL2RPCVersion.strIDL2RPCVersion_len = 0;                 102
                                                                               103
                                                                               104
        free (pSInput->strInterfaceVersion.strInterfaceVersion_val);          105
        pSInput->strInterfaceVersion.strInterfaceVersion_val = NULL;          106
        pSInput->strInterfaceVersion.strInterfaceVersion_len = 0;             107
                                                                               108
                                                                               109
        free (SOutput.strIDL2RPCVersion.strIDL2RPCVersion_val);               110
        SOutput.strIDL2RPCVersion.strIDL2RPCVersion_val = NULL;               111
        SOutput.strIDL2RPCVersion.strIDL2RPCVersion_len = 0;                  112
                                                                               113
                                                                               114
```

```
       free (SOutput.strInterfaceVersion.strInterfaceVersion_val);       115
       SOutput.strInterfaceVersion.strInterfaceVersion_val = NULL;       116
       SOutput.strInterfaceVersion.strInterfaceVersion_len = 0;          117
                                                                          118
   }                                                                      119
   // ===========================================================        120
   /// Return the return value                                           121
   *pSOutput = SOutput;                                                   122
       return (bool_t)1;                                                  123
}                                                                         124
```

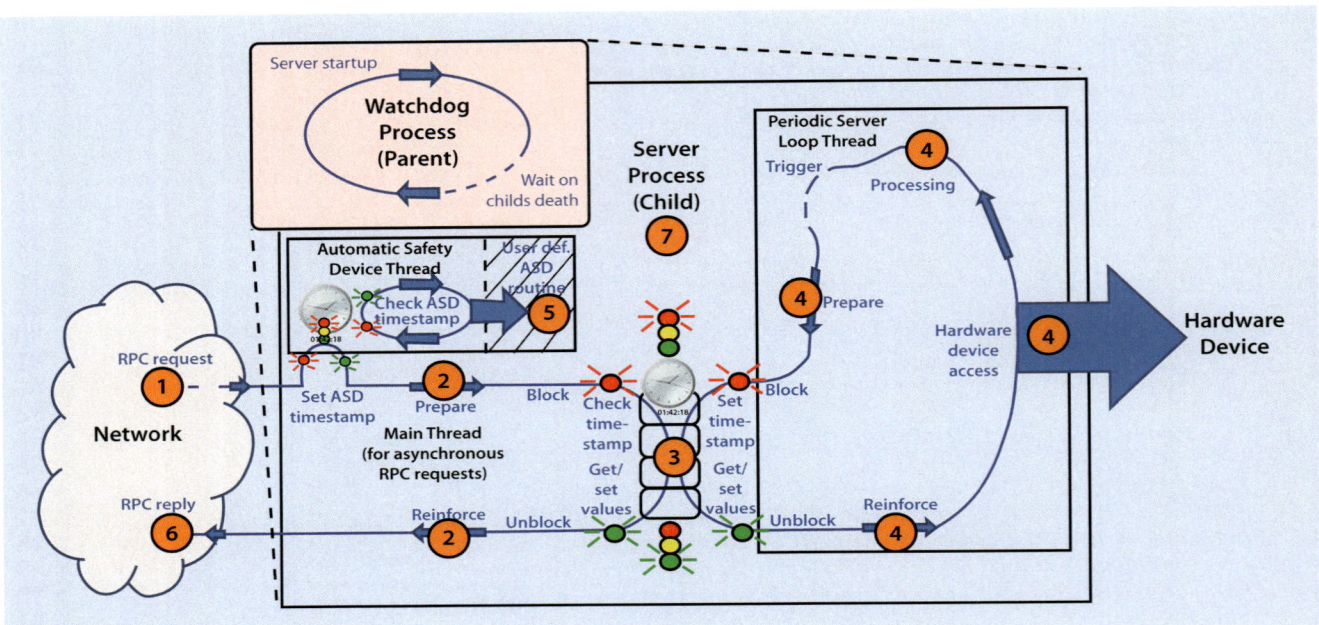

Fig. 3.22   The server safety techniques working together for an internal stability avoiding possible error points (marked with the included numbers)

Server Safety

The «idl2rpc.pl» generator also creates several possibilities of ensuring internal server safety. ☐ Figure 3.22 shows all the integrated techniques which work together. With such a setup, a requesting client always gets an answer immediately even if a connected hardware or processing event in the periodic server loop hangs. In detail the following techniques are used:

— A startup control
— A consequent processing separation with threads and semaphores
— A server-side watchdog process
— A server-side «automatic safety device» («ASD» argument of the «idl2rpc.pl» call)

Possible Error Sources

In principle, the safety techniques try to detect error states, which can happen at specific error points (see ☐ Fig. 3.22), and perform a suitable safety strategy to prevent damage to the server and connected devices. The possible error states are:

Client	Server
1. Request cannot be delivered	
If the server is not accessible or the network is down, the client ends up in an exception immediately; if the network does not work properly, the client ends up in an exception after the standard behavior of retransmissions	The server does not get notice of a connection request. After the expiration of the «ASD» time, the automatic safety device fires the «ASD» alerts (see ◘ Fig. 3.26)
2. Problems in main thread	
If the request never comes back («while(1)» loop), the client timeout expires and throws an exception; if the server crashes, the client ends up in an exception immediately	If the request loops endlessly («`while(1)`» loop), the server becomes unreachable (other threads are not influenced) and the automatic safety device will fire (e.g., the measuring of the downtime and a forced exit of the server for a following restart are possible here in the application code); if the server crashes, the watchdog restarts it immediately
3. Problems in critical section	
As in 2	If the request loops endlessly («`while(1)`» loop), the server becomes unreachable (critical for periodic threads, which will be blocked as well), and the automatic safety device will fire (e.g., the measuring of the downtime and a forced exit of the server for a following restart are possible here in the application code); if the server crashes, the watchdog restarts it immediately
4. Problems in periodic server loop thread	
Client does not get a notice of reply behavior; a check with the client timestamp will result in «not ok» as function return	If a section in the periodic server loops endlessly («`while(1)`» loop), the update of variables is not processed, and the device does not receive orders (it can be checked with the update of the «ASD» timestamp as well; e.g., the measuring of the down time and a forced exit of the server for a following restart are possible here in the application code); if the server crashes, the watchdog restarts it immediately
5. Problems in user-defined automatic safety device routine	
Client does not get notice of it	The safety device is not working correctly after it has been fired, but all of the other threads are completely functional (some checks in the periodic server loop thread can also keep the safe states); if the server crashes, the watchdog restarts it immediately
6. Reply cannot be delivered	
If the request never comes back with a reply, the client timeout expires and throws an exception; if the server crashes, the client ends up in an exception immediately	The server behavior is not influenced; if the server crashes, the watchdog restarts it immediately
7. General crash	
If the server crashes, the client class throws an exception immediately if a communication is currently active	If the server crashes, the watchdog restarts it immediately

3

Server Safety: StartUp Control

During the startup of the server, it checks if the same server is already running on the same machine. For this, it uses the «portmapper» entry in combination with the pinging mechanism. Because each server registers itself at the «portmapper», it is a central register for processes on the same machine. The existence of the port is also a clear sign to check if the server is already activated or if another server has occupied the port preferred. But checking the registration and the port is only half the job. Under certain circumstances, it is possible that servers have crashed, but have not cleared their registration. Therefore, a ping is sent to the server to check its reaction time. These checks are done using the client class also in the server. During the startup of the server, it uses an instance of the client to perform a ping to check if the server is already started or whether another process is using the port in question on the same computer. If none is found, the server starts.

Server Safety: Multi-threading

The next step to establish server safety in the remote functions themselves is to design a non-blocking behavior. Here the use of threads is very convenient. All the necessary parts are offered by the module file «<INTERFACEFILE>_simple_thread.cpp/.hpp».[49] They separate sequential behavior into (apparently) parallel working tasks (controlled by the scheduling mechanisms of the thread control in the operating system). This means that for a safe setup, the periodic loop should always be implemented as a separate thread (by using the «TLT» calling parameter as input for the generation). The advantage is that the asynchronous client requests, which are processed in the main thread, are not influenced by longer-lasting or broken activities in the periodic thread. It is a good mechanism to separate RPC requests from the periodic accesses to a hardware device as long as mutually blocking critical sections are reduced to the real change of common variables and do not include the hardware communication. Therefore, it does not matter how much time the device control needs or even if it hangs. The client will always be served immediately. It is like a «timing isolation» on software level, so that disturbances cannot diffuse through this barrier. To protect the mutual accesses, semaphores are the suggested mechanism (see ▶ Sect. 2.6.4).

Server Safety: Semaphores

A proper solution is offered by the semaphore-protected variables (see also ▶ Sect. 2.6.4) from the generated file «<INTERFACEFILE>_simple_semvar .cpp/.hpp».[50] By using such protected variables, it is safe to share variables between the periodic function «_vPeriodicServerActivity» and the asynchronously called RPC functions. Semaphore-protected variables only work for single values in the stack. Areas in the heap to which a pointer points are not directly protected. An explicit blocking and unblocking must be used for the protection of dynamic memory. The semaphore methods «vBlock(...)» and «vUnblock(...)» must be used. The same mechanism can be used if more than one variable needs to be changed at once, for example, to trigger activities of a state machine in the periodic loop. But the blocking and unblocking must be undertaken with great care, as forgetting to unblock a semaphore leads to a deadlock, as no further call will succeed anymore. This means that the blocked critical section should be limited to a function body and a very short code section there (e.g., just the copying of critical variables).

Server Safety: Time Limitations

To avoid long periods within critical sections, only the setting and getting of values should be protected. All of the preparations (like allocating memory, opening of file, checking method arguments, etc.), the processing of variable contents, and the reinforcement of values for following actions should be done outside. This means that the critical section is only occupied for a very short time. The replying function should also be kept as short as possible to reduce server occupation time, so that many parallel clients can compete successfully for server time. Large calculations or long-lasting hardware communications can be triggered by an RPC

---

49 *<INTERFACEFILE>* is replaced by the filename of the interface specification file, e.g., if the interface specification is in the file «*test*.idl», the resulting thread class is in the file «*test_*simple_thread.cpp» and «*test*_simple_thread.hpp».

50 See previous footnote.

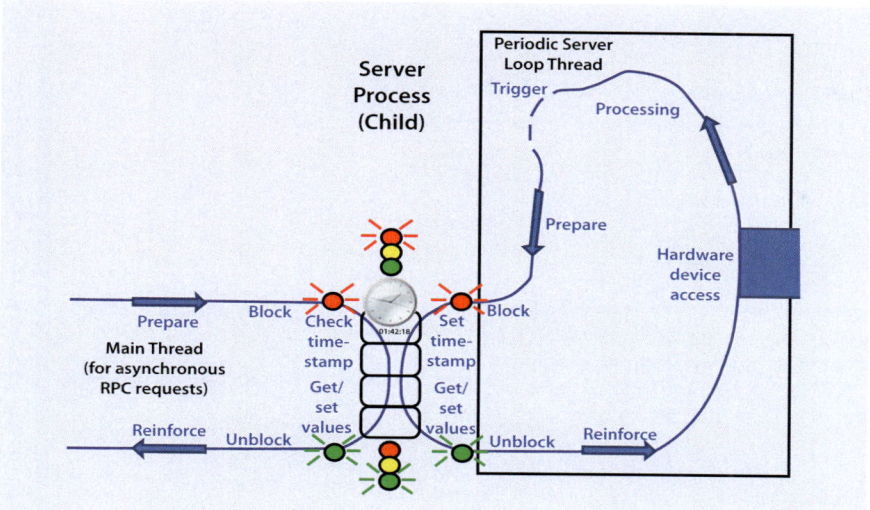

Fig. 3.23   Consequent processing separation with threads and semaphores

function, but should be processed in the periodic server loop or in another individual application thread.

Example 3.20 and Example 3.21 show the concept of semaphore-protected code sequences for the periodic server loop thread and the RPC main thread, written for the monitoring and control of the Holzworth synthesizer HS9000, used for the first-stage local oscillator of the VLBI twin telescopes of the Wettzell observatory. The periodic loop function in Example 3.20 uses a clear design to command the synthesizer with a state machine. It only protects the copying of the private member variables of the controlling class to local copies, which can be used without protection later. Errors while blocking and unblocking of semaphores are critical errors which require a restart of the server to reset them. The next step is reading the data from the hardware to have the latest values available. Using these values and the commanded order, the real hardware commanding can be implemented as a state machine with «switch/case» states. The processing in the state machine updates changed values from the hardware after the commanding. Finally, the new values must be copied from the local values into the member attributes of the controlling class to have them available for the RPC requests. This copying is protected as a critical section. Finally, the RPC requests are quite simple and need less time, as shown in Example 3.21. They only protect the copying between the member values and the input or return values of the RPC call. This code is a clean, safe, and less blocking solution for the items discussed in ◘ Figs. 3.8 and 3.23.

Server Safety: Demonstration

■■ **Example 3.20**
Excerpt of the periodic code, controlling and monitoring the Holzworth synthesizer HS9000, using a clear design for the semaphore protected sequences with short blocking times while copying the data

```
// ...                                                                          1
                                                                                2
/*******************************************************************            3
 *   class       rxmon_server                                                   4
 *   function    _vPeriodicServerActivity                                       5
 *******************************************************************/            6
/*!          Periodic tasks done at the loop time (defined by user)             7
 *   \param    —                                                                8
 *   \return   —                                                                9
 *******************************************************************/            10
```

```cpp
/*  author      Alexander Neidhardt
 *  date        14.05.2007
 *  revision    —
 *  info        Part of the idl2rpc.pl — generator!
 ********************************************************************/
void rxmon_server::_vPeriodicServerActivity ()
{
    // USERDEFMETHODBEG: Userdefined method body
    unsigned long ulBlockHandle = 0;
    bool bCommandOrdered = false;

    // ========================================================
    /// <b> Copy all private variables to a local copy </b>
    // ========================================================

    /// Block memory with semaphore
    try
    {
        priv_cProtectPrivateVariablesSemaphore.vBlock (ulBlockHandle);
    }
    catch (...)
    {
        std::cerr << "[ERROR] _vPeriodicServerActivity: Can't block semaphore \n";
        priv_uiFatalError = FATALERROR_SEMAPHORE;
        return;
    }

    /// Copy all private variables to a local copy
    // General server admistration ———————————————————
    unsigned int       uiFatalError = priv_uiFatalError;
    unsigned int       uiInternalError = priv_uiInternalError;
    unsigned long      ulInternalErrorTimestampInUnixtime =
        priv_ulInternalErrorTimestampInUnixtime;
    // priv_cProtectPrivateVariablesSemaphore ==> no local copy allowed
    // Configurations ————————————————————————————————
    // priv_CConfiguration    ==> no local copy required (just used in _vInitServerActivity)
    // priv_strConfigFilepath ==> no local copy required (just used in _vInitServerActivity)
    // Connected sensors ——————————————————————————————
    // priv_CLOSynthesizer    ==> no local copy required (just used here in
    //     _vPeriodicServerActivity)
    // priv_CReceiver         ==> no local copy required (just used here in
    //     _vPeriodicServerActivity)
    // Control ————————————————————————————————————————
    unsigned int       uiCommandId = priv_uiCommandId;
    if (priv_uiCommandId != COMMANDID_NOCOMMAND)
    {
        bCommandOrdered = true;
    }
    // Status data from the LO synthesizer ————————————
    unsigned short     usNumberOfChannels = priv_usNumberOfChannels;
    double             dCurrentChannelFrequenciesMHz[4];
    double             dCurrentChannelMaxFrequenciesMHz[4];
    double             dCurrentChannelMinFrequenciesMHz[4];
    double             dCurrentChannelPowerDBm[4];
    double             dCurrentChannelMaxPowerDBm[4];
    double             dCurrentChannelMinPowerDBm[4];
    double             dCurrentChannelPhaseDeg[4];
    double             dCurrentChannelMaxPhaseDeg[4];
    double             dCurrentChannelMinPhaseDeg[4];
    double             dCurrentChannelTemperatureDegC[4];
    bool               bCurrentChannelPowerOn[4];
    for (unsigned int iIndex = 0; iIndex < 4; ++iIndex)
```

```
{                                                                                70
    dCurrentChannelFrequenciesMHz[iIndex] = priv_dCurrentChannelFrequenciesMHz[iIndex];    71
    dCurrentChannelMaxFrequenciesMHz[iIndex] = priv_dCurrentChannelMaxFrequenciesMHz[      72
        iIndex];
    dCurrentChannelMinFrequenciesMHz[iIndex] = priv_dCurrentChannelMinFrequenciesMHz[      73
        iIndex];
    dCurrentChannelPowerDBm[iIndex] = priv_dCurrentChannelPowerDBm[iIndex];                74
    dCurrentChannelMaxPowerDBm[iIndex] = priv_dCurrentChannelMaxPowerDBm[iIndex];          75
    dCurrentChannelMinPowerDBm[iIndex] = priv_dCurrentChannelMinPowerDBm[iIndex];          76
    dCurrentChannelPhaseDeg[iIndex] = priv_dCurrentChannelPhaseDeg[iIndex];                77
    dCurrentChannelMaxPhaseDeg[iIndex] = priv_dCurrentChannelMaxPhaseDeg[iIndex];          78
    dCurrentChannelMinPhaseDeg[iIndex] = priv_dCurrentChannelMinPhaseDeg[iIndex];          79
    dCurrentChannelTemperatureDegC[iIndex] = priv_dCurrentChannelTemperatureDegC[iIndex];  80
    bCurrentChannelPowerOn[iIndex] = priv_bCurrentChannelPowerOn[iIndex];                  81
}                                                                                82
// Values for the LO synthesizer ——————————————                                  83
double              dChannelFrequencyMHz = priv_dChannelFrequencyMHz;            84
double              dChannelPowerDBm = priv_dChannelPowerDBm;                    85
// Status data from the receiver ——————————————                                  86
// Values for the receiver ——————————————                                        87
                                                                                 88
/// Unblock semaphore                                                            89
try                                                                              90
{                                                                                91
    priv_cProtectPrivateVariablesSemaphore.vUnblock (ulBlockHandle);            92
}                                                                                93
catch (...)                                                                      94
{                                                                                95
    std::cerr << "[ERROR] _vPeriodicServerActivity: Can't block semaphore \n";  96
    priv_uiFatalError = FATALERROR_SEMAPHORE;                                    97
    return;                                                                      98
}                                                                                99
                                                                                 100
// ========================================================                      101
/// <b> Read all values from the devices </b>                                    102
// ========================================================                      103
                                                                                 104
/// Check fatal error state                                                      105
if (uiFatalError)                                                                106
{                                                                                107
    std::cerr << "[ERROR] _vPeriodicServerActivity: Still in fatal error state\n";  108
    sleep (1);                                                                   109
    switch (uiFatalError)                                                        110
    {                                                                            111
        case FATALERROR_CONFIGFILE:                                             112
        case FATALERROR_SEMAPHORE:                                              113
        case FATALERROR_LOSYNTH:                                                114
        case FATALERROR_RECEIVER:                                               115
        {                                                                        116
            exit (0);                                                            117
        }                                                                        118
        default:                                                                 119
        {                                                                        120
            std::cerr << "[ERROR] _vPeriodicServerActivity: undefined fatal error\n";  121
            exit (0);                                                            122
        }                                                                        123
    }                                                                            124
}                                                                                125
                                                                                 126
/// Read status from LO synthesizer                                              127
try                                                                              128
{                                                                                129
    std::string strUni;                                                          130
    for (unsigned short usChannelNumber = 1;                                    131
        usChannelNumber <= usNumberOfChannels;                                  132
        ++usChannelNumber)                                                      133
```

```cpp
        {
            /// Read temperature
            if (priv_CLOSynthesizer.usGetChannelTemperature (usChannelNumber,
                                        dCurrentChannelTemperatureDegC[
                                              usChannelNumber −1],
                                        strUni) == synth_hs9000::
                                              SYNTH_HS9000_NOK)
            {
                std::cerr << "[ERROR] _vInitServerActivity: Cannot read temperature of channel
                            "
                            << usChannelNumber << "\n";
                goto Ret_vPeriodicServerActivitySemUnblock;
            }
        }
    }
    catch (...)
    {
        std::cerr << "[ERROR] _vInitServerActivity: Cannot read temperature of channel\n";
        goto Ret_vPeriodicServerActivitySemUnblock;
    }

    // ========================================================
    /// <b> Command devices and read changed values </b>
    // ========================================================

    /// Command LO synthesizer
    try
    {
        /// Reset internal error if new command arrived
        if (uiCommandId != COMMANDID_NOCOMMAND)
        {
            uiInternalError = INTERNALERROR_OK;
            ulInternalErrorTimestampInUnixtime = 0;
        }
        switch (uiCommandId)
        {
            case COMMANDID_NOCOMMAND:
            {
                break;
            }
            case COMMANDID_SETLOCALOSCILLATOR1:
            {
                /// Set local oscillator 1
                bool bPowerStatus;
                unsigned int uiError;
                if ((uiError = uiSetLocalOscillator (1,
                                            dChannelFrequencyMHz,
                                            dChannelPowerDBm,
                                            bPowerStatus)) != INTERNALERROR_OK)
                {
                    uiInternalError = uiError;
                    ulInternalErrorTimestampInUnixtime = time(NULL);
                    goto Ret_vPeriodicServerActivitySemUnblock;
                }
                dCurrentChannelFrequenciesMHz[0] = dChannelFrequencyMHz;
                dCurrentChannelPowerDBm[0] = dChannelPowerDBm;
                bCurrentChannelPowerOn[0] = bPowerStatus;
                break;
            }
```

134
135
136
137
138
139
140
141
142
143
144
145
146
147
148
149
150
151
152
153
154
155
156
157
158
159
160
161
162
163
164
165
166
167
168
169
170
171
172
173
174
175
176
177
178
179
180
181
182
183
184
185
186
187
188
189

```
case COMMANDID_SETLOCALOSCILLATOR2:                                          190
{                                                                            191
    /// Set local oscillator 2                                               192
    bool bPowerStatus;                                                       193
    unsigned int uiError;                                                    194
    if ((uiError = uiSetLocalOscillator (2,                                  195
                                         dChannelFrequencyMHz,               196
                                         dChannelPowerDBm,                   197
                                         bPowerStatus)) != INTERNALERROR_OK) 198
    {                                                                        199
        uiInternalError = uiError;                                           200
        ulInternalErrorTimestampInUnixtime = time(NULL);                     201
        goto Ret_vPeriodicServerActivitySemUnblock;                          202
    }                                                                        203
    dCurrentChannelFrequenciesMHz[1] = dChannelFrequencyMHz;                 204
    dCurrentChannelPowerDBm[1] = dChannelPowerDBm;                           205
    bCurrentChannelPowerOn[1] = bPowerStatus;                                206
    break;                                                                   207
}                                                                            208
case COMMANDID_SETLOCALOSCILLATOR3:                                          209
{                                                                            210
    /// Set local oscillator 3                                               211
    bool bPowerStatus;                                                       212
    unsigned int uiError;                                                    213
    if ((uiError = uiSetLocalOscillator (3,                                  214
                                         dChannelFrequencyMHz,               215
                                         dChannelPowerDBm,                   216
                                         bPowerStatus)) != INTERNALERROR_OK) 217
    {                                                                        218
        uiInternalError = uiError;                                           219
        ulInternalErrorTimestampInUnixtime = time(NULL);                     220
        goto Ret_vPeriodicServerActivitySemUnblock;                          221
    }                                                                        222
    dCurrentChannelFrequenciesMHz[2] = dChannelFrequencyMHz;                 223
    dCurrentChannelPowerDBm[2] = dChannelPowerDBm;                           224
    bCurrentChannelPowerOn[2] = bPowerStatus;                                225
    break;                                                                   226
}                                                                            227
case COMMANDID_SETLOCALOSCILLATOR4:                                          228
{                                                                            229
    /// Set local oscillator 4                                               230
    bool bPowerStatus;                                                       231
    unsigned int uiError;                                                    232
    if ((uiError = uiSetLocalOscillator (4,                                  233
                                         dChannelFrequencyMHz,               234
                                         dChannelPowerDBm,                   235
                                         bPowerStatus)) != INTERNALERROR_OK) 236
    {                                                                        237
        uiInternalError = uiError;                                           238
        ulInternalErrorTimestampInUnixtime = time(NULL);                     239
        goto Ret_vPeriodicServerActivitySemUnblock;                          240
    }                                                                        241
    dCurrentChannelFrequenciesMHz[3] = dChannelFrequencyMHz;                 242
    dCurrentChannelPowerDBm[3] = dChannelPowerDBm;                           243
    bCurrentChannelPowerOn[3] = bPowerStatus;                                244
    break;                                                                   245
}                                                                            246
default:                                                                     247
{                                                                            248
    break;                                                                   249
}                                                                            250
}                                                                            251
}                                                                            252
```

```cpp
      catch (...)
      {
          std::cerr << "[ERROR] _vPeriodicServerActivity: Can't command ACU \n";
          uiFatalError = FATALERROR_MEMORY;
          goto Ret_vPeriodicServerActivitySemUnblock;
      }

Ret_vPeriodicServerActivitySemUnblock:
      /// Reset command identification
      uiCommandId = COMMANDID_NOCOMMAND;
      /// Reset internal error after defined time
      if (uiInternalError &&
            ((time(NULL) - ulInternalErrorTimestampInUnixtime) > 60))
      {
          uiInternalError = INTERNALERROR_OK;
          ulInternalErrorTimestampInUnixtime = 0;
      }


      // =======================================================
      /// <b> Copy all local copies of values to the private variables </b>
      // =======================================================


      /// Block memory with semaphore
      try
      {
          priv_cProtectPrivateVariablesSemaphore.vBlock (ulBlockHandle);
      }
      catch (...)
      {
          std::cerr << "[ERROR] _vPeriodicServerActivity: Can't block semaphore \n";
          priv_uiFatalError = FATALERROR_SEMAPHORE;
          return;
      }


      /// Copy all local copies of values to the private variables
      // General server admistration ───────────────────
      priv_uiFatalError = uiFatalError;
      priv_uiInternalError = uiInternalError;
      priv_ulInternalErrorTimestampInUnixtime = ulInternalErrorTimestampInUnixtime;
      // priv_cProtectPrivateVariablesSemaphore ==> no local copy allowed
      // Configurations ───────────────────────
      // priv_CConfiguration    ==> no local copy required (just used in _vInitServerActivity)
      // priv_strConfigFilepath ==> no local copy required (just used in _vInitServerActivity)
      // Connected sensors ─────────────────────
      // priv_CLOSynthesizer    ==> no local copy required (just used here in
      //    _vPeriodicServerActivity)
      // priv_CReceiver         ==> no local copy required (just used here in
      //    _vPeriodicServerActivity)
      // Control ────────────────────────────
      if (bCommandOrdered)
      {
          priv_uiCommandId = uiCommandId;
      }
      // Status data from the LO synthesizer ──────────
      for (unsigned int iIndex = 0; iIndex < 4; ++iIndex)
      {
          priv_dCurrentChannelFrequenciesMHz[iIndex] = dCurrentChannelFrequenciesMHz[iIndex];
          priv_dCurrentChannelMaxFrequenciesMHz[iIndex] = dCurrentChannelMaxFrequenciesMHz[
              iIndex];
          priv_dCurrentChannelMinFrequenciesMHz[iIndex] = dCurrentChannelMinFrequenciesMHz[
              iIndex];
          priv_dCurrentChannelPowerDBm[iIndex] = dCurrentChannelPowerDBm[iIndex];
          priv_dCurrentChannelMaxPowerDBm[iIndex] = dCurrentChannelMaxPowerDBm[iIndex];
          priv_dCurrentChannelMinPowerDBm[iIndex] = dCurrentChannelMinPowerDBm[iIndex];
          priv_dCurrentChannelPhaseDeg[iIndex] = dCurrentChannelPhaseDeg[iIndex];
```

```
        priv_dCurrentChannelMaxPhaseDeg[iIndex] = dCurrentChannelMaxPhaseDeg[iIndex];     314
        priv_dCurrentChannelMinPhaseDeg[iIndex] = dCurrentChannelMinPhaseDeg[iIndex];     315
        priv_dCurrentChannelTemperatureDegC[iIndex] = dCurrentChannelTemperatureDegC[iIndex];  316
        priv_bCurrentChannelPowerOn[iIndex] = bCurrentChannelPowerOn[iIndex];             317
    }                                                                                     318
    // Values for the LO synthesizer ─────────────────────                               319
    if (bCommandOrdered)                                                                  320
    {                                                                                     321
        priv_dChannelFrequencyMHz = dChannelFrequencyMHz;                                 322
        priv_dChannelPowerDBm = dChannelPowerDBm;                                         323
    }                                                                                     324
    // Status data from the receiver ─────────────────────                               325
    // Values for the receiver ─────────────────────                                     326
    if (bCommandOrdered)                                                                  327
    {                                                                                     328
        bCommandOrdered = false;                                                          329
    }                                                                                     330
                                                                                          331
    /// Unblock semaphore                                                                 332
    try                                                                                   333
    {                                                                                     334
        priv_cProtectPrivateVariablesSemaphore.vUnblock (ulBlockHandle);                  335
    }                                                                                     336
    catch (...)                                                                           337
    {                                                                                     338
        std::cerr << "[ERROR] _vPeriodicServerActivity: Can't block semaphore \n";        339
        priv_uiFatalError = FATALERROR_SEMAPHORE;                                         340
        return;                                                                           341
    }                                                                                     342
    // USERDEFMETHODEND                                                                   343
}                                                                                         344
                                                                                          345
// ...                                                                                    346
```

■■ Example 3.21

Excerpt of the RPC code, manipulating the Holzworth synthesizer HS9000, using semaphore protected sequences to inject new commands and values and to read current values, prepared by the periodic loop

```
// ...                                                                                    1
                                                                                          2
/*******************************************************************                      3
 *  class       rxmon_server                                                              4
 *  function    usGetStatusOfLocalOscillator1                                             5
 *******************************************************************/                      6
/*!            Generated interface method. See interface definition file! (defined by user)  7
 *******************************************************************/                      8
/*  author      Alexander Neidhardt                                                       9
 *  date        14.05.2007                                                                10
 *  revision    —                                                                         11
 *  info        Part of the idl2rpc.pl — generator!                                       12
 *******************************************************************/                      13
unsigned short rxmon_server::usGetStatusOfLocalOscillator1 (double &                      14
    dCurrentChannelFrequencyMHz, double & dCurrentChannelPowerDBm, double &
    dCurrentChannelPhaseDeg, double & dCurrentChannelTemperatureDegC, unsigned short &
    usChannelIsSwitchedOn) throw (_interface_throw)
{                                                                                         15
    // USERDEFMETHODBEG: Userdefined method body                                          16
    unsigned long ulBlockHandle = 0;                                                      17
    unsigned short usRetVal = RXMON_OK;                                                   18
                                                                                          19
```

```
    // Set default return values
    dCurrentChannelFrequencyMHz = 0.0;
    dCurrentChannelPowerDBm = 0.0;
    dCurrentChannelPhaseDeg = 0.0;
    dCurrentChannelTemperatureDegC = 0.0;
    usChannelIsSwitchedOn = RXMON_RFSPOWERIS_OFF;

    /// Block memory with semaphore
    try
    {
        priv_cProtectPrivateVariablesSemaphore.vBlock (ulBlockHandle);
    }
    catch (...)
    {
        std::cerr << "[ERROR] usGetStatusOfLocalOscillator1: Can't block semaphore \n";
        priv_uiFatalError = FATALERROR_SEMAPHORE;
        return RXMON_NOK;
    }

    /// Check fatal state and check if a previous command is undone
    if (priv_uiFatalError)
    {
        usRetVal = RXMON_NOK;
        goto ReturnSemaphore;
    }

    /// Set return values
    dCurrentChannelFrequencyMHz = priv_dCurrentChannelFrequenciesMHz[0];
    dCurrentChannelPowerDBm = priv_dCurrentChannelPowerDBm[0];
    dCurrentChannelPhaseDeg = priv_dCurrentChannelPhaseDeg[0];
    dCurrentChannelTemperatureDegC = priv_dCurrentChannelTemperatureDegC[0];
    if (priv_bCurrentChannelPowerOn[0])
    {
        usChannelIsSwitchedOn = RXMON_RFSPOWERIS_ON;
    }
    else
    {
        usChannelIsSwitchedOn = RXMON_RFSPOWERIS_OFF;
    }

    /// Check if doubler mode is used and divide by 2
    if (priv_bUseFrequencyDoublerMode == true)
    {
        dCurrentChannelFrequencyMHz = dCurrentChannelFrequencyMHz * 2.0;
    }

ReturnSemaphore:
    /// Unblock memory with semaphore
    try
    {
        priv_cProtectPrivateVariablesSemaphore.vUnblock (ulBlockHandle);
    }
    catch (...)
    {
        std::cerr << "[ERROR] usGetStatusOfLocalOscillator1: Can't block semaphore \n";
        priv_uiFatalError = FATALERROR_SEMAPHORE;
        usRetVal = RXMON_NOK;
    }

    return usRetVal;
    // USERDEFMETHODEND
}

// ...
```

20
21
22
23
24
25
26
27
28
29
30
31
32
33
34
35
36
37
38
39
40
41
42
43
44
45
46
47
48
49
50
51
52
53
54
55
56
57
58
59
60
61
62
63
64
65
66
67
68
69
70
71
72
73
74
75
76
77
78
79
80
81
82
83
84

```
/***********************************************************************    85
 *   class      rxmon_server                                              86
 *   function   usSetLocalOscillator1                                     87
 ***********************************************************************/   88
/*!          Generated interface method. See interface definition file! (defined by user)  89
 ***********************************************************************/   90
/*   author    Alexander Neidhardt                                        91
 *   date      14.05.2007                                                 92
 *   revision  —                                                         93
 *   info      Part of the idl2rpc.pl — generator!                       94
 ***********************************************************************/   95
unsigned short rxmon_server::usSetLocalOscillator1 (double   dCurrentChannelFrequencyMHz,   96
    double  dCurrentChannelPowerDBm) throw (_interface_throw)
{                                                                          97
    // USERDEFMETHODBEG: Userdefined method body                          98
    unsigned long ulBlockHandle = 0;                                      99
    unsigned short usRetVal = RXMON_OK;                                  100
                                                                         101
    /// Block memory with semaphore                                     102
    try                                                                  103
    {                                                                    104
        priv_cProtectPrivateVariablesSemaphore.vBlock (ulBlockHandle);  105
    }                                                                    106
    catch (...)                                                          107
    {                                                                    108
        std::cerr << "[ERROR] usSetLocalOscillator1: Can't block semaphore \n";  109
        priv_uiFatalError = FATALERROR_SEMAPHORE;                       110
        return RXMON_NOK;                                                111
    }                                                                    112
                                                                         113
    /// Check fatal state and check if a previous command is undone     114
    if (priv_uiFatalError ||                                            115
            priv_uiCommandId != COMMANDID_NOCOMMAND)                     116
    {                                                                    117
        usRetVal = RXMON_NOK;                                            118
        goto ReturnSemaphore;                                           119
    }                                                                    120
                                                                         121
    /// Check if doubler mode is used and divide by 2                   122
    if (priv_bUseFrequencyDoublerMode == true)                          123
    {                                                                    124
        dCurrentChannelFrequencyMHz = dCurrentChannelFrequencyMHz/2.0;  125
    }                                                                    126
                                                                         127
    /// Check input variables                                           128
    if (dCurrentChannelFrequencyMHz < priv_dCurrentChannelMinFrequenciesMHz[0] ||  129
        dCurrentChannelFrequencyMHz > priv_dCurrentChannelMaxFrequenciesMHz[0])    130
    {                                                                    131
        usRetVal = RXMON_NOK;                                            132
        goto ReturnSemaphore;                                           133
    }                                                                    134
                                                                         135
    /// Set processing varibales for periodic loop                      136
    priv_dChannelFrequencyMHz = dCurrentChannelFrequencyMHz;            137
    priv_dChannelPowerDBm = dCurrentChannelPowerDBm;                    138
    priv_uiCommandId = COMMANDID_SETLOCALOSCILLATOR1;                   139
                                                                         140
ReturnSemaphore:                                                         141
    /// Unblock memory with semaphore                                   142
    try                                                                  143
    {                                                                    144
        priv_cProtectPrivateVariablesSemaphore.vUnblock (ulBlockHandle);  145
    }                                                                    146
```

```
    catch  (...)                                                       147
    {                                                                  148
        std::cerr << "[ERROR]  usSetLocalOscillator1:  Can't block semaphore \n";  149
        priv_uiFatalError = FATALERROR_SEMAPHORE;                      150
        usRetVal = RXMON_NOK;                                          151
    }                                                                  152
                                                                       153
    return  usRetVal;                                                  154
    //  USERDEFMETHODEND                                               155
}                                                                      156
                                                                       157
//  ...                                                                158
```

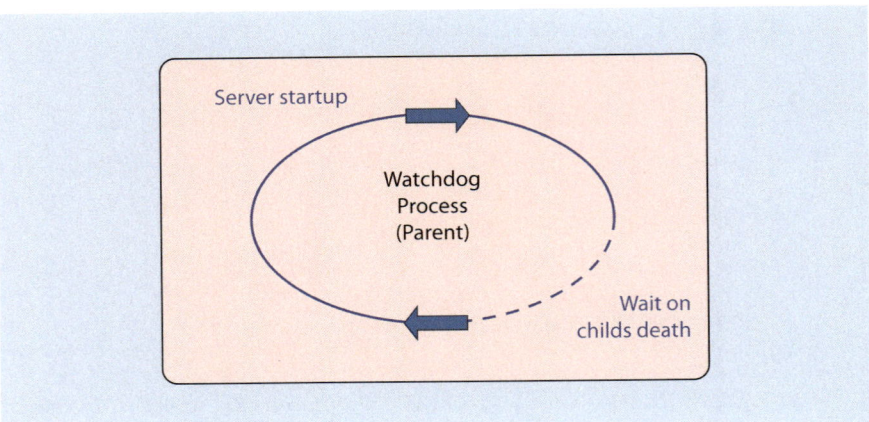

🔲 **Fig. 3.24**   A watchdog parent process

Server Safety: Watchdog

Another disturbing situation for clients is if the server crashes and is not available anymore. Such situations are detected by the clients when the request fails and timeouts expire. But originally, the server has no mechanism to clear the situation. This problem can be solved using another parallel process, the watchdog (see also ▶ Sect. 3.3.1). After the server is started, two processes become active: the actual RPC server and its parent process. The process which was originally started forks another child process. The parent process is then responsible for controlling the existence of the child process (see 🔲 Fig. 3.24), which serves the actual RPC requests from the clients. In principle, the parent process just waits for the «death» of the child process. After such an unauthorized shutdown of the child (e.g., because of crashes), the parent watchdog detects it, performs all the «on exit» activities necessary, closes all sockets, unregisters from the «portmapper», and restarts the child again within a few microseconds. During the start, all of the startup sequences are processed in the same way as during the usual first start. The watchdog keeps the replying child server alive and is accessible even when there are shutdowns (possibly caused by programming errors).[51]

Server Safety: Persistence and Idempotence

The principle of persistent and idempotent activities is very important for a correct restart of the RPC server. An idempotent function can be called several times without any negative consequences or wrong states at the server (see also ▶ Sect. 3.3.2). The state before, after, and even on error does not change the state of the server. Therefore, a crash within such a function is not critical, as it can be called again later without any precautions. If the server state is changed by a function, for

---

51 An instructed shutdown which forces the system to stop must call «_v<INTERFACE>Stop ServerLoop()» to ensure that the control flow should exit without an automated restart. For a sophisticated management of the server, the operating system signals to «kill» or «terminate» a process are rearranged, so that they can handle a safe and instructed shutdown of the program from the command line with «Ctrl-C».

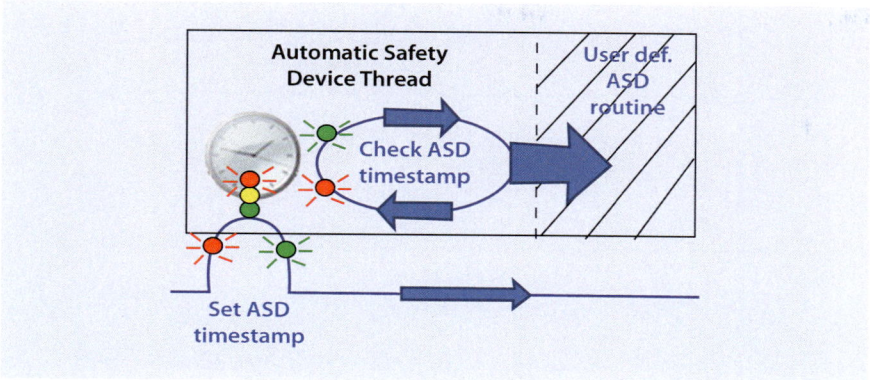

Fig. 3.25 The automatic safety device to listen for frequent requests from the clients

example, because data are written to a file, it is necessary for the server to implement a stable state again after a crash. Ideally the called function is processed completely or not at all. This is quite a difficult problem and implements a transaction concept (see also ▶ Sect. 4.6).[52] The server must persistently save consistent state information to the file system. During the restart, it evaluates this information and runs a recovery process in the «_vInitServerActivity()». Here it must complete the started activity or undo it according to the information. For example, if data are written to a file, the already written bytes must be removed again. Currently, the «idl2rpc.pl» does not offer supporting mechanisms for this task, so that the application engineer must take care of it himself.

Networks can often break down during an active request. For servers which control hardware, this is a critical situation, as there is no responsible client available anymore. The server must run autonomously and must make decisions. But in order to do that, it must first of all detect such a situation. But in client-server architectures, the server does not know anything about the client and its state. Only the client has all the information about the errors of the connected server. To give the server the option to detect situations where it should run autonomously, the principle of an automatic safety device is used again. It is similar to the «deadman control» in trains, which stops the train automatically after something critical has happened to the driver. The critical situation of the driver is detected with a simple procedure: the driver has to regularly push a particular button to signal that he is still alive and able to operate the train. If the frequent signals from the button do not come, the train performs an emergency stop. A similar solution is also integrated into the «idl2rpc.pl» generated server code (if it is parametrized with the calling argument «ASD»; see ▪ Fig. 3.25).[53] In principle, the automatic safety device implements a «heartbeat» control for clients, which means that clients must periodically send a heartbeat request to show that they are still responsible for the server. The same technique can be used to detect a hanging periodic loop.

The automatic safety device is organized as an additional thread in the server. It defines a certain time within which clients must touch the server frequently. It is implemented in a similar way to an «egg timer». The timer (implemented with semaphores in the parallel, multi-threading system) is started during the startup of the server. Each remotely called function sets it back to the original waiting time. If there are no longer any requests, the timer expires. Then the automatic safety device thread calls a specific method «_vAutomaticSafetyDeviceAlert ()»

Server Safety: Automatic Safety Device

Server Safety: Parallel Safety Device Thread

---

52 A transaction is a sequence of functionalities which are processed completely in the correct order, or not at all.

53 If specific methods do not need to be controlled by the automatic safety device, the extension «!ASD» behind the function name can be used in the IDL file to disable the automatic safety control for that function (e.g., « vGetState!ASD (...)»).

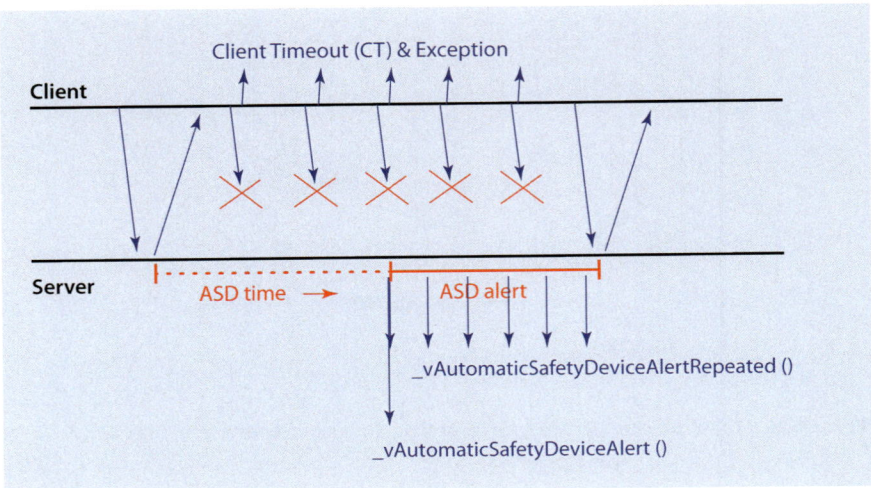

Fig. 3.26 The behavior of the automatic safety device during communication blackouts to responsible clients

once, and as long as there are no more client requests, it calls the method «`_vAutomaticSafetyDeviceAlertRepeated ()`» each second (see ◘ Fig. 3.26). Within these two methods, the application developer can start actions to control the system autonomously or to bring it into a stable state (e.g., by shutting down) without any danger to human beings or the hardware.

All of these safety techniques allow a stable processing of the RPC client-server-communication. Another important issue is the guarantee of a secure communication, which means that no one can attack, control, or destroy the server and consecutive devices.

Security

Server Security in the LAN and WAN

It is important for communication systems that they can be used in the LAN, as well as in the WAN (see also ▶ Sect. 3.3.1). While risks for misuse in LAN s are manageable, risks from attackers and spies from the WAN are serious. Therefore, it is important to protect the server machines in particular from such destructive risks, which might at best just crash the server but can also mean a complete loss of control with the risk of putting human beings or the systems themselves in danger. ▶ Section 5.5 will discuss security and the achievement of a trusted computing system in more detail. The following sections touch on «idl2rpc.pl» relevant parts of security.

Server Security: Authentication

For the protection of the server, it is important that callers of RPC s are in fact the parties they claim to be (Brookshear 2012, *l.c. page 179*). This means that users must be authenticated at the server to get the authorization for activities at the server. This authorization can be very detailed, so that, for example, specific users are only allowed to process specific functions. Nevertheless, the first step is to identify a user. A simple way to do this is to use a username and a password. This is certainly the most commonly used technique. It can be implemented with the techniques offered by Sun RPC (see the security item about RPC in ▶ Sect. 3.3.2 starting). These techniques solve the authentication. But if higher role-based or task-based techniques are available, additional methods are possible in the higher levels of the application programming. Therefore, the «idl2rpc.pl» generated code also allows the transfer of user signatures and resulting user credentials in addition to the Sun RPC techniques (see ▶ Sect. 3.3.2). If the generation is parametrized with the argument «AUTH» of the generator call, the opening of the generated interface with «`_usOpenInterface (...)`» can be extended with the arguments for a username and a password. After the connection is physically established to the server, the server stub then activates the method «`_usAuthenticate (...)`», which gets the transferred username and password and returns an integer number as further user identification for the credentials. The function itself must be filled by the application programmer, so that he can use any authentication technique he wants. It is only then that a user identifier different from «0» is returned to the client. This identifier is then used automatically for all further RPC calls to identify the user.

Using the returned authentication identifier for each received RPC call, the server can identify the sender. This can be used to define further criteria for the activation of specific functions. Usually there are different classes of users in a system. Some are just allowed to read data, others are allowed to change them, and still others have complete administration rights on the machine (superusers or «root» users), so that they are even allowed to change configurations. This classification is also possible for RPCs. The user identification number and a generated remote method number (from the enumeration type «MethodEnum» in the server class) are used as arguments for the additional function «_usAuthorize (...)». This function is called by the server stub each time before the real remote function is activated. The remote method identifier is automatically derived by the server stub and handed over to this authorization method. Inside this method, the application engineer can write his own code to permit or block remote functions, which are identified with the function enumeration value for specific users, who are identified with the user identification number (credentials). If the «_usAuthorize (...)» returns with «0», the access for the current function is denied. Otherwise it is granted. A similar technique can also be used explicitly in the remote methods themselves if specific code parts of the method are not granted for some users. The generated code around this authentication and authorization only organizes the workflow, but does not define the access management. This is precisely what makes it open for all possible and available authentication techniques, which are then part of the application design (e.g., by using the toolbox component «simple_role_manager» from ▶ Sect. 3.2).

A problem with the above access management is that the standard Sun RPC mechanism does not offer techniques to encrypt data transfers. Passwords, credentials, etc., are transferred in a readable form. Additionally, the identity of the client is only defined by the username and password, which is a little weak. A remedy for that problem is provided by «Secure RPC» from Oracle. It provides a public key cryptography (see also ▶ Sect. 5.5.2) (McClure et al. 2006, *l.c. page 173*). Secure RPC uses the DES mechanism (see also ▶ Sect. 3.3.2) in combination with the Diffie-Hellman authentication mechanism[54] (see ▶ Sect. 5.5.2). It authenticates the host and the user who is making a request for a service (for the following, see (Oracle n.d., *l.c. page 299 ff.*)). Here the client and the server use their private keys (secret keys) to create a common key from the public keys. The common key, which is encrypted with the DES, is then used for the communication between client and server. The public and private keys are managed with an Network Information System (NIS), where the private key is only known to the user. The cryptography also uses synchronized clocks on the client and server machines. The sender encrypts its current time and sends it to the receiver, where it is decrypted and compared to the local time. An authentication then follows the following steps (Oracle n.d., *l.c. page 301 ff.*):

1. *Entering a password*: Secure RPC uses the «keylogin» command to request a password from the user. The password is then used to decrypt the private key, which is passed to the key server on the local machine.
2. *Generating the common key for the communication*: When the user initiates a transaction between a client and a server, the key server creates a random conversation key. This key is used to encrypt the timestamp of the client. Additionally, it uses the public key of the server and the private key of the client to encrypt the random conversation key.
3. *Contacting the server*: With the first transmission, a credential and a verifier are included as parameters. The credential contains the client's network name, the encrypted conversation key, and an encrypted time window, which defines the maximum allowed time delay between the clocks of the client and the server machine. The verifier contains the encrypted timestamp and an

Server Security: Authorization

Server Security: Secure RPC

---

54 In combination with RPC known as «AUTH_DH».

3

encrypted verifier of the specified window, which is decremented by 1, to reduce the chance of fake credentials.

4. *Receiving the contacting message by the server*: After receiving the contact attempt, the client's public key is requested from the local NIS. Together with the private key of the server, it can calculate the common key. This is used to decrypt the received conversation key. This decrypted conversation key is then used to decrypt the received timestamp. The server compares the timestamp with its own time.

5. *Storing information for the following communication*: The server stores the decrypted timestamp, together with the client's network name, the conversation key, and the time window at a specific index position in a credential table. The stored timestamp is used to identify replays, as timestamps must be chronological. All data are used for the authentication during the following communication.

6. *Return to the client*: The server returns a verifier with the index of the record in the credential table and the encrypted client's timestamp minus 1.

7. *Establishing of the communication by the client*: The client uses the returned timestamp to identify the server, as only the server knows the sent client's timestamp.

8. *Further communication transactions*: Each following transaction contains the record index to the credential table together with a new encrypted timestamp. The server always replies with the client's timestamp, reduced by 1.

This mechanism securely identifies the authenticity of the client and the server, as only they are able to create the necessary keys with their private keys. All messages which do not fit into the transaction schema are automatically rejected. This secure RPC mechanism is very reliable, but requires a lot of overheads on the communicating machines. Another problem is that the payload of the messages is not encrypted. A third problem is that Secure RPC is not available for all Linux/Unix platforms (McClure et al. 2006, *l.c. page 327*). A possible solution is an SSH tunnel. This topic is touched in more detail in ▶ Sect. 5.5. Nevertheless, the method demonstrates how complex it can become to explicitly identify authorized users and access.

All of the modules and components in the code toolbox, and especially the generative programming approach (as used for the IPC) build the background for further, rapid developments of qualitative, well-tested software systems, which have to use extended communications between widely distributed single processes of a distributed system.

3.4   A Rudimentary Middleware for Controlling of Distributed Systems

Principles

The abstraction and individual parametrization of communication tasks are essential for modern distributed systems (see ▶ Sect. 3.3.2). They offer a consistent and high-level access point to the network communication system but hide the communication details, so that the use is similar to local function calls. Additional services manage the distributed tasks and support and simplify the application engineering. Simple activation mechanisms of remote procedures (see ▶ Sect. 3.3.2), marshalling of data representations between different machines (see ▶ Sect. 3.3.2), controlled safety of the communication (see ▶ Sect. 3.3.3), secure data transfers (see ▶ Sect. 5.5), and a simple way to find relevant communication partners (see ▶ Sect. 3.3.2 about the «portmapper» as a very simple solution)[55] are the most important

---

55   For larger systems, those port mapping services can be extended with a naming and interface repository service, which collects references to possible communication partners (such as the alias or IP address, the port, and the program number) to offer them as a centralized information service, similar to the «yellow pages» in telecommunications. But for the usual tasks to design distributed and remotely accessible control, as in an observatory, the techniques described are sufficient.

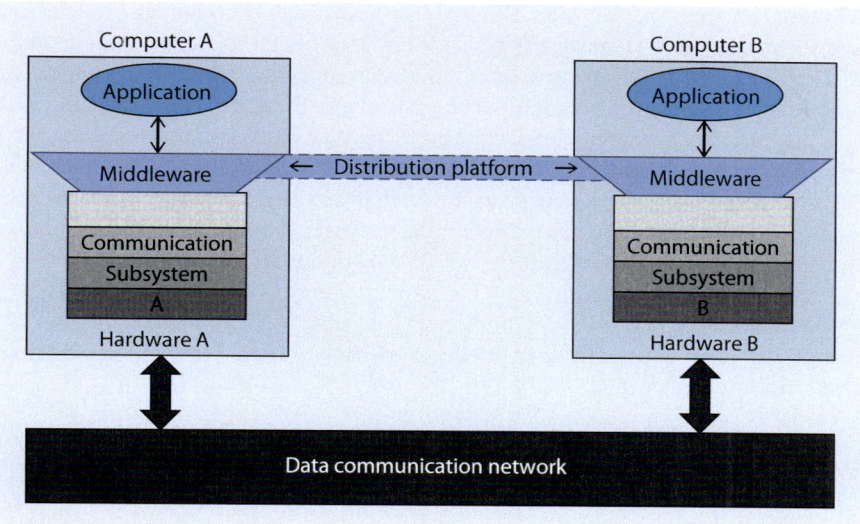

techniques. Together they offer everything required for a distribution platform as used for communication middleware systems (Neidhardt 2005, *l.c. page 39 ff.*).

> *Definition DEF3.12. of «communication middleware»* (see (Puder and Römer 2001, *l.c. page 2*))
> A communication middleware is a distribution platform for communication tasks, which offers generative and homogenizing services and programming tools to implement an abstraction for the communication between an application and the communication network.

Communication Middleware

In such a middleware system, the communication subsystem, as described for the TCP/IP stack in ■ Fig. 3.3, is extended by an additional layer as a service access point for the application. This additional layer provides the same access functions for the hardware to homogenize the different implementations of the communication subsystem of different operating systems and platform hardwares (see ■ Fig. 3.27). Therefore, the «idl2rpc.pl» generated code also creates an operating system module «<*INTERFACEFILE*>_os.cpp/.hpp»[56] with operating system specific code.[57] The new layer seems like a unique platform which automatically handles communications if required.

Distribution Platform

Even if pure middleware solutions, such as the Common Object Request Broker Architecture (CORBA), no longer play a big role today, the principles are still used and active in the background of cloud computing, grid computing (see ▶ Sect. 3.3.2), and so on. But a big challenge is how to abstract interfaces for different servers in a system, so that they can be arbitrarily combined, for example, for control tasks. One possibility is to define one interface specification each server must at least fulfill. But in some circumstances, this is not really satisfactory, as different servers require different

Command Line Language

---

56   <*INTERFACEFILE*> is replaced by the filename of the interface specification file, e.g., if the interface specification is in the file «*test*.idl», the resulting operating system specific code is in the files «*test*_os.cpp» and «*test*_os.hpp».

57   The current version is only usable under Linux and Unix systems, but are prepared for further operating system.

functions for their specific tasks. But with the knowledge from domain engineering and the definition of languages, it is also possible to design and implement an automatically generated middleware communication language. The generator can not only generate the communication relevant parts from the interface specification with IDL (e.g., as from Example 3.22), it can also write code for a command line interpreter (e.g., see the code fragment of a generated interpreter in Example 3.23), which converts textual RPC instructions (e.g., as shown in Example 3.24) into the real RPC activations. This interpreter can be part of the client class and can be used during runtime to control the client and therefore the server activities using text commands.

■■ **Example 3.22**
Sample IDL file with really used methods to control radio telescopes to demonstrate the interpreter idea

```
interface test                                                                1
{                                                                             2
                                                                             3
    // ==================================================================    4
    // Send a single chat message strChatMessage as continuous text and define if    5
    // the message should be added to the log file (usMsgToLogFlag = 1)       6
    // Returned error code is 0 for OK and 1 for ERROR                        7
    // ==================================================================    8
    unsigned int uiSendChatMsg (in string strChatMessage,                    9
                            in unsigned short usMsgToLogFlag);              10
                                                                            11
                                                                            12
    // ==================================================================   13
    // Send a checklist as table ppstrChecklist with the columns             14
    //       Question      |      Value request      | Answer possibilities  15
    // Any defined question |   An individual value   | OK, NOK              16
    // and with an additional note strAdditionalNotes as continuous text     17
    // plus the real name strOperatorName of the responsible operstor        18
    // Returned error code is 0 for OK and 1 for ERROR                       19
    // ==================================================================   20
    unsigned int uiSendChecklist (in string strChecklistName,               21
                            in string strOperatorName,                      22
                            in string ppstrChecklist <><>,                  23
                            in string strAdditionalNotes);                  24
};
```

■■ **Example 3.23**
Code fragment of a possible interpreter method in the client class, which is generated by «idl2rpc.pl» to convert textual RPC instructions into real RPCs

```
// ...                                                                        1
// Enumeration of method identifiers in client class declaration             2
//     enum MethodEnum {MID_ulPing = 1,                                       3
//                   MID_vResetServer = 2,                                    4
//                   MID_usCheckIDLVersion = 3,                               5
//                   MID_usAuthenticate = 4,                                  6
//                   MID_usLogOut = 5,                                        7
//                   MID_usCommandHelp = 6,                                   8
//                   MIDuiSendChatMsg,                                        9
//                   MIDuiSendChecklist                                      10
//                   };                                                      11
//                                                                          12
// ...                                                                       13
//                                                                          14
unsigned int test_client::uiInterprete (std::string strCommand, std::string & strAnswer)    15
```

```
{                                                                             16
    unsigned int uiRetVal = INTERPRETATION_OK; /// Return value of interpreter method    17
    enum MethodEnum EMethodID = (enum MethodEnum) 0; /// EMethodID = Identifier of remote  18
        function
    std::string strCommandCopy = strCommand.c_str(); /// strCommandCopy = Copy of entered  19
        command
    std::string strEmptyString;   /// strEmptyString = Empty string, used to clear other   20
        strings threadsafe
    std::string strCurrentToken; /// strCurrentToken = Currently split token                21
    std::string strFunctionName; /// strFunctionName = Name of the remote function in the   22
        command
    std::string strValue; /// strValue = Used for back conversions from numbers             23
                                                                                            24
    // Init return value                                                                    25
    strAnswer = strEmptyString.c_str();                                                     26
                                                                                            27
    // Clean white spaces to get a standard format                                          28
    if (usCleanupWhiteSpaces (strCommandCopy)) {uiRetVal = INTERPRETATION_NOK; goto         29
        uiInterprete_Ret;}
                                                                                            30
    // Cut of function name from command as token and convert it into enum representation   31
    if (usGetFunctionName (strCommandCopy, strCurrentToken)) {uiRetVal = INTERPRETATION_NOK; 32
        goto uiInterprete_Ret;}
    if (usConvertFunctionNameToEnum (strCurrentToken, EMethodID)) {uiRetVal =               33
        INTERPRETATION_NOK; goto uiInterprete_Ret;}
    strFunctionName = strCurrentToken.c_str();                                              34
                                                                                            35
    // Process the different remote methods                                                 36
    switch (EMethodID)                                                                      37
    {                                                                                       38
        case MID_ulPing:                                                                    39
        {                                                                                   40
            // ...                                                                           41
            break;                                                                          42
        }                                                                                   43
        case MID_vResetServer:                                                              44
        {                                                                                   45
            // ...                                                                           46
            break;                                                                          47
        }                                                                                   48
        case MID_usCheckIDLVersion:                                                         49
        {                                                                                   50
            // ...                                                                           51
            break;                                                                          52
        }                                                                                   53
        case MID_usAuthenticate:                                                            54
        {                                                                                   55
            // ...                                                                           56
            break;                                                                          57
        }                                                                                   58
        case MID_usLogOut:                                                                  59
        {                                                                                   60
            // ...                                                                           61
            break;                                                                          62
        }                                                                                   63
        case MID_usCommandHelp:                                                             64
        {                                                                                   65
            // ...                                                                           66
            break;                                                                          67
        }                                                                                   68
```

```
case MIDuiSendChatMsg:                                                    69
{                                                                         70
    // Convert input arguments                                           71
    if (usGetArgument (strCommandCopy, strCurrentToken)) {uiRetVal =      72
        INTERPRETATION_NOK; goto uiInterprete_Ret;}
    std::string strChatMessage = strCurrentToken.c_str();                 73
    if (usGetArgument (strCommandCopy, strCurrentToken)) {uiRetVal =      74
        INTERPRETATION_NOK; goto uiInterprete_Ret;}
    unsigned short usMsgToLogFlag = 0;                                    75
    if (usConvertTokenToUnsignedShort (strCurrentToken, usMsgToLogFlag)) {uiRetVal =   76
        INTERPRETATION_NOK; goto uiInterprete_Ret;}
    // Call remote procedure                                             77
    unsigned int uiFuncRetVal = 0;                                       78
                                                                         79
    try                                                                  80
    {                                                                    81
        uiFuncRetVal = uiSendChatMsg (strChatMessage, usMsgToLogFlag);   82
    }                                                                    83
    catch (...)                                                         84
    {                                                                    85
        uiRetVal = ANSWER_NOK;                                          86
        goto uiInterprete_Ret;                                          87
    }                                                                   88
    // Clear input values                                              89
    // Convert output arguments and create answer                      90
    strAnswer = strFunctionName.c_str();                               91
    strAnswer += "!";                                                  92
    if (usConvertUnsignedIntegerToString (uiFuncRetVal, strValue)) {uiRetVal =
        ANSWER_NOK; goto uiInterprete_Ret;}
    strAnswer += strValue.c_str();                                     93
    strAnswer += ";\n";                                                94
    break;                                                             95
}                                                                      96
case MIDuiSendChecklist:                                               97
{                                                                      98
    // Convert input arguments                                         99
    if (usGetArgument (strCommandCopy, strCurrentToken)) {uiRetVal =    100
        INTERPRETATION_NOK; goto uiInterprete_Ret;}
    std::string strChecklistName = strCurrentToken.c_str();            101
    if (usGetArgument (strCommandCopy, strCurrentToken)) {uiRetVal =    102
        INTERPRETATION_NOK; goto uiInterprete_Ret;}
    std::string strOperatorName = strCurrentToken.c_str();             103
    if (usGetArgument (strCommandCopy, strCurrentToken)) {uiRetVal =    104
        INTERPRETATION_NOK; goto uiInterprete_Ret;}
    std::string ** ppstrChecklist_val = NULL;                          105
    unsigned int _ppstrChecklist_len = 0;                              106
    unsigned int _ppstrChecklist_len1 = 0;                             107
    unsigned short * pstrLengths = NULL;                               108
    if (usConvertTokenToStringArray (strCurrentToken, 2/*Dimension*/, (void *&)  109
        ppstrChecklist_val, pstrLengths)) {uiRetVal = INTERPRETATION_NOK; goto
        uiInterprete_Ret;}
    _ppstrChecklist_len = pstrLengths[0];                              110
    _ppstrChecklist_len1 = pstrLengths[1];                             111
    delete [] pstrLengths;                                             112
    if (usGetArgument (strCommandCopy, strCurrentToken)) {uiRetVal =    113
        INTERPRETATION_NOK; goto uiInterprete_Ret;}
    std::string strAdditionalNotes = strCurrentToken.c_str();          114
    // Call remote procedure                                           115
    unsigned int uiFuncRetVal = 0;                                     116
                                                                       117
    try                                                                118
    {                                                                  119
        uiFuncRetVal = uiSendChecklist (strChecklistName, strOperatorName,
            ppstrChecklist_val, _ppstrChecklist_len, _ppstrChecklist_len1,
            strAdditionalNotes);
    }                                                                  120
```

```
        catch  (...)                                                         121
        {                                                                    122
            uiRetVal = ANSWER_NOK;                                           123
            goto uiInterprete_Ret;                                           124
        }                                                                    125
        // Clear input values                                               126
        if (ppstrChecklist_val != NULL)                                      127
        {                                                                    128
            for (unsigned int uiIndex; uiIndex < _ppstrChecklist_len; uiIndex++)  129
            {                                                                130
                delete [] ppstrChecklist_val[uiIndex];                       131
                ppstrChecklist_val[uiIndex] = NULL;                          132
            }                                                                133
            delete [] ppstrChecklist_val;                                    134
            ppstrChecklist_val = NULL;                                       135
        }                                                                    136
        // Convert output arguments and create answer                       137
        strAnswer = strFunctionName.c_str();                                 138
        strAnswer += "!";                                                    139
        if (usConvertUnsignedIntegerToString (uiFuncRetVal, strValue)) {uiRetVal =  140
            ANSWER_NOK; goto uiInterprete_Ret;}
        strAnswer += strValue.c_str();                                       141
        strAnswer += ";\n";                                                  142
        break;                                                               143
        }                                                                    144
    }                                                                        145
uiInterprete_Ret:                                                            146
    if (uiRetVal == INTERPRETATION_NOK)                                      147
    {                                                                        148
        strAnswer = "INTERPRETER ERROR! Token=";                            149
        strAnswer += strCurrentToken.c_str();                                150
        strAnswer += ";\n";                                                  151
    }                                                                        152
    if (uiRetVal == ANSWER_NOK)                                              153
    {                                                                        154
        strAnswer = "ANSWER ERROR!;\n";                                      155
    }                                                                        156
    return uiRetVal;                                                         157
}                                                                            158
```

■ ■ **Example 3.24**

Interpretable textual RPC instruction, which can be handed to the interpreter of
the client class

```
Command examples:                                                            1
==================                                                           2
uiSendChatMsg ("Hello, world!", 0);                                          3
uiSendChecklist ("Before", "Alexander", <[["Antenna"][""]["OK"]][["WX"]["Sunny"]["OK"]][["  4
    System"][""]["OK"]][["Pointing"]["casa"]["OK"]]>, "");
                                                                             5
Array notation:                                                              6
==============                                                               7
e.g. a two-dimensional (4x3) string array (string matrix)                    8
<                                                                            9
   [["Antenna"] [""]      ["OK"]]                                            10
   [["WX"]      ["Sunny"]["OK"]]                                             11
   [["System"]  [""]      ["OK"]]                                            12
   [["Pointing"]["casa"] ["OK"]]                                             13
>                                                                            14
e.g. a two-dimensional (3x3) number array (matrix)                           15
```

```
<                                                                        16
   [[1][2][3]]                                                          17
   [[4][5][6]]                                                          18
   [[7][8][9]]                                                          19
>                                                                        20
e.g. a three-dimensional (3x3x2) number array                           21
<                                                                        22
   [[[1] [2] [3]]                                                        23
    [[4] [5] [6]]                                                        24
    [[7] [8] [9]]]                                                       25
   [[[10][11][12]]                                                       26
    [[13][14][15]]                                                       27
    [[16][17][18]]]                                                      28
>                                                                        29
```

Command Line Scanner

The interpreter of the new generated interface uses methods from a general parser which help to find specific tokens, such as the function name or individual arguments.[58] Further parser methods are used to convert between string and value representations, so that a string can be converted into an integer or float representation. Those converters check the individual value for syntax correctness, as shown for floating point numbers in ◘ Fig. 3.7. The check of arrays with dynamic dimensions is a little more difficult. But a syntax structure as shown in Example 3.24 defines a clear pattern for the input of dynamic arrays. In this case, it is only necessary to also check that the number of elements per rows is equal and so on. As unlimited dimensions are possible with this notation (in real life, a realization up to three dimensions is usually enough, as arrays with more dimensions are confusing), it is necessary to count the elements per row of each dimension. The routines described build up a sophisticated scanner, which already does some parser work.

Command Line Parser

These checks and conversions are then part of the runtime environment of the client object. As shown in Example 3.23, the interpreter takes an instruction and searches for a remote function name, which is converted into an enumeration value to be used within a «switch» condition. This «switch» statement selects the individual interpreter code, generated by «idl2rpc.pl». After the selection of the individual interpretation, the input arguments are read and converted (interspersed output arguments are symbolized with an empty space between the argument commas[59]). The derived arguments are used for the call of the RPC method. After the request and the successful reply, the return and output values are converted to an answer string of the style «`<functionname>!<argument1>,<argument2>,...,<argumentn>;`». This string is handed back to the caller. A comparison of this workflow shows that the structure is again very similar to the design of compilers shown in ► Sect. 3.3.2. It is a parser which starts semantic actions and generation activities.

Command Line Help

As the interpreter is generated individually for specific interface definitions in the IDL file, it is even possible to generate an additional function which offers a command line help. Using the remote function definitions (and maybe added code attributes as comments per argument and function name), the generator can generate a help output, which can be printed to the command line each time the help function is called with the name of a remote function. If no function is specified, help can print a complete overview of all available remote functions.

---

58 In the example shown, the functions get a string, from which they extract the searched value at the beginning of the string. They cut off the token found and return the rest of the string for further extractions.

59 It is beneficial if the interface arguments of a remote function are arranged with the order: only input, input/output, and only output arguments.

The interpreter method is also added to an additional abstract class,[60] which is equal for all generated interfaces. This class provides assistance in using the methods which are common for all remote clients and part of the abstract object within a centralized administration object. To do this, an object instance of the individual class must be created, which can then be converted («casted») into the abstract class. All the standard methods, such as pinging, can still be used, so that broadcast pings can easily be implemented. Only if specific methods of the specific interface class are activated can the object be cast back to implementation object. Using this technique, the interpreter can also be called from the abstract object. But as it is implemented by the individual object which was cast into the abstract class, the individual function is activated from the abstract object reference. This avoids moving the interpreter functionality to the server, as this would require the conversion of all remote calls into ASCII representations with all their disadvantages (see ▶ Sect. 3.3.1) and a loss of the advantage of binary data representation for the communication. Another advantage of client abstraction is that the client user must be in possession of the actual client class interface to connect to the server. If the interpretation is in the server, each command is sent to the server directly, which offers the possibility of blocking him with permanent requests, which would result in a «Denial-of-Service» attack (see ▶ Sect. 5.5.2).

Command Line Abstraction

The described middleware command line interpretation is a key technology for further developments of schedulers (see ▶ Sect. 4.4). The interpretation of textual instructions provides a simple runtime interpreter, which makes it possible to write simple, sequential scripts. If those instructions are timed in a scheduler for control systems, frequent tasks can be organized. But loops and conditions are missing which must be added by the scheduler. The greatest advantage is that by adding a new interface method, the command line interpretation is directly available after the generation, so that error-prone and exhaustive manually written command line interpreters are not required anymore. Similar technologies are only known in interpreter languages. Python, for example, offers the method «eval», which argument is parsed and evaluated as a Python expression (Python Software Foundation n.d.). It makes it possible to use textual notations to evaluate Python expressions during runtime.

Key Technology for a Command Interpretation

3.5 Summary

3.5.1 What Did We Learn?

Software is decomposed using modules and components. These elements can use generative ways to support parametrized applications. We saw different techniques to increase the adaptable modularity with generative techniques like template patterns, template metaprogramming, or completely generated modules. These techniques split a development into domain and application engineering. Most of the engineers can use predefined, parametrized domains to implement their own applications. This method reduces development work and increases stability.

Generative programming is ideal for tasks with similar characteristics, which solve a specific class of programming problems. One of these classes is the communication between two communication partners over the network. We saw the classic socket communication using the network protocol stack. The disadvantage is that the complete communication protocol has to be written manually. Pitfalls

---

60 An abstract class is a class which only declares the methods, but does not define them with bodies. Derived classes must then implement all of these class methods. Abstract classes can be used to cast between different occurrences of related classes, e.g., if hardware devices with the same interface but with different implementations for different hardware are used together from one controlling object.

are data representations, bit orders, and the message and dataflow correctness. Regular expressions help to check the correct flow of text message. Regular expressions implement simple grammars, so that the use of generative programming requires knowledge of language theory.

The Remote Procedure Calls use the technique of generative programming. They standardize the communication and also all related tasks. The results are remote function calls which look similar to local calls and implement the communication implicitly. The Remote Procedure Call shown uses an External Data Representation to convert data types, for example, the floating point representations, between different platforms. Another aspect of Remote Procedure Calls is that they have to implement a call semantics which results in clear states before, during, and after the call on the client and on the server side, even if there are error situations.

The chapter also explained how the generative programming can be adapted to improve the generation of Remote Procedure Calls for one's own requirements. To do this, it is important to know the basics of language theory. Formal grammars can be used to implement individual interpreters. The different languages can also be characterized with the classification in the Chomsky hierarchy.

Compilers and interpreters are software components which implement the conversion of languages to a specific program code or directly to machine code. We saw the different phases of a compiler which analyze the original source in the defined domain language and generate optimized source code as output. This output includes concepts to guarantee safety and security. One technique used is a watchdog. Another is the automatic safety device. The result is a new middleware which can be used as distribution platform for standardized Remote Procedure Calls.

3.5.2 How Did We Use the Contents Learned?

The theory shown was used to implement an individual interprocess communication with the Sun RPC technique. We created an individual interpreter written in Perl, which is able to read an interface definition and which converts it to individual communication code for a client-server application.

We used the theory from the chapter about computer languages to design an individual domain language, the IDL. Using Perl, an individual scanner and parser were implemented, which are able to use the domain language for a specific parametrization of a communication workflow. The implementation contains several solutions for an individual middleware.

3.6 Questions

1. What is the difference between a module and a component?
2. How does template metaprogramming produce a faster code? What will take longer instead?
3. For what is a domain engineer responsible?
4. The TCP/IP protocol stack uses four layers. Which?
5. What does it mean to bind a socket?
6. How many bytes are required to represent the decimal number 1024?
7. Which decimal number is given by the binary representation 11010011 (show calculation)?
8. What do you have to change if you have to convert a 4-byte number from big endian to little endian?
9. Explain each with one sentence: alphabet, syntax, grammar, and semantic.
10. Draw a graphic with the sets of the Chomsky hierarchy and name the language classifications and types.

11. Given is the following simple grammar to check simple sentences:

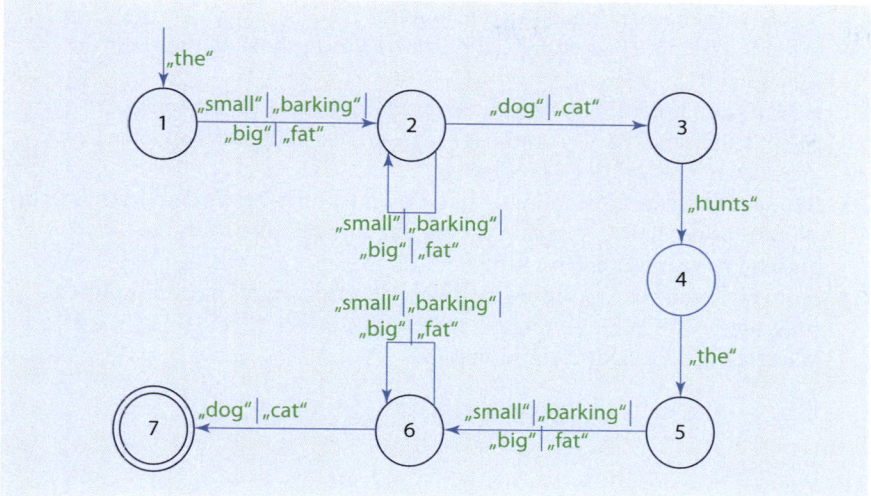

(a) Create three sentences which are accepted by the grammar.
(b) Check the following sentences if they can be produced with the grammar:
«The dog hunts the big fat cat»
«The small dog hunts the small cat»
«The small barking dog hunts the big fat cat»
«The small small small small dog hunts the barking fat dog»
(c) Extend the state chart, so that the following sentences are additionally accepted:
«The nice cat hunts the fat dog»
«The nice cat hunts mice»
«The cat likes me»

12. Explain the difference between grammatically correct and semantically correct for the grammar in the previous question using the example sentences «the big barking dog hunts the small big cat» and «the big fat dog hunts the small barking cat»!
13. Given is the following grammar which accepts simple mathematical expressions:
Write down the single steps from the start symbol «E» to the final expression «(a + b) * (b + c)».

ETF-grammar:
Non-terminal symbols: E, T, F
Terminal symbols: +, *, (,), a, b, c, ..., z
Start symbol: E
Productio rules: E → T | E+T
T → F | T*F
F → (E) | a | b | c | d | ... | z

14. Which phases are part of a compilation workflow (draw the flow chart)?
15. What does IEEE 754 describe? Explain the basic structure!
16. All floating point representations deal with some problems. Which do you know?
17. What is the difference between ASCII and a binary representation of an integer number?

18. What happens if you calculate the mathematical difference of the ASCII characters «a» minus «A» and if you add the result to the ASCII code of «I» before you interpret it again as a character?
19. What is a Remote Procedure Call? Draw a sketch showing the general functionality?
20. Which call semantics do you know? Explain them with one sentence!
21. Safety and security are essential for high-quality software. Compare them!
22. What does a «watchdog» process?
23. Remember semaphore. Why should the semaphore-blocked code section be as short as possible, for example, only to copy variables from the function arguments to internal structures?
24. For which issue can an automatic safety device be used in communication processes?
25. What is the idea behind a middleware?

Controlling a Laser Ranging System

4.1   Principles of Laser Ranging Systems – 248

4.2   The Laser Ranging System as an Autonomous
        Production Cell – 253
4.2.1   Distributed Hardware Control – 254
4.2.2   The Construction of the Autonomous Production Cell – 277

4.3   User Interfacing – 283

4.4   Autonomous Coordination Cell – 300

4.5   Autonomous Hardware Control Cells – 332

4.6   Autonomous Data Management Cell – 338

4.7   Autonomous System Monitoring and Safety Cell – 369

4.8   Summary – 390
4.8.1   What Did We Learn? – 390
4.8.2   How Did We Use the Contents Learned? – 391

4.9   Questions – 391

© Springer International Publishing Switzerland 2017
A. Neidhardt, *Applied Computer Science for GGOS Observatories*, Springer Textbooks in Earth Sciences,
Geography and Environment, DOI 10.1007/978-3-319-40139-3_4

4

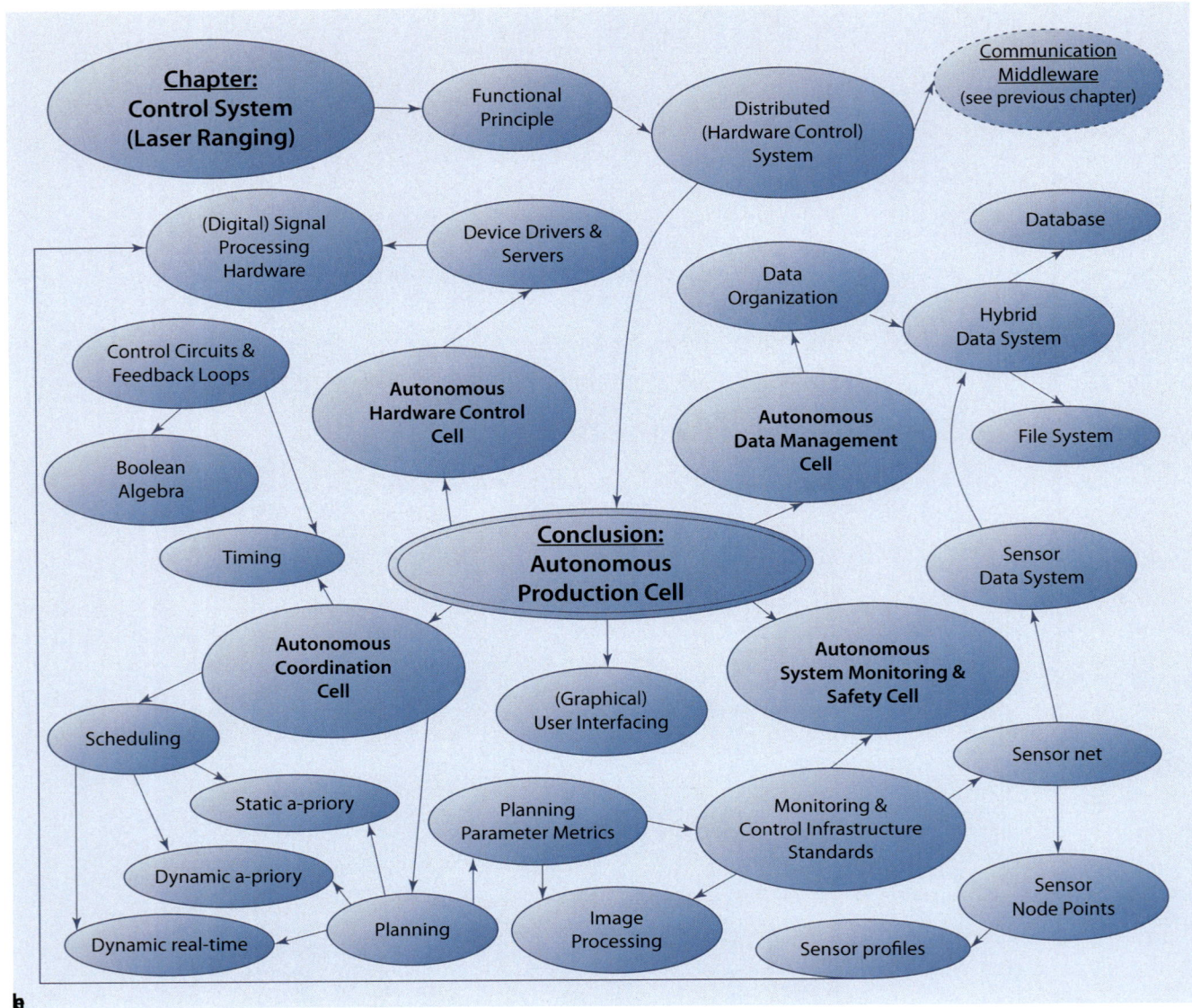

4.1   Principles of Laser Ranging Systems

Example Environments: Laser Ranging
and Radio Interferometry

The use of the software techniques described in the previous chapters is shown within two example environments. The first example field is laser ranging. The second one is the geodetic use of radio interferometry (see next ► Chap. 5). The type of laser ranging in the field described is the geodetic space technique of SLR and LLR. SLR measures the distances to satellites in orbits of a height from a few hundred kilometers (Low Earth Orbit (LEO)) to about 40,000 kilometers (High Earth Orbit (HEO)) with a pulsed laser. Using the same technique, LLR measures the distances to (passive) reflector targets on the moon. Even though the following chapter is not an explanation of the laser ranging technique, but rather a view of it from the perspective of computer science and design, it is necessary to explain some basic principles (compare (Seeber 1988, *l.c. page 320 ff.*) for the following blocks).

Technical Principles

Laser ranging is based on two segments: the ground and the space segment. The ground segment consists of several, worldwide distributed ground stations with a laser ranging system. The space segment consists of satellites in different orbits, nearby the moon, and the moon itself. Distance ranging measures the delay of a short laser pulse between the ground station and the satellite or the moon. Therefore, the target must consist of a suitable (retro-)reflector or a reflector field (geodetic satellites like LAGEOS with its radius of 30 cm are passive satellites in the form of a «retroreflector ball»), which reflects the laser pulse coming from the sending equipment of the ground station back to the receiving equipment of the same ground station. The delay between the time epoch when the laser pulse is leaving the telescope (electronic start pulse) and the time epoch when the laser pulse is received again (electronic stop pulse) is measured with an electronic time-interval counter (the event timer). The delay measured between the start and stop pulse is double the elapsed time the laser pulse needed for the distance between the ground station and orbit target.[1] Using the speed of light, it is possible to calculate the distance between both. In this context, laser ranging is a two-way method, as the laser passes almost the «same» way twice (to and from the target).[2] This helps to reduce transients, such as delay influences from the atmosphere.

To implement the laser ranging to satellites, a number of technical hardware pieces must be arranged to cooperate on the ground. A laser ranging system consists of the hardware components shown in ▣ Fig. 4.1. A laser (as an acronym for Light Amplification by Stimulated Emission of Radiation (LASER)) produces short, bright, coherent light pulses (with a length of about 50 picoseconds). Each pulse produces an electronic start signal when it passes the transmit section of the «transmit-receive unit». The pulses are then guided toward an optical telescope using optical mirrors, which have a special coating for the laser light frequencies (usually 532 nanometers («green» light) and a second frequency with 1,064 nanometers (infrared light), which is the double of 532 nanometers because of the amplification). The optical system of the transmit-receive unit also offers adjustment possibilities for the beam, such as the divergency parameter, filters, or the field of view. The telescope, which is usually protected by a dome, follows the satellite orbit which was previously calculated from orbit predictions for the different satellites by the ranging control and data system.

Interaction of the Hardware

There are two types of laser ranging telescopes: biaxial and monoaxial ones. Biaxial mounts use one smaller sending and one bigger receiving telescope on two different optical axes. This simplifies the beam guidance through the system, as sending and receiving beams are automatically separated. Monoaxial mounts use just one telescope and the same axis for sending and receiving. Monoaxial telescopes simplify a precise ranging to distant targets like the moon because an adaption of the convergence or a mounting correction is not so relevant as for a mounting of two independent telescopes for sending and receiving. But it requires a separation of the sending and the receiving beam in the transmit and receive unit. Usually this can be solved with a rotating disk, where a mirror and a transmission area are arranged concentrically. The rotation speed of the mirror is precisely matched with the laser fire rate (from several Hertz, e.g., 10 Hertz to a few kilohertz) and the expected return time of the pulses from the satellite. The strong sending beam (with an energy of a few hundred millijoules per pulse) is then sent to the telescope, and while rotating and opening the air gap, the receiving beam can cross the transmit-receive unit to the

Telescope and Reception

---

1   Further missions with transponder systems (active laser reply systems) offer further possibilities for the space segment because larger distances can be measured.

2   It is not exactly the same way, because, for example, the satellite moves ahead on its orbit, while the laser pulse is on the way from and back to the ground station. Therefore, the length of the sending path is slightly different to the length of the receiving path. Usually this is not relevant for the pointing accuracy to satellites. But it must be implied for lunar ranging to get the correct travel times of the laser for the correct range gate calculation, which defines the time when returns are expected.

4

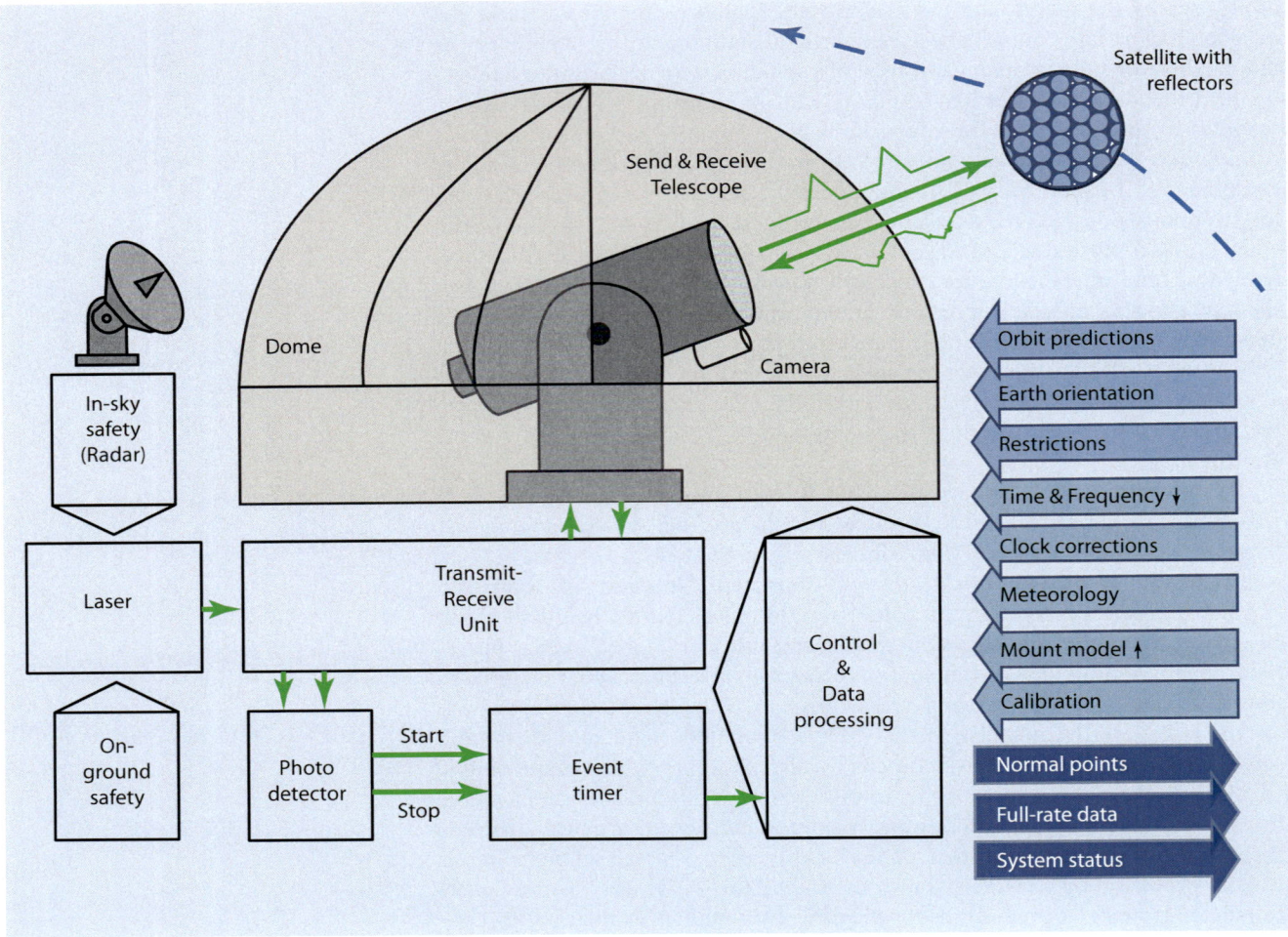

◻ Fig. 4.1   The interaction of hardware in a laser ranging system

Data for Laser Ranging

detectors. The detectors are single-photon detectors, which work as electron multipliers. For LEO satellites up to orbits such as those from the LAGEOS, a type of Micro-channel Plate (MCP) is normally used. Higher satellites are tracked with an Avalanche Photo Diode (APD). They produce an electronic signal as an event for each returning photon (except during a short dead time after each detection for regeneration). Photons of outgoing and incoming laser pulses can produce electronic pulses as start and stop signals for a time-interval counter, which is used as an event timer with an accuracy of about 4 picoseconds. It measures the delay between start and stop pulses and returns it as residual (reference value minus actual value) to the data acquisition and processing system. This finally calculates the Normal Points (NP) as representative, thinned, actual supporting points of the measured orbit.

The ranging system needs some input data to produce the output of NP or full-rate data (the complete, standardized measurement set) in the Consolidated Laser Ranging Data Format (CRD) (Ricklefs and Moore 2009) and system status information in the EUROLAS status format. The input data can be split into three different categories:

— *External data from external partners:* First of all, the laser ranging system needs orbit predictions. In the simplest case, these can be ephemerides or two-line elements. But then the orbits themselves must be integrated and perturbations or influences must be included. Nowadays, the predictions can be downloaded with all of these influences from central services in the CPF (Ricklefs 2006). Another data set from an external partner is the EarthOrientation Parameters (EOP). They describe the orientation and variation of the rotation pole of the earth and the variation of the rotation speed per day. During the observation, additional permission or restriction information must be used for specific satellites to protect them from the destructive impacts of laser energy. Those data sets are offered in different formats from the satellite operating organizations.

— *System external data from other local systems at the observatory*: Most of the hardware parts need a connection to the time as time epoch (exact timestamp) for events and also to a stable frequency. This input comes from a timing laboratory at the observatory, which consists of a simple time receiver or, ideally, of an atomic time standard (atomic clock) and is also used for other systems as common clock. The cables used and the time distribution equipment normally delay the signals, so a time correction parameter is used as additional input. Another very important input set from the observatory is the meteorological information, consisting of air temperature, humidity, and pressure.

— *System data derived from the system itself*: The ranging system also produces data, which are then used again to improve the data quality. Calibration data are frequently collected. They represent the runtime statistics of laser pulses on a clearly defined path through the system on which the regular pulses to the satellites are also guided. These delay lines represent the runtime behavior of the laser light through the optical system. These data can be collected in real-time or as pre- and post-measurements before and after a real satellite observation. Besides the calibration data, another data set is the mount model correction, which describes and fits errors of the mechanical mounting system of the telescope. It is derived from astronomic position observations of stars which are well distributed over the visible sky. Typical parameters are position and orientation errors of the vertical and horizontal axis, which are then corrected with a model calculation added to the pointing directions for the telescopes.

In-Sky Laser Ranging Safety with Radars

Because the laser used is not eye safe and uses short pulses with high energy per square centimeter over distances of more than 50 kilometer, it is necessary to protect human beings on the ground and also in airplanes in the sky from accidental eye contact. While it is difficult to install no-fly zones in a suitable diameter around and up to a certain height above the observatory, in-sky safety is mostly realized with a local radar system (from Radio Detection and Ranging (RADAR)). It points in the direction of the laser and «envelops» the dangerous beam with a protection zone of microwaves with a diameter of about six degree. If an object is detected within this zone, the safety system sets a hardware interlock to the laser, which switches the transmission of laser light off within a suitable and safe time interval. Currently, there are some plans to replace the radar systems with other techniques, such as eye safe laser beams or other optical systems. But they are still at the development stage and not yet productive. Another problem with the new systems is that they have difficulty in supporting the protection for more than 50 kilometer along the beam, as low elevation positions cross larger distances through the atmosphere where airplanes fly.

In-Sky and On-Ground Laser Ranging Safety with Additional Data

In addition to the hard interlock systems, the operator can be supported by live information about current air traffic above and around the observatory. A suitable option is the use of a virtual radar box, which decodes the Automatic Dependent Surveillance Broadcast (ADS-B) secondary radar messages from the airplanes, which update position data, aircraft identification, and so on, every second. There are also web pages offering this semi-real-time information printed on maps (such as Flightradar24 (Flightradar24 AB n.d.)). But there are no guarantees for real-time and safety issues, so that they are just a nice add-on for the planning of satellite tracks by the operator. In some countries, it is possible to get live data from the air traffic control network by paying service fees. These data are offered with guarantees and can be used in a better way for software protection zones around the airplanes. Then the protection zone is dependent on the update rate of the position data and its accuracy, so that the local system can guarantee that the laser will be switched off during the time an airplane needs to pass through the area protected by the zone. The operator always carries some responsibility for in-sky safety, so that optical control with a camera is essential for example to protect paragliders, which cannot be seen even by a radar system. In addition to in-sky safety, it is also necessary to implement on-ground safety. This can easily be done by locking the doors of dangerous zones against any unauthorized entrance. Other techniques use hardware interlock switches at the doors to protect people from contact with the laser energy.

Workflow of Observations: Before the Observation

The control of an actual observation uses the previously mentioned elements of a laser ranging system. The first step in a control workflow is the acquisition of prediction data. These data are used to calculate whole orbits and to derive visibility times of the satellites for a given ground station (topocentric position). Within these visibility times, way points for those satellite passages are calculated. These preparations are usually made automatically in the background.

Workflow of Observations: During the Observation

The operator then selects an individual satellite depending on what the priority and the current meteorological situation is. The passage data is used for further interpolations to calculate the real-time way points and delays of the orbit track. The additional input data, such as refraction corrections with meteorological information, is included in the calculated satellite positions within the passage. These adapted positions are then used for the telescope's autotracking and for the range gate calculation (the time interval to listen for return events around the expected pulse return time from the satellite) in the event timer. The telescope automatically starts to follow the target if it appears in the sky. The dome and the additional radar system for in-sky safety follow wherever the telescope is pointing. If the telescope is on target, the operator can switch on the laser and can adjust the pointing direction of the telescope in steps of arc seconds along and across the satellite movement (paddle settings). He can verify the cloud coverage condition in the direction of the telescope axis, using the image from an installed camera. He can also check the settings of the transmit-receive system, as the field of view or the divergency can be adjusted to optimize the return rate or other receiving parameters. The detected returns from the event timer are plotted as residuals over the epoch times on the x-axis. If the satellite target is hit, significantly more events are detected at the expected pulse return time within the gate interval. Therefore significantly more residuals appear around zero on the y-axis. If such a hit series is detected by the operator (or by an automated algorithm), the paddle adjustment of the telescope is stopped. The satellite is then observed until a defined number of returns is received, for example 1,200 returns. If too many returns saturate the detector, the operator has the option to further reduce the return energy with filters or apertures in the transmit-receive unit. During the observation time, the data acquisition system saves all residual points and additional settings to an observation data file for the satellite passage.

Workflow of Observations: After the Observation

After collecting sufficient hits, the operator stops the observation, which stops the tracking and all other processes. After a successful observation, the operator runs a local data analysis, to calculate the NPs, which are representative, statistical points for the observed passage. He validates the results and sends the final data sets to the external data centers. Then the operator can select another satellite and repeat the same steps. During the whole process of an observation, it is quite important for the operator to control the correct functionality of the system, the system quality, and the safety (a visual check of the sky, the interlock states, and so on).

The whole observation process has to deal with influences from the system and from effects on the path of the laser beam outside the system. After leaving the telescope, the laser beam passes through the atmosphere and space to the satellite or lunar target.[3] At the target, the light is reflected back in the direction it came from. But the returning signal pulse has some deformations because of perturbations from the atmosphere, the coverage of different return signals from different reflectors at the satellite target, and the relative movement of the ground station and satellite (Seeber 1988, *l.c. page 329*). Temporary variations of the optical path in the ground station also influence the signal quality. For this reason, the beam path is calibrated frequently with a pre- and post-method before and after an observation or using a real-time calibration (as described above). A refraction correction improves the pointing angles and the calculated satellite positions during the local analysis, using the current meteorological data. A regular local and manually operated quality control process checks the beam quality and its position on the beam path of the system. Another quality control process is carried out by the analysis centers, which check the data quality for irregularities and outliers.

Perturbations, Calibration, and Quality Control

All the data are collected, analyzed, and published by the ILRS. The main goals of this international cooperation and the use of laser ranging are the determination and update of variations of satellite orbits, Earth and sea tides, geocentric station coordinates, tectonic plate movement, Earth gravity, Earth orientation parameters (polar motion and Earth rotation variations), tide friction, dynamics of the Earth-moon system, and so on.

Goal of Observations

Taking a look on ◘ Fig. 4.1 about the laser ranging principles and the observation workflow, it is clear that a laser ranging system consists of several hardware components, which must work together to solve one common task, managed by a control system and an operator. As most of the components are distributed over the whole ranging system and are usually connected via a communication network, a suitable solution to design the control software can be based on the idea of distributed systems (see the definition in ▸ Sect. 3.3.2). The following sections show and demonstrate general ideas and a possible implementation version, using the example of the software for the laser ranging system of the Geodetic Observatory Wettzell.

4.2 The Laser Ranging System as an Autonomous Production Cell

As shown in the previous section, a laser ranging system is an interacting formation of several hardware devices. They all need to be represented by information technology structures which communicate with each other. The hardware devices described are complex control elements consisting of several more or less intelligent control units. Usually those intelligent units and hardware components are produced by external companies. They are integrated into the digital control software design as more or less autonomous software units. They interact as a distributed system. The goal is to design and construct them to be as autonomous and automatic as possible. An idea on how to do this comes from the automation theory for industrial manufacturing plants and is called autonomous production cells. For the control of laser ranging systems, this concept can be enlarged to a concept of interacting autonomous control cells.

---

3   Lunar targets are reflector fields on the moon, positioned during three different Apollo missions, 11, 14, and 15, and by two Russian unmanned missions Lunokhod 1 and 2.

4.2.1 Distributed Hardware Control

Intelligent Control Units

As described, a laser ranging system consists of several, interacting hardware components. Each component is a hardware device or software process which fulfills a specific task together with its controlling components. These implement a more or less intelligent control loop in hardware and/or software. Therefore, each component can be seen as an intelligent control unit. The structure of a solved control loop, partly in software and partly in hardware, to run a hardware device, is shown in ◘ Fig. 4.2 (combined from (Langmann 2010, *l.c. page 166 ff.*), (Langmann 2010, *l.c. page 103 ff.*), (Langmann 2010, *l.c. page 141 ff.*), and (Favre-Bulle 2004, *l.c. page 22*)). It controls a physical system with an actuator system. The actuator is gated from a controller, processing signals and logical control. It uses input from a sensor system, which measures parameters from the physically controlled system. It is a classic feedback control loop. To command the intelligent control unit from outside and to request parameters and control states, the controller is connected to a communication interface. Each hardware device in the laser ranging system is similar to this structure, which can be split into information technology engineering parts and elements of the physical, (electro-)technical engineering. The separation happens somewhere in the actuators and sensors, where a conversion between physical indicators and digital representations is made.

Actuators

Digital representations of reference values change into electrical control signals for an electromechanical device in the actuators. But the structure of actuators can be quite different. A possible design is to convert the digital levels into analog usable current and voltage signals on a servo drive board, using a digital-to-analog converter.[4] The servo performs an electrical matching between the lower power levels of the electronic circuit and the higher power levels for the control of the actuating element. This element is variable for different actuator categories, for example, for electronic actuators of power electronics, electromagnetic actuators for motors and electromagnets, fluid-technical actuators as outlets or pneumatic/hydraulic engines, etc. ((Langmann 2010, *l.c. page 141 ff.*) and (Favre-Bulle 2004, *l.c. page 42 ff.*)).

Sensors

There are also sensors which take physical indicators and convert them into digitally usable representations for a control loop (more about the sensor technology can be found in ► Sect. 4.7 about the system monitoring). In principle, they build up a test circuit for the physical value. With this circuit in the sensor element, electrical power levels are produced. For a clean conversion into digital numbers, they must be amplified and filtered within a measurement amplifier unit (transducer). Then an analog-to-digital converter[5] can convert the levels into binary signals. Sensors can be classified into sensors for forces, mass, torque, pressure, distance, position, angle, speed, angular speed, acceleration, temperature, humidity, liquid capacity, fill level, flow-through, chemical indications, electromagnetic indications, optics, visual characteristics, and so on ((Langmann 2010, *l.c. page 103 ff.*) and (Favre-Bulle 2004, *l.c. page 20 ff.*)). This means that sensors can have

---

4   Digital-to-analog converters read a binary word of bits (usually 8 to 32 bits) and produce an analog signal representing the bit value. A possible way to implement this is to create a circuit for a summation of weighted currents. The resolution is defined by the number of bits. The precision is defined by the discretization error, the variations of the reference voltage, and the resistor values in the summation circuit. The goal is a monotone behavior of the analog current according to the digital input (Langmann 2010, *l.c. page 98*).

5   Analog-to-digital converters quantize time-discrete analog values to binary coded bit words (usually 8 to 32 bits). The quantization steps and therefore the resolution depend on the maximum analog value divided by 2 to the power of used bits. It is possible to convert actual and current analog values or integrated values over a defined time interval. The whole process is an iterative sampling and digitization of the analog signal bit by bit when the signal is stable after a setting time (Langmann 2010, *l.c. page 99 ff.*).

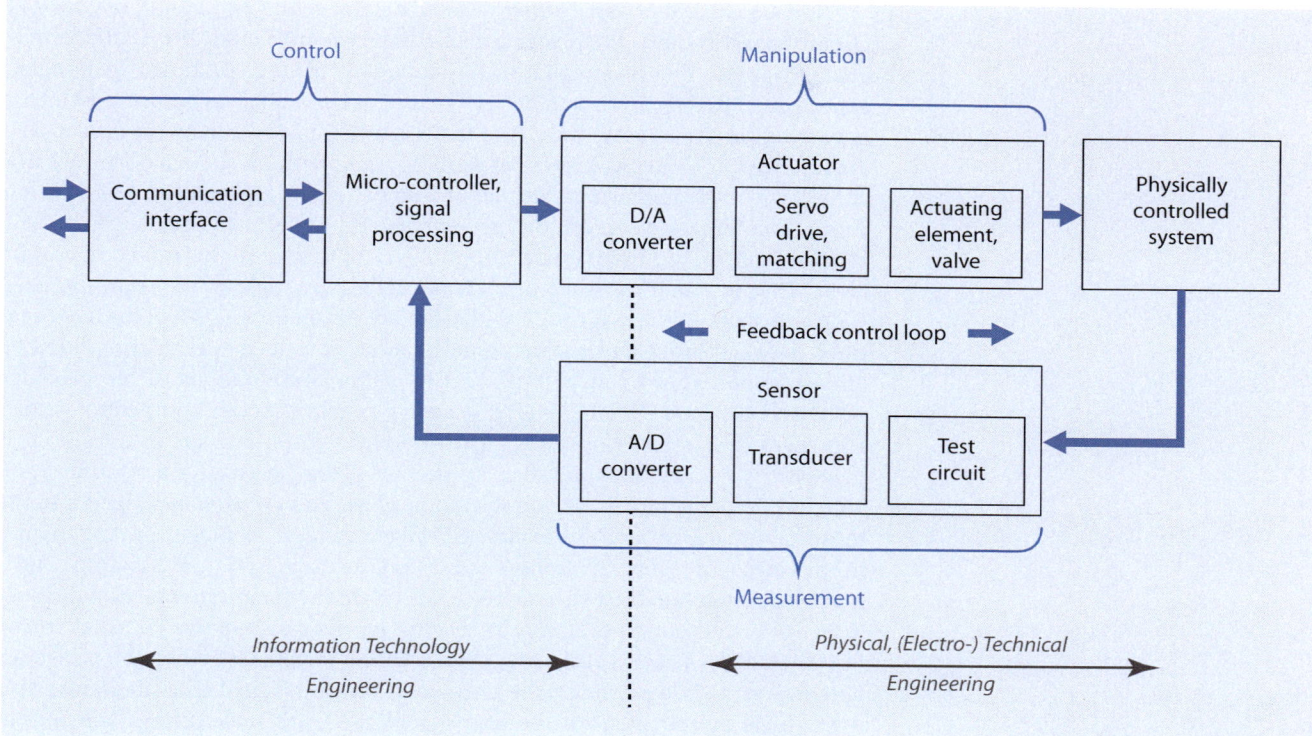

different complexity levels, and the boundaries of how the control intelligence is distributed are blurred (e.g., already a sensor or actuator can use a microcontroller, to solve its task).

But higher, more sophisticated applications require a separate signal processing unit where the actual control is processed. Currently, a lot of control units use programmable controller boards to solve their controlling task. The process of controlling means a continuous, clock-triggered, sensor-based monitoring of an actual, physical indicator value, which must be adjusted to a reference value by commanding an actuator. The algorithmic controller program or the controlling signal processing circuit continuously compares the actual value with the reference and uses different transfer functions to set input values of the actuator (Langmann 2010, *l.c. page 165 f.*). At this complexity level, this is a data- or signal-based solution. The activity sequence combines input signals or data from the sensor to derive output signals or data for the actuators. It only reacts to current states of the controlled system without considering its previous history. In some cases, the control algorithm can also use a rule-based realization at this level (e.g., fuzzy logic) where predefined rules and limits determine the output from a given input scenario (Favre-Bulle 2004, *l.c. page 59*). Usually standard controller units are used, such as proportional controller, integral controller, proportional-integral controller, proportional-derivative controller, and so on (Langmann 2010, *l.c. page 176 ff.*). The whole system implements a closed feedback control loop.[6]

Feedback Control Loop

---

6   Closed-loop control systems use a direct, «instantaneous» feedback mechanism, while open-loop or feed-forward control systems do not. They only command an actuator and read the feedback from the controlled system a bit later (Favre-Bulle 2004, *l.c. page 59 f.*).

4

Signal Processing Circuit with Logic Gates

A basic controller can consist of only a static signal processing circuit on a circuit board. It offers the possibility of a binary control after the digitization of analog signals. This means that binary input signals are combined with logical gates using Boolean algebra to derive binary control signals as output which trigger actions in the actuator line. The results are measured again by the sensors and are reported in binary as new input signals (Langmann 2010, *l.c. page 163 f.*). The circuit can be created using the rules of Boolean algebra (see ◘ Fig. 4.3), which is a set of two binary elements, 0 (low voltage) and 1 (high voltage), together with at least one of the binary operations AND or OR and the unary operation NOT. The combination of elements from the algebra follows the common laws, such as distributivity, associativity, commutativity, and so on. All of the combinations between input and output signals can be described with truth tables. All circuit combinations can be built just by using combinations of the gates for AND and NOT or OR and NOT. It is even possible to create comparator, adder, and memory circuits (flip-flops) with the basic gates. For larger circuits, it is necessary to reduce the number of gates used by processing a simplification which is, in principle, only the application of the rules of Boolean algebra to the terms of a standardized (or normalized) form of a logic formula (e.g., for all lines in the truth table where the output is 1, a conjunction (AND combination) of the input lines is created; all conjunctive clauses are then combined with a disjunction (OR combination), which results in the disjunctive normal form; (see ◘ Fig. 4.4). There is also a graphical solution for the simplification of circuits with at least four input lines: the Karnaugh-Veitch diagram (KV diagrams). It is a graphical arrangement of the terms from the truth table, using their output values. The KV diagram contains all of these outputs, so that those with a value of 1 can be used to build blocks of 16, 8, 4, 2, and 1 elements. The maximum possible number of elements must always be combined. The reduction of the inputs per block (where only those inputs are kept which do not appear also in negated form in the block) is the simplification (see also ◘ Fig. 4.4) (Rembold et al. 1991, *l.c. page 61 ff.*). This reduction is quite important to keep binary circuits small. But fixed soldered circuit boards are very static. A more flexible way to use the binary signal-based control gates is Field Programmable Array (FPGA)s, which emulate logic circuits on a programmable field for the truth tables on an integrated circuit. Generally, the signal processing circuit on circuit boards can control physical devices very fast. Therefore, they are usually used for safety issues, such as interlock mechanisms.

Microcontroller

A better way to implement the logic for control loops is the usage of microcontrollers. They are formed with highly integrated and, therefore, very compact, logic circuits. Microcontrollers, such as the ATmega series from Atmel (Atmel Corporation 2012), which is currently used for the open-source electronics prototyping platform Arduino (Arduino team n.d.), are very compact and modern solutions of a classic John von Neumann architecture (see ◘ Fig. 4.5, partly adopted from (Atmel Corporation 2012, *l.c. page 5*)). John von Neumann and his colleagues defined a universal architecture for a data processing computer system in the mid-1940s.[7] This architecture is still valid and implemented in all modern computer systems. It defines four main elements (Langmann 2010, *l.c. page 57 f.*):

– *Input and output system*: Direct contact between the machine and the user happens via information exchange and representation with an input and output system. These are the peripheral devices to enter data and programs as input to the system and to retrieve results as output from the system.

---

7   Similar basic ideas for a processing unit with an arithmetical unit and memory and input/output units were already described by Charles Babbage in the 1820s/1830s for a difference engine (a mechanical calculator) and later for the analytical engine (Rembold et al. 1991, *l.c. page 28*).

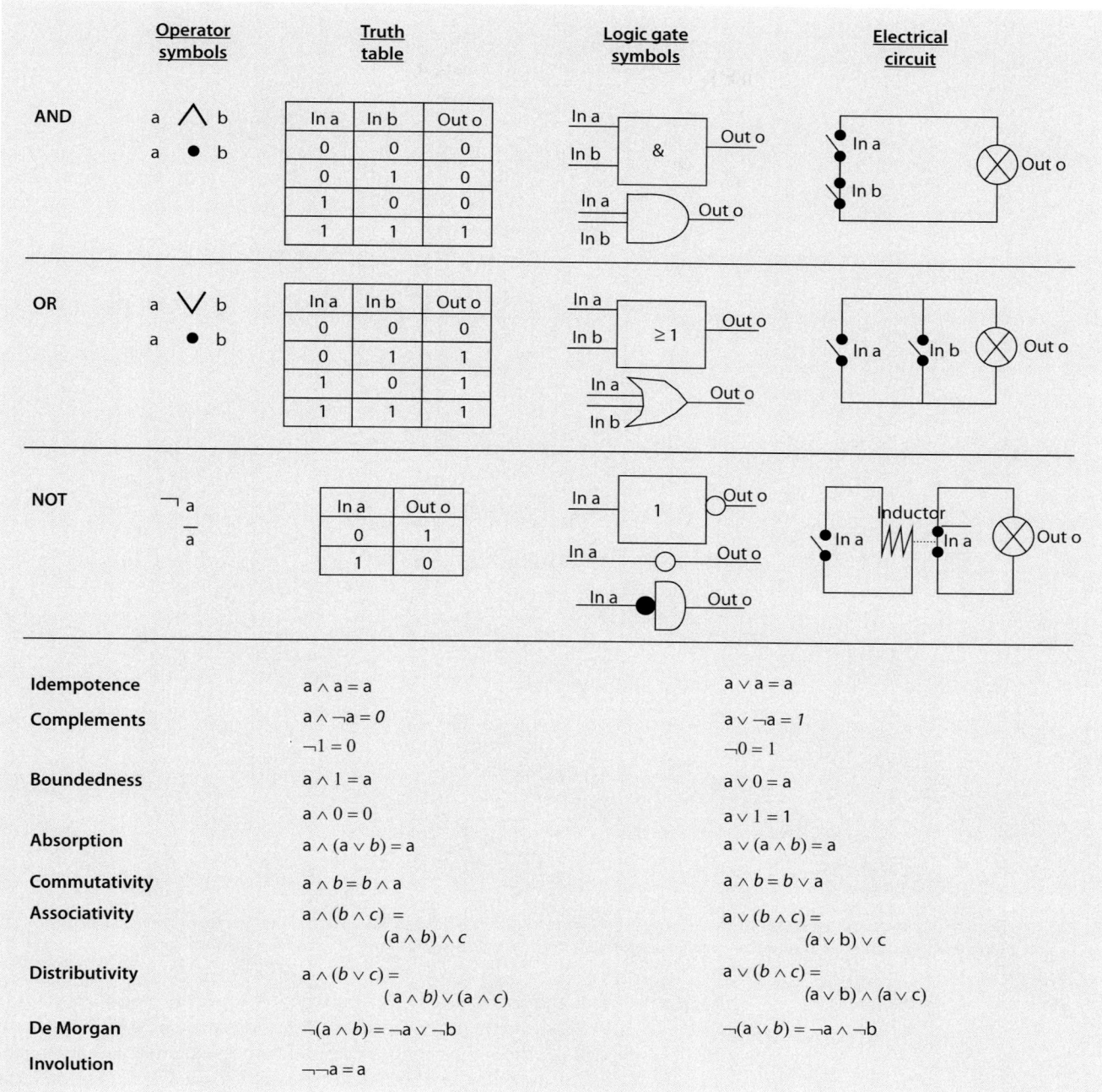

	Operator symbols	Truth table			Logic gate symbols	Electrical circuit

AND

In a	In b	Out o
0	0	0
0	1	0
1	0	0
1	1	1

OR

In a	In b	Out o
0	0	0
0	1	1
1	0	1
1	1	1

NOT

In a	Out o
0	1
1	0

Idempotence	$a \wedge a = a$	$a \vee a = a$
Complements	$a \wedge \neg a = 0$	$a \vee \neg a = 1$
	$\neg 1 = 0$	$\neg 0 = 1$
Boundedness	$a \wedge 1 = a$	$a \vee 0 = a$
	$a \wedge 0 = 0$	$a \vee 1 = 1$
Absorption	$a \wedge (a \vee b) = a$	$a \vee (a \wedge b) = a$
Commutativity	$a \wedge b = b \wedge a$	$a \wedge b = b \vee a$
Associativity	$a \wedge (b \wedge c) = (a \wedge b) \wedge c$	$a \vee (b \wedge c) = (a \vee b) \vee c$
Distributivity	$a \wedge (b \vee c) = (a \wedge b) \vee (a \wedge c)$	$a \vee (b \wedge c) = (a \vee b) \wedge (a \vee c)$
De Morgan	$\neg(a \wedge b) = \neg a \vee \neg b$	$\neg(a \vee b) = \neg a \wedge \neg b$
Involution	$\neg\neg a = a$	

Fig. 4.3 Boolean algebra and its laws

- *Memory*: the memory saves raw and processed data together with the programs (the representations of algorithms) to process the data in binary coded form.
- *Control unit*: The control unit controls the operations of the arithmetic logic unit. It fetches the binary data and instructions from the memory («FETCH»), decodes the contained instructions («DECODE»), controls and executes the processing of the instruction in the arithmetic logic unit («EXECUTE»), and writes the results back to the memory («WRITEBACK») (Rembold et al. 1991, *l.c. page 308 ff.*).
- *Arithmetic logic unit*: the arithmetic logic unit processes all arithmetical (mathematical) and logical (Boolean) operations.

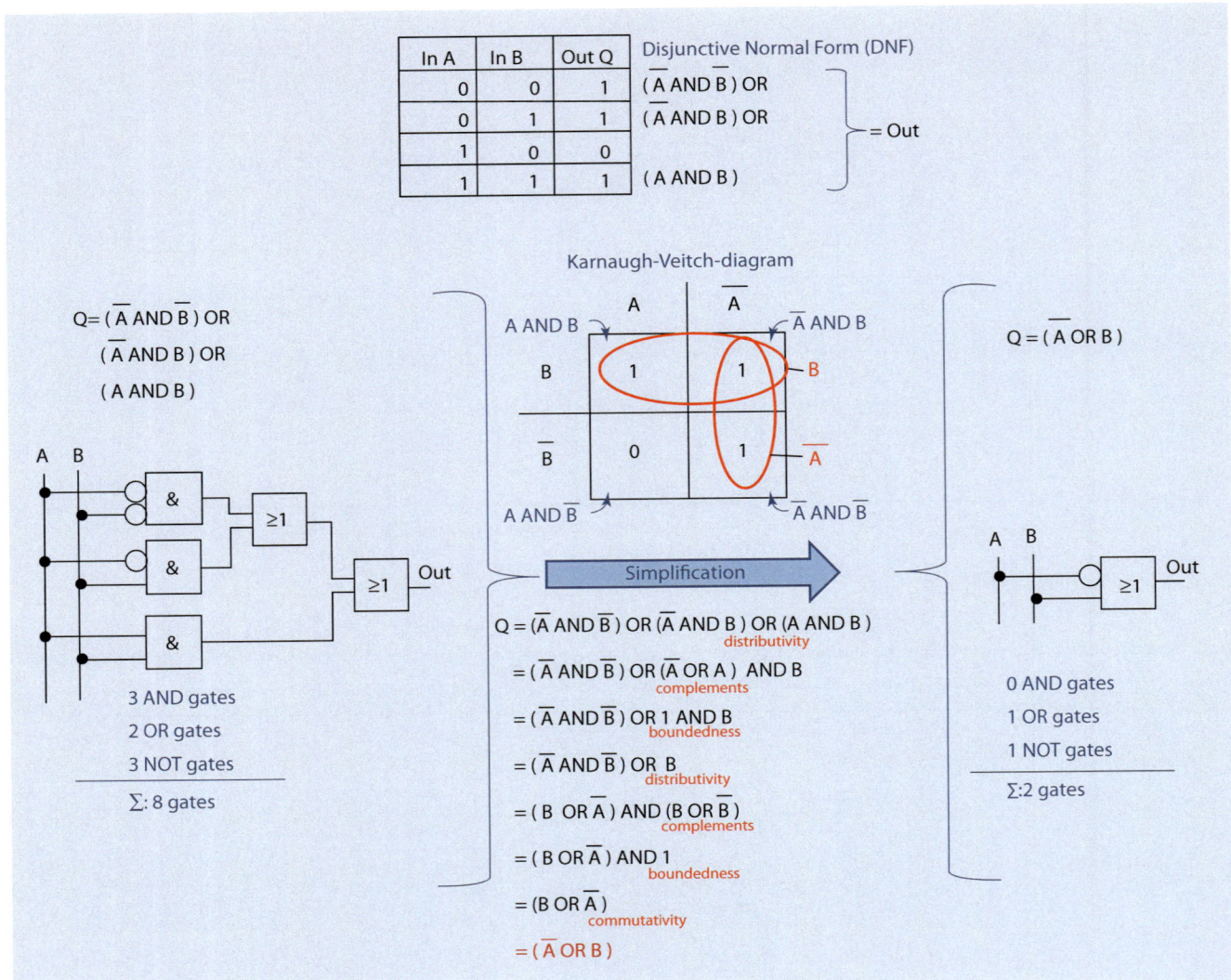

In A	In B	Out Q	Disjunctive Normal Form (DNF)
0	0	1	($\overline{A}$ AND $\overline{B}$) OR
0	1	1	($\overline{A}$ AND B) OR
1	0	0	
1	1	1	(A AND B)

= Out

Karnaugh-Veitch-diagram

$Q = (\overline{A}$ AND $\overline{B})$ OR
$(\overline{A}$ AND B) OR
(A AND B)

3 AND gates
2 OR gates
3 NOT gates

Σ: 8 gates

$Q = (\overline{A}$ OR B)

Simplification

$Q = (\overline{A}$ AND $\overline{B})$ OR $(\overline{A}$ AND B) OR (A AND B)
      distributivity
$= (\overline{A}$ AND $\overline{B})$ OR $(\overline{A}$ OR A) AND B
      complements
$= (\overline{A}$ AND $\overline{B})$ OR 1 AND B
      boundedness
$= (\overline{A}$ AND $\overline{B})$ OR B
      distributivity
$= (B$ OR $\overline{A})$ AND (B OR $\overline{B}$)
      complements
$= (B$ OR $\overline{A})$ AND 1
      boundedness
$= (B$ OR $\overline{A})$
      commutativity
$= (\overline{A}$ OR B)

0 AND gates
1 OR gates
1 NOT gates

Σ:2 gates

Fig. 4.4   The use of a Karnaugh-Veitch diagram in comparison with the formal mathematical method with the rules of Boolean algebra to reduce the number of logic gates for a controller circuit, starting from the disjunctive normal form derived from the truth table

The control unit and arithmetic logic unit together form the microprocessor or CPU. The CPU and the memory together form the processor core (Langmann 2010, *l.c. page 57 f.*). The components are nowadays implemented with integrated circuits. A microcontroller is a controller board with a CPU, memory, and required periphery. Peripheral elements are a clock and timers, digital and/or analog input and output ports (with converter circuits), and serial or bus interfaces. The memory can be an Electrically Erasable Programmable Read-Only Memory (EEPROM) for more persistent, non-volatile data, which can be electrically deleted. Nowadays, flash memories are also very popular. These are compact, low-energy, flash EEPROM. A possibility for volatile storage is the Static Random Access Memory (SRAM) (see also the ATmega block diagram in ▪ Fig. 4.5). As microcontrollers understand programs in binary form, they can be programmed with the usual programming languages to solve the controlling tasks. For example, in the case of the Arduino board, a C/C++ dialect is used to write and run structured and OOP programs. Additionally, digital signal processors are optimized and specialized microcontrollers for digital controlling tasks (Favre-Bulle 2004, *l.c. page 168*).

Microcontrollers are very sophisticated computing machines. The small boards and prototyping platforms, such as the Arduino board, offer tremendous

Computer

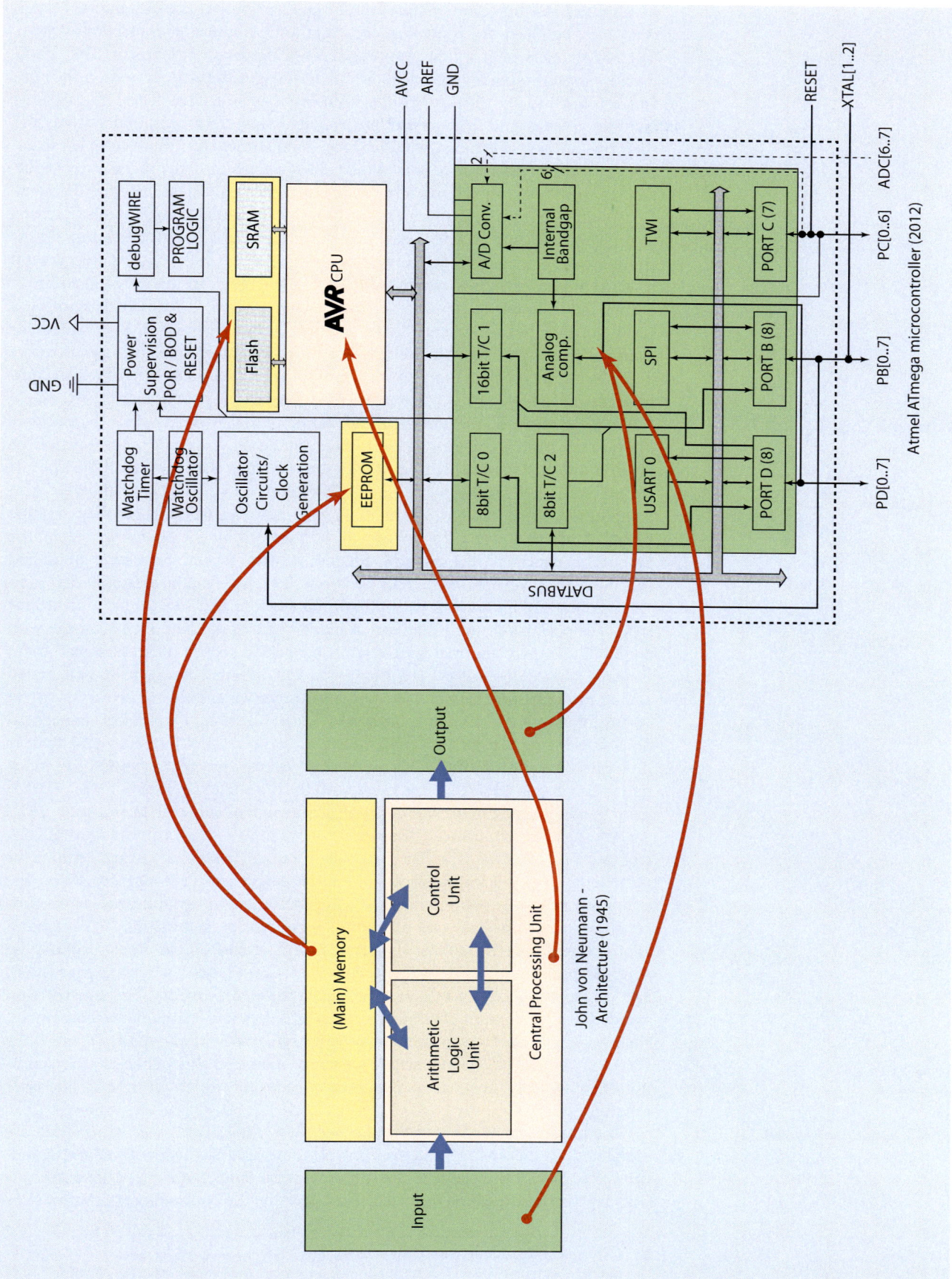

possibilities. As they can operate several input and output lines of different types, they can be used to control more than one physically controlled system. Microcontrollers offer a first grouping level of hardware equipment. They allow higher programming languages to describe the algorithms processed on the controllers. But usually they don't offer high-level software interfaces, multitasking facilities, queues, and so on, which are available in modern computer systems and their operating systems. A similar but more sophisticated and more complex structure is the use of microprocessors in the field of electronic data processing computers. In principle, those modern computer systems (including the Personal Computer (PC)) are just more advanced microcontroller boards (and therefore, again, von Neumann architectures) with more sophisticated input and output possibilities, human user interfaces, a powerful (maybe multi-core) CPU; several internal, higher-level bus systems, different data storage media for data, and programs and different fast and large main memory types. To run such systems, a rudimentary type of software, the Basic Input Output System (BIOS) is loaded, for example, from an EEPROM. Then the operating system is loaded from a hard drive or flash card memory in the form of precompiled machine code.[8] It offers all common functionalities of a PC (see also ▶ Sect. 2.6.4).

Programmable Logic Controller

While a standard PC is universal and, therefore, ideally prepared for the end-user or office environment, hardware controllers require more specific processing for the reading, processing, and writing of digital and analog data. One of those specific controllers is a Programmable Logic Controller (PLC), which is available in a compact or a modular form. The modular variants are very popular, as they combine a central unit (with the CPU board and the memory for the operating system, the processing data, and the running programs) with several peripheral units and functional units on a bus system as backbone. The units can be arranged very individually and are mountable with plug-in connectors and sockets on a connector strip. The peripheral units offer digital input and output components usually with 8–32 bits. Junctors and matching circuits adapt and reduce different signal levels, voltages, and interferences. The analog input and output components use integrated analog-to-digital and digital-to-analog converters with the usual word format of 10–16 bits, different sampling rates, and adaptable integration times. Functional units extend the central unit with counters, servo drivers for servo and stepper motors, feedback loop controllers, numerical control units for machines, and much more (Langmann 2010, *l.c. page 156 ff.*). All the units together offer an ideal mechanism to bundle and group different controlling loops as shown in ▫ Fig. 4.6.

Industrial PC

Another very sophisticated and feasible solution for this combination of control loops is industrial PCs. They are very robust PCs with additional options for peripheral units and the possibility of interrupting a priori running processes. The operating system supports guarantees for such interrupt routines to enable processing in real-time (see one of the following blocks about timing) (Langmann 2010, *l.c. page 58*). The robust chassis of an industrial PC thoroughly protects the components from dirt, humidity, corrosion, Radio Frequency Interference (RFI), and other influences encountered in industrial environments. The construction avoids sensitive components, such as fans, and uses low-maintenance equipment, such as solid-state disks as memory medium to avoid rotating parts (Favre-Bulle 2004, *l.c. page 168 f.*). As industrial PCs can be managed almost like standard computers or servers, their use is ideal for scientific control environments like laser ranging.

Automatic Sequence Control

The programmable units control several lower-level controller loops. They use a higher-level automatic sequence control. Compared to the low-level and signal-driven feedback controllers, they use a program to implement a feed-forward controller without immediate feedback. They describe the controlling task with a flow

---

8   Early operating systems were written in machine code (assembly language used in combination with assemblers as a form of simple compiler), while nowadays they are programmed in higher-level languages such as C (e.g., the Linux kernel is written in C).

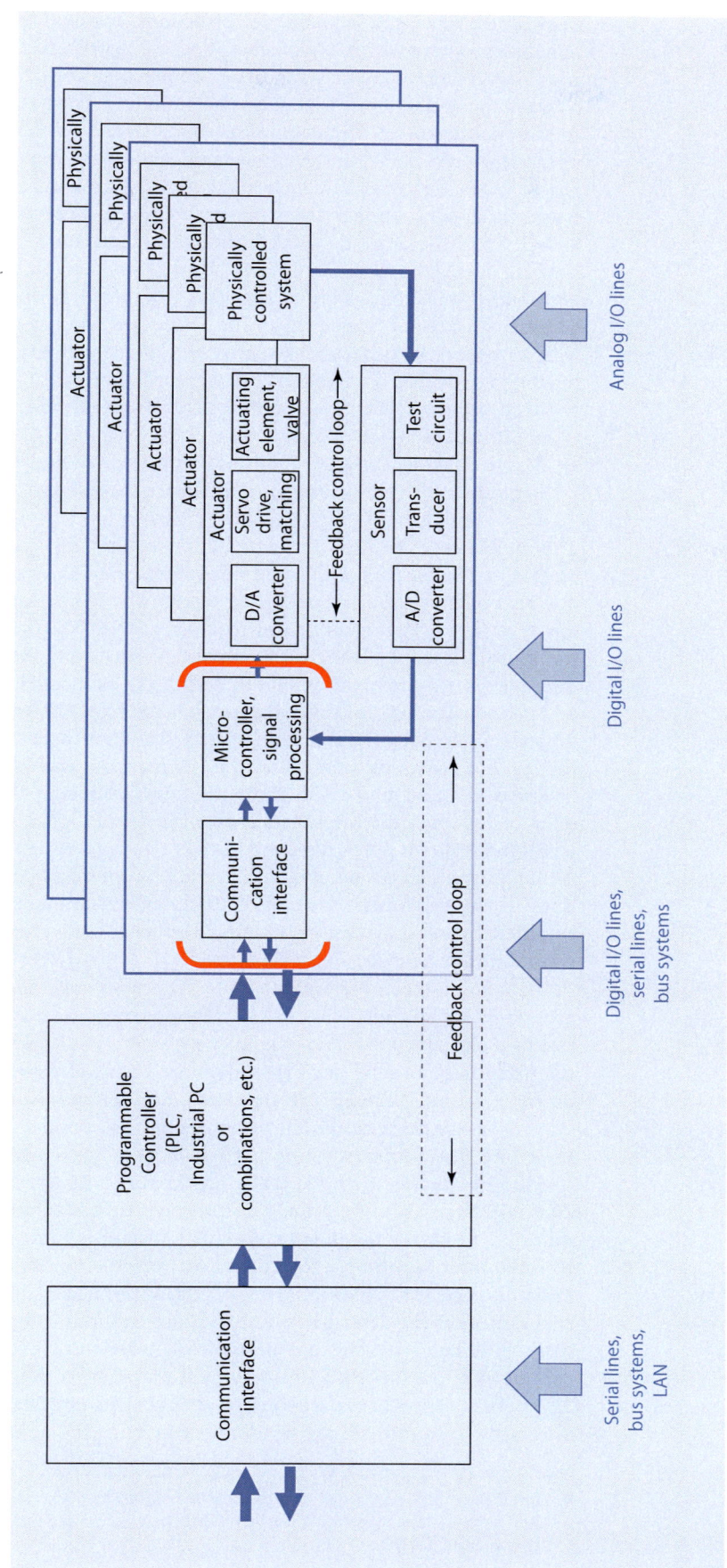

Fig. 4.6 The next level of control loops combines and groups several hardware control units in the form of sophisticated programmable logic controllers and industrial PCs as extension to intelligent hardware control units in ◼ Fig. 4.2

chart for the order and sequence of instructions. The flow chart is a graphical notation for a program which implements a specific algorithm. It is more or less a «state graph» with different node symbols and connecting labeled and non-labeled edges between the nodes. It offers single instructions (represented as rectangular blocks), conditional statements (represented as rhombs), and combinations for conditional instructions, alternative instructions, and conditional loops (Langmann 2010, *l.c. page 213 ff.*). But generally, such charts just describe state machines of different types, as already familiar from the processing of programming languages (see ▶ Sect. 3.3.1). Therefore, the same techniques and diagrams, for example, the known state graphs, can be used. But usually the process and controlling state machines are higher-level machines with a sophisticated memory which uses an arbitrary addressing of memory elements.[9] The results are sequential finite state machines with a memory loop-back represented by computer programs. They can be separated into two types (Langmann 2010, *l.c. page 219 f.*):

— *Mealy state machines*: the resulting control signals depend on input signals and the states in the memory.
— *Moore state machines*: the resulting control signals are only defined by the individual states in the memory which are the results of the inputs.

Communication Interface Layers

This higher-level control and automation equipment is usually produced and offered by the vendors of the controlled hardware. For example, a telescope vendor also offers the basic controlling equipment which might consist of several low-level servo drives, controlled by an industrial PC. The antenna control unit for radio telescopes from the company Vertex Antennentechnik GmbH, for example, uses a combination of a PLC with the connected hardware servo drives and an industrial PC for the higher-level tasks. The problem is that different vendors use different communication interfaces and protocols. As shown in ▣ Fig. 4.6, the level of communication interfaces increases at each step where the control becomes more sophisticated. While the most physically controlled systems just offer analog lines, the converters enable the use of digital lines with fixed voltage levels and just two states: high and low/zero voltage. After the microcontrollers, it is then possible to use parallel, serial, and bus interfaces besides the single digital lines. These are combinations of digital lines, which make it possible to transfer complete bytes controlled by a hardware and/or software protocol and a special communication device.

Different Physical Communication Lines

Parallel connectors combine eight data lines together with eight electrical ground lines and control lines to transfer one byte in parallel. Serial lines use one transmit and one receive line together with different control lines and an electrical ground line to send one byte sequentially and synchronously triggered bit by bit (an example is the RS-232 standard). Bus communications are special serial systems where more communication partners are connected sequentially on the same line. The communication partners on the bus are then addressed and use specific bus access methods (an example is the RS-485 standard) (Langmann 2010, *l.c. page 312*). Field buses are special, robust solutions, which guarantee minimal response times over specified distances, and are protected against RFI and offer high safety for critical manufacturing environments (Langmann 2010, *l.c. page 356*). Additional units, such as those available in the PLC or industrial PCs, offer the described point-to-point communications but also add a network interface with the use of network protocols as TCP/IP. Programming communication sockets with individual messages is then possible (see ▶ Sect. 3.3.1). Most of the hardware devices offer sockets or serial connections. But the interfaces and protocols are vendor specific (e.g., the dome control for the laser

---

9   But these machines can also theoretically be reduced to a Turing machine with just one memory tape, on which a read/write head can access and move by one step (see ▶ Sect. 3.3.3).

ranging system at the Wettzell observatory operates on a serial line, the telescope uses sockets, and the radar uses different serial lines with a specific protocol). Some of the connection lines are additionally protected with «galvanic isolators» to separate the different grounds (very important for sensible high-frequency processing boards, as used in the event timer) and to reduce influences from peak voltages (lightning protection). Sometimes it is also necessary to use additional line drivers, which stabilize the voltages and currents on a physical line, so that longer distances can be reached, or to stabilize the communication in RFI-polluted environments.

This situation of individual implementations for communication tasks is not really desirable to guarantee connectivity, stability, synchronization, and so on. Usually several external devices are connected to a centralized controller computer. But this solution is a very tight coupling because each device requires a special handling for the timing and synchronization, which might also be repercussive to the main control if the devices do not respond anymore. As shown for safety in ▶ Sect. 3.3.3, a lot of techniques are necessary to implement basic and reliable communication safety. The use of techniques from the already described middleware in ▶ Sect. 3.4 is one suitable solution to avoid these disadvantages. Each hardware device is represented by an RPC server as a sophisticated software representation for a device driver (see ▫ Fig. 4.7). Each server then implements something like a virtual software plug-in connector, which can be connected to the controller middleware network, building something like a virtual connector strip. This setup implements systems with software that are already very feasible for modular PLCs on the lower levels. The server must build up another abstraction level. Just as the PLC does not implement or even take care of controlling the implementations of lower-level proportional controllers or proportional-integral controllers, the higher-level servers in the RPC network should not be forced to take care of controlling the hardware in the server. To implement this, the new pluggable hardware device server uses three layers: the hardware driver module or component, the periodic scheduler loop, and the RPC interface. Using the «idl2rpc.pl» generator, the safety arrangements are created directly besides the skeleton for the periodic control loop and the other functions of the server.

The hardware driver module or component implements access and communication to the hardware controller, which is, for example, the PLC or the industrial PC. It combines low-level communication functionalities, such as complete communication message flows with compact functions which are usable from the periodic scheduler loop. The hardware driver module also implements a simple timing and safety environment which is just enough to run the communication. In principle, the module hides vendor-specific communication messages, protocols, and hardware connectors. To implement its tasks, the module can include further modules from the code toolbox if necessary (e.g., for socket or serial communication). A sample excerpt for an implementation of a hardware driver component in C++ to operate the Antenna Control Unit (ACU) for the controlling of radio telescopes of the company Vertex Antennentechnik GmbH is shown in Example 4.1. The function «ESetPositionInAzElOfACU» shown processes the complete communication in order to command the radio telescope to a specific position, given as floating point arguments for the azimuth and elevation angles in degree. It checks the arguments, the communication, and the ACU status. The next step sets the right state of the ACU if necessary. Then the binary ACU command which was defined by the vendor is composed. The assembled command is then sent via a socket connection to the ACU industrial PC in the telescope tower, which replies with an acknowledgment message. The receiving is handled by another function «bReceiveAcknowledgeFromACU». Then the positioning function waits until the telescope is really moving. All these steps build the workflow to control and automate the movement of the telescope to a specific position.

RPC Server as Device Driver

Hardware Driver Module or Component

4

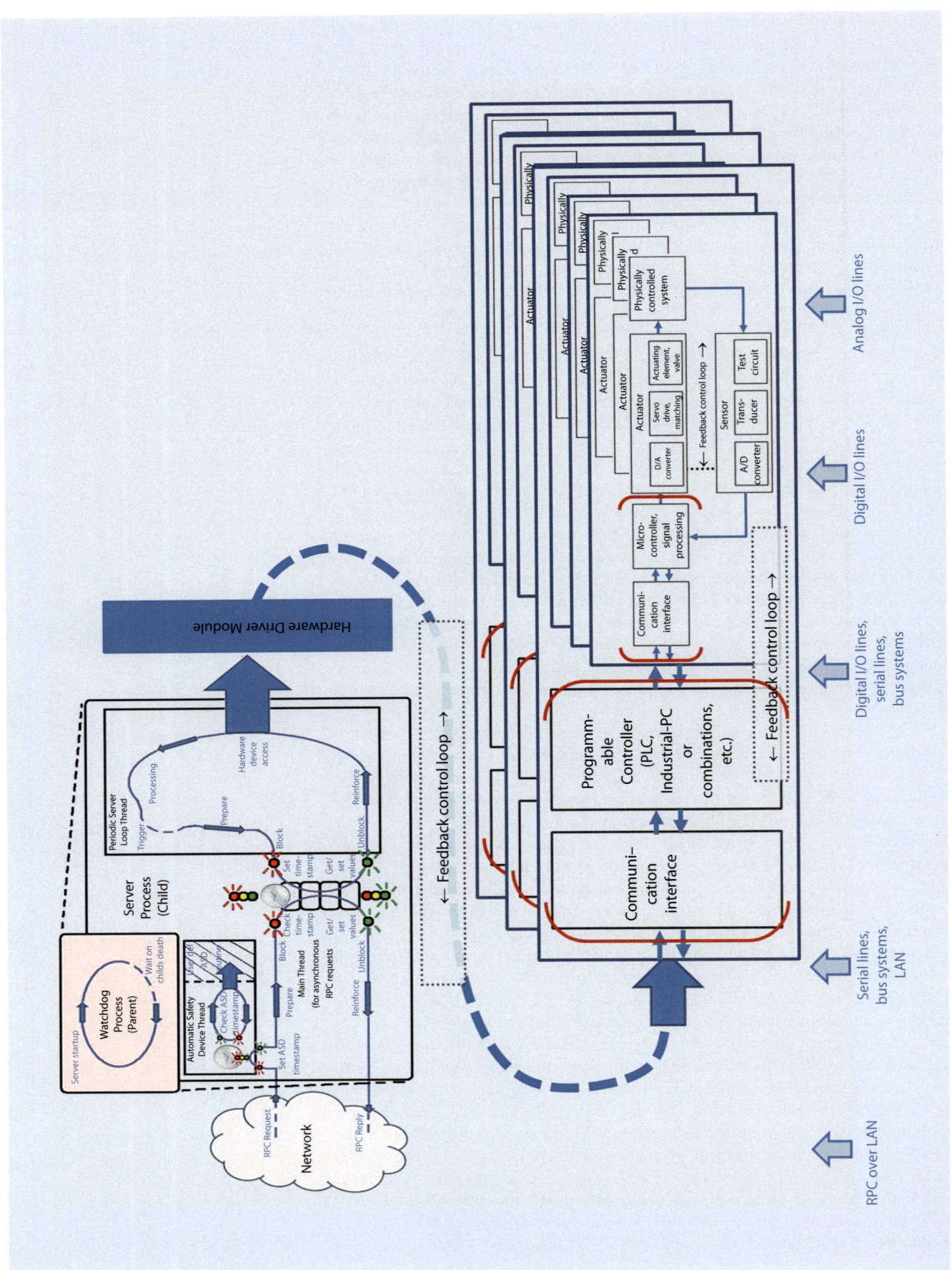

▪ ▪ Example 4.1

Excerpt of the hardware driver module to operate the ACU for the controlling of
radio telescopes of the company Vertex Antennentechnik GmbH

```
// ...                                                                1

class VertexAntennaControlUnit                                       2
{                                                                    3
    private:                                                         4
        bool bDebugOn;                                               5
        unsigned int uiCommunicationTimeoutSec;                      6
        SimpleSocketType SClientSocket;                              7
        std::string strIPAddress;                                    8
        unsigned int uiPort;                                         9
        double dAzimuthMinimum;                                      10
        double dAzimuthMaximum;                                      11
        double dElevationMinimum;                                    12
        double dElevationMaximum;                                    13
        std::string strCommandedSourceName;                          14
    protected:                                                       15
        union ChecksumUnion                                          16
        {                                                            17
            int iChecksum;                                           18
            char acChecksum[4];                                      19
        };                                                           20
        union MessageLengthUnion                                     21
        {                                                            22
            unsigned short usLength;                                 23
            char acLength[2];                                        24
        };                                                           25
        union IntegerConversionUnion                                 26
        {                                                            27
            unsigned short usInteger[2];                             28
            unsigned int uiInteger;                                  29
            int iInteger;                                            30
            char acArray[4];                                         31
        };                                                           32
        union DoubleConversionUnion                                  33
        {                                                            34
            double dDouble;                                          35
            char acArray[8];                                         36
        };                                                           37
    protected:                                                       38
        // ...                                                       39
        void vCalculateChecksum (char * acBuffer,                    40
                                 unsigned int uiBufferLength,        41
                                 unsigned int uiStartIndex,          42
                                 char acChecksumArray[2]) const;     43
        unsigned short usIsConnectionOpen () const;                  44
        bool bReceiveAcknowledgeFromACU ();                          45
        // ...                                                       46
                                                                     47
                                                                     48
    public:                                                          49
        // ...                                                       50
        RetValueEnumType ESetPositionInAzElOfACU (double dAzimuthDegree, double   51
            dElevationDegree);

        // ...                                                       52
};                                                                   53
                                                                     54
// ...                                                               55
                                                                     56
```

```
/********************************************************************
 *   class       VertexAntennaControlUnit
 *   function  bReceiveAcknowledgeFromACU
 ********************************************************************/
/*!            Receive ACU acknowledge message from ACU
 *   \param      —
 *   \return    bool <— Answer (true = acknowledge received, false = not received)
 ********************************************************************/
/*   author    Alexander Neidhardt
 *   date       15.06.2012
 *   revision —
 *   info       —
 ********************************************************************/
bool VertexAntennaControlUnit :: bReceiveAcknowledgeFromACU ()
{
    char acBuffer[1024];
    unsigned long ulBufferLength = 1024;

    /// Check if connection opened
    if (!usIsConnectionOpen ())
    {
        return false;
    }

    /// Clean buffer
    vCleanBuffer (acBuffer, 1024);

    /// Receive acknowledge and check for correctnes
    if (uiReceiveWithTimeout (&SClientSocket,
                              acBuffer,
                              &ulBufferLength,
                              uiCommunicationTimeoutSec))
    {
        return false;
    }
    if (bDebugOn)
    {
        printf (" >>> bReceiveAcknowledgeFromACU: ");
        vPrintMessageBuffer (acBuffer, ulBufferLength, 'R');
    }
    if (!(acBuffer[0] == 0x06 && acBuffer[1] == '\0'))
    {
        return false;
    }

    return true;
}

/********************************************************************
 *   class       VertexAntennaControlUnit
 *   function  ESetPositionInAzElOfACU
 ********************************************************************/
/*!            Set position in azimuth [degree] and elevation [degree]
 *   \param      double dAzimuthDegree —> Azimuth position in degree
 *   \param      double dElevationDegre —> Elevation position in degree
 *   \return    VertexAntennaControlUnit :: RetValueEnumType <— Error (ACUOK, ACUNOK)
 ********************************************************************/
/*   author    Alexander Neidhardt
 *   date       15.06.2012
 *   revision —
 *   info       —
 ********************************************************************/
```

57
58
59
60
61
62
63
64
65
66
67
68
69
70
71
72
73
74
75
76
77
78
79
80
81
82
83
84
85
86
87
88
89
90
91
92
93
94
95
96
97
98
99
100
101
102
103
104
105
106
107
108
109
110
111
112
113
114
115
116
117
118
119

```
VertexAntennaControlUnit::RetValueEnumType VertexAntennaControlUnit::ESetPositionInAzElOfACU (    120
    double dAzimuthDegree,                                                                          121
        double dElevationDegree)                                                                    122
{                                                                                                   
    int iAzimuth = 0;                                                                               123
    int iElevation = 0;                                                                             124
    char acBuffer[1024];                                                                            125
    char acChecksumArray[2] = { '\0', '\0' };                                                       126
    char acArray[4] = { '\0', '\0', '\0', '\0' };                                                   127
    double dPreviousCommandedAzimuthDegree = 0.0;                                                   128
    double dPreviousCommandedElevationDegree = 0.0;                                                 129
    VertexAntennaControlUnit::TimeStructType STime;                                                 130
    VertexAntennaControlUnit::PositionStructType SPosition;                                         131
    VertexAntennaControlUnit::StatusStructType SStatus;                                             132
    VertexAntennaControlUnit::TiltStructType STilt;                                                 133
    VertexAntennaControlUnit::SubreflectorStructType SSubrefelctor;                                 134
    unsigned long ulStartTime = 0;                                                                  135
                                                                                                    136
    /// Check input                                                                                 137
    if (dAzimuthDegree < dAzimuthMinimum || dAzimuthDegree > dAzimuthMaximum ||                     138
            dElevationDegree < dElevationMinimum || dElevationDegree > dElevationMaximum)           139
    {                                                                                               140
        return ACUNOK;                                                                              141
    }                                                                                               142
    iAzimuth   = (int)(dAzimuthDegree * 1000000.0);                                                 143
    iElevation = (int)(dElevationDegree * 1000000.0);                                               144
                                                                                                    145
    /// Check if connection opened                                                                  146
    if (!usIsConnectionOpen ())                                                                     147
    {                                                                                               148
        return ACUNOK;                                                                              149
    }                                                                                               150
                                                                                                    151
    /// Check ACU status                                                                            152
    if (ERequestStatusOfACU (STime, SPosition, SStatus, STilt, SSubrefelctor))                      153
    {                                                                                               154
        return ACUNOK;                                                                              155
    }                                                                                               156
    if (SStatus.EAzimuthStatusMode != ACUMODE_PRESET ||                                             157
            SStatus.EElevationStatusMode != ACUMODE_PRESET)                                         158
    {                                                                                               159
        if (ESetOperationModeOfACU (VertexAntennaControlUnit::ACUMODE_PRESET))                      160
        {                                                                                           161
            return ACUNOK;                                                                          162
        }                                                                                           163
    }                                                                                               164
    dPreviousCommandedAzimuthDegree = SPosition.dCommandedAzimuthDegree;                            165
    dPreviousCommandedElevationDegree = SPosition.dCommandedElevationDegree;                        166
                                                                                                    167
    if (((int)(fabs(SPosition.dAzimuthDegree   - dAzimuthDegree) * 1000.0)) == 0 &&                 168
            ((int)(fabs(SPosition.dElevationDegree - dElevationDegree) * 1000.0)) == 0)             169
    {                                                                                               170
        // Nothing to do                                                                            171
        return ACUOK;                                                                               172
    }                                                                                               173
                                                                                                    174
    /// Clean buffer                                                                                175
    vCleanBuffer (acBuffer, 1024);                                                                  176
                                                                                                    177
    /// Prepare buffer                                                                              178
    acBuffer[0] = (char)0x02;   // STX                                                              179
    acBuffer[1] = (char)0x50;   // Message ID                                                       180
    acBuffer[2] = (char)0x13;   // Size byte 1                                                      181
    acBuffer[3] = (char)0x00;   // Size byte 2                                                      182
    vConvertSignedIntegerToSignedIntegerArray (iAzimuth, acArray);                                  183
    acBuffer[4] = acArray[0];   // Azimuth position byte 1                                          184
```

```
acBuffer[5]  = acArray[1];   // Azimuth position byte 2          185
acBuffer[6]  = acArray[2];   // Azimuth position byte 3          186
acBuffer[7]  = acArray[3];   // Azimuth position byte 4          187
vConvertSignedIntegerToSignedIntegerArray (iElevation, acArray); 188
acBuffer[8]  = acArray[0];   // Azimuth position byte 1          189
acBuffer[9]  = acArray[1];   // Azimuth position byte 2          190
acBuffer[10] = acArray[2];   // Azimuth position byte 3          191
acBuffer[11] = acArray[3];   // Azimuth position byte 4          192
acBuffer[12] = (char)0x00;   // not used => empty                193
acBuffer[13] = (char)0x00;   // not used => empty                194
acBuffer[14] = (char)0x00;   // not used => empty                195
acBuffer[15] = (char)0x00;   // not used => empty                196
/// Calculate checksum                                           197
vCalculateChecksum (acBuffer, 19, 1, acChecksumArray);           198
/// Finalize buffer                                              199
acBuffer[16] = acChecksumArray[0]; /// Checksum byte 1           200
acBuffer[17] = acChecksumArray[1]; /// Checksum byte 2           201
acBuffer[18] = (char)0x03; // ETX                                202
                                                                 203
/// Send buffer to ACU                                           204
if (bDebugOn)                                                    205
{                                                                206
    printf (" >>> ESetPositionInAzElOfACU: ");                  207
    vPrintMessageBuffer (acBuffer, 19, 'S');                    208
}                                                                209
if (uiSend (&SClientSocket, acBuffer, 19))                       210
{                                                                211
    return ACUNOK;                                               212
}                                                                213
                                                                 214
                                                                 215
/// Receive acknowledge message from ACU                         216
if (!bReceiveAcknowledgeFromACU ())                              217
{                                                                218
    return ACUNOK;                                               219
}                                                                220
                                                                 221
/// Wait until movement                                          222
vSleep (200);                                                    223
ulStartTime = time(NULL);                                        224
while (1 == 1)                                                   225
{                                                                226
    if (time(NULL) − ulStartTime > uiCommunicationTimeoutSec)    227
    {                                                            228
        return ACUNOK;                                           229
    }                                                            230
    if (ERequestStatusOfACU (STime, SPosition, SStatus, STilt, SSubrefelctor)) 231
    {                                                            232
        return ACUNOK;                                           233
    }                                                            234
                                                                 235
    if (((int)(fabs(SPosition.dCommandedAzimuthDegree   − dAzimuthDegree  ) * 1000.0)) ==
        0 ||
            ((int)(fabs(SPosition.dCommandedElevationDegree − dElevationDegree) * 1000.0)) 236
                == 0)
    {                                                            237
        break;                                                   238
    }                                                            239
}                                                                240
                                                                 241
strCommandedSourceName = "Az/El Pos";                            242
                                                                 243
return ACUOK;                                                    244
}                                                                245
                                                                 246
// ...                                                           247
```

The periodic control loop as the second level of a hardware-driving RPC server is the heart of the software controller. It is the place where functions of the hardware driver module or component are called. This is one of the main design decisions. No RPC function which arrives asynchronously activates a hardware control or communication function directly. This avoids negative feedback or backing up of delay effects from the lower hardware interface to higher levels of control at the RPC client. Another reason is that the periodic loop acts as a synchronous scheduler. All asynchronously incoming requests can be ordered, sequentialized, and synchronized with periodic controlling tasks. Therefore, the periodic loop is the place where workflow errors and other faults can be managed, so that the higher level of the control in the RPC client does not have to be responsible for the precisely ordered arrival time of hardware commands. The timing depends on the current state of the periodic state machine, derived from the feedback of the hardware. Usually access to the hardware is also a critical section (see ▶ Sect. 2.6.4). Communication lines like serial connections are just single access points, which cannot be operated in parallel with different handlers. The hardware can usually only process the communication protocol sequence for one command at a time. As the RPC calls arrive asynchronously from different clients, this might lead to conflicts if incompatible calls are commanded. With the design restriction that only the periodic loop has access to the real hardware, the critical section around the hardware communication, which is usually very time-consuming, has no effect and cannot be protected with semaphores.

Periodic Control Loop for Devices

To increase the performance of the server, it is recommended that threads are used. Therefore, semaphore-protected shared variables must always be used between the thread for the asynchronous RPC and the activities in the periodic loop. The RPC functions only order activities or read the values via shared variables. The periodic loop takes the information about activities and the incoming inputs to schedule control tasks for the hardware. Output variables are then filled after the activities for the feedback to the client. The semaphore variables act here as something similar to a software-based «timing isolation» between the software for the hardware control and the virtual plug-in interface (see ▶ Sect. 2.6.4). The periodic loop provides the necessary abstraction for controlling hardware on higher tasks. Example 4.2 shows a possible realization of a periodic loop in the form of a finite state machine (the control task can be realized with different strategies, but experience has shown that finite state machines with sophisticated memory are ideal at this level). The code is taken from the hardware server to control the event timer and shows the three different states: idle, ranging, and calibrating. They are commanded with an RPC function, where the semaphore-protected «uiMode» is changed. After the change, «uiMode» is checked in a semaphore-protected way and selects the state which is activated each time the periodic loop is started, as long as «uiMode» is changed again in the periodic loop (e.g., for the calibration state in the example) or by an RPC function.

Periodic Control Loop Thread

▪▪ Example 4.2
The periodic control loop to control the real-time industrial PC of the eventtimer over a specific hardware driver module, which realizes the socket communication.

```
/*****************************************************************
 *   class         eventtimer_server
 *   function      _vPeriodicServerActivity
 *****************************************************************/
/*!           Periodic  tasks  done  at  the  loop  time  (defined  by  user)
 *   \param    —
 *   \return   —
 *****************************************************************/
/*   author    Alexander  Neidhardt
 *   date      14.05.2007
 *   revision  —
 *   info      Part  of  the  idl2rpc.pl  —  generator!
 *****************************************************************/
```

```
void eventtimer_server::_vPeriodicServerActivity ()                        14
{                                                                          15
    // USERDEFMETHODBEG: Userdefined method body                          16
    unsigned long ulBlockHandle = 0;                                      17
    unsigned short usCalibFinished = 0;                                   18
                                                                          19
    /// Check error states                                               20
    if (usFatalError || uiAlert.CGetSemVal() == ET_NOK)                  21
    {                                                                     22
        sleep (5);                                                        23
        usReset ();                                                       24
        return;                                                           25
    }                                                                     26
                                                                          27
    /// Check, if eventtimer is connected                                28
    if (pET == NULL)                                                      29
    {                                                                     30
        usFatalError = 1;                                                 31
        return;                                                           32
    }                                                                     33
                                                                          34
    /// Process finite state machine for eventtimer control              35
    /// Re-connections to the eventtimer hardware are done implicitly    36
    switch (uiMode.CGetSemVal())                                         37
    {                                                                     38
        case ET_MODE_IDLE:                                               39
            /// Idle mode => do nothing                                  40
            break;                                                        41
        case ET_MODE_CALIB:                                             42
            /// Calibration mode (started with calibration RPC function) 43
            try                                                           44
            {                                                             45
                if (usETRangingActive.CGetSemVal())                      46
                {                                                         47
                    /// Do calibration                                   48
                    cEventtimerSocketSemaphore.vBlock(ulBlockHandle);    49
                    /// Read data                                        50
                    (void) pET->ReadTimer();                             51
                    /// Stop calibration mode if eventimer is finish     52
                    if (!pET->iCalibrationFinished())                    53
                    {                                                     54
                        usCalibFinished = 1;                             55
                    }                                                     56
                    else                                                 57
                    {                                                     58
                        usCalibFinished = 0;                             59
                    }                                                     60
                    cEventtimerSocketSemaphore.vUnblock(ulBlockHandle);  61
                    if (usCalibFinished)                                 62
                    {                                                     63
                        if (usSetIdleMode())                             64
                        {                                                 65
                            uiAlert = ET_NOK;                            66
                        }                                                 67
                    }                                                     68
                }                                                         69
            }                                                             70
            catch (...)                                                  71
            {                                                             72
                usFatalError = 1;                                        73
            }                                                             74
            break;                                                        75
```

```
     case ET_MODE_RANGE:                                                  76
         /// Range mode (started with range RPC function)                 77
         try                                                              78
         {                                                                79
             if (usETRangingActive.CGetSemVal())                          80
             {                                                            81
                 /// Do ranging                                           82
                 cEventtimerSocketSemaphore.vBlock(ulBlockHandle);        83
                 /// Read data                                            84
                 (void) pET->ReadTimer();                                 85
                 cEventtimerSocketSemaphore.vUnblock(ulBlockHandle);      86
                 /// Range mode is stopped from external with RPC function 87
             }                                                            88
         }                                                                89
         catch (...)                                                      90
         {                                                                91
             usFatalError = 1;                                            92
         }                                                                93
         break;                                                           94
     default:                                                             95
         usFatalError = 1;                                                96
}                                                                         97
                                                                          98
/// Space for further states of a state machine for the data migration   99
/// ...                                                                   100
                                                                          101
// USERDEFMETHODEND                                                       102
}                                                                         103
```

The definition of the RPC interface finally offers the implementation of the software plug-in connector. It is a standardized access point to the server. All asynchronous function bodies must be filled there with their functionalities. These are set and get methods of variables and contents which control the finite state machine in the periodic loop. The RPC functions feed it with data which should be sent to the hardware and they take data which comes from the device. As described, the RPC functions do not contact the hardware device directly and use the shared variables for the data transfer to the periodic loop. A sample interface in the IDL for the «idl2rpc.pl» generator is shown in Example 4.3, which is taken from the dome control of the laser ranging system. It defines all possible remote methods to control the dome or to influence its behavior. An example for a suitable implementation of an RPC function can be found in ► Sect. 3.3.3, which shows the software integration of the Holzworth synthesizer.

RPC Device Interface

■■ Example 4.3
An IDL file for the RPC interface to control the dome of the laser ranging system.

```
// ********************************************************************   1
// * Description:                                                        2
// * This is the RPC interface for the dome controller from Baader       3
// * Planetarium GmbH. It bases on an existing C-code for WLRS-dome.     4
// * Some additional functionality is added given by possibilities with  5
// * RPC techniques.                                                     6
// ********************************************************************   7
                                                                         8
// ********************************************************************   9
// Constants                                                             10
// ********************************************************************   11
```

```
// Define operation modes
const OPMODE_MANUAL          = 0;
const OPMODE_AUTOTELESTEP    = 1;
const OPMODE_AUTOTELECONT    = 2;
const OPMODE_AUTOSATSTEP     = 3;
const OPMODE_AUTOSATCONT     = 4;
const OPMODE_EXTTRIGGERED    = 5;
// Define shut positions
const SHUTPOS_CLOSE  = 0;
const SHUTPOS_MOVING = 1;
const SHUTPOS_OPEN   = 2;
// Define simulation states
const SIMMODE_OFF = 0;
const SIMMODE_ON  = 1;
// Define sun avoid states
const SUNAVOIDMODE_OFF = 0;
const SUNAVOIDMODE_ON  = 1;
const INSUN   = 0;
const OUTSUN  = 1;
// Define sat list load state
const SATLISTUNLOADED = 0;
const SATLISTLOADED   = 1;
// Define errors
const ERROR_OK          = 0;
const ERROR_DOME        = 1;
const ERROR_SERVER      = 2;
const ERROR_OUTOFORDER  = 3;
// Define serial device settings
const TTY_BITSPERSEC_50        = 50;
const TTY_BITSPERSEC_75        = 75;
const TTY_BITSPERSEC_110       = 110;
const TTY_BITSPERSEC_134       = 134;
const TTY_BITSPERSEC_150       = 150;
const TTY_BITSPERSEC_200       = 200;
const TTY_BITSPERSEC_300       = 300;
const TTY_BITSPERSEC_600       = 600;
const TTY_BITSPERSEC_1200      = 1200;
const TTY_BITSPERSEC_1800      = 1800;
const TTY_BITSPERSEC_2400      = 2400;
const TTY_BITSPERSEC_4800      = 4800;
const TTY_BITSPERSEC_9600      = 9600;
const TTY_BITSPERSEC_19200     = 19200;
const TTY_BITSPERSEC_57600     = 57600;
const TTY_BITSPERSEC_115200    = 115200;
const TTY_CHARSIZE_5      = 5;
const TTY_CHARSIZE_6      = 6;
const TTY_CHARSIZE_7      = 7;
const TTY_CHARSIZE_8      = 8;
const TTY_PARITY_NO       = 0;
const TTY_PARITY_ODD      = 1;
const TTY_PARITY_EVEN     = 2;
const TTY_STOPBITS_1      = 1;
const TTY_STOPBITS_2      = 2;
const TTY_FLOWCONTROL_NO  = 0;
const TTY_FLOWCONTROL_HW  = 1;
const TTY_FLOWCONTROL_SW  = 2;

// ****************************************************************
// Interface definition
// ****************************************************************
interface domectrl #Dome20090409001 {

    // ————————————————————————————————————————
    // Set/get settings
    // ————————————————————————————————————————
    // Set IP of telescope control RPC-server
    unsigned short usSetIPOfTelescopeController (in string strIP);
```

```
// Get IP of telescope control RPC—server
unsigned short usGetIPOfTelescopeController (out string strIP);
// Set serial device settings
unsigned short usSetSerialDevice (in string strName,
                                  in unsigned short usBitsPerSec,
                                  in unsigned short usCharSize,
                                  in unsigned short usParity,
                                  in unsigned short usStopBits,
                                  in unsigned short usFlowControl);
// Get serial device settings
unsigned short usGetSerialDevice (out string strName,
                                  out unsigned short usBitsPerSec,
                                  out unsigned short usCharSize,
                                  out unsigned short usParity,
                                  out unsigned short usStopBits,
                                  out unsigned short usFlowControl);
// Set azimuth aberration when using step mode
unsigned short usSetStepModeAberration (in unsigned short usAberrationDegree);
// Get azimuth aberration when using step mode
unsigned short usGetStepModeAberration (out unsigned short usAberrationDegree);
// Set starting sequence which is done after server starts
unsigned short usSetStartSequence (in string strStartSequence);
// Get starting sequence which is done after server starts
unsigned short usGetStartSequence (out string strStartSequence);
// Set azimuth tolerance of pointing
unsigned short usSetAzimuthTolerance (in unsigned short usAzimuthTolerance);
// Get azimuth tolerance of pointing
unsigned short usGetAzimuthTolerance (out unsigned short usAzimuthTolerance);
// Set azimuth offset of pointing
unsigned short usSetAzimuthOffset (in short sAzimuthOffset);
// Get azimuth offset of pointing
unsigned short usGetAzimuthOffset (out short sAzimuthOffset);


// ————————————————————————————————
// Simple ASCII—calls to do everything by string—orders
// ————————————————————————————————
unsigned short usASCIIOrder (in string strOrder);
unsigned short usASCIIState (out string strState);


// ————————————————————————————————
// Post single orders
// ————————————————————————————————
// Reset dome system
unsigned short usReset ();
// Open shut
unsigned short usOpenShut ();
// Close shut
unsigned short usCloseShut ();
// Move azimuth
unsigned short usMoveAzimuth (in unsigned short usAzimuthDegree);
// Move azimuth
unsigned short usExtTriggeredAzimuth (in unsigned short usAzimuthDegree);
// Set satellite passage
unsigned short usUploadSatPassage (in unsigned long ulUnixEpoch <>,
                                   in double dAzimuth <>);
// Set satellite passage
unsigned short usClearSatPassage ();
// Get state
unsigned short usGetState (out unsigned short usOperationMode,
                           out unsigned short usSimulationMode,
                           out unsigned short usAzimuthPosition,
                           out unsigned short usShutState,
                           out unsigned short usError,
                           out unsigned short usSunAvoidState,
                           out unsigned short usIsInSunAvoid,
                           out unsigned short usSatListLoaded);
```

```
// Get state GUI                                                               145
unsigned short usGetStateGUI!ASD (out unsigned short usOperationMode,          146
                            out unsigned short usSimulationMode,               147
                            out unsigned short usAzimuthPosition,              148
                            out unsigned short usShutState,                    149
                            out unsigned short usError,                        150
                            out unsigned short usSunAvoidState,                151
                            out unsigned short usIsInSunAvoid,                 152
                            out unsigned short usSatListLoaded);               153
// Get operation mode                                                          154
unsigned short usGetStateOperationMode (out unsigned short usOperationMode);   155
// Get simulation mode                                                         156
unsigned short usGetStateSimulationMode (out unsigned short usSimulationMode); 157
// Get azimuth position                                                        158
unsigned short usGetStateAzimuthPosition (out unsigned short usAzimuthPosition); 159
// Get shut state                                                              160
unsigned short usGetStateShutState (out unsigned short usShutState);           161
// Get error state                                                             162
unsigned short usGetStateErrorState (out unsigned short usError);              163
// Get sun avoidance state                                                     164
unsigned short usGetStateSunAvoidanceState (out unsigned short usSunAvoidState, 165
                            out unsigned short usIsInSunAvoid);                166
// Get satlist upload state                                                    167
unsigned short usGetStateSatListLoaded (out unsigned short usSatListLoaded);   168
                                                                               169
// ──────────────────────────────────                                         170
// Set simulation modes                                                        171
// ──────────────────────────────────                                         172
// Set simulation mode for serial device                                      173
unsigned short usTurnOnSimulationMode ();                                      174
// Reset simulation mode for serial device                                    175
unsigned short usTurnOffSimulationMode ();                                     176
                                                                               177
                                                                               178
// ──────────────────────────────────                                         179
// Set sun avoidance modes                                                     180
// ──────────────────────────────────                                         181
// Set sun avoidance mode                                                      182
unsigned short usTurnOnSunAvoidanceMode ();                                    183
// Reset sun avoidance mode                                                    184
unsigned short usTurnOffSunAvoidanceMode ();                                   185
                                                                               186
                                                                               187
// ──────────────────────────────────                                         188
// Set operation modes                                                         189
// ──────────────────────────────────                                         190
// Set manual mode                                                             191
unsigned short usSetManualMode ();                                             192
// Set extarnal triggered mode                                                 193
unsigned short usSetExtTriggeredMode ();                                       194
// Set automatic mode using satellite table with movement by steps            195
unsigned short usSetAutoSatStepMode ();                                        196
// Set automatic mode using satellite table with continual movement           197
unsigned short usSetAutoSatContMode ();                                        198
// Set automatic mode using telescope positions with movement by steps        199
unsigned short usSetAutoTeleStepMode ();                                       200
// Set automatic mode using telescope positions with continual movement       201
unsigned short usSetAutoTeleContMode ();
};
```

Distributed Hardware Control

Each hardware device in a laser ranging system (as described in ▶ Sect. 4.1 and explained in ◼ Fig. 4.1) is implemented in the three-tier form described. Therefore, all devices can be accessed remotely via the RPC middleware. It does not matter where they are or even how many hardware devices must be controlled. It is only necessary that an RPC server is installed and running somewhere on a computer

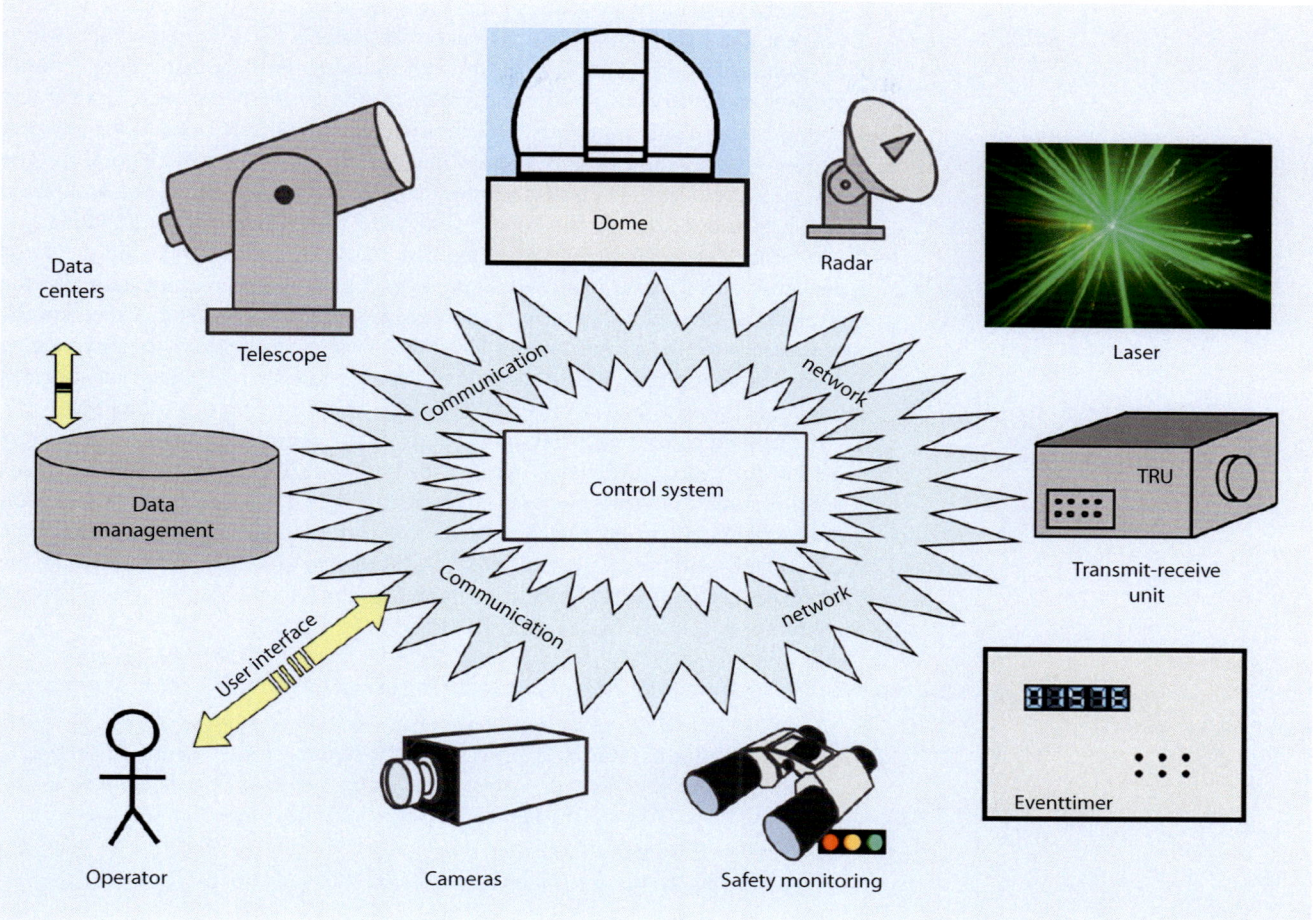

Fig. 4.8 Laser ranging as a distributed system

which is plugged to the hardware or which can access the device. Accessibility, availability, safety, and so on are implemented by the communication backbone of the RPC middleware. Additionally, all other communication partners, such as databases, processes, and centralized controlling processes, are managed in the same way. They are encapsulated by an RPC server, which translates the asynchronous, remote requests into local requests in the periodic loop.[10] All these devices together form a distributed system for hardware control (see ◻ Fig. 4.8), which offers different layers of complexity of hardware, controlling mechanisms, and timing.

The different layers of hardware complexity have already been described in the previous sections and can also be seen in the incremental way that communication interfaces become more complex from the hardware to the network of the distributed system (see also ◻ Fig. 4.7). The change in the complexity of controlling mechanisms was also mentioned briefly. While signal-based controlling is used in the logic circuits of the low-level hardware feedback loops, data-driven decisions can be made in microcontrollers. Also

Controlling Layers from Signal Circuits to Genetic Algorithms

---

10 Dependent on the task and situation, e.g., in combination with a database, it might be possible to loosen the strict separation between asynchronous RPC functions and the control loop. Databases have a very complex transaction system, which allows multitasking requests, so that there is not necessarily a need for external semaphore protection. It is only important that the response from large table sets from the database do not back up to the requester. It is an individual decision. The separation described is only obligatory for real hardware connections.

rule-based control loops, for example, those with fuzzy logic,[11] are located here. With the use of processors, the controlling tasks can be described in the form of programs which offer additional possibilities. The control with industrial PCs and similar higher-level controlling equipment can additionally include the use of an expert system,[12] a neuronal network,[13] or a genetic algorithm.[14] All these techniques make it possible for a central controller for the whole system to implement a self-learning, neural algorithm. The results are more or less intelligent machines, designed to be more tolerant of inaccuracies, uncertainty, and factoids. It is a change from hard computing to soft computing (see (Favre-Bulle 2004, *l.c. page 59*) and (Langmann 2010, *l.c. page 484 ff.*)). But the use is dependent on the use case. In most cases predefined, a priori state machines also solve the controlling, as they can be optimized to perform a specific task. To do this, expert knowledge is usually required. Therefore, all currently known control systems for laser ranging just use hard computing. But in the case of dynamic satellite scheduling, event-driven controlling of systems, and so on, there might also be potential for soft computing.

Timing

The layered concept also has another advantage: it directly supports the correct timing of the tasks. At the layer close to the hardware, reactions are very fast. This is ideal for safety systems, where interlocks must be propagated within reaction times far below microseconds. The event counting in the event timer must also be implemented with four picoseconds resolution, so that LEO satellites can be tracked properly. This is the field of hard real-time scenarios (for the following consideration, also see (Favre-Bulle 2004, *l.c. page 172 ff.*)).

Real-Time Systems

> *Definition DEF4.1. of «real-time systems»* (see Favre-Bulle 2004, *l.c. page 173*)
> Real-time systems must read, process, and write data correctly until predefined deadlines (absolute time) or within predefined time intervals (relative time). The result is a deterministic behavior on the time line. The real-time system guarantees this behavior reliably.

Timing Layers from Real-Time to Normal Time

Usually real-time systems are event, time, or interrupt triggered, work on small portions of data per time unit, and work very closely with hardware (see also ▶ Sect. 4.5). If the time criteria are very strict in resolution or in reliability, it is hard real-time. It can be implemented on tightly coupled systems with hardware boards as well as with PLCs or industrial PCs with real-time operating systems. But principally, it defines the field of hardware control. The synchronization is implemented with special clock modules, using a time (Pulse Per Second (PPS)) and frequency distribution network, or over Ethernet using the new Precision Time Protocol (PTP), which enables a synchronization in nanoseconds if the network components implement PTP in their hardware.[15] The hardware driver module or

---

11　Fuzzy logic makes decisions based on a probabilistic logic with memberships of fuzzy sets. But here the thresholds are not fixed. The classification of a membership can fluctuate between all scales of «completely related» and «not related». The individual memberships to a set are defined by membership functions (Langmann 2010, *l.c. page 485 ff.*).

12　Expert systems use the knowledge of experts coded in a computer program to make more or less complex decisions (Langmann 2010, *l.c. page 490 f.*).

13　Neuronal networks imitate processing of information in a human brain. They use nodes, the neurons, with edges in between. These connections have alterable weights. Each node sums up the different inputs (the attributes) and propagates the result by using a nonlinear activation function if the input result reaches a defined threshold (the bias). This construct allows a dynamic adaption of an algorithm, which means «learning» (Langmann 2010, *l.c. page 492 f.*).

14　Genetic algorithms use neuronal networks and process a search for an optimal solution with a construct of an artificial evolution. Each state is a population of qualifications to solve a problem (the fitness). The algorithm modifies the processing and tests again the fitness of the «offspring». Then it selects the solution which is fitting best and modifies the solution again (Langmann 2010, *l.c. page 492 f.*).

15　PTP is also part of the White Rabbit Solution developed by the European Organization for Nuclear Research CERN and partners to synchronize networks over long fiber lines.

component and the periodic control loop of an RPC control server operate with weak real-time in an order of microseconds to milliseconds. Delayed reactions or failure to meet the deadlines do not lead to fatal or critical system states which compromise whole systems or human beings. But longer lasting or added delays might result in critical situations. For this weaker timing, the complete computer can be synchronized to a clock server using Network Time Protocol (NTP), which offers the required order of magnitude.[16] The weak real-time systems also use cyclic interrupt generation together with a time measurement of the computer time to control periodic tasks (as given for the periodic loop). Another important part is the watchdog, whose timing is independent from the real-time process (see, e.g., the automatic safety device of the generated code for RPC server in ▶ Sect. 3.3.3). While the periodic loop is synchronous, the RPC calls are asynchronous. Their reaction times can range between several 100 milliseconds to a few seconds. Usually the message transfer over the network works quite fast (which of course depends on network topologies and distances). But in the case of errors, timeouts are used which destroy reliable timing. The timing can be seen here as being as accurate as from a human operator which defines the normal time of processing. This means that the centralized control system offers reaction times in terms of seconds in the same way as an operator. Other tasks, such as the graphical interface on a very remote machine, data transfer of the results from a satellite track, or the data preparation for the tracking of the following week, require fewer timing constraints. The usual ones are for a few seconds, minutes, or even days and weeks. In summary, this means that a layered timing implementation can be established over the whole system in a similar way as for the controlling. The control elements are assigned to a special timing situation in the whole system, which they guarantee with different hardware solutions. But the assignment is variable.

In brief, this idea of distributed systems offers a lot of possibilities for a flexible design of control systems with several, different hardware pieces, which must be controlled. This is also ideal if a new system needs to be integrated into an existing one or if a smooth transition between two systems should be performed. Therefore, it was easy to marry the existing, old laser ranging system with the newly developed one of the described style at the Geodetic Observatory Wettzell. But even though this is already quite a useful setup, the design does not go far enough. Ideas for an additional step come from the automation of industrial manufacturing processes and are known as autonomous production and manufacturing cells.

4.2.2 The Construction of the Autonomous Production Cell

Laser ranging systems and also all individual systems of space geodetic techniques are, in principle, production facilities. While industrial manufacturing plants produce hardware products, the scientific measurement systems produce data. To do this, they use hardware machines which must be coordinated. Usually well-trained and highly educated personnel are involved in planning, controlling, interacting with, and repairing these data production systems. Increasing requirements on the production of data (e.g., more satellites need to be observed by more stations to produce more data in less time) are similar to industrial processes. Therefore, it is a good idea to take solutions and ideas from the commercial manufacturing of individual, customer-adapted, fast available products.

Adaption from the Industry

A way to adapt to given economic requirements is to increase the autonomy of single system or production equipment. The idea behind this is to enable systems to solve complex working processes with a maximum degree of independence and autonomy. The process should be operative, reliable, and without interruptions and

Autonomous Production and Manufacturing Cells

---

16 NTP is a standard to synchronize computer clocks over packet-based computer networks with an accuracy of a few hundred microseconds.

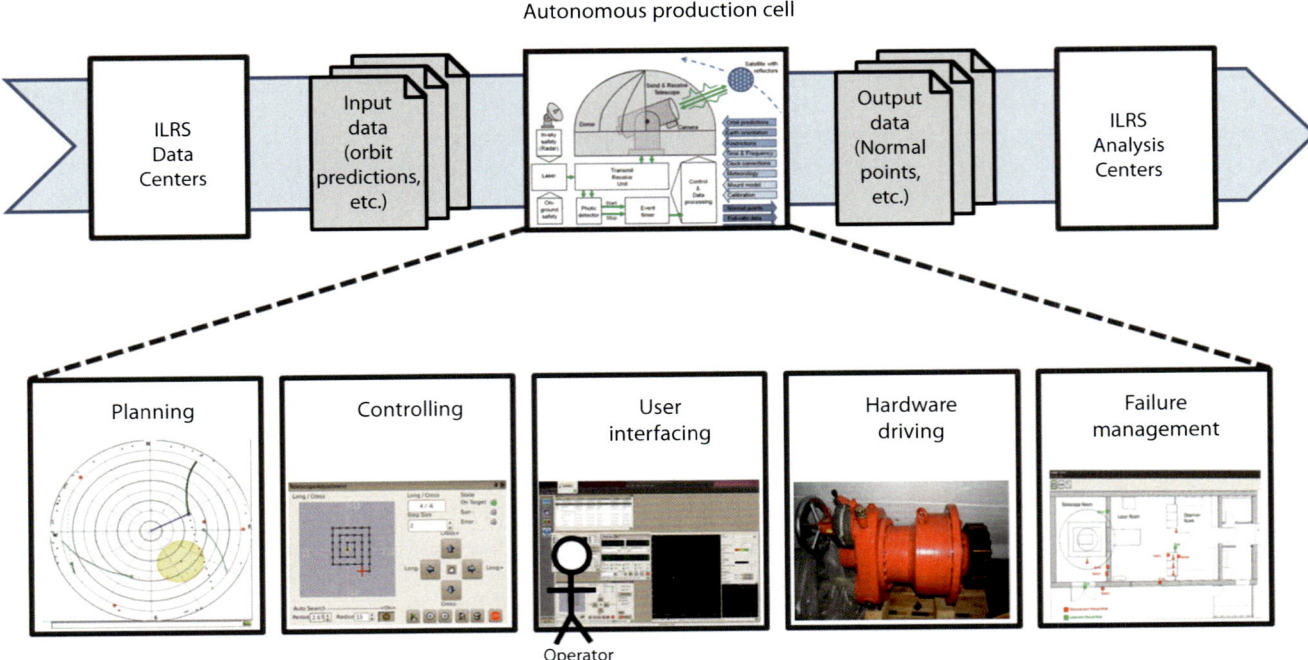

Fig. 4.9 The autonomous production cell of a laser ranging system in analogy to the industrial autonomous manufacturing cells (analog to (Pfeifer and Schmitt 2006, *l.c. page 9*))

disturbances over a specific, longer time period. Autonomous production and manufacturing cells are the result. A system like a laser ranging system can also be seen as an autonomous production facility. In analogy to the industrial cells, the autonomous functionalities of such a cell are planning, controlling, user interfacing, hardware driving, and managing failure (see ◘ Fig. 4.9, analog to (Pfeifer and Schmitt 2006, *l.c. page 9*)). The input is the required external information and data, such as orbit predictions, which are used to produce the output data, such as the distance measurements (Pfeifer and Schmitt 2006, *l.c. page 3 f.*).

Autonomous Production Cell

Definition DEF4.2 of «autonomous production cell» (adapted from (Pfeifer and Schmitt 2006, *l.c. page 12 f.*))

An autonomous production cell solves the autonomous functionalities, planning, controlling, user interfacing, hardware driving, and failure management, and offers the following performance characteristics:

— The production solves complex processing sequences as far as possible autonomously and with failure resistance because of extended functionalities and capabilities.
— The extended functionalities and capabilities are possible because of the interlocking of planning, processing, and monitoring at each cell.
— The operator has a complete set of functionalities in a professional way for the implementation and monitoring of the processing on the location of the cell.
— The controlling operator and the system together form an autonomous union, which acts as a highly autonomous unit.

Autonomous Functionalities

In the case of a laser ranging system, the functionalities and interactions between an operator and the autonomous production cell are mainly defined with the interface of the centralized coordination facility. The part of the «planning» are the tasks to calculate, to derive, and to plan the observation flow with satellite selection, satellite hopping, and tracking prioritization. «Controlling» means synchronizing and controlling all the related hardware devices. Commands are sent and the eradication of the tasks is frequently requested. This creates an overall central feedback control loop. It is mandatory that all equipment uses standards, as described in the previous chapters to support the simplification of stable and simple collaborations (Pfeifer and Schmitt 2006, *l.c. page 8*). This use of standards and the integration of a reliable information and communication infrastructure, such as the middleware described (see ▶ Sect. 3.4), are indispensable elements to create a sophisticated sensor/actuator network and also to connect production cells to each other (Pfeifer and Schmitt 2006, *l.c. page 12*). The «user interfacing» enables the operator to get complete knowledge of the current processing states in the form of a graphically pre-selected display. He can interact with the cell to adapt parameters or to solve problems. «Driving hardware» is the real control of devices which are all connected units designed to solve a common task. Finally, «failure management» is quite important to process preventive safety in combination with complex technical systems. Large sets of combined robot systems in workshops of factories can be controlled with such a system. It is also an equivalent approach for the sensitive and individual instruments which are parts of a laser ranging system. But a key role is played by the integration of a detailed monitoring and diagnosis functionality to establish the basis for autonomous fault correction and autonomous error recovery to a certain degree (Pfeifer and Schmitt 2006, *l.c. page 12 f.*), which is finally dependent on the stability of the controlled system. All of these elements can work with different degrees of autonomy and automation. But they are essential to allow a higher degree of automation.

But what does this autonomy and this automation mean? Autonomy of a production system, as already implied, means the ability to solve complex processing tasks with a maximum degree of independence and autonomy over a longer time period and without failures. This can be obtained with the intensification of functionalities per cell, so that it can solve an extended set of tasks and is no longer restricted to a limited one. It combines planning, monitoring, and controlling in the location of the cell and is highly resistant and tolerant against failures. The system is also able to react to failures itself. The goal is to reduce the number of user interactions. But if interactions are necessary, they are optimized. The system offers all functionalities transparently and processes state changes autonomously. An intervention is always possible. The user or operator is released from routine jobs and has more potential for other tasks, such as maintenance and development (Pfeifer and Schmitt 2006, *l.c. page 3*). This means a change in the abilities of an operator from manual, continuous activities to more cognitive tasks (Pfeifer and Schmitt 2006, *l.c. page 19*).

Autonomy and Automation

This change also influences the situation of employees. While researchers and engineers currently spend up to 60% of their time operating just one system in shifts, this is not possible anymore if more than one system needs to be operated at one observatory. Often a separation between operators and technical staff occurs. But this is not beneficial as pure operators, for example, in the case of student operators, are usually not able or willing to understand the big picture and complexity of a system. Failures cannot be interpreted, and the operation stops until the technical staff fix the problem. With autonomous production cells, a level of automation can be reached to reduce operation shifts much more efficiently (in combination with remotely shared observations), while keeping well-educated staff and bringing the manpower back to the engineering activities. More systems can be operated by one person. The systems can react more flexibly. A good example of the power of automation is shown by optical astronomic telescopes, where, for example, at the Liverpool telescope, the number of scientific publications could be increased by a factor of about 16 in about five years of automation, while keeping the downtime at about 4% per year (Neidhardt et al. 2012a, *l.c. page 273 ff.*). Therefore, the prognosis is that automation offers a lot of new possibilities. Usually it will not mean that scientists lose their jobs. Nevertheless, the decision depends on the management. But it will change the daily duties from routine tasks to more scientific and maintenance activities. But one limitation must be added: the degree of automation is dependent on the capacity of system safety to protect humans and the system itself. Therefore, a huge «showstopper» for the laser ranging automation

Situation of Employees

Self-Similar Decomposition

might be in-sky safety, where human beings must be protected autonomously from a high-energy laser beam. This is a real challenge.

In combination with the idea of a distributed hardware system (see ▶ Sect. 3.3.2), the central coordination facility is supported by several intelligent subsystems, the distributed hardware controllers. A closer look shows that all of this supporting equipment, which is represented by a server, is, in principle, smaller mappings of the structure of the autonomous production cell, represented by a central coordination facility and its autonomous functionalities of planning, controlling, user interfacing, hardware driving, and failure management. While the central coordination facility applies the autonomous functionalities to coordinate several connected device facilities in the form of subsystems, the device controlling servers apply similar functionalities reduced to their tasks to run the connected hardware. This means the complex central autonomous functionalities can be split into device-relevant parts, so that they can be shifted to the next level of intelligent control cells. The whole autonomous production cell can be split into similar smaller, interacting autonomous control cells, which are copies of the central coordination. The tight and detailed monitoring and diagnosis at this new cell level again play a key role. This means that each cell needs all the necessary information about the state and defective or error-prone elements in its administrative control field. This requires additional sensors to monitor the behavior of the hardware (e.g., an electric eye which detects the minimum elevation limit of the telescope tube to switch off the laser). In addition to the functionality of supporting the hardware server, it can also feed a parallel, centralized system monitoring and safety system, which collects safety data for the whole autonomous production cell. By doing this decomposition, the whole system of an autonomous production cell is reduced to several smaller autonomous control cells.

Autonomous Control Cells

> *Definition DEF4.3 of «autonomous control cells»*
> Autonomous control cells selectively solve the autonomous functionalities of planning, controlling, user interfacing, hardware driving, and failure management for specific tasks of the whole production cell. They decompose a complex autonomous production cell into similar smaller, manageable units, while keeping the performance characteristics of the whole cell for their smaller tasks. Each cell is something like a copy of the central controlling facility and keeps itself alive and stable by implementing the autonomous functionalities.

Types of Autonomous Control Cells

Generally, the system can now be separated into two different types of autonomous control cells, which are similar in behavior and differ in the level of controlling:

— *Autonomous coordination cell*: It is the main access point to the autonomous production cell for the operator. Therefore, it provides a sophisticated graphical user interface (GUI) for the general control of the whole production cell. It is the main and central control feedback loop. All the information is concentrated here. An access to such cell is enough to run the complete system. It derives the observation schedule dynamically or statically. It runs the operations by activating, controlling, and monitoring the related devices. It monitors and evaluates the whole state of the system, and it runs central failure recovery scenarios or stops the processing after an irreparable error to inform the operator.

— *Autonomous hardware control cells*: These cells control the hardware and are implemented in the same way as the RPC servers (see ▣ Fig. 4.7). They can take more or less intelligent scheduling and control tasks from the central coordination cell. They prepare all the required setups for their specific devices and tasks. For tight-coupled tasks, they can act as tie points, so that they control other device cells on lower hierarchy levels besides their own hardware (as shown in ▣ Fig. 4.7). They offer detailed user interfaces for administrative user interactions and keep their own administrative area alive with a local failure recovery, which is fitted to the specific hardware task.

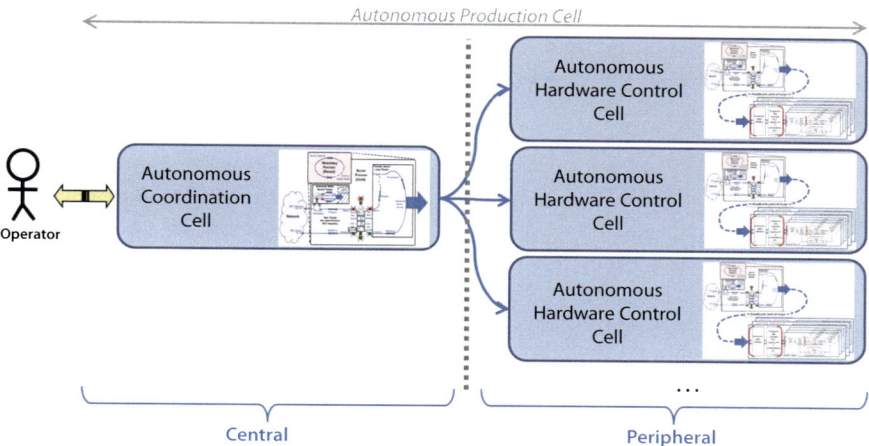

Fig. 4.10 The basic two-layer model of an autonomous production cell on the basis of defined autonomous control cells (Adapted from Pfeifer and Schmitt 2006, *l.c. page 26*)

With these autonomous control cell types, it is possible to design a production cell as a two-layer system (see ◘ Fig. 4.10, adapted from (Pfeifer and Schmitt 2006, *l.c. page 26*)). As already indicated in the first design of the distributed system in ◘ Fig. 4.8, the central control system controls the distributed hardware. The central control system is now represented by an autonomous coordination cell, and the different lower-level hardware devices are represented by autonomous hardware control cells. But not all cells define all the autonomous functionalities, so that they can process and schedule their tasks independently and autonomously.

The practical implementation of the autonomous cells has shown that there are situations which require an autonomous device control cell to coordinate other autonomous device control cells. This makes sense if timing constraints for the propagation of data in tight-coupled hardware systems are critical or if status data from one device is not required by the whole system and is just relevant for another device (see also ► Sect. 4.5 about timing constraints for data samples). An example is telescope control, where position data must be propagated fast enough to the radar. The telescope control also propagates positions to the dome and requests the position status from it. The telescope, radar, and dome must always point in the same direction. With such tightly coupled systems, autonomous device control cells can also coordinate other device cells and become a mixture of both the previously defined cell types. The result is a multilayered model of an autonomous production cell. Each node of this hierarchy is responsible for the following cell components of the next level and coordinates all autonomous functionalities for these cells. The periodic control loop of this coordinating cell must frequently request the states, error situations, and processing information from the devices of the following lower level. Then it can interpret the situation, using logic compositions (like fuzzy logic), to make decisions and derive its current, internal state.

The basic structure of autonomous coordination cells, autonomous hardware control cells, and a mixture of them is additionally extended by (almost) complete, parallel-working, additional autonomous production cells, which support the tasks of the main production cell. They solve their own tasks independently, and they create a two-layered or multilayered architecture with coordination and device control. Without these parallel production cells, the main autonomous production cell cannot work correctly. An example of such a supporting cell structure is data management. Several data collecting, processing, preparing, and delivering tasks are coordinated independently from the rest of the actual measurement system and tasks. Data managing can be implemented with a similar structure like a complete autonomous production cell, where the different tasks are done by autonomous control cells. The existence of such parallel production cells is essential for the main task. They take over workload which is not intrinsically tied to the core

Two-Layer Model of an Autonomous Production Cell

Multilayer Model of an Autonomous Production Cell

Supporting Production Cell Structures

task of the central production cell but necessary for the achievement of its tasks. Therefore, they usually implement tasks which can be shared among different production cells, such as data management, or which implement a second, parallel safety setup. To include such parallel subsystems in the rest of the architecture, it is quite suitable for all cells to just be represented by their communication interface on the basis of RPC, which acts as a plug-in connector to the communication network. Therefore, the appearance of such a parallel production cell is like each other cell in the cell structure. Hidden behind it is a main structure and layered design which incorporates smaller versions of the main shape, so that something like the fractal property of self-similarity applies.[17]

Types of Supporting Production Cells

Using this design of self-similarity, the following supporting, parallel production cells can be found for a laser ranging system:

- *Autonomous data management cell*: The autonomous production cell is supported by a centralized knowledge base and information storage (Pfeifer and Schmitt 2006, *l.c. page 10*). It feeds the central coordination cell with the conditioned input data loaded from the data centers and takes the output data to send them to the data analysis centers. In principle, it absorbs all of the controlling inconvenient communication activities with the outer world. By doing this, it unburdens the main task of observation from that distracting and resource intensive work. It also organizes the knowledge base and respectively stores archive information, so that fast access is given. It conditions the data for controlling the system, makes it accessible over the common infrastructure, and receives the resulting measurement data to prepare them for the data transfer to the outer world.

- *Autonomous system monitoring and safety cell*: The autonomous system monitoring and safety cell is a parallel instance to monitor and control critical situations which require interlocks or failure recoveries. The main focus lies on failure management. Intrinsically, the functionality is not necessary, as all devices implement their own safety system, which can be propagated via the interfaces to the higher controlling instances. But a parallel and independent safety system is anyway desirable as a redundant instance, which additionally takes care of critical tasks. Therefore, it also runs on different hardware and uses a different software and monitoring structure. With this system, hardware interlock mechanisms are conjoined, so that their states are accessible over the common infrastructure. The failure states, events, or logic values can be logically combined to get a parallel «big picture» of the failure state of the whole system. It is like a system-wide watchdog.

Multilayered Autonomous Production Cell of a Laser Ranging System

All the elements described together offer the design of the multilayered autonomous production cell of a laser ranging system (see ■ Fig. 4.11). In this design, it is useful to distribute the autonomous cells and supporting production cells to different hardware machines, according to their resource hunger. Therefore, general coordination and data acquisition during an observation are on one machine: the control server. The tightly coupled elements for tracking a satellite with mechanical devices, like the telescope, the dome, and the radar, build the second server: the telescope server. The calculation-intensive and therefore resource-intensive data management is on a third server: the data management server. Finally, system monitoring and safety should be as independent as possible to offer an additional «watchdog» instance to the whole system. Therefore, it is located in another server: the safety monitoring server. All the elements are connected via Ethernet and the communication infrastructure, defined by the middleware. Each accessible cell is implemented as an RPC server. Each cell of a higher level in the control hierarchy includes the RPC clients and coordinates them.

---

17 Fractals show repeating patterns that display at every scale, for example, known from the Mandelbrot set.

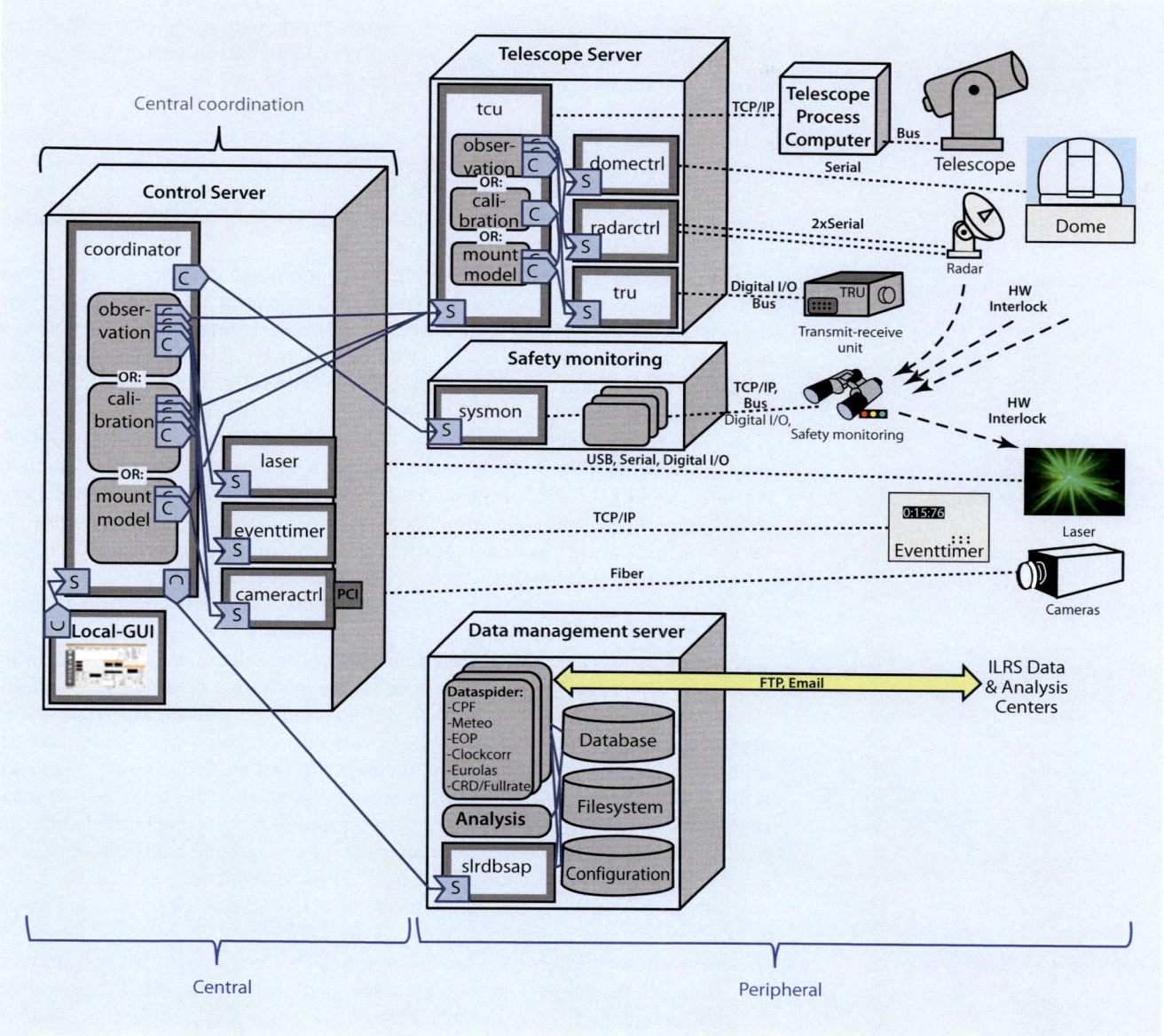

Fig. 4.11   The resulting multilayer model of the autonomous production cell of a laser ranging system on the basis of distributed autonomous control cells

Before the individual cell types can be described, it is appropriate to show and explain the background and design decisions of user interfaces, which are the access points for users to the different cells.

4.3   User Interfacing

Nowadays, user interfaces are GUI clients of autonomous control cells. As they are the usual and often only way to interact with the system, GUIs are quite important.

    The simplest way to interact with a computer program is the command line interface in a shell, which is usually part of the operating system. The communication is based on text messages, which can be read over the standard input («`stdin`» in C and «`std::cin`» in C++). Messages are printed by the program to the standard output («`stdout`» in C and «`std::cout`» in C++) or to a special error channel («`stderr`» in C and «`std::cerr`» in C++) and appear as messages in the shell.

Shell Interface

Escape Sequences and Ncurses

There are not many ways to make the outputs more graphical at this level. One possibility is the use of ANSI escape sequences to beautify the output. The escape sequences are sequences of nonprinted characters, which are interpreted by the shell to change the attributes for the output (e.g., «`\033[22;32m Green Text \033[22;39m`» to print the green characters «`Green Text`») (ASCII-Table.com n.d.). For example, the generator «idl2rpc.pl» uses these sequences to highlight error messages in red and successful processing steps in green. Another possibility for a more graphical style in shells is the use of libraries like «ncurses» (Ncurses team n.d.), which emulate the output terminal, so that background colors, multiple highlights, function-key mapping, and so on can be defined.

Linux GUI System X11

While all servers should be tested with such command line clients, they are not suitable for large, interactive user interfaces. The appropriate way is to use the window system of the operating system with a graphical user interface (GUI). It makes it possible to issue commands with input devices like a mouse to click graphical buttons or to drag and drop icons. This plays an important role in interacting with a computer and its programs. For Linux systems, the basic graphical user interface is the X11 system. Within this system, it is possible to create windows and other graphical elements, like buttons, boxes, frames, menus, and so on, which are then administrated by the window manager of the operating system. The window manager defines the general style of the graphical user interface. It is then responsible for converting mouse signals and positions to events in the specific program which is responsible for handling the event (Brookshear 2012, *l.c. page 116 f.*).

wxWidgets Framework

But the programming of X11 applications is quite bulky. There are several frameworks which offer software components to write desktop or mobile applications much more conveniently. The methods of the components hide all the «housekeeping» work to provide a well-known behavior of graphical applications. One of these sophisticated toolkits is «wxWidgets».[18] wxWidgets[19] offers a multi-platform interface, so that applications can run on different platforms with only a few changes. The appearance of the windows provides a native look and feel, as it uses the native window manager attributes of the underlying operating system. wxWidgets benefits from the heavy use of OOP concepts. Therefore, it offers classes for several types of graphical objects, can handle events from the computer environment with well-defined class behaviors, and runs as open-source project under the LGPL. This flexibility has won wide industry support, so that wxWidgets is used in all kinds of software developments and for all fields of applications (Smart et al. 2006, *l.c. page 1 ff.*).

wxWidgets Architecture

Internally, wxWidgets has a layered architecture. On the operating system layer, for example, Linux or Unix, it uses the platform-specific GUI Application Programming Interface (API), such as the «Xlib» for the X11 system. The following «wxWidget port» layer hides the system-specific interfaces and creates an adapter to its functionality. Such a port is necessary for each operating system (e.g., Windows™, Linux, Unix, Mac OS X, OS/2, and Palm OS are supported) and its related window system API. The final layer defined by the wxWidgets API implements a common, platform-independent access point to the window functionalities (Smart et al. 2006, *l.c. page 8*). The internal code is additionally separated into six characteristics (Smart et al. 2006, *l.c. page 12 f.*):

- *Common code*: These code parts are used in all wxWidgets ports and are, therefore, common to the wxWidgets framework. They implement common types and base classes.
- *Generic code*: These code parts contain all advanced, platform-independent widgets.

---

18   There are also several other frameworks, for example, the «Qt» or «GTK+», which offer similar possibilities. But they had some limitations in licensing (e.g., no open-source licensing) and supporting by user groups in the days as the main development projects at the Wettzell observatory started. Therefore, the focus here is laid on the «wxWidgets» framework. Nevertheless, all the general principles discussed can also be applied to the other frameworks.

19   «Widgets» are the elements of a graphical user interface.

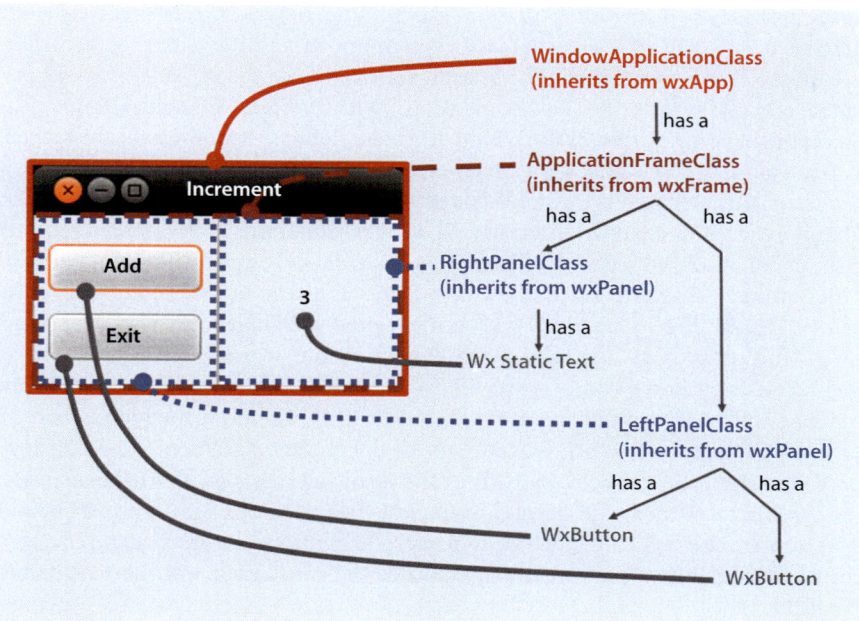

■ **Fig. 4.12** The composition of a wxWidgets window with different graphical elements, which is again reflected in the structure and composition of C++ class objects

- *wxUniversal*: These code parts contain all native widgets for platforms which don't have them.
- *Platform-specific code*: It contains classes which use the native functions from the window system of the operating system.
- *Contributed code*: These are useful classes which combine existing elements or define new, additional elements. These code parts are, among other benefits of wxWidgets, the reason it was, and still is, so popular for industrial control applications, as a lot of industrial displays are already developed and available here.[20]
- *Third-party code*: Independent libraries used by wxWidgets are located here. Those libraries, like the libraries to read image formats, are essential elements in the whole code.

The programming is typically object-oriented. The example window in ■ Fig. 4.12 demonstrates the relation of graphical window elements to according class objects. The whole program to create this window can be found in Example 4.4.[21] ■ Figure 4.12 shows that the implementation of windows is a hierarchical composition, where graphical elements contain other graphical elements. This finds its analogy in the class hierarchy. The classes for the graphical elements contain instances of other elements as attributes. The creation and final definition of the elements, with its behavior, forms, colors, and positions are then done in the constructor of the classes which contain them. This implements a classic «has a» or

Window Designing

20 But wxWidgets is no framework for safety-critical systems, as its sources do not manage pointers and memory in a safe way, for example, memory allocation running out of memory is not checked with «try/catch» blocks. This enables a better readability, but dramatically reduces safety. In the case of autonomous production cells, graphical user interfaces are pure presentation layers without any logic. Logic and safety issues are usually completely implemented in the coordination cell and related sub-cells. Thus, crashes of the user interface are not safety critical.
21 For a better illustration of the relations between the different classes, all methods are «inline» code (the method body is directly in the class declaration, except the event handler «vOnAddButtonHandler», which uses methods from another class and needs a forward declaration), and the class components are not separated into different module files.

«uses a» relation in the OOP design. The graphical class objects themselves are derived from basic wxWidgets objects, which implement the native implementation of the final appearance of the graphical elements on the screen.[22] This inheritance complies with the «is a» relation of the object-oriented design. The hierarchical composition of the windows finally defines the design on the screen.

wxWidgets Program Elements and Flow

In a similar way to the generated server code of the «idl2rpc.pl» generator, a program with wxWidgets must follow special structures defined by the framework. Therefore, a wxWidgets program has no «main» function, as the startup requires a set of functions which should be hidden to allow a better transparency for the programmers. Instead of the «main», the macros «DECLARE_APP» and «IMPLEMENT_APP» are used. They initialize the wxWidgets-internal data structures and create an instance of the application window (the instance of «WindowApplicationClass» in Example 4.4). Then wxWidgets calls the method «OnInit» of the application window. This method is the place where an instance of the frame, which will contain all the graphical elements, is created. To do this, a dynamic memory element of the frame class is allocated. This activates the constructor of the frame class (in Example 4.4: «ApplicationFrameClass»). This constructor allocates further dynamic elements of the next hierarchy level (in Example 4.4: two panels) which activates the constructors of these elements. After the allocation of all objects has been successful, «OnInit» shows the frame on the screen and returns with the state «true». After that, wxWidgets starts with an event loop which waits for events (mouse clicks, keyboard inputs, etc.). Each time an event arrives, it is dispatched to an appropriate event handler (Smart et al. 2006, *l.c. page 23*).

wxWidgets Event-Driven Programming

The heart of each window system framework is the handling of events. The main idea behind it is that an event loop waits for input events from the user or from other components (such as a window refresh event). Sophisticated management of the events gives the user the illusion that several things are happening simultaneously in the window. The implementation behind can cause phenomenal complexity. The main way to implement such event-driven programming under wxWidgets is the creation of an event table for each reactive class (for illustration, see the implementation of the class «LeftPanelClass» in the Example 4.4). To do this, it is necessary to declare a class which is directly or indirectly derived from the event handler class «wxEvtHandler». Then it is necessary to add an event handler method for each event to the derived class that should be caught. Such methods are, in principle, usual member methods, which have a fixed head structure (e.g., see the event handler «vOnAddButtonHandler» of the Example 4.4). Those functions are called if an event of a specific type arrives. The functions contain the code which should be processed in case the related event arrives. The assignment of specific events (e.g., the event after a button is clicked) is done with an event table. To implement such a table, it is necessary to define event identifiers, which are constant numbers declared as constants or enumerations. The event table must be declared in the class for which it should be active. This is done by the macro «DECLARE_EVENT_TABLE». Finally, the event table can be implemented in the source file between the macros «BEGIN_EVENT_TABLE» and «END_EVENT_TABLE».[23] Each table entry defines an appropriate macro (e.g., «EVT_MENU», «EVT_SIZE», or «EVT_BUTTON») which maps an event identifier to a related event handler method. Event identifiers are assigned to the reactive, graphical element. Event handler functions are previously defined. In principle, this technique is an application of function pointers where the addresses to functions of the same calling structure are arranged (Smart et al. 2006, *l.c. page 25 ff.*). The whole handling behind it is organized by wxWidgets.

---

22 Help about the available and possible elements and their member methods can be found on web pages and Wikis such as ► http://docs.wxwidgets.org/2.8/wx_contents.html.

23 Compare this similar structure of begin and end tags with the generated server code of the «idl2rpc.pl» generator, where user-specific code and definitions are also identified with similar labels, but not with macros.

■ ■ Example 4.4

Composition of a wxWidgets window with different graphical elements and according event handlers.

```cpp
#include <wx/wx.h>                                                      1
#include <wx/panel.h>                                                   2
                                                                        3
// *********************************************                        4
/// Left-panel class with two buttons in a column                       5
// *********************************************                        6
/// Declare event identifiers                                           7
const int ID_ADD  = 101; // Event identifier for add-button             8
const int ID_EXIT = 102; // Event identifier for exit-button            9
                                                                       10
/// Declare and define class LeftPanelClass                            11
class LeftPanelClass : public wxPanel                                  12
{                                                                      13
    public:                                                            14
        wxButton * pCAddButton;                                        15
        wxButton * pCExitButton;                                       16
        wxPanel *  pCParentPanel;                                      17
        int iCounter;                                                  18
    public:                                                            19
        /// Constructor, which creates all elements of the             20
        /// left panel                                                 21
        LeftPanelClass (wxPanel * pCNewParentPanel) throw (int) :      22
            wxPanel (pCNewParentPanel, wxID_ANY, wxPoint(0,-1),        23
                     wxSize(100,100), wxBORDER_SUNKEN)                  24
        {                                                              25
            iCounter = 0;                                              26
            pCParentPanel = pCNewParentPanel;                          27
            try                                                        28
            {                                                          29
                pCAddButton = new wxButton (this, ID_ADD, wxT("Add"),  30
                                            wxPoint(5,15));             31
            }                                                          32
            catch (...)                                                33
            {                                                          34
                pCAddButton = NULL;                                    35
                throw 1;                                               36
            }                                                          37
            try                                                        38
            {                                                          39
                pCExitButton = new wxButton (this, ID_EXIT, wxT("Exit"), 40
                                             wxPoint(5,55));            41
            }                                                          42
            catch (...)                                                43
            {                                                          44
                pCExitButton = NULL;                                   45
                throw 1;                                               46
            }                                                          47
        }                                                              48
                                                                       49
        /// Destructor                                                 50
        virtual ~LeftPanelClass ()                                     51
        {                                                              52
            delete pCExitButton;                                       53
            pCExitButton = NULL;                                       54
            delete pCAddButton;                                        55
            pCAddButton = NULL;                                        56
        }                                                              57
                                                                       58
```

```
private:
    /// Event handler to handle click events on add
    /// button
    void vOnAddButtonHandler (wxCommandEvent & WXUNUSED(CEvent));

    /// Event handler to handle click events on exit
    /// button
    void vOnExitButtonHandler (wxCommandEvent & WXUNUSED(CEvent))
    {
        Close(true);
        exit (0);
    }

    /// Declare wxWidgets event table for class
    DECLARE_EVENT_TABLE()
};

/// Define event table for class LeftPanelClass
BEGIN_EVENT_TABLE (LeftPanelClass, wxPanel)
    EVT_BUTTON   (ID_ADD,    LeftPanelClass::vOnAddButtonHandler)
    EVT_BUTTON   (ID_EXIT,   LeftPanelClass::vOnExitButtonHandler)
END_EVENT_TABLE ()

// *********************************************
/// Right—panel class with the text output
// *********************************************
class RightPanelClass : public wxPanel
{
    public:
        wxStaticText * pCTextOutput;
    public:
        /// Constructor, which creates all elements of the
        /// right panel
        RightPanelClass (wxPanel * pCNewParentPanel) throw (int) :
            wxPanel (pCNewParentPanel, wxID_ANY, wxPoint(100,−1),
                    wxSize(100,100), wxBORDER_SUNKEN)
        {
            try
            {
                pCTextOutput = new wxStaticText (this, −1, wxT("0"),
                                            wxPoint (40, 40));
            }
            catch (...)
            {
                pCTextOutput = NULL;
                throw 1;
            }
        }

        /// Destructor
        virtual ~RightPanelClass ()
        {
        }
};

// *********************************************
/// Frame class with the two panels in a row
// *********************************************
class ApplicationFrameClass : public wxFrame
{
    public:
        wxPanel          * pCMainPanel;  /// Main panel, holding all elements
        LeftPanelClass   * pCLeftPanel;  /// Left panel in the main panel
        RightPanelClass  * pCRightPanel; /// Right panel in the main panel
```

59
60
61
62
63
64
65
66
67
68
69
70
71
72
73
74
75
76
77
78
79
80
81
82
83
84
85
86
87
88
89
90
91
92
93
94
95
96
97
98
99
100
101
102
103
104
105
106
107
108
109
110
111
112
113
114
115
116
117
118
119
120
121
122

```
public:
    /// Constructor, which creates all elements of the
    /// right panel
    ApplicationFrameClass (const wxString & strTitle) throw (int) :
        wxFrame (NULL, wxID_ANY, strTitle, wxDefaultPosition,
                wxSize(200,100))
    {
        pCMainPanel = NULL;
        pCLeftPanel = NULL;
        pCRightPanel = NULL;

        /// Create main panel, holding all elements
        try
        {
            pCMainPanel = new wxPanel (this, wxID_ANY);
        }
        catch (...)
        {
            pCMainPanel = NULL;
            throw 1;
        }

        /// Create left panel
        try
        {
            pCLeftPanel = new LeftPanelClass (pCMainPanel);
        }
        catch (int)
        {
            delete pCLeftPanel;
            pCLeftPanel = NULL;
            delete pCMainPanel;
            pCMainPanel = NULL;
            throw 1;
        }
        catch (...)
        {
            pCLeftPanel = NULL;
            delete pCMainPanel;
            pCMainPanel = NULL;
            throw 1;
        }
        /// Create right panel
        try
        {
            pCRightPanel = new RightPanelClass (pCMainPanel);
        }
        catch (int)
        {
            delete pCRightPanel;
            pCRightPanel = NULL;
            delete pCLeftPanel;
            pCLeftPanel = NULL;
            delete pCMainPanel;
            pCMainPanel = NULL;
            throw 1;
        }
```

```
        catch  (...)
        {
            pCRightPanel = NULL;
            delete pCLeftPanel;
            pCLeftPanel = NULL;
            delete pCMainPanel;
            pCMainPanel = NULL;
            throw 1;
        }

        /// Center main panel
        this->Centre();
    }

    /// Destructor
    virtual ~ApplicationFrameClass ()
    {
        delete pCRightPanel;
        pCRightPanel = NULL;
        delete pCLeftPanel;
        pCLeftPanel = NULL;
        delete pCMainPanel;
        pCMainPanel = NULL;
    }
};


// *********************************************
/// Application class with the frame
// *********************************************
class WindowApplicationClass : public wxApp
{
    public:
        /// Startup method, which is automatically called
        /// on application startup (method is mandatory)
        virtual bool OnInit ()
        {
            ApplicationFrameClass * pcFrame = NULL; /// Frame with main panel

            /// Create frame with main panel in application window
            /// and show it
            /// (use the converter method wxT to convert standard
            /// strings into wxWidgets specific strings)
            try
            {
                pcFrame = new ApplicationFrameClass (wxT("Increment"));
            }
            catch (int)
            {
                delete pcFrame;
                pcFrame = NULL;
                return false;
            }
            catch (...)
            {
                pcFrame = NULL;
            }
            pcFrame->Show (true);

            /// Return to start the event loop
            return true;
        }
};
```

181
182
183
184
185
186
187
188
189
190
191
192
193
194
195
196
197
198
199
200
201
202
203
204
205
206
207
208
209
210
211
212
213
214
215
216
217
218
219
220
221
222
223
224
225
226
227
228
229
230
231
232
233
234
235
236
237
238
239
240
241
242

```
// ******************************************
/// Main program
// ******************************************
/// Implement the application
DECLARE_APP ( WindowApplicationClass )

/// Implement the wxWidgets main
IMPLEMENT_APP ( WindowApplicationClass )

// ******************************************
/// Definition of the event handler, which
/// uses a member function of another class
// ******************************************
/// Event handler to handle click events on add
/// button
void LeftPanelClass::vOnAddButtonHandler (wxCommandEvent & WXUNUSED( CEvent ))
{
    ApplicationFrameClass * pcFrame = NULL;

    /// Increase counter
    ++iCounter;

    try
    {
        /// Fetch wxFrame pointer
        pcFrame = (ApplicationFrameClass *) pCParentPanel->GetParent();
        /// Write text into text field of right panel,
        /// while using the pointer dereferencing chain,
        /// starting with the frame
        pcFrame->pCRightPanel->pCTextOutput->SetLabel (wxString::Format(wxT("%d"),
                                                       iCounter));
    }
    catch (...)
    {
        --iCounter;
    }
}
```

243
244
245
246
247
248
249
250
251
252
253
254
255
256
257
258
259
260
261
262
263
264
265
266
267
268
269
270
271
272
273
274
275
276
277
278
279
280

After the compilation and linking of the wxWidgets libraries (in the example shown with «g++ `wx-config --cxxflags --version=2.8 --libs core,base' <file>.cpp»[24]), the final application can be started like the other applications and will appear in the designed graphical form. Using such a programming framework, it is possible to design almost any complex structure. But the implementation of these more complex structures in the way described can be exhausting. Generative programming is again the solution. Rapid Application Development (RAD) tools, such as the open-source development «wxFormBuilder»[25] or other proprietary licensed tools, offer generators with a graphical design board, which use a drawing of the final design with available graphical elements to generate the skeleton of the final window program. The programmer only has to fill the method bodies and can, therefore, reduce the development times. A problem is that most of the RAD tools do not use already existing code, so that each new generation requires a manual compilation of the existing elements into the newly generated source. Nevertheless, it is a quick way to design confined window elements. Another way to design new applications rapidly is to use the industrial controls or other classes in the contributed code or from external libraries.

wxWidgets Rapid Application Development

Graphical user interfaces to control a laser ranging system are very complex systems. They combine drawings, several input types, different standard and

wxWidgets Advanced

---

24 The example uses wxWidgets version 2.8. The latest stable version is 3.0.3.
25 Available at ► http://sourceforge.net/projects/wxformbuilder/.

custom dialogs, image outputs, data structures, streams, and numerous events like sizers, buttons, text field inputs, and so on (see (Smart et al. 2006)). As most of the activities appear at the same time, the user interfaces are multi-threading programs. This means that the problem of critical sections is omnipresent. The output of a string to a text widget, as used in the previous example, can become a critical section if several, parallel tasks need to do the same thing at the same time. Therefore, it is necessary to use protections. wxWidgets offers «mutex» classes for those cases (see also ► Sect. 2.6.4). As with semaphores, it is valuable to keep the protected section as small and local as possible. Another problem is the entanglement of application-wide events, which again can trigger other events triggering further events and so on. In principle, events are like jump points. The workflow is not sequential anymore, but jumps between the event handlers. Careful design is necessary, which is also required for the memory management, so that deleting orders do not result in application crashes after an application is quit. Internationally used applications must also support different character code sets, so that ASCII and ANSI are not always the ideal solution. Character codes on the basis of UTF are better. But strings must then be converted properly. Designing and implementing of such huge applications are a very complex task, which can be supported effectively by the practices employed to develop clean scientific code, as described in ► Chap. 2.

Thin Clients and Fat Servers

Nevertheless, another design criterion is very important for the design of graphical user interfaces for autonomous control and production cells: strict separation between «cell coordination and control code» and «cell presentation and user interface code». While the processing logic is concentrated in the servers, where all arithmetic, logic, administrative, planning, controlling, scheduling, and failure managing tasks are processed, the client is just a presentation layer. It holds no essential logic for the system, but only for the processing which is necessary for graphical presentation, visualization, user interaction, announcing, and reporting (Langmann 2010, *l.c. page 48*). This concept splits the components into «thin clients» and «fat servers». The glue between both is the communication infrastructure code with its communication interfaces. But why is this design, as shown in ◘ Fig. 4.13, so important? A server of an autonomous production cell must be able to solve its tasks without the graphical presentation and even without any interaction from outside. This design is quite important, as graphical user interfaces often do not support a higher level of code safety. But if GUIs do not share critical control tasks, which are all in the server code, the requirements can be reduced for the presentation layer. If processing logic were in the presentation client, this design criterion would not be possible. Another difficulty would arise if several users want to monitor the same system from different locations. This is quite valuable, as, for example, a supervisor can log in

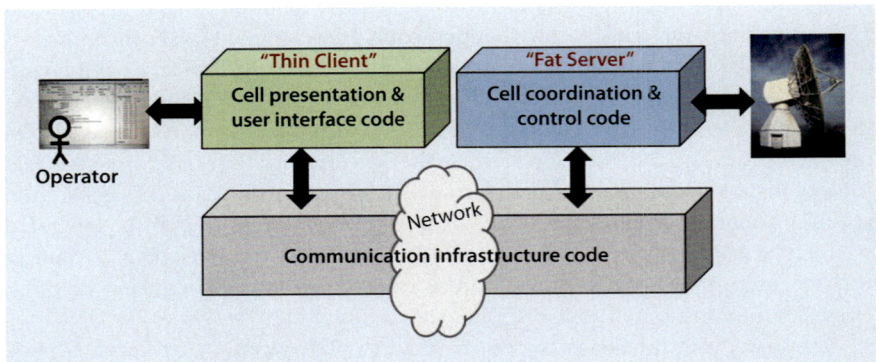

◘ **Fig. 4.13** The use of «thin clients» and «fat servers» for the example of an autonomous production cell for the control of radio telescopes

and use another instance of the presentation layer to guide an operator or to run recovery activities after a failure situation. But if processing logic is enmeshed in presentation code, the different clients compete with their local states and work-flows for the server. As a consequence, misleading requests from the different clients could disturb the system, as no client knows the state and current process-ing of the others. How then should the server decide which orders are the right one? The thin client and fat server concept is essential for the realization of autonomous control cells, as then all the logic for a specific task is in the server. Each further hierarchy level, such as a graphical interface, or a following coordi-nation node can build on this and is only able to manipulate the internal states of the state machine of the fat server. This offers the highest degree of flexibility in a controlling system.

Strict separation also enables and simplifies the replacement of the GUI frame-work or the implementation type of the user interface. This is quite useful to adapt to technical evolution. But it also makes it possible to support other access points over the usual web and a browser. With thin clients, it is even possible to imple-ment this with dynamic HTML pages, so that often no programming language like JavaScript is necessary. Those active elements in a browser are in some cases not secure, as it is necessary to open firewalls (see ▶ Sect. 5.5.3 about network security and firewalls), which opens a backdoor for viruses. A simple solution is the use of dynamically generated web pages. When controlling autonomous production cells, it is possible to add Common Gateway Interface (CGI) scripts or programs to the CGI directory of a web server, like the Apache web server. CGI scripts are small programs, which can be called by the web server if a dedicated web path is requested by a browser over HTTP. The browser can even send input data to the script. It adds the «+» separated arguments list at the end of the web page URL. Another possibility is to send larger data sets as input streams to the CGI script. The script or program reads the calling arguments or the larger data sets from standard input, processes the request, and prints an ordinary HTML page to the standard output, beginning with the text «`Content-type: text/html\n\n`». In autonomous control cells, the CGI program is just an RPC client, requesting information from and sending inputs to the RPC server of the coordination cell. The client then works as a converter, which converts the RPC infrastructure into the web infrastructure.

Experience has shown that a very suitable way to design the graphical elements is with standard HTML elements in a template file. It contains the HTML code with some replacement tags, which can be searched and replaced by actual values for the finally transmitted web page. This template file can be read by the CGI pro-gram after the RPC requests have returned the requested data. All the templates included are searched using pattern matching which is nothing more than another application of regular expressions (see regular languages in ▶ Sect. 3.3.1) to replace them with the actual values from the RPC reply. Orders to the RPC server are interpreted by the CGI program, taking the input arguments, so that they can be converted into adequate RPC order requests. Possible sophisticated and useful designs are shown in ▪ Fig. 4.14, where the alternative HTML GUI of the dome control cell is presented. The relevant HTML template file can be found in the fol-lowing Example 4.5. The tags included are style settings, value outputs, and order instructions. Style settings, such as «`[LEDOrder]`» are matched and replaced with style class definitions from the HTML style definition section at the beginning of the template page, according to the replied states from the autonomous cell. Value outputs are taken from RPC replies and replace value tags, like «`[DisplayAzimuth]`». Finally order instruction tags, such as «`[ButtonActionMoveFastPlus]`», dynamically define changing orders (e.g., if a specific relative displacement is added to a current azimuth position) and change the URL of a request and its argument list in an HTML button.

Alternative GUI: HTML

HTML Templates

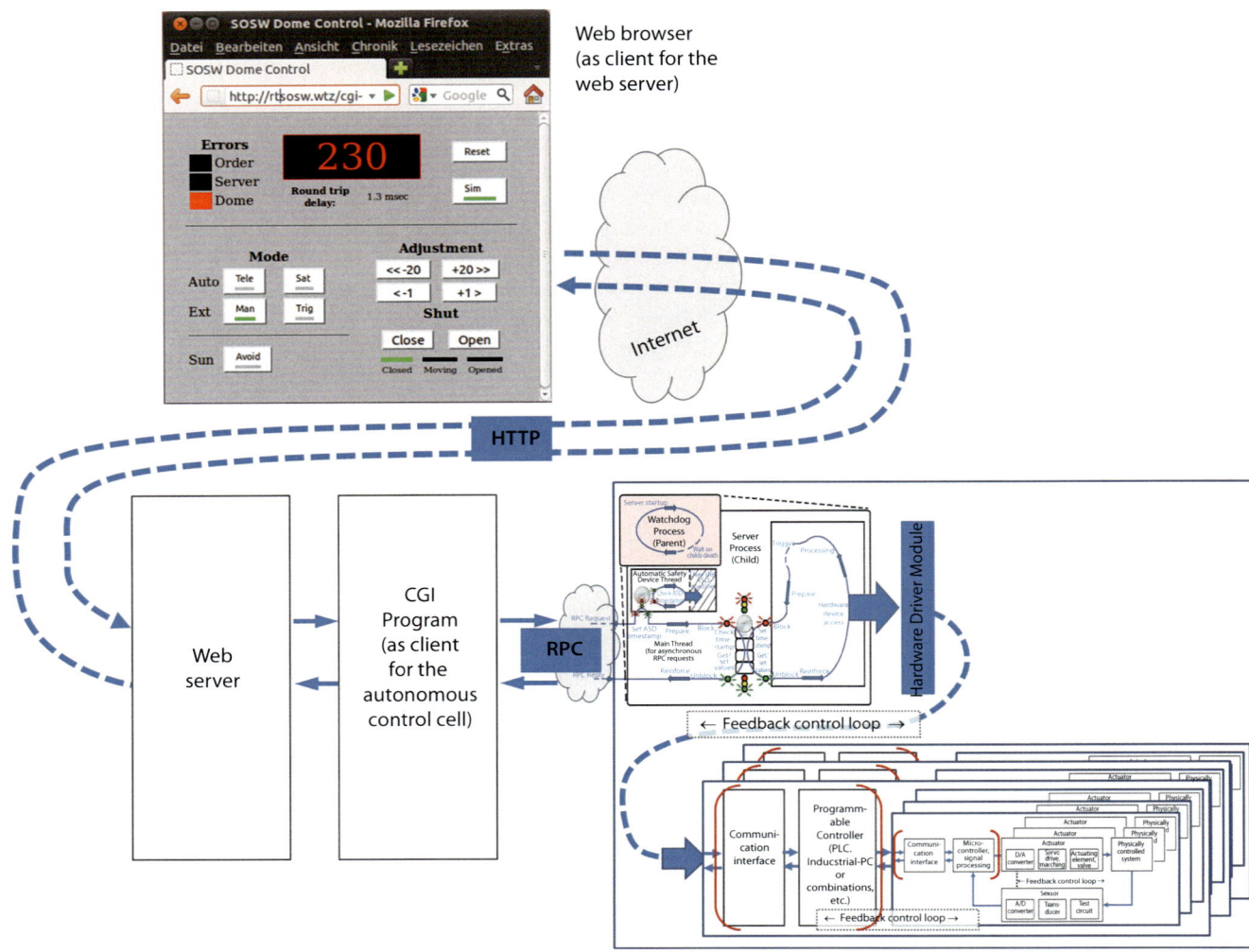

Fig. 4.14   An alternative GUI for the dome control cell on the basis of standard HTML in a web browser, using CGI programs in a web server as extension to the RPC server in ■ Fig. 4.7

■ ■ **Example 4.5**

HTML-templates as basis for an alternative dome GUI.

```
<!DOCTYPE HTML PUBLIC "−//W3C//DTD HTML 3.2 Final//DE">          1
<HTML>                                                           2
 <HEAD>                                                          3
  <TITLE>SOSW Dome Control</TITLE>                               4
  <meta http−equiv="Content−Type" content="text/html"; charset="iso−8859−1">   5
  <meta http−equiv="copyright" content="WETTZELL.BKG">          6
  <META HTTP−EQUIV="Refresh" CONTENT="1; URL=http://dome.wtz/cgi−slruserbin/domectrlhtmlc.cgi"   7
     >
  <style type="text/css">                                       8
  body { background−color:#C0C0C0; color: #000000; margin:20px; font−size:11pt }   9
  p,h1,h2,h3,h4,h5,ul,ol,li,div,td,th,tr,address,blockquote,center,nobr,b,i,caption   10
      { background−color:#C0C0C0; color: #000000; font−size:11pt }   11
  button { background−color:#FFFFFF;   color:#000000 }           12
  td.buttontext { background−color:#FFFFFF; font−size:8pt }      13
```

```
     td.roundtriptext { font-size:8pt }
     td.display { background-color:#FF0000; font-size:32pt }
     td.ledblack { background-color:#000000 }
     td.ledgreen { background-color:#00FF00 }
     td.ledred { background-color:#FF0000 }
     td.ledgrey { background-color:#C0C0C0 }
     center.display { background-color:#000000; color:#FF0000; font-size:32pt }
     center.roundtriptext { font-size:8pt }
     b.roundtriptext { font-size:8pt }
</style>
</HEAD>
<BODY bgcolor="#C0C0C0">

   [BEGControlUnit]

  <TABLE>
     <TR>
        <TD width="115">
           <CENTER>
              <b> Errors </b>
           <TABLE>
              <colgroup>
                 <col width="25">
                 <col width="50">
              </colgroup>
              <TR>
                 <TD class="[LEDOrder]">
                     
                 </TD>
                 <TD>
                    Order
                 </TD>
              </TR>
              <TR>
                 <TD class="[LEDServer]">
                     
                 </TD>
                 <TD>
                    Server
                 </TD>
              </TR>
              <TR>
                 <TD class="[LEDDome]">
                     
                 </TD>
                 <TD>
                    Dome
                 </TD>
              </TR>
           </TABLE>
           </CENTER>
        </TD>
        <TD width="20">
            
        </TD>
        <TD width="240">
           <TABLE>
              <TR>
                 <TD width="240" class="display" align="right">
                 <CENTER class="display">
                    [DisplayAzimuth]
                 </CENTER>
                 </TD>
              </TR>
```

14
15
16
17
18
19
20
21
22
23
24
25
26
27
28
29
30
31
32
33
34
35
36
37
38
39
40
41
42
43
44
45
46
47
48
49
50
51
52
53
54
55
56
57
58
59
60
61
62
63
64
65
66
67
68
69
70
71
72
73
74
75
76
77

```
                <TR>
                    <td>
                        <table>
                            <TD width="120" class="roundtriptext">
                                <CENTER class="roundtriptext">
                                <b class="roundtriptext"> Round  trip  delay: </b>
                                </CENTER>
                            </TD>
                            <TD width="120" class="roundtriptext">
                                <CENTER class="roundtriptext">
                                [DisplayAnswer]
                                </CENTER>
                            </TD>
                        </table>
                    </td>
                </TR>
            </TABLE>
        </TD>
    <TD width="20">
         
    </TD>
    <TD width="115">
        <center>
            <button name="reset" type="button" onClick="self.location.href='http://rtsosw.
                wtz/cgi-slruserbin/domectrlhtmlc.cgi?reset'"> <table width="40"> <tr><td
                class="buttontext">Reset</td></tr> </table></button><br> <br>
            <button name="simulation" type="button" onClick="self.location.href='http://
                rtsosw.wtz/cgi-slruserbin/domectrlhtmlc.cgi?sim'"><table width="40"><tr><
                td class="buttontext">Sim</td></tr><tr><td height="4" class="[LEDSim]"></
                td></tr></table></button>
        </center>
    </TD>
</TR>
</TABLE>

<TABLE>
    <colgroup>
        <col width="500">
    </colgroup>
    <TR>
        <TD>
            <hr>
        </TD>
    </TR>
</TABLE>
<table>
    <tr>
        <td width="222">
            <center>
            <table>
                <tr>
                    <td colspan="5">
                        <CENTER>
                            <b> Mode </b>
                        </CENTER>
                    </td>
                </tr>
                <tr>
                    <td width="25">
                        Auto
                    </td>
```

78
79
80
81
82
83
84
85
86
87
88
89
90
91
92
93
94
95
96
97
98
99
100
101

102

103
104
105
106
107
108
109
110
111
112
113
114
115
116
117
118
119
120
121
122
123
124
125
126
127
128
129
130
131
132
133

```html
<td>                                                                          134
  <button name="autotelestep" type="button" onClick="self.location.href='http   135
     ://rtsosw.wtz/cgi-slruserbin/domectrlhtmlc.cgi?autotelecont '"><table
     witdh="40"><tr><td class="buttontext">Tele</td></tr><tr><td height="4"
     class="[LEDAutoTeleCont]"></td></tr></table></button>
</td>                                                                         136
<td width="5">                                                                137
</td>                                                                         138
<td width="25">                                                               139
  <button name="autosatcont" type="button" onClick="self.location.href='http://  140
     rtsosw.wtz/cgi-slruserbin/domectrlhtmlc.cgi?autosatcont '"><table witdh="
     40"><tr><td class="buttontext">Sat </td></tr><tr><td height="4"
     class="[LEDAutoSatCont]"></td></tr></table></button>
</td>                                                                         141
<td width="25">                                                               142
  <center>                                                                    143
    <table>                                                                   144
      <TR>                                                                    145
        <TD width="25" class="[LEDSatListLoaded]">                            146
                                                                         147
        </TD>                                                                 148
      </TR>                                                                   149
    </table>                                                                  150
  </center>                                                                   151
</td>                                                                         152
</tr>                                                                         153
<tr>                                                                          154
  <td width="25">                                                             155
    Ext                                                                       156
  </td>                                                                       157
  <td width="25">                                                             158
    <button name="man" type="button" onClick="self.location.href='http://rtsosw.  159
       wtz/cgi-slruserbin/domectrlhtmlc.cgi?man'"><table witdh="40"><tr><td
       class="buttontext">Man</td></tr><tr><td height="4" class="[LEDManual]"><
       /td></tr></table></button>
  </td>                                                                       160
  <td width="5">                                                              161
  </td>                                                                       162
  <td width="25">                                                             163
    <button name="exttrig" type="button" onClick="self.location.href='http://    164
       rtsosw.wtz/cgi-slruserbin/domectrlhtmlc.cgi?exttrig '"><table witdh="40">
       <tr><td class="buttontext">Trig</td></tr><tr><td height="4" class="[
       LEDExtTrig]"></td></tr></table></button>
  </td>                                                                       165
  <td width="25">                                                             166
  </td>                                                                       167
</tr>                                                                         168
<tr>                                                                          169
  <td colspan="5">                                                            170
    <hr>                                                                      171
  </td>                                                                       172
</tr>                                                                         173
<tr>                                                                          174
  <td width="25">                                                             175
    Sun                                                                       176
  </td>                                                                       177
  <td width="25">                                                             178
    <button name="Avoid" type="button" onClick="self.location.href='http://rtsosw  179
       .wtz/cgi-slruserbin/domectrlhtmlc.cgi?sunavoid '"><table witdh="40"><tr><
       td class="buttontext">Avoid</td></tr><tr><td height="4" class="[
       LEDSunAvoid]"></td></tr></table></button>
  </td>                                                                       180
```

```
                    <td colspan="2">                                    181
                        <center>                                        182
                          <table>                                       183
                            <TR>                                        184
                              <TD width="20" class="[LEDInSun]">        185
                                                                   186
                              </TD>                                     187
                            </TR>                                       188
                          </table>                                      189
                        </center>                                       190
                    </td>                                               191
                </tr>                                                   192
              </table>                                                  193
            </center>                                                   194
          </td>                                                         195
          <td width="6">                                               196
                                                                   197
          </td>                                                         198
          <td width="222">                                             199
            <CENTER>                                                    200
            <TABLE>                                                     201
              <tr>                                                      202
                  <td colspan="2">                                      203
                    <CENTER>                                            204
                      <b> Adjustment </b>                               205
                    </CENTER>                                           206
                  </td>                                                 207
              </tr>                                                     208
              <TR>                                                      209
                  <TD>                                                  210
                    <button name="minusfast" type="button" onClick="self.location.href='http   211
                        ://rtsosw.wtz/cgi-slruserbin/domectrlhtmlc.cgi?[
                        ButtonActionMoveFastMinus]'"> <table width="40"> << -20 </table></
                        button>
                  </TD>                                                 212
                  <TD>                                                  213
                    <button name="plusfast" type="button" onClick="self.location.href='http   214
                        ://rtsosw.wtz/cgi-slruserbin/domectrlhtmlc.cgi?[
                        ButtonActionMoveFastPlus]'"> <table width="40">  +20 >>  </table></
                        button>
                  </TD>                                                 215
              </TR>                                                     216
              <TR>                                                      217
                  <TD>                                                  218
                    <button name="minusslow" type="button" onClick="self.location.href='http   219
                        ://rtsosw.wtz/cgi-slruserbin/domectrlhtmlc.cgi?[
                        ButtonActionMoveSlowMinus]'"> <table width="40">  < -1  </table></
                        button>
                  </TD>                                                 220
                  <TD>                                                  221
                    <button name="plusschnell" type="button" onClick="self.location.href='   222
                        http://rtsosw.wtz/cgi-slruserbin/domectrlhtmlc.cgi?[
                        ButtonActionMoveSlowPlus]'"> <table width="40">  +1 >  </table></
                        button>
                  </TD>                                                 223
              </TR>                                                     224
              <tr>                                                      225
                  <td colspan="2">                                      226
                    <CENTER>                                            227
                      <b> Shut </b>                                     228
                    </CENTER>                                           229
                  </td>                                                 230
```

```
            <tr>                                                              231
                <td colspan="2">                                             232
                 <CENTER>                                                     233
                 <TABLE>                                                      234
                    <TR>                                                      235
                        <TD>                                                  236
                            <TABLE>                                           237
                                <colgroup>                                    238
                                    <col width="35">                          239
                                    <col width="10">                          240
                                    <col width="35">                          241
                                </colgroup>                                   242
                        <TR>                                                  243
                            <TD>                                              244
                                    <button name="openshut" type="button" onClick="self.    245
                                    location.href='http://rtsosw.wtz/cgi-slruserbin/
                                    domectrlhtmlc.cgi?closeshut'"> <table width="20">
                                    Close </table></button>
                            </TD>                                             246
                            <TD>                                              247
                                                                         248
                            </TD>                                             249
                            <TD>                                              250
                                    <button name="closeshut" type="button" onClick="self.   251
                                    location.href='http://rtsosw.wtz/cgi-slruserbin/
                                    domectrlhtmlc.cgi?openshut'"> <table width="20">   Open
                                    </table></button>
                            </TD>                                             252
                        </TR>                                                 253
                 </TABLE>                                                     254
                        </TD>                                                 255
                    </TR>                                                     256
                    <TR>                                                      257
                        <TD>                                                  258
                            <TABLE>                                           259
                                <TR height="5">                               260
                                    <TD width="25" class="[LEDShutClosed]">   261
                                    </TD>                                     262
                                    <TD width="5">                            263
                                    </TD>                                     264
                                    <TD width="25" class="[LEDShutMoving]">   265
                                    </TD>                                     266
                                    <TD width="5">                            267
                                    </TD>                                     268
                                    <TD width="25" class="[LEDShutOpened]">   269
                                    </TD>                                     270
                                </TR>                                         271
                                <TR>                                          272
                                    <TD> <font size="-3"> Closed </font>      273
                                    </TD>                                     274
                                    <TD width="5">                            275
                                    </TD>                                     276
                                    <TD> <font size="-3"> Moving </font>      277
                                    </TD>                                     278
                                    <TD width="5">                            279
                                    </TD>                                     280
                                    <TD> <font size="-3"> Opened </font>      281
                                    </TD>                                     282
                                </TR>                                         283
```

```
                              </TABLE>                            284
                            </TD>                                 285
                          </TR>                                   286
                        </TABLE>                                  287
                       </CENTER>                                  288
                      </td>                                       289
                  </tr>                                           290
                </TABLE>                                          291
               </CENTER>                                          292
             </td>                                                293
          </tr>                                                   294
      </table>                                                    295
      [ENDControlUnit]                                            296
                                                                  297
    </BODY>                                                       298
    </HTML>                                                       299
```

HTML Limitations

The conversion approach to the web infrastructure with browsers and web servers is very suitable, but it also has its limitations. Automatic updates can only be processed in second intervals with standard HTML (see the meta-information line «<META HTTP-EQUIV="Refresh" CONTENT="1; URL=..."»). With the intermediate processing steps in the web server, it is not ideal for direct control. Another disadvantage is that the complete HTML page must always be transferred to show the graphics. This reduces performance significantly. But the concept offers a very cheap, simple, and comprehensive way to implement graphical outputs without any knowledge of GUI programming.

Having discussed these different ways to interact with the different autonomous cells of the system, it is now time to take a deeper look at the different types of control and production cells and their implementations.

4.4   Autonomous Coordination Cell

Topology

The autonomous coordination cell is the heart of the autonomous production cell and, therefore, of the whole system. The coordination cell is the center of a star topology. All operational accesses from the users arrive here at this point, where the main control loop for the system is located. The system alive state with all the required parameters must be available at this point, to which all status information, error situations, and current processing details must be propagated. Only administrative accesses to the real hardware devices can be separated, as they use individual functionalities of the individual hardware control cells. For error tracking and configuration, it is suitable to directly communicate with the end devices. Therefore, such accesses are done using the special GUI of the specific hardware cell.

Coordination Functionalities

Autonomous coordination cells have four different categories of interface functionalities in this context (see ◘ Fig. 4.15):

- *Internal functionalities of the autonomous coordination cell itself*: The coordination cell acts like every other autonomous cell in the control network. From an external view, it is no difference to a hardware control cell, which in the given case controls more than one connected hardware device. It offers its own functionalities to control its internal behavior and to set and get parameters. These functionalities are offered as ordinary RPC functions to the outer world of clients. A typical example would be the request for error numbers, current states of the internal state machine, or other status information of the coordination cell.

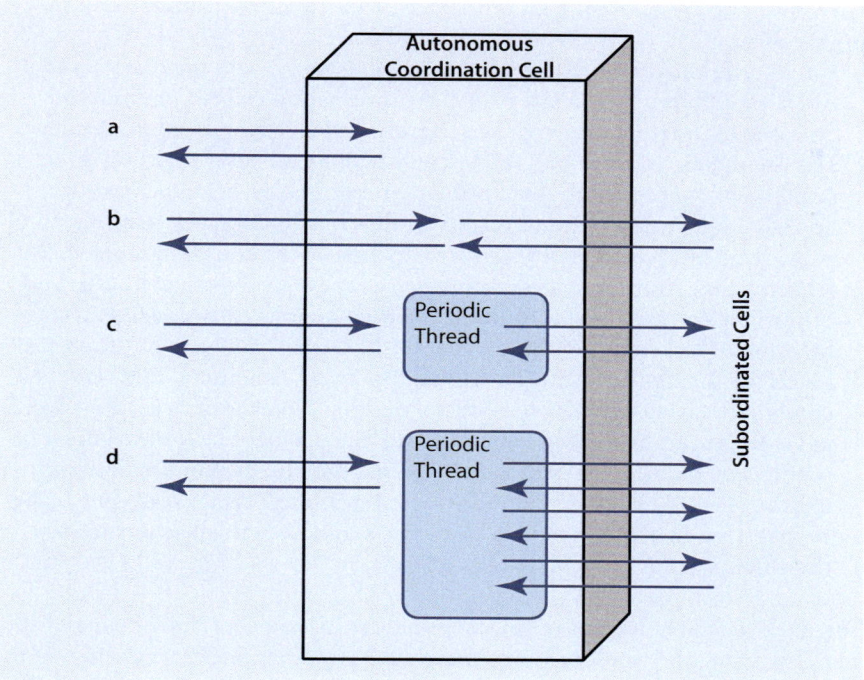

◘ Fig. 4.15 The different categories of interface functions of an autonomous coordination cell: **a** access to internal data, methods, and states of functionalities of the autonomous coordination cell; **b** directly propagated functionalities from the subordinated cells; **c** internally managed, propagated functionalities from the subordinated cells; and **d** internally managed, combined functionalities of propagated and internal data, methods, and states

- *Synchronous, directly propagated functionalities from the subordinated cells*: As the coordination cell multiplexes the different functionalities of the subordinated cells, it requires a way to easily forward requests, parameters, and configurations to the other cells and use data, states, and methods from the other cells. The first approach is to directly propagate these functionalities from the subordinated cells without any local buffering or management in the coordination cell. Usually the hardware cells offer all possible control mechanisms for the controlled hardware or devices. Not all of these functions are always relevant for the higher control layers. This means that the designer of the coordination cell decides which functions, data, and states are also usable via the coordination cell interface. He also has to select which parameters and configurations of the sub-cells should be set from the coordination cell. This selection helps to keep the central interface compact and prevents the diversification of interfaces and their parametrization for a huge amount of very individual sub-cells. A typical example of directly propagated functionalities is the functions which stop or reset hardware immediately, because the commands directly clamp down on the subordinated cells and hardware.
- *Asynchronous, internally managed, buffered, temporally separated, propagated functionalities from the autonomous subordinated cells*: Another category follows the same ideas to offer a forwarding of requests, parameters, and configurations to other cells and use data, states, and methods from the other cells. But by doing this, it temporally and logically separates these functionalities, so that there is no direct access and propagation to the final hardware. Requests are locally buffered in the coordination cell. Decisions are made and the according functionality is called asynchronously

from the request. Usually all propagated RPC functions should be of this category.

- *Combined functionalities from the subordinated cells and the coordination cell itself*: Often it is necessary for the coordination cell to combine and coordinate functionalities of the subordinated cells, for example, if data sets are requested from one subordinated cell (e.g., from a hardware) to hand them over to another subordinated cell. The function behaves like an individual, internal functionality of the coordination cell itself, as it provides a completely new interface function of the coordination cell. But its activation really activates a sequence of calls to subordinated cells and implements a combined activity. Using the previous categories, each combination of tasks can be handed over to the highest instance of the control hierarchy. Using the locally combined functions, functional units can be combined, which simplify the control design. Typical examples are when passage data are requested from the data management cell, which must be sent to several other devices which process the observation, such as moving the telescope, dome, and radar system. Another typical example is the propagation of error situations from one sub-device to all others to stop the whole system.

Technical Realization of the Functionalities

The internal functionalities of the coordination cell are very individual and specific. Therefore, they must be programmed like any other interface function of the RPC interface. During runtime of the cell, these functions are then fixed in their behavior. But this static management of hard-coded interface functions is not necessarily useful for the connection to individual sub-cells and, therefore, for the forwarded operations. If variable devices of the same type need to be coordinated, the fixed, static management restricts a dynamic behavior, as only those device configurations are possible which were statically defined during the programming.[26] More flexible and dynamic management techniques are required to enable dynamic changes during runtime and which use a dynamic memory management. They can be partly created by generative programming. Therefore, the functionalities of the coordination cell can technically be implemented in three different ways:

1. *Static communication management* (see ◘ Fig. 4.16): The RPC server class can include all devices as member variables. This can be programmed manually. Each hardware control cell gets its client representation in the form of such a member variable. It is even possible to control several parallel devices of the same type, as they use the same client class, but are labeled with different identifiers for the member variables.[27] Methods which directly propagate and forward data to the connected sub-device are separately defined in the IDL definition of the coordination cell, so that the bodies are generated during the build process of the cell. The bodies of these functions must be filled by the application developer. The same must be done for combined functionalities. As everything needs to be created manually, it is a very error-prone task (e.g., with a reset, all client connections must be manually closed and reopened again). This does not improve if a separate module is developed as part of the code toolbox, because then the manually changeable parts are just transferred to another management unit.

---

26 The classic way to hard code the connections is used in the NASA Field System, which is used to operate radio telescopes. The system limits the number of supported data acquisition systems of the type Mark5. Therefore, it is necessary to create workarounds, such as the use of a parallel field system if more than the allowed acquisition systems are used.

27 But as described, the number of devices of the same type is fixed during runtime, and the use of additional devices of the same type requires code changes.

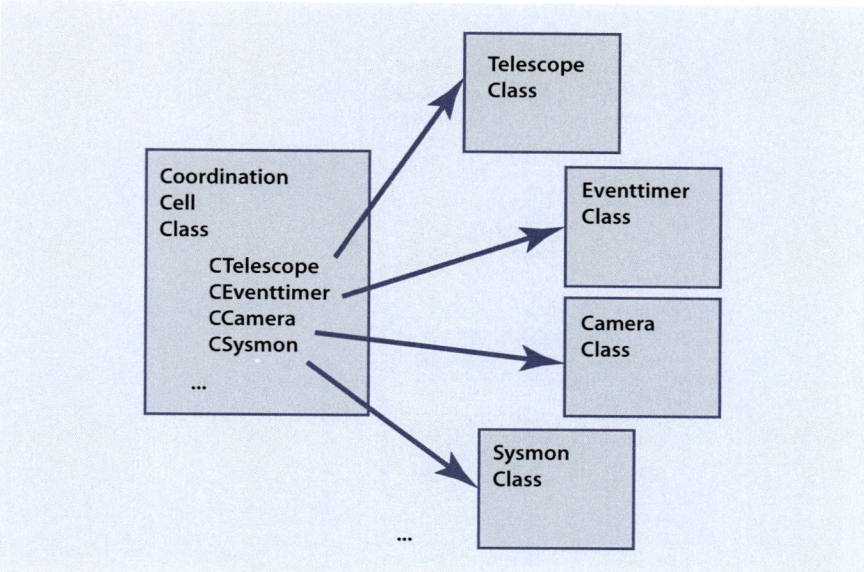

Fig. 4.16 The structure of static communication management for the subordinated cells in the autonomous coordination cell

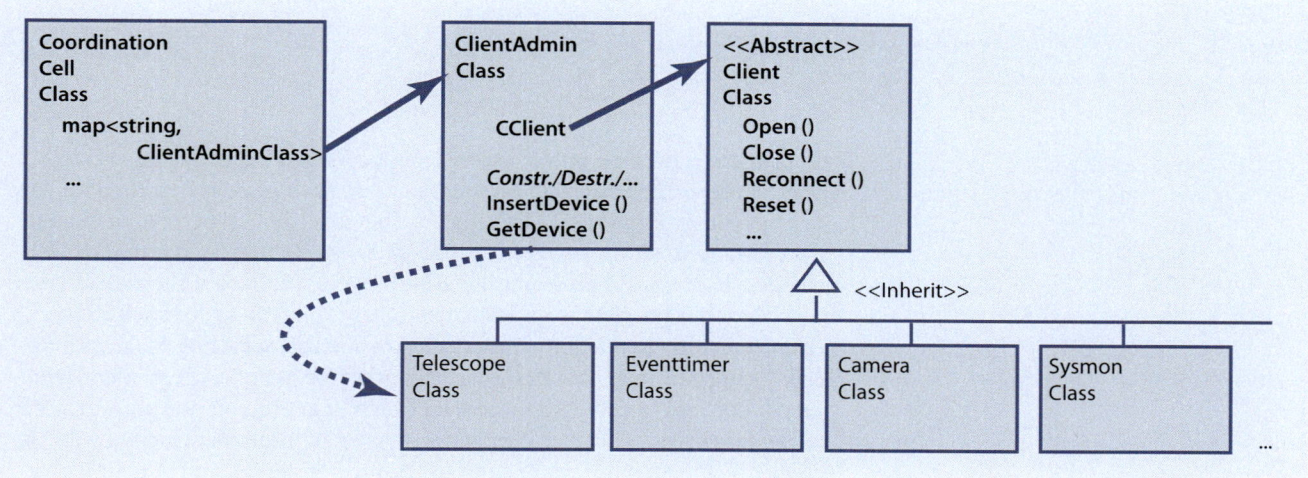

Fig. 4.17 The structure of a dynamic communication management for the subordinated cells in the autonomous coordination cell using maps

2. *Dynamic connection management* (see **Fig. 4.17**): Another possibility is to use an implementation for connections which can be changed during runtime. This is possible with dynamic heap elements and the OOP paradigm. The key is STL maps. Maps are hash tables (see ▶ Sect. 2.7.2) where values are assigned to key values. The keys can be efficiently used for the indexing of the values. As maps are template classes, they can deal with any type of variables. Therefore, they can be used to assign the client classes of the connected sub-cells to a unique, identifying name (e.g., a standard STL string). The only mechanism necessary is that all client classes must be inherited from one abstract interface class to which they must be converted («cast»[28]) before inserting. The abstract class holds declarations of the methods supported by

---

28 Casting is a programming mechanism where a specific pointer or value of a particular type is interpreted as (or converted to) a pointer or value of another type.

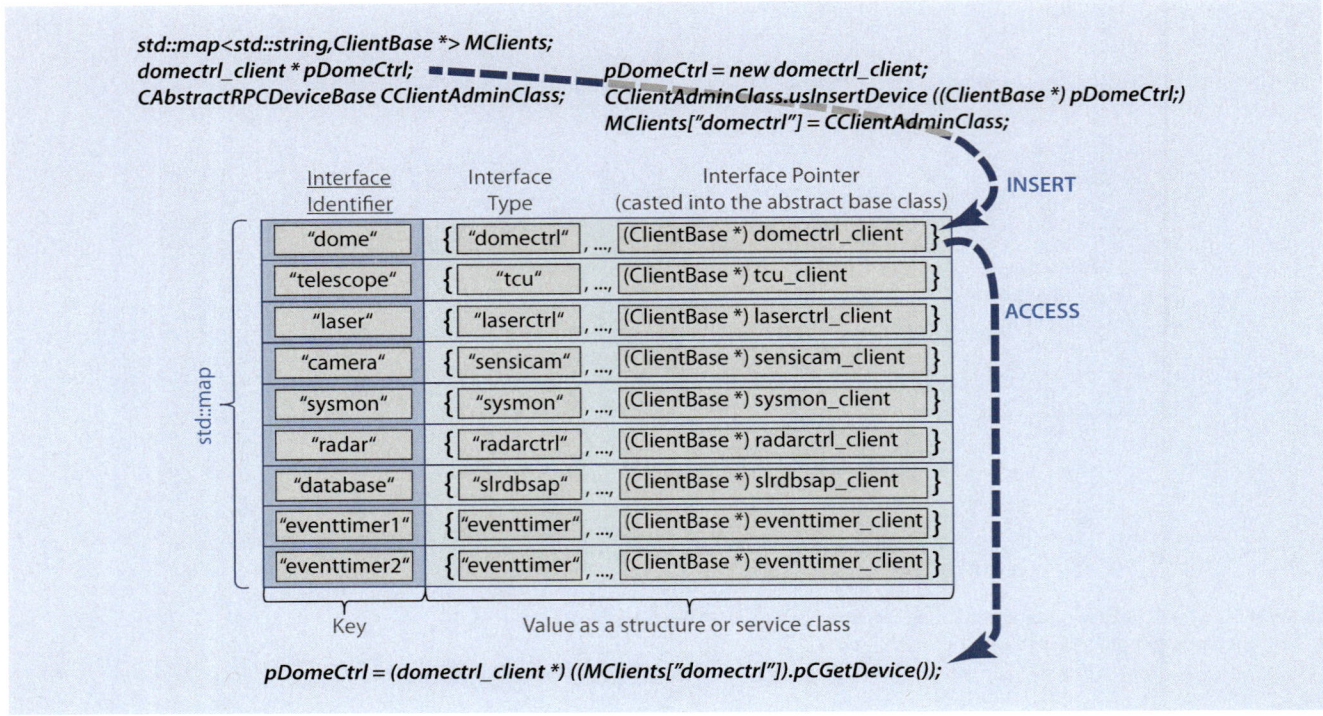

```
std::map<std::string,ClientBase *> MClients;
domectrl_client * pDomeCtrl;              pDomeCtrl = new domectrl_client;
CAbstractRPCDeviceBase CClientAdminClass; CClientAdminClass.usInsertDevice ((ClientBase *) pDomeCtrl;)
                                          MClients["domectrl"] = CClientAdminClass;
```

Interface Identifier	Interface Type	Interface Pointer (casted into the abstract base class)	
"dome"	{ "domectrl" , ...,	(ClientBase *) domectrl_client	}
"telescope"	{ "tcu" , ...,	(ClientBase *) tcu_client	}
"laser"	{ "laserctrl" , ...,	(ClientBase *) laserctrl_client	}
"camera"	{ "sensicam" , ...,	(ClientBase *) sensicam_client	}
"sysmon"	{ "sysmon" , ...,	(ClientBase *) sysmon_client	}
"radar"	{ "radarctrl" , ...,	(ClientBase *) radarctrl_client	}
"database"	{ "slrdbsap" , ...,	(ClientBase *) slrdbsap_client	}
"eventtimer1"	{ "eventtimer" , ...,	(ClientBase *) eventtimer_client	}
"eventtimer2"	{ "eventtimer" , ...,	(ClientBase *) eventtimer_client	}

INSERT

ACCESS

Key — Value as a structure or service class

```
pDomeCtrl = (domectrl_client *) ((MClients["domectrl"]).pCGetDevice());
```

Fig. 4.18 The use of STL maps to enable a dynamic connection management during runtime

any client. In the case of the «idl2rpc.pl»-generated clients, such a class is offered under the name «ClientBase». As each pointer to a client object[29] can be cast into this base class, it is possible to administrate several connections, indexed within a client map (see ◘ Fig. 4.18). The declaration of all clients which should be used must be known in the class of the coordination cell. Then it is possible to allocate dynamic heap memory for such a client object and open a new connection to a suitable server of the according cell during runtime. This new object can then be assigned to an identifying search key and stored within the (STL) map. If another connection to a cell of the same type is used, it is only necessary to allocate new memory for the client object, open the connection to the further device, and store it with a different identifier in the map. Even if the functionality is still hard-coded, the use of dynamic memory offers new possibilities for the flexible management of connections to cells of the same type during runtime. Additionally, service methods can easily be created, which open, reopen, or close all stored client connections at once. This can be done with map iterators and a loop (see Example 4.6), in which, for example, an open function (defined in the abstract interface class) can be called iteratively for all clients, which must be implemented by each client object or within an additional administration class.[30]

---

29 A restriction is that the clients need to be generated with the same version of the «idlerpc.pl» generator.

30 The services and maintenance of heap elements are usually much better if the dynamic heap elements are members of an additional administration class which adds classifying information, an error management, or the implicit memory management to the pure memory allocation (see the abstract interface device class «CAbstractRPCDeviceBase» in ◘ Fig. 4.18).

▪▪ Example 4.6

Iteration over all interfaces in a map to re-open all of them at once.

```
/********************************************************************
    class       RPCDeviceAdmin
    function    usOpenInterfaces
 ********************************************************************/
/*!             Open interfaces and reset control values
 *  \param      —
 *  \return     unsigned short <— Errorcode (see enum ErrorCode)
 ********************************************************************/
/*  author      Alexander Neidhardt
    date        23.04.2008
    revision    —
    info        CAbstractRPCDeviceBase is a helper class, containing the RPC
                client, with service routines for clients
 ********************************************************************/
unsigned short RPCDeviceAdmin::usOpenInterfaces ()
{
        /*! <b> Variables </b> */
        unsigned short usRetVal = SLRCTRLERR_OK;
            /*! usRetVal = return value of function */
        std::map<std::string, CAbstractRPCDeviceBase*>::iterator CDeviceInterator;
            /*! CDeviceInterator = Iterator to walk through client devices */

        /*! <b> Operations </b> */
        // Reopen all interfaces in the map within a loop using the iterator
        for (CDeviceInterator = m_CRPCConnectionList.begin();
            CDeviceInterator != m_CRPCConnectionList.end();
            ++CDeviceInterator)
        {
            if (!((*(CDeviceInterator->second)).usReopenInterface()))
            {
                usRetVal = SLRCTRLERR_INIT;
            }
        }

        return usRetVal;
}
```

```
1
2
3
4
5
6
7
8
9
10
11
12
13
14
15
16
17
18
19
20
21
22
23
24
25
26
27
28
29
30
31
32
33
34
35
36
```

3. *Dynamic connection management with a generated interpreter* (see ◘ Fig. 4.19):
   At this point, generative programming is again quite helpful. It can be used to
   automatically generate the forwarding functionalities which can be
   parametrized for a specific situation. Additionally, the STL maps and assisting
   administration classes can be generated, which contain all the necessary
   management methods (e.g., open all, close all and reset all sub-cells and
   sub-devices, etc.). Additionally, the interpretation of commands, as described
   in ▶ Sect. 3.4, can be used to simplify the use of propagated functionalities. A
   possible way to enable this is to extend the «idl2rpc.pl» generator, so that it can
   interpret additional IDL files of the sub-cells with a selection of subordinated
   RPC functions for the forwarding of functionalities. This technique reduces
   manual code writing and reduces errors. Only some combined functionalities
   need to be manually implemented. Together with the use of maps, this
   approach makes it possible to dynamically add connections to subordinated
   cells or delete them again during runtime, as long as the interface class is
   known during compilation and linking.

The structure of the generated coordination cell with the dynamic connection
management is shown in ◘ Fig. 4.20. It is a standard RPC server with all functionalities like a watchdog process, an automatic safety device, a separate control loop

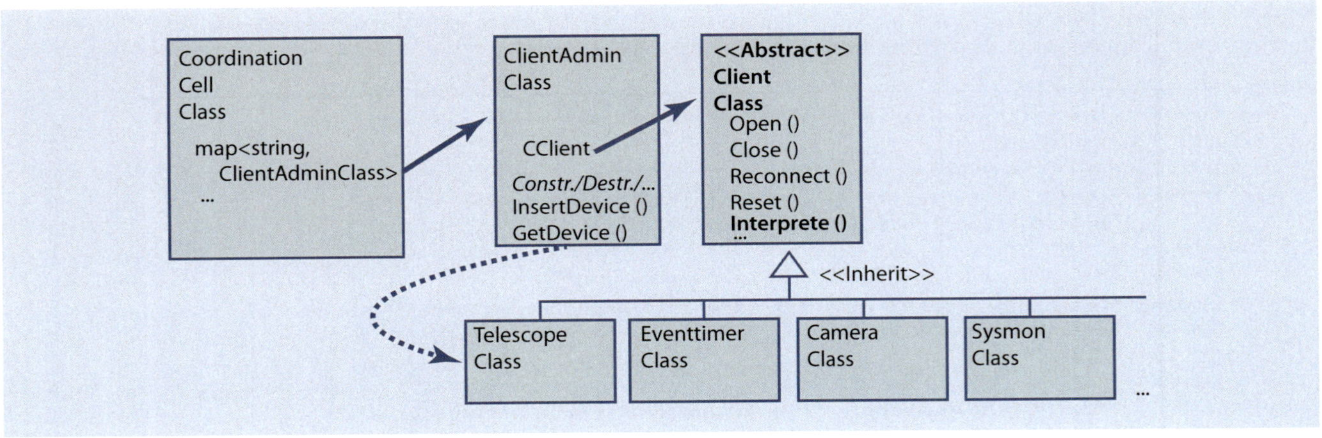

Fig. 4.19 The structure of a dynamic communication management with a generated interpreter function for the subordinated cells in the autonomous coordination cell, using maps and generative programming

Client Selection for the Data Propagation

thread, and the semaphore-protected variables, which uses the map for the management and control of dynamic RPC connections to serve subordinated cells.

The newly generated remote functions operate with different clients, which they have to select for specific data propagation. On the one hand, this selection and the according activation of forwarded functions from sub-cells can be done by adding the key of the client to the map as an additional parameter to the function. To run it, it is then necessary to enter the right key, which is checked in the (generated) forwarding function. Everything is then used in the form of RPC calls, and the conversion between the abstract base class and the specific client class object is done within this forwarding function. Another possibility is the use of the command line functionality described in ▶ Sect. 3.4 starting. With this technique, the key can be added at the beginning of the function name in the ASCII command, similar to a namespace scope (e.g., «domecontrol::usOpenShut()»).[31] Both techniques make it possible to propagate information and to start RPC functions on different hierarchy levels from the central coordination cell. The information is propagated step by step through the hierarchy and is implicitly copied at each level to call the next remote function. But the command line technique offers one additional possibility. It does not necessarily require the remote functions to be repeated at each level of the hierarchy. As each client implements the «uiInterprete» (see Example 3.23), it is available on each hierarchy level. This allows nested calls, as the interpreter function is also a regular RPC function if it is implemented on the server side as an RPC function. Therefore, within an interpreter function of a coordination cell, it is possible to select a sub-cell and call the interpreter function of this again.[32] This allows calls like:

```
tcu::uiInterprete
  ("
      domectrl::uiInterprete
        ("
            this::usOpenShut()
        ")
  ");[33]
```

---

31 Internal functions in the connector class can be characterized with «this», as in C++ or without any key scope.

32 Cyclic/recursive calls can be avoided if it is not allowed that a method calls itself again (e.g., «this::uiInterpreter» or just «uiInterpreter» is not allowed as input of «uiInterpreter»).

33 For the given nested call, a shortcut is conceivable, like «tcu::domectrl::usOpenShut()», which is then automatically translated into the above-nested combination. The commas without a following argument symbolize an output argument of the called function. In the case of «uiInterprete», it is the answer string.

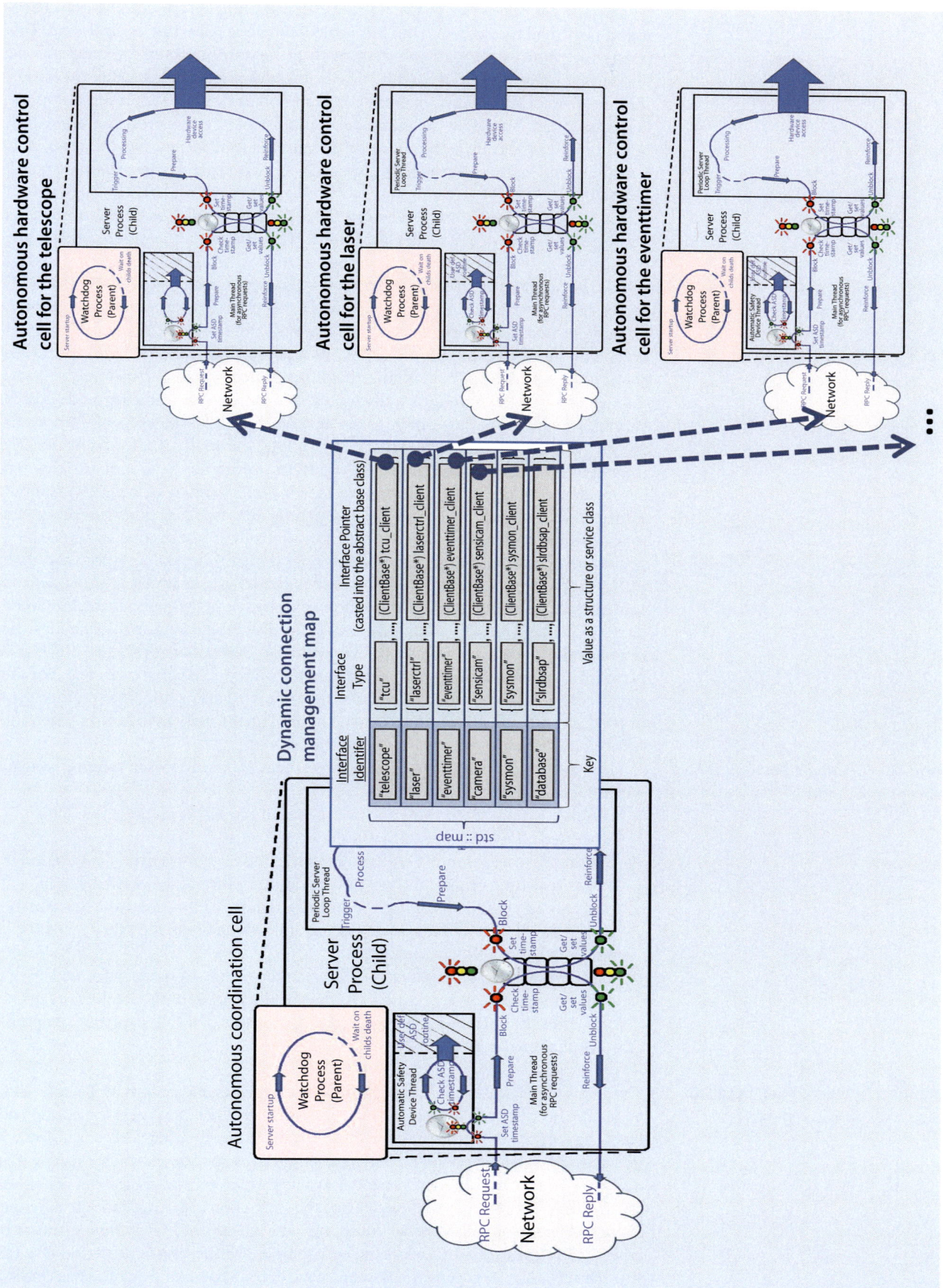

■ **Fig. 4.20** The structure of the autonomous coordination cell, using the map for the dynamic management of RPC connections in ■ Fig. 4.18 using the architecture of RPC server with safety techniques from ■ Fig. 3.22

4

It calls the interpreter of the telescope control unit («tcu» class) with a command for the dome control, which is a sub-cell of the telescope control unit. This possibility reduces the inflationary growth of forwarding interface functions on higher hierarchy levels. But it deals with the disadvantage that instead of the binary XDR sets with original data of complete precision, now ASCII representations are communicated. Therefore, to automatically transfer the same precision of floating point numbers, significantly, more bytes must be transferred. Nevertheless, it is a suitable and useful mechanism (especially if a command line schedule should be used).

Separation of the Workflows

In any case, a separation between the different parallel workflow threads is necessary. At least one thread serves the RPC requests and another controls the connected sub-cells and devices described in ▶ Sect. 3.3.3. The propagation of data should be implemented with semaphore-protected variables between the RPC server and the periodic loop. The RPC functions of the coordination cell set local, semaphore-protected variables and activate flags checked by the periodic loop. It organizes the calls of the sub-cell functions and propagates the values to them. The answers are saved again in similar semaphore-protected values. Also frequent status requests of RPC functions of the connected cells result in data stored in such variables. From there they can be taken again by the RPC functions of the coordination cell. This mechanism is not really simple, especially if a function directly returns a result in accordance with the input data which has been sent. But this can be prevented by the design of the interface if the activation of a task with input values and the requesting of results are separated into two functions (usually given for hardware controlling tasks such as for controlling a telescope, where, e.g., a positioning command is released, which only returns if it was accepted, and then the new status of the telescope controller is requested separately). This cumbersome mechanism avoids the propagation of time delays throughout the hierarchy levels of control, which is quite important in distributed systems. It handles the remote calls asynchronously to the control loop and serializes them within this periodic loop to control the connected hardware device. In some cases (e.g., for the requests to a database where specific data is returned according to an input selection), it is quite difficult or unacceptable to implement such a separation. Then the sub-cell must guarantee specific answer times.

Autonomous Functionalities: Rough-Cut Planning of Main Tasks

The planning must organize the three main tasks of a laser ranging system which recur regularly:

— Observation and determination of the mount model corrections (which is processed about every second week).
— Calibration measurement (internally in the optical system over a calibration line as real-time or pre- and post-observation calibration), which is made continuously while observing. Additional, regular calibrations to external targets in the dome and outside on the ground, which are made every few days, offer further stability parameters.
— Laser ranging observation (to satellites, to the moon, to spacecrafts, and maybe to space debris according to the visibility times of the satellite, its priority, and the current meteorological conditions; a possible sequence of operations is shown in ◘ Fig. 4.21).

Autonomous Functionalities: Detailed Planning of Actions

The planning must also coordinate the different activities and actions of the main task in detail. This detailed planning results in a sequence of actions processed during one task. The goal of this planning is to use the system optimally. Therefore, it is necessary to evaluate and rate a possible plan according to one specific plan. This means it is necessary to generate different variants and alternatives of a plan for further activities, starting from the current situation. During the generation of the variants, the different restrictive, external, and internal constraints are included. One or more evaluation criteria can then be used to select the best solution for the next time period (Pfeifer and Schmitt 2006, *l.c. page 42 ff.*). But this requires continuous and quantifiable counts, measures and tolerances, and, therefore, a suitable evaluation metrics (Pfeifer and

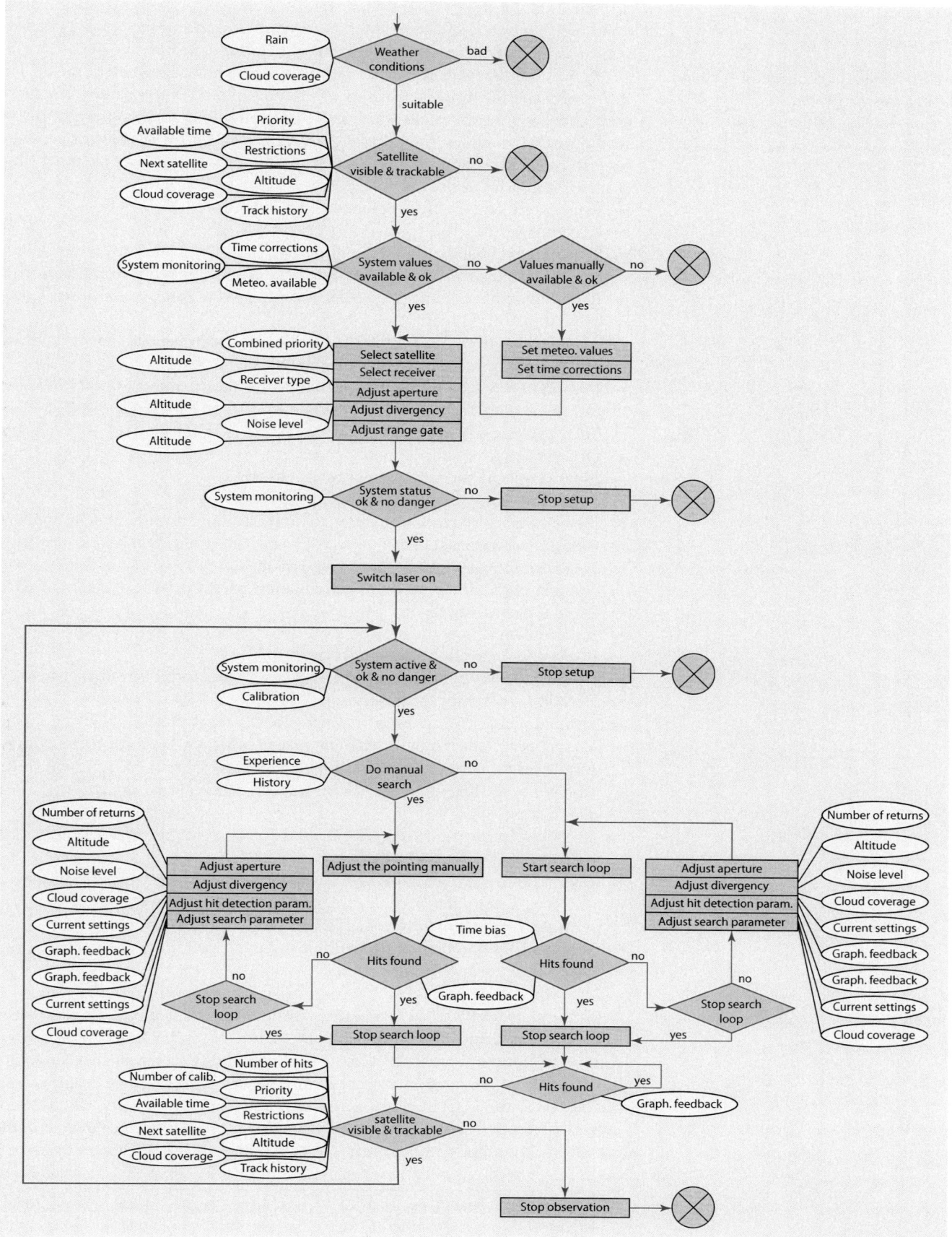

Fig. 4.21 The sequence of steps during a regular laser ranging observation including required decisions (represented as *rhomb*), tasks (represented as *rectangle*), and related information (represented as *ovals*)

Schmitt 2006, *l.c. page 47*). This also requires continuous monitoring of the most important health and quality relevant parameters throughout the whole workflow (which can also be achieved with a separate monitoring system) (see ▸ Sect. 4.7).

Autonomous Functionalities: Planning Evaluation Metric Factors

As with the code testing metrics, it is necessary to find characteristic factors for the metrics and the related evaluation of a given situation or simulation. The characteristics are numeric values which make it possible to evaluate a given situation with quantitative values (Wooldridge 2009, *l.c. page 38 ff.*). The different parameters can be weighted and result in a quality criterion for a generated planning situation. Examples for such factors might be:

- Rise, transit, and set times of the satellites (targets)
- Satellite priority
- Satellite orbit height
- Satellite category (such as GNSS, altimetry satellites, pure geodetic satellites, etc.)
- Required number of hits per satellite passage (and maybe per real-time calibration)
- Achieved number of hits per satellite passage (and maybe per real-time calibration)
- Maximum elevation reached by the satellite
- Minimum elevation which must be reached by the satellite at transit time (maybe in combination with a horizon mask[34])
- Total duration of the passage
- Observation time already spent per passage
- Elapsed waiting time per passage to get observed
- Minimum observation time per satellite passage
- Meteorological situation consisting of humidity, rain, or snow and air pressure
- Cloud coverage situation
- Total visibility time of satellite during passage
- Number of recurrences within a specific time period (maybe in eight hours)
- Sunlit times during night
- Sun avoidance times during day
- In-sky and on-ground safety parameters
- External satellite restrictions, like the go/no-go flag, forbidden segments, or elevation constraints for specific satellites
- Minimum interleaving times between satellite observations
- Minimum distances moved, accelerations, or other costs (similar to boundary conditions for machines in workshops (Pfeifer and Schmitt 2006, *l.c. page 35*))
- Number of satellites observed during a specific time period (maybe in eight hours)
- Number of prioritized satellites observed during a specific time period (maybe in eight hours)
- Distribution of observations over a whole satellite passage to get results for the rise, transit, and set times
- Ideal sky coverage of passages as percentage distribution
- Simultaneous observation from different ground stations or asynchronous observation from different stations

Image Processing for Metric Parameters

To derive some of these metric factors, it is necessary to evaluate graphical representations in the form of images. The evaluation of images is one of the central abilities for autonomous and automated systems, in the same way that the ability to see is one of the most important human senses. Two intuitive examples can directly be found in the field of automated laser ranging systems: cloud detection, as mentioned above, and the detection of star misalignments due to mounting errors in a mount model observation.[35] Therefore, it is necessary to make a

---

34 A horizon mask defines different elevation values for the different azimuth positions, below which it is not possible to observe.

35 Ideas from the image processing field might also be interesting for hit detection in ranging returns, which can be plotted as residuals of reference return times minus measured

short introduction to image processing. Image processing can be completely or partly assigned to autonomous sub-cells for the camera management. Nevertheless, the results are quite important for several coordination tasks and should include additional information, for example, outdoor temperatures from the meteorological sensors to calibrate infrared images which are not directly available at the sub-cell. Image processing can be explained in the context of the coordination cell.

Digital images or photographies are usually raster pictures. These are, in principle, matrices with two dimensions in the form of two-dimensional arrays, where each element represents one image point (pixel). On the one hand, the quality of a digital image is represented by the resolution used, which is the size of the raster matrix used as a display window of a real world object (giving spatial discretization). On the other hand, the quantization of the real world color into discrete equivalents also determines the image quality. Each pixel of a digital color image is usually represented by three values from 0 to 255 for the intensities in the red, green, and blue channel. These are the base colors of the Red, Green and Blue (RGB) color space. Gray images use only one value from 0 (black) to 255 (white), where each number represents one gray scale or brightness scale. In binary images, which are often used as bit masks, each pixel is only represented by two values: black and white (Kopp 1997, *l.c. page 33 f.*).

Digital Images

The digital images taken by a digital camera are streamed as binary sequences directly from the camera or are represented in different file formats (e.g., JPG and GIF are very well-known representatives of such formats). It is necessary to read and interpret the digital or binary formats which need to be offered by suitable graphic libraries. As the retrieved digital images are two-dimensional, discrete functions, they can be statistically characterized. Besides the statistical moments,[36] like the mean value of the gray scales (representing the mean brightness of the image) and the variance (a statistical, central moment representing the contrast of the image), also histograms[37] with the quantitative number of appearances of the different gray values in the whole image (or just in a part of it) are quite important for the characterization of images (Kopp 1997, *l.c. page 37 f.*). As in most cases of image processing, it is often enough to handle gray-scale images (e.g., brightness distributions from temperature scales in Infrared (IR) images from the sky). For cloud coverage detection, the raw data for the images come either from cameras for the human visible spectrum (sky imager), from IR cameras,[38] from multispectral cameras, from Light detection and ranging (LIDAR) systems,[39] or limited from Water Vapor Radiometer (WVR),[40] laser ceilometer,[41] or Nubiscope.[42] They usually show intensities as «brightness» distributions.

Digital Image Formats, Characteristics, and Sources

To separate clouds from free sky in such images, it is necessary to do a segmentation.

Segmentation

---

actual return times over the relevant start time, the start epoch. The resulting image can be treated like a standard image for pattern matching.

36 Moments are representative measures to quantify a distribution.

37 A histogram is a visual impression of the distribution of statistic data sets. It splits the set into a specific number of sections of equal sizes, for which the number of values belonging to each individual section is counted. These counts are then assigned to the section which represents the distribution of the specific value set for a specific accumulation time.

38 IR cameras use the principle of a photo camera for the IR spectrum.

39 LIDAR systems are laser ranging systems which use the RADAR principle in the optical spectrum usually to measure atmospheric parameters.

40 A WVR is a passive sensor which detects the water vapor concentration from the received radiation of water molecules along a detection beam (Geodätisches Observatorium Wettzell n.d.). As a complete sky scan would last too long, it would be just able to scan areas around the satellite positions or along the passage, depending on the integration times.

41 A laser-ceilometer is an instrument to measure the cloud heights using the backscatter profile of a laser beam returned from the clouds.

42 A Nubiscope uses the emitting IR rays from clouds to get the regional distribution of the radiation temperature (Feister 2013).

Elementary Segmentation

> *Definition DEF4.4 of «elementary segmentation»* (adapted from (Kopp 1997, *l.c. page 110*))
> An elementary segmentation method processes images in a way which clearly separates objects from each other or from the background to detect them for a following manual or automated analysis. The elementary segmentation only uses the scale of gray values as a one-dimensional feature space.

Image Textures

For a suitable segmentation of clouds, it is necessary to manage inhomogeneous textures. Textures are properties of an area or surface which are more complicated than simple, single gray levels in the interior of a contour. The problem is how to characterize such complex textures, so that different areas can be segmented successfully. Usually it is necessary to use higher statistical moments of second order (variance) or third order (skewness) to identify them (Niemann 1990, *l.c. page 110*). These are usually combined in a multidimensional feature space, together with gray-scale gradients, regional contrasts, geometrical characteristics (e.g., linear arrangements), or topological attributes (e.g., two- or three-dimensional surface properties) (Kopp 1997, *l.c. page 114*). Multispectral information can support this multidimensional pattern classification.

Classification

> *Definition DEF4.5 of «classification»* (adapted from (Kopp 1997, *l.c. page 110*) and (Kopp 1997, *l.c. page 114*))
> A classification is a sophisticated form of a segmentation which uses a multidimensional feature space of different classifiers as an extended information basis to come to better results for the identification of objects. As most features are unique to the specific classification task, it is necessary to select suitable classifiers, for example, by using sample images for the object class which should be detected.

Threshold Segmentation

But in the given use case of the separation between clouds and free sky, a simpler approach can be used. Taking a look into scientific papers (e.g., (Li et al. 2011, *l.c. page 1287*)) shows that thresholding techniques are the basis for cloud detection at present. Thresholding is a technique of elementary segmentation. It is a binarization of gray-scale images using a fixed threshold where all gray values lower than this threshold become black and all greater ones become white (Kopp 1997, *l.c. page 111 f.*). This means that gray-scale images must be used (e.g., from a black-and-white camera or an IR camera) or that RGB images must be transformed into single-channel feature images. While a simple threshold for the gray values already produces suitable results for IR images (see the example in ◘ Fig. 4.22 which demonstrates the segmentation with different threshold values of 40, 80, 120, 160, 200, 255[43]), different approaches are commonly used in the field of cloud detection in RGB images. One is the use of a ratio between the red and the blue channel. Pixels with a red/blue ratio less than a given threshold are labeled here as sky (Li et al. 2011, *l.c. page 1287*). Another technique uses saturation of the image and, therefore, the color «purity». The RGB image space must be converted into intensity, hue, and saturation space. A threshold for the saturation component of this space can then be used after this transformation, because clear sky shows high saturation values (Souza-Echer et al. 2006, *l.c. page 439 f.*). More sophisticated methods use the geometrical features in the three-dimensional RGB space. Because clouds and sky populate different areas in the RGB space, it is possible to set a threshold for the Euclidean geometric distance between these regions (Luiz et al. 2010, *l.c. page 1504 ff.*).

---

43 The sky image is artificially composed from different other sky images, using Adobe® Photoshop® Elements, so that an almost equivalent IR image is the result. All effects and editing in the following images are demonstrated with the utilities of Photoshop® Elements.

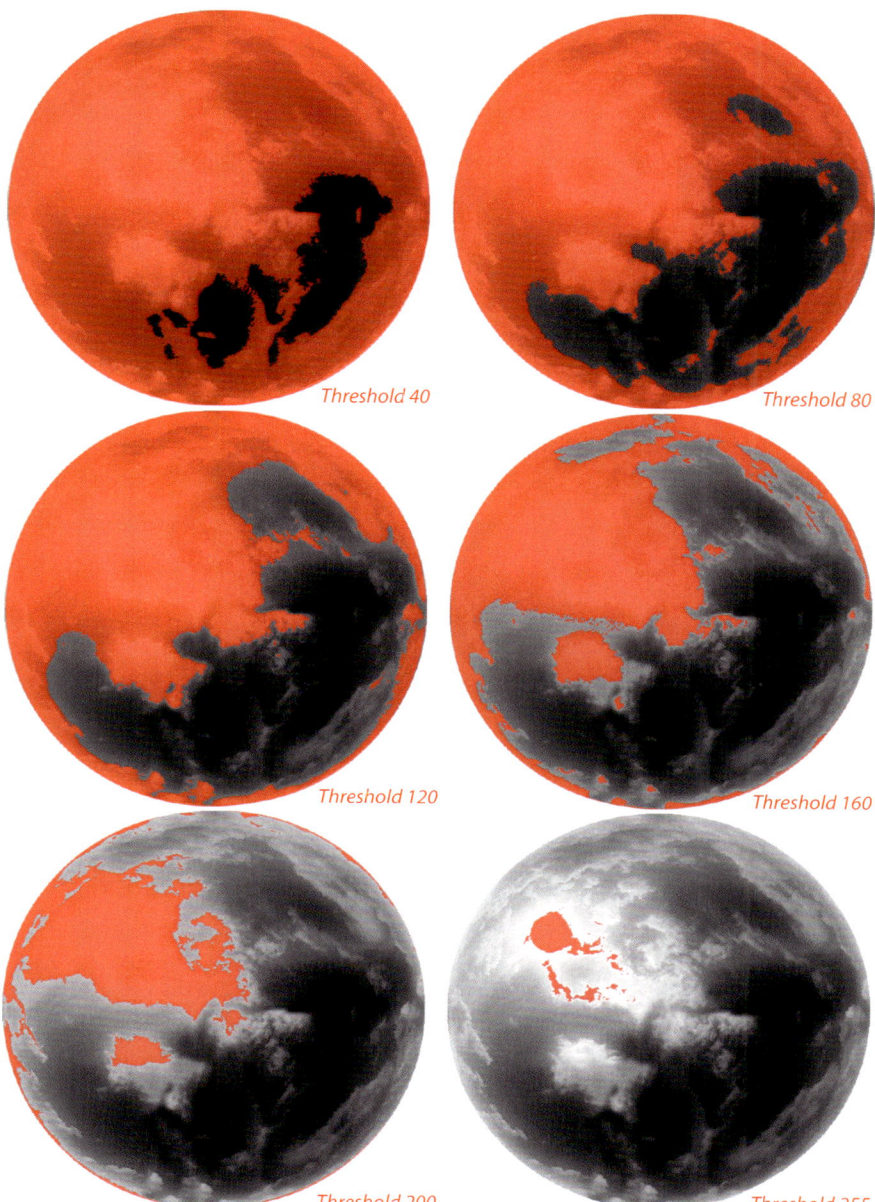

Threshold 40

Threshold 80

Threshold 120

Threshold 160

Threshold 200

Threshold 255

◘ Fig. 4.22 Image segmentation of an example IR sky image, using a binary conversion of gray scales with different fixed global thresholds, which allow the detection of clear sky (area outside the segmentation in the *upper left image*), different coverage situations, and the position of the sun (the matched area in the *lower right image*)

This global thresholding is only effective if the two detectable objects (clouds and sky) can be easily separated in the histogram, because of its bimodal distribution.[44] For clouds and sky, this is not always easily possible, because the gray-scale distribution depends on the camera type and resolution, the time of day, when the picture was taken, the general brightness, the inhomogeneity of the illumination over the whole image, as the sun illuminates the sky only from one side, and so on. Therefore, for example, dark areas in dark clouds can result in the same gray level as the dark areas of the blue sky in images in ordinary gray-scale images taken in the visible part of the light spectrum. Therefore, it is important to adapt the threshold to the different situations, so that the classification requires additional information and adequate

Camera Selection

---

44 A bimodal distribution shows a frequency distribution which has two accumulations of gray levels (showing two significant extrema), so that they can be split easily with a threshold between these accumulations.

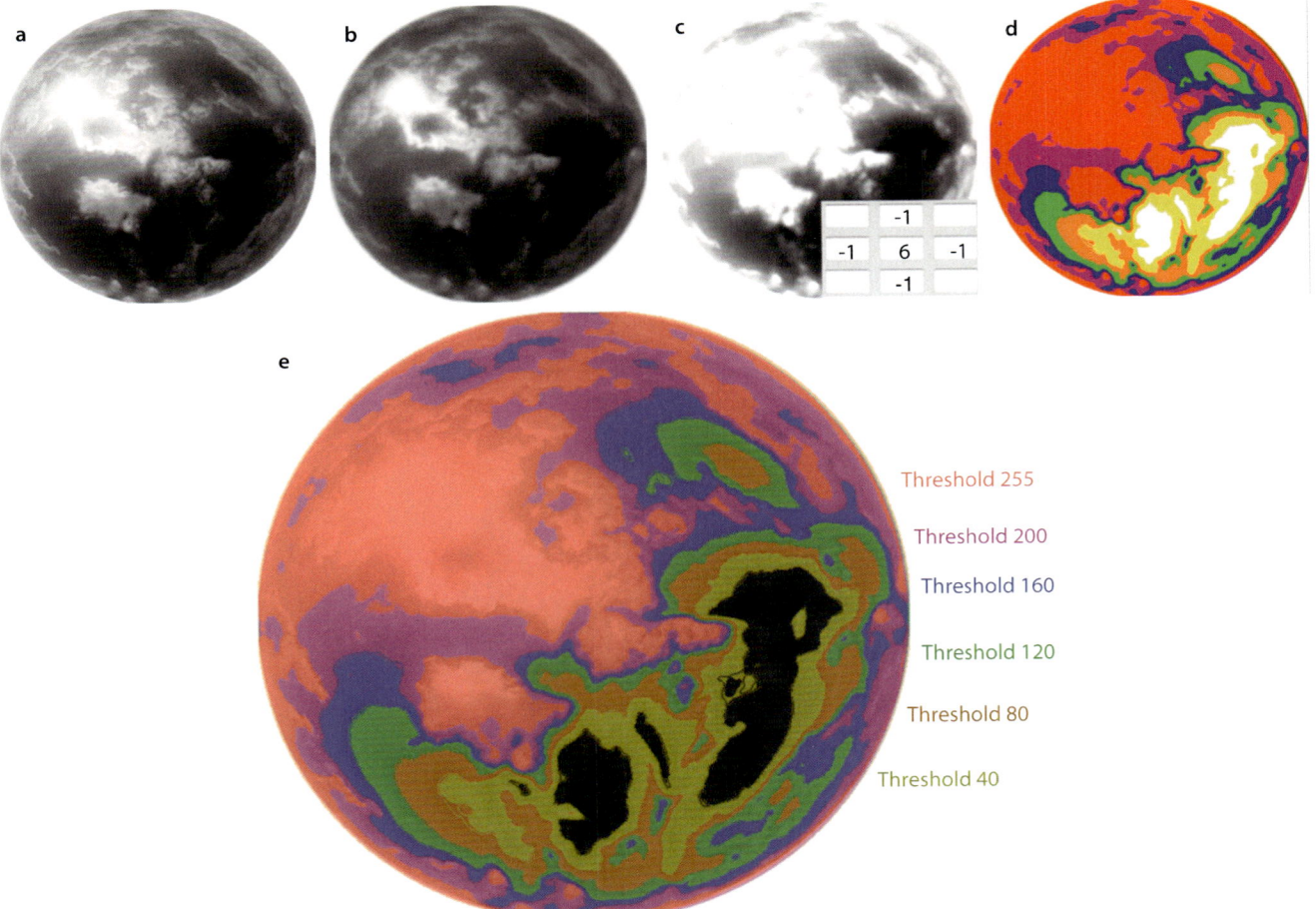

Fig. 4.23 Image segmentation (result image in **e** of an example sky image **a**, using a filter queue of a Gaussian blur filter **b**, an individual morphologic filter mask **c**, binarizations with different fixed thresholds **d** colorize different threshold levels with different colors **e** shows the negated binary image with all colorized thresholds

Processing Chain

classifiers. IR images which show temperatures can offer better results, because clouds are warmer than clear sky, so clouds appear brighter. This means that the selection of the camera influences the following processing chain.

While image taking with a camera is the first step in the processing, further processing steps are usually sequentialized to improve images and do the final segmentation. The low-level conditioning can be done with local operators or filters, working on the bit-mapped graphics.[45] Usually the first step is a normalization of brightness and contrast using the mean and variance values of the image (Niemann 1990, *l.c. page 112*). Ideally, the normalization is done using reference images and parameters for the current light situation of the sky according to the mood of the light for a specific day in the year and time of day. This partly avoids erroneous adaptions of the original image, which may lead to segmentations of clear sky in a completely cloudy sky image, because of wrongly normalized gray scales. The normalization is followed by different filters to improve the image according to the segmentation or classification method used. A sample for gray-scale sky images is shown in ▣ Fig. 4.23, where a Gaussian blur filter is used as a first step to smooth the contours (Kopp 1997, *l.c.*

---

45 Filters are local operators on regions of an image which calculate a new value for a dedicated pixel by using the sum of weighted values in a mask around that pixel. The mask is moved over the whole image until each value is calculated newly. Very well-known filters are low-pass (reduces the high frequent parts and results in a blurring effect) or high-pass (enforces the high frequent parts, so that the fast gradients and, therefore, the edges become visible) filters (Kopp 1997, *l.c. page 92 ff.*).

page 99 f.) in combination with another, specific, individual morphologic filter to whiten the bright cloud regions. The filters are followed by thresholding with different values to classify the regions.[46] The final result of this sample processing chain is shown in part **e** of ◘ Fig. 4.23, where the negated results of the different segmented binary masks were superimposed on a final cloud detection image. The red and violet areas represent cloudy regions, where no satellite tracking is possible. The blue, green, and orange areas represent light misty areas, where low and medium orbit satellites and only partly high orbit satellites can be tracked. Finally, the yellow and remaining parts represent free sky, where all satellites could be observed.

If a bimodal distribution is not clear for the whole picture (see ◘ Fig. 4.24a), a simple technique enables the use of the thresholding method anyway: a local threshold for a reduced mask which is shifted over the image. A derived threshold for the whole image can be modified by a weighted, local mean value, or a completely separate threshold can be calculated. The principle is shown in ◘ Fig. 4.24 with the already normalized and filtered sample sky image. A local bimodality check offers good results for local thresholds (see ◘ Fig. 4.24b). Together with a segmentation criterion for blue sky (mainly low gray levels in the mask, like in ◘ Fig. 4.24c) and

Local Thresholding

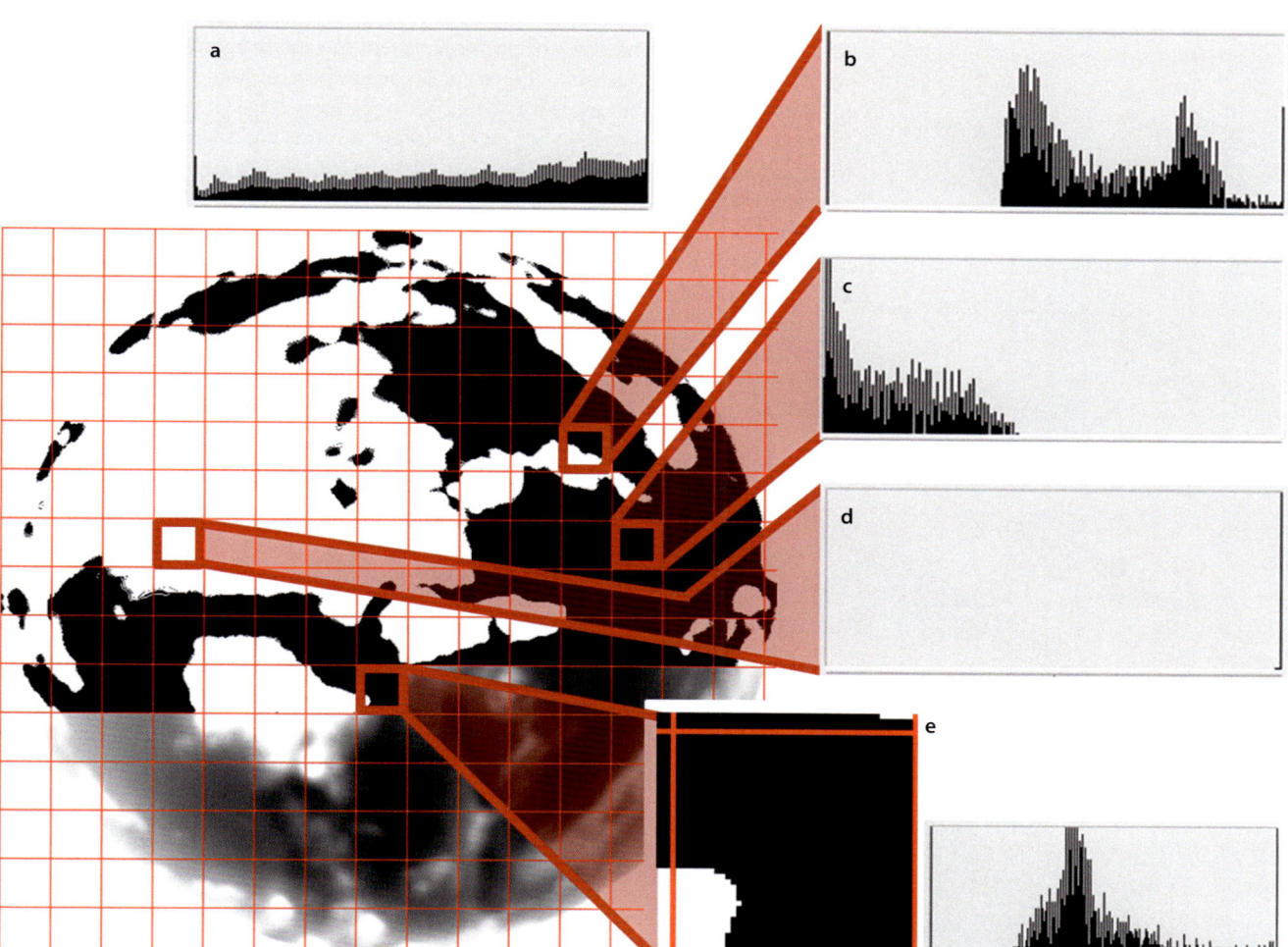

◘ **Fig. 4.24** Image segmentation of a sky image using a binary conversion with local thresholding (instead of a global; **a** with a bimodal distribution check **b**), a histogram analysis of clear sky based on dark regions **c** or of covered sky based on bright regions **d**, and a connecting boundary criterion at the edges of the local thresholding mask **e**

---

46 The already mentioned fuzzy logic can be helpful to adapt more variable threshold values for specific day and image situations.

the criterion for clouds (mainly high gray levels or pure white, like in ◘ Fig. 4.24d), local thresholds can be derived, which offer a better segmentation than the global thresholding (Kopp 1997, *l.c. page 110 ff.*). But as local thresholds just recognize the local area within the mask, the final binarization of neighboring regions can jump in scale. Therefore, a connecting boundary criterion is required (see ◘ Fig. 4.24e), which optimizes the threshold so that the outermost pixel edge of a currently binarized mask fits to the outermost pixel edge of an already converted neighbor area. This can easily be done after a local threshold is derived, for example, as a mean value of the local mask. If the trend in the mask is more toward darkness, the threshold value can be shifted toward the low-levels, until more than 90–95% of the boundary edge pixel of the current mask has the same binary value for black as the previous one. If the trend in the mask is more toward brightness, a similar approach can be performed for the higher levels. Other algorithms can try to shift in both directions to find the optimum. The combination of different segmentation criteria offers a very good performance for most of the different sample images.

Binary Protection Mask and Observation Planning

The segmented binary image selects similar areas. The inverted image directly shows regions where clouds prevent observation (see also ◘ Fig. 4.23). Other regions where no observation is possible or allowed are given by the horizon mask (defined by azimuth direction specific, minimum elevation values), the sun avoidance zone (a circular area around the sun position, where too much received signal strength would destroy the detectors), elliptic masks around the airplanes (protection zones for laser hazards to humans in airplanes, which become larger if the planes are closer to the laser ranging system in higher elevations or lower flight heights and which are more extended into the flight direction of the plane, according to the speed of the aircraft), and restriction zones from the satellite operators (defined by go/no-go flags, restricted orbit segments, or minimum/maximum elevation restrictions). All zones together show a binary mask where the observation is protected or impossible (see ◘ Fig. 4.25). As the mask is defined in the image coordinate system, which describes a section of the monitor (viewport), it is necessary to transform it into the real world coordinates which represent the same area in the real world image (window). With

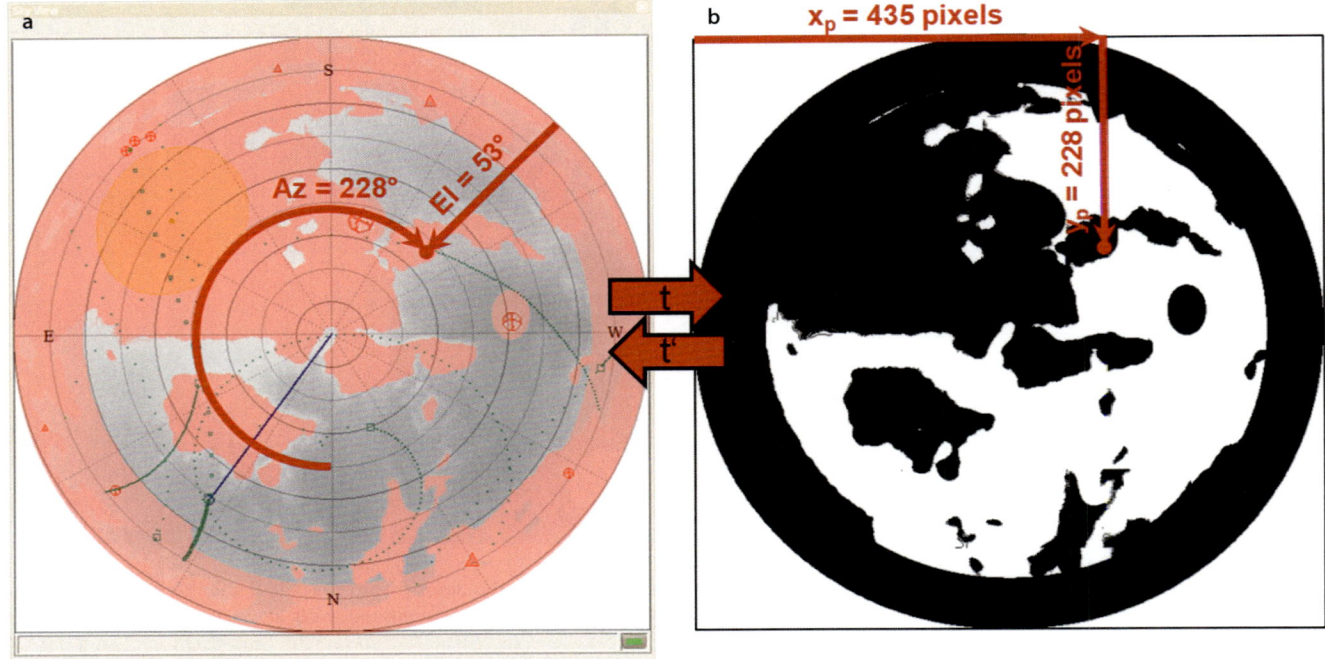

◘ **Fig. 4.25** The sky plot with all the necessary information about areas which cannot be observed **a** and the resulting, binary protection mask, including the sun position, the cloud coverage mask from a local thresholding processing, the horizon mask, and the airplane masks **b**. Pixel positions of the binary mask can be transformed into position angles for observations and vice versa, using the window/viewport transformation

the window/viewport transformation and its inverse, it is possible to map both coordinate systems and to convert pixel positions of the image into azimuth and elevation positions of the passages and vice versa (compare also (Zavodnik and kopp 1995, *l.c. page 23 f.*)). Therefore, it is quite simple to check if a specific position is currently observable. The specific topocentric coordinate must only be transformed into the image coordinate system, where the binary value defines whether the position can be observed or not. This information can be used for planning an observation schedule for satellite passages in an autonomous coordination cell.

But image processing is not only powerful for the central planning of observations of satellite passages, the segmentation and classification of objects in images is also suitable for the observation of the mount model corrections. As shown in (Ettl 2006), it is possible to support a workflow to derive mount model correction parameters with image processing techniques. Usually several different star positions which are distributed over the whole visible sky are observed. The human operator uses a sight device which consists of a small hair cross, which is projected into a highly sensitive camera through the beam path. The hair cross position defines the actual pointing direction. If the telescope points to a star, the hair cross can be used to find the offsets between the internal pointing direction and the real position of the star. The direction is changed in small steps, until the hair cross is ideally in the center of the star projection. Here it is necessary for the operator to compensate for the permanent jittering of the star position given by the atmospheric fluctuations (seeing effect). The offsets derived are then used for a fitting calculation, which results in the different fitting parameters for the different mounting corrections according to a given mount model. This process of finding the star and hair cross positions can be managed by image processing. First of all, the image of the hair cross must be segmented and classified (see ◼ Fig. 4.26a taken from (Ettl 2006, *l.c. page 24*)). Knowing the orientation of the orthogonal axes of the hair cross makes it possible to calculate the pixel position of the intersection. This position corresponds via a predefined, heuristically derived window/viewport transformation with the real world offsets in arc seconds. After determining the position of the hair cross, the different star positions can be pointed. After a normalization and segmentation of the star image, it is possible to calculate the graphical center of the star area (see ◼ Fig. 4.26b taken from (Ettl 2006, *l.c. page 43*)). To compensate the seeing effect mentioned before, it is necessary to accumulate different images taken over a specific time interval,[47] or to increase

Automated Observation of Mount Model Corrections

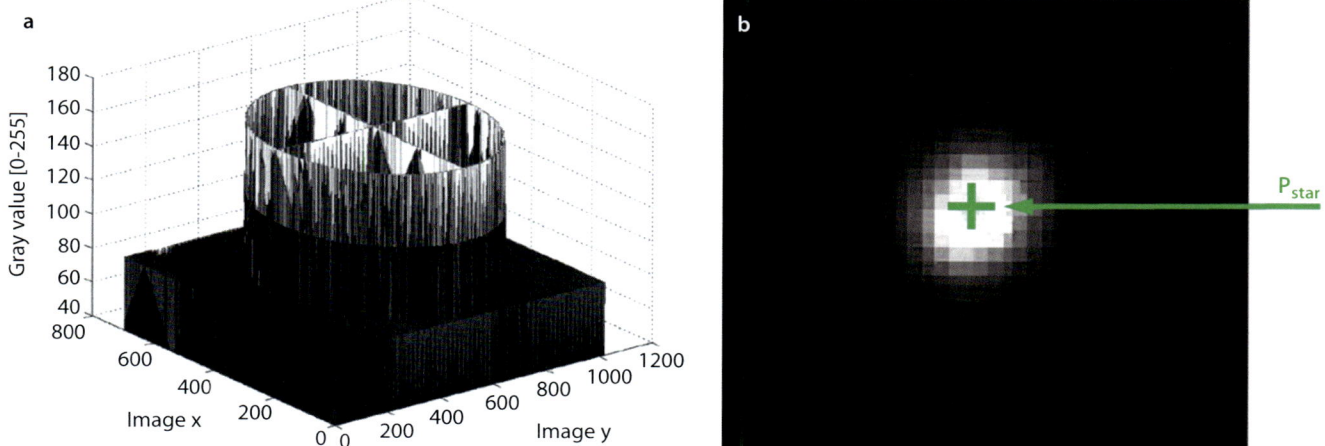

◼ **Fig. 4.26** Classification of a hair cross projection in an image, taken via the telescope beam path **a** and of the graphical center of area detection of a star image **b**, both needed for an automated detection of pointing errors during a mount model observation (Taken from Ettl (2006) *l.c. page 24*)

47 Hobby astronomers use a cheap but effective method to improve their images: lucky imaging. It is an image stabilization by selecting and combining a number of optimal single images according to a mathematical criteria using computer-based image processing.

the exposure time. Having the star position and the hair cross position makes it possible to calculate the offsets in pixels, which can be transformed into real world offsets with the specific window/viewport transformation.

Image Processing Tools and Libraries

The wide field of image processing offers many possibilities. In this age of machines which are able to see and classify things in a similar way to humans, image processing is quite essential. Several available programs, tools, and libraries offer support for normalization, filtering, segmentation, and classification of objects in images. Projects at the Geodetic Observatory Wettzell made good progress with the licensed software HALCON from the company MVTec (see ▶ http://www.mvtec.com/halcon/). It offers a broad field of matching techniques for different applications, such as remote sensing, positioning, security, and so on.

Autonomous Functionalities: Planning Analogies to Operating Systems

Using some of the metric factors, it is possible to rate one resulting plan against others. The goal is an optimal assignment of observation time to the different satellites for one telescope system. This describes an analogous situation to processor assignment strategies in operating systems. Different applications or orders (the processes in operating systems analogous to satellite passages in laser ranging systems) compete for an operator station (the CPU in computers analog to the telescope, optics, and detection system in the laser ranging). Only one application can be processed per time unit, which is the cycle time Δt of a processor. The rest of the applications wait in something like a waiting queue for future processing time. The planning system (the scheduler in operating systems analog to the coordination cell in the laser ranging system) must try to comply with the given boundary conditions and requirements (compare (Hofmann 1991, *l.c. page 107 f.*) and ▶ Sect. 2.6.4) and must prevent orphans (computer processes which never get processing time). Equivalent to that scenario of an operating system for computers, it is possible to define a minimum observing time Δt (e.g., three minutes are suitable as a heuristic, empirical value), which defines the cycle time of observations. The other parallel observable passages must wait during this time interval for future observation time, and a coordinating scheduler must try to avoid passages which are starving. Even some of the evaluation factors found can have analogies in the processor assignment strategies. For example, the satellite priority is equivalent to the process priority. Interrupts can be implemented in a similar way. The total length of the passage is similar to the required processor time for the whole process. The time already spent waiting for an observation can be compared to the time a process took in the waiting queue. The observation time already spent is equivalent to the processor time of a process. Fortunately, the arrival times of observations are known in advance. But unfortunately, environmental conditions like the meteorological situation or the sky coverage are restricting factors. These are comparable to asynchronous, human interactions with programs in a computer system, which force prioritizing of specific tasks compared to others. Differences in the analogy are that the complete passage does not have to be observed, that the passage rise and set times define fixed deadlines, and that the success of the observation is not defined by the fact that the passage was observed at all, but that there were useful returns as hits.

Autonomous Functionalities: Planning Assignment Strategies

Knowing about these analogies, a good approach for a planning strategy is to look for techniques already used in operating systems (see also ▶ Sect. 2.6.4). Different operational or statistical, analytic approaches can be found, which can be adapted to satellite observing systems with the restrictions of the passage time constraints. A simple strategy is «first come, first served», where the first passage found will be observed until it is successful. This strategy is not so useful, as it has a strong tendency to starvation of single passages. A similar strategy is that of «shortest job first», even though it favors the short passages where only a few minutes are possible to get enough ranging hits. This means that preemptive or time-slicing techniques are required, as «preemptive shortest job first» (or «weighted shortest job first) or «round robin». At each start of a new observation of a passage, the satellite with the shortest remaining observation time is selected. This favors long-lasting, high orbit passages to be mostly observed at the end of their transit. Therefore, a round robin technique (see also ▶ Sect. 2.6.4) is a good choice, splitting the possible observation time into time slices. These time intervals are assigned to the passages which are observable at this time (compare also (Hofmann 1991, *l.c. page 130 f.*)). While in the usual operating systems, it is desirable to split the processor time uniformly between the competing processes and the priority of the processes usually plays a minor role, satellite observation prioritizes satellites much more, so that the selection for the next assignment is mainly dependent on the priority. These priorities can differ along the passage time and should be dynamically changed during the observation and according to the time already spent on observation (compare (Tanenbaum 1995, *l.c. page 80 ff.*)). This avoids starvation of the other observations with lower priorities. The environmental parameters are additional restricting factors. Therefore, it is necessary to evaluate the situation for each time slice separately according to these prioritizing parameters.

Different more or less complex planning strategies are conceivable, which are all based on time slicing, but consider dynamic parameters differently:

Autonomous Functionalities: Planning Strategies

1. *Static a priori planning*: This planning is comparable to the cooperative, non-preemptive scheduling in an operating system. Different variants of plans are generated for the following time period (up to maybe a few hours). Depending on the evaluation scheme, the best variant is chosen for the following time period. This plan is then static for this period, and each satellite returns the observation control cooperatively back to the scheduling coordination cell at the predefined return time, or the coordination cell takes the control back preemptively. An example of such static scheduling is shown in ◘ Fig. 4.27, where the common satellite table in the HTML format is used to indicate the schedule plan for the coming eight hours. A disadvantage of this method is that it cannot include or react to faster changing metric parameters, such as meteorological conditions. Nevertheless, a big advantage is its simplicity.

2. *Dynamic a priori planning*: This planning is more or less comparable to the preemptive scheduling in an operating system. As with static planning, a schedule is chosen according to the metric factors for the next time period. But in short, regular time intervals (for the laser ranging an empiric time interval might be three minutes), the situation is evaluated again. Changing parameters can then change the evaluation result, so that the schedule is updated. A current observation will be revoked preemptively, and the coordination cell selects the highest rated passage for the next observation for the following tie period.

3. *Dynamic real-time planning*: The dynamic real-time planning is more or less event or interrupt driven and can be compared to a scheduling in a real-time operating system. The scheduler continuously checks the different conditions and event states in a loop, according to deadlines or driving metric factors. If a changed condition leads to a rearrangement of the current processing hierarchy of the possible passages, it revokes the current observed passage and hands over to the currently better rated possibility. The evaluation of the framework requirements is important here, as they are the general drivers for the system. Examples of these are the minimal observation time per passage or the interleaving time to change between two satellites, so that the telescope does not permanently move between different satellites if priorities permanently toggle. The algorithms to make decisions are very similar to the previous planning strategies, with the difference that they are performed

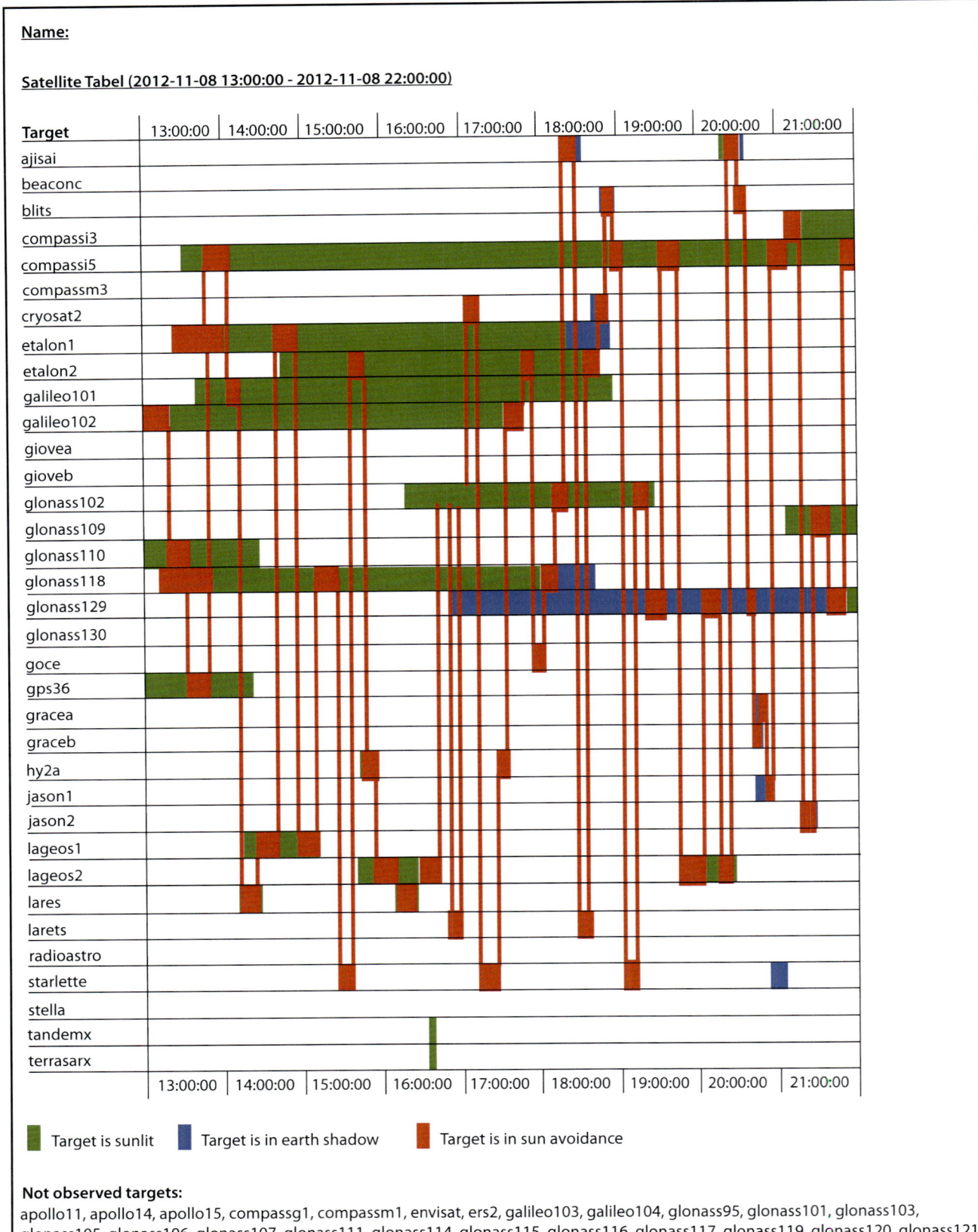

Fig. 4.27 The satellite passage table in the current HTML version and an indicated scheduling plan, as a result of a static a-priori-planning

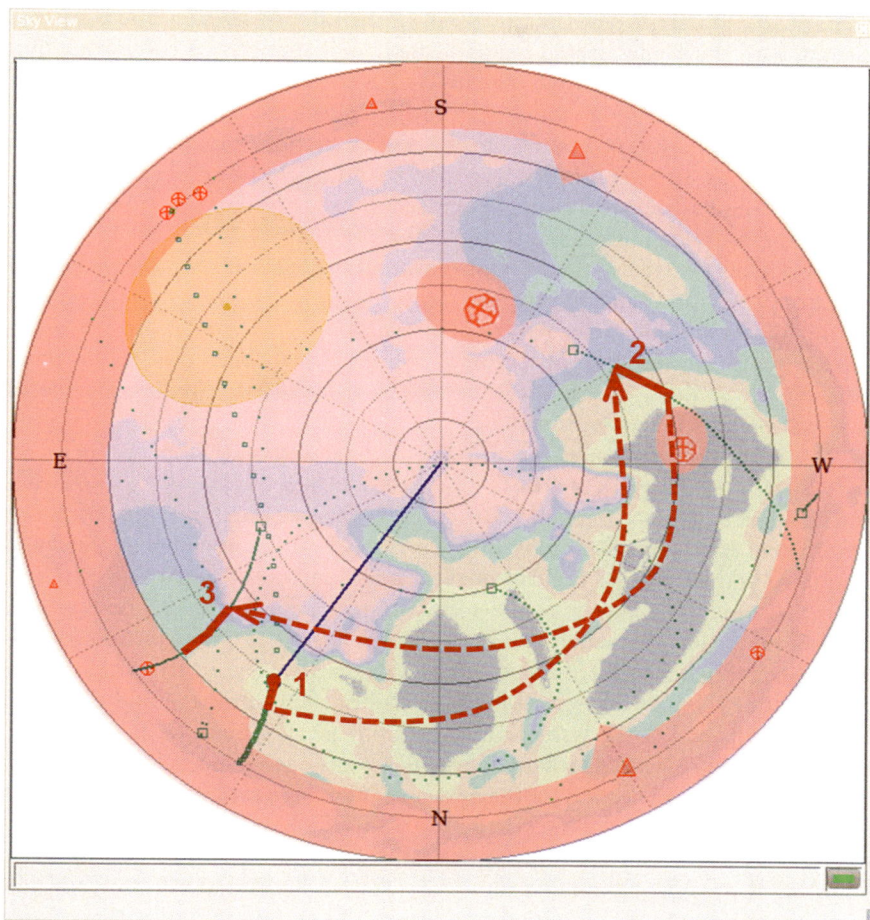

Fig. 4.28 A sky plot with an indicated cloud detection mask, the horizon mask, the airplane protection areas, and the possibly resulting, dynamic real-time planning of satellite observations

permanently and that, therefore, plans change frequently according to real-time parameters. The current decision, based on the current situation, cannot always be the best if a complete time period is regarded. Weighting and controlling parameters can improve this situation. Real-time planning is still the most flexible technique for an agile, autonomous system, because it can immediately react to changing conditions. An example real-time scheduling scenario for a following time period is shown in ◘ Fig. 4.28, which includes a current passages situation, the meteorological, in-sky and on-ground safety restrictions.

As already noted, all strategies must deal with the prioritization which is a result of the evaluation of static and dynamic priorities to select the optimal passage which should be observed during the next time slot Δt. It is derived from the evaluation of the metric factors. Each factor must be represented as a count or measure which can be weighted (the representation as a function set of fuzzy logic is also possible). The weights are then the «adjustable screws» to optimize the parametrization of observation plans. The sum of all weighted counts defines the priority level used for the selection. Analogous to autonomous production cells in factories, it is possible to use negotiation mechanisms to come to a suitable selection. Here the different orders (in the case of a laser ranging system, the passages) compete against each other with their calculated priorities, bidding for observation times with their weighted evaluation factors. These mechanisms are known as the auction (Pfeifer and Schmitt 2006,

Autonomous Functionalities: Planning
Priority Evaluation

l.c. page 30). Different types can be used. In the English auction, the winner is the one with the highest-rated factors. The Dutch auction is a reverse auction and starts with a specific, high starting value, which is reduced by negatively rated values until the first bidder with fitting factors is found. Others are then ignored even if they have a better optimum. The auctions can have one or more bidding rounds. In satellite planning, one round might be enough, so that the first-price sealed-bid auction might be used. It uses no subsequent rounds and selects the most fitting bidder within one round (Woolcridge 2009, *l.c. page 295 f.*). This is possible if the weighting factors are optimized to the real world selection and optimization process. Only the winning passage is processed during the next time slice. After each time slice, the evaluation process is started again for the next observation interval. If another process becomes more highly ranked, the observation time is revoked and the newly selected observation gets time for processing.

To find an optimal planning represented by the weighting factors, it is necessary to train the system. One possibility here is to find out about the knowledge and habits of the human operators with a questionnaire checking human decisions for different scheduling situations. The planning model can then be weighted and compared to the statistical evaluation of the questionnaire. It might also be necessary to add additional metric parameters if human operators use different criteria. While doing this optimization, it is possible to adjust the planning process to get a mean weighting behavior. It is also suitable to consider not only the behavior for the next time slice but to also rate the situation for a longer time period according to good utilization of the system time, most observation times on highly prioritized satellites, or a good distribution over all the passages. The result is a two-stage method, based on a selection for the next time interval in combination with a general evaluation of the resulting scheduling plan for a longer time period of a few hours. It is possible to create different variants of a plan for the next few hours by adjusting the weighting parameters within defined limits and selecting the best solution according to the utilization parameters. This optimizes in an n-dimensional parameter space to a local or even absolute optimum. A post-evaluation, according to hit statistics or ranging quality parameters, is critical, as these depend on highly unpredictable, indescribable, and hard-to-measure, external parameters. Therefore, it could be easy to adjust wrongly. Using tangible parameters for the adjustment is much better.

Autonomous Functionalities: Planning Adjustment

Autonomous Functionalities: Intelligent Planning Systems

Adjusting the planning parametrization is something like training the system. For this, intelligent, learning, and trainable algorithms like neuronal nets or genetic algorithms are predestined. Artificial neuronal nets simulate the structures and, therefore, imitate the behavior of the human brain. A huge number of simple processing elements (neurons) are connected via a huge number of synapses to code and learn internal optimization knowledge for a real-time reaction. Simple neurons add up the values of the input synapses and create one output, usually via a nonlinear activation function (a threshold or sigmoid function (S-function)). The output is again connected to the next level of neurons and so on. It is not a simple approach to select the suitable topology, layers, weighting, activation functions, learning rules, feedback structures, and so on, to define the properties of the net, the neurons, and the system dynamic for a suitable application (Langmann 2010, *l.c. page 491 ff.*). In the field of the optimization of passage schedules, genetic algorithms might be more suitable. As already mentioned, one way is to create different versions of a plan, which is optimized according to the utilization parameters. The next optimization uses these settings and manipulates the parameters. This is precisely what is described by the theory of artificial evolution. Each selection step creates a new set of possible solutions (a new population of new «individuals»). They code different sets of optimization parameters defined by the weights (the chromosomes). One of the populations is then selected according to «fitness» criteria. The next selection process uses the previous chromosomes and performs a modification in the form of a hybridization (change of parameters according to a predefined rule) or a mutation (random adaption of parameters) (Langmann 2010, *l.c. page 495 f.*). This mutation and selection process can be continued as a trial-and-error process, until the most fitting solution is found related to a suitable stop criterion, and it can be continued throughout the operation of the coordination cell.

The planning ends in a sequence of RPC calls to the subordinated control cells. These calls can be hard coded. But using the command line functionality from ▸ Sect. 3.4 with the extension from this section, it is possible to describe the whole observation schedule with detailed ASCII orders. These commands can even be assigned to timestamps, defining when the individual RPC call must be processed. Then they are interpreted and processed in the periodic controlling loop of the coordination cell at the defined times. Such a text-based, but hard coded command interpretation technique is already used for the processing of VLBI schedules in the Standard Notation for Astronomical Procedures (SNAP), which are used by the NASA Field System.[48] The SNAP commands are used to control and monitor software and hardware states. They are command line instructions which are interpreted by the NASA Field System. Basically, two types can be used: real commands and time-flow statements. While commands control the hardware, the time-flow statements control the timing of commands during a schedule plan (with a resolution of seconds). This is important to trigger activities on specific timestamps, to sequence delayed events, and to organize a time list for periodic executions. Each SNAP command can be a single, programmed action, which is started after the interpretation of the command. It can also be a sequence of such actions grouped into SNAP procedures. These procedures are not interruptible and are quasi-«atomic» for the control system (Field System Team 1997b, *l.c. page ARCH-2 f.*). The code Example 4.7 shows an excerpt of a sample SNAP file for an VLBI-observation of the INTENSIVE-type observation on January 2, 2012. Single commands without equal signs are calls of procedures without arguments (like «preob», which calls the procedure with a sequence of checking commands required before each observation). Commands with equal signs are procedure calls, to which the following argument list separated by commas is handed over (e.g., «disk_record=on», which calls the command «disk_record» to hand over the argument «off» or «on», which starts or stops the recording of data). Commands with a question mark after the equal sign request status information or return values, which are then printed to a log output. Commands which should be directly sent to and processed at a subordinated device use the name of the device in front of the command, separated by an additional equal sign (e.g., «mk5=bank_set=b», which changes the hard drive bank at the data recording system «Mark 5» to use the second bank «b» of hard drives). Commands can be triggered with timestamps of the format «!2013.002.18:29:50»,[49] which defines the year, day of the year, hour, minute, and second) or can only be arranged in sequence, so that the next command is always processed when the previous one has been completed (see Field System Team 1997b, *l.c. page SUBR-2 ff.*).

Autonomous Functionalities: Controlling

---

48 The commands from the field system are programmed manually and not generated.
49 A higher time resolution can be optimal for laser ranging to offer a better granularity for RPC calls and the related generated text commands.

■ ■ **Example 4.7**

First sequence of a VLBI SNAP file for the 20 m radio telescope Wettzell with the ASCII description of the VLBI schedule for an INTENSIVE observation on January 2nd, 2013.

```
"  I13002      2013  WETTZELL  V  Wz                                           1
"  V  WETTZELL  AZEL    .0000  180.0     0    251.5    831.0    90.0    0    2.0    89.0  20.0    2
Wz  33                                                                        3
"  Wz  WETTZELL   4075539.89941    931735.27025    4801629.35185  72247801     4
"  33      WETTZELL  Mark5A                                                    5
"  drudg  version  2008Oct08  compiled  under  FS   9.10.04                    6
"  Rack=Mark4        Recorder  1=Mark5A       Recorder  2=none                 7
scan_name=002−1830,i13002,Wz,134,134                                          8
source=1357+769,135755.37,764321.1,2000.0,ccw                                9
ready_disk                                                                    10
setupsx                                                                       11
!2013.002.18:29:50                                                           12
preob                                                                        13
!2013.002.18:30:00                                                           14
disk_pos                                                                     15
disk_record=on                                                              16
disk_record                                                                 17
data_valid=on                                                               18
midob                                                                       19
!2013.002.18:32:14                                                           20
data_valid=off                                                             21
disk_record=off                                                           22
disk_pos                                                                     23
postob                                                                       24
scan_name=002−1833,i13002,Wz,64,64                                          25
source=OJ287,085448.87,200630.6,2000.0,ccw                                 26
checkmk5                                                                     27
setupsx                                                                     28
!2013.002.18:33:05                                                           29
...                                                                          30
```

Autonomous Functionalities: Controlling Tasks

A similar text schedule as for VLBI can also be achieved using the automatically generated command line functionality of the middleware. Then the coordination cell can be split into two modes: a manual mode where the operator controls all parameters manually within the limitations of the autonomous hardware sub-cells and an autonomous mode where the system takes control mainly with periodically generated schedules, for example, in text form. The tasks of this autonomous controlling mode can be split into different sections for each passage observation (with some analogies to the VLBI approach):

1. *Setup (e.g., «setupsx» in the VLBI schedule)*: The setup task requests the passage data with the orbit supporting points from the database[50] and hands it over to the different sub-cells. It sets all starting values for the transmit-receive unit and selects the detector (e.g., APD or MCP). It also sets the correction parameters, such as the clock delays or meteorological refraction corrections. Another very important task of the setup is to request the configuration parameter or files from the central data management cell and forward them to

---

50 Experience has shown that not all orbit points of a complete passage should be transferred during the setup to reduce the transfer and processing times and, therefore, the startup times. A reduction is possible, because usually an observation of one satellite is limited in time if an observation interleaving between different satellites is used. While usually short passages of LEO satellites can be transferred completely, a maximum passage section of one hour for HEO satellites and of three hours for moon orbits are suitable.

all sub-cells after a parameter change or a restart of the system. This ensures that the complete description of the configuration is always available in the central data storage and that all devices run with these configurations and settings. Configuration errors are directly alerted via the GUI to the operator.

2. *Pre-observation («preob» in the VLBI schedule)*: The pre-observation controls and monitors all devices via the sub-cells if they are ready and if all states and parameters are correct (call of the RPC status functions which all cells must implement). If the system uses a pre-calibration, it is also operated here. Then it starts the tracking with a start function call to all subordinated cells.

3. *Observation («midob» in the VLBI schedule)*: After the pre-observation, all sub-cells and the connected devices start with the tracking autonomously. This means that, for example, the telescope calculates the intermediate supporting points of the orbit in azimuth and elevation in the required time resolution and points to the satellite. It automatically follows the passage. All other devices also start with their local control. The event timer starts with the timing and ranging according to the set time/range table. It is also responsible for hit detection. The coordination cell then continuously requests the status information from all sub-cells to check their correctness and operability. If there are alerts or errors, it immediately propagates these to all connected devices, stops the observation, and returns with a specified notification for the user. If critical alerts happen, the schedule is blocked until a human interaction occurs. During the observation, the coordination cell also requests the return, residual, and hit statistics from the event timer to offer it for plotting in the graphical user interface and to record the observation data. Using these statistics, it adapts the parameters of the transmit-receive unit and controls the satellite search loop. This loop controls the long- and cross-paddle settings, which set pointing offsets in arcseconds at the telescope. It has been shown by the different operators that an ideal search algorithm starts at the predicted satellite position and changes the offsets with a bigger step size in the form of a spiral around this position, until the first hit sequence is detected (see ▢ Fig. 4.29). Then it is suitable to reduce the step size and move back and forth along the previous track, while checking the return statistics. If the peak is found, a search loop with small step size around this position can be started. During this search, it

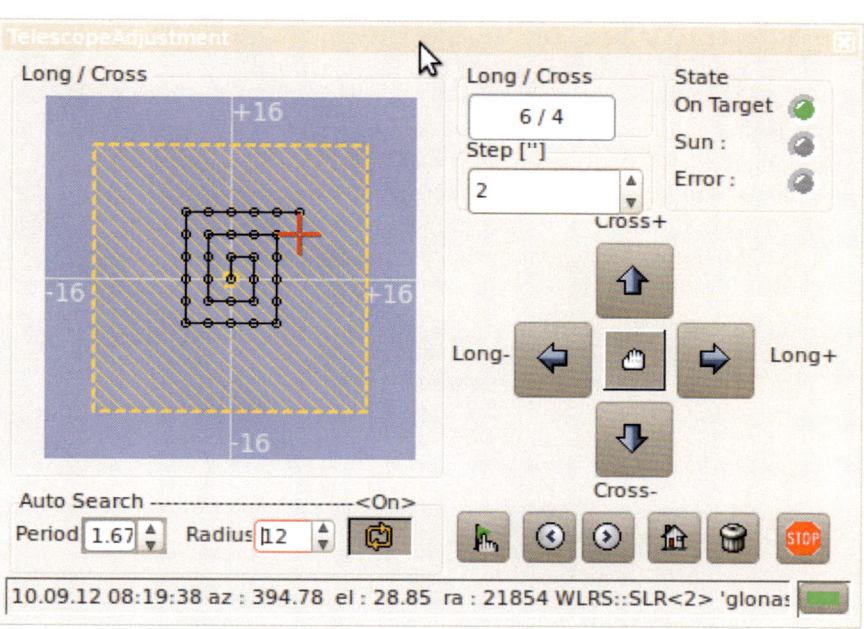

▢ **Fig. 4.29** A possible search spiral in the telescope control window to find the actual satellite position around the predicted orbit position

is necessary to include the current cloud coverage conditions along the orbit path, as the return statistics are forged by the blockages of returns by clouds in the beam. After finding the satellite and while hits are continuously detected, the coordination cell just checks the quality and performs some small adaptations to the position offsets or to the optics. If the hits are lost, the search loop must start again, preferably around the position which has already been found. If no new satellite contact can be found within a specific time period, the search spiral is reset and starts again at the predicted orbit position for the current time. Here it is also possible to use a genetic algorithm or even better a very flexibly adaptable neuronal net to adapt to and to train the best behavior for the search, using different real-time feedback parameters. If the system uses real-time calibration, it is processed by the event timer during this section, and the results are handed to the recording system. In the a priori scheduling, the observation is interrupted at the scheduled time or if an operator switches to manual mode. In real-time scheduling, the observation can be interrupted at any time if the planning decides to switch to another satellite because of changed parameter priorities. Then the post-observation is started.

4. *Post-observation («postob» in the VLBI schedule)*: The post-observation task stops the tracking at all subordinated cells, performs a final quality and system check, and saves the data for further on-site analysis. All observation data files are copied to the data management cell if the passage is completely over. If the satellite is still visible and only part of the passage has been observed, the observation file is kept locally at the control cell, so that the next observation sections can be appended. After the setting of the satellite, the recorded data file is then copied to the data management cell. If the system uses pre- and post-calibration, an additional calibration is measured here and added to the observation data before it is sent to the data management cell. If the observation schedule uses a fast interleaving or satellite hopping, the post-calibration can be skipped because the pre-calibration of a following satellite can be used as post-calibration of the previous satellite.

5. *On-site-analysis*: On-site-analysis can be processed in parallel with the observation. To reduce mutual interference between the analysis and the observation process, the analysis is processed on the hardware of the data management cell. It calculates the NP and prepares everything for transfer to the data centers. Then the analyzed passage is listed, to be manually checked by the operator if the quality of the NP is given. This manual interaction should remain as a manual task, so that the final instance is the operator. He can request different reports about the analysis and also about the logged scheduling information. If he accepts the analysis results, it is marked for sending, which is periodically processed by the data management cell. The observed and analyzed passage is then added to the weekly report (see ◘ Fig. 4.30). If the operator does not accept the analysis results, the resulting data file is hosted in a quarantine folder for further manual analyses.

Weekly pass report from 01.11.12 to 07.11.12 for 8834

Nr	Satellite	Detector	Rms	Returns	Note
1	stella_8834_20121101115159	MCP 1	11.00	810	Good/ToSend
2	etalon1_8834_20121101121940	APD 2	43.90	1738	Good/ToSend
3	glonass118_8834_20121101120705	APD 2	25.30	1802	Good/ToSend
4	hy2a_8834_20121101163208	MCP 1	10	1789	Good/ToSend
5	terrasarx_8834_20121101165657	MCP 1	8.10	954	Good/ToSend
6	tandemx_8834_20121101165657	MCP 1	9.00	565	Good/ToSend
7	larets_8834_20121101172019	MCP 1	11.10	626	Good/ToSend
8	lageos2_8834_20121101165151	MCP 1	12.90	151	Good/ToSend
9	cryosat2_8834_20121101175923	MCP 1	8.70	76	Good/ToSend
10	galileo101_8834_20121101120023	APD 2	21.60	4504	Good/ToSend
11	etalon2_8834_20121101133155	APD 2	41.80	4980	Good/ToSend

◘ **Fig. 4.30** First part of the weekly report with all successful and sent passage observations

The handling of the other main tasks (pure calibration and observation of the mount model corrections) is similar to the sections described. It just includes and uses different sub-cells with different, adapted parameters and status information for the specific tasks. For example, for the mount model, it is necessary to select a certain number of suitable, visible, and bright reference stars. Then the tracking of the stars is performed manually or with support from image processing utilities. It can follow a predefined, a priori scheduling plan, which is executed instruction by instruction until the final star is observed. As with satellite tracking, quality parameters are used to define a successful observation of a specific star. After the complete schedule is executed, the fitting is performed as an analysis section. The resulting fit is accepted if further quality parameters show a suitable quality.

Controlling such a complex system also requires possibilities for human interactions and a suitable graphical representation[51] of the system states and current activities. First of all, it is necessary for the operator to be able to change between auto-tracking mode and manual mode, so that he can always acquire full control of the coordination cell and, therefore, of the whole system. Additionally, two different main roles for local users can be determined for other user roles (see also ► Sect. 5.5.2): administrators/supervisors («admin») and operators («oper»). Administrators have full access rights to all the functionalities of the whole system. They can change system parameters and are able to run all functionalities manually. Therefore, they need direct access to all connected hardware cells. This becomes manifest in a special GUI (see ◘ Fig. 4.31), where all individual windows for the sub-cells are included and can be shown independently. The regular operator, which can be a single account for a general operator role or in the form of personalized accounts for each human operator, does not need all of these functionalities. He just needs all control panels required to run the main tasks and to control the main functionalities of the system. Therefore, he does not need direct access to the sub-cells (e.g., he does not necessarily need the option to move the telescope manually to a specific azimuth/elevation position). All operator functions are located at the coordination cell and are forwarded to the subordinated hardware cells after a check to see whether they are currently allowed and can be operated in the current state. Additionally, the operator needs an overview display to get the «big picture» of the system for visual cross-checks. All of these elements are implemented in the operator view of the GUI (see ◘ Fig. 4.32). Because of the defined and implemented separation of a «thin-client» and «fat-server» model (see ► Sect. 4.3), several graphical user interfaces and different roles can interact with the system in parallel. If this is implemented, a suitable user access and role management is required (see ► Sect. 5.5.2).

Additionally, a third role can be identified: the developer or programmer («prog»). His role exists in parallel with the others and is only used for integration tests of the software on the system hardware. The real development happens on different machines. Programmer interactions interfere with all other activities, as they also include the starting of test servers for the different cells, and so on. Therefore, they must act very carefully. These three types of user roles find their reflection in the user accounts of the operating system.

As the coordination cell is the central representation of and access point to the whole system, it also offers the interface to the outer world. All functionalities which are implemented as RPC functions can be called in a network of other observation systems from other sites. This is precisely what creates the type of interaction needed for a globally organized observation system, consisting of different, centrally coordinated space techniques in the form of multi-agent systems (see also ► Sect. 5.1).

Autonomous Functionalities: Controlling Tasks for the Mount Model

Autonomous Functionalities: User Interfacing and Main Roles

Developer Role

Autonomous Functionalities: User Interfacing to the Outer World

---

51 Special thanks should be extended here to Andreas Leidig for his development of several graphical windows for the laser ranging system at the Geodetic Observatory Wettzell, so that the illustrations in the different figures were possible, using real implementations of wxWidgets GUIs!

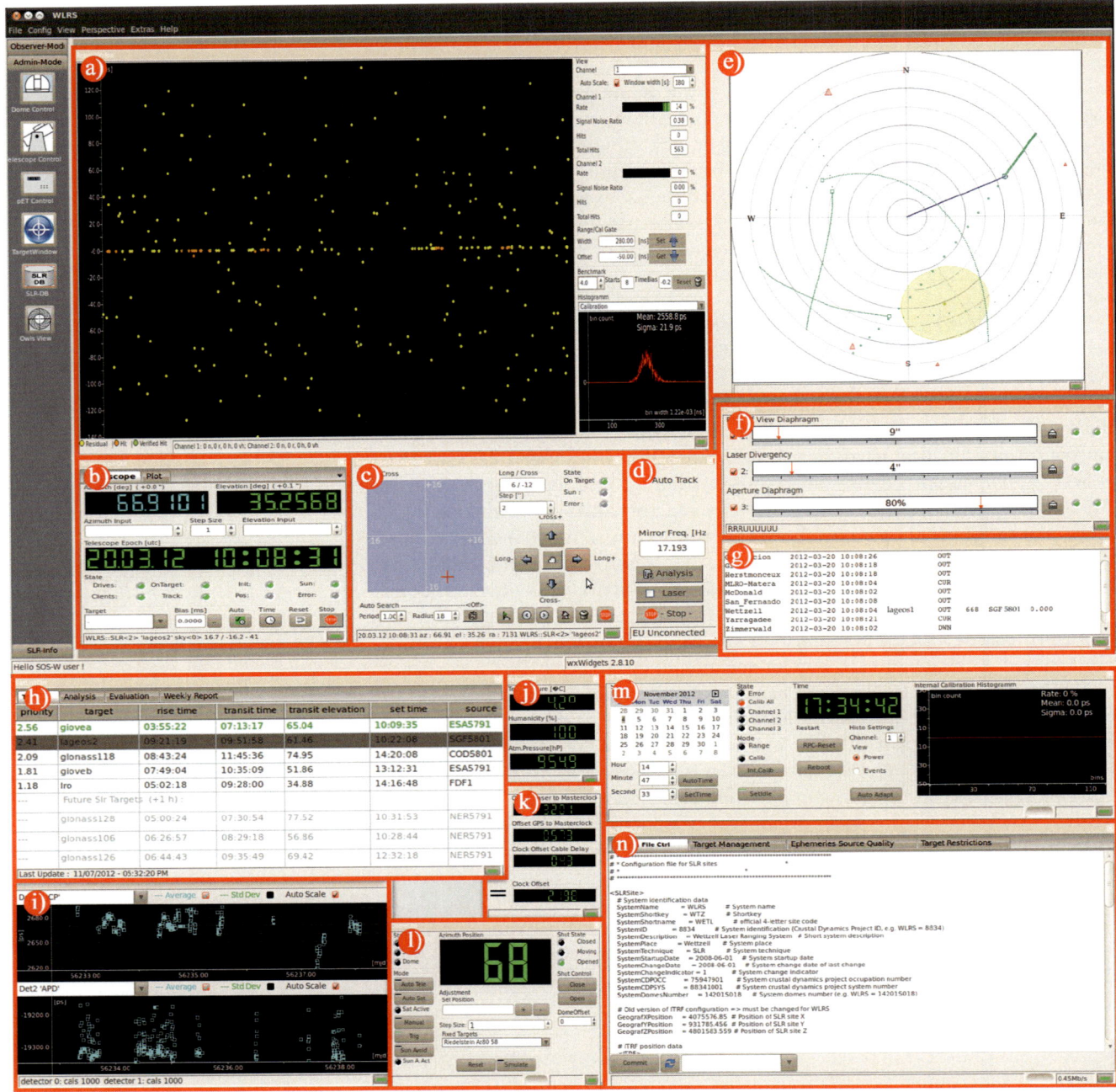

Fig. 4.31 The graphical user interface with the administrator view including several detailed interfaces for the coordinated hardware control cells (*top window*: **a** the residual plot and return/hit statistics, **b** the telescope control, **c** the tracking/search panel, **d** a test scheduler, **e** the sky plot, **f** the transmit-receive unit control, **g** the EUROLAS status (*lower window*), **h** the target and analysis list, **i** the calibration quality plots, **j** the meteorology information, **k** the time correction information, **l** the dome control, **m** the event-timer control panel, and **n** the configuration file editor of the data management cell)

Autonomous Functionalities: Hardware Driving

Unlike all the other autonomous cells, hardware driving of the coordination cell is only implemented implicitly. As no direct hardware is connected to the coordination cell, the managed sub-cells are treated as intelligent hardware systems, which are commanded and controlled according to the planning. The real hardware is then autonomously controlled within these sub-cells. But as the coordination cell organizes all the connected cells via a map, it is the right place for an additional naming service for all connected devices. It extends the functionality of the «portmapper» process of RPC, which should only be used locally on one system because of security issues. The naming service is just a general RPC

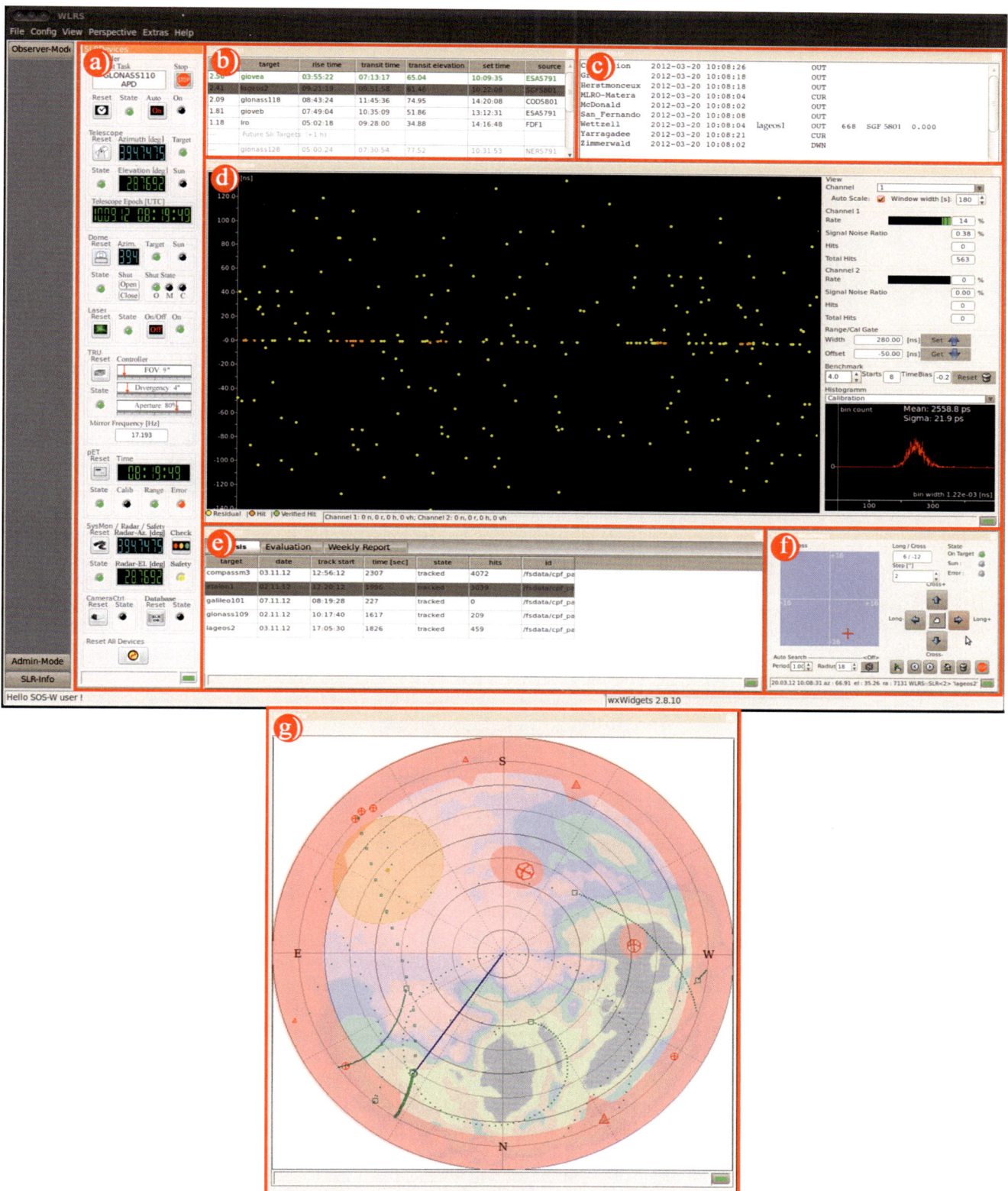

Fig. 4.32 A design sketch of the graphical user interface with the operator view including only the main parts to run the coordination cell (*top window*: **a** the hardware control panel to control the main functions of the connected hardware cells, **b** the target list, **c** the EUROLAS status, **d** the residual plot and return/hit statistics, **e** the analysis display, and **f** the tracking/search panel (*lower window*): **g** the sky plot with the protection zones and cloud coverage)

functionality which is offered by each coordination cell to return the direct access data to all sub-cells, like the IP address, port number, the interface name and version, the interface generator version, and the SSH access information if required.

The final, but very essential task of the coordination cell is failure management. As already described in the controlling section, the central coordination cell periodically requests different status information from the connected sub-cells. The data from the system safety and monitoring cell (see ▶ Sect. 4.7) are important here. All of these data are evaluated against predefined thresholds or fuzzy sets to derive the final, overall system state. It is used to derive decisions with a form of state machine (e.g., see ▶ Sect. 3.3.1) which may lead to interruptions, automated restarts of parts or of the whole system, and maybe complete emergency stop with alert messages, requesting human interaction. As the independent, distributed hardware cells do not know anything about each other, all relevant status information must be propagated to the central coordination cell, which makes decisions and initiates resulting activities in the other cells. For example, if the dome has a problem with following the position of the telescope and the azimuth values differ, the coordination cell can reset the dome server, and if this is not successful, it can stop the whole processing. It is important in this field that all devices, and especially the coordination cell, must implement a suitable recovery system, which continues the processing after a restart at the point where the workflow was still correct. Additionally, all activities and decisions are logged together with the schedule in a log file, which is the logbook of the system.

Several mechanisms can be implemented to enable suitable logging. The usual way is to use «syslog». Kernel and services write all of their log messages to the «syslog» daemon, which is a standard component of the Linux systems. The generated log files then appear in the folder «*/var/log*» and contain messages with different warn levels from «*LOG_DEBUG*» (lowest priority for debugging use) to «*LOG_EMERG*» (highest priority for panic conditions).[52] The problem with this solution is that it becomes confusing if huge amounts of log data are written from different servers. It is also located on a single computer and does not support network-wide logging. Searching specific logs is stressful as it only supports the file search (Müller 1999).

A solution to all of these problems is offered by «rsyslog». It adds additional features, such as sophisticated filter mechanisms, network-wide logging over TCP to other logging computers, and the possibility of including database systems as «PostgreSQL» (see also ▶ Sect. 4.6 about databases). The addition of database functionalities allows fast search mechanisms and quick access to the log messages, which can also be presented via web browser applications. «rsyslog» is also a simple option, as it supports all of the known «syslog» options and only adds additional functionalities (Müller 1999). «rsyslog» is quite an important way to centralize the error management of distributed systems, so that all systems send their log messages not only to a local but also to a central data logger, which then makes decisions on the information received and notifies people observatory-wide.

Another way of logging is given by one's own «proprietary» logging systems and libraries. In daily work, they are quite useful for debugging issues. One suitable mechanism is offered by «logog», a «logger optimized for games» (Gigantic Software n.d.). «logog is a portable C++ library to facilitate logging of real-time events in performance-oriented applications, such as games. It is especially appropriate for projects that have constrained memory and constrained CPU requirements» (Gigantic Software n.d.). It offers additional flexibility for the development. Therefore, the systems at Wettzell support a generalized log library «xlog», where the user can select between the different logging mechanisms. A similar approach of a proprietary logger is realized in the NASA Field System, where all VLBI activities are logged to a session log file per participating station, which is then transferred to the correlation and analysis centers using File Transfer Protocol (FTP) after the observation session (as

Autonomous Functionalities: Failure Management

Logbook «Syslog»

Logbook «Rsyslog»

Logbook as a Proprietary Solution

---

52   For more information, see the manual pages of Linux with the command «*man syslog*».

sample, see Example 4.8 with the logbook for the intensive session from the 20 meter radio telescope Wettzell, scheduled with the SNAP file from Example 4.7).

■ ■ Example 4.8

Selection of a sequence of a VLBI log file of the 20 m radio telescope Wettzell for the INTENSIVE observation from January 2nd, 2013 («=>»-entries are additional comments to explain the sections; the output corresponds to the call of the procedure 'preob' in the according SNAP file of Example 4.7).

```
  ...                                                                              1
  2013.002.18:16:52.06:!2013.002.18:29:50                                          2
  => Check  error  states                                                          3
  2013.002.18:17:13.23?ERROR  ch  −309  v5  alarm  is  on                          4
  2013.002.18:17:34.39?ERROR  ch  −309  v5  alarm  is  on                          5
  2013.002.18:17:55.54?ERROR  ch  −309  v5  alarm  is  on                          6
                                                                                   7
  => Request  and  check  the  cable  delay  measurement                           8
  2013.002.18:18:29.45;cable                                                       9
  2013.002.18:18:29.77/cable/3.69106E−04                                          10
                                                                                  11
  => Request  and  check  the  time  offset  of  the  formatter  to  the  GPS−time 12
  2013.002.18:18:32.89;counter2                                                   13
  2013.002.18:18:34.15/gps−fmout/2.687E−05                                        14
                                                                                  15
  => Request  meteorological  data                                                16
  2013.002.18:18:45.14;wx                                                         17
  2013.002.18:18:45.14/wx/ 2.0,955.8,76.3                                         18
                                                                                  19
  => Request  values  and  status  of  the  cryogenic  dewar  with  the  amplifiers 20
  2013.002.18:18:51.27;rxdew                                                      21
  2013.002.18:18:51.27&rxdew/rx=17,*,*,*,*,*,*                                    22
  2013.002.18:18:51.27&rxdew/rx                                                   23
  2013.002.18:18:51.27&rxdew/rx=1e,*,*,*,*,*,*                                    24
  2013.002.18:18:51.27&rxdew/rx                                                   25
  2013.002.18:18:51.27&rxdew/rx=1f,*,*,*,*,*,*                                    26
  2013.002.18:18:51.27&rxdew/rx                                                   27
  2013.002.18:18:51.36/rx/17(PRES),on,a,on,on,on,off,locked,1.903                 28
  2013.002.18:18:51.44/rx/1E(20K),on,a,on,on,on,off,locked,17.16                  29
  2013.002.18:18:51.53/rx/1F(70K),on,a,on,on,on,off,locked,59.06                  30
                                                                                  31
  => Check  if  the  antenna  is  already  on  the  correct  position  of  the  source 32
  2013.002.18:18:57.22;onsource                                                   33
  2013.002.18:18:57.31/onsource/TRACKING                                          34
  2013.002.18:19:00.50;status                                                     35
  2013.002.18:29:50.01:preob                                                      36
  2013.002.18:29:50.01&preob/onsource                                             37
                                                                                  38
  => Measure  the  system  temperatures                                           39
  2013.002.18:29:50.01&preob/caltsys                                              40
  2013.002.18:29:50.01&preob/check=*                                              41
  2013.002.18:29:50.10/onsource/TRACKING                                          42
  2013.002.18:29:50.10&caltsys/tpi=formvc,formif                                  43
  2013.002.18:29:50.10&caltsys/ifd=max,max,*,*                                    44
  2013.002.18:29:50.10&caltsys/if3=max,*,*,*,*,*                                  45
  2013.002.18:29:50.10&caltsys/!+2s                                              46
  2013.002.18:29:50.10&caltsys/tpzero=formvc,formif                              47
  2013.002.18:29:50.10&caltsys/ifd=old,old,*,*                                   48
  2013.002.18:29:50.10&caltsys/if3=old,*,*,*,*,*                                 49
```

```
2013.002.18:29:50.10& caltsys/calon                                                    50
2013.002.18:29:50.10& caltsys/!+2s                                                     51
2013.002.18:29:50.10& caltsys/tpical=formvc,formif                                     52
2013.002.18:29:50.10& caltsys/caloff                                                   53
2013.002.18:29:50.10& caltsys/tpdiff=formvc,formif                                     54
2013.002.18:29:50.10& caltsys/caltemp=formvc,formif                                    55
2013.002.18:29:50.10& caltsys/tsys=formvc,formif                                       56
2013.002.18:29:50.56/tpi/1d,7557,2u,4956,3u,5485,4u,4293,i1,6701                       57
2013.002.18:29:50.56/tpi/9u,8141,au,9839,bu,12794,cu,9890,du,11285,eu,20608,i2,29272   58
2013.002.18:29:50.56/tpi/5u,7672,6u,7788,7u,6887,8d,6260,i3,10159                      59
2013.002.18:29:53.07/tpzero/1d,594,2u,947,3u,731,4u,1030,i1,95                         60
2013.002.18:29:53.07/tpzero/9u,613,au,918,bu,686,cu,724,du,549,eu,3799,i2,878          61
2013.002.18:29:53.07/tpzero/5u,1410,6u,644,7u,911,8d,932,i3,55                         62
2013.002.18:29:53.14& calon/"turn cal on                                               63
2013.002.18:29:53.14& calon/rx=*,*,*,*,*,*,on                                          64
2013.002.18:29:55.65/tpical/1d,11341,2u,7024,3u,8233,4u,6443,i1,10674                  65
2013.002.18:29:55.65/tpical/9u,11216,au,13421,bu,17190,cu,13047,du,14986,eu,25944      66
2013.002.18:29:55.65/tpical/i2,33938                                                   67
2013.002.18:29:55.65/tpical/5u,12475,6u,13211,7u,11672,8d,10443,i3,18005               68
2013.002.18:29:55.65& caloff/"turn cal off                                             69
2013.002.18:29:55.65& caloff/rx=*,*,*,*,*,*,off                                        70
2013.002.18:29:55.71/tpdiff/1d,3784,2u,2068,3u,2748,4u,2150,i1,3973                    71
2013.002.18:29:55.71/tpdiff/9u,3075,au,3582,bu,4396,cu,3157,du,3701,eu,5336,i2,4666    72
2013.002.18:29:55.71/tpdiff/5u,4803,6u,5423,7u,4785,8d,4183,i3,7846                    73
2013.002.18:29:55.71/caltemp/1d,20.800,2u,20.800,3u,20.800,4u,20.800,i1,20.800         74
2013.002.18:29:55.71/caltemp/9u,21.400,au,21.400,bu,21.400,cu,21.400,du,21.400,eu,21.400  75
2013.002.18:29:55.71/caltemp/i2,21.400                                                 76
2013.002.18:29:55.71/caltemp/5u,20.800,6u,20.800,7u,20.800,8d,20.800,i3,20.800         77
2013.002.18:29:55.71/tsys/1d,38.3,2u,40.3,3u,36.0,4u,31.6,i1,34.6                      78
2013.002.18:29:55.71/tsys/9u,52.4,au,53.3,bu,58.9,cu,62.1,du,62.1,eu,67.4,i2,130.2     79
2013.002.18:29:55.71/tsys/5u,27.1,6u,27.4,7u,26.0,8d,26.5,i3,26.8                      80
2013.002.18:29:55.71/!2013.002.18:30:00                                                81
...                                                                                    82
```

While former systems concentrated all controlling parts within one central control system, the distributed approach allows a structural design separation of the different subsystems. The coordination cell as the brain of the whole system then relies on the controlling functionalities of the connected sub-cells and can just concentrate on the main tasks, as described. Therefore, the autonomous hardware cells, which finally do the actual controlling of the hardware, play an important role in the whole system.

4.5   Autonomous Hardware Control Cells

General Structure

The different autonomous hardware control cells have at least one directly connected piece of hardware which they control and operate, as described in ▶ Sect. 4.2.1 about distributed control. They consist of a control server together with a watchdog process. Each control server is split into at least three different threads: the RPC serving thread to serve the asynchronous RPC requests, the control loop thread to run the hardware, and a safety thread to handle internal safety issues, such as excessively long time delays in the processing loops. Additionally, the developer of a server can also implement several individual threads, for example, to manage the timing of the parallel communication using different communication lines to a specific hardware. The communication between the different threads is implemented with semaphore-protected variables. The hardware itself is a more or less intelligent device consisting of simple digital circuits, microcontrollers, or also self-contained PC systems with their own operating systems (partly with real-time abilities). The internal constitution of each server using the structure of a sophisticated device driver with a standardized, generative interface was shown in ▢ Fig. 4.7. While the control of hardware using different control and feedback loops was described there, the following sections describe the specifics of the cells in general and not the individual setup of each cell.

As the data management cell and the system monitoring and safety cell can be separated from the lower-level hardware cells, because they implement completely parallel distributed, supporting production cells with a higher-level coordination and control for safety reasons, the remaining hardware cells are:

Hardware Control Cells

- Telescope and mounting control
- Dome control
- Radar control
- Laser control
- Optics, transmitting and receiving unit control
- Timing system and event-timer control
- Camera control

The planning within these control cells is minimal. It is just limited to local, periodic activities, operated in the periodic loop, and their timing. Usually the planning in the hardware control cells is not really dynamic and is defined by the programmer of the server code during the code development. An exception is the telescope control, as it is more or less a coordination cell as well. As shown in the multilayer model in ▢ Fig. 4.11, it not only controls its own telescope hardware but also coordinates tightly coupled subsystems for a faster propagation of position data. With this characteristic trait, it also implements approaches described for coordination cells (see previous ▶ Sect. 4.4 starting). Therefore, it has a more dynamic planning (similar to the described coordination cell) for the different modes for the observation and determination of the mount model corrections, the calibration measurement, and the laser ranging observations.

Autonomous Functionalities: Planning

The controlling is the main part of each hardware control cell. The controlling is processed in the periodic control loop of each server, as shown in Example 4.2 or in Example 3.20, where it is also described in detail. But a short look should be taken at the timing behavior of the controlling. Usually the time requirements (see also ▶ Sect. 4.2.1) can be split into hard real-time (failures to comply with deadlines result in unacceptable, critical system states), weak real-time (the system should comply with deadlines and should avoid delays, but they do not result in unacceptable system states), stable real-time (weak real-time with deadlines) (see (Vigenschow 2005, *l.c. page 223*)), and normal time (no constraints according to the time scheduling with the system time of the standard operating system). In this context, each hardware control cell is the separation point of hard real-time at the local hardware devices and stable time at the coordination cell. In other words, the hardware control cell is usually the conjunction between hard real-time and stable time, which requires at least an internal implementation of (weak) real-time abilities. In most cases, weak real-time implementations are enough for hardware control cells because the hardware control and feedback loop encapsulate the hard real-time constraints. It is often possible to keep the hard real-time constraints as local as possible in the lower control layers by changing the hardware parameters (e.g., if the update rate for the RADAR system cannot be guaranteed, so that aircrafts can safely be protected around the laser beam, it might be possible to extend the RADAR beam pattern to a wider angle, while losing only as much energy as needed for the detection to still be possible for the given application[53]). Nevertheless, most of the hardware controlling cells should be able to comply with weak deadlines and with periodic updates or reactions, and therefore, it is necessary to know something about the update rate, its

Autonomous Functionalities: Controlling

---

53 Another possibility is to increase the protection zone around an aircraft so that a detection and the consecutive protection from laser contact are possible within the longer, weaker time period.

consequences, and the propagation of delays, which can be derived from a theorem of the signal processing.

Nyquist-Shannon Sampling Theorem

> *Definition DEF4.6 of «Nyquist-Shannon sampling theorem»* (see (Hermes 2005, *l.c. page 150 f.*) and (Yadav 2008, *l.c. page 266*))
> A continuous, band-limited signal with a minimal frequency of $f_{min} = 0\,Hz$ and a maximal frequency f_{max} can be sampled alias-free and without any information losses into an approximated discrete signal over time if the sampling rate $f_{rate} > 2f_{max}$ (Nyquist rate), where $f_{max} = f_{nyquist} \leq f_{rate}/2$ is the Nyquist frequency and the sampling interval is defined by $t_{rate} = 1/f_{rate}$

Autonomous Functionalities: Update Constraints for Controlling Data

As each cell (especially the hardware control cells) and even each data updating component in the system work like a sampling unit of data from equipment which is subordinated to it, it needs to be able to deal with the above sampling theorem.[54] This means if, for example, the RADAR hardware periodically requires the position data from the responsible hardware control cell every 50 milliseconds, the superordinate cell (in the design described this is the telescope control cell) must update the position information at the RADAR server with more than 25 milliseconds. This is why the telescope control cell builds an additional coordination instance, so that the timing for position updates is possible. Additionally, the internal processing in the RADAR control cell must also be performed within this time interval. Ideally, this is solved by an additional, manually written thread, parallel to the status-requesting control loop thread. The additional thread acts like a server for the RADAR hardware, so that the current value can almost immediately be forwarded in answer to a hardware request. This implementation complies with the response requirements.

Autonomous Functionalities: Control Data Propagation

But this sampling rate is not only relevant for the update timing of controlling data in the data flow from higher levels to the hardware and back. It also defines the maximum rate at which it is possible to have the status data from the hardware available on the higher controlling and coordination levels. If an ergonomic response time of at least 1 second is required in the GUI, the responsible coordination cell must request data more than every half second. The next level has to offer the data faster than 250 milliseconds and so on. This allows a rough estimation of the timing constraints for a layered system. In the case of the design from ◘ Fig. 4.11, where the longest path through the layers for the main components has five steps,[55] the hardware control cells should implement a timing with something like about 100 milliseconds to the higher layers. The hardware and the internal control of the hardware control cell should comply with more than about 50 milliseconds time intervals. This estimation is just for the informal status propagation and does not consider stronger constraints, as described before, for example, for the update rate of position data. In summary, this means that data propagation should always be limited, where this is possible to avoid time constraints. In the laser ranging system, this can mostly be done if each hardware control cell gets the necessary orbit data directly at the beginning and calculates the necessary values for the tracking within a short setup phase itself. Then each cell involved can control the hardware according to the

---

54 The components act as a sampling unit, because one design aim separates the asynchronous RPC part from the synchronous control loop part in the server. If this design is waived, the timing for update requests is equal throughout the whole system hierarchy. But then also all delays from the lower hardware elements are forwarded in the same way. The separation avoids this, but pays for it with sampling constraints.

55 For example, the GUI, the coordination cell server, the telescope control server, the RADAR control server, and the RADAR hardware

positions calculated over time without any additional data propagation between the systems. This is also one reason for independent, autonomous control cells. Only for components which rely on each other because of safety issues (as, e.g., the RADAR should always point in the direction to which the telescope actually points[56]), a data propagation with the according timing constraints is a must. This is also one important design driver for the design in ◘ Fig. 4.11.

The hardware driver also acts as an abstraction level. This means that all the parameters do not have to be set from outside, so that the operator would have to know all of the specifics of one local hardware cell. For example, it is possible to set the startup values of the optics in the transmit-receive unit if just the detector type, such as MCP or APD, and the height of the satellite in the categories, LEO, Medium Earth Orbit (MEO) (or «Lageos» type), or HEO, are known. For the Wettzell laser ranging system, this means that for MCP the detector is of the type PMT210 (of the Photec company). If all LEO satellite is observed with this detector, the field of view can automatically be opened to 15 arcseconds, and the divergence can be set to 12 arcseconds, while for MEO satellites with the same detector, the divergence can be set to 4 arcseconds.[57] This example shows that the operator response can be reduced in some cases. During the observation, he only needs to know which parameters he must increase or decrease in specific situations, for example, if too many returns appear and could saturate the detector. He also does not need to know the absolute values. It is enough to offer different steps, which are internally converted into the actual parameter settings. This abstraction simplifies the handling and is one important task of the hardware control cell, as all the necessary information is only available there. This abstraction even simplifies automation, because decisions of higher-level coordination cells can be reduced if hardware control cells hide the very specific, individual division parameters. Nevertheless, there should be an option to also control the actual parameters directly, for example, by an administrator.

This abstraction possibility is also used for the information produced by the hardware control cells. One example here is the event-timer control. The responsible hardware control cell has the only access to the hardware for which data propagation can be challenging. For laser ranging systems with a higher laser pulse rate, for example, of kilohertz and more, there are so many events from returning photons and, therefore, so many resulting residuals[58] that a filtering must avoid unnecessary data traffic, for example, to the following coordination cells or the GUI. The event-timer control cell does such an abstraction by creating histograms and calculating statistics parameters per defined time interval, which are used as reduced representatives equivalent to the raw data. Only the reduced information about the statistics is then forwarded to the higher control layers. Internally it must be able to cache and save the data stream to the file system, which may be packed to reduce the data amount. The handling of these files is also part of the event timer control cell tasks. Another statistics calculation is the evaluation of returns to detect hits, as all data is available at the event-timer cell without additional propagation. The orbit ranges are calculated during the setup phase, and the returns and statistics are available during the observation. In Wettzell, the hit detection is implemented with a version of a double histogram analysis in combination with an analysis of the hit progression and history together with a time bias correction. The histogram analysis uses two histograms, which are shifted against each other by half of a bin width in the axis of the residuals, while accumulating data over

Autonomous Functionalities: Control Abstraction

Autonomous Functionalities: Information Abstraction

---

56 The dome control and the control of the transmit-receive unit are also directly connected with the telescope control using direct data propagation because both implement safety. The transmit-receive unit can block the beam with a shutter, and the dome can turn the dome window away from the telescope beam if any danger is detected for humans or the system.
57 The APD at Wettzell is the SAP500 of the company Laser Components, which is used for all HEO satellites with the field of view of nine arcseconds and the divergence of four arcseconds.
58 Residuals are the result of a subtraction of predicted reference ranges to the orbit and the measured, actual ranges to the satellite.

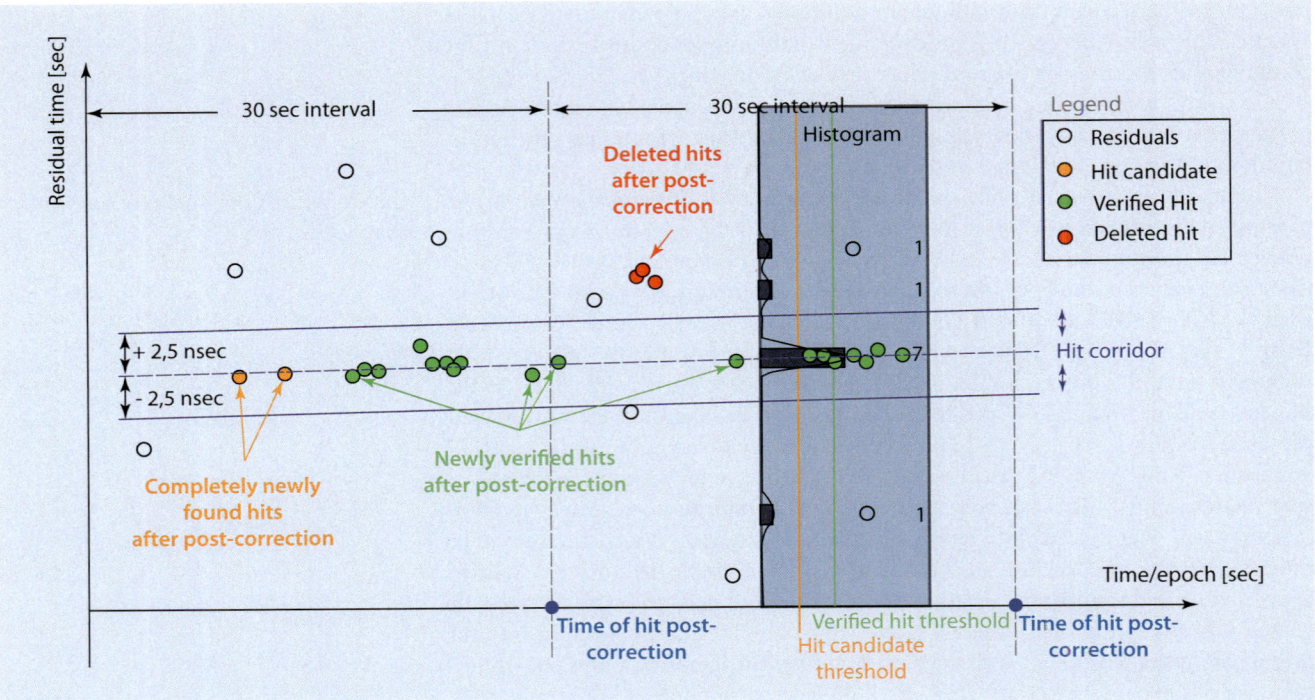

Fig. 4.33 Illustration of a hit detection using histograms and a post-correction with an interpolated hit corridor (Inspired by Hiener (2006) *l.c. page 61*; reproduced by permission of Michael Hiener)

3–5 seconds to optimize the adjustment around the peak number of returns. The bin with the peak value of the counter is a hit candidate. A predefined and adjustable threshold for the number of required numbers of verified and unverified hits in the bin segments classifies the candidates found. The additional progress detection uses the information already derived to adapt the search for future hits within a corridor along histogram bins containing hits which have already been found. The algorithm also looks back at the historic residuals already received in order to do a post-processing for previous values. The added time bias detection and correction search for trends in the hit data and calculate an interpolation between the hit points, which are used to correct the trend found. The frame parameters of this hit detection algorithm are the width of the bins, the time interval used to accumulate bin residuals, the accepted thresholds for verified and unverified hits, and the interpolation function type for the hit track. The hit detection described improved on a simple histogram detection algorithm, which did not use the knowledge acquired for improved future detection and for post-detection of historic values (see also the illustration in **Fig. 4.33**) (Hiener 2006, *l.c. page 50 ff.*). Other (e.g., graphical or image processing) algorithms might also be conceivable, but one or the other variant of the double histogram detection offers very good results. Each detection requires knowledge of the current ranging situation and returns. Therefore, the hit detection is ideally located at the event-timer control cell, so that it also processes higher analysis tasks as well.

As already shown in **Fig. 4.31**, each hardware cell can be controlled directly by an administrator. This offers more possibilities for parametrization of the control cell. If additionally the GUI of each cell is developed individually as a separate object and directly assigned to the code of the hardware cell, it can be developed completely independently and in parallel during the software development process. The individual hardware GUI object can then be easily integrated into the administrator view of the coordination cell. In this context, each hardware cell offers its own user interface for administrative needs (see **Fig. 4.34**).

Autonomous Functionalities: User Interfacing

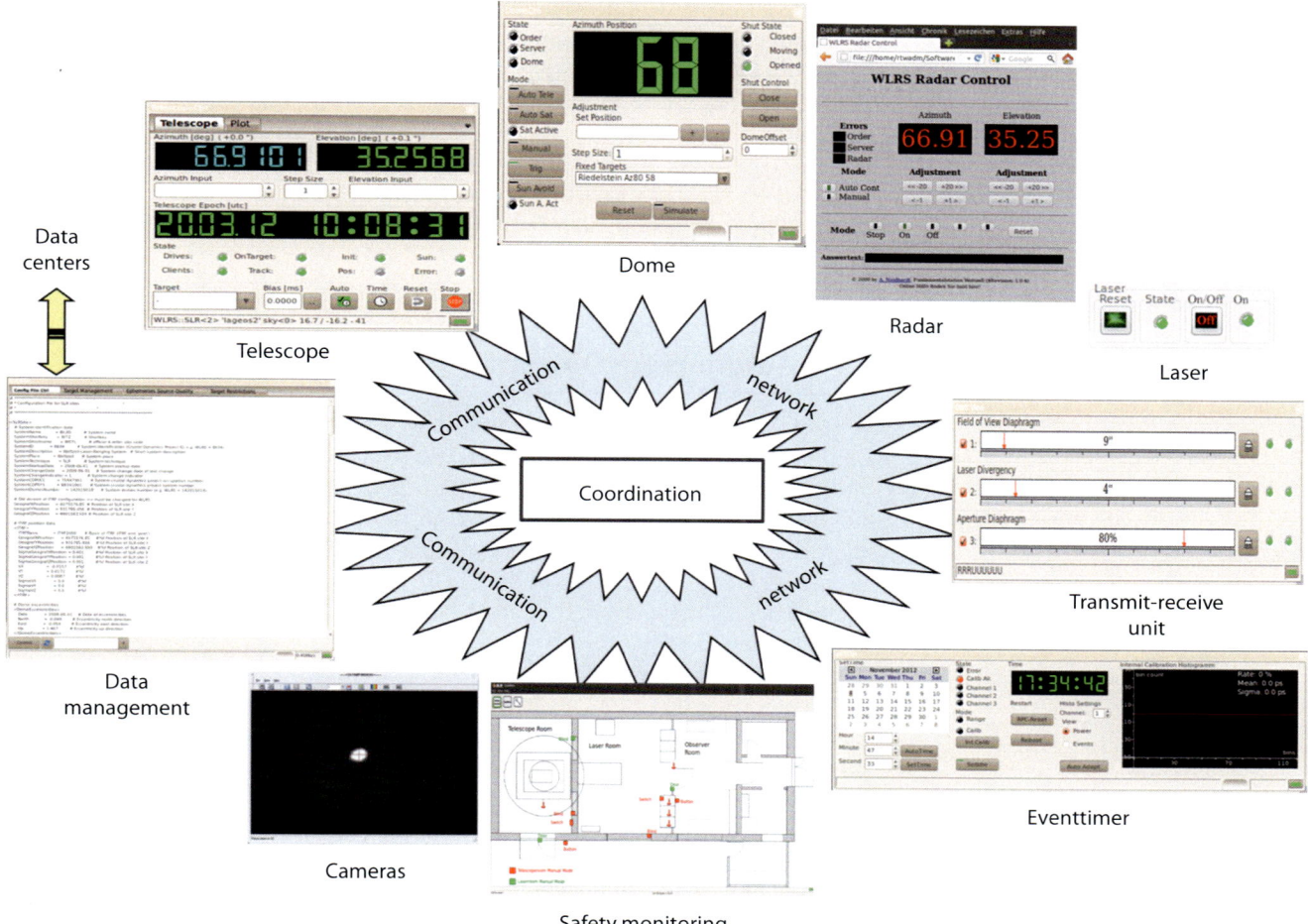

Fig. 4.34 The different GUIs of the different connected sub-cells including the autonomous hardware control cells, which are coordinated by the coordination cell as distributed system like in ▶ Fig. 4.8

In any case, the main task of the hardware control cell is the driving of the connected hardware. As already described in ▶ Sect. 4.2.1 about the functionality of a device driver implementation using an RPC server, each hardware controlling cell uses the implemented hardware driver module or component to access and control the connected hardware. This module implements the hardware-specific communication and control and offers suitable interface access points which can be used in the control loop to request status information and to send control messages.

Autonomous Functionalities: Hardware Driving

Because each hardware control cell is a completely autonomous cell, it must fulfill the local safety implementations, as described in ▶ Sect. 4.2.1. Together with the automatic safety device, described in ▶ Sect. 3.3.3, suitable failure management is possible. If a failure appears, it is then evaluated to get at least the failure levels «alert», «error», or «warning». Alerts are system-critical situations which cannot be solved autonomously by the cell. They must be propagated through the whole system and lead to a complete stop of all other cells until there is a human interaction. They usually do not happen if the system is used within the limits of the task described or if no defective equipment is used. Errors are dangerous situations which must be fixed locally. One single error does not compromise the system and can usually be fixed autonomously (see also (Vigenschow 2005, *l.c.* page 224)). But if more than one error appears, it might result in a critical state, which would result in an alert (evaluations with fuzzy techniques might be quite suitable for the estimation of danger levels). Warnings are more or less the first

Autonomous Functionalities: Failure Management

indication for the operator that something might be wrong or that an error or alert could be the result if a potentially dangerous situation develops.

To offer possibilities for testing, for bug fixing, and for the isolation of errors, the implementation of autonomous hardware control cells should also include a simulation mode. It should simulate real behavior as well as possible without using or controlling the real hardware. This makes it possible to perform integration tests on the real production system without the hardware and separates the controlling code from the hardware with the hardware driver module or component for tests.

Autonomous Functionalities: Simulation of Failures and Functionalities

4.6   Autonomous Data Management Cell

File Systems

A centralized knowledge base and information storage is essential for each autonomous production cell, as it administrates the required data and makes it accessible in a fast and sophisticated way (Pfeifer and Schmitt 2006, *l.c. page 10*). While the data is offered to the communication infrastructure in the usual RPC way, there are several ways to administrate information in the data storage system. The usual way is the organization of data as files in a hierarchical directory tree. This method is well known from the personal computer at home, where the documents are organized like this. This structure of a file system is just a virtual organization generated by the operating system, as the actual data is saved block by block (e.g., with a block size of 4 kilobyte (Schicker 1996, *l.c. page 15*)) and sequentially on the hard drive or another storage medium. Only the file system of the operating system creates the logical structure of files and folders as presentation layer to enable a simpler usage.

File System Operations

Therefore, the operating system offers an interface library:

- To use directory paths
- To get the content of folders
- To manipulate the file or folder names
- To delete or move files
- To read from, write, or append to files

File Manipulations

Each file has a filename and usually an extension using the ASCII alphabet. The internal structure of a file is a sequential order of binary values or characters. Directories are also named using the ASCII alphabet. This means that directory trees of files can be navigated using tree paths and a lexical order or search criterion on file or directory names. But the files must be read sequentially, using a file pointer which is shifted byte by byte through the file. An exception is if the position of the required value is known as a relative number of bytes counted from the beginning of the file. Then the file pointer can be directly shifted to that position with a «seeking» function. This can be used in addressable file structures on the basis of records with equal sizes. For example, if each record has a length of 100 bytes, it is possible to calculate the position of the 5th record with «5*100=500», so that the file pointer can be directly positioned to that address. A problem occurs if not all actual data sets have exactly the same record size. On the one hand, if it is smaller, the rest of the record space is wasted in favor of quick access. On the other hand, if the size is bigger, the content must be cut. Another significant problem is the insertion of new data records between existing ones in ordered record sequences. This is quite difficult and slow, as the correct insert position must be found. Then the rest of the sequential file must be buffered, so that it can be attached again after the new values are written to the found position (Schicker 1996, *l.c. page 16 ff.*).

Already the previous simple example has illustrated one basic principle in combination with data organization: if a fast access is implemented, it is usually necessary to use more memory space (e.g., because of the structure or, as shown later, for additional information used for administration). Less memory space usually means that the access times increase, because of missing administrative structures. The goal of each data management design is to find the optimum between access times and required space. This optimum depends on the administrated data, their natural structure, and the possibility of converting them into more manageable organizations. Another principle can be identified: the implementation of a quick access is always based on an algorithm to calculate addresses for a direct access (for simple, fixed-sized records, a simple calculation «Position = RecordNumber * RecordSize» can be enough). If it is possible to find an algorithm where the position of a record can be calculated, the access time can be reduced with a direct addressing mechanism.

Memory Space Versus Access Speed

To allow such direct addressing mechanisms in an extended way, records assign a search criterion to the user data. Records can be split into a search key (or just key), which is used to calculate the position of the record and its associated data list. A simple example is the organization of a personal organizer or filofax, which arranges appointments in a calendar system. If the record size is again fixed, each address of an appointment can be calculated with the formula «Position = StartAddressOfTheYear + (DayOfTheYear-1) * RecordSize». The «DayOfTheYear» can be calculated with a time and date library function from the Gregorian date, and the «StartAddressOfTheYear» can be calculated from a start year in combination with the fixed space needed per year (Schicker 1996, *l.c. page 18 ff.*). Ideally, the mapping between key values and address space is distinct (= definitely). But previous chapters have already discussed a technique if this is not possible: the hash method (e.g., see ▶ Sect. 2.7.2). But the file system is too bulky for the use of those algorithms in an efficient way. Therefore, the usual way is to read the complete file and organize it in the dynamic heap memory with an adequate structure which allows a sophisticated use of addressable memory.

Direct Addressing

To enable an efficient and fast reading of files, file formats structure the content in a suitable way. One format has already been explained: the use of records with fixed sizes. The records themselves are structured again with values of a fixed size. This is quite straight-forward and builds up something like a table (two-dimensional array) in the heap memory. This is usually possible for binary files of numbers, where the data values are saved with their binary representations which are of fixed sizes (e.g., a double value usually has 8 bytes[59]). The required time for reading such a file depends on the size of a read block. Byte-by-byte operations are always very slow, while the reading of a block size which is used by the operating system is optimal. Instead of fixed-sized elements, it is possible to use reserved signs as separators between the data values, for example, a tabulator space or a comma sign. The existence of these signs within the textual user data is forbidden, as it would compromise the reading. The reading program can then assign the values between two reserved signs to specific positions, for example, in a table. A more sophisticated and more flexible way is to add additional information which classifies the assigned user data. This means that each data element is extended with an additional classifier or identifier. This additional information is in the form of meta tags, which can be used to interpret the values and order them in the heap structure. With this metadata, it is even possible to organize the file in a similar way to how it would be organized in the heap, so that relations between data become clear or that closely

File Format

---

59 But such binary files have to deal with representation differences as already discussed in previous chapters, e.g., ◻ Figs. 3.5 and 3.10.

related data can be composed within block structures. The result is a data format which follows grammar rules. This means that the file contents are interpreted with a parser, similar to the ones described for DSLs or programming languages. Simple, fixed-sized records, binary file structures, or plain textual structures follow regular grammars and can be parsed with regular expressions, while formats with blocks and nested or hierarchical tree structures use context-free languages. Therefore, it is possible to parse the files using file format grammars which are defined with replacement rules as shown in ◘ Sect. 3.3.3.

The ILRS defined such standard file formats for their products for the orbit prediction data (CPF) and for the resulting observation data (CRD). For example, the CPF (Ricklefs 2006, see also the excerpt of a sample file in Example 4.9) defines a line-based structure where each line starts with a unique identifier. The file is generally split into a header part and a body with the prediction records. The format uses no parentheses and there are no relations between the lines. Header lines begin with «H» and a number. Record types in the body can be identified by their numbers with the labeling method of 10, 20, 30, and so on. Each line defines one textual record with several, individual values, separated with white spaces (e.g., for record type «10»: a direction flag, the Modified Julian Date, the seconds of the day (UTC), a leap second flag, and the geocentric X, Y, and Z position in meters (Ricklefs 2006, *l.c. page 16*)). The whole format follows a regular grammar and can be defined with regular expressions (see the regular expressions for the whole grammar of the CPF in Perl style in Example 4.10; the part for record type «10» can be found in variable «$Record10»). These can be used to check the correctness of the files (see the Perl code snippet to check the format correctness in Example 4.10), which is not only required for the reading but also for the quality control of incoming files from the data centers.[60] Therefore, such a correctness test (e.g., with a Perl script as shown in Example 4.11) is done for each file which is imported to the autonomous data management cell. This ensures that data is correct, reliable, and trustworthy for the rest of the autonomous production cell.

File Format Example: ILRS CPF

▪▪ **Example 4.9**

Excerpt of the prediction file 'glonass102_cpf_120528_6511.hts' for the laser ranging in CPF format.

```
H1  CPF   1   HTS  2012    5  28  12    6511  glonass102              NONE                                    1
H2      606201  9102          29670  2012  05  28    0   0   0  2012  06  01  22  15    0      900  1  1    0  0  1    2
H9                                                                                                           3
10  0  56074    81000.00000    0   -17643284.005    16983200.872    -6945524.610                            4
10  0  56074    81900.00000    0   -18037108.261    17556170.419    -3782857.538                            5
10  0  56074    82800.00000    0   -18095751.191    17890690.412     -546226.344                            6
10  0  56074    83700.00000    0   -17852717.400    17939761.533     2701085.905                            7
10  0  56074    84600.00000    0   -17351724.526    17665869.527     5895583.127                            8
10  0  56074    85500.00000    0   -16644668.395    17042766.012     8974824.246                            9
10  0  56075        0.00000    0   -15789255.050    16056781.111    11878668.190                           10
10  0  56075      900.00000    0   -14846403.076    14707610.853    14550468.902                           11
10  0  56075     1800.00000    0   -13877529.312    13008545.531    16938194.916                           12
10  0  56075     2700.00000    0   -12941835.628    10986129.688    18995450.198                           13
                                                                                                           14
...                                                                                                        15
                                                                                                           16
10  0  56079    78300.00000    0    22668326.820   -10270025.720    -5837582.257                           17
10  0  56079    79200.00000    0    21787114.811    -9963089.403    -8907160.198                           18
10  0  56079    80100.00000    0    20637346.358    -9372896.660   -11804738.514                           19
99                                                                                                         20
```

---

60 Experience has shown that files on the data centers are sometimes corrupted.

■■ Example 4.10

Regular expressions for the whole grammar of the CPF in Perl style.

```
# Define the elements of CPF-lines (Consolidated Laser Ranging Prediction Format Version 1.01      1
    17.02.2006)
my $HeaderH1 = 'H1\s+CPF\s+(-?\d{1,2})\s+([a-zA-Z0-9]{1,4})\s+(-?\d{1,4})\s+(-?\d{1,2})\s+(-?\     2
    d{1,2})\s+(-?\d{1,2})\s+(-?\d{1,5})\s+([a-zA-Z0-9]{1,10})\s+([a-zA-Z0-9\s]*).*';
my $HeaderH2 = 'H2\s+(-?\d{1,8})\s+(-?\d{1,4})\s+(-?\d{1,8})\s+(-?\d{1,4})\s+(-?\d{1,2})\s     3
    +(-?\d{1,2})\s+(-?\d{1,2})\s+(-?\d{1,2})\s+(-?\d{1,2})\s+(-?\d{1,4})\s+(-?\d{1,2})\s+(-?\
    d{1,2})\s+(-?\d{1,2})\s+(-?\d{1,2})\s+(-?\d{1,2})\s+(-?\d{1,5})\s+(\d{1})\s+(\d{1})\s
    +(-?\d{1,2})\s+(\d{1})\s+(-?\d{0,1}).*';
my $HeaderH3 = 'H3\s+(-?\d{1,5})\s+(-?\d{1,5})\s+(-?\d{1,5})\s+(-?\d{1,5})\s     4
    +(-?\d{1,5})\s+(-?\d{1,5})\s+(-?\d{1,5})\s+(-?\d{1,5}).*';
my $HeaderH4 = 'H4\s+(-?\d{1,12})\.?(-?\d{0,5})\s+(-?\d{1,10}\.?\d{0,4})\s+(-?\d{1,11}\.?\d     5
    {0,2})\s+(-?\d{1,11}\.?\d{0,2}).*';
my $HeaderH5 = 'H5\s+(-?\d{1,7})\.?(-?\d{0,4}).*';                                                   6
my $HeaderH9 = 'H9.*';                                                                               7
my $Record10 = '10\s+(\d{1})\s+(-?\d{1,5})\s+(-?\d{1,13}\.?\d{0,6})\s+(-?\d{1,2})\s+(-?\d     8
    {1,17}\.?\d{0,3})\s+(-?\d{1,17}\.?\d{0,3})\s+(-?\d{1,17}\.?\d{0,3}).*';
my $Record20 = '20\s+([a-zA-Z0-9]{1,4})\s+(\d{1})\s+(-?\d{1,19}\.?\d{0,6})\s+(-?\d{1,19}\.?\d     9
    {0,6})\s+(-?\d{1,19}\.?\d{0,6}).*';
my $Record30 = '30\s+([a-zA-Z0-9]{1})\s+(-?\d{1,18}\.?\d{0,6})\s+(-?\d{1,18}\.?\d{0,6})\s+(-?\     10
    d{1,18}\.?\d{0,6})\s+(-?\d{1,5}\.?\d{0,1}).*';
my $Record40 = '40\s+(-?\d{1,6}\.?\d{0,3}).*';                                                       11
my $Record50 = '50\s+(\d{1})\s+(-?\d{1,5})\s+(-?\d{1,13}\.?\d{0,6})\s+([a-zA-Z0-9]{1,10})\s     12
    +(-?\d{1,17}\.?\d{0,3})\s+(-?\d{1,17}\.?\d{0,3})\s+(-?\d{1,17}\.?\d{0,3}).*';
my $Record60 = '60\s+(-?\d{1,5})\s+(-?\d{1,13}\.?\d{0,6})\s+(-?\d{1,17}\.?\d{0,12})\s+(-?\d     13
    {1,17}\.?\d{0,12})\s+(-?\d{1,17}\.?\d{0,12})\s+(-?\d{1,17}\.?\d{0,12}).*';
my $Record70 = '70\s+(-?\d{1,5})\s+(-?\d{1,6})\s+(-?\d{1,8}\.?\d{0,5})\s+(-?\d{1,8}\.?\d{0,5})     14
    \s+(-?\d{1,10}\.?\d{0,6}).*';
my $Record99 = '99.*';                                                                              15
my $Record00 = '00.*';                                                                             16
```

■■ Example 4.11

Perl script snippet, which parses a CPF file to do a format correctness check.

```
# >------------------------------------------------------------<       1
# > Subroutine:  ValidateCPFFile                              <         2
# >              Check the correctness of a CPF-file          <         3
# > Parameter:   $Filepath --> Filepath to CPF-file          <         4
# > Return:      <-- Result of check (0 = ok, 1 = not ok)     <         5
# > Author:      A. Neidhardt                                 <         6
# > Date:        09.01.2007                                   <         7
# > Revision:    —                                            <         8
# > Info:        —                                            <         9
# >------------------------------------------------------------<       10
sub ValidateCPFFile {                                                  11
    my ($Filepath)=@_;                                                 12
    my $Line;                                                          13
    my $Error = 0;                                                     14
    my $ErrorMsg = "OK";                                               15
    my @FileContainsHeaderTags = (0,0,0,0,0,0); # H1, H9, H2, H3, H4, H5 --> each tag can   16
        appear once
    my $FileContainsRecordTag = 0;   # one of 10, 20, 30, 40, 50, 60, 70 is included once or   17
        more times
    my $FileContainsTimeRecord = 0;   # A record with time information is included   18
    my $FileContainsLastRecord = 0; # 99 is included => file is complete                19
    my ($StartingYear, $StartingMonth, $StartingDay, $StartingHour, $StartingMinute,    20
        $StartingSecond);
                                 # Starting date and time of ephemeris records given in   21
                                     header
```

```perl
my $StartingMJD;                    # Starting time of ephemeris records given in header as    22
    MJD
my ($EndingYear, $EndingMonth, $EndingDay, $EndingHour, $EndingMinute, $EndingSecond);       23
                                    # Ending date and time of ephemeris records given in      24
                                        header
my $EndingMJD;                      # Ending time of ephemeris records given in header as MJD  25
my $EntriesDeltaTime;               # Time between CPF-nodes in file given in header           26
my $RecordMJD;                      # Node time days given as MJD                              27
my $RecordSeconds;                  # Node time seconds of day given as MJD                    28
my $RecordStartingMJD;              # Complete starting time of ephemeris records given as MJD 29
my $RecordEndingMJD;                # Complete ending time of ephemeris records given as MJD   30
my $DummyValue;                     # Value to handle unneeded return values                   31
my $FilenameInfoSatellitename;      # Value derived from filepath to check internals (         32
    satellitename)
my $FilenameInfoEphemerisSequenceNumber; # Value derived from filepath to check internals     33
    (sequence number)
my $FilenameInfoEphemerisSourceID;  # Value derived from filepath to check internals (         34
    ephemeris source ID)
my $Satellitename;                  # Value derived from header h1 to check internals (        35
    satellitename)
my $EphemerisSequenceNumber;        # Value derived from header h1 to check internals (        36
    sequence number)
my $EphemerisSourceID;              # Value derived from header h1 to check internals (        37
    ephemeris source ID)
# Comment tags are not relevant                                                               38
                                                                                              39
# Open file (old style open, so that older perl versions can handle it)                       40
if (!($Filepath =~ /([a-zA-Z0-9]+)_cpf_\d+_(\d+)\.([a-zA-Z0-9]+).*$/)) {                      41
    return (2,"Filename is wrong");;                                                          42
}                                                                                             43
open (FileHandle,"< $Filepath") or return (1,"Open not possible");                            44
                                                                                              45
# Get filepath information                                                                    46
($FilenameInfoSatellitename, $FilenameInfoEphemerisSequenceNumber,                            47
    $FilenameInfoEphemerisSourceID) =
        ($Filepath =~ /([a-zA-Z0-9]+)_cpf_\d+_(\d+)\.([a-zA-Z0-9]+).*$/);                     48
$FilenameInfoSatellitename = lc($FilenameInfoSatellitename);                                  49
$FilenameInfoEphemerisSourceID = uc($FilenameInfoEphemerisSourceID);                          50
                                                                                              51
# Read lines                                                                                  52
while (($Line = <FileHandle>)) {                                                              53
    if ($FileContainsHeaderTags[1] != 1) {                                                    54
        # Header                                                                              55
        # Header H1                                                                           56
        if ($Line =~ /^$HeaderH1$/){                                                          57
            if ($FileContainsHeaderTags[0] == 1) {                                            58
                $Error = 1;                                                                   59
                $ErrorMsg = "Header H1 found again";                                          60
                goto ValidateCPFFile_Return;                                                  61
            }                                                                                 62
            $FileContainsHeaderTags[0] = 1;                                                   63
            ($Error,$ErrorMsg) = GetHeaderInfoH1 ($Line,                                      64
                                                  \$DummyValue,                               65
                                                  \$EphemerisSourceID,                        66
                                                  \$DummyValue,\$DummyValue,\$DummyValue,\    67
                                                      $DummyValue,
                                                  \$EphemerisSequenceNumber,\                 68
                                                      $Satellitename,
                                                  \$DummyValue);                              69
            if ($Error) {                                                                     70
                $Error = 2;                                                                   71
                goto ValidateCPFFile_Return;                                                  72
            }                                                                                 73
            if (!($EphemerisSourceID =~ /$FilenameInfoEphemerisSourceID/) ||                  74
                !($EphemerisSequenceNumber =~ /$FilenameInfoEphemerisSequenceNumber/) ||      75
                !($Satellitename =~ /$FilenameInfoSatellitename/)) {                          76
```

```
        $Error = 2;                                                          77
        $ErrorMsg = "Filename doesn't fit to internal data";                 78
        goto ValidateCPFFile_Return;                                         79
    }                                                                        80
    next;                                                                    81
}                                                                            82
# Header H2                                                                  83
if (($Line =~ /^$HeaderH2$/)){                                               84
    if ($FileContainsHeaderTags[2] == 1) {                                   85
        $Error = 1;                                                          86
        $ErrorMsg = "Header H2 found again";                                 87
        goto ValidateCPFFile_Return;                                         88
    }                                                                        89
    ($Error, $ErrorMsg) = GetHeaderInfoH2 ($Line,                           90
                                                                             91
                          \$StartingYear ,\ $StartingMonth ,\               92
                            $StartingDay ,\ $StartingHour ,\
                            $StartingMinute ,\ $StartingSecond ,
                          \$EndingYear ,\ $EndingMonth ,\ $EndingDay ,\      93
                            $EndingHour ,\ $EndingMinute ,\
                            $EndingSecond ,
                          \$EntriesDeltaTime ,                               94
                          \$DummyValue ,\ $DummyValue ,\ $DummyValue ,\      95
                            $DummyValue ,\ $DummyValue );
    if ($Error) {                                                            96
        $Error = 2;                                                          97
        goto ValidateCPFFile_Return;                                         98
    }                                                                        99
    Timecalc::GRD2MJD ($StartingYear , $StartingMonth , $StartingDay , $StartingHour ,  100
        $StartingMinute , $StartingSecond ,
                      \$StartingMJD );                                       101
    Timecalc::GRD2MJD ($EndingYear , $EndingMonth , $EndingDay , $EndingHour ,  102
        $EndingMinute , $EndingSecond ,
                      \$EndingMJD );                                         103
    $FileContainsHeaderTags[2] = 1;                                          104
    next;                                                                    105
}                                                                            106
# Header H3                                                                  107
if (($Line =~ /^$HeaderH3$/)){                                               108
    if ($FileContainsHeaderTags[3] == 1) {                                   109
        $Error = 1;                                                          110
        $ErrorMsg = "Header H3 found again";                                 111
        goto ValidateCPFFile_Return;                                         112
    }                                                                        113
    $FileContainsHeaderTags[3] = 1;                                          114
    next;                                                                    115
}                                                                            116
# Header H4                                                                  117
if (($Line =~ /^$HeaderH4$/)){                                               118
    if ($FileContainsHeaderTags[4] == 1) {                                   119
        $Error = 1;                                                          120
        $ErrorMsg = "Header H4 found again";                                 121
        goto ValidateCPFFile_Return;                                         122
    }                                                                        123
    $FileContainsHeaderTags[4] = 1;                                          124
    next;                                                                    125
}                                                                            126
# Header H5                                                                  127
if (($Line =~ /^$HeaderH5$/)){                                               128
    if ($FileContainsHeaderTags[5] == 1) {                                   129
        $Error = 1;                                                          130
        $ErrorMsg = "Header H5 found again";                                 131
        goto ValidateCPFFile_Return;                                         132
    }                                                                        133
    $FileContainsHeaderTags[5] = 1;                                          134
    next;                                                                    135
```

```
            }
            # Header H9 — final tag                                          136
            if (($Line =~ /^$HeaderH9$/)){                                   137
                if ($FileContainsHeaderTags[1] == 1) {                       138
                    $Error = 1;                                              139
                    $ErrorMsg = "Header H9 found again";                     140
                    goto ValidateCPFFile_Return;                            141
                }                                                            142
                $FileContainsHeaderTags[1] = 1;                             143
                next;                                                        144
            }                                                                145
            # Record 00 (comment)                                           146
            if (($Line =~ /^$Record00$/)){                                   147
                next;                                                        148
            }                                                                149
            $Error = 1;                                                      150
            $ErrorMsg = "Unknown tag in header: ".$Line;                    151
            goto ValidateCPFFile_Return;                                    152
        }                                                                    153
        else {                                                               154
            # Record                                                         155
            if ($FileContainsLastRecord == 1) {                             156
                $Error = 1;                                                  157
                $ErrorMsg = "Record after final record tag: ".$Line;       158
                goto ValidateCPFFile_Return;                                159
            }                                                                160
            # Record 00 (comment)                                           161
            if (($Line =~ /^$Record00$/)){                                   162
                next;                                                        163
            }                                                                164
            # Record 10 with time information                               165
            if (($Line =~ /^$Record10$/)){                                   166
                ($Error,$ErrorMsg) = GetRecordInfo10 ($Line,               167
                                          \$DummyValue,                      168
                                          \$RecordMJD,\$RecordSeconds,      169
                                          \$DummyValue,\$DummyValue,\$DummyValue,\  170
                                              $DummyValue);                  171

                if ($Error) {                                               172
                    goto ValidateCPFFile_Return;                            173
                }                                                            174
                if ($FileContainsRecordTag == 0) {                         175
                    $RecordStartingMJD = $RecordMJD + ($RecordSeconds/86400.0);  176
                }                                                            177
                else {                                                       178
                    $RecordEndingMJD = $RecordMJD + ($RecordSeconds/86400.0);    179
                }                                                            180
                $FileContainsRecordTag = 1;                                 181
                $FileContainsTimeRecord = 1;                                182
                next;                                                        183
            }                                                                184
            # Record 50 with time information                               185
            if (($Line =~ /^$Record50$/)){                                   186
                ($Error,$ErrorMsg) = GetRecordInfo50 ($Line,               187
                                          \$DummyValue,                      188
                                          \$RecordMJD,\$RecordSeconds,      189
                                          \$DummyValue,\$DummyValue,\$DummyValue,\  190
                                              $DummyValue);                  

                if ($Error) {                                               191
                    goto ValidateCPFFile_Return;                            192
                }                                                            193
                if ($FileContainsRecordTag == 0) {                         194
                    $RecordStartingMJD = $RecordMJD + ($RecordSeconds/86400.0);  195
                }                                                            196
                else {                                                       197
                    $RecordEndingMJD = $RecordMJD + ($RecordSeconds/86400.0);    198
                }                                                            199
```

```
                $FileContainsRecordTag = 1;                                  200
                $FileContainsTimeRecord = 1;                                 201
                next;                                                        202
            }                                                                203
            # Record 60 with time information                               204
            if (($Line =~ /^$Record60$/)){                                   205
                ($Error,$ErrorMsg) = GetRecordInfo60 ($Line,                 206
                                        \$RecordMJD,\$RecordSeconds,         207
                                        \$DummyValue,\$DummyValue,\$DummyValue,\  208
                                        $DummyValue);
                if ($Error) {                                                209
                    goto ValidateCPFFile_Return;                             210
                }                                                            211
                if ($FileContainsRecordTag == 0) {                          212
                    $RecordStartingMJD = $RecordMJD + ($RecordSeconds/86400.0);  213
                }                                                            214
                else {                                                       215
                    $RecordEndingMJD = $RecordMJD + ($RecordSeconds/86400.0);  216
                }                                                            217
                $FileContainsRecordTag = 1;                                  218
                $FileContainsTimeRecord = 1;                                 219
                next;                                                        220
            }                                                                221
            # Record 70 with time information                               222
            if (($Line =~ /^$Record70$/)){                                   223
                ($Error,$ErrorMsg) = GetRecordInfo70 ($Line,                 224
                                        \$RecordMJD,\$RecordSeconds,         225
                                        \$DummyValue,\$DummyValue,\$DummyValue);  226
                if ($Error) {                                                227
                    goto ValidateCPFFile_Return;                             228
                }                                                            229
                if ($FileContainsRecordTag == 0) {                          230
                    $RecordStartingMJD = $RecordMJD + ($RecordSeconds/86400.0);  231
                }                                                            232
                else {                                                       233
                    $RecordEndingMJD = $RecordMJD + ($RecordSeconds/86400.0);  234
                }                                                            235
                $FileContainsRecordTag = 1;                                  236
                $FileContainsTimeRecord = 1;                                 237
                next;                                                        238
            }                                                                239
            # Record 20, 30, 40, 00 (comment) without time information     240
            if (($Line =~ /^$Record20$/) ||                                  241
                ($Line =~ /^$Record30$/) ||                                  242
                ($Line =~ /^$Record40$/) ||                                  243
                ($Line =~ /^$Record00$/)){                                   244
                $FileContainsRecordTag = 1;                                  245
                next;                                                        246
            }                                                                247
            # Header 99 — final tag                                         248
            if (($Line =~ /^$Record99$/) ){                                  249
                $FileContainsRecordTag = 1;                                  250
                $FileContainsLastRecord = 1;                                 251
                next;                                                        252
            }                                                                253
            $Error = 1;                                                      254
            $ErrorMsg = "Unknown tag in record: ".$Line;                    255
            goto ValidateCPFFile_Return;                                     256
        }                                                                    257
    }                                                                        258
                                                                             259
    # Check if header end with endtag                                       260
    if ($FileContainsHeaderTags[0] != 1 ||                                   261
        $FileContainsHeaderTags[1] != 1) {                                   262
```

```
            $Error = 1;                                                     263
            $ErrorMsg = "Header not complete";                             264
            goto ValidateCPFFile_Return;                                   265
        }                                                                   266
                                                                            267
        # Check if records end with endtag                                 268
        if ($FileContainsRecordTag != 1 ||                                 269
            $FileContainsLastRecord != 1) {                                270
            $Error = 1;                                                     271
            $ErrorMsg = "Records not complete";                           272
            goto ValidateCPFFile_Return;                                   273
        }                                                                   274
                                                                            275
        # Check semantic if header dates are correct for record dates     276
        if ($FileContainsTimeRecord == 1) {                               277
            if ($RecordStartingMJD > $StartingMJD &&                      278
                abs ($StartingMJD-$RecordStartingMJD) > ($EntriesDeltaTime/86400.0)) {   279
                $Error = 1;                                                 280
                $ErrorMsg = "Records starts too late (header info doesn't match with records)";   281
                goto ValidateCPFFile_Return;                               282
            }                                                               283
            if ($RecordEndingMJD < $EndingMJD &&                          284
                abs ($EndingMJD-$RecordEndingMJD) > (2*($EntriesDeltaTime/86400.0))) {   285
                $Error = 1;                                                 286
                $ErrorMsg = "Records ends too early (header info doesn't match with records)";   287
                goto ValidateCPFFile_Return;                               288
            }                                                               289
        }                                                                   290
                                                                            291
    ValidateCPFFile_Return:                                                292
        close (FileHandle);                                               293
                                                                            294
        return ($Error, $ErrorMsg);                                       295
}                                                                           296
```

File Format Example: XML

Another quite well-known format for files is nowadays described with the Extensible Markup Language (XML). XML is a description language for document structures and data representations. It is not a fixed format but more a structure to define individual formats. The background is the addition of meta-information. This appears in the form of markups, which add a label and relation information to a data value. A similar style is already familiar from HTML. But while HTML has fixed markups, XML makes it possible for a format developer to define his own markups with their own meanings. Therefore, it is more a metalanguage. The important basis is that each document is defined by a document definition (e.g., with the Document Type Definition (DTD)) or a scheme description, which is, in principle, a mapping of the underlying grammar. They define the replacement rules and, therefore, structures of the individual document. It is possible to check automatically if a document is «well formed», which means if it follows the correct implementation of the syntax. XML makes it possible to define namespaces for the data identifiers. Values can be linked with XLink or can be addressed with XPath, as they define a hierarchical tree structure. With the available parsing tools, such as Document Object Model (DOM) or the simpler, event-driven parser Simple API for XML (SAX) with its activation of callback functions for each correctly detected markup, XML is a very powerful document metalanguage. This explains why it is still so popular for web services (about XML, see also (Kazakos et al. 2002, *l.c. page 7 ff.*) and (Neidhardt 2005, *l.c. page 131 ff.*)). Experience at the Geodetic Observatory Wettzell has shown that the XML structures are too bulky for engineers of other disciplines if data management is not their daily business. Therefore, the use of XML was reduced. But some of the tools are able to produce documents in XML. If the programs or outputs are used for other observatories, the inclusion of XML is a suitable feature. One of these programs is the orbit and passage calculation for observable satellites, which uses the CPF files to create observatory-specific observation schedule and tracking files. An excerpt output in XML from the orbit calculation for the Wettzell site is shown in Example 4.9 for the CPF input file «`glonass102_cpf_120528_6511.hts`» (it dispenses with a document definition in this case). The according file with the quick info about the passage with rise, transit, and set times (valid times) and the start and end times for the periods, when the satellite is in the state «sunlit»[61] or in the state «sun avoidance»,[62] is shown in Example 4.13.

■ ■ **Example 4.12**

Excerpt of the calculated passage file `glonass102_6511_hts_20120529080636.pass` for the laser ranging in XML format.

```
<PASSAGE>                                                                    1
    <HEADER>                                                                 2
        <SIC> 9102 </SIC>                                                    3
        <EPHEMERIS_SEQUENCE> 6511 </EPHEMERIS_SEQUENCE>                      4
        <EPHEMERIS_SOURCE> HTS </EPHEMERIS_SOURCE>                           5
    </HEADER>                                                                6
    <BODY>                                                                   7
        <IRVPASS EPOCH="2012.05.29 08:06:36" MJD="56076.3379166667" X="23250218.350278" Y="   8
            -4967483.386056" Z="-9099762.035900" VX="1241.999528" VY="-263.784252" VZ="
            3326.644995" AZ="204.266198" EL="2.001533" RA="24407416.561233" SUNLIT="1" SUNAVOID=
            "0"/>
        <IRVPASS EPOCH="2012.05.29 08:06:37" MJD="56076.3379282407" X="23251460.111494" Y="   9
            -4967747.214059" Z="-9096435.280961" VX="1241.522899" VY="-263.871741" VZ="
            3326.864871" AZ="204.268601" EL="2.009284" RA="24406561.347256" SUNLIT="1" SUNAVOID=
            "0"/>
        <IRVPASS EPOCH="2012.05.29 08:06:38" MJD="56076.3379398148" X="23252701.396063" Y="  10
            -4968011.129514" Z="-9093108.306185" VX="1241.046233" VY="-263.959156" VZ="
            3327.084667" AZ="204.271005" EL="2.017035" RA="24405706.144262" SUNLIT="1" SUNAVOID=
            "0"/>
        <IRVPASS EPOCH="2012.05.29 08:06:39" MJD="56076.3379513889" X="23253942.203947" Y="  11
            -4968275.132347" Z="-9089781.111654" VX="1240.569530" VY="-264.046496" VZ="
            3327.304382" AZ="204.273409" EL="2.024786" RA="24404850.952273" SUNLIT="1" SUNAVOID=
            "0"/>
        <IRVPASS EPOCH="2012.05.29 08:06:40" MJD="56076.3379629630" X="23255182.535111" Y="  12
            -4968539.222482" Z="-9086453.697448" VX="1240.092791" VY="-264.133762" VZ="
            3327.524017" AZ="204.275815" EL="2.032537" RA="24403995.771309" SUNLIT="1" SUNAVOID=
            "0"/>
        <IRVPASS EPOCH="2012.05.29 08:06:41" MJD="56076.3379745370" X="23256422.389517" Y="  13
            -4968803.399846" Z="-9083126.063647" VX="1239.616016" VY="-264.220953" VZ="
            3327.743572" AZ="204.278220" EL="2.040288" RA="24403140.601389" SUNLIT="1" SUNAVOID=
            "0"/>
        <IRVPASS EPOCH="2012.05.29 08:06:42" MJD="56076.3379861111" X="23257661.767131" Y="  14
            -4969067.664364" Z="-9079798.210332" VX="1239.139204" VY="-264.308070" VZ="
            3327.963046" AZ="204.280627" EL="2.048040" RA="24402285.442534" SUNLIT="1" SUNAVOID=
            "0"/>
```

61 Sunlit means that the sun illuminates the satellite.
62 Sun avoidance means that the satellite is in front or close to the direction of the sun where a tracking is not possible without damaging the sensitive detector and optics system.

```
<IRVPASS EPOCH="2012.05.29 08:06:43" MJD="56076.3379976852" X="23258900.667914" Y="        15
    −4969332.015961" Z="−9076470.137582" VX="1238.662357" VY="−264.395112" VZ="
    3328.182440" AZ="204.283034" EL="2.055792" RA="24401430.294765" SUNLIT="1" SUNAVOID=
    "0"/>
<IRVPASS EPOCH="2012.05.29 08:06:44" MJD="56076.3380092593" X="23260139.091832" Y="        16
    −4969596.454563" Z="−9073141.845478" VX="1238.185473" VY="−264.482080" VZ="
    3328.401754" AZ="204.285442" EL="2.063544" RA="24400575.158101" SUNLIT="1" SUNAVOID=
    "0"/>
<IRVPASS EPOCH="2012.05.29 08:06:45" MJD="56076.3380208333" X="23261377.038847" Y="        17
    −4969860.980096" Z="−9069813.334101" VX="1237.708552" VY="−264.568972" VZ="
    3328.620987" AZ="204.287850" EL="2.071297" RA="24399720.032563" SUNLIT="1" SUNAVOID=
    "0"/>
                                                                                              18
                                                                                              19
...                                                                                           20
                                                                                              21
<IRVPASS EPOCH="2012.05.29 14:18:40" MJD="56076.5962962963" X="3280246.813589" Y="
    25279649.914337" Z="1920509.506736" VX="273.367458" VY="237.781824" VZ="−3587.007855
    " AZ="102.732469" EL="2.017590" RA="24530659.398245" SUNLIT="1" SUNAVOID="0"/>
<IRVPASS EPOCH="2012.05.29 14:18:41" MJD="56076.5963078704" X="3280520.167959" Y="           22
    25279887.441833" Z="1916922.475978" VX="273.341268" VY="237.273167" VZ="−3587.053649
    " AZ="102.738280" EL="2.011684" RA="24531307.856177" SUNLIT="1" SUNAVOID="0"/>
<IRVPASS EPOCH="2012.05.29 14:18:42" MJD="56076.5963194444" X="3280793.496099" Y="           23
    25280124.460670" Z="1913335.399468" VX="273.314999" VY="236.764508" VZ="−3587.099357
    " AZ="102.744092" EL="2.005778" RA="24531956.328191" SUNLIT="1" SUNAVOID="0"/>
                                                                                              24
</BODY>
</PASSAGE>                                                                                    25
```

■ ■ **Example 4.13**

Excerpt of the quick info file 'glonass102_6511_hts_20120529080636.passinfo' for the calculated passage file 'glonass102_6511_hts_20120529080636.pass' in XML format.

```
<PASSAGEINFO>                                                                                 1
  <HEADER>                                                                                     2
    <SIC> 9102 </SIC>                                                                          3
    <EPHEMERIS_SEQUENCE> 6511 </EPHEMERIS_SEQUENCE>                                            4
    <UNIXEPOCH_OF_EPHEMERIS> 1338206400 </UNIXEPOCH_OF_EPHEMERIS>                              5
    <EPHEMERIS_SOURCE> HTS </EPHEMERIS_SOURCE>                                                 6
    <DELTATIME> 1.000000 </DELTATIME>                                                          7
  </HEADER>                                                                                     8
  <BODY>                                                                                        9
    <VALIDFROM EPOCH="2012.05.29 08:06:36" UnixEpoch="1338278796" AZ="204.266198" EL="       10
        2.001533" RA="24407416.561233" />
    <TRANSIT EPOCH="2012.05.29 10:53:41" UnixEpoch="1338288821" AZ="317.273887" EL="         11
        75.004242" RA="19285284.061568" />
    <VALIDTO EPOCH="2012.05.29 14:18:42" UnixEpoch="1338301122" AZ="102.744092" EL="         12
        2.005778" RA="24531956.328191" />
    <SUNLITFROM EPOCH="2012.05.29 08:06:36" UnixEpoch="1338278796" AZ="204.266198" EL="      13
        2.001533" RA="24407416.561233" />
    <SUNLITTO EPOCH="2012.05.29 14:18:42" UnixEpoch="1338301122" AZ="102.744092" EL="        14
        2.005778" RA="24531956.328191" />
  </BODY>                                                                                      15
</PASSAGEINFO>                                                                                 16
```

File Format Example: Proprietary Configuration File Format

As XML is too confusing for developers at observatories, but as the nested tree structures are quite suitable to combine data sets, an XML-alike proprietary format was developed at the Wettzell observatory. As different data sets must be compiled for the system configuration of the autonomous production cell, the format is used especially for the configuration files. It also uses something like markups. But the implemented identifiers are much simpler than the XML ones. There are only two types: block definitions for the beginning and ending of a value grouping and the value definitions. A restriction to simplify the format is that each data or block definition must use a separate line and that each configuration file must at least consist of one root block definition. Block definitions which are similar to XML, for example, for the value grouping «SLRSite», the styles «<SLRSite>» for the beginning and «</SLRSite>» for the ending of the block are used. But these definitions do not have arguments. They are just grouping tags. The data definition lines for a single configuration parameter are between the grouping tags and consist of an identifier, followed by an equal sign and the value. Optionally it is possible to add comments, which begin with «#» and end at the end of the same line. To allow a rudimentary type checking (similar to the grammar check of floating points in ◘ Fig. 3.7), it is possible to add format tags, such as «%f» or «%i» directly behind the «#» sign of a following comment. The structure allows a simple use in text editors, as unnecessary formal design elements are avoided for the developer. A sample of such a configuration file is shown in Example 4.14.

It defines a hierarchical tree structure where the connection nodes are the block identifiers and where the final level carries the leaves given by the value settings. To read the proprietary configuration files, the module «simple_structured_conf» is offered as part of the code toolbox. It reads a complete file and creates a syntax tree with all the values in the dynamic heap memory. After that, each single element can be addressed directly, using the path through the tree (e.g., «SLRSite_0.CPF_0.CPFSpider_0.CPFServer_1.FtpServer» addresses the value «edc.dgfi.badw.de» in Example 4.14; the numbers define the ordered number of the node on its level). Another address mode uses the relative position of a value, counted from the beginning (e.g., «SLRSite.CPF.CPFSpider.CPFServer.FtpServer_1» addresses the same value as before). It is even possible to use the «*» (asterisk) in the tree path to find all nodes with a specific path combination. The tool can cut out complete block sections from one configuration file to save them as a reduced new one (useful if several autonomous control cells are served from one central configuration file for the whole autonomous production cell). All of these access methods are coded within the parser module from the toolbox. All configuration files of all autonomous cells use this style. Therefore, it is possible to host a central configuration in the data management cell, from which all individual cells fetch their local, individual configuration section. This is an essential mechanism to allow a central overview of all configuration settings in the data management cell, so that an administrator just has to make his changes there.

Toolbox Module for Configurations

▪▪ Example 4.14

Excerpt of the configuration file for the Wettzell Laser Ranging System in the XML-alike proprietary configuration format.

```
# ********************************************************************    1
# * Configuration file for SLR sites                           *    2
# *                                                            *    3
# ********************************************************************    4
                                                                      5
<SLRSite>                                                             6
    # System identification data                                     7
    SystemName          = WLRS          # System name                8
    SystemShortkey      = WTZ           # Shortkey                    9
    SystemShortname     = WETL          # official 4-letter site code 10
    SystemID            = 8834          # System identification (Crustal Dynamics Project  11
          ID, e.g. WLRS = 8834)
```

```
SystemDescription        = Wettzell Laser Ranging System    # Short system description    12
SystemPlace              = Wettzell           # System place                              13
SystemTechnique          = SLR                # System technique                          14
SystemStartupDate        = 2008-06-01         # System startup date                       15
SystemChangeDate         = 2008-06-01         # System change date of last change         16
SystemChangeIndicator    = 1                  # System change indicator                   17
SystemCDPOCC             = 75947901           # System crustal dynamics project occupation number  18
SystemCDPSYS             = 88341001           # System crustal dynamics project system number      19
SystemDomesNumber        = 14201S018          # System domes number (e.g. WLRS = 14201S018)        20
                                                                                          21
# ITRF position data                                                                      22
<ITRF>                                                                                    23
    ITRFBasis            = ITRF2000           # Basis of ITRF (ITRF and 'year')           24
    GeografXPosition     = 4075576.85         #%f Position of SLR site X                  25
    GeografYPosition     = 931785.456         #%f Position of SLR site Y                  26
    GeografZPosition     = 4801583.559        #%f Position of SLR site Z                  27
    SigmaGeografXPosition = 0.001             #%f Position of SLR site X                  28
    SigmaGeografYPosition = 0.001             #%f Position of SLR site Y                  29
    SigmaGeografZPosition = 0.001             #%f Position of SLR site Z                  30
    VX                   = -0.0157            #%f                                         31
    VY                   = 0.0172             #%f                                         32
    VZ                   = 0.0087             #%f                                         33
    SigmaVX              = 0.0                #%f                                         34
    SigmaVY              = 0.0                #%f                                         35
    SigmaVZ              = 0.0                #%f                                         36
</ITRF>                                                                                   37
                                                                                          38
# ...                                                                                     39
                                                                                          40
# CPF definitions                                                                         41
<CPF>                                                                                     42
    DatabaseName         = slr                     # Database name                       43
    DatabaseHost         = wlrsdbold.wlrs          # Database host                       44
    DatabaseUsername     = xxxxxx                  # Database user                       45
    DatabasePassword     = xxxxxx                  # Database password                   46
    LocalCPFFilesPath    = /fsdata/cpf_predicts/   # Directory for CPF files             47
    <CPFSpider>                                                                           48
        WebLogFilepath = /slruser/data/systeminfos/cpfspider.html # File where html log is  49
            saved to
        MultiServerLogic = OR                      # Mode when more than one server      50
            is used (AND, OR (=on error))
        <CPFServer>                                                                       51
          FtpServer = cddis.gsfc.nasa.gov               # Server IP                      52
          Username  = xxxxx                        # FTP user                            53
          Password  = xxxxx                        # Password of FTP user                54
          Mode      = ascii                        # Transfer mode                       55
          RemoteCPFFilesPath = ./pub/slr/cpf_predicts/  # Relative directory to CPF      56
              files
          Filemask  = [a-zA-Z0-9]+_cpf_\d{6}.+     # Mask to identify searched           57
              files
        </CPFServer>                                                                      58
        <CPFServer>                                                                       59
          FtpServer = edc.dgfi.badw.de                  # Server IP DGFI                 60
          Username  = xxxxx                        # FTP user                            61
          Password  = xxxxx                        # Password of FTP user                62
          Mode      = ascii                        # Transfer mode                       63
          RemoteCPFFilesPath = ./pub/slr/cpf_predicts/     # Relative directory to CPF   64
              files
          Filemask  = [a-zA-Z0-9]+_cpf_\d{6}.+     # Mask to identify searched           65
              files
        </CPFServer>                                                                      66
    </CPFSpider>                                                                          67
                                                                                          68
# ...                                                                                     69
                                                                                          70
</SLRSite>                                                                                71
```

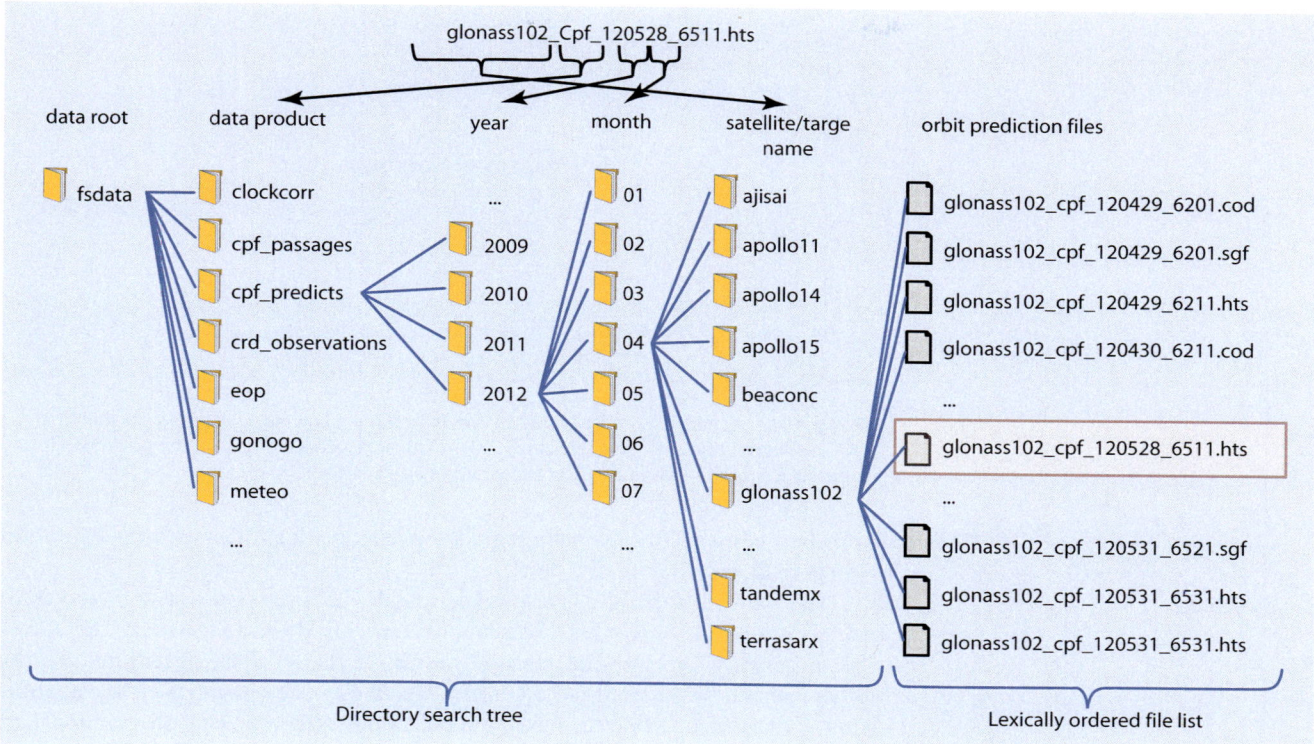

□ **Fig. 4.35** The organization of files with laser ranging prediction in a suitable search tree structure on the basis of the directory hierarchy and (ordered) file lists

The formats[63] described in combination with suitable directory structures allow the organization of laser ranging data in the regular file systems in an appropriate way. Single-level data structures, such as unsorted lists,[64] ordered lists,[65] ordered or non-ordered tables,[66] or even linked lists,[67] and multi-level structures, such as binary trees or general trees[68] with search paths, are possible (Schicker 1996, *l.c. page 21 ff.*). Hierarchical file systems enable such tree structures in a restricted way, as the search trees are defined by the directory names (see □ Fig. 4.35 for the organization of files with the laser ranging predictions). The leaves of these trees, which are the (lexically) ordered files,[69] can only be

Single-level and Multi-level Data Structures

---

63 Nowadays, Network Common Data Format (NetCDF) is also a very popular file format for scientific data, which also uses the XDR for the extended data representation (Unidata n.d.). But currently, it plays no big role in geodetic space techniques.

64 The individual entries have no order. No meta-information is used. The elements do not know anything about their previous and following elements. A search is only sequential and new elements are always added at the end.

65 The individual entries are ordered and, therefore, allow search mechanisms like the use of a binary search tree (see the following section). The elements do not know anything about their previous and following elements. New elements must be added at the right position in the ordered list.

66 The individual entries are records with a key element and according values (each row represents a dedicated record). The elements can be addressed with their index (address) or using the key element which can be mapped to the index. The elements do not know anything about their previous and following elements. New elements can be added somewhere according to the position found by the key algorithm. Fixed element sizes are usually used. Records which are not completely filled waste memory space.

67 Same as ordered lists but each element contains a pointer to the address of the following and/or previous element. Therefore, the size of the records is arbitrary, and the elements can be located somewhere in the memory using the addresses stored in the records.

68 Additional meta-elements are used to split a sequential list into a hierarchically searchable structure.

69 The order is generated only by the program which reads the directory content. The storage of the files has not to be ordered.

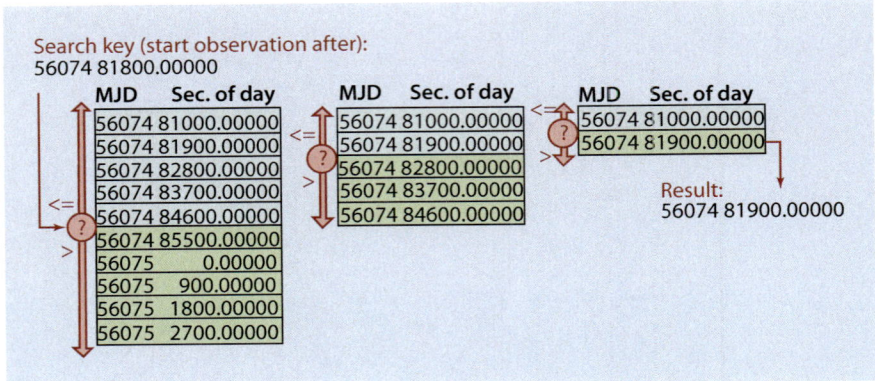

Fig. 4.36 Application of the binary search, to find the following supporting point after a specific observation start time (Modified Julian Date and seconds of day) in the prediction list from the file for the «glonass102» of the previous examples

matched sequentially against a search key (e.g., print all files with the extension «.hts» in ◘ Fig. 4.35).

Binary Search

The content of the files is also sequential. But it is possible that it is read completely into single- and multi-level data structures in the heap memory. This allows fast accesses with array indexes and hashes. It also enables algorithms like the binary search for ordered record entries (see ◘ Fig. 4.36 which shows the application of the search to find the next valid orbit point after a specific observation start time). The binary search splits a sequentially ordered list into two, ideally equal-sized sets of values and compares the search string with the list values at the position where it is cut. If the value is (e.g., lexically) smaller or equal to the last value of the first set, the first block is taken to process the same algorithm of splitting and comparing again. This iterative processing is continued until only one element is left, which is the found value, or no match was found. The efficiency of this method is tremendous: for example, for one million values, a sequential search would require in mean 500,000 search accesses, while the binary search would get the result within 20 iterations, as 1,000,000 is about 2^{20} (Schicker 1996, *l.c. page 19 f.*).

Limits of Directories and File Formats

The limits of these organizations are reached if data needs to be combined or selected from different, separate data files. This is, for example, necessary to find all currently visible satellites to plan the next tracking observation. Up to a certain degree, it is possible to use the selection with the directory search path. But the final level with the files must be searched sequentially. It is possible to differentiate the directory search path much more (e.g., with time keys into directory levels for the year, the month, the day, the hour, the minute, the second, the ephemeris source, etc.), so that at least the path for one particular search criterion directly ends at a single, matching file. But the problem is that the structure then only supports this very specific search path. If another criterion is used (e.g., a request for all currently used ephemeris source centers), the whole directory structure must be searched sequentially as long as no multiple structures with duplicated (or linked) files are created. But this redundancy and housekeeping would increase the required memory and the costs for updates and changes as all redundant files would need to be touched. Another limitation is given in a multi-user access. The file system structures only offer very limited control to ensure that parallel read and write accesses are safe (see the problem of critical section from ▶ Sect. 2.6.4 which is a similar class of problems).

Use of Files for Specific Laser Ranging Data

Data management which is only based on file systems is not really ideal for the administration of data sets in an autonomous data management cell. It is very suitable for data which are sequential in their structures, occupy larger volumes of memory, are limited to one file, exist for a longer time without changes, and require just one search path or ideally no search at all. This means they are ideal for the general organization of the prediction files, passage data files, observation files, or configuration files. But additional data about the satellites, the passages, the data centers, and so on, which are used for the planning of the scheduling and which must be combined and compiled, require better organization. Another task is to organize the files in the file system with additional metadata to allow faster, extended search requests, so that the relevant files can be found easily. Another aspect is that it should also manage the multiple accesses from different clients with its parallel read and write tasks. This can be done ideally by a database management system for databases.

> *Definition DEF4.7 of «database»* (see (Schicker 1996, *l.c. page 46*))
> A database is a collection of file sets (data stock) with logical dependencies and administrated by a DBMS including a standardized access interface for the user requests.

Database

The use of a DBMS with its user interface and the internal data stock gives a different view of data as being possible with single files in hierarchical file systems (see ◘ Fig. 4.37). But there are different database models. One still very popular

Relational Database Concept

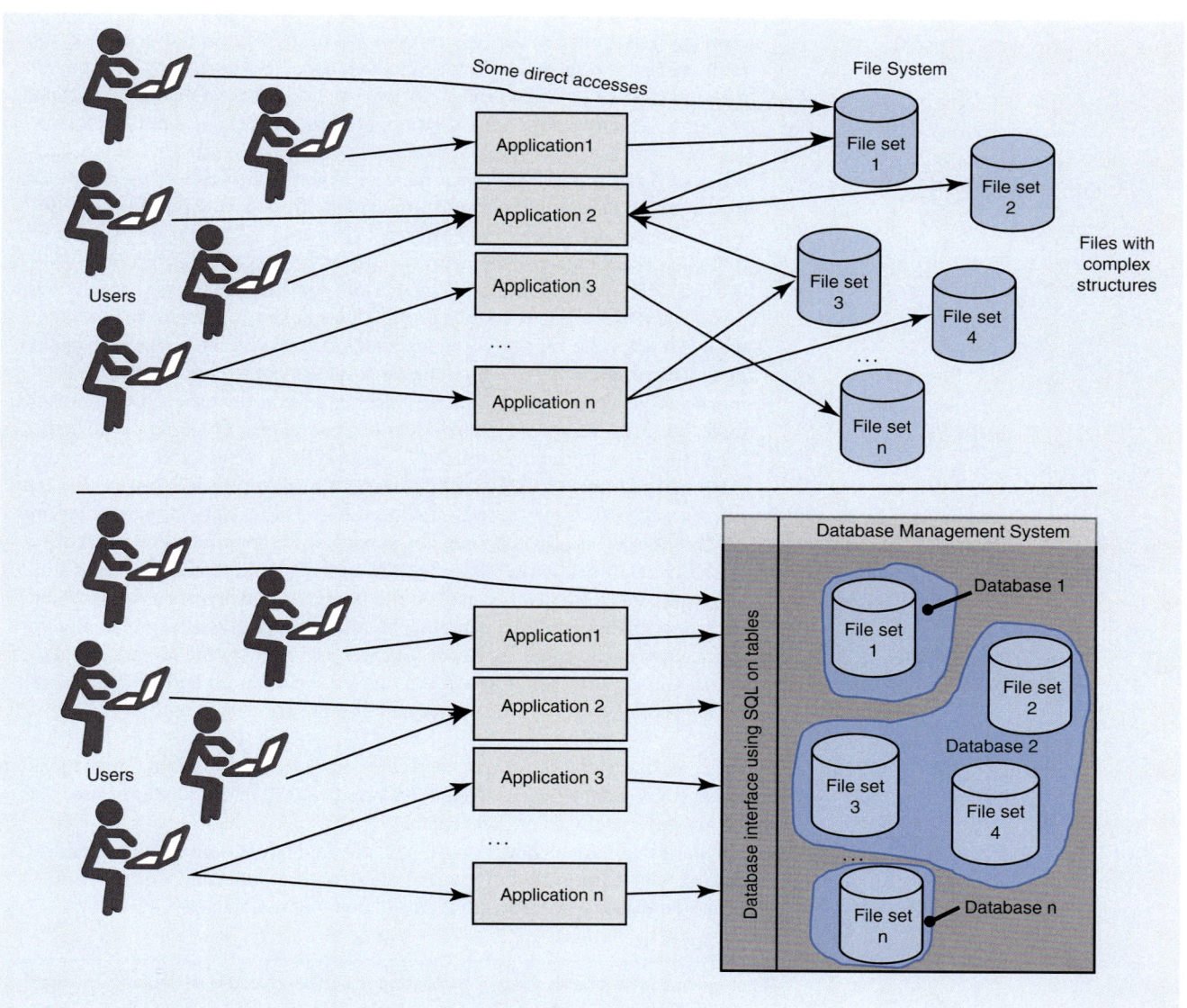

◘ **Fig. 4.37** The difference between file systems and a database management system with databases and an application interface (Adapted from Schicker (1996) *l.c. page 45 ff.*)

design uses the relational database model.[70] It organizes the data in the form of tables. Each table has a headline where each column gets a unique name. This defines the attributes of the table. Each of these attributes has an assigned type and can define restrictions (e.g., that an entry is not allowed to be empty or that it must be equal to a set of predefined values). One or a set of specific attributes can be used to define a non-empty key index, the primary key. The body of the table contains the records. Each line with its data tuples is one record. The records are also not ordered, and the same primary key is not permitted for two different records. Each record is filled with values, where each attribute is a container for exactly one value ((Schicker 1996, *l.c. page 57*) and (Schicker 1996, *l.c. page 63 ff.*)). Using the primary keys, it is possible to define relations between the different tables. The primary key of one table is used as a foreign key in another table. This means that the foreign key points to a value in the related table, which is selectable with its primary key (Schicker 1996, *l.c. page 74 ff.*). It is possible to create the following relation cardinalities with this method (Schicker 1996, *l.c. page 137 ff.*):

- *1:1/0 relationship*[71]: A record in one table has exactly one relational record in another (1:1). Usually it would be possible to combine both tables into one. But for reasons of clarity, the relation is a structuring possibility. But it makes more sense for attributes which do not exist in all records (e.g., the stellar aberration correction in meters in the record type «30» of the CPF would be such a case, as the record is only defined for some satellites). To avoid empty elements in the records, a suitable solution is to create an additional table for the optional elements. This forms the 1:1/0 relation, as it can exist (1:1) or not (1:0).

- *1:n relationship*: If a record has several related other records, then it is a 1:n relation. An example is the assignment of targets to a satellite, for example, the moon carries several targets («n» targets), like Apollo 11, Apollo 14, Apollo 15, and so on, but each target only belongs to the moon («1» carrier). The 1:n relationship described is realized with two tables, in the example described, one for the satellites (in the example the moon) and one for the targets (here the reflectors). The targets table then contains the foreign key as a link to the satellite, and each record in the targets table defines exactly one satellite, which is identified by its primary key to which the foreign keys of the targets point.[72]

- *m:n relationship*: If a record in one table has several related records in another table, but these can also be dependent on several other records in the first table, then it is an m:n relationship. An example is the assignment of ephemeris source centers to CPF files for a target. On the one hand, one center can calculate the orbits for several satellites. On the other hand, the orbit for one satellite (target) is also calculated by several, different ephemeris source centers. This relationship requires an additional relation table, as the cross-linking is not possible with just two tables. The relation table defines a primary key for each m:n relation and just contains the foreign keys of the two related tables (here a key to the ephemeris source center and a key to the target). In this additional table, it is then possible to assign the different entries of the targets with optional entries in the source center table and vice-versa.

Database Principles

This means that (relational) databases offer several administrative principles on data sets ((Schicker 1996, *l.c. page 50 ff.*) and (Saake et al. 2008, *l.c. page 7 ff.*)), which are quite useful to organize data:

- Databases contain a collection of logically dependent data sets in several, related tables, integrated in internal file structures and independent from other databases with different logically dependent data sets.

---

70   Nowadays the relational model is extended and partly replaced with an object-oriented model. It organizes table relations similar to the attribute inheritance in OOP languages.

71   This relationship is also named as 1:c, where c is 1 or 0.

72   The CPF does not define satellites anymore, but only targets in the file. This means that the target name is the main key for almost all relationships.

- Well-designed databases organize data with reduced redundancy. Constraints for key relations between tables force an automated data propagation.
- Data sets in the form of tables which are accessible via the user interface are independent from the actual, internal, physical storage, which is encapsulated by the management system.[73]
- The system offers individual user views on the data and defines relationships between the tables, using key values as links.
- The data in the databases are secure and protected from unauthorized access by mechanisms of authentication and authorization to differentiated, single-data tables.
- The management keeps the physical and logic integrity, consistency, and correctness of the data by the use of constraints for data checks (e.g., value intervals or allowed value lists) and can partly check the semantic integrity.
- Databases allow operations on the data, like creating, inserting, updating, deleting (data mutations), and searching/selecting (queries), on the basis of mathematical set operations from the relational algebra (see ▫ Fig. 4.38 (Schicker 1996, *l.c. page 82 ff.*)). The database management must optimize the complexity, redundancy, access times, and memory consumption. Each operation is one atomic process.

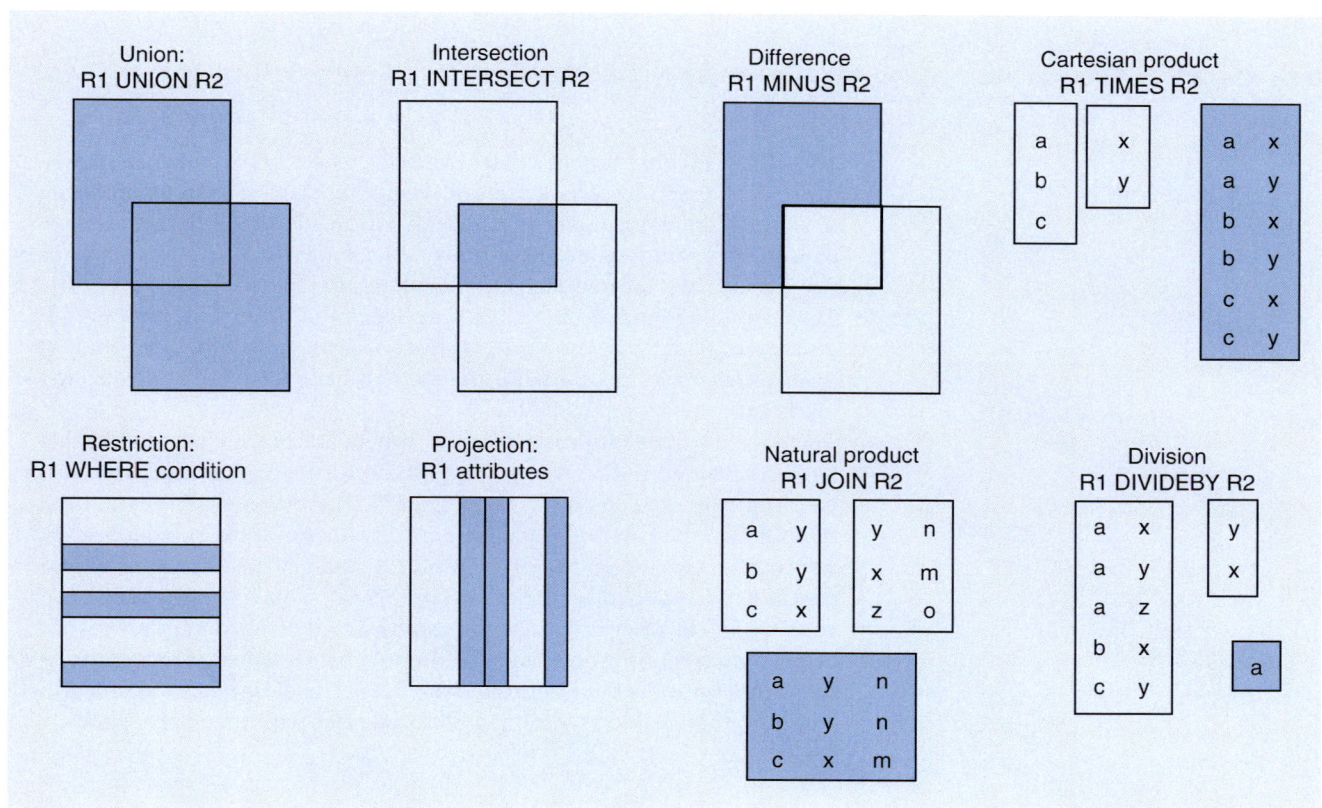

▫ **Fig. 4.38** Possible mathematical set operations of the relational algebra applied to tables of databases ((Schicker 1996, *l.c. page 82 ff.*))

---

73 The physical mapping to the file system is usually done with Index Sequential Access Method (ISAM) files. These files contain referenced blocks of a size used by the operating system for one read/write access. Multilayered header blocks refer to other header blocks or data blocks, so that they split the data into ordered lists, which can be identified by the first element key and which are addressed by the block address of this first element. The result is a tree-like structure, which allows accesses with a minimum number of physical read/write actions (Schicker 1996, *l.c. page 28 ff.*).

4

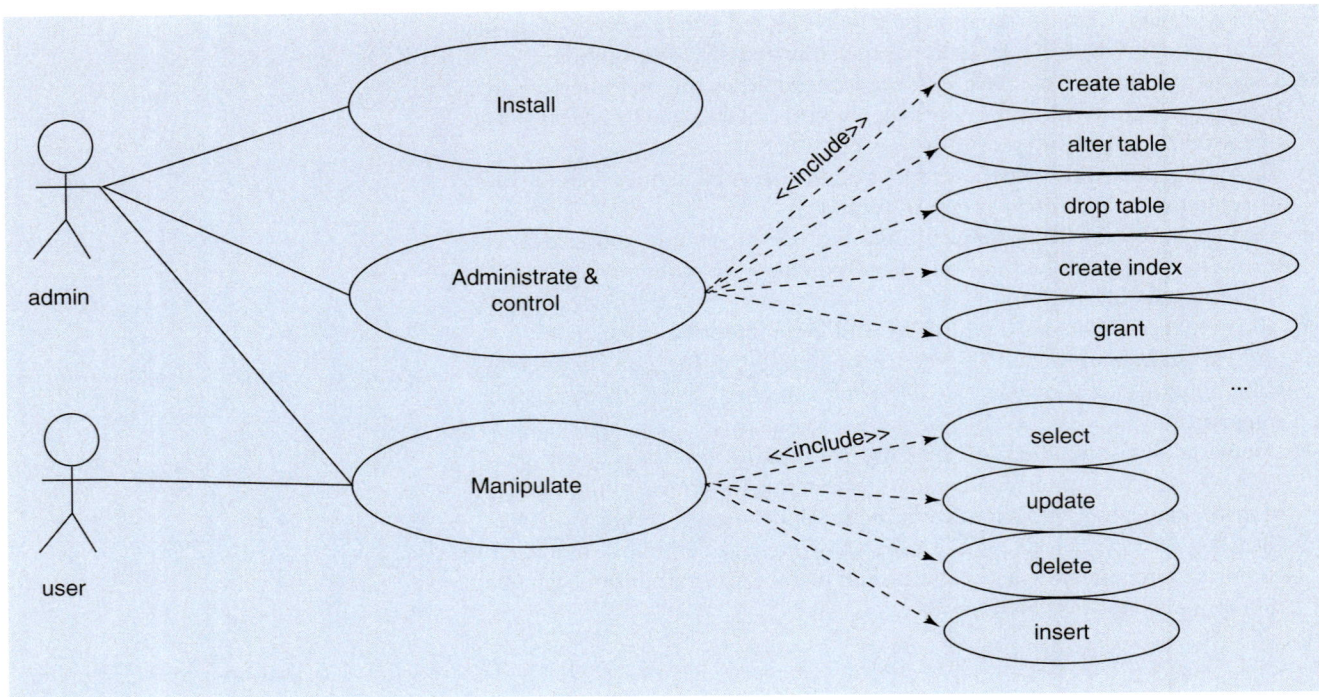

◘ Fig. 4.39 The user roles in a database system with the according operations in the form of a UML use case diagram (see also ▶ Sect. 2.5.1)

— The management system of a database enables safe concurrent accesses to the same data sets by several users. It synchronizes queries, sequentializes operations, and implements transactions, which guarantee that sequences of concurrent operations are either processed completely as an atomic process or not at all to keep the integrity and consistency of the data.

— The data storage and all operations are reliable and stable. This must even be guaranteed with a fail-safe scenario after a system crash, which also prevents inconsistencies in the data sets if transactions have been processed during the blackout.

— The databases are administrable, and system states are always checkable by a database administrator, so that the filling degree, the used volume or user statistics, and so on can be generated. Therefore, user roles can also be split into the two types of an «administrator» and the standard «user», which have specific operation possibilities and get specific rights granted (see ◘ Fig. 4.39 (Schicker 1996, *l.c. page 55 f.*)). This diversification is also suitable for the laser ranging database, as usually all program processes of the autonomous production cell do not change database structures and only work with the data in these structures. The design itself is predefined and only changeable by a human administrator with the appropriate knowledge.

Database Design: Normalization

The designing of the structure and relations in a database is a complex task. The design has to follow specific constraints to avoid unnecessary redundancies or unnecessary memory consumption and to offer multiple combination possibilities with given relations between the tables. The way to do that is to process a normalization on the tables to convert the tables to «normal forms». There are different definitions of normal forms for databases, but the most popular and in most cases sufficient types are the first three, defined by Boyce/Codd (see ◘ Fig. 4.40 and (Schicker 1996, *l.c. page 120 ff.*), (Warner 2007, *l.c. page 129 ff.*), and (Saake et al. 2008, *l.c. page 175 ff.*)).

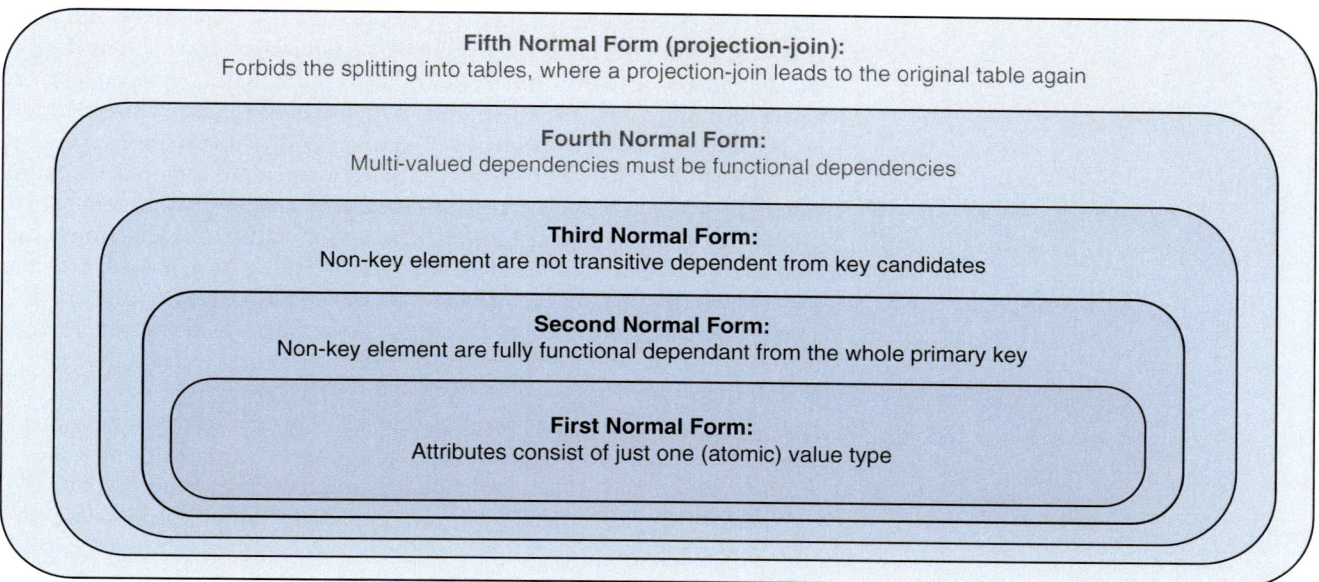

Fig. 4.40 The most important normal forms for relational databases

The goal of these formalized definitions is to reduce redundancies and to optimize relations. For practical use, experience has shown that the following recipe is a suitable way to reach this goal of normalized data tables for the limited use cases of a geodetic observatory:

Database Design: Recipe to Avoid Redundancies

1. Take enough suitable sample data which gives a good impression of the complete data stock.
2. Group data to logical units according to their similar properties. Such collections form the tables. To do this, search for values which can be used as topics for others. This creates a functional dependency. Usually there will be several key candidates. Ideally, those taken should form a general topic (like a headline of the table), so that most of the other values can be subordinated to them (like the subtitles of the table contents). These topic values are then primary keys.
3. Arrange the values so that each value gets its own column with a headline attribute, so that the usual tables are formed (first normal form).
4. Avoid combined keys. This can be done by creating an additional attribute column, which then acts as a key. If it is not possible to use single key attributes, check if each other element subordinates to the complete key as topic and not just to parts of the key. Hints are redundant values in a different column which is not dependent on other key candidates, but on the primary key or parts of it. If such a functional dependency on a key part is found, split the table into parts, so that the values which are just dependent on a part of the key form a new table. The primary key of this new table is the key part from the previous table (second normal form).
5. Usually key candidates are subordinated to the primary keys. But other values can be subordinated to those keys as well. Hints are again redundant values in a column, which is not only subordinated to the primary key but also to another key candidate. Check if the primary key is really the main topic for the data table. If it is, take the key candidate and its dependent values and form a new table, where the key candidate is a new primary key. This key can then be used as a foreign key column in the original table (third normal form).
6. Check all further redundancies and try to reduce them by grouping the parts to new tables.

7. Take all the elements which are partly empty and create new tables with three columns: a new primary key, the optionally empty value, and a foreign key, which points back to the primary key of the original table (1:c dependencies).

8. Check all tables for relations to others. Is there a need to relate tables to others because of a causal dependency? Causal dependencies can be found if queries combine table values in an additional form. If it is required, check if a «1:n» or a «m:n» relationship is given. Then add an additional column with a foreign key to that one table (1) which has a relationship to several others (n). If an «m:n» relation is given, add a new relation table with a new primary key and two foreign keys, one as a link to the primary key of the first and one as a pointer to the primary key of the second table. Check if an attribute is more dependent on the relationship than on a single table, and rearrange it to the new relation table.

9. Look through the tables and relations and add the required types and constraints to each attribute (e.g., not null, foreign key to table with a dedicated primary key, limited value selection, limits for an interval value, cascading values to propagate changes via relations, etc.) (PostgreSQL Global Development Group 2003, *l.c. page 68 ff.*).

10. Simulate the most common operations for inserts, updates, deletes, and selects. Check which tables are touched if the operations work smoothly with the designed structure and if the values are propagated via the cascades to the related tables.

Database Design: ERM

The result of the above process is the final database design. For a better human interpretation, it can also be drawn using the Entity Relationship Model (ERM) diagrams. Throughout the history of these graphs, several slightly different notations with some changes in the graphical representations were established. For example, the Chen notation of an ERM diagram (see ▢ Fig. 4.41, which shows the relation between «tar_targets» and «tar_passive_artificial_satellite», which can also be found in ▢ Fig. 4.42 in another Wettzell specific notation explained in the next text block) defines the data values as entities, which are combined to entity types (the tables of a relational database) according to similar properties. The entity types are then graphically represented with rectangles with a label identifying the entities. The attributes (headlines of the table columns) of the entity types are graphically represented as ovals with an attribute name as a label and connected by a line to the corresponding entity type (table). The names of primary keys are underlined. The relations between the entities are usually represented as rhombuses, labeled with the relationship type. Each rhombus connects the corresponding entity types with a line and defines the relationship cardinality (1:c, 1:n, n:m) at

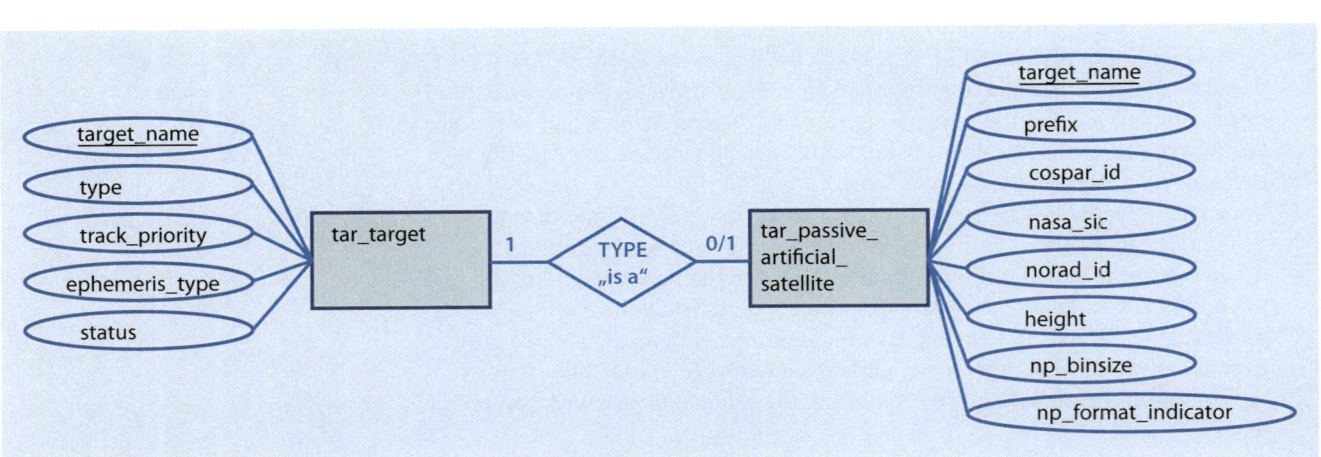

▢ **Fig. 4.41**   Example for the description of the relation between the table «tar_targets» and the table «tar_passive_artificial_satellite» in the Entity-Relationship Model (ERM) in Chen notation (the relation can be found again in ▢ Fig. 4.42 in another, Wettzell specific, notation)

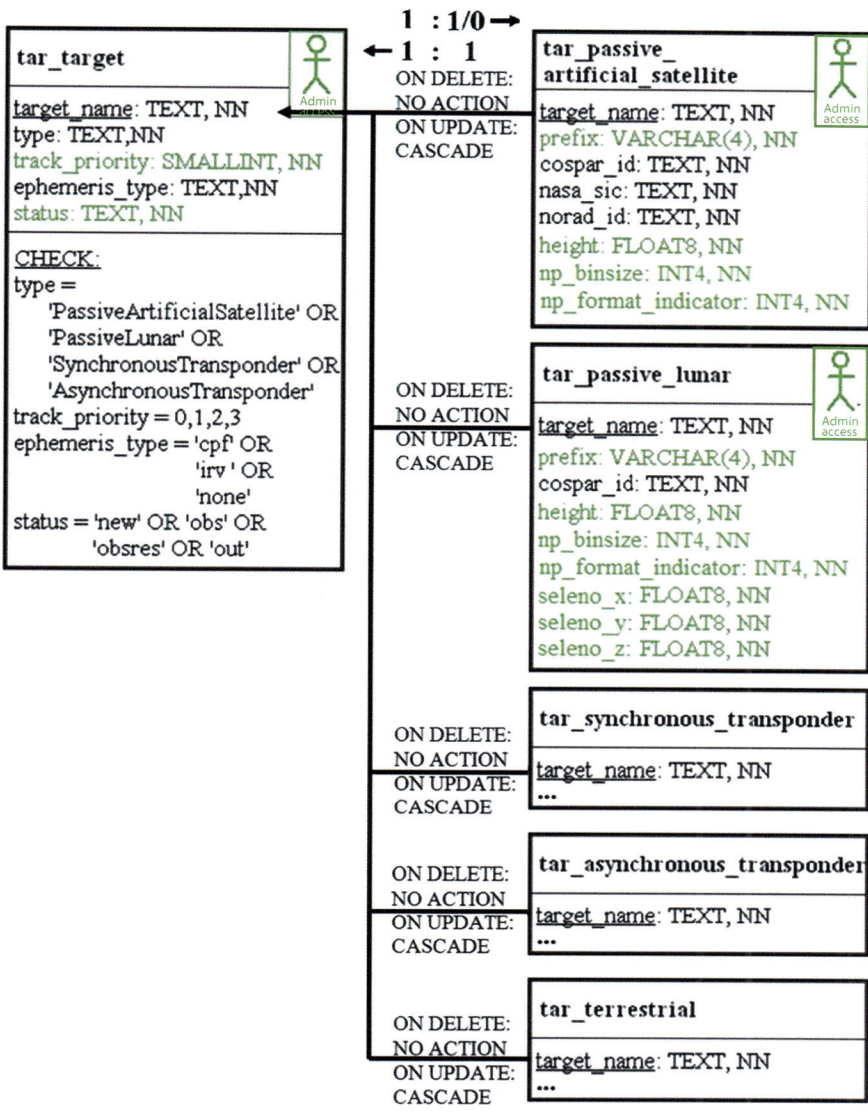

Fig. 4.42 Excerpt of the design of the target administration in the laser ranging database of the laser ranging systems at Wettzell

the ends of the connection lines. The result is a graphical net between entities (Saake et al. 2008, *l.c. page 59 ff.*).

In the practical field, another proprietary graphical representation (which is similar to those from Microsoft® Access™) is used at the Wettzell observatory, as it is much more intuitive. Similar to the ERM, it gives a graphical representation of the database structure. But unlike the ERM, it represents a table as a box with the table name and a list of the attributes of the table (column headlines). Each attribute identifier of the list also defines its type and individual constraints (e.g., «NN» for «not null» and, therefore, for the constraint that the element is not allowed to be empty). It can be colored to show if the value needs to be set by a user or if it is set automatically by the autonomous data management cell. Other constraints (allowed values, limits, etc.) are also defined in a second section of the table box, so that a complete description of the table is visible in the graphical design. Relations are then lines or arrows which directly connect foreign key columns with primary key columns. The lines are labeled with the cascading actions to check the referential integrity, which defines if a data update or delete should be propagated between the related tables via the relationship. For samples, see the

Database Design: Proprietary Documentation Style

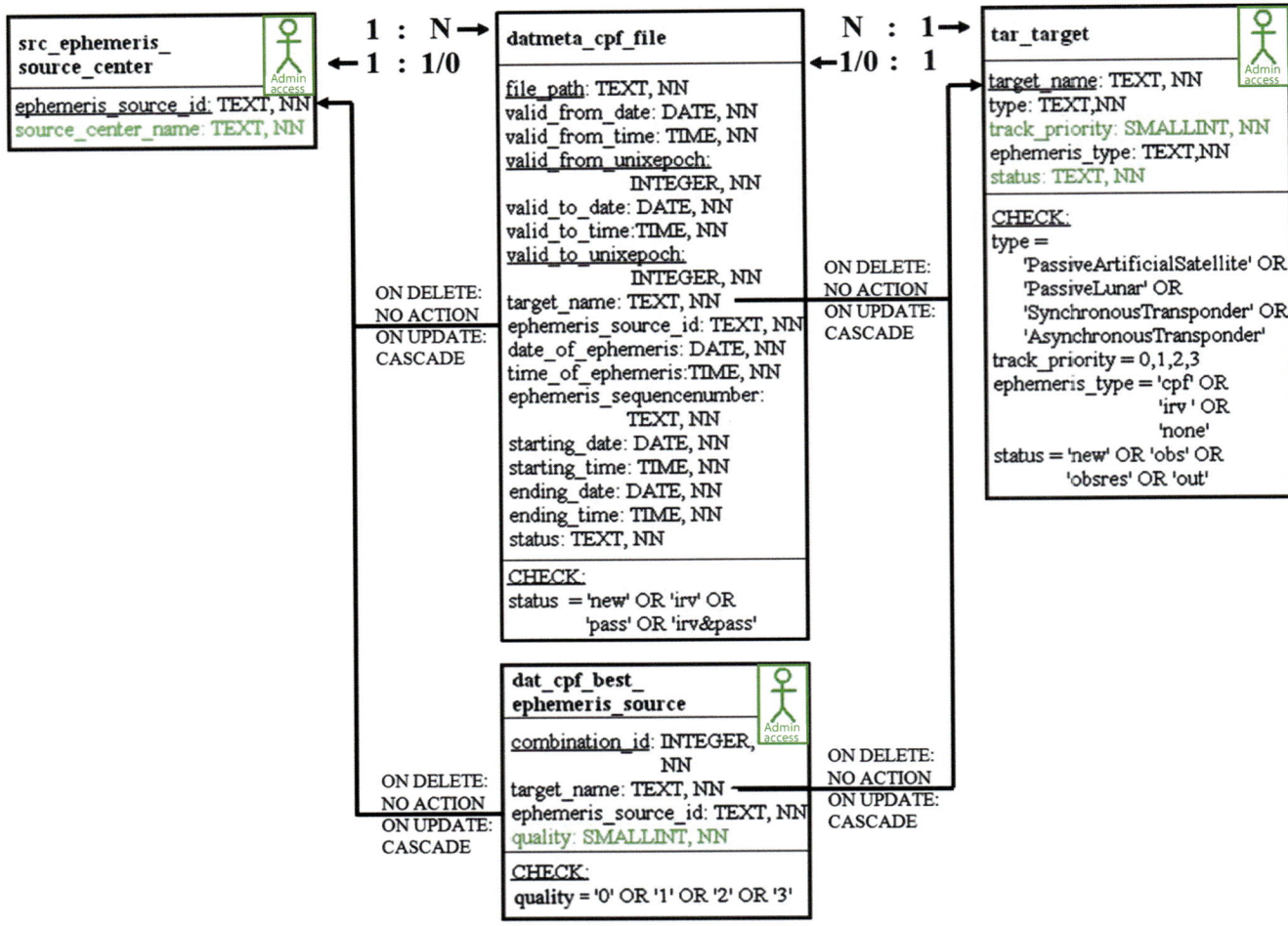

Fig. 4.43 Excerpt of the design of the CPF file administration in the laser ranging database of the laser ranging systems at Wettzell

two excerpts from the design of the laser ranging database of the Wettzell observatory for the administration of targets in ◘ Fig. 4.42 and of CPF prediction files in ◘ Fig. 4.43.

Database Interface: SQL

The implementation of the database can only be done via the database interface using a standardized language: SQL. SQL consists of three parts: a data definition language (e.g., to create, alter, and drop tables), a data manipulation language (e.g., to insert, select, update, and delete records), and a data control language (e.g., to grant rights, commit, revoke, or rollback transactions). It is an ISO standard with different releases. If a relational database system conforms with the standard, it must at least implement the Core SQL part of the standard from 1999. It is defined by a grammar (as shown in ▶ Sect. 3.3.3). SQL is not case sensitive. But a common notation means that reserved words are written in upper case. SQL allows identifiers in the usual way and defines data types, for example, FLOAT, REAL, INTEGER, VARCHAR (for varying character strings), DATE, TIME. The input of values must fit to SQL literals, for example, numerical values can be entered in the usual way, while text inputs or dates must be quoted with apostrophes. The whole instruction follows an implementation of set operations. Therefore, SQL statements are more descriptions of what should be done and not how it should be processed (Warner 2007, *l.c. page 159 ff.*). A sample of how to create the tables «tar_target» and «tar_passive_artificial_satellite» of ◘ Fig. 4.42, how to insert values to the table «tar_passive_artificial_satellite», and how to drop the tables again is shown in Example 4.15 (see also (PostgreSQL Global Development Group 2003, *l.c. page 89 ff.*).

■ ■ **Example 4.15**

SQL statements to create tables, fill values into one table and drop the table again.

```
—— ********************************************************
—— Create tar_target table
—— ********************************************************
CREATE TABLE tar_target (
target_name      TEXT       NOT NULL,  —— name of the target (header H1 or filename)
type             TEXT       NOT NULL,  —— type of target (e.g. artiificial, natural, terrestrial)
track_priority   SMALLINT NOT NULL,   —— observation track priority (high=3,medium=2,low=1,off
      =0)
ephemeris_type   TEXT       NOT NULL,  —— type of ephemeris (e.g. cpf)
status           TEXT       NOT NULL,  —— status of satellie: new, obs, out
PRIMARY KEY (target_name),
CHECK (type = 'PassiveArtificialSatellite' OR type = 'PassiveLunar' OR
       type = 'SynchronousTransponder' OR type = 'AsynchronousTransponder' OR
       type = 'Terrestrial'),
CHECK (track_priority = 0 OR track_priority = 1 OR track_priority = 2 OR
       track_priority = 3),
CHECK (ephemeris_type = 'cpf' OR ephemeris_type = 'irv' OR ephemeris_type = 'none'),
CHECK (status = 'new' OR status = 'obs' OR status = 'obsres' OR status = 'out')
);
—— Grant rights for dedicated operations to the users of the database
grant select on tar_target to slr;
grant select on tar_target to slradm;
grant update on tar_target to slradm;


—— ********************************************************
—— Create tar_passive_artificial_satellite table
—— ********************************************************
CREATE TABLE tar_passive_artificial_satellite (
target_name         TEXT        NOT NULL, —— name of the target (header H1 or filename)
prefix              VARCHAR(5) NOT NULL, —— Shortcut of satellite
cospar_id           TEXT        NOT NULL, —— COSPAR identifier
nasa_sic            TEXT        NOT NULL, —— NASA system integration code (= Goddard Space
      Craft Number)
norad_id            TEXT        NOT NULL, —— NORAD identifier
height              FLOAT8      NOT NULL, —— Mean distance to earth center [km]
np_binsize          INT4        NOT NULL, —— Length of normal point window
np_format_indicator INT4        NOT NULL, —— Normal point format indicator according to ILRS
      Normal Point Format
PRIMARY KEY (target_name),
FOREIGN KEY (target_name) REFERENCES tar_target (target_name)
                    ON DELETE NO ACTION ON UPDATE CASCADE

—— Grant rights for dedicated operations to the users of the database
grant select on tar_passive_artificial_satellite to slr;
grant select on tar_passive_artificial_satellite to slradm;
grant update on tar_passive_artificial_satellite to slradm;

—— ...


—— ********************************************************
—— Insert values into tar_passive_artificial_satellite table
—— ********************************************************
INSERT INTO tar_passive_artificial_satellite VALUES ('ajisai', 'AJI', '8606101', '1500', '
      16908', 30, 5, 1485);
INSERT INTO tar_passive_artificial_satellite VALUES ('beaconc', 'BEC', '6503201', '317', '
      1328', 15, 3, 927);
INSERT INTO tar_passive_artificial_satellite VALUES ('etalon1', 'ETA1', '8900103', '525', '
      19751', 300, 9, 19105);
INSERT INTO tar_passive_artificial_satellite VALUES ('etalon2', 'ETA2', '8903903', '4146', '
      20026', 300, 9, 19105);
```

1
2
3
4
5
6
7
8
9
10
11
12
13
14
15
16
17
18
19
20
21
22
23
24
25
26
27
28
29
30
31
32
33
34
35
36
37
38
39
40
41
42
43
44
45
46
47
48
49
50
51
52
53

```
|  — ...                                                                    54
|                                                                          55
|                                                                          56
|  — *****************************************************               57
|  — Drop tables                                                          58
|  — *****************************************************               59
| DROP TABLE tar_passive_artificial_satellite;                           60
| DROP TABLE tar_target;                                                  61
|                                                                         62
|  — ...                                                                   63
```

Database Interface: SQL in the Code

Usually SQL is used in combination with a command line interface to the database management system, which is offered by the vendor of the database management system. But SQL instructions can also be used within source code in another programming language (e.g., C++) in the form of «embedded SQL» or in combination with an interface module from the vendor of the database system. The last option is offered as call-level interface, where the user of the interface must convert the database table structures into suitable objects of the programming language (Saake et al. 2008, *l.c. page 416 f.*). Such an interface is, for example, provided by the database management system «PostgreSQL» in the form of the library «libpq». PostgreSQL is an object-relational database management system and the most sophisticated open-source system on the market (PostgreSQL Global Development Group 2003, *l.c. page 20)*. The interactions with the database system are organized with a client-server architecture. A client sends requests, which the database system server processes. For this «libpq» offers functions to connect to the database server («PQconnectdb»), to execute an order at the server («PQexec»), to read the results of orders («PQgetvalue»), to clear results again («PQclear»), and to close the connection («PQfinish»). The library offers the option to process the orders asynchronously (PostgreSQL Global Development Group 2003, *l.c. page 325 ff.)*. To simplify the handling of a database connection, an additional driver module («simple_psqlquery.cpp/.hpp») (see the modules in ▶ Sect. 3.2) was written as a high-level interface at the Wettzell observatory. It hides the specific parts and offers sophisticated converter functions (see also (Saake et al. 2008, *l.c. page 438 ff.)*), which understand SQL instructions and return two-dimensional tables, using memory from the heap. This is the main part of the «hardware» driving module of the autonomous data management cell. Sample code for the use of the driver is shown in Example 4.16.

■ ■ **Example 4.16**

Use of SQL statements to process a join operation between two tables in C++ source code in combination with the driver module «simple_psqlquery.cpp/.hpp» and the underlaying 'libpq' of PostgreSQL.

```
#include "simple_psqlquery.hpp"                                           1
                                                                          2
// ...                                                                     3
                                                                          4
PostgreSQLDB CDatabase;       // Connection to database system PostgreSQL 5
                                                                          6
// ...                                                                     7
                                                                          8
// Open connection to database system                                     9
if (CDatabase.uiDBOpen (strDatabaseHost, strDatabasePort, strDatabaseName,10
                  strDatabaseUsername, strDatabasePassword, strTimeout)) 11
{                                                                        12
    uiRetVal = DBSAP_NOK;                                               13
    goto slrdbsap_server_uiGetPassiveArtificialSatellites_Return;       14
}                                                                        15
```

```
// ...

std::string strQuery = ""; // Query text
std::string * pstrColNamesOfPassiveArtificialSatellites = NULL;        // Table headline
std::string ** ppstrAnswerTableOfPassiveArtificialSatellites = NULL;  // Table with values
unsigned long ulNumOfRowsOfPassiveArtificialSatellites = 0;           // Number of rows
unsigned long ulNumOfColsOfPassiveArtificialSatellites = 0;           // Number of columns

// Do query to get already set information of passive targets
strQuery = "SELECT t.target_name,t.ephemeris_type,t.track_priority,t.status,";
strQuery += "p.prefix,p.cospar_id,p.nasa_sic,p.norad_id,p.height,p.np_binsize,";
strQuery += "p.np_format_indicator ";
strQuery += "FROM tar_target t, tar_passive_artificial_satellite p ";
strQuery += "WHERE t.target_name=p.target_name ";
strQuery += "ORDER BY t.status, t.target_name;";
if (CDatabase.uiDBQuery (strQuery,
                         pstrColNamesOfPassiveArtificialSatellites,
                         ppstrAnswerTableOfPassiveArtificialSatellites,
                         ulNumOfRowsOfPassiveArtificialSatellites,
                         ulNumOfColsOfPassiveArtificialSatellites))
{
    uiRetVal = DBSAP_NOK;
    goto slrdbsap_server_uiGetPassiveArtificialSatellites_Return;
}
if (ulNumOfRowsOfPassiveArtificialSatellites == 0)
{
    uiRetVal = DBSAP_NOK;
    goto slrdbsap_server_uiGetPassiveArtificialSatellites_Return;
}

// Process two-dimensional array with table content
// and delete dynamic array again
// ...

slrdbsap_server_uiGetPassiveArtificialSatellites_Return:
// Error processing
// ...

// Close connection to database system
(void) CDatabase.uiDBClose ();

// ...
```

```
16
17
18
19
20
21
22
23
24
25
26
27
28
29
30
31
32
33
34
35
36
37
38
39
40
41
42
43
44
45
46
47
48
49
50
51
52
53
54
55
56
57
58
```

Databases must also deal with some limitations. One problem is that it is not always easy to reorganize data as related table structures. Another problem is that a query can have to follow several relation dependencies, which make it quite processing intensive and, therefore, sometimes slow for large databases. In fact, one positive feature can become in some cases a disadvantage: data cannot be read directly. It is always necessary to use the database interface and the DBMS in the background to read data. Therefore, it is not easy to access data with a simple program like an editor, as is usual with simple files. In practice, this is a real disadvantage for engineers who have no experience with databases, so additional GUIs are necessary. Finally, another disadvantage is that the saving of sequential or binary data sets, which just makes sense as a complete block, is not ideal and often slow with databases.[74]

Limits of Database

Nevertheless, databases offer considerable advantages for the scheduling of laser ranging passages, where a number of tables must be combined using several relationships. On the one hand, the beneficial value of such an administration

Hybrid Data Management

---

74 The use of BLOBs in modern databases is not really a fast or useful solution.

4

system is also clear if several parallel processes are used to fetch, analyze, or send data in parallel. Transactions can be a way to coordinate and synchronize activities. On the other hand, file systems have huge advantages for sequential files, such as those given for orbit supporting points, observation points, and so on. Therefore, it is valuable to combine both techniques in the autonomous data management cell. This means that a database is used to organize, administrate, and synchronize files. Metadata about the files and their contents can also be used to allow fast mathematical set operations for queries to find and select suitable combinations for the observation strategies. The combined file system contains the real data as sequential files. This can also be used to separate data according to their update and request rate or their expiration time: data which are requested several times (e.g., passage information of satellites for one observation station which are used to schedule the observation dynamically[75]), which require several updates (e.g., passages can change if newer, more accurate orbits are available), or which live for just a short period are located in the database. Data which are not changed at all (e.g., orbit ephemeris), which are just requested once (detailed positions of a satellite during a passage), or which are saved and archived unchanged (observation results) are stored in the file system. The result is a hybrid data management cell, which is represented by an RPC server for the communication infrastructure (see ◘ Fig. 4.44).

Autonomous Planning

With this knowledge, it is now possible to take a look at the autonomous functionalities of the autonomous data management cell. The first part is the autonomous planning. The main purpose of the data management cell is to support the rest of the system with all the required data. It is the connection point to the outer world of data centers like the European Data Center (EDC) or the Crustal Dynamics Data Information System (CDDIS). In this environment, the data management cell is more complex than a simple autonomous control cell. It is more like a complete autonomous production cell. The planning of this cell organizes the processing of data fetching, preparing, evaluating, and sending with parallel but partly synchronized tasks at the right time. At Wettzell, those jobs are done by parallel processes, triggered, for example, as a «cron job»[76]:

— The data fetching is done by data spiders, which are programs «crawling» through file structures of the data centers, using FTP to search for new files or directories, which are then automatically copied into the local file system structures and registered at the database. Input data are:
— CPF prediction files
— Files with EOP (which are inserted into a separate database table, as the predicted values per day are daily updated with the actual values after they are determined by the external analysis)
— Clock correction parameters with runtime values between the time standard and the laser ranging system
— Meteorological data (also inserted into the database to allow dedicated select statements)
— Observation files from the autonomous production cell
— Status information from other laser ranging systems (EUROLAS status,[77] which is fetched over a TCP/IP socket)

---

75 At the Wettzell observatory, this is defined as the «satellite observation alert list».

76 Cron jobs are timed processes started by the operating system and defined in a «crontab» file. The only difficulty here is to prevent one job being started several times. But this can be done with operating system scripts, which check if a process with the same name is already running. A better way is to trigger the external programs in the periodic control loop, so that a pipelining of tasks can be arranged.

77 It is a simple table with updated information about sites, like status, last observation, hits, and used ephemeris sources.

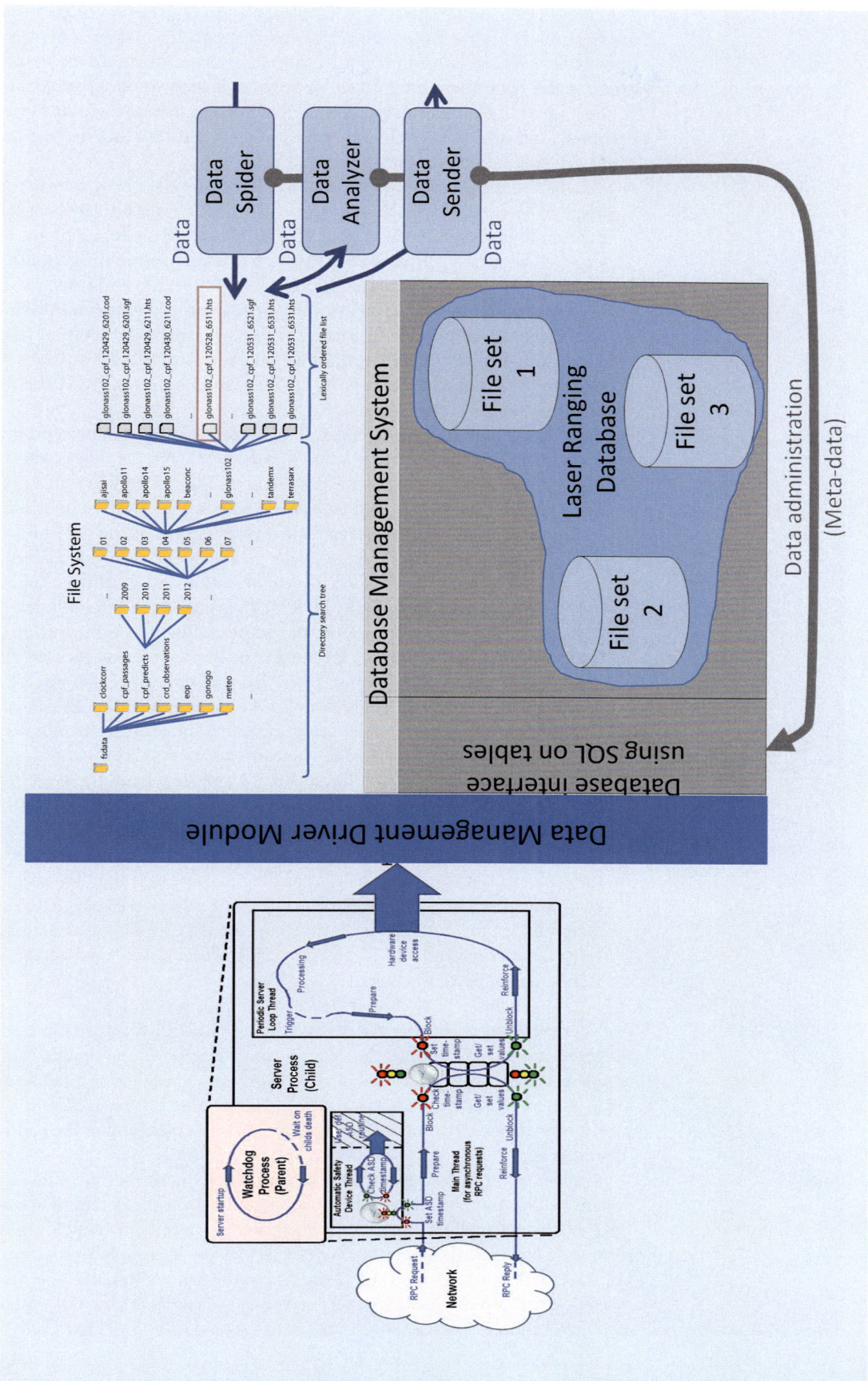

■ **Fig. 4.44** The autonomous data management cell as a hybrid data organization and represented as an RPC server in the autonomous production cell using the server safety techniques from ■ Fig. 3.22

The integrity of the files herein is handled via the registration of the files in the database. Only after a file is completely and successfully fetched, it is registered at the database. Unregistered files, for example, because of a system crash, do not exist for the data management cell, even though they could be saved in the file system. The processing of these missing files restarts during the next cycle, where eventually files which have already been saved are replaced and the registration is processed successfully to start the next steps.

— The data preparation is done by the «cpf2irv» converter program (which is inspired by the CPF sample software of the ILRS). It interpolates the orbit prediction points from the registered CPF files to get sampling points in steps of one second. Then it does the transformation from the geocentric to a topocentric view for the laser ranging system position and converts the X/Y/Z representation into an azimuth/elevation representation. Additionally, it calculates the passage information with rise, transit, and set times and the positions when the satellite is illuminated by the sun (sunlit) or when it is too close to the sun direction (sunavoid). The resulting passage and orbit files are registered again at the database.

— The observation evaluation (obseval) is done with the evaluation programs from the RGO with the observation files fetched from the production cell and registered at the database. The evaluation is a call sequence of external RGO programs: «solve1» and «solve2» to solve a least squares solution for the residuals and perform a range reduction, «gauss3» to process a Gaussian fit to binned data, «normpt» to create NPs for the quicklook transmission, and «npcheck» to check the NP for a variety of station and satellite parameters. All of these programs are originally FORTRAN sources. Additionally, the output files are created or converted from the output of the RGO observation evaluation software which produces the current observation result files for full-rate data and NP in the CRD format.[78] Another output is an HTML report, which describes the results and offers feedback to the operator. The resulting files are then registered again at the database, including the statistics parameters.

— The sending is done by an FTP program, which checks the database for registered CRD files, which have also got a permit from the system and/or the operator. He can manually check the results and gives a permit if the statistical results are good enough for a transfer to the data centers. The data cell starts this sending asynchronously from time to time. Then the observation is marked as sent at the database and the planning loop ends for those data. The EUROLAS status is also sent by the data management cell, as the data server hardware is also active if the computer of the central coordination cell is restarted. Therefore, it can forward the correct status messages all the time.

Another planning activity is to calculate the first priority for a satellite as a sum of weighted terms according to the general satellite priority, the height, the passage length, and meteorology. The calculation is performed each time a satellite passage list is loaded via the user interface. This priority value can be used as the starting priority for the observation scheduling, which is the planning of the central coordination cell.

Autonomous Controlling

The controlling processes all of the operations described on files in the file system, which are administrated and synchronized via the database management system. The controlling here is not hardware management, which reads values from a device and makes decisions for that device. It is more or less the service implementation to offer the data to the rest of the production cell. The controlling takes care to keep the integrity and correctness of the data. The only access point

---

78   Previous formats are CSTG for the NP and the ILRS Full-Rate Data Format V2 (MERIT-II) for the full-rate data with all the observation information necessary (ILRS n.d.).

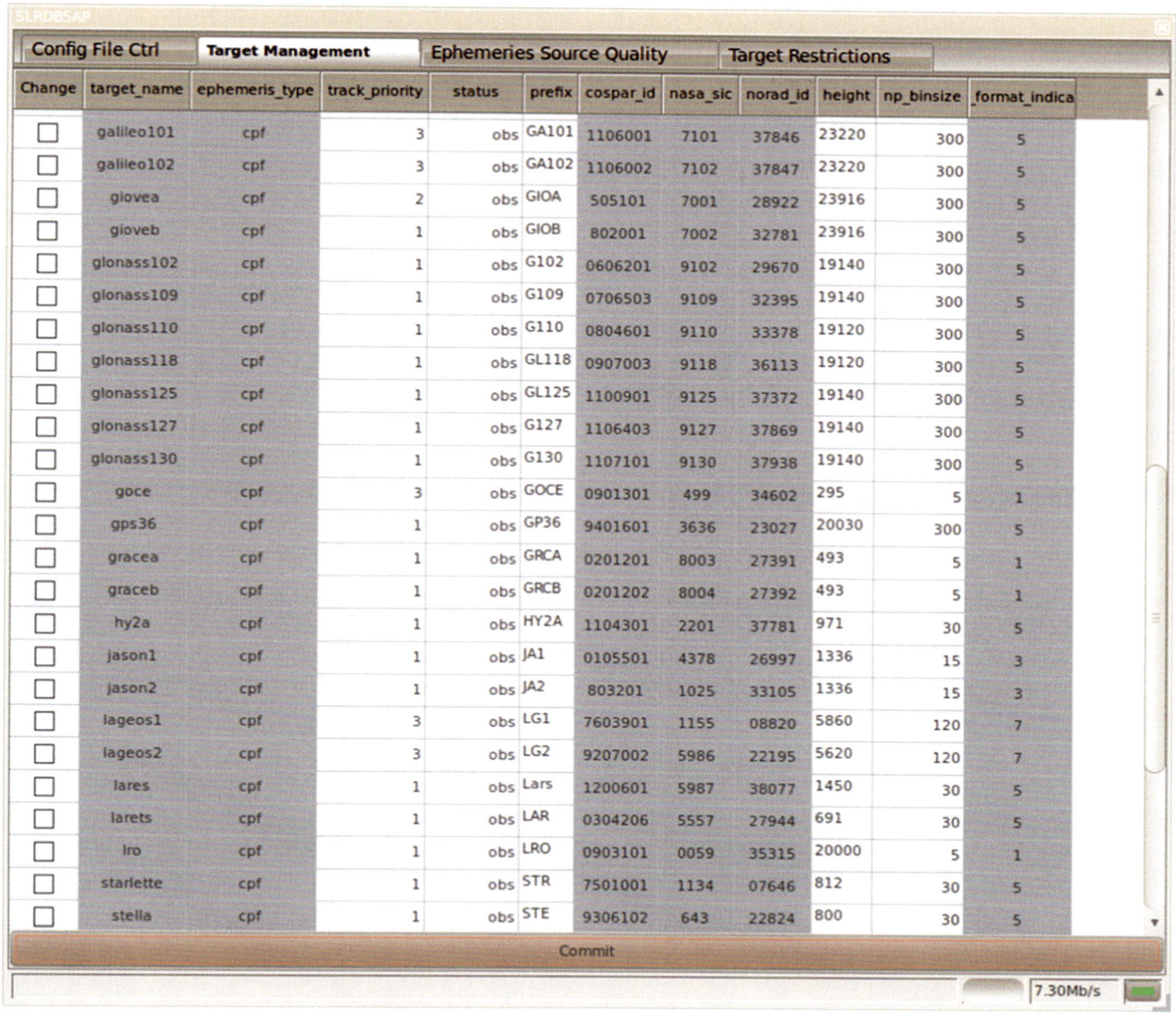

Change	target_name	ephemeris_type	track_priority	status	prefix	cospar_id	nasa_sic	norad_id	height	np_binsize	format_indica
☐	galileo101	cpf	3	obs	GA101	1106001	7101	37846	23220	300	5
☐	galileo102	cpf	3	obs	GA102	1106002	7102	37847	23220	300	5
☐	giovea	cpf	2	obs	GIOA	505101	7001	28922	23916	300	5
☐	gioveb	cpf	1	obs	GIOB	802001	7002	32781	23916	300	5
☐	glonass102	cpf	1	obs	G102	0606201	9102	29670	19140	300	5
☐	glonass109	cpf	1	obs	G109	0706503	9109	32395	19140	300	5
☐	glonass110	cpf	1	obs	G110	0804601	9110	33378	19120	300	5
☐	glonass118	cpf	1	obs	GL118	0907003	9118	36113	19120	300	5
☐	glonass125	cpf	1	obs	GL125	1100901	9125	37372	19140	300	5
☐	glonass127	cpf	1	obs	G127	1106403	9127	37869	19140	300	5
☐	glonass130	cpf	1	obs	G130	1107101	9130	37938	19140	300	5
☐	goce	cpf	3	obs	GOCE	0901301	499	34602	295	5	1
☐	gps36	cpf	1	obs	GP36	9401601	3636	23027	20030	300	5
☐	gracea	cpf	1	obs	GRCA	0201201	8003	27391	493	5	1
☐	graceb	cpf	1	obs	GRCB	0201202	8004	27392	493	5	1
☐	hy2a	cpf	1	obs	HY2A	1104301	2201	37781	971	30	5
☐	jason1	cpf	1	obs	JA1	0105501	4378	26997	1336	15	3
☐	jason2	cpf	1	obs	JA2	803201	1025	33105	1336	15	3
☐	lageos1	cpf	3	obs	LG1	7603901	1155	08820	5860	120	7
☐	lageos2	cpf	3	obs	LG2	9207002	5986	22195	5620	120	7
☐	lares	cpf	1	obs	Lars	1200601	5987	38077	1450	30	5
☐	larets	cpf	1	obs	LAR	0304206	5557	27944	691	30	5
☐	lro	cpf	1	obs	LRO	0903101	0059	35315	20000	5	1
☐	starlette	cpf	1	obs	STR	7501001	1134	07646	812	30	5
☐	stella	cpf	1	obs	STE	9306102	643	22824	800	30	5

Commit

7.30Mb/s

Fig. 4.45 The observed satellite tab of the administrators GUI for the laser ranging data in the autonomous data management cell

for the user is the interface offered. The controlling also helps against the limits of the constraints in the database design, which, for example, do not check that a registered target is either a passive artificial satellite, a passive lunar, or one of the transponder reflectors.

The interfacing is implemented as an RPC server. The user is either the autonomous production cell or an operator. Both have to use the interface to get, change, or insert data. The operator has the option to change some dates, such as the satellite priorities or the ranking of ephemeris source centers. The administrator interface is a GUI, which organizes and combines table structures as a user view onto the database and configuration file (see ▪ Figs. 4.45 and 4.46). Therefore, the operator and the usual system administrator no longer have to edit directly in the database tables, which could easily lead to inconsistencies. Another graphical interface for operators is given by a web server which allows access to the observation evaluation reports and to other HTML-based outputs (e.g., a documentation archive, a wiki, etc.).

Autonomous User Interfacing

The hardware driving is the use of the previously described implementations (e.g., the embedded SQL or the «libpq») in combination with the usual file

Autonomous Hardware Driving

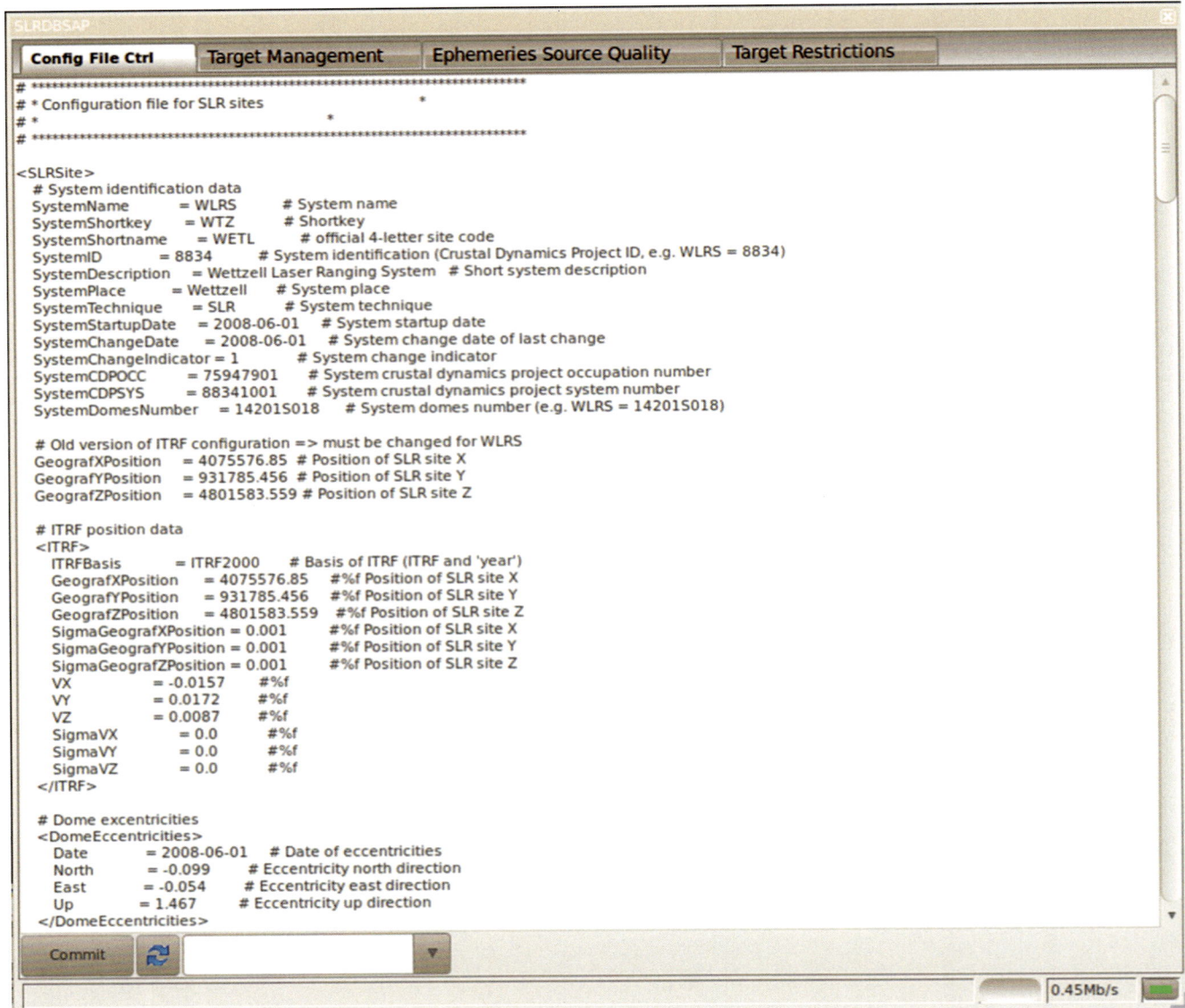

Fig. 4.46 The configuration file editor tab of the administrators GUI for the laser ranging data in the autonomous data management cell

Autonomous Failure Management

operations (open, read, write, close, rename, remove from the C standard) and directory operations (offered in operating system libraries) to give access to the hybrid data system. The «hardware» in this case is the server of the database management system and the file system.

Failure management is more or less completely handed over to the database management system with its recovery mechanisms. The database acts as a central authority, where all the data must be registered. Unregistered data does not exist for the system. The rest is organized with stable scripts and programs which avoid multiple activations. The direct access point of the RPC server is protected and monitored by the watchdog of the «idl2rpc.pl» mechanism. The database propagates information about bad results of requests or inconsistencies to the requesting cell or operator.

The data from the data management cell is mainly used by the central coordination cell, which requests required data, plans future observations, and propagates the relevant data to the subsystems of hardware control cells.

4.7   Autonomous System Monitoring and Safety Cell

A separate, additional safety system is essential for each control system (see also ▶ Sect. 3.3.3 about safety basics) as it reduces common-mode failures. A common-mode failure is given if a single-point failure influences different parts of the system, so that it runs into a critical state, but cannot react to it because the single-point failure also blocks the failure exception processing. Therefore, no safety relevant parts are allowed to run on the same hardware, which is also used for the rest of the controlling. A safe system is possible if no errors appear as long as the system is used in the defined specifications for the task for which it was designed. A single-point failure in such a safe system does not result in a critical state for the whole system. The system must be able to react to the appearance of the failure within a defined time limit. Therefore, additional, separated channels are required for safety relevant parts and for controlling components to offer the necessary redundancy. Each channel collects data or information to derive logical decisions about the safety status of the system. It is more or less a connection of sensor data in series, as given for emergency stop switches. An error detected in the channel affects the whole channel (Vigenschow 2005, *l.c. page 224*). As the data has already been collected and evaluated via the channels and the safety system, it can also be used as an additional monitoring system for the controlled system.

Safe Systems with Separate Channels

As with the supporting data management cell, the ideas can be used not only for the monitoring and safety evaluation of laser ranging systems, but they are also relevant for other space geodetic techniques and even as a general data acquisition system. The ideas are discussed by a VGOS Monitor and Control Infrastructure (MCI) group with the intention of promoting uniformity of MCI throughout the next-generation international observing network. This section is more or less a copy of the suggestions from the working document of this group.[79] To allow a possibly open use for almost any case in the field of monitoring, the key ideas which are also the main drivers for an autonomous system monitoring and safety cell in a laser ranging system are:

MCI Standardization

— The architecture and software should be open-source and open-ended, so that stations are free to use, modify, and adapt it for local needs.
— The architecture should be hierarchically extendable, so that one MCI node collects MCI data from different sensors, but acts itself as a sensor for further MCI nodes in the higher hierarchy.
— The architecture should be self-identifying in order to facilitate straightforward expansion of the station's MCI and promote IVS network uniformity.
— Data logging should be cast into the form of a data management with safety, completeness, integrity, and different logging and sampling rates.

With these ideas in mind, a nodal structure with a standardized setup and an MCI RPC interface is designed. Each MCI software for a sensor and each node must implement it to comply with the international network. As the system monitoring and safety cell implements a parallel system of different channels, which acquire additional data from the system hardware to drive safety decisions and to log the monitored data, the whole system is something like a copy of the design of the observing system itself, but with fewer main controlling parts. The result is a hierarchical structure, which uses similar ideas as explained for example in ◘ Fig. 4.7. The result is a layered structure, shown in ◘ Fig. 4.47.

Cell Structure

On layer 1 are all the sensors (see ▶ Sect. 4.2.1 about the definition of sensors) of different types such as temperature sensors, sensors for front end monitoring, meteorological sensors, emergency switches, and so on, which offer different kinds

Hardware Layers

79   Many thanks must here be given to the MCI group, mainly consisting of Chris Beaudoin, Ed Himwich, Martin Ettl, and Matthias Schönberger, for the fruitful discussions and the acceptance of the design issues. The working document was written together with these authors.

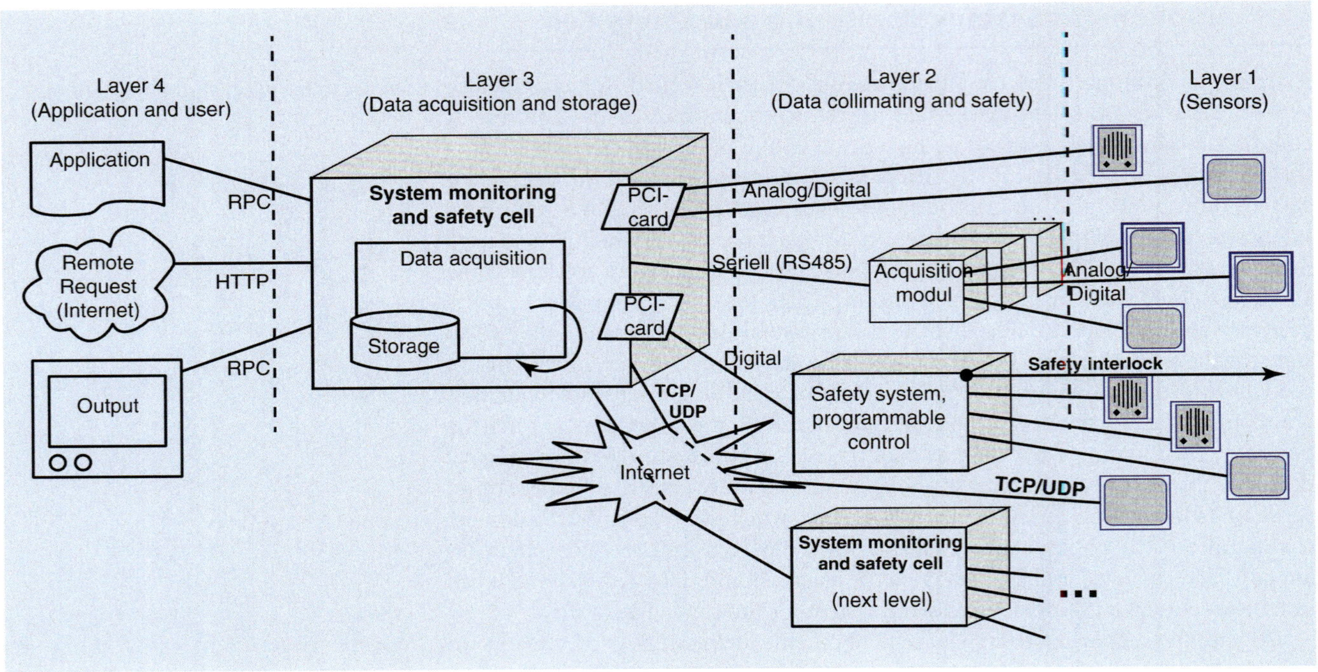

Fig. 4.47 The hardware structure of a system monitoring and safety cell with the four levels (*from right to left*): the sensors, the data «collimating»/gathering and hard safety, the data acquisition and storage, and the application and user layer

of data from different types of hardware. The sensors are connected with analog or digital lines in combination with PC cards, serial lines (e.g., RS-232 or RS-485),[80] and Ethernet over a LAN or any other connection (e.g., over the Universal Serial Bus (USB)), which can be read with software in the system monitoring and safety cell. Layer 2 is an optional data collimation/gathering and safety layer. It helps to combine several sensors on the level of hardware boards with analog and/or digital lines to sample data with higher rates or to implement an interlock system. The interlock for RADAR detections, emergency switches, or safety switches of the laser ranging system are arranged and combined here. They must react immediately to critical situations for humans or the system in real-time within a very strict deadline. Layer 3 is a computer (ideally, a fanless, low-energy PC), which implements the data acquisition, preparation, and presentation. In principle, this is the coordination layer for the system monitoring and safety production cell. All software parts for the sensors, recorded by the node, reside here. Finally, layer 4 implements the use of the data in the observation system, for presentation, further processing, additional controlling and coordination, automation, and distribution to international analysis centers. Therefore, the access point is implemented with RPC or also as HTML presentation via HTTP. Layer 4 can also be another monitoring and safety cell, which collects data from different other monitoring and safety cells via Ethernet. This makes it possible to build up hierarchical systems, which propagate their data throughout the different layers to a final controlling and interpretation instances. These can be high-level system monitoring tools such as Nagios®, Pandora FMS, or Zabbix (e.g., see ▶ Sect. 4.7).

Software Structure   The paper of the MCI group suggests the following very open and integrative method for the internal architecture of a node (see ◲ Fig. 4.48). The main part of the cell is comparable to the data storage and management system (see also ▶ Sect. 4.6), because the central component is again a modern DBMS like PostgreSQL, which allows many concurrent accesses and is scalable to different hardware platforms.

---

80   RS232 and RS485 are industrial standards for serial communications, defining the hardware and the endpoints.

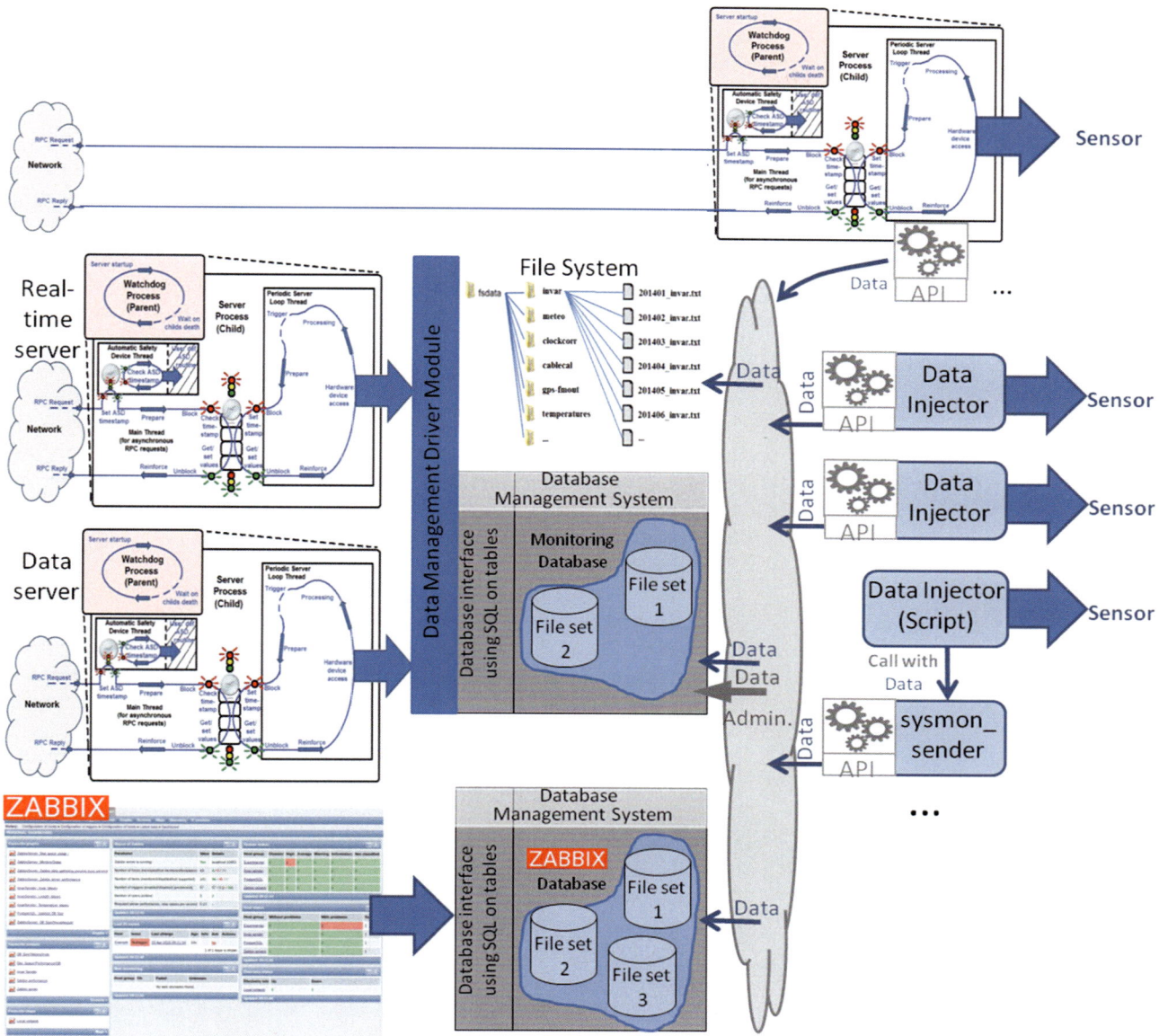

Fig. 4.48 The software structure of a system monitoring and safety cell with the hybrid data management of a database and a file system, the real-time server for current values and the data server for historic data (left side). On the right side is an illustration of the sensor control points with RPC-based sensor control point servers and (simple) data injectors, which directly include the API module or use an external program «sysmon_sender», offered by the monitoring program suite. The lower part demonstrates a beneficial extension: the Zabbix suite for presentation and on-call services

Similar to the data management cell of the observation control system, it is extended by a (additionally included) file system server. This again makes it possible to build up hybrid storage capabilities, for example, to save configuration files or sequential data files (e.g., with higher sampled data sets or archived, historic data). These files can be requested using remote file access techniques, like FTP, Secure File Transfer Protocol (SFTP), Network File System (NFS), or maybe Samba. For standard users, read access should be enough here. As the system monitoring requires fast access to current data and also to data sets for a short time period, parts of the latest data are also stored in tables in the database. A parallel service server can periodically archive data automatically from the database tables to a central folder of the file system in a directory structure of years, months, and days. The database is local and can only be directly accessed from there. The data are available outside using servers with standardized interfaces. Two basic versions of RPC interfaces are planned (see ◘ Fig. 4.48 and compare it to ◘ Fig. 4.44 about the autonomous data management cell to see the similar structures): one fast for the monitoring of current, real-time data values with timestamps and warning levels (real-time server) and the other to offer selected sets of historic data (data server). The historic values can only be selected from data sets still available in the data tables. Data which has already been archived to the file system are not available anymore using the RPC interface and can only be downloaded as complete files. But the data sets from the tables can be individually requested for different time intervals, for example, to support plotting graphs. But retrieving huge blocks of historic data from the database is time-consuming, which blocks other requesters but does not influence the real-time channel for the current values.

Sensor Control Points and Data Injection

The programs which send data into the system are sensor control points and have to be written individually for specific hardware. They are comparable to the data spiders in the data management cell (see ▶ Sect. 4.6). They are individual programs dealing with the driver hardware. The programs are responsible for sampling, requesting, controlling, warning, and time tagging. They can operate and manage one single device or several external hardware sensors as subsystems (see ◻ Fig. 4.48). They can inject data into the data storage system using a standardized software component in the form of an API (e.g., an extended and more specialized version of «simple_psqlquery») (see ▶ Sect. 3.2) for the database and possible additional data storage (like the file system) and presentation units (like Zabbix) (see ▶ Sect. 4.7). But it is also possible to write individual formats of files to specific paths in the file system. Three types of sensor control points are possible:

— *Data injectors*: These are simple, single, and separate programs which control and read the hardware. They sample the data and process the first simple logical activities on them. Then they inject the current data into the data management system of the system monitoring via the standardized software API, which implements a simple connection to the database and according storage systems. They use no watchdog process and are not directly controllable from remote. These injector programs allow an easy startup, because they only include the API modules and use the existing functions from there to register or inject data. It is even possible to use scripts because the monitoring suite offers a separate program «sysmon_sender», which can be activated using a system call. The program arguments of this call make it possible to register or deregister sensors using a configuration file (see Example 4.17). It can also be used to send sensor data into the database and the file system using the program arguments (if one single sensor value is injected) or a separate file with a table of sensor values (if a set of sensor values is injected).[81] The use of scripts is quite suitable to connect to other monitoring systems, such as: the Australian Open-MoniCA system, used for the Australia Telescope Compact Array for VLBI[82] (see ▶ https://www.narrabri.atnf.csiro.au/monitor/monica/ and http://code.google.com/p/open-monica/); the Goddard Mission Services Evolution Center (GMSEC) middleware, which is currently under development by the NASA Goddard Space Flight Center and will probably be used for the new US VGOS sites[83]; The developments for the NASA Field System using Telegraf - InfluxDB - Grafana (TIG) (see ▶ http://www.haystack.mit.edu/workshop/TOW2017/files/Seminars/horsley-tig-notes-2017.pdf and https://grafana.com/grafana); Some individual developments like the "VLBI Monitor"

---

81   More information can be found in the descriptions of the implementation of the MCI system monitoring at the Geodetic Observatory Wettzell.

82   Open-MoniCA is developed by the Commonwealth Scientific and Industrial Research Organisation (CSIRO) as a versatile real-time monitor and control system, written in Java. The server makes it possible to collect data from an unlimited range of devices. It logs the data to disk. Control loops allow the implementation of feedback structures to the devices. A graphical client program allows the inspection and plotting of real-time and historical data (CSIRO n.d.).

83   GMSEC is a solution which is applicable to current and future NASA missions, using standardized interfaces, a middleware, which implements a software bus, and a flexible design for the integration of individual hardware components (National Aeronautics and Space Administration (NASA), Goddard Space Flight Center n.d.).

of Dr. P. Schellart (see https://gitlab.science.ru.nl/radiolab/vlbi_monitor); Other industrial monitoring systems, like Nagios® Pandora FMS, or Zabbix, etc.

— *Sensor control point servers*: These are complete servers in the sense of the described hardware control servers or cells which control the hardware via a separate thread with a control loop and offer a standardized interface on the basis of RPC to the external world. In a similar way to the simple data injectors, they write the current data values directly into the data management system of the system monitoring via the standardized software API for database connections. But unlike these simple programs, they allow direct remote access to the sensor and, therefore, also a sophisticated remote configuration and commanding. They use all the mechanisms of the generative technique, described in ▶ Sect. 3.3.3 (e.g., watchdogs, automatic safety devices, threads, semaphores, etc.). Therefore, they can also be individually extended with controlling functionalities besides the raw monitoring abilities to command specific checks, changes, or an order to the hardware. This can be used to acknowledge error situations or handle a clearance after a critical situation.

— *System monitoring and safety cells*: The hierarchical idea described also compels the design to offer the possibility that each system monitoring and safety cell can be treated like a sensor control point by a following monitoring cell on a higher hierarchy level. This cell can then request all values from the subordinated cell at once with restrictions in timing, as discussed in ▶ Sect. 4.5.

Configuration

During the very first startup, sensor data injectors of the sensor control points must register themselves at the data storage system. After that, the servers can get their startup configuration from the data storage system, which they must request during the start. Another possibility is to use a local configuration file, which updates the registered one automatically during the startup of the server. Both techniques guarantee that the server and the data storage system use the same configuration. The registration is done by sending the content of the configuration file to the local data management. The processes of the sensor control points are then started automatically in the startup time of the computer with the system monitoring and safety cell. The elementary structure of such a configuration file for the data injectors and sensor control point servers is shown in the Example 4.17. The structure for other, subordinated system monitoring and safety cells is shown in Example 4.18. All configuration files can be read and administrated with the configuration parser «simple_structured_config» from ▶ Sect. 3.2. The specific identifier tags in the configuration file are mandatory. Besides these, individual others tags are allowed. The configurations are also stored in the database, acting as a centralized data stock for them. Local changes in the servers must be registered again. All other changes can be propagated from remote via the data storage system. In principle, the database registration offers a sophisticated method similar to the hash maps in ▶ Sect. 4.4.

▪▪ Example 4.17

The configuration file for the simple data injectors and sensor control point servers in the XML-alike proprietary configuration format.

```
<MCISensorControlPoint>                                                          1
   ControlPointID = Wz_Invar_1 # Unique identifier (name or number) for a sensor 2
                     # in a system => this will be converted to the MCI          3
                     # identifier of <SCPID>(<IPADR>,<PORT>)                      4
   ControlPointType = IDL2RPC  # Type of this control point, e.g. IDL2RPC or PROPRIETARY 5
   ControlPointPort = 50500    # Access port of the SCP for direct requests       6
   # Individual control point configuration values e.g. device settings for all sensors 7
   <MCISensor>                                                                    8
       # Connected hardware sensor                                               9
       SensorID = Wz_TempCenter_1   # Unique identifier (name or number) for a sensor 10
                     # in a system                                               11
```

```
    SensorName = TempCenter        # Identifying name of a sensor controlled by SCP        12
    SensorType = Temperature sensor # Type of sensor e.g. temperature sensor               13
    SensorUnit = řC                # Unit of the measured value                            14
    SensorManufacturer = Company   # Manufacturer of the sensor                            15
    SensorModel = AX510Temp        # Model number etc. of the sensor                       16
    SensorPosition = Midway in azimuth axis # Descriptive position explanation or          17
                                   # geometric location                                    18
    SensorUpdateInterval = 180s    # Time steps between each value update (~ rate)          19
                                   # in seconds [s] or microseconds [us]                   20
    SensorResolution = 0.05        # Resolution of sensor in the above unit                21
                                   # (also defines the valid decimal places)               22
    SensorDataAvailabilityTime = 1d # Direct availability of data from database            23
                                   # in days [d] or seconds [s]                            24
    SensorMinLimit = −20           # Lower limit of representable values                   25
    SensorMaxLimit = 50            # Upper limit of representable values                   26
    SensorMinWarningLimit = −15    # Lower than this value throws warning state (or 'off') 27
    SensorMaxWarningLimit = 25     # Greater than this value throws warning state          28
                                   # (or 'off')                                            29
    SensorMinAlertLimit = −18      # Lower than this value throws alert state (or 'off')   30
    SensorMaxAlertLimit = 30       # Greater than this value throws alert state (or 'off') 31
    SensorFlagProvider = yes       # Flag that server collects HW data and offers them     32
    SensorFlagConsumer = no        # Flag that data can be sent to the server              33
    SensorFlagCommandable = no     # Flag that server offers a command line funct.         34
    SensorFlagManageable = no      # Flag that server offers additional RPC funct.         35
    SensorDataArchiveDirectory = /archive/MCI/ # Directory for standard archive service    36
                                   # or empty                                              37
    SensorPropArchiveDirectory = # Directory for proprietary SCP data archiving            38
                                   # or empty                                              39
    # Individual sensor configuration values e.g. sensor specific device settings          40
  </MCISensor>                                                                              41
  <MCISensor>                                                                               42
    ...                                                                                     43
  </MCISensor>                                                                              44
  <MCISensorSubnode>                                                                        45
    # Connected MCI subnode                                                                 46
  </MCISensorSubnode>                                                                        47
    ...                                                                                     48
</MCISensorControlPoint>                                                                    49
```

■ ■ Example 4.18

The configuration file for system monitoring and safety sub-cells in the XML-alike proprietary configuration format.

```
<MCISensorControlPoint>                                                                     1
  ControlPointID = Wz_RequSubnodeCabine_1 # Unique identifier (name or number)             2
                                   # for a sensor in a system                              3
                                   # => this will be converted to the MCI                  4
                                   # identifier of <SCPID>(<IPADR>,<PORT>)                  5
  ControlPointType = IDL2RPC     # Type of this control point, e.g. IDL2RPC or PROPRIETARY  6
  ControlPointPort = 50520       # Access port of the SCP for direct requests              7
  # Individual control point configuration values e.g. device settings for all sensors     8
  <MCISensorSubnode>                                                                        9
    # Connected MCI subnode                                                                 10
    SubnodeIP = 192.168.178.200    # IP−address of subnode                                 11
    SubnodeRealtimePort = 50550    # Real−time port of subnode                             12
    SubnodeUpdateInterval = 180s # Update interval for data from the subnode               13
                                   # (also propagated as concatenation to the higher       14
                                   # hierarchy levels e.g. 180s<<180s)                      15
  </MCISensorSubnode>                                                                       16
</MCISensorControlPoint>                                                                    17
```

The real-time and the data service server of each system monitoring and safety cell is also registered at the local data management system. This allows a remote configurability and centralizes all configuration information at the data management of the cell. The structure of the server configuration is shown in Example 4.19.

■■ Example 4.19

The configuration file for the real-time and data server in the XML-alike proprietary configuration format.

```
<MCINode>                                                                     1
  NodeID = Wz_SubnodeCabine_1  # Unique identifier (name or number) for a sensor  2
                              # in a system => this will be converted to the MCI  3
                              # identifier of <SCPID>(<IPADR>,<PORT>)          4
  NodeType = IDL2RPC  # Type of this control point, e.g. IDL2RPC or PROPRIETARY  5
  NodeRealtimePort = 50600    # Access port of the SCP for direct requests   6
  NodeSelectPort = 50601      # Access port of the SCP for history requests   7
  # Individual control point configuration values e.g. device settings for all sensors  8
</MCINode>                                                                    9
```

The identification of sensors or system monitoring and safety cells in a hierarchical sensor network is generated automatically using a concatenation of the different identifiers (keys) from the individual configuration files in a similar way as the maps described in ▶ Sect. 4.4. The result is a similar key path as described there. Individual remote functions can be activated with such a path on sensor control point servers. As data injectors just send data to the data management system, they cannot be used in this way. But after the injection into the data storage system, the real-time and the data servers offer the possibility of using the standardized way via RPC. With a small extension to the path structure, it is even possible to directly request sensor data without calling an individual RPC function explicitly. Each path identifier is extended with the IP address of the machine where it is available and the IP port of the MCI access server. The path can implicitly be split into connection open and request functions, which follow the path hop by hop to the final sensor point, from which the data is finally requested.[84] To separate this path structure from the one already described, it uses «<<» instead of «::». The implementation can either be integrated into «idl2rpc.pl» or implemented as a separate module which converts the new data path into a call path of the «idl2rpc.pl» style, which can then be interpreted by the generated interpreter function (see ▶ Sect. 4.4).

Identification in a Network

The following example can demonstrate the idea. A proprietary data injector only propagates data to the data storage system and will be registered with the combined «ControlPointID = Wz_Invar» and «SensorID = Wz_TempCenter» on the local data system, for example:

Network Identification Example

```
Wz_Invar<<
    Wz_TempCenter
```

A sensor control point server also offers data to the external world via RPC. It is registered with a combination of the «ControlPointID = Wz_InvarSCP» and its IP address and port in brackets together with the «SensorID = Wz_TempCenter», so that the connection information for a direct remote data access is defined, for example:

```
Wz_InvarSCP(10.0.0.10,50500)<<
    Wz_TempCenter
```

---

84 The basic idea comes from the Simple Network Management Protocol (SNMP), where a similar path structure can be used to identify sensors and where standardized functions can be used to get data sets.

As the data is also offered by a real-time and data server, the concatenation is extended with the information about one of these access servers, for example, for the real-time server:

```
Wz_SubcellCabineRealtime(10.0.0.10,40600)<<
    Wz_InvarSCP(10.0.0.10,50500)<<
        Wz_TempCenter
```

or, for example, for the data server:

```
Wz_SubcellCabineData(10.0.0.10,40700)<<
    Wz_InvarSCP(10.0.0.10,50500)<<
        Wz_TempCenter
```

At the next hierarchy level, another sensor control point server represents the subordinated system monitoring and safety cell. It follows the style already described and extends the existing address with an additional path element, for example:

```
Wz_SubcellCabineRealtimeSCP(193.174.168.50,50123)<<
    Wz_SubnodeCabine(10.0.0.10,50600)<<
        Wz_InvarSCP(10.0.0.10,50500)<<
            Wz_TempCenter
```

The data from this sub-cell is then also offered via the local data servers of the higher-level system monitoring and safety cell. Therefore, the path is again extended with this information, for example, for the real-time server:

```
Wz_SubcellRTWRealtime(193.174.168.50,40200)<<
    Wz_SubcellCabineRealtimeSCP(193.174.168.50,50123)<<
        Wz_SubnodeCabine(10.0.0.10,50600)<<
            Wz_InvarSCP(10.0.0.10,50500)<<
                Wz_TempCenter
```

and so on.[85]

Database Tables

All of these structures are organized using the API as a standardized library interface[86] for the data storage system. All functionalities and administrative activities are hidden behind this interface module and library. It uses similar functions to the defined RPC access, which is generated with «idl2rpc.pl» plus additional, useful service functions and methods which combine and simplify the interaction between data sender and storage system. The according data tables in the database system which administrates the metadata from the configuration and also the temporarily stored real data are kept very simple. ◘ Figure 4.49 shows a possible table structure in the database for the administrative metadata tables (see ◘ Fig. 4.49a) and for the data tables (see ◘ Fig. 4.49b) with their attributes.[87]

Database Metadata Tables

The administrative data tables contain redundant information, once in configuration file text form and once in separated easily accessible data sets of the database. Changes can only be propagated in the form of configuration files. The information is then split up into the according tables, which might also contain local modifications, for example, the locally used sensor update interval in hierarchical systems.

Database Data Tables

---

85 If the identifiers are unique within the whole path structure, it is even possible to use a shortcut leaving all connection information which is anyway registered at the data management system.

86 A library is a partly compiled object file, which can be used as a shared object (the usual file extension is «.so»). The project which wants to use it only has to include the according header files in its source code and can directly link the shared object during linking of the executable.

87 The degree of normalization of such tables is not so relevant because the complete administration is automated, and inserts, updates, and deletes always concern a whole record.

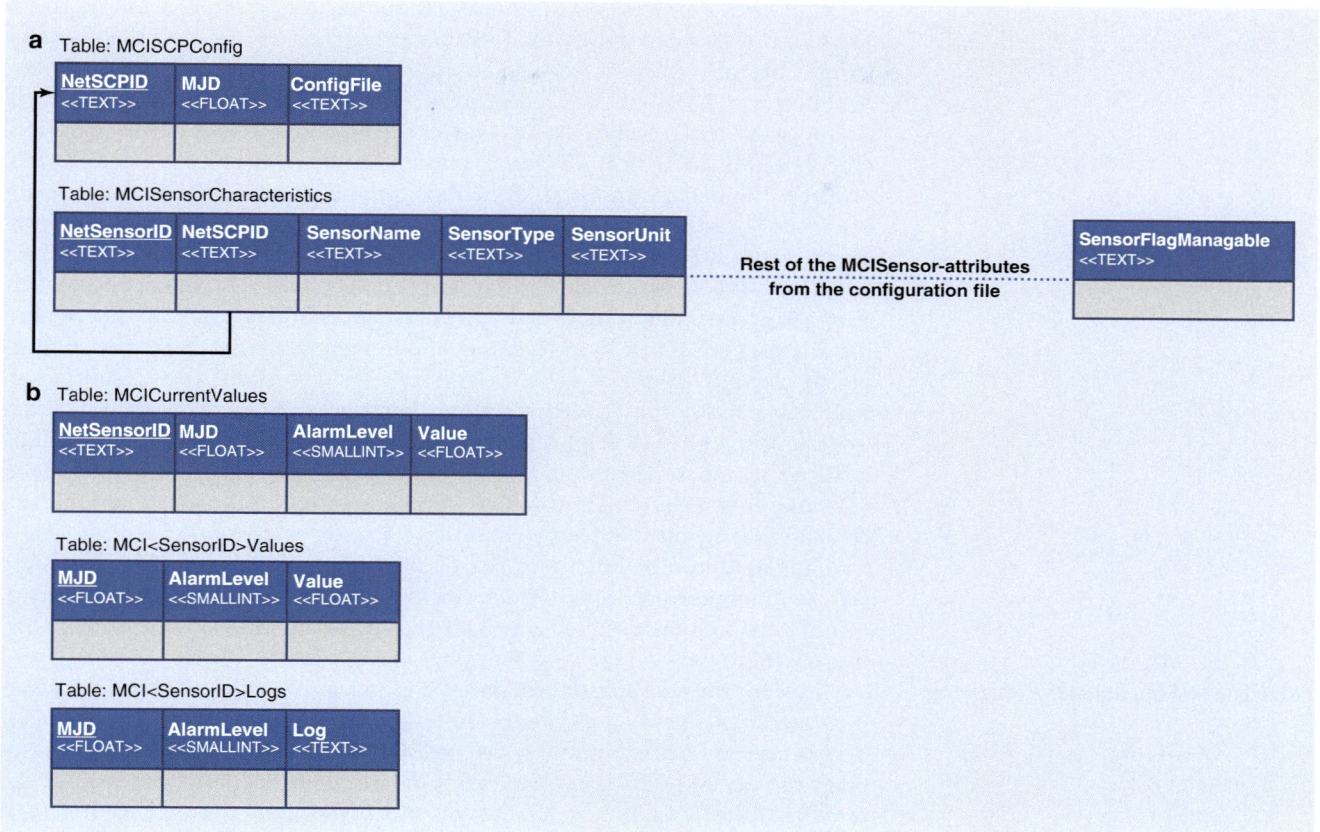

Fig. 4.49 The table structure of the database for a system monitoring and safety cell with the administrative metadata tables (part **a**) and for the data tables (part **b**) with their attributes

The data tables are for the current time-tagged values, the historic time-tagged values in a specific table per sensor, and the time-tagged log information in a specific table per ASCII logbook sensor.[88] Each table contains the time in Modified Julian Date (MJD) with a minimum precision of milliseconds. This double value offers the possibility of a simple ordering of the records. The values themselves are saved in IEEE double-precision format (see also ► Sect. 3.3.2). The values can also be sent or requested as ASCII text. Each conversion from binary to ASCII prints the complete double-precision number with all decimal places. Each record also already contains a flag for alarm levels, where 0 = no alarm, 1 = sensor control point alarm, 2 = warning, and 3 = alert (compare also ► Sect. 4.5). The alarm levels are defined within the sensor control points and their configurations and can be derived with a fuzzy set (see ► Sect. 4.2.1), or with fixed thresholds and logic combinations. If additional specifications are necessary, they can be coded with alarm values greater than three. The alarm levels can also be used for logging mechanisms (see ► Sect. 4.4). The alarm messages can also be used to send text messages to mobile phones. Regular telephones can be called using Voice over IP (VoIP). Suitable devices or gateways use the Session Initiation Protocol (SIP) to make phone calls and to play voice messages generated from text files.[89]

---

88 The system should also be able to deal with text information, so that centralized logbooks can be implemented and not only binary system values can be used.

89 A mini-computer with Voyage-Linux is used at the Wettzell observatory to run the open-source software «Asterisk», which is a communication framework and can be used to make phone calls (see ► http://www.asterisk.org).

Comparison to Industrial Standard: OPC UA

In industrial process management, similar techniques have been known for years and are de facto standards. Therefore, a short comparison is suitable at this point. Currently, the best known and most promising standard is OPC Unified Architecture (OPC UA),[90] which is the International Electrotechnical Commission (IEC) standard 62,541. Compared to the classic OPC definition, OPC UA adds a new TCP-/IP-based communication stack and extended security features. The major advantage of the new communication stack is that it is not Microsoft®-related anymore, so that it can be used in portable implementations. The methods are defined within OPC UA and can be implemented in different languages. This is the middleware layer of OPC UA. OPC UA also defines a standardized access point to read and write current data («Data Access»), to request historic data («Historic Data Access»), and to request alarms and events according to changed priorities («Alarm and Events» or «Alarm and Conditions»), which also allows subscriptions, so that clients get notified, when critical situations appear. OPC UA defines nodes and hierarchical namespaces which can be browsed (compare (Enste and Müller 2007, *l.c. page 142 ff.*) and in general (Mahnke et al. 2009)). All in all, OPC UA shows that similar ideas, design drivers, and architectural implementations are already successfully used for industrial automation. The functionalities of an autonomous monitoring and safety cell try to implement similar structures which are independent from expensive, commercial architectures and which are conditioned for the monitoring of scientific observatories.

Autonomous Functionalities: Planning

Knowing the background structures, it is now possible to consider the autonomous functionalities of the system monitoring and safety cell. First of all, the planning is very asynchronous and distributed between the individual sensor control point servers and data injectors. They can use individual strategies to plan the activities. Usually the timing and the sequence of the tasks are hard coded. Central activities (e.g., the writing of data backups) are planned by the cell administrator and defined as cron jobs or as individual processes using configuration files. As the structure is very similar to the data management cell, it is possible to use similar planning mechanisms (see ▶ Sect. 4.6).

Autonomous Functionalities: Controlling

The controlling is split into two parts: the part which is developed by the programmers of the data injectors and the centralized data management. The controlling processes all of the described operations on database tables and files. The controlling here is similar to the data management cell and is not pure hardware management. It is more or less the service implementation to offer the data to the rest of the production cell, which takes care of data integrity and correctness. A very important task is the management of the alarm levels in the sensor control points. But besides the hardware interlock mechanisms in the second layer, no additional actions are started or commanded if levels climb over defined limits. The alarm levels are only passive signals, which can be requested by higher coordination instruments via the user interface.

Autonomous Functionalities: User Interfacing

The user interface consists of two parts: the standardized interface definition of RPC functions and separate GUI implementations, such as with Zabbix. The first version of the standardized interface was defined by the MCI group and can be found in Example 4.20 as a collection of functions in the IDL format for «idl2rpc. pl». While the RPC interface is standardized to offer a unique access point, the GUI can be designed very individually. The style can be adapted to individual tasks and wishes. Different possibilities are shown in ▫ Fig. 4.50. The simplest style is a text output, organized with tabs, where each tab represents one sensor control point

---

90 Originally «OPC» stood for «OLE for Process Control», where Microsoft®'s Object Linking and Embedding (OLE) technology was used. As nowadays OPC is not tied to this environment, OPC has no specific meaning anymore, but is still used for the foundation name.

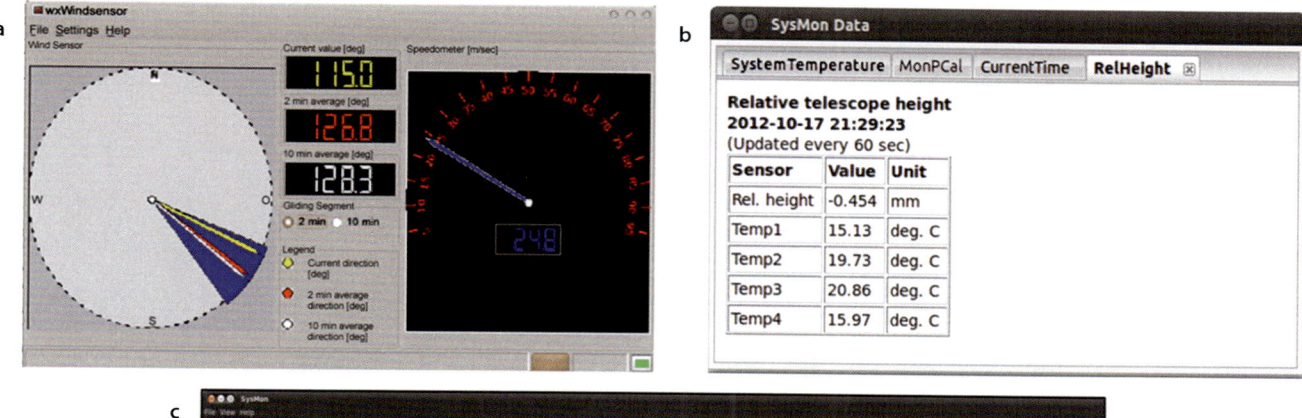

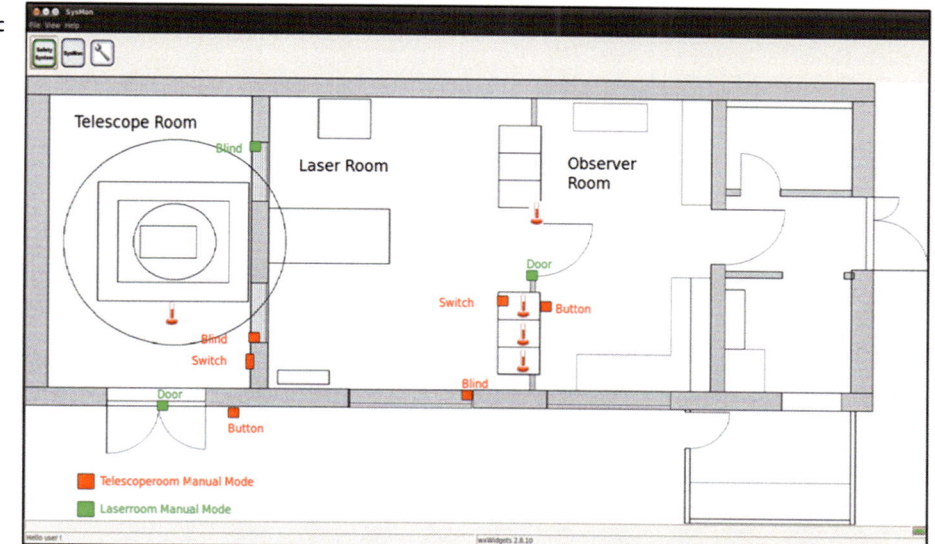

■ **Fig. 4.50** Different system monitoring GUIs: the wind sensor display at the German Antarctic Receiving Station O'Higgins **a**; the system monitoring parameters at the Wettzell radio telescope, organized as text and HTML output in tabs **b**; and the GUI for the ground safety system of the laser ranging system on a touch screen, showing the warning levels at the specific place of the sensors in a map of the building **c**

(see ■ Fig. 4.50b). Using an HTML output, it is already possible to colorize different alarm levels using a font or background color. But even more graphically oriented designs are possible. The wind sensor is a very eye-catching display (see ■ Fig. 4.50a), which shows wind direction, wind speed, and different average calculations. As almost all graphical styles can be arranged, it is even possible to show the alarm levels on the position of the sensors in a map of a building where they are triggered. If this is presented on a touch screen, where the operator can get the current alarm information by just touching the position of the sensor, it is a very ergonomic and interactive presentation of system monitoring and safety parameters (see ■ Fig. 4.50c). Another possibility is the installation of a local web server (e.g., Apache) on the system monitoring and safety cell. This web server can then be used to offer configuration and data access via HTML web pages. Some other graphical or higher-level monitoring tools (like Zabbix) can be included at this point as well.

Zabbix is a complete monitoring system, which is available under the conditions of open-source code from ▶ http://www.zabbix.com/. It is mainly developed for the management of servers, virtual machines and network devices. Nevertheless, it can also be used to define individual metrics to support other devices from which data should be gathered. Besides collecting of data, it offers numerous visualization

Zabbix

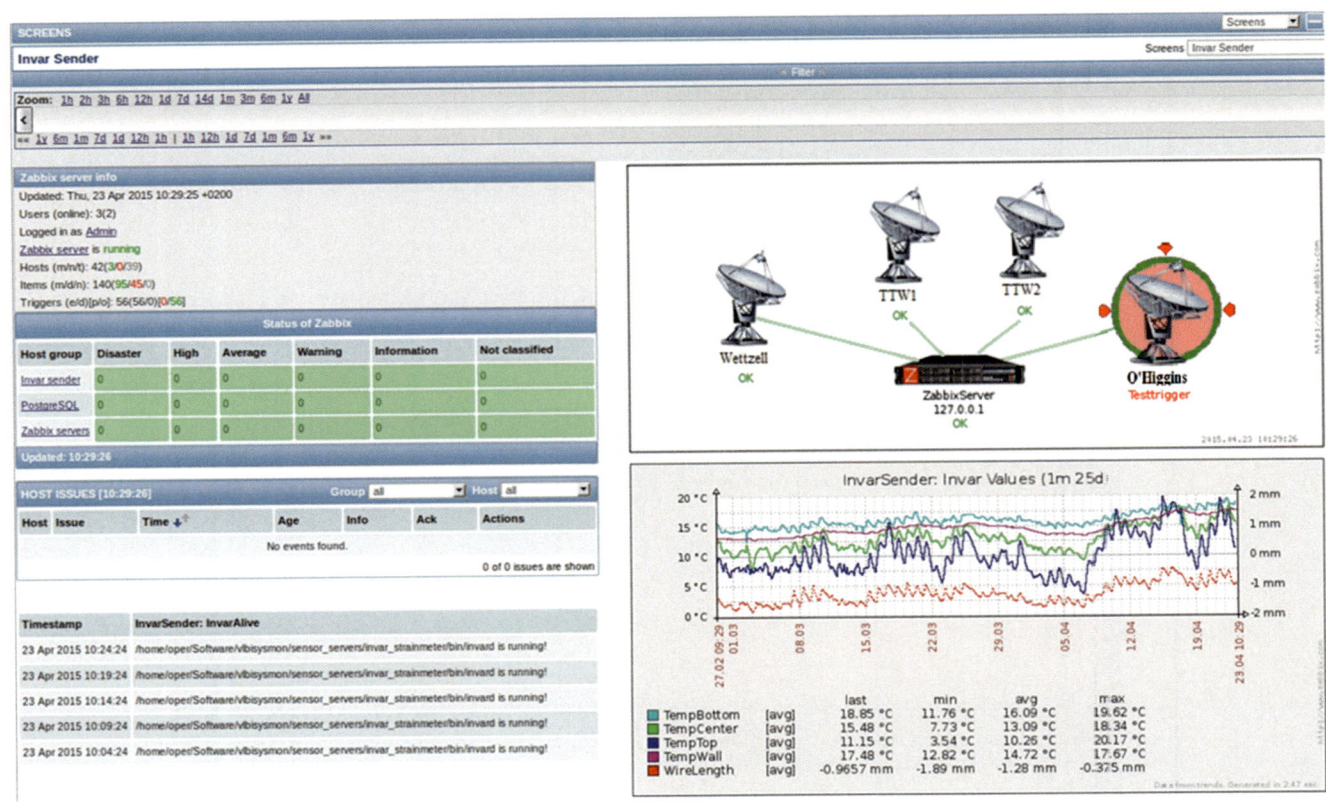

Fig. 4.51 A possible Zabbix web page screen to present the states of the different radio telescopes with their subsystems (O'Higgins was triggered manually to show an error)

features, like graphs, maps, screens, or overviews. Flexible trigger definitions additionally make it possible to analyze the data for the purpose of alerts. In the case of an alert, Zabbix can send emails or even short messages to mobile phones. In addition, the Zabbix community already offers many tools for monitoring computers, server hardware, network devices, and web infrastructures (Zabbix LLC n.d.). Because Zabbix is a pure monitoring system without the aspect of direct interaction with the devices, it does not completely fulfill the requirements for an autonomous safety cell, as described in the context of this book. Nevertheless, the cell can benefit tremendously from the powerful abilities of Zabbix, especially for presentation tasks. This was why it was included in the general implementation of the system monitoring suite and the API at the Wettzell observatory (see ▣ Fig. 4.48[91]). A possible graphical feedback in the form of the Zabbix monitoring web page is shown in ▣ Fig. 4.51. The big advantage of including Zabbix is that all the data of the system monitoring and safety cell can easily be injected to other Zabbix servers using the Proxy concept of Zabbix, which makes it possible to propagate data. Another big advantage is that with Zabbix, each monitoring node directly offers a web-based GUI, which can be used to inform operators about the states of the connected systems in different information ranges. The MCI hardware and software parts in the background ensure activities to keep the system safe and stable or to switch off in critical situations.

---

91 Special thanks should be given here to Matthias Schoenberger and Katharina Kirschbauer for the implementation of hardware and software solutions for the Wettzell system monitoring with Zabbix.

■ ■ Example 4.20

A first version of the standardized RPC interface from the MCI group.

```
// Function return values
const MCISCP_RET_OK  = 0;   // Function returns ok
const MCISCP_RET_NOK = 1;   // Function returns not ok

// Warning levels
const const MCISENSOR_WARNINGLEVEL_OK = 0;   // No warning
const const MCISENSOR_WARNINGLEVEL_SCPALARM = 1;   // Sensor control point server alarm
const const MCISENSOR_WARNINGLEVEL_WARNING = 2;   // Value is in warning interval
const const MCISENSOR_WARNINGLEVEL_ALERT = 3;     // Value reached alert interval

// Interface function definition
interface <SensorControlPointName> #MCIVer1.0_20120417001
{
    // ###################################################################
    // GET MONITORING DATA
    // ###################################################################
    /*********************************************************************
        function   usGetCurrentDataText
    *********************************************************************/
    /*!            Return the content of the database table for current values
                   in form of the original database table
                   with the possibility to select dedicated net sensors by ID
                   Returns: NetSensorID | MJD | AlarmLevel | Value
        \param     in pstrNetSensorIDSelect<> --> Array to select net sensors or empty
        \param     out string pstrSensorValueColumnNames<> <-- The table column head lines as
            array
        \param     out string ppstrSensorValueTable <><> <-- 2-dim. Array with the table content
            of the
                                                request
        \return    unsigned short <-- Error code (0 = ok, >0 = error)
    *********************************************************************/
    /*  author    MCI colaboration group
        date      17.04.2012
        revision  —
        info      —
    *********************************************************************/
    unsigned short usGetCurrentDataText (in pstrNetSensorIDSelect<>,
                                         out string pstrSensorValueColumnNames<>,
                                         out string ppstrSensorValueTable<><>);
    /*********************************************************************
        function   usGetCurrentDataMJD
    *********************************************************************/
    /*!            Return the content of the database table for current values
                   in form of separated column vectors and in Modified Julian Date (MJD)
                   with the possibility to select dedicated net sensors by ID
                   (each row over the column vectors is one sensor entry)
        \param     in pstrNetSensorIDSelect<> <--> Array to select net sensors or empty,
                                            returns the resulting, found net sensors
        \param     out double pdTimeMJD<> <-- Resulting array with the record times in MJD
        \param     out unsigned int puiAlarmLevel<> <-- Resulting array with alarm levels
        \param     out double pdValue<> <-- Resulting array with values
        \return    unsigned short <-- Error code (0 = ok, >0 = error)
    *********************************************************************/
    /*  author    MCI colaboration group
        date      17.04.2012
        revision  —
        info      —
    *********************************************************************/
    unsigned short usGetCurrentDataMJD (inout pstrNetSensorID<>,
                                        out double pdTimeMJD<>,
                                        out unsigned int puiAlarmLevel<>,
                                        out double pdValue<>);
```

```
/*****************************************************************
    function    usGetCurrentDataTextSinceMJD
 *****************************************************************/
/*!              Same as usGetCurrentDataText including a selection for dates since a
                 dedicated time, so that only values are transferred which have been updated
                 during this time period
    \param       in pstrNetSensorIDSelect<> -> Array to select net sensors or empty
    \param       in double dSinceTimeMJD -> Time since when the data should be returned
    \param       out string pstrSensorValueColumnNames<> <- The table column head lines as
        array
    \param       out string ppstrSensorValueTable<><> <- 2-dim. Array with the table content
        of the
                                               request
    \return     unsigned short <- Error code (0 = ok, >0 = error)
 *****************************************************************/
/*  author     MCI colaboration group
    date        17.04.2012
    revision    —
    info        —
 *****************************************************************/
unsigned short usGetCurrentDataTextSinceMJD (in pstrNetSensorIDSelect<>,
                                             in double dSinceTimeMJD,
                                             out string pstrSensorValueColumnNames<>,
                                             out string ppstrSensorValueTable<><>);
/*****************************************************************
    function    usGetDataFromToMJDText
 *****************************************************************/
/*!              Transfer complete tables of data of one sensor with selection of start
                 time and an end time
    \param       in strNetSensorIDSelect -> Select net sensors
    \param       in double dStarttimeMJD -> Time since when the data should be returned
    \param       in double dEndtimeMJD -> Time to which data should be returned
    \param       in unsigned short usTableSelector -> Select table (0=Value&Log, 1=Value, 2=
        Log)
    \param       out string pstrSensorValueColumnNames<> <- The table column head lines as
        array
    \param       out string ppstrSensorValueTable<><> <- 2-dim. Array with the table content
                                               request for data with the structure
                                               NetSensorID | MJD | AlarmLevel | Value
    \param       out string pstrSensorLogColumnNames <> <- The table column head lines as
        array
    \param       out string ppstrSensorLogTable <><> <- 2-dim. Array with the table content
        of the
                                               request for logs with the structure
                                               NetSensorID | MJD | AlarmLevel |
                                                   LogText
    \return     unsigned short <- Error code (0 = ok, >0 = error)
 *****************************************************************/
/*  author     MCI colaboration group
    date        17.04.2012
    revision    —
    info        —
 *****************************************************************/
unsigned short usGetDataFromToMJDText (in strSelectNetSensorID,
                                       in double dStarttimeMJD,
                                       in double dEndtimeMJD,
                                       in unsigned short usTableSelector,
                                       out string pstrSensorValueColumnNames<>,
                                       out string ppstrSensorValueTable<><>,
                                       out string pstrSensorLogColumnNames<>,
                                       out string ppstrSensorLogTable<><>);

// ##############################################################################
// SET MONITORING DATA
// ##############################################################################
```

```
/*****************************************************************         120
      function    usSetDataText                                           121
 *****************************************************************/        122
/*!             Insert a data table into system with the structure        123
               NetSensorID | MJD | AlarmLevel | Value                     124
     \param      in string ppstrSensorValueTable<><> -> 2-dim. Array with the  table content  125
        of the                                                            
                                                     insert              126
     \return     unsigned short <- Error code (0 = ok, >0 = error)        127
 *****************************************************************/        128
/*   author    MCI colaboration group                                    129
     date       17.04.2012                                               130
     revision  -                                                          131
     info      -                                                          132
 *****************************************************************/        133
unsigned short usSetDataText (in string ppstrSensorValueTable <><>);      134
/*****************************************************************         135
      function    usSetSingleDataMJD                                      136
 *****************************************************************/        137
/*!             Insert a single data set into system                      138
     \param      in string strNetSensorID -> Net sensor identifier        139
     \param      in double dTimeMJD -> Time in MJD when the value was recorded  140
     \param      in unsigned int uiAlarmLevel -> Alarm level of value     141
     \param      in double dValue -> The value itself                     142
     \return     unsigned short <- Error code (0 = ok, >0 = error)        143
 *****************************************************************/        144
/*   author    MCI colaboration group                                    145
     date       17.04.2012                                               146
     revision  -                                                          147
     info      -                                                          148
 *****************************************************************/        149
unsigned short usSetSingleDataMJD (in string strNetSensorID,              150
                                   in double dTimeMJD,                    151
                                   in unsigned int uiAlarmLevel,          152
                                   in double dValue);                     153
/*****************************************************************         154
      function    usSetDataMJD                                            155
 *****************************************************************/        156
/*!             Insert a data sets into system                            157
     \param      in string pstrNetSensorID <> -> Net sensor identifiers   158
     \param      in double pdTimeMJD <> -> Times in MJD when the values were recorded  159
     \param      in unsigned int puiAlarmLevel <> -> Alarm levels of value 160
     \param      in double pdValue <> -> The values itself                161
     \return     unsigned short <- Error code (0 = ok, >0 = error)        162
 *****************************************************************/        163
/*   author    MCI colaboration group                                    164
     date       17.04.2012                                               165
     revision  -                                                          166
     info      -                                                          167
 *****************************************************************/        168
unsigned short usSetDataMJD (in string pstrNetSensorID <>,                169
                             in double pdTimeMJD <>,                      170
                             in unsigned int puiAlarmLevel <>,            171
                             in double pdValue <>);                       172
                                                                          173
// ###############################################################        174
// SET LOG DATA                                                           175
// ###############################################################        176
/*****************************************************************         177
      function    usSetLog                                                178
 *****************************************************************/        179
/*!             Insert a log table into system with the structure         180
               NetSensorID | MJD | AlarmLevel | LogText                   181
     \param      in string ppstrSensorLogTable <><> -> 2-dim. Array with the table content of  182
        the
```

```
                                                       insert
      \return    unsigned short <- Error code (0 = ok, >0 = error)
      **********************************************************************/
/*    author    MCI colaboration group
      date       17.04.2012
      revision   —
      info       —
      **********************************************************************/
unsigned short usSetLog (in string ppstrSensorLogTable <><>);
/**********************************************************************
      function   usSetSingleLogMJD
      **********************************************************************/
/*!              Insert a single log set into system
      \param     in string strNetSensorID -> Net sensor identifier
      \param     in double dTimeMJD -> Time in MJD when the value was recorded
      \param     in unsigned int uiAlarmLevel -> Alarm level of value
      \param     in string strLogText -> The log text itself
      \return    unsigned short <- Error code (0 = ok, >0 = error)
      **********************************************************************/
/*    author    MCI colaboration group
      date       17.04.2012
      revision   —
      info       —
      **********************************************************************/
unsigned short usSetSingleLogMJD (in string strNetSensorID,
                                  in double dTimeMJD,
                                  in unsigned int uiAlarmLevel,
                                  in string strLogText);
/**********************************************************************
      function   usSetLogMJD
      **********************************************************************/
/*!              Insert a log sets into system
      \param     in string pstrNetSensorID <> -> Net sensor identifiers
      \param     in double pdTimeMJD <> -> Times in MJD when the values were recorded
      \param     in unsigned int puiAlarmLevel <> -> Alarm levels of value
      \param     in string pstrLogText <> -> The log texts itself
      \return    unsigned short <- Error code (0 = ok, >0 = error)
      **********************************************************************/
/*    author    MCI colaboration group
      date       17.04.2012
      revision   —
      info       —
      **********************************************************************/
unsigned short usSetLogMJD (in string pstrNetSensorID <>,
                            in double pdTimeMJD <>,
                            in unsigned int puiAlarmLevel <>,
                            in string pstrLogText <>);

// #######################################################################
// SET CONFIGURATION TO ADMINISTRATE SENSOR CONTROL POINT
// #######################################################################
/**********************************************************************
      function   usSetSingleConfigurationMJD
      **********************************************************************/
/*!              Insert a single configuration into system
      \param     in string strNetSensorControlPointID -> Net sensor control point identifier
      \param     in double dTimeMJD -> Valid-from time of configuration
      \param     in string strConfigFileContent -> Configuration content
      \return    unsigned short <- Error code (0 = ok, >0 = error)
      **********************************************************************/
/*    author    MCI colaboration group
      date       17.04.2012
      revision   —
      info       —
```

```
*********************************************************************/    247
unsigned short usSetSingleConfigurationMJD (in string strNetSensorControlPointID ,    248
                                     in double dTimeMJD ,    249
                                     in string strConfigFileContent );    250
/*********************************************************************    251
      function   usSetConfigurationMJD    252
 *********************************************************************/    253
/*!           Insert configurations into system    254
    \param     in string pstrNetSensorControlPointID <> -> Net sensor control point    255
          identifiers
    \param     in double pdTimeMJD <> -> Valid-from times of configuration    256
    \param     in string pstrConfigFileContent <> -> Configuration contents    257
    \return    unsigned short <- Error code (0 = ok, >0 = error )    258
 *********************************************************************/    259
/*   author     MCI colaboration group    260
     date        17.04.2012    261
     revision   -    262
     info       -    263
 *********************************************************************/    264
unsigned short usSetConfigurationMJD (in string pstrNetSensorControlPointID <>,    265
                                in double pdTimeMJD <>    266
                                in string pstrConfigFileContent <>);    267
                                                                          268
// ########################################################################    269
// GET CONFIGURATION TO ADMINISTRATE SENSOR CONTROL POINT    270
// ########################################################################    271
/*********************************************************************    272
      function   usGetConfigurationMJD    273
 *********************************************************************/    274
/*!           Return configurations from system with selection of sensors    275
    \param     inout string pstrNetSensorControlPointID <> <-> Net sensor control point    276
          identifiers
                                                    and selector    277
    \param     out double pdTimeMJD <> <- Valid-from times of configuration    278
    \param     out string pstrConfigFileContent <> <- Configuration contents    279
    \return    unsigned short <- Error code (0 = ok, >0 = error )    280
 *********************************************************************/    281
/*   author     MCI colaboration group    282
     date        17.04.2012    283
     revision   -    284
     info       -    285
 *********************************************************************/    286
unsigned short usGetConfigurationMJD (inout string pstrNetSensorControlPointID <>,    287
                                out double pdTimeMJD <>,    288
                                out string pstrConfigFileContent <>);    289
/*********************************************************************    290
      function   usGetConfigurationSinceMJD    291
 *********************************************************************/    292
/*!           Return configurations from system with selection of sensors (similar to    293
          usGetConfigurationMJD) with an additional selector , to just get recently    294
                changed
          configurations    295
    \param     inout string pstrNetSensorControlPointID <> <-> Net sensor control point    296
          identifiers
                                                    and selector    297
    \param     in double dSinceTimeMJD -> Select from this time on    298
    \param     out double pdTimeMJD <> <- Valid-from times of configuration    299
    \param     out string pstrConfigFileContent <> <- Configuration contents    300
    \return    unsigned short <- Error code (0 = ok, >0 = error )    301
 *********************************************************************/    302
/*   author     MCI colaboration group    303
     date        17.04.2012    304
     revision   -    305
     info       -    306
```

```
*****************************************************************/    307
unsigned  short  usGetConfigurationSinceMJD ( inout  string  pstrNetSensorControlPointID <>,    308
                                     in  double  dSinceTimeMJD ,    309
                                     out  double  pdTimeMJD <>,    310
                                     out  string  pstrConfigFileContent <>);    311
                                                                     312
// ####################################################################    313
// OTHER ADMINISTRATION FUNCTIONS    314
// ####################################################################    315
/*************************************************************    316
    function    usGetRegisteredNetSensorIDs    317
    *************************************************************/    318
/*!            Return a list of registered net sensors    319
    \param     out string pstrNetSensorID <> <— List of sensor identifiers    320
    \return    unsigned short <— Error code (0 = ok, >0 = error)    321
    *************************************************************/    322
/*  author    MCI colaboration group    323
    date       17.04.2012    324
    revision   —    325
    info       —    326
    *************************************************************/    327
unsigned  short  usGetRegisteredNetSensorIDs (out string pstrNetSensorID <>);    328
                                                                     329
// ####################################################################    330
// Individual , proprietary functions    331
// ####################################################################    332
// <your personal functions>    333
};    334
```

Autonomous Functionalities: Hardware Driving

The functionality to drive the hardware has already been explained in the previous general explanations about system monitoring and safety cells. The hardware drivers are modules or components in the data injectors or in the sensor control point servers. They include the individual methods and libraries from the vendors of the different hardware or run specific protocols over communication lines to interact with the hardware. The sensor control points are, in principle, the local hardware control cells and can be treated like these (see ▶ Sect. 4.5). But usually the sensors of the system monitoring and safety cell deal with a not too complex hardware of analog, digital, or binary sensors, which are based on the elementary sensor techniques.

Sensors

The general definition of sensors was already introduced in ▶ Sect. 4.2.1. At this point, some details should be touched on again. Generally, sensors can be split into two types: active and passive. Active sensors are activated by external forces, energies, or exciter, using the thermo, photo, piezo, or electrodynamic effect. Passive sensors use their own separate power supply and detect changes of properties in resistors, capacities, inductors, or inductive reactance (Schmid et al. 2011, *l.c. page 224*). Therefore, they are usually minimally invasive depending on their installation in existing systems. For example, sensors which are based on inductance are ideal for measuring the voltages and currents of existing, already installed power and signal cables, as it is only necessary to strap an inductive sleeve around the cable. This minimally invasive concept is a good way to install additional sensors on existing systems, so that they can be upgraded to (semi-) automated systems.

Characteristic Sensor Profiles

Sensors can be characterized with their signal properties. The best case would be a proportional (or linear) behavior, which means if the measured value changes linear, the resulting signal also changes linear (maybe with a fixed factor) (see ▣ Fig. 4.52c). But usually this is not the case. The resulting signals can differ from the ideal proportional relation to the measured value and, for example, describe a

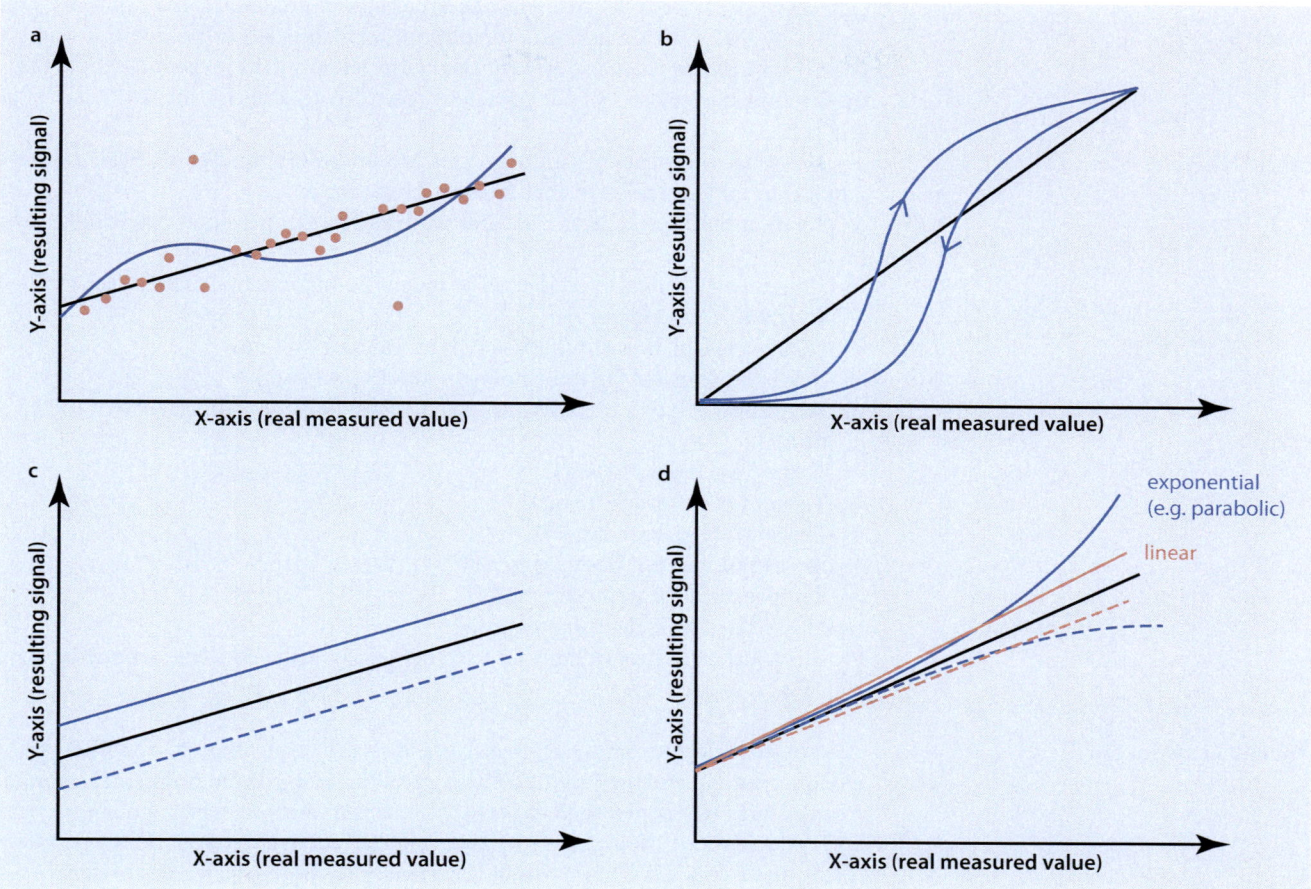

■ **Fig. 4.52** The different accuracy properties of sensors: **a** deviations from the ideal linear, proportional line or outliers (*dots* with large distances), **b** different signal curves for increasing and decreasing signals in the form of a hysteresis, **c** positive and negative (*dashed line*) linear offsets, and **d** positive or negative (*dashed lines*), linear, or exponential drifts ((Schmid et al. 2011) *l.c. page 225*)

signal swinging around the head curve[92] of the sensor (see the swinging solid line in ■ Fig. 4.52a). It is also possible that the sensor produces some outliers (see the dots also in ■ Fig. 4.52a; e.g., signals from ultrasonic sensors jitter, as the resulting distances depend on the quality and angle of the reflecting surface). Other resulting signals show a hysteresis, so that the measured signal differs from the characteristic line in a way which is different for increasing and decreasing signals (see ■ Fig. 4.52b; e.g., mechanical instruments often have such reversal errors). Very often signals have an offset or a drift (see ■ Fig. 4.52d). The offset is defined by a factor by which the signal is modified. Usually most sensors show a temperature drift, so that the measured signal curve depends on external temperatures. The drift, which is a variable factor on the reference curve, can be linear or exponential (typical examples are time standards: cesium atomic standards show a linear drift over time, while hydrogen masers show a parabolic drift over time). All the inaccuracies together define the maximum error limit (Schmid et al. 2011, *l.c. page 225*). Because of all these inaccuracies, it is necessary to calibrate sensors against a reference sensor. This must be done regularly to also check changes over time. Other huge influences are interferences from galvanic cross-talking of different, parallel signal lines, inductive reactances of magnetic fields in the environment, or

92   A head curve is a graphical description of two physical indicators from a (electrical) component which are in relation to each other.

4

capacitive cross-feeds because of different electrical potentials. They must be prevented by filters with a galvanic separation, with shielded cables, with a suitable connection to the electrical ground, or with better suited communication techniques and cumulative signal repeaters (see also (Schmid et al. 2011, *l.c. page 259 f.*)).

Sensor Measures

The sensors can be split into various types (see (Schmid et al. 2011, *l.c. page 226 ff.*)) to measure relative, incremental, or absolute:

- Positions and tracks (e.g., coded measuring scales, fiber optic gyros, light barriers)
- Angles (e.g., optical-incremental or magnetic-incremental angular sensors and angular encoders)
- Distances (e.g., ultrasonic sensors, laser ranging systems)
- Velocities (e.g., RADAR technique, inductive sensors)
- Expansion (e.g., strainmeter, invar cable with fast linear displacement transducer)
- Forces (e.g., piezo sensor)
- Torques (e.g., torque meter)
- Pressures (e.g., barometer)
- Accelerations (e.g., accelerometer)
- Temperatures (e.g., thermometer)
- Humidities (e.g., hygrometer)
- Electrical signals as voltages or currents (e.g., voltage converter, inductive or capacitive probes)

Additional Sensor Net

A set of well selected, minimally invasive sensors can be positioned in addition to the already existing systems (used for the controlling loops of the mechanical components at large telescopes, such as servo systems with tachometers, angular encoders, and end switches). The sensor signals are collected and analyzed by the system monitoring and safety cells. To do this, it is necessary to convert all signals to digital. For the sampling rate, it is important to determine the necessary Nyquist frequency (see ▶ Sect. 4.5) of the signal to be measured to get the required sampling interval. Another important factor is defined by the binary resolution, which is given by the number of used bits for the amplitude of a signal. It defines the number of intervals between the maximum and minimum signal amplitude. After all the signals have been digitized, a logic comparison and combination are possible. Safety-relevant parts are directly connected with a serial hardware logic (e.g., a special safety-certified PLC) and produce interlock signals which are directly sent to switches in the relevant controllers to stop the processing immediately. All other signals are just collected and evaluated to assess the alarm levels. The whole system is specially designed for an individual observing system and can be split into a hierarchical structure, so that different cells are responsible for different areas in the telescope system or building. An example of a possible distribution of sensors over a radio telescope is indicated in ◻ Fig. 4.53 or can be found in the suggestions of the MCI proposals.

Sensor Data Categories

The resulting data sets are not only interesting for local issues. They can partly account for improved data analysis of the observation data. Another quite important use is the maintenance, where, for example, higher sampled data offer information about possible error sources. Therefore, the monitoring data can be split into three different basic types to which the different sensors offer their data:

1. *Data for science and analysis* with mostly lower sampling rates of several seconds; scheduled or predefined data sets; for example, meteorology, WVR, clock offsets, etc.

2. *Data for system operations* with medium sampling rates of seconds and sub-seconds similar to human reaction times (except the directly used interlock lines, which must switch according to the defined reaction and protection times); permanently available for the system control and coordination as system health states; for example, power voltages and currents, wind speed and direction, emergency stops and switches, rack temperatures, etc.

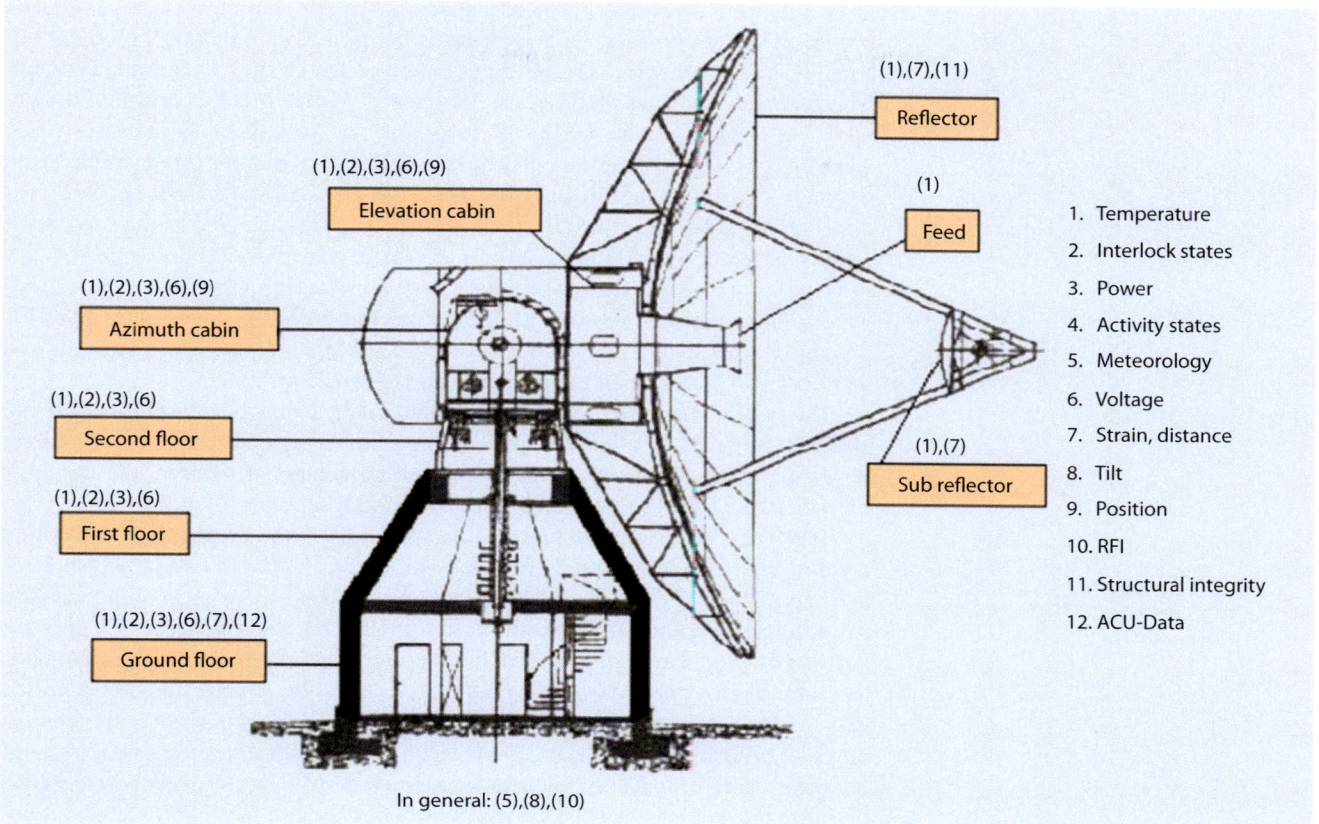

Fig. 4.53 The distribution of different, additional sensors of the system monitoring and safety cell, indicated for the Wettzell radio telescope. A similar plan can also be drawn for the laser ranging systems

3. *Data for diagnostics* with higher sampling rates of milliseconds and microseconds; requested on demand according to maintenance requirements and sensor control point possibilities; for example, servo currents, contouring errors, relative rotational movement, etc.

Failure management uses the different mechanisms offered by «idl2rpc.pl» servers. Simple data injectors must themselves deal with failure scenarios. The central data management system of each system monitoring and safety cell must check if the registered sensor control points send in data at the proposed time intervals. If they do not, an alarm level must be set. The different failure states (also of the real sensor values) are directly converted into these alarm levels, which represent alarms with the values 0 = no alarm, 1 = sensor control point alarm, 2 = warning, or 3 = alert, as described before. To follow the definitions of a safe device, it is necessary to react to alarms in a defined time with a defined reaction which must be guaranteed (Vigenschow 2005, *l.c. page 224*). Fast reactions are always important when human beings could be in danger. Therefore, the in-sky and on-ground safety is an essential part of the system monitoring and safety cell. It is also an essential part of the whole laser ranging system, as it cannot be automated without a well-functioning and safe in-sky and on-ground safety system. While other passive systems do not have such problems, the active laser ranging system with a laser beam that is not eye safe strongly requires such a protection technique. As described, it is implemented as a hardware-based interlock mechanism and directly controls the dangerous laser. In-sky safety is supported by a combined system of a RADAR,[93] of RADAR data

Autonomous Functionalities: Failure Management

---

93 RADARs are a problem at geodetic core sites, as they interfere with other techniques, such as VLBI. But currently, they are the only system which guarantees the protection of aircrafts.

from air traffic control, of received data from aircraft transponders, and of additional LIDAR derivatives. One problem is hang- and paragliders, which are not detected by the systems. Currently, they are protected by the local human operator, who manually switches the laser off if a glider is visible in a live image of the sky taken by a video camera on the telescope mount. Automated systems must also solve this problem to autonomously detect such targets. But this is not simple, so an operator will still be required at these active systems. All the techniques used produce interlock signals with different tolerance times and can be used for the planning of observation schedules. On-ground safety is given using door and window sensors, scanners, and other detectors and with protecting areas. In any case, hardware solutions are usually very fault-tolerant. But adequate checks must offer the possibility of evaluating the system's correct functionality. Regular alarm tests and practices also help to train the operators and to check the system.

Other Fields for a Safety System

The system described is a very well-suited data acquisition system, which can also be configured for other tasks where a monitoring or a safety system is required, for example, in computer rooms of computing centers. Therefore, the design is even planned for these facilities at the Geodetic Observatory Wettzell.

Together with the other autonomous control cells – the autonomous coordination cell, the different autonomous hardware control cells, and the autonomous data management cell – they build up an autonomous production cell. As most of the activities run automatically and autonomously here, this offers a suitable design which does not necessarily require an operator in front of the console screen of the system. Therefore, the autonomous production cells offer a new operation strategy for observatories, which uses a human remote control and shared observations between different observing systems. It has become possible for one operator to control several systems (even for different space techniques) from one remote operator workplace. This can optimize the work at the observatories and can be used to increase the observations output significantly, with fewer assignments for operator staff, who can then perform the increasing number of maintenance tasks in an increasing network of systems. This is the starting point for remotely controlled multi-agent systems.

4.8   Summary

4.8.1   What Did We Learn?

The chapter focuses on a design for the controlling of a laser ranging system. It gives a short explanation about the technical aspects and backgrounds of laser ranging and describes the required data and the final product of such a system. A mapping of the real structures to software creates a distributed system. The main focus is an intelligent controlling of hardware which has to cooperate to produce the final product.

We learned about the different layers of feedback loops which use data from sensors to make decisions for the commanding of actuators. To deal with such inputs from digital input lines, the chapter discussed the Boolean algebra and the combination of logic gates to implement one's own simplified but static controller circuits. The next higher level is John von Neumann computers which implement central processing units. These units can be programmed to process data. Finally, every hardware can be represented in the form of standardized software stubs for the hardware functionality with specific interfaces for the middleware used.

Ideas are on the way to replace them by other techniques, such as combined missile and aircraft tracking systems, on the basis of optical and IR sensors. In any case, the best solution would be to use lasers with an eye-safe beam on longer distances than about 50 kilometer. Then only the unsafe area lower than 50 kilometer needs to be protected with the combined techniques described.

The way of controlling advised is to implement autonomous production cells, which are similar to implementations in industrial manufacturing. They provide techniques for planning, controlling, interfacing, hardware driving, and failure managing of production processes. Using this strategy, we identified different tasks which can be implemented as individual autonomous software cells. They all are coordinated by a central coordination cell.

A short excursus explained the way to implement user interfaces with different techniques from simple shell interfaces to sophisticated graphical designs. The background is the method of «thin clients» and «fat servers» which is used by each autonomous cell.

The following sections explained the different autonomous cells and the theory which is required. The coordination cell again uses the technique of metrics to implement static and dynamic planning and scheduling components. As many inputs come from optical systems, image processing is a key technology which makes it possible to find segmentations and classifications of objects in images to improve the planning and scheduling of observation tasks. The observation itself is split into specific procedures. The procedures use the functionalities of separate autonomous hardware cells which control the individual hardware. Each step processed is logged in a log file.

The data are organized in an autonomous data management cell using a hybrid data management. Fast-changing data, or data which have to be selected for individual tasks, are organized in a database. Therefore, the concepts of databases were explained. Sequential data are stored in files in the file system of the operating system. The combination of both techniques makes quick accesses possible. Searching algorithms on the basis of keys are combined with fast, block-oriented file operations.

The chapter ends with an explanation of a parallel system monitoring, which is used to fulfill safety criteria and gives a detailed overview of additional sensors, sensor control points, and a possible implementation of a monitoring system.

4.8.2 How Did We Use the Contents Learned?

The theory explained was necessary to design and implement real control systems which are used at the Geodetic Observatory Wettzell to control the laser ranging systems. Each component is explained in detail and can be found in real in the system developed at Wettzell.

The knowledge of databases was used to show examples for data selections in the real system. The code snippets are part of the official software at Wettzell.

4.9 Questions

1. Explain the working principle of a laser ranging system, and name the relevant components for a representation of hardware in a distributed software system.
2. Which external data do you need to run a laser ranging system?
3. Which hardware architecture is implemented in every modern computer? Draw the scheme, name the units, and explain them (each in one sentence).
4. What are the four operations (steps) of a CPU workflow? Explain each step with one sentence.
5. What is an assembler? What disadvantages can you find in combination with simple assembly languages?
6. What is specific for closed-loop control systems?

7. ◘ Figure 4.4 shows the use of a Karnaugh-Veitch diagram to reduce the number of logic gates for a truth table with two inputs. Think about what has to be changed to deal with three or four inputs. Draw examples of such extended diagrams.

8. A half adder 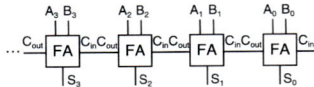 is a logic gate to sum up two single binary digits. It has two outputs: the sum (S) and the carry (C). Create the truth table for the sum and for the carry.

9. If you want to sum up two n-bit binary numbers, you have to use a series of full adder (ripple-carry adder).

Full adders can add two single bits plus the carry bit from the previous addition of a lower bit. What do you have to do to get a full adder?

10. Simplify the Following Truth Table:

A	B	C	D	OUT
0	0	0	0	0
0	0	0	1	1
0	0	1	0	0
0	0	1	1	0
0	1	0	0	0
0	1	0	1	0
0	1	1	0	1
0	1	1	1	1
1	0	0	0	0
1	0	0	1	1
1	0	1	0	0
1	0	1	1	0
1	1	0	0	0
1	1	0	1	0
1	1	1	0	1
1	1	1	1	1

- Create the disjunctive normal form (DNF) from the truth table.
- Use a Karnaugh-Veitch diagram for the simplification.
- Use the formal mathematical method with the rules of Boolean algebra to reduce the number of gates.
- Draw the gates diagram for the simplified solution.

11. What is the difference between unsorted, ordered, and linked lists? What would you expect? Which list is better suited for inserts, updates, or deletes of entries?

12. Describe the method of a binary search. Does it also work for keys using character types?

13. What is the secret behind ISAM files?

14. Which three elements from the definition form a database?

15. What is a transaction and why is it so important?

16. For which type of data are databases ideal? Where would you better use sequential files instead?

17. Which operations with data sets do you know from relational algebra applied to tables of databases?
18. What is an ERM?
19. What is a threshold used for image processing?
20. An autonomous production cell converts starting products to finished goods or, in analogy to that, input data to output data products. Five functionalities support this. Which?
21. Which sensor types do you know to measure relative, incremental, or absolute?
22. Classify the typical accuracy properties of sensors.

Controlling a VLBI System Remotely

5.1   Autonomous Production Cells as Parts of Multi-agent
        Systems – 396

5.2   Principles of Very Long Baseline Interferometry – 400

5.3   The Used Control System for VLBI: The NASA Field
        System – 409

5.4   Extend Existing Control Systems with Multi-agent
        Abilities – 415

5.5   Security for the Controlled Systems – 431
5.5.1   Security for Internal Local Access to a Computer – 433
5.5.2   Security for External Accesses to a Computer – 438
5.5.3   Security for a Complete Distributed System – 457
5.5.4   Security for a Complete Observatory – 466
5.5.5   On-the-Fly Management of Temporary SSH Tunnels – 469

5.6   Summary – 481
5.6.1   What Did We Learn? – 481
5.6.2   How Did We Use the Contents Learned? – 482

5.7   Questions – 482

© Springer International Publishing Switzerland 2017
A. Neidhardt, *Applied Computer Science for GGOS Observatories*, Springer Textbooks in Earth Sciences,
Geography and Environment, DOI 10.1007/978-3-319-40139-3_5

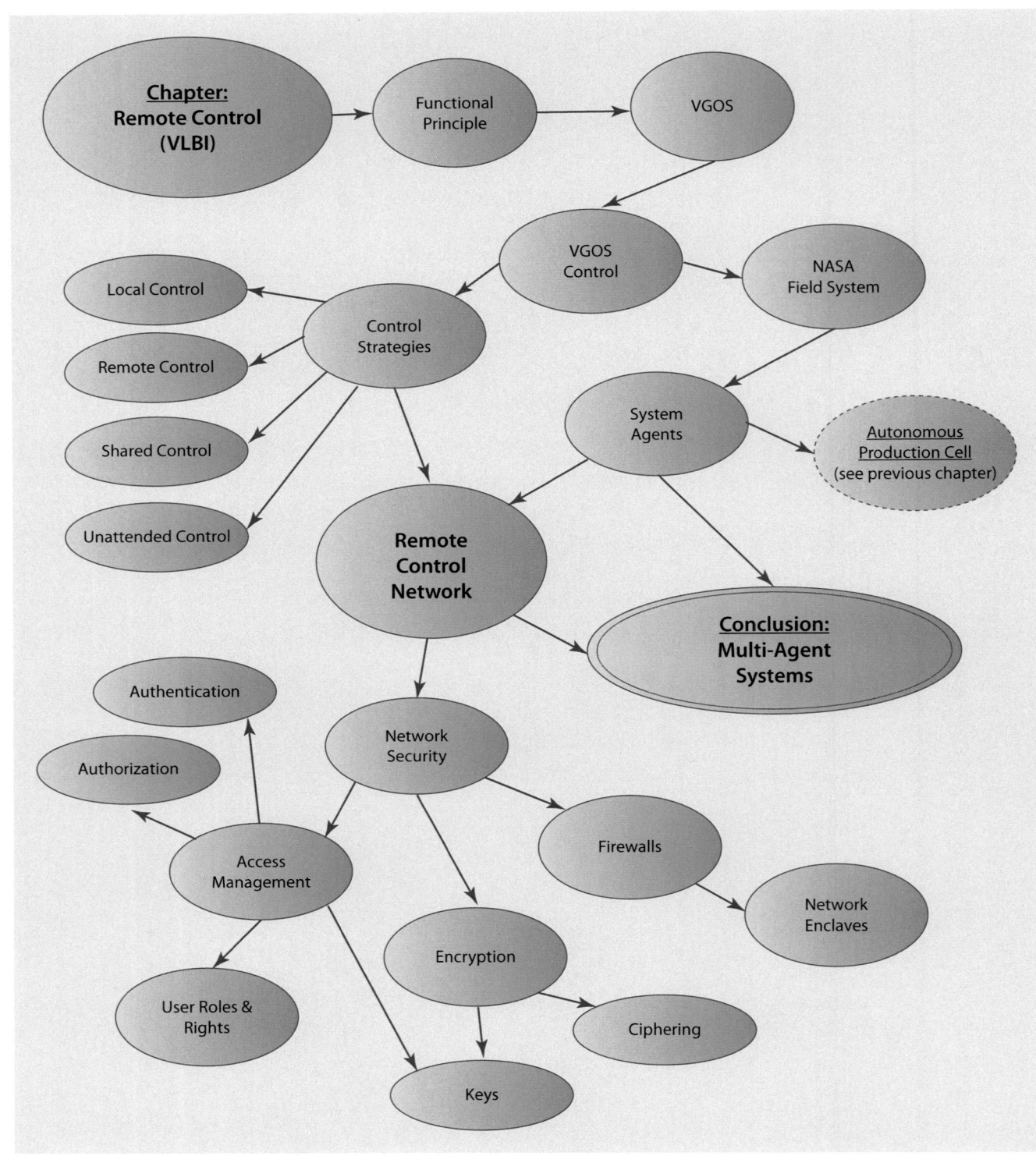

5.1　Autonomous Production Cells as Parts of Multi-agent Systems

Remote Control Idea

The idea behind controlling systems remotely is quite comprehensible with the techniques previously described to control systems locally. As all devices are represented by individual autonomous cells, which offer a dedicated RPC interface as connection to the Internet, they can be used from anywhere. This means it is quite easy to take a local GUI client for an autonomous coordination cell and use it miles away on a separate computer to control the cell remotely (see the illustration in ◻ Fig. 5.1). As long as the systems detect and react to errors autonomously within the specified time, it results in a risk only slightly greater than for local control.

Remote control of industrial systems is no new idea. But it is becoming increasingly interesting, as nowadays the bandwidth and stability of communication networks allow safe control structures. Technical restrictions or barriers no longer exist. Therefore, different industries have started extending their remote networks to control complete plants over large distances from an Remote Operations Center (ROC). As the technique is very safe, it does not matter anymore whether a technician controls a system next door, or over large distances through the Internet or mobile communication systems. An example is the company Linde in Basel, Switzerland. The company uses an ROC in Leuna, Germany, to control different air separation facilities around Europe. Because it is possible to compare production capacities, using remote control, it results in a better control and optimization of the production. An additional advantage is given with the possibility for remote assistance and guidance for local teams. Professional solutions are provided by industrial companies like Siemens, which offers complete systems in the SIMATIC system line for peripheral automation (Golem.de n.d.). But also with open solutions in scientific environments, it has become increasingly possible to create autonomous systems, which can be controlled remotely, as shown in the previous chapter. A good example of such a university development is «The Liverpool Telescope», operated robotically by John Moores University, Liverpool, since 2004. It is a 2.0 meter, unmanned, fully robotic telescope at the Observatorio del Roque de Los Muchachos on the island of La Palma (Canary Island), Spain. The team working with Prof. Iain Steele has complete access to the telescope from the remote location of Liverpool. Locally, only one member of the on-call service staff is available, who does weekly maintenance and who can be called on critical errors. The statistics show that downtimes are very rare and that the scientific output has increased significantly (JMU, Liverpool n.d.). The key techniques for such distributed control systems can be found in multi-agent systems and in suitable mechanisms to guarantee access and network security to prevent hacking attacks. But before these security issues are discussed, the ideas behind multi-agent systems should be shown.

Industrial Remote Control

> *Definition DEF5.1. of «agent»* (see (Wooldridge 2009, *l.c. page 21*) and (Pfeifer and Schmitt 2006, *l.c. page 27*))
> Agents are intelligent and autonomous software components, which are situated in complex environments and which are capable of making decisions about actions to meet their delegated objectives.

Agent

Multi-agent systems are based on autonomous agents to control a dedicated environment. Autonomy in this field means that a software or computing entity can be used to delegate goals using orders. The entity then decides itself how best to accomplish its goals and can negotiate with the applicant about the orders. The entity has a sufficient overview of the controlled environment to make decisions about taking orders. To do that, it gets feedback from the environment via different sensors. This feedback information is perceived and then evaluated to come to a suitable decision. This can be used either for the negotiation for new orders and/or for controlling the environmental parameters using effectors and actuators. This autonomy can be adjusted, so that responsibility is taken to higher levels or back to people if they want to make more beneficial decisions, if

Agent Autonomy

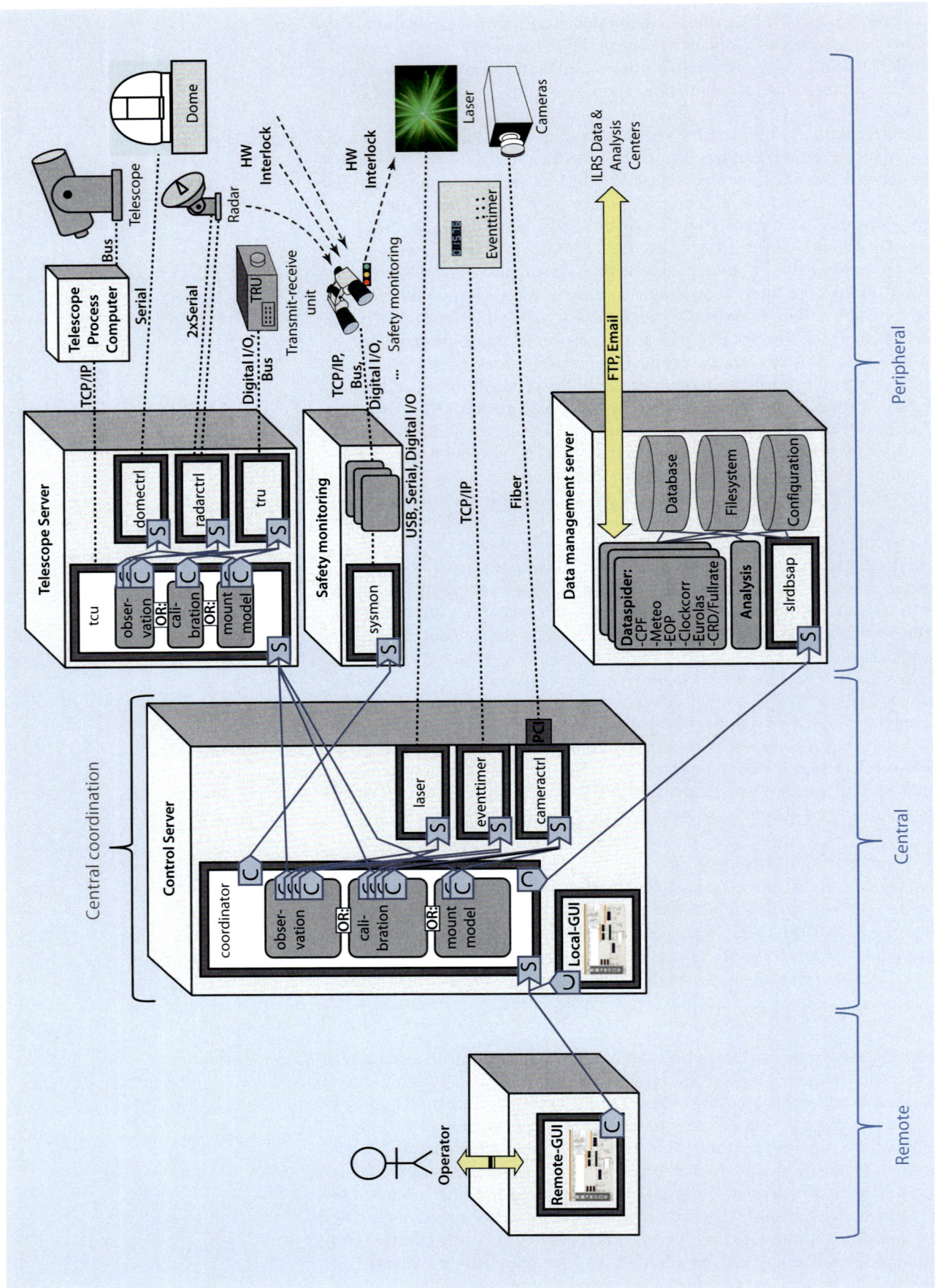

Fig. 5.1   Extending the local control of systems using the same type of local GUI client on a remote machine

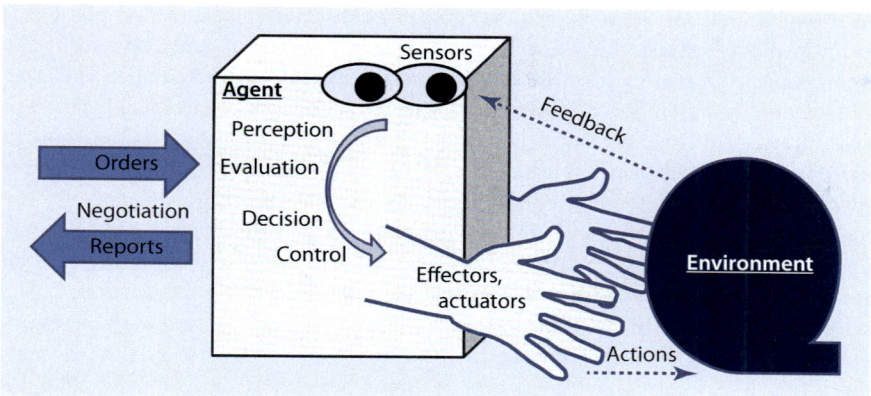

Fig. 5.2 Sketch of an agent which controls its environment autonomously. It gets feedback from the environment via sensors. This perception is evaluated to make decisions which result in the control of actuators to perform dedicated actions, influencing the environment again. The delegated objectives are also influenced by orders from the external world, which are negotiated. The agent replies with reports about the processing of the orders (Inspired by Wooldridge (2009) *l.c. page 22;* ©2009 John Wiley & Sons Ltd.)

the decision might cause harm, or if the agent capabilities are not sufficient. A basic sketch of an agent in this case is drawn in ◘ Fig. 5.2 (Wooldridge 2009, *l.c. page 21 ff.*). Compared to the descriptions of autonomous production cells, starting in ▶ Sect. 4.2.2, it can be seen that each autonomous production cell offers all the abilities which an agent needs. The autonomous coordination cell, as representative of the whole autonomous production cell of an observing system, can be used as an agent for the requirements of an observation.[1]

Multi-agent systems represent each control task with these intelligent software units. Agents for control systems must be reactive, proactive, and social. This means that they must react autonomously to changing environmental parameters. They must use control loops in separate threads which permanently check the environment states, evaluate them, come to decisions, and control actions. Threaded, nonterminating, continuous control loops enable these immediate, frequent reactions and allow long-term decisions. The agents must also be proactive, which means that they activate tasks autonomously to fulfill their delegated goals. This is given by the planning authority in autonomous production cells. Additionally, each agent is accessible over a communication network, and itself controls sub-cells via this network. Therefore, each agent is quite social. In terms of the control system described, the social competence is not defined by the hierarchical control structures in the observing system to the sub-cells. The social character of such multi-agent systems is given by the possibilities offered to the external world of cooperating on shared tasks or negotiating about orders (see (Wooldridge 2009, *l.c. page 25 and 29 ff.*)). The main focus in the field described is on the network between the autonomous coordination cells of the different observing systems of space geodetic techniques.

Multi-agent systems use an underlying communication infrastructure to which each agent is connected and which they can use to receive orders and to send reports about the results and current activities. Unlike other meshed topologies for trainable systems with artificial intelligence or mobile agents,[2] the

Agent Properties

Agent Networks

---

1   Each autonomous control cell could be seen as a single agent. But as their main focus is not on coordination tasks, and as they are very specialized, the autonomous coordination cell can be seen as the agent for the observing system.
2   Mobile agents are usually used to solve huge calculations, where the agents migrate over the network to the place with the best conditions for a dedicated task, taking the algorithms and the data used with them.

topology to control networks of observing systems can be organized as hierarchical, top-down order request and response systems. This is already offered by the communication platform (middleware) and its features generated with «idl2rpc.pl» (see ▶ Sect. 3.4). Each agent is then a server which processes local tasks, represents an observing system, and offers functionality to the rest of the multi-agent system. These functionalities are registered at superordinated coordination agents in the network, where each observing system agent must sign on and off. The agent interfaces are represented by an abstract interface definition used to generate an abstract base class, which makes it possible to use one abstract method set for communication with each individual agent and their administration. Using this, agents can take over individual pieces of the whole global task of the network (task sharing). They can cooperate to solve the global, delegated goal of the network with which they share the results of their work (Pfeifer and Schmitt 2006, *l.c. page 28 ff.*) and (Wooldridge 2009, *l.c. page 154 ff.*). For this cooperation, agents must define pre- and post-conditions for their own jobs. Then they must evaluate incoming orders, which are parts of the decomposed, global problem or task. These tasks are announced by superordinate applicants. Here it is possible for the applicants to bid for processing time and therefore for observation time,[3] or for the agents of the observing systems to bid for assignments of tasks. In both cases, it is necessary to perform an award (local at the agents or at the higher-level coordination agents), which can follow similar rules, as already defined for the decision-making process in autonomous coordination cells, using auctions, for example (see ▶ Sect. 4.4). After the evaluation of the suitable metrics parameters (e.g., free operation times, current states, etc.), a distribution of the common task can be planned, and all agents in the multi-agent system must be informed about their dedicated tasks and rough schedules (compare (Wooldridge 2009, *l.c. page 157 ff.*)). This is a similar solution to job controlling in large industrial plants, which adapts to individual capacities, as described above.

Even in less complex, self-adapting, and self-configuring high-end networks, multi-agent systems offer advantages. As mentioned for industrial processes, they offer stable, autonomous control, which can be used for remote access and which deals with communication blackouts, because of the capacity of its agents to make decisions. They offer standardized connection points to the systems which can be shared so that new observation strategies can be performed to increase the observation, data, and scientific output. Ideally, they can optimize the capacities to increase the efficiency of global observing networks. This is also a challenge for the GGOS, which designs a network of distributed observatories, which cooperate together as if they were one single system. But where could such an approach be tested better than with a system which already requires global cooperation? This is where VLBI plays a pioneering role.

5.2   Principles of Very Long Baseline Interferometry

Principles of VLBI

VLBI is a passive measuring technique using noise sources in space. In principle, any noise source, such as the pseudo-random noise of a transmitting GPS satellite, or noise of natural radio sources, such as compact radio stars or quasars,[4] can be used. The geodetic VLBI mostly uses well-selected quasars. Quasars are cores of galaxies at a distance of several billion light years up to the border of the currently visible universe. They continuously emit radiation throughout the electromagnetic spectrum, including radio waves which travel to the Earth, where they can be received with radio antennas. Because of their huge distance, the quasars selected for geodetic observations can be used as a realization of a quasi-inertial reference frame with

---

3   A similar strategy is used at the Liverpool robotic telescope to share observation times between the astronomical institutes.

4   The red shifts of the quasars indicate cosmological distances, and their emission must be powered by mass flows into massive black holes (Plag and Pearlman 2009, *l.c. page 259*).

high precision for global surveying tasks, as they are stable enough in position. But as the signals travel such large distances, their signal strength measured in Jansky

$$\text{where} \qquad 1\,\text{Jansky} = 10^{-26}\,\frac{\text{Watt}}{\text{Hertz} \times \text{Meter}^2}$$

is very low (sources with a spectral flux density of only a few Jansky are used in geodesy: usually 0.5–1 Jansky (Schüler et al. 2015, *l.c. page 18777*), max 10 Jansky (Kilger 1990, *l.c. page 44*)). Therefore, the size of the antenna dish, the optimal signal receiving, the amplification with cryogenic low-noise amplifiers, and the integration time for which an antenna points to the same position in space are essential. The incoming signals form quasi-plane wave fronts, which appear with a slightly time difference (max. 20 milliseconds) at different places on Earth, according to their relative position to the radio source. If at least two radio telescopes at different positions on the Earth point in the same quasar direction and receive the same noise signals from there, the time delay can be derived if the signals from the different telescopes are correlated[5] with each other. Each signal path is phase locked to a local, highly precise hydrogen maser frequency standard as oscillator. The connection to UTC allows a synchronization of the generated PPS signals for the correlation. Additionally, synchronously derived phase calibration tones are recorded with the noise signals from the quasars to optimize dispersion over different frequencies. This intercontinental observation technique is almost weather-independent and also works if the signals are overlaid by the usual other noise patterns (e.g., RFI), as long as these are just local interferences, do not saturate the whole signal path, or change the signal dispersion. Limiting factors are time distribution, atmospheric refraction effects especially from the troposphere (density changes in water vapor (Kilger 1990, *l.c. page 46*)), and geometric, structural effects at the reference point[6] (see (Seeber 1988, *l.c. page 394 ff.*)). These limiting factors must be estimated or corrected with suitable models during correlation and analysis. The observation principle of the VLBI technique is also pointed out in ◘ Fig. 5.3, together with the equipment used.

To enable the above observation principle, several components at the location of the radio telescope and also at the correlation center are involved (see also ◘ Fig. 5.3 (inspired by (Kilger 1990, *l.c. page 45*) and (Clark 2006))). In geodetic systems, the radio signals from the quasars are received with a dual-frequency feed-horn mainly located in the focus point of a Cassegrain or offset antenna (for the possible and used radio antenna dish types, see ◘ Fig. 5.4). The frequencies used are 2.1–2.4 gigahertz (S-band) and 8.1–8.9 gigahertz (X-band) (Seeber 1988, *l.c. page 399 ff.*). To amplify the weak signals, low-noise amplifiers are used within a cryogenic, evacuated pressure box (dewar or cryostat), which is cooled to about 15–20 Kelvin with helium gas expansion by a two-stage «cold-head»[7] (compare (Colucci 1999, *l.c. page 13 and 25*) and (Clark 2006)), which is connected in a closed helium circuit to a helium compressor to build a «refrigerator system». They amplify the signal by about 25–35 decibels. The point where the phase calibration is injected is also in front of the low-noise amplifiers. Another injection is generated by the calibrated noise diode, which can be switched on and off to calibrate the system temperature (Kilger 1990, *l.c. page 47*).

The now amplified signals are bandpass filtered and down-converted to Intermediate Frequency (IF) bands using a heterodyne conversion with local oscillator

Technical VLBI Aspects: Reception

Technical VLBI Aspects: Down-Conversion and Sampling

---

5 During the correlation process, the different signals are shifted along the time axis until they are «congruent» and offer a maximum correlation amplitude.

6 The geodetic, geometric reference point is defined by the axes intersection of the azimuth and elevation axis. But physically the reference point is in the feed-horn where the radio signals are focused and converted to electrical signals transmitted on a wave guide. Both points are in relation to each other.

7 A «cold-head» is an expander using a compression and expansion space, and a displacer to cool down to a few Kelvin in a cryocooler system.

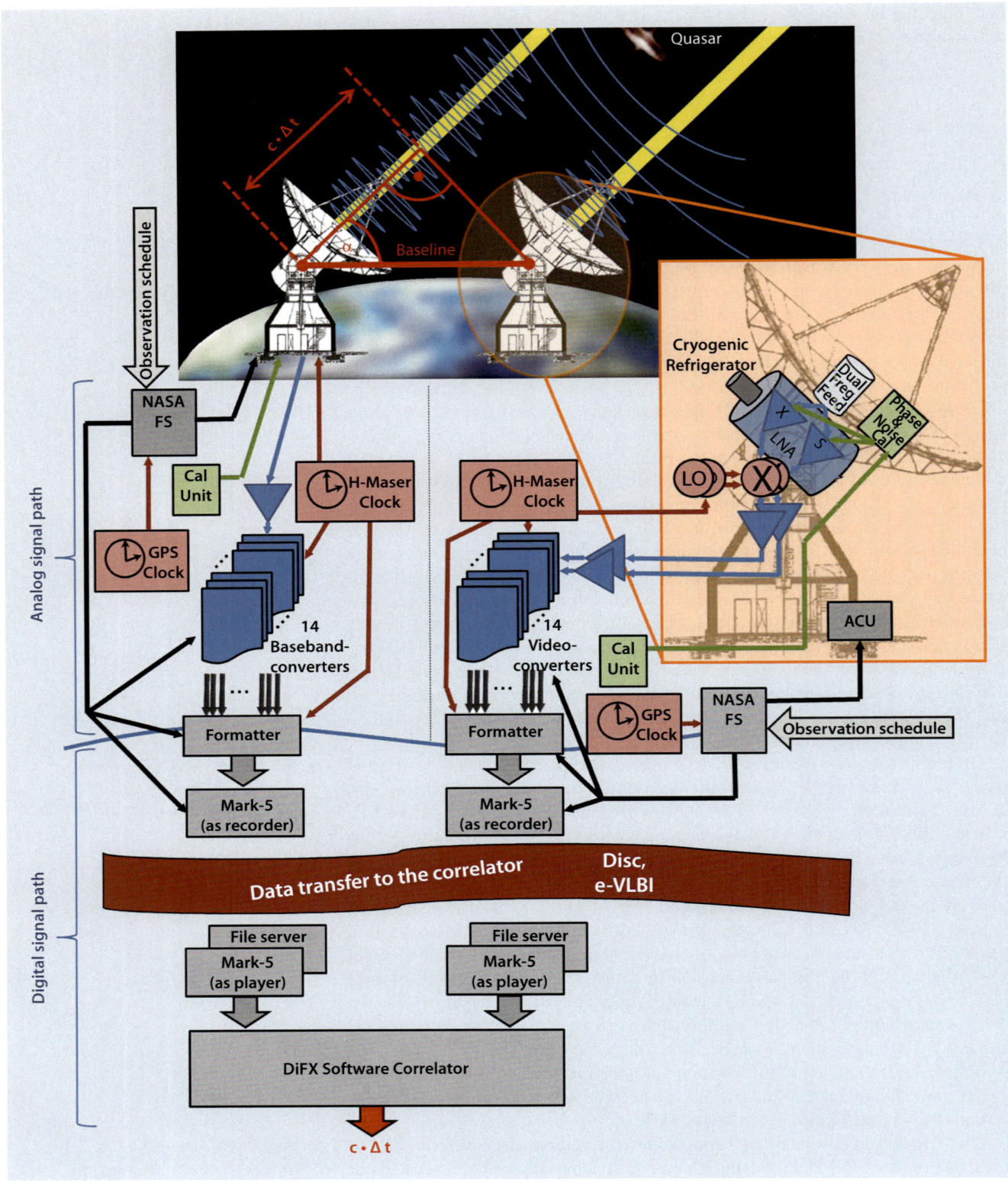

Fig. 5.3   The observation principle of VLBI and the components involved

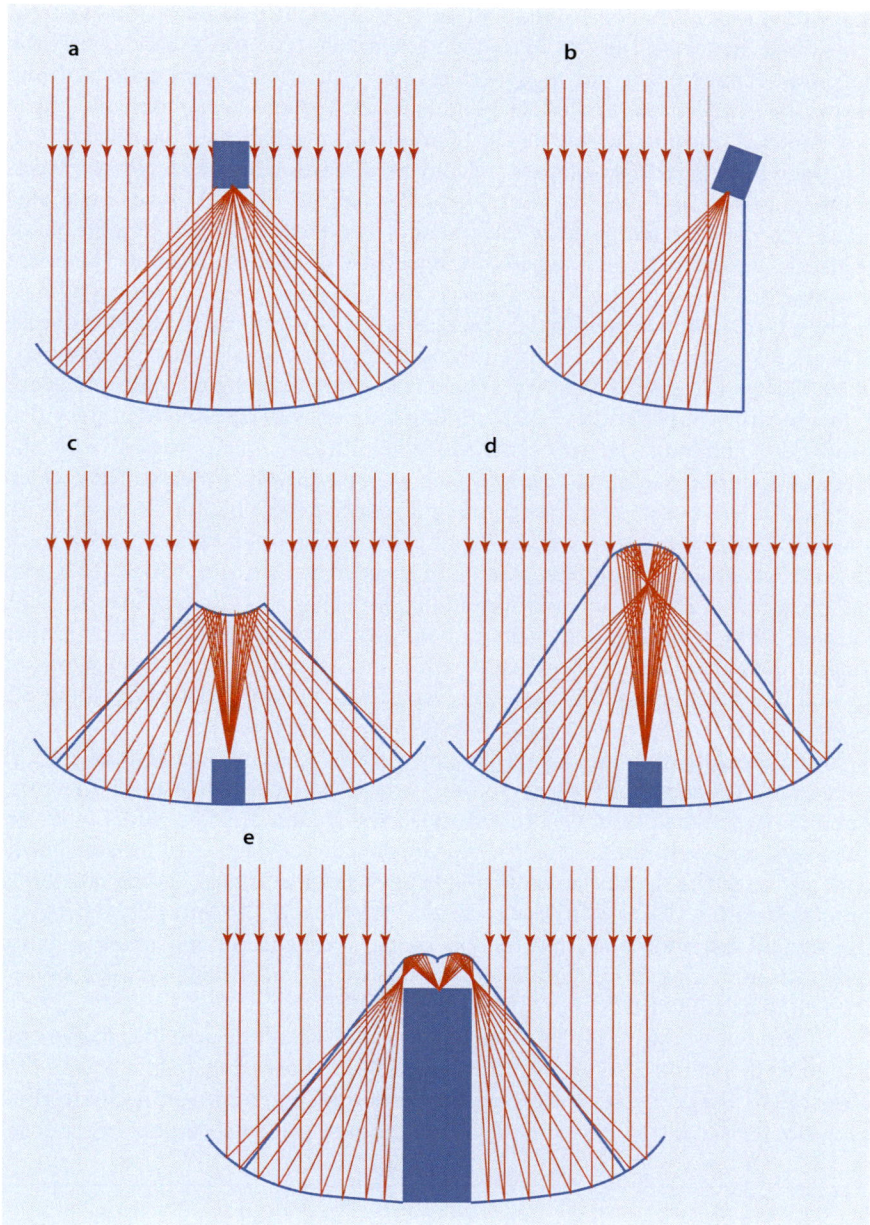

◘ **Fig. 5.4**   The different, possible, and commonly used antenna dish types: **a** prime-focus antenna, **b** offset antenna, **c** Cassegrain antenna (e.g., 20 m Radio Telescope Wettzell), **d** Gregorian antenna, **e** ring-focus antenna (e.g., 13.2 m VGOS TWIN Telescope Wettzell)

frequencies of 8,080 (X-band) and 2,020 megahertz (S-band), multiplied by a 5 megahertz signal, which is phase locked to the local hydrogen maser as a frequency standard. The IF reduces transmission losses on the coaxial cable from the receiver in the antenna to the data acquisition system in the control room (Takahashi et al. 2000, *l.c. page 34*). After the down-conversion, two IF bands (S with about 300 megahertz bandwidth, X with about 500 megahertz bandwidth, and an extended X wide band from 500 to 800 megahertz bandwidth, which can be selected with a local oscillator there) are available and transmitted to the data acquisition rack in the control room of the telescope. The data acquisition system[8] takes the IF signals and selects 16 frequency windows (10 from X-band, where the upper side bands of the video converters 1–8 are used in combination with the lower side bands of the converters 1 and 8; 6 from S-band, where only the upper side bands of the video converters 9–14 are used) of a dedicated bandwidth of 2, 4, 8, or 16 megahertz. These bands are converted to basebands with video or baseband converters and their local oscillator frequencies. A total power integrator makes it possible to measure the integrated power of each band. The single bands are then sampled with 1 bit, where the signal strength is lost and only the phase information is preserved (clipping)[9] to get digital data streams, which are formatted by a formatter, which adds a frame header (e.g., with a size of 32 bytes in case of Mark 5B) with time and control information to the digital signal payload (e.g., of 10,000 bytes in case of Mark 5B). These frames are then digitally recorded with a data recorder like Mark 5.[10] The sampling frequency is dependent on the video bandwidth and must be twice the video bandwidth in accordance with the Nyquist-Shannon sampling theorem (see ▶ Sect. 4.5). This means if the video bandwidth is 4 megahertz, the sampling frequency must be 8 megahertz. With 1-bit sampling, two states must be represented per sampling point, so that each channel has a binary sampling rate of 16 megabit per second, which leads to a sampling rate of 256 megabit per second for all 16 channels (Takahashi et al. 2000, *l.c. page 54 f.*). Currently 4 megahertz bandwidth (respectively, 128 megabit per second data rate) and 8 megahertz bandwidth (respectively, 256 megabit per second data rate) are used for geodetic experiments (compare (Colucci 1999, *l.c. page 13 and 25*), (Clark 2006) and (Kilger 1990, *l.c. page 48 f.*)).

Then the data disks can be shipped by a courier service or sent with e-VLBI techniques[11] through the Internet to the correlation center (correlator). There the same Mark 5 systems are used to download the recorded data streams from the shipped modules. In the case of e-VLBI, the data directly arrive on file server systems. After all the streams are available from the different sites, the correlation is processed with a DiFX[12] software correlator. It produces the time delays between the signals and additional information (such as calibration and phase stability information or correlation amplitudes) (compare (Colucci 1999, *l.c. page 13 and 25*) and (Clark 2006)).

Unlike the workflow of laser ranging systems, where the activities in the first line must be coordinated locally and limited to the observing system within one observatory, the principle of VLBI requires a coordination of two or more telescope sites. But the radio telescope and the technique used are not so dependent on local

Technical VLBI Aspects: Shipment and Correlation

VLBI Observation Workflow

---

8 The standard equipment is the Mark-III rack with the IF distributor, the 16 video or baseband converters (usually 14 are used for geodetic experiments), and the formatter. «The IF distributor provides two totally independent channels (IFI and IF2) for distribution of IF signals to the video converters. Each channel includes a switch for selecting one of two IF inputs, a step attenuator, a power divider [or filters] for splitting the IF into two different frequency bands (upper and lower sub-channels [, upper sub-channels from 216 to 504 Megahertz and lower sub-channels from 96 to 224 Megahertz (Kilger 1990, l.c. page 48)]), bandpass filters, and eight-way power dividers to send the IF signals to the converters». ((Field System Team 1997a) *l.c. page MKIII-8*) The old Mark-IV formatters are nowadays upgraded to VLBI Standard Interface (VSI) formatters. In modern systems, such as VGOS compatible systems, the whole rack is replaced by a Digital Back End (DBE) or Digital Baseband Converter (DBBC), solving the same task digitally on an FPGA.

9 For astronomical observations, a 2-bit sampling can be selected.

10 The standard equipment are Mark 5 systems of different versions like Mark 5A, Mark 5B, or Mark 5C, which use a stack of eight hard drives to record the data with high data rates of one Gigabit per second and more.

11 Increasing network capacities, new technologies, and higher transfer rates make the direct transmission of large data sets through the Internet possible.

12 DiFX means digital correlation where the Fourier transformation (symbolized by the «F») is done before the cross-correlation, symbolized by the «X». Software correlators replaced the hardware correlators, using hardware boards with integrated circuits to process the correlation, as nowadays FPGAs can be used to set up programmable correlation arrays.

environmental situations or local changes, for example, the weather situation. The control flow does not have to adapt so flexibly to such constraints. Therefore, the coordination is currently organized using schedule files, containing a sequence of time-tagged commands which control the individual systems (see the SNAP files already mentioned in ▶ Sect. 4.4). While the antennas and their control are quite individual, most of the components used in a VLBI system are standardized. The systems use the same or compatible receiving systems, data acquisition racks, formatters, data recording systems (Mark 5), and the NASA Field System for their control.[13] By limiting the number of variables, it becomes much easier to concentrate on the main coordination and controlling parts in the control software. But the tricky part is again the generation of the schedule for the workflows at the different observatories. It is centralized and done by the human schedulers with specific programs, which include the telescope specifics, such as the type of cable wrap and the resulting slewing range, the speed of the slewing to combine large but slow telescopes with smaller but faster ones, and so on. It also defines the same setups for the frequency channels. These schedules are then transformed with the local parameters (using the Field System program «drudg»), for example, the topocentric coordinates and the attenuations, so that they can be started within the NASA Field System. The observation then runs almost completely automatically. It is split into different scans, which are the time intervals for which the signals are recorded. Each of these observations has a «preob», «midob», and «postob» section, which defines activities before, during, and after a scan recording (see also ▶ Sect. 4.4). These activities are combined from Field System-specific commands and station-defined procedures, which are station-specific programs or combinations of Field System commands. All of these activities and also the parameters used, and the analysis and system health-relevant parameters, are recorded into an observation-specific log file (see the explanations in ▶ Sect. 4.4), which is sent with the data to the correlation and analysis centers.

The results can be used in the field of geodesy and geodynamics. The time delay can be transformed into distances using the relation to the speed of light. Therefore, it is possible to derive the length and relative changes of baselines between the telescope reference points with very high precision. This can be used to monitor plate tectonics between continents. Because of the angular and directional accuracy of radio interferometry over long baselines, it is also possible to monitor the EOP of polar motion and Earth rotation. Regular INTENSIVE observation sessions are especially installed to monitor these parameters. Another result is the monitoring of the quasar positions and therefore of the International Celestial Reference Frame (ICRF), in which all sources are registered with their right ascension and declination coordinates. As the same technique is also used for radio astronomic research, it is possible to participate in those astronomical programs with S- and X-band. One result is the monitoring of the source structure of the radio sources used (compare (Seeber 1988, *l.c. page 399 ff.*)).

To achieve precise results, it is necessary to evaluate the antenna performance continuously. In the first phase, this concerns the pointing of the antenna, which means the ability to point to several positions in the sky under different conditions, so that the source is always centered in the beam with minimal aberrations from the peak signal of the source. Usually the pointing is measured with different measurement slices through the source to detect the signal peak.[14] The actual direction angles are then stored and compared to the calculated angles of expected position, where an aberration of lower than 0.2 of the beam width is quite convenient (about an arcminute). Another test can be combined with the pointing test to sources with known, suitable, sufficiently strong radio transmission power, which defines the flux

VLBI Observation Results

System Efficiency and Calibration

---

13   In laser ranging systems almost every component is individual and must be controlled individually.

14   Special functions in the control system, like the NASA Field System, support a pointing test, e.g., «fivept» in the Field System.

density of the source. By performing a pointing on and off a source, it is possible to derive the increase of the noise, measuring the different power levels. Using these power levels, where $P_{\text{offsource}}$ is the power of the IF off source, P_{onsource} is the IF power on source, and P_{zero} is the zero level power (the signal bias, which is a nonzero output if nothing is connected) and the Source Flux Density (SFD) (from an astronomical table), it is possible to calculate the System Equivalent Flux Density (SEFD) with

$$\text{SEFD} = \frac{P_{\text{offsource}} - P_{\text{zero}}}{P_{\text{onsource}} - P_{\text{zero}}} SFD$$

It is a number defining the general system sensitivity of the VLBI system. The smaller the value of the SEFD, the better. For example, the 20 meter radio telescope Wettzell has an SEFD of about 500–700 Jansky for the X-band. This sensitivity is influenced by the antenna dish size, the antenna efficiency, and the receiver system temperature, which together influence the Signal to Noise Ratio (SNR). The system temperature is the noise signal, added to the received signal by the equipment used, the spillover and scattering, the atmospheric emission, the receiver noise temperature, and so on. With a well-calibrated noise signal from a noise diode of temperature T_{cal}, it is possible to find the effective system temperature at any time, measuring the output voltages of the square-law detector (power meter) if the noise diode is switched on (V_{on}) and off (V_{off}) in combination with the output bias voltage (V_0), which can be found if no input signal is connected. The system temperature measured is

$$T_{\text{sys}} = \frac{(V_{\text{off}} - V_0) T_{\text{cal}}}{V_{\text{on}} - V_{\text{off}}}$$

and should be far lower than 100 Kelvin (for more information see (Shaffer and Vandenberg 1993, *l.c. page 3 ff.*) and (Colucci 1999, *l.c. page 15 ff.*)).

System Time

Generally, VLBI is an interferometric technique, which measures relative time delay differences between two telescopes. Therefore, the technique is very independent from hard timing constraints according to the time epoch, which is offered by a GPS time receiver, and at least one Cesium frequency standard. An absolute time of a few microseconds is usually enough as it is just necessary for the pointing, where the position angles must be derived from the celestial system via the right Earth rotation parameter.

Phase Calibration

Nevertheless, the system is dependent on a (short-term) stable frequency from the hydrogen maser and from the phase of the local oscillators and video/baseband converters. To measure and record these phases, it is useful to inject a sharp peak phase calibration pulse tone (shorter than 50 picoseconds) with a rate of 1 microsecond into the noise signal in front of the low-noise amplifiers. It is generated from a 5 megahertz signal from the hydrogen maser, which triggers a tunnel diode, so that 5 megahertz pulses are generated, from which only every fifth one is selected. This forms a frequency comb with spectral lines at each megahertz. By mixing the signal with the local oscillator frequencies of the video or baseband converters, the tones appear in the basebands at 10 kilohertz, 1.01 megahertz, 2.01 megahertz, and so on (see (Vandenberg et al. 1993, *l.c. page 7 ff.*)). The phase calibration is used for the correction of the frequency-dependent dispersion during the correlation, as it is also recorded with the received signal.

Cable Delay

In parallel with the 5 megahertz signal for phase calibration, a 5 kilohertz signal is sent. It forms the uplink line for a cable delay measurement. The receiver unit multiplies the two signals. The resulting product of 4.995 megahertz is then sent back to a ground unit. There the returning downlink signal is mixed back to a 5 megahertz signal, by multiplying it again by 5 kilohertz. As the ground unit now has the up- and downlink signal, it can perform a phase comparison. The result is the relative delay change on the cable, which also transports the 5 megahertz signal for phase calibration. Therefore, it is also a correction value for the phase calibration tones, according to propagation changes in the cables, for example, because of a turning cable wrap, which changes the cable radii. Another cause is changing temperatures. For example, a temperature change of 5 degree celsius results in a delay change of about 200 picoseconds over the standard coaxial cables used (1 centimeter change causes a change of 20 microseconds in a scaled output of the cable delay unit[15]). The cable calibration measurement is recorded once per observation scan and is one of the most important calibrations (see (Vandenberg 1993, *l.c. page 4 ff.*) and (Vandenberg et al. 1993, *l.c. page 3*)). The maser, which feeds all of these calibration systems, is finally also monitored against a master clock and UTC.

Additionally, the surface meteorological data are used to calibrate the VLBI system during an observation. The temperature, pressure, and humidity are used during analysis as input parameters to a model that is used to compute the excess delay due to the neutral atmosphere, where pressure is the most important influence (accuracy must be higher than 0.1 millibar) (see (Vandenberg 1993, *l.c. page 2 ff.*)).

Meteorology and Calibration

The technique described has been used and well known since the 1970s/1980s for geodetic tasks. Since then, the RFI at the S-band (around 2 gigahertz) has increased, because of mobile phone and wireless network transmitters. The antennas are slow, badly distributed over the Earth, and unmanned operations are almost impossible because of the lack of automation on site and in the whole processing pipeline (Niell et al. 2005, *l.c. page 2*). To meet future geodetic requirements, the Working Group 3 of the IVS developed the vision VLBI2010 (now known as VGOS) about the requirements for new VLBI systems. The goals for the next-generation VLBI system are essentially an accuracy of 1 mm position[16] and 1 mm/year velocity measurement (in the terrestrial reference frame), a continuous measurement for EOP, and a rapid generation and real-time distribution of the IVS products (Niell et al. 2005, *l.c. page 3 f.*). The resulting design drivers and following suggestions for new sites are (see (Niell et al. 2005, *l.c. page 4 ff.*) and (Schüler et al. 2015, *l.c. page 18767 ff.*)):

Future Techniques: VGOS

— *Small, low-cost, fast-moving antennas:* An optimum between the effective active dish surface and the possibility of moving very fast between sources, while keeping the mechanical stress low, is the setup of about 12 meter dishes with more than 60% efficiency and a slew rate higher than 5 degree per second (e.g., 12 degree per second in azimuth and 6 degree per second in elevation). The stiff back-structure should keep mechanical errors very low. The ring-focus design seems to be ideal for these small antenna dishes because it optimizes the receiving characteristics (Schüler et al. 2015, *l.c. page 18771*), even if there are some spillover effects at high elevation angles.

— *Optimum and practical observing frequencies:* The RFI in the S-band forced the use of new observation bands, supported by new possibilities like the determination of group delays over much broader bands and the optimal selection of observing channels. It is recommended that broadband-receiving systems are used for a continuous coverage between about 1–14 gigahertz with a system temperature $T_{sys} = 45$ Kelvin. Cooled broadband feeds (like the Eleven feed or the Quad Ridge Feed Horn (QRFH); see (Beaudoin and Whittier n.d.)) and up-down-converters with tunable first local oscillators to select individual Nyquist zones (Schüler et al. 2015, *l.c. page 18777 ff.*) are required.

— *Inclusion of existing antennas:* The existing, larger antennas should be integrated into the new strategies, as they cannot be completely replaced by the new, smaller antennas as the sensitivity is mainly influenced by the antenna dish size, as explained above. The existing systems can be upgraded to the new frequency bands and can offer new ways to analyze the influences of the source structures if they are used for the source structures monitoring. Then they close the gaps between geodesy, astrometry, and astronomy.

— *Modernization of VLBI data acquisition systems for higher stability and reliability, wider bandwidth, and lower cost:* The new systems should use

---

15 1 microsecond = 2.5 picosecond one-way delay.
16 Currently a vertical position accuracy of about three millimeters is possible, because of the limiting factors of the per-observation delay measurement error (Niell et al. 2005, *l.c. page 4*).

digitized systems as early as possible. They should channelize the signals into several frequency segments directly selected from the front end bandwidth without a conversion to IF bands. The phase and cable calibration system must also offer higher accuracy to detect hidden delays. The data rates should be increased from 256 or 512 megabit per second nowadays to 2–4 gigabit per second (with future increases to 8, 16, or 32 gigabit per second), using a Polyphase Filter Bank (PFB) mode, which, for example, samples 16 continuously arranged 32 megahertz channels per radio frequency band. Direct sampling of a complete 1 gigahertz wide band is another option. A DBBC of the newest generation combined with a modern formatter, such as the FILA10G device,[17] supports these modes (HATLab n.d.).

— *Transmission of data via high-speed network (e-VLBI):* The new 20–40 sites should have more than one system per site and can use the new possibilities of fiber-optic networks to transfer the data with the high speed of 10 gigabit per second and more, directly to the correlator. High-speed data recorders with removable disks (like the Mark6) or with static data raids (like the Flexbuff system) are the basis for these transfers and recording scenarios.

— *New observing strategies:* The availability of a global array of small fast-moving antennas will allow many more observations per day, for example, one observation scan per minute. With two or more antennas at one site, it is possible to increase the observing density to 24 hours a day, 7 days a week. The combination of satellite observations with quasar observations may improve the results. Strategies where the different antennas at one site support different programs at the same time, or are used to optimize the tropospheric limitations, are also conceivable.

— *Automation of observations and remote monitoring:* To improve the productiveness of future observation strategies, it is necessary to automatically tailor the observation schedules to the different purposes of observing programs and sessions, while taking into account the telescope arrays at the location of the observation site. It must be possible to read quality and benchmark parameters from the telescopes and from the real-time correlation process to flexibly add and subtract stations, sources, or observation frequencies remotely during an observation session. Automated diagnostic procedures and notification mechanisms with self-organizing, autonomous systems increase the safety of personnel staff, heavy equipment, and within the whole geodetic network. The shipping and data transfer of the recorded data according to network transfer capacities are also an essential element in this case.

— *Possible correlator upgrades:* A near-real-time correlation with distributed correlation technologies can distribute the load of the high data rates over several correlation centers. Software correlators on fast, multiprocessor, high-end systems offer high flexibilities.

— *Advances in models and strategies for data analysis:* The robustness and reliability of the solutions must be improved. Intra-VLBI and intra-technique solutions offer a complete approach. The different software solutions must be benchmarked and compared.

— *Automation of data processing:* The data chain must be automated where states and alarm levels are automatically announced and reported to those responsible. The systems must automatically check the quality of the resulting product, which is offered in real-time to the scientific communities.

Reading the ideas described for future observing systems, the technique of multi-agent systems with unmanned, autonomous production cells, which offer stable, safe, and notifying operations, explicitly supports all mechanisms of automation, autonomy, and remote access and control, like the automation of observations and remote monitoring and the automation of data processing. «The development of

---

17   FILA10G is a double port 10 gigabit per second VLBI-Ethernet interface, formatting raw VLBI for the recording on high-speed recording systems.

these automated systems and procedures is particularly important for the timely production of EOP results, but will also significantly add to the robustness and productiveness of the system and lead to higher quality geodetic products» (Niell et al. 2005, *l.c. page 15*). While it is generally necessary to implement correlator agents, which support distributed correlation and negotiation of correlation orders in more or less intelligent order assignment systems, the following sections focus on the aspects for the observation sites. Here one aspect must still be solved: how already existing systems can be integrated into the net of multi-agents without rigorous changes of the whole technique and extensive new developments.

5.3 The Used Control System for VLBI: The NASA Field System

The control system used for geodetic VLBI radio telescopes is the NASA Field System.[18] This software was originally written in FORTRAN, but greatly extended and converted into C code. Most of the documentation and also the architectural structures were created in the early 1990s. Nevertheless, similar structures can be found described in ► Chap. 4. But it is not defined as a distributed system, so that it is organized as a more centralized architecture, where a central coordination and controlling instance «boss» controls the peripheral hardware equipment, using program stubs for each piece of hardware. Unlike the distributed system in ► Chap. 4, all software components are located on one central PC. As they are not distributed over different computers, the communication between them can be organized with IPC on the basis of shared memory (see ► Sect. 3.3.1). The programs themselves are more or less loosely coupled, parallel running programs controlled by the central control program «boss», which share date via the shared memory. This memory is protected by an individual semaphore implementation within the NASA Field System. The main controlling architecture is shown in ◘ Fig. 5.5 (inspired by (Field System Team 1997b, *l.c. page ARCH-9 ff.*)).

NASA Field System

The NASA Field System can be split into six main layers, which are similar to elements in the control structure in ► Chap. 4 (something like a horizontal structure with the hardware on the right and the user interaction on the left):

Control Layers

1. Programs for hardware control (hardware driving)
2. Programs for module checking (monitoring)
3. Programs for the SNAP command interpretation
4. Programs for Command Processing and Control (coordination: «boss» or, e.g., the Antenna Calibration Data Acquisition «aquir»)
5. Programs for error reporting
6. Programs for user interfacing

Furthermore, the programs can be split into two categories, according to where the code is developed: the general Field System programs (offered by the developers at the NASA, Goddard Space Flight Center, and the company NVI) and the station code, which must be developed by the station staff. All general controlling, analyzing, calibrating, interfacing, and hardware adapting programs, which can be used in the same way at the different telescopes, belong to the first category. The station-specific code is the «glue» between the general programs and the local hardware and software. Hooks allow a station antenna control (with the program «antcn»), specific control programs, station module checking (with the program «cheks»), station SNAP commands (with the program «stqkr»), and station error reporting (with the program «sterp»). The NASA

Station-Specific Parts

---

18 At this point, special thanks should be extended to Ed Himwich from the Goddard Space Flight Center, NASA, NVI Inc. for his support, explanations, openness for new developments, and directness in responses during the development of new remote control components for the NASA Field System.

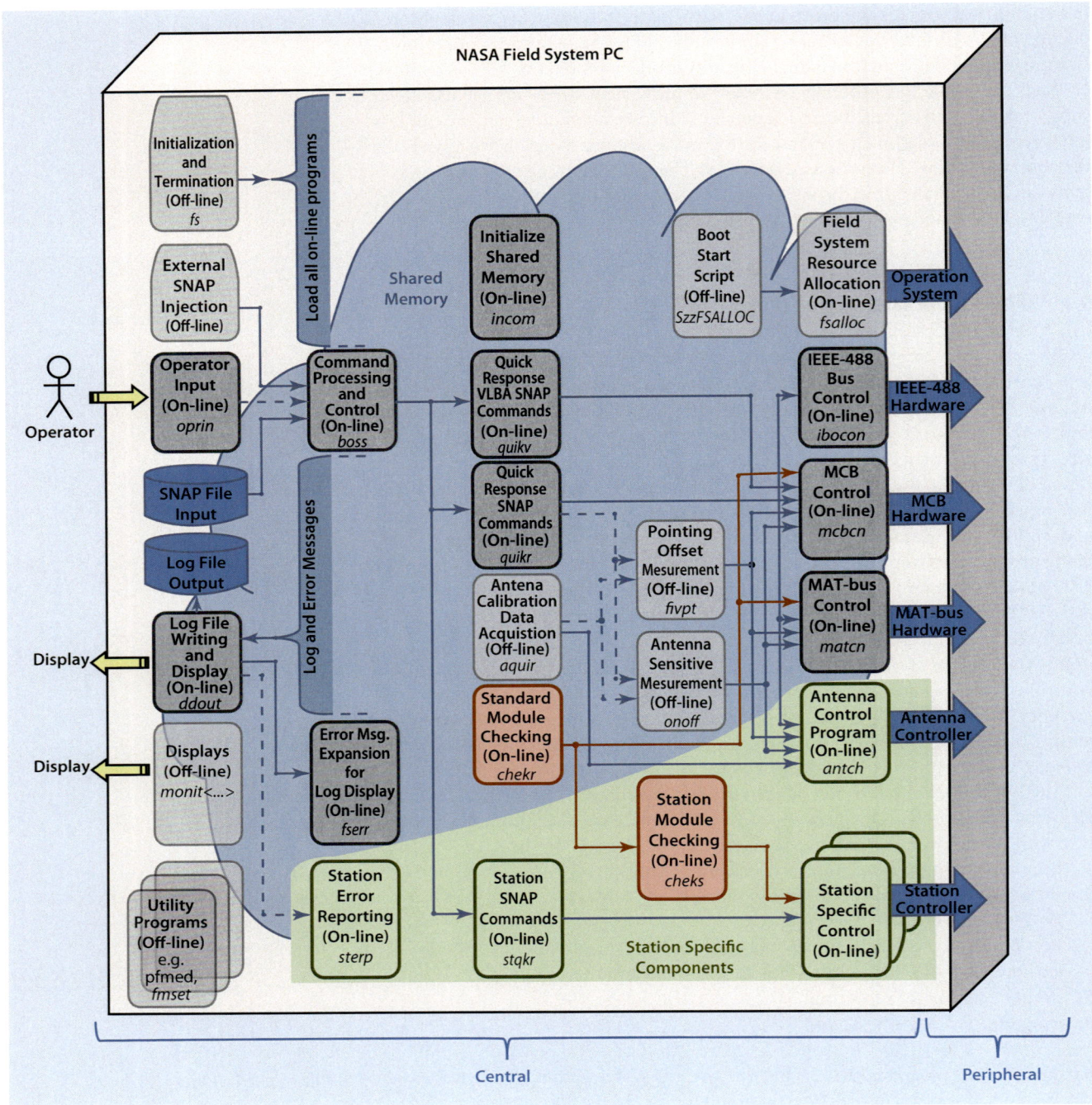

Fig. 5.5 The structure of the NASA Field System components and internal controlling and commanding techniques

NASA Field System Tasks

software contains skeletons with sample code for the station-specific parts, which must then be extended by the local staff (Field System Team 1997b, *l.c. page ARCH-3 f.*).

During runtime, some of the programs must be active all the time (on-line), and some can be optionally started as additional utilities (off-line) to process the two main tasks of the software, namely, the Antenna Calibration Data Acquisition and the real observation operation. The system operates either the calibration or the observation (also compare the similar tasks of a SLR system in ▪ Fig. 4.11). The tools offered can be used with any antenna type as long as the antenna control unit supports the antenna commanding required. They automate the measuring of calibration pointing offsets and sensitivity parameters for the antenna performance. Additional programs fit the pointing data to a mount model. All required online programs for the observation are started and stopped by the initialization and termination program «fs», while all required resources, such as the shared memory, are generated with a start script, which runs during booting of the system (Field System Team 1997b, *l.c. page ARCH-4 f.*).

The NASA Field System uses three different times: the formatter[19] time, the computer time, and the artificial Field System time. The first one is the time which is reported from the formatter hardware, which is assumed to be UTC. The operator must set this time correctly to a real epoch, while the 1 pulse-per-second signal from the maser keeps it accurate. The computer time comes from the computer clock, which has a rate and an offset. The Field System time is generated from the computer time, using an error model which describes the errors of the computer clock (Field System Team 1997b, *l.c. page ARCH-8 f.*). This generation is an artifact from the days before the NTP was available and accurate enough.

Used Time

The general, not station-specific Field System programs in detail are (see (Field System Team 1997b, *l.c. page ARCH-4 ff.*) and (Field System Team 1997b, *l.c. page ARCH-15 ff.*)):

NASA Field System Programs

- *Startup, initialization, and termination programs:* The required Field System resources, e.g., shared memory, message queues, and semaphores, are allocated at boot time by a special start script «SzzFSALLOC». Then they are available and can be accessed by attaching a function «`setup_ids()`» to the source code.
  - *Field System Resource Allocation «fsalloc»:* This program is used during the computer boot-up. It is started within the «SzzFSALLOC» script and allocates all system resources like message queues (for the class input/output emulation), semaphores, and shared memory.
  - *Initialization and Termination «fs»:* «fs» starts all online programs (active, synchronous programs required for the control) and takes care of their correct termination when the control is closed again. The operator starts this program manually in a terminal window to start the control software and initialize all synchronous programs.
  - *Initialize Shared Memory «incom»:* After starting the Field System programs with «fs», the shared memory is initialized with proper values by «incom». It reads all configuration files to get the relevant data for the initialization of the shared memory.
- *Bus control programs:* The bus control programs provide an abstraction of the bus communication to the controlling and coordinating instances. They drive the bus hardware and send the sequences of requests in the right order, which is not interruptible. They use a queuing system for the incoming commands from the higher-level programs sent via the IPC using an emulation for the class Input/Output of the Hewlett-Packard Real-Time Executive operating system[20] (Field System Team 1997b, *l.c. page ARCH-6 f.*).
  - *IEEE-488 Bus Control «ibcon»:* The General Purpose Interface Bus (GPIB) (or IEEE–488 Bus) hardware is controlled with this program. It offers a suitable communication interface to the higher-level control programs to send requests via and receive the responses from the bus hardware to which the specific hardware devices are connected.
  - *Microprocessor ASCII Transceiver (MAT) Bus Control «matcn»:* The MAT Bus hardware is controlled with this program and provides a communication interface abstraction for the higher-level programs.
  - *Monitor and Control Bus (MCB) Control «mcbcn»:* The Very Long Baseline Array (VLBA) uses this special MCB, which is controlled with «mcbcn», which also provides an abstract communication interface for the according hardware.

---

19 The formatter is the digitizer of system.
20 The original NASA Field System was written for the Hewlett-Packard Real-Time Executive operating system, which required a special structure and design of the control system. To enable a further use, it was ported to the UNIX System V, which offered similar structures for shared memory, IPC, program scheduling, semaphores, suspending and resuming, etc., and which could be converted easily to more modern Linux operating systems such as Debian. It was only necessary to emulate some of the original Hewlett-Packard Real-Time Executive features (Field System Team 1997b, *l.c. page ARCH-26 ff.*).

— *Antenna Control Program skeleton «antcn»:* This station-specific program is the abstraction for the antenna controller hardware (usually an industrial PC, a PLC, or another processing unit). The program implements the state machine to command radio source positions or position offsets and to determine if the antenna is tracking the position commanded. The programs to calibrate and measure the antenna performance also use these states. As the antenna is very individual at the different sites, the «antcn» skeleton must be completed in station-specific software.

— *SNAP command interpretation programs:* The parsing and interpretation of SNAP commands are separated from the central coordination of the Command Processing and Control program «boss».

— *Quick Response SNAP commands «quikr»:* «quikr» is the second level of coordination and control. It is controlled by the Command Processing and Control program «boss» to parse and interpret the SNAP commands for the data acquisition system. It directly commands the bus control programs to interact with the hardware. The commanding is flexible enough to enable the creation of new SNAP subroutines.

— *Quick Response VLBA SNAP commands «quikv»:* In principle, «quikv» is a copy of «quikr» for the SNAP commands, required for the VLBA. It is completely written in C code.

— *Coordination programs:*

— *Command Processing and Control «boss»:* This program is the central coordination of the whole system. It processes SNAP commands from the operator or from an input file and schedules the interpretation programs. It also controls the startup and initialization of the hardware/bus driving programs during the start of the Field System, where it uses the different configuration files to parametrize the subordinated programs.

— *Antenna Calibration Data Acquisition «aquir»:* It automates the measurement of the pointing and antenna sensitivity with the Pointing Offset Measurement «fivpt» and the Antenna-Sensitive Measurement «onoff». It reads a list of sources and operations from the configuration files. The «aquir» program then repetitively cycles through these sources and according activities.

– *Pointing Offset Measurement «fivpt»:* This program can be used to derive the offsets for the azimuth and elevation axis between the actual and the expected position of a radio source in local topocentric coordinates. It measures the power level received at a set of discrete offsets along slices through the expected position. The acquired data are fitted to a Gaussian curve with an offset and a slope, where the calculated peak of the Gaussian fit defines the actual position. Adjustable SNAP arguments are used to customize the behavior.

– *Antenna-Sensitive Measurement «onoff»:* This program is used to measure the single-dish sensitivity of the receiving system. It measures the power on and off the source to calculate the SEFD, source-to-calibration signal ratio, or aperture efficiency. The power values for this calibration are obtained from up to two detectors in the back end system. The behavior of the program can be customized with the SNAP command arguments provided.

— *Status checking programs:* The checking programs perform the monitoring of the control system (system monitoring).

— *Standard Module Checking «chekr»:* The checking program uses defined checklists to check the system states of the different online programs. It is a parallel scheduling loop, which runs repetitively every 20 seconds.[21] If there are discrepancies, it sends error messages to the logging and user interfacing programs. During each run, it also activates the station-specific checking programs.

---

21 With autonomous systems, it might take too long if system states are only checked each 20 seconds. Therefore, it is necessary to implement additional, independent monitoring tools or to extend the existing ones.

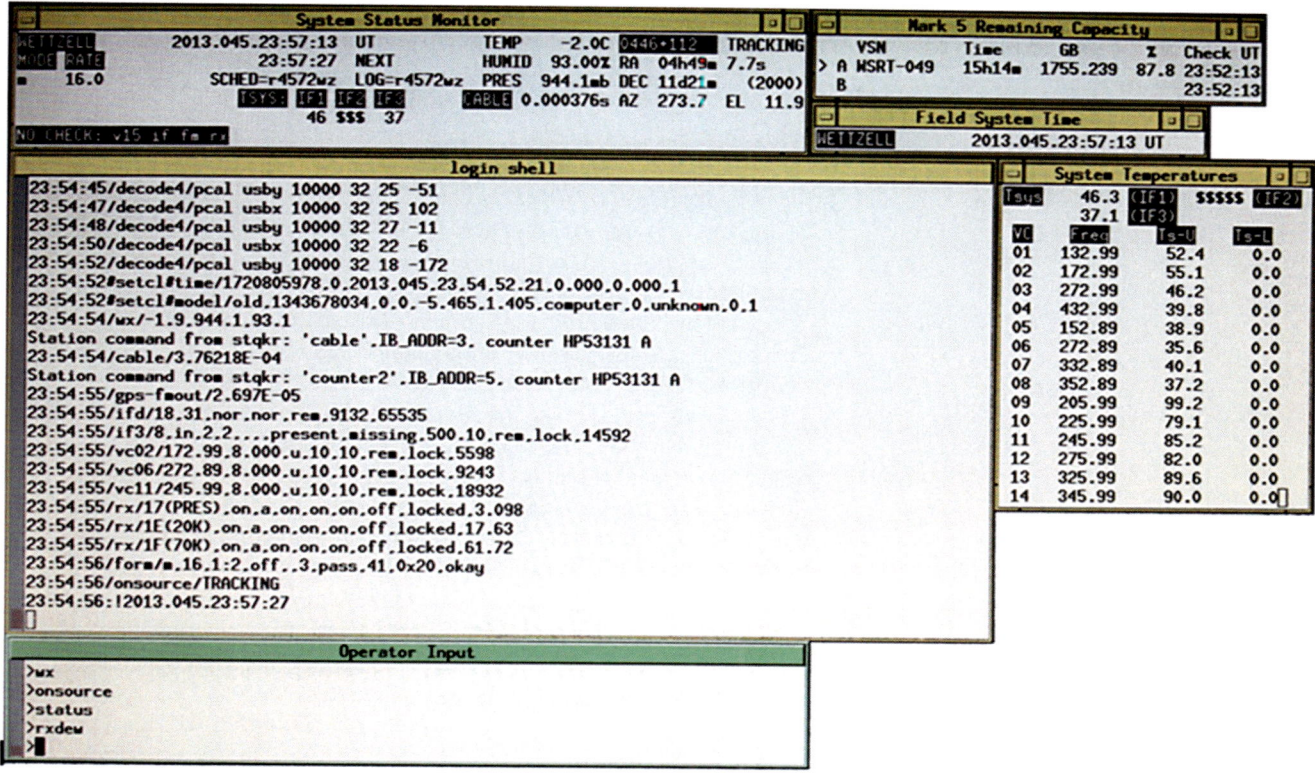

Fig. 5.6 The graphical user interface of the NASA Field System with the System Status Display («monit2»), the Mark 5 Remaining Capacity Display, the Field System Time Display («monit1»), the login shell with the log file and standard output, the System Temperatures Display («monit3»), and the Operator Input Display («oprin»)

— *Logging and user interfacing programs:* The user interfacing is solved with the graphical X Window System of the Linux operating system. It allows a more or less graphical output (see ▫ Fig. 5.6), as described in ▶ Sect. 4.3 (Field System Team 1997b, *l.c. page ARCH-7*).

— *Operator Input «oprin»:* All command line inputs of SNAP commands from the operator are handled with this program. The commands received are transmitted to the Command Processing and Control program «boss».

— *SNAP File Input:* The usual way to control the system during an observation is the interpretation and processing of schedule files. These files (usually in the standard schedule file format of the NASA sked program with the file extension «.skd» (Vandenberg 1997) or as VLBI EXperiment format (VEX) file output (Whitney et al. n.d.) from the NRAO scheduling program with the file extension «.vex»[22] (Vandenberg 2000, *l.c. page DRUDG-1*)) are generated by the schedulers of the IVS. They can be downloaded from IVS servers. At the location of the telescopes, they are then prepared and converted into SNAP command files for the location of the telescope, using the utility program «drudg» for the Experiment Preparation Drudge Work. This generation produces the according SNAP files, containing a time-tagged sequence of observation SNAP commands (see also ▶ Sect. 4.4). The SNAP files can be started with the SNAP command «schedule»,[23] which then automatically starts the required activities at the predefined time.

---

22   The VEX2 format is currently under specification. Information can be found under
   ▶ https://safe.nrao.edu/wiki/bin/view/VLBA/Vex2community.

23   For example, «schedule=i14167wz, #1», where «i14167wz» is the experiment name and «#1» is the line number at which the processing should start.

— *External SNAP Injection:* Another possibility is a programming interface with the class input/output communication to inject SNAP commands directly to the Command Processing and Control program «boss». It is possible to use its message queue to directly send SNAP commands, without using the standard Operator Input «oprin». All the functions required are available in the source code of the Field System.

— *Log File Writing and Display «ddout»:* The «ddout» writes log files with the status and error information and displays these to the user. Ideally, all outputs are sent directly to this logging facility, so that the log file is a complete overview of the processing (similar to a flight data recorder system in aircrafts). The program itself is a classical server, which waits on incoming log messages, which are queued and sequentially printed to the output while timestamps are assigned to each message.

— *Displays «monit(...)»:* The monitor programs offer a time, system status, system temperature, and Mark 5 display, where the specific system information from the shared memory is presented to the operator.

— *Error message expansion for log display «fserr»:* The Log File Writing and Display program «ddout» uses this program to look up error messages, which correspond to the error codes, consisting of two characters and an error number. This lookup table is defined in one of the configuration files.

— *Help Display Shell Script «helpsh»:* Help texts and information about SNAP commands can be shown to the operator using this script, which prints the according help text files to the operator screen.

— *Utility programs (off-line programs):* Off-line programs are all additional programs which can be started individually and asynchronously to process specific tasks.

— *Programs for observations:* There are several programs which support the operator in his daily work to prepare, run, and evaluate the observations with the NASA Field System, which are mainly:

  – The Experiment Preparation Drudge Work program «drudg» for the conversion of schedules to station-specific SNAP files
  – The Procedure File Manager and Editor «pfmed» to list, edit, and manipulate the generated SNAP files
  – The Mark IV/VLBA Formatter Time Set program «fmset» to set the time of the formatter
  – The Field System Time Model Control «setcl» to control and monitor the Field System time used for the controlling
  – The Log Examination program «logex» to extract specific commands from a log and produce printer plots

— *Programs for the antenna performance analysis:* The Field System also offers adequate utility programs for the evaluation of the pointing measurements, which are mainly

  – The Pointing Data Extraction program «xtrac» to extract pointing data from the log file
  – The Pointing Data Editing tool «sigma», used for an automated editing of pointing data to extract outliers
  – The Pointing Data Analysis program «error» to fit pointing data to a model
  – The Pointing Data Residual Plot program «resid» to generate printer plots of the pointing residuals

— Even though the structure of the VLBI control system is very similar to structures of autonomous production cells, it does not directly fit into this design. It does not support the required communication interfaces, and it is supplied by an external vendor, so that it is difficult to adapt its code without an administrative overhead. Nevertheless, it is quite useful to integrate and use the already existing programs described and their communication structures, especially as they are the standard for all IVS radio telescope observations. A complete rewriting is difficult, is expensive, and would separate an observatory from the rest of the VLBI community. Nevertheless, it is necessary to enable the possibilities of the new agent system structures. This can only be implemented with extensions, which are add-ons to the existing programs of the NASA Field System.

5.4 Extend Existing Control Systems with Multi-agent Abilities

To extend an existing legacy system to get a modern multi-agent architecture requires different techniques, which have already been described in the previous chapters, for example, in ▶ Sect. 2.4 about the use of legacy code and in ▶ Sect. 2.6.2 about the architecture of software tests, to monitor and bypass other systems. For the implementation of tests, it is necessary that the software module under test is not dependent on other programs or on hardware. The solution is the implementation of stubs which act like proxies of the dependent hardware with the expected behavior, but with more interaction and manipulation options. This mechanism can be used to replace the existing control programs for hardware in the NASA Field System with other components, which offer the functionality to include them in the newer distributed control structures with newer RPC communication interfaces. Comparable to the test setup in ▪ Fig. 2.20, it creates the «lower parenthesis» of interaction to the existing Field System at the level which represents the equipment controlled by the Field System. External programs can also monitor these lower-level components, for example, with higher monitoring rates, in parallel with the existing methods of the Field System. The «higher-level parenthesis» of interaction manages and controls the NASA Field System with all its connected components from the perspective of the user interactions and external inputs. The equivalent representations in test scenarios are mocks, where tests need to be able to control software and treat it with external input to test dedicated, single functionalities of the program. These «test spy» techniques of mocks and stubs can be adapted to extract data from and to inject input into existing control systems, such as the NASA Field System, and to bypass monitoring information to higher control levels, such as multi-agent systems (see ▪ Fig. 5.7 and compare it to ▪ Fig. 2.20).

Equivalence to Software Test Systems

The equivalent to mock objects in the case of the extension of the given NASA Field System is a remote control RPC server. It implements all functionalities described in ▶ Sect. 3.3, is generated with «idl2rpc.pl», and treats the existing Field System like external hardware which has to be controlled within the periodic loop. A minimally invasive concept uses only the access points which are already offered by the Field System and which are covered by an abstracting Field System monitor layer (see ▪ Fig. 5.8 and the code snippet in Example 5.1 to get the meteorology and to send SNAP commands with existing functionalities). To inject control input, it uses:

Field System Monitor

- The external SNAP command injection
- The SNAP files

To get feedback from the Field System, the monitor layer uses:
- The log files which are continuously read, so that always only the new lines are processed (similar to the Linux command «tail»)
- The shared memory from which the current control values (e.g., the pointing position, current time, meteorological information) can be retrieved
- The text sent by the Field System programs to standard output (for this, it is necessary to obtain the information using a child process, which runs the third-party program and sends the outputs of the program via IPC to a monitoring parent process, as shown with the code snippet in Example 5.8 of ▶ Sect. 5.5.5 about logging of a SSH client)

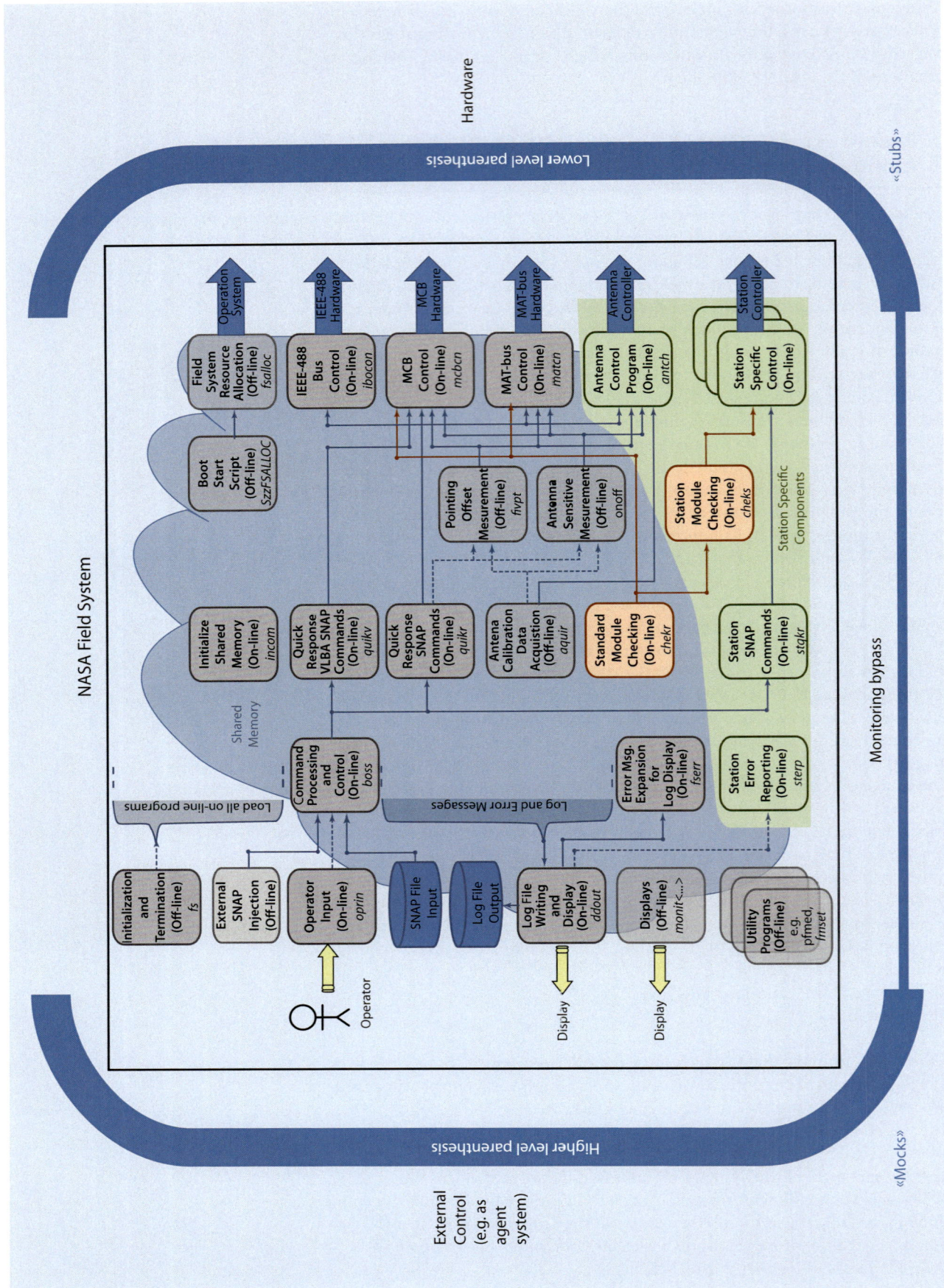

Fig. 5.7   The control «parentheses» around the existing programs of the NASA Field System with a bypass to monitor data from the hardware stubs to the higher-level agents

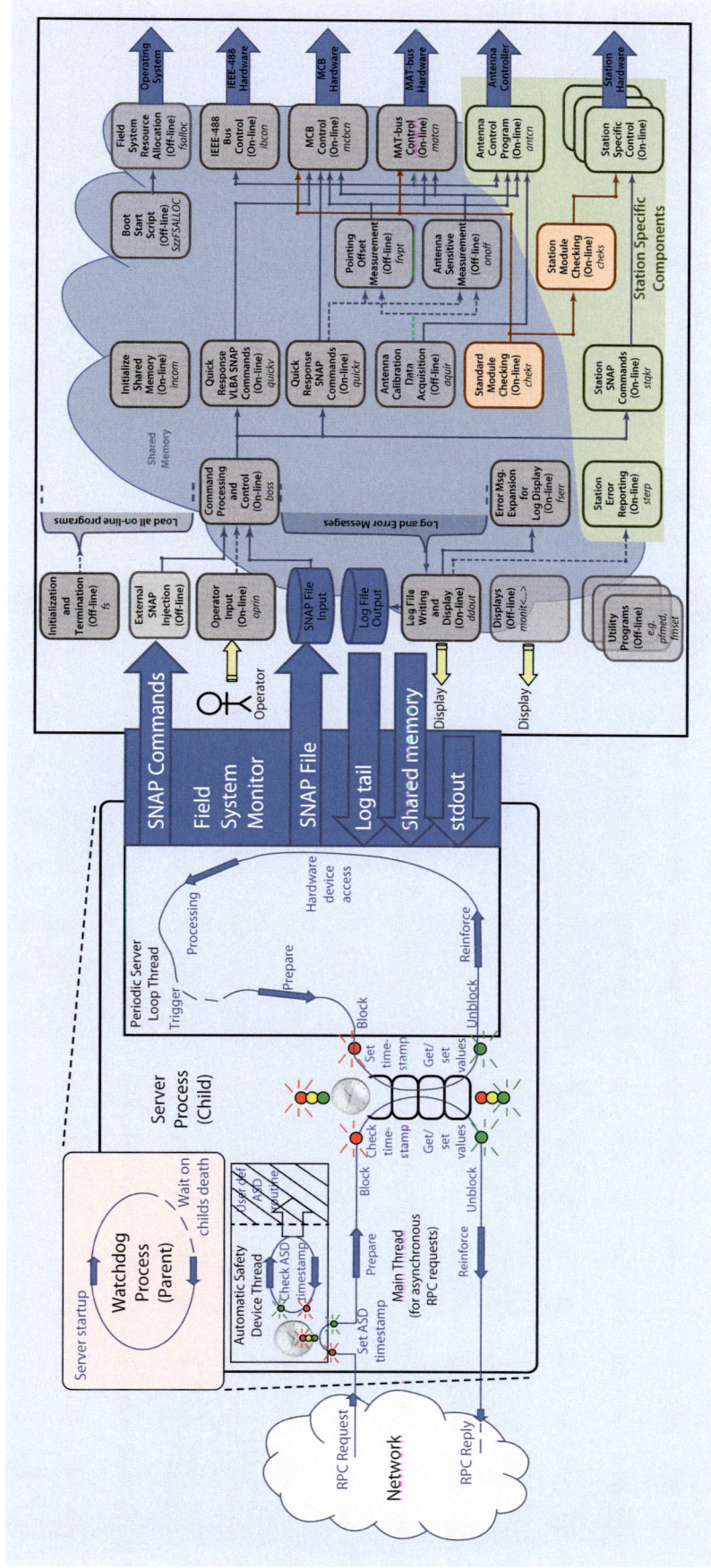

Fig. 5.8 The RPC remote control server with its server safety techniques from ▫ Fig. 3.22 using an additional Field System monitor layer as a higher-level bracket to send in input data and retrieve the feedback information

■ ■ **Example 5.1**

The Field System monitor as interface to the NASA Field System.

```
extern struct fscom *shm_addr;                                      1
                                                                    2
static struct fscom *fs;                                            3
                                                                    4
static unsigned short usSetupIDSActivated;                          5
                                                                    6
/* ... */                                                           7
                                                                    8
/**                                                                 9
 * Checks whether the field system is running                      10
 *                                                                  11
 * @return 1 if the field system is running; 0 otherwise           12
 *                                                                  13
 * @author    Helge Rottmann                                       14
 * @date      14.1.2009                                            15
 * @version   1.0                                                  16
 */                                                                17
unsigned short usIsFieldSystemRunning()                            18
{                                                                  19
    if (usSetupIDSActivated == 0)                                  20
    {                                                              21
        setup_ids();                                               22
        usSetupIDSActivated = 1;                                   23
    }                                                              24
                                                                   25
    if (nsem_test(NSEMNAME) == 1)                                  26
    {                                                              27
        return 1;                                                  28
    }                                                              29
                                                                   30
    return 0;                                                      31
}                                                                  32
                                                                   33
/**                                                                34
 * Initializes the field system shared memory pointer.             35
 * This method should be called by every function that tries to obtain   36
 * information from the shm_addr pointer.                          37
 *                                                                  38
 * @return 0 in case of success                                    39
 * @return 1 in case the field system is not running               40
 *                                                                  41
 * @author    Helge Rottmann                                       42
 * @date      14.1.2009                                            43
 * @version   1.0                                                  44
 */                                                                45
unsigned short usInitSHM()                                         46
{                                                                  47
    if (usSetupIDSActivated == 0)                                  48
    {                                                              49
        setup_ids();                                               50
        usSetupIDSActivated = 1;                                   51
    }                                                              52
    if (!usIsFieldSystemRunning())                                 53
    {                                                              54
        return 1;                                                  55
    }                                                              56
    if (fs == NULL)                                                57
    {                                                              58
        fs = shm_addr;                                             59
        if (nsem_test(NSEMNAME) != 1)                              60
        {                                                          61
            return 1;                                              62
        }                                                          63
```

```
        }

    return 0;
}


/* ... */

/**
 * Injects a SNAP command via boss.
 *
 * @param command the SNAP command
 * @return 0 in case of success
 * @return 1 in case the field system is not running
 *
 * @author    Helge Rottmann
 * @date      29.4.2009
 * @version   1.0
 */
unsigned short usInjectSnapCommand(const char *command)
{

    int iLength;

    /* check if the field system is running and initialize the shared memory */
    if (usInitSHM())
    {
        return 1;
    }

    iLength = (int)strlen(command);

    cls_snd( &(shm_addr->iclopr), (char*)command, iLength, 0, 0);
    skd_run("boss ", 'n', ip);
    return 0;

}

/* ... */

/**
 * Returns the current air humidity
 * value stored in the field system shared memory.
 *
 * @param[out] fHumiWX the current air humidity
 * @return 1 in case the field system is not running; 0 otherwise
 *
 * @author    Helge Rottmann
 * @date      14.1.2009
 * @version   1.0
 *
 */
unsigned short usGetSHMHumiWX(float *fHumiWX)
{
    if (usInitSHM())
    {
        return 1;
    }

    *fHumiWX = shm_addr->humiwx;

    return 0;
}

/* ... */
```

Field System Monitor Restrictions

This concept has to deal with some restrictions, which also limit the possibilities of the agent system. It is only possible to control the whole system with the ASCII-based SNAP commands. All data sets which are not available in the shared memory are not directly accessible and can only be requested in parallel with class Input/Output communication. Another possibility during observation sessions is parsing log files to search for specific values. But the timing of updates is here defined by the status checking programs. The update rate of values is a general restriction, because it is defined within the Field System or in the command sequences of SNAP files and procedures. But higher control levels usually require their own timing. Therefore, the stubs replace the modules for the real hardware bus control in the bus control programs and create bypasses to the real system data from the bus or act like a «sniffer» program directly on the bus communication. But this produces a lot of development work or unmanageable dependencies between external and local code. With the Field System, it is mostly unnecessary, as functionalities which are not relevant for safety can be solved with regularly scheduled, timed Field System SNAP commands. The stub technique will focus on safety relevant, station-specific components.

Autonomous Field System Hardware Control Cells

One component in this field is the antenna control program «antcn» in combination with the station-specific ACU. This control element can be designed as an autonomous control cell. The «antcn» program is then nothing more than a client program, which connects to the ACU server. The server offers the control functionality via RPCs and is implemented as an autonomous hardware control cell (see ◻ Fig. 5.9 and also ▶ Sect. 4.5 about autonomous hardware control cells). The NASA Field System with its existing libraries and modules must here be treated as legacy code in the programming language C, into which new code in C++ with the RPC client functionality has to be integrated. It is necessary to use an adequate adapter module, as described in ▶ Sect. 2.4. Each RPC function of the client class, which should be used in «antcn», gets a stub function in the adapter module. Each of these functions uses converted versions of the original arguments plus a pointer to the client object, which is cast to «void *».[24] Instead of a module-global, static client object, the version with the cast return pointer allows the use of the module in one and the same program in parallel, where the pointer is something like an object descriptor. The new converter module uses the internal preprocessor definition for «external "C"», as described in ▶ Sect. 2.4 and is then translated with the C++ compiler, but linked with the C modules using the C compiler.

Optimization for C++

Instead of an explicit converter for a dedicated, connected hardware device, another possibility is to use a specialized Field System monitor as a converter for the connection to the NASA Field System. The monitor is written in C and uses the above «external "C"» nomenclature. Then the rest can be written in C++, while only the monitor is kept in C. This technique reduces the cost of new developments if usually C++ is used. Nevertheless, both techniques work fine and only depend on whether the whole project is written more in C or C++.

Bypass

This setup for legacy code allows several, parallel clients in the existing Field System and also outside, which implement a communication bypass to the antenna beside the Field System structures. This allows the direct control and checking of emergency and safety relevant functionalities of the antenna (e.g., a direct stop command to all axes without using all layers of the Field System). Such bypass functionalities should exist for all of these safety relevant parts, so that an additional safety control loop can be implemented. An additional advantage is that the actual, real data from the antenna can be directly forwarded to the graphical display used by a remote operator.

System Health and Human Safety

The importance of human and system safety increases a great deal if no local operator is taking care of the system and its surroundings. The short distance of a local operator provides a lot of options to solve critical situations for human beings and the system's health itself, as a local operator can use all of his senses in direct

---

24 This cast can easily be hidden if a new type definition is realized for the pointer, e.g., «typedef void * ACUDescriptor;».

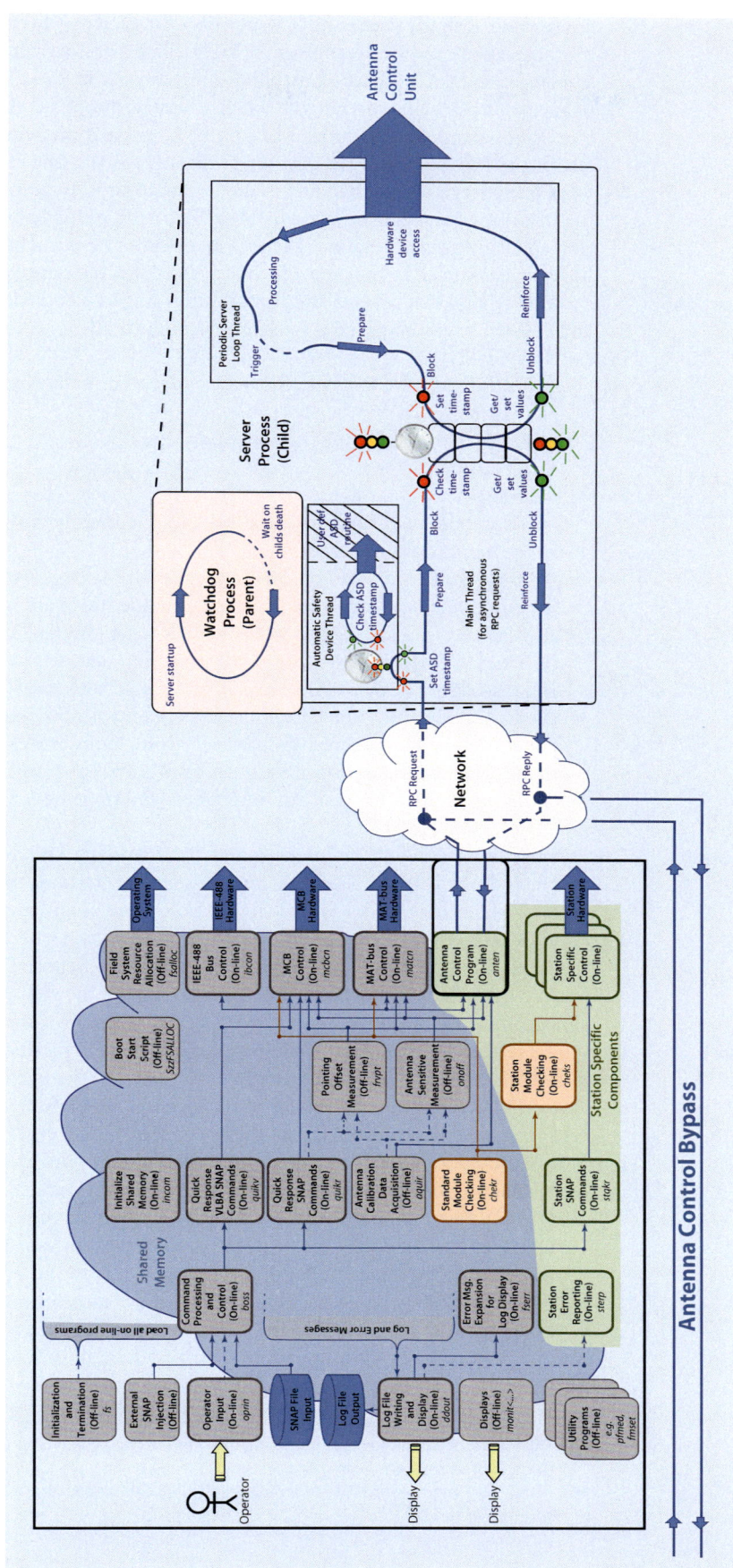

Fig. 5.9 The use of hardware control cells with RPC technique from ◘ Fig. 3.22 as a lower-level bracket for station-specific code like the antenna control program

connection with the system he is responsible for. One real-life example can be given for the 20 meter radio telescope of the Wettzell observatory. The servo control of the axes uses several feedback controlling mechanisms in the PLC which shows the behavior of an oscillation ringing when a new position is commanded. But defective, aging components can lead to a build-up oscillation, which shakes the whole telescope. This can result in disastrous damage, as the huge mass of the movable telescope parts is accelerated and braked again in short time intervals. As this shaking produces very loud noises, a local operator can immediately react to this situation and can stop the antenna. The remote operator does not have this possibility, as his view is reduced to the data which are offered by the remote control server.

Additional Aspects for Remote Control

This means that several additional things must be considered when operating large systems remotely (see (Lovell and Neidhardt 2013, *l.c. page slide 21*)):
— *Accessibility:*
  — Nothing can be touched, moved, or turned directly (e.g., changing hard drive modules for the recording of VLBI data).
  — Cables cannot be plugged off and on directly.
  — Keyboards on location are not directly accessible.
— *Visibility:*
  — Telescopes and their surroundings are not directly visible.
  — Sky conditions are unratable.
  — Racks and hardware equipment are not visible (e.g., lights on devices).
  — Computer monitors, desktops, and module stores are not directly visible.
— *Redundancies:* Single systems offer several safety bottlenecks, which can only be avoided with enough redundancy of equipment, which is «hot pluggable» («hot standby»)[25] or which is available as «cold standby».[26]
— *Mental detachment:* The lack of direct contact with the real hardware device suggests a bigger distance, so that everything looks as if it is in a computer game. This mental detachment must be avoided by those responsible.
— *Maintainability:* The local and remote control must be easily maintainable.

Importance of System Monitoring

A consequence is that maintenance of the hardware and software is becoming increasingly important if such huge, but sensitive systems are operated remotely or even unattended. This means that remote control and automation should not entail a reduction of personnel. It should just reduce operator shifts, so that a realignment of jobs and their description is a consequence, where regular maintenance with clearly defined checklists must be done by well-educated and responsible staff.

Importance of Parallel Remote Equipment

Nevertheless, the best maintained system can produce failure situations. Therefore, the remote control extension must emulate the senses of the remote operator on location. But it is not necessary to really rebuild the human senses. Local sensors must just be able to detect the same error situation, so that the mechanisms of the autonomous production cell can react to them. With the previously described error situation of servo oscillations with the subsequent shaking of the telescope, inertial measurement units with suitable data analysis, or the evaluation of the motor currents, might also detect the problem. Another example is the autonomous reaction on wind speed if it exceeds defined limits, so that the antenna can automatically be moved to its stow position. This means that a remotely controlled system necessarily requires a suitable, flexible, extended, parallel system monitoring, which replaces the senses of a local operator to protect human beings and the system itself.

---

25   «Hot pluggable» means that hardware or software can be exchanged without influencing the rest of the system during operations.

26   «Cold standby» means that hardware or software can be exchanged after the equipment is shut down and off.

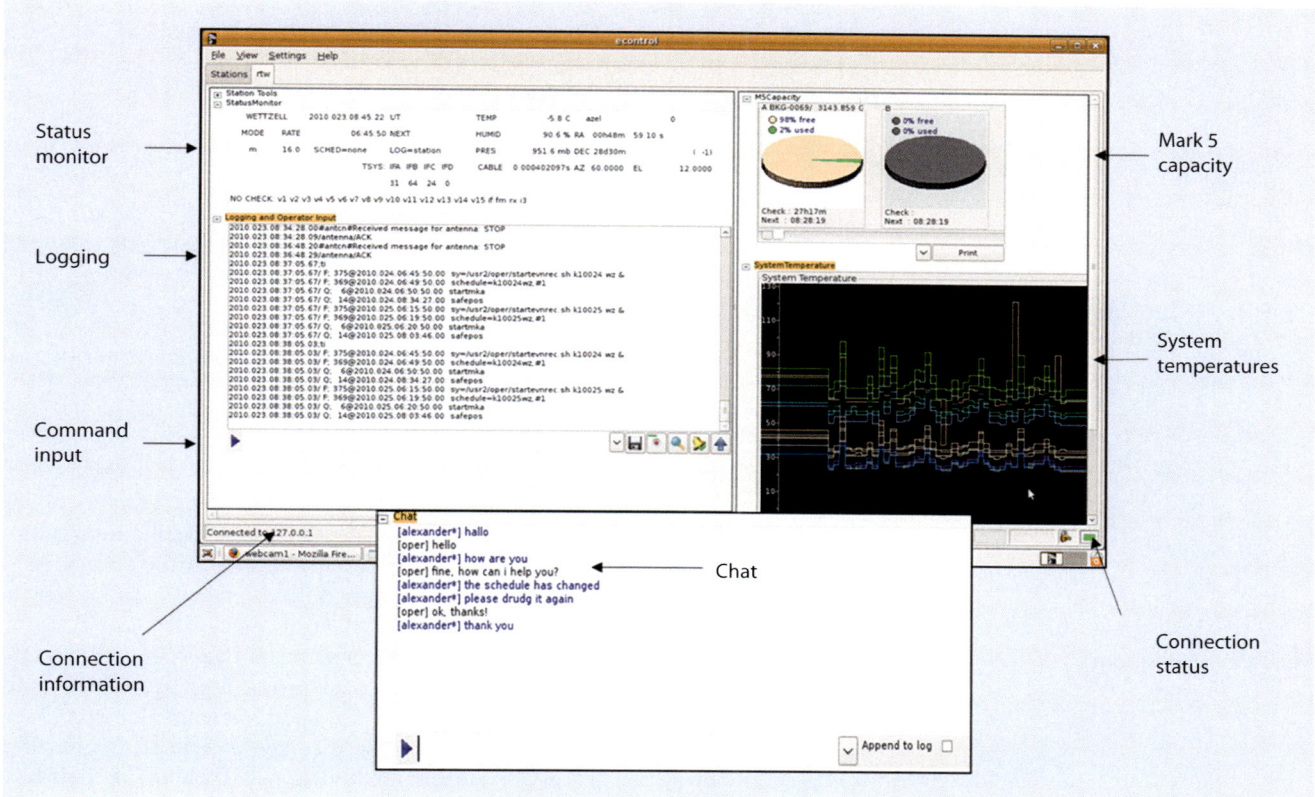

Status monitor

Logging

Command input

Connection information

Mark 5 capacity

System temperatures

Chat

Connection status

Fig. 5.10 The new remote GUI of the NASA Field System agent with the mixture of selectable inputs and outputs in classic and modern, graphical style

Here the system monitoring and safety cell can be used (see ▶ Sect. 4.7) where the data for system operations and diagnostics are most important for this task. These data are collected in parallel with the Field System control and are available at the remote control server, where they can be used to make autonomous decisions, as described for the autonomous production cells in the previous chapter.

Using the RPC architecture of «idl2rpc.pl» (see ▶ Sect. 3.3), the NASA Field System is treated like complex, controllable hardware, offering a standardized interface plug, which can be accessed remotely. This offers all possibilities for multi-agent systems and for a remote GUI (see ■ Fig. 5.10). The principle again is to use «fat servers», where all autonomous activities are processed, in combination with «thin clients», which only present the current system values or receive events from the operator which are commanded to the server (see the descriptions in ▶ Sect. 4.3).

The idea behind the implementation of new operator interfaces to already existing systems should be to keep the look of the already known and well-established GUI, as the users are familiar with it (e.g., a new GUI for the NASA Field System should have the basic elements of the original GUI). New features and more graphical-oriented presentations can be added, so that the operator can select between the classic style and the new one (see ■ Fig. 5.11). Experience has shown that this approach is widely accepted by the user group. For the example of the remote control of radio telescopes with the NASA Field System, the existing windows outputs and inputs from ■ Fig. 5.6 are copied to the new «wxWidgets» GUI (see ▶ Sect. 4.3 for information about «wxWidgets») and supplemented by more sophisticated outputs and inputs in additional tabs. Then the operator can decide which style he wants to use by selecting the related tab. These additional,

Graphical User Interface

Look and Feel with New Graphic Extensions

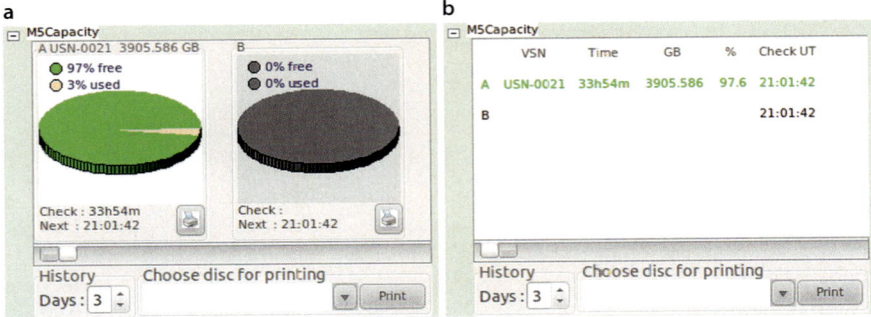

Fig. 5.11 An example of the combination of **a** modern, graphical displays with **b** classic, text-based, look-and-feel outputs on the example of the «Mark 5 Capacity» status. The operator can select the individual view

graphical outputs contain pie charts for the Mark 5 recording system capacity, X/Y-plots for the system temperatures with selectable system channels, or a functionality to select log lines, using text pattern matching to extract specific information from the logs. An additional part in the new user interface is the emulation of alarm tones, so that different alarm levels produce different sound outputs over the sound card hardware of the client computer.

Functional Extensions

The additional coordination layer, which implements the access point to the autonomous production cell of a VLBI system, can also be used to add new functionalities to the existing possibilities of the official Field System. Value-added services, which facilitate the work of an operator, are ideal in this field. In combination with the graphical abilities of wxWidgets for the client GUI and the high-level coordination at the server, it is possible to implement

- Optical presentations of system and connection states
- Pattern matching and graphically highlighted and/or acoustical output of logging information (like errors)
- Chat dialogs where operators can talk with each other
- Print services
- Graphical checklists for operator activities and checks before, during, and after an observation session
- A general antenna control dialog
- Web cam picture presentations
- Complete HTML presentations within the wxWidgets GUI

GUI Variants

In particular, the possibility of displaying HTML directly in a wxWidgets window in the GUI reduces the costs of switching the presentation layer to a different type, such as a browser-based version, as described in ► Sect. 4.3 about alternative GUI variants. This also enables the use of mobile devices, like smartphones or tablet PCs. Another advantage here is that the server can define the layout of the presentation if it sends the template at the beginning of the session into which the regularly updated data are plotted. It is also very beneficial, if station-specific outputs are presented in the graphical client, such as those required for station-specific system monitoring data when the station programmers don't want to use the extensive framework of wxWidgets. In this case, the server just has to offer the option to transfer such HTML pages, which are then automatically presented in separated HTML tabs of the GUI. As most station programmers can simply arrange HTML code with their system data, they do not have to know anything about the graphical programming but get very attractive graphical outputs in the user interface. This technique can also be used to automatically present HTML help pages, which are converted from text files. A disadvantage of this technique is that large parts of the wxWdigets GUI are nothing more than a lightweight web browser, which can be reached much easier with

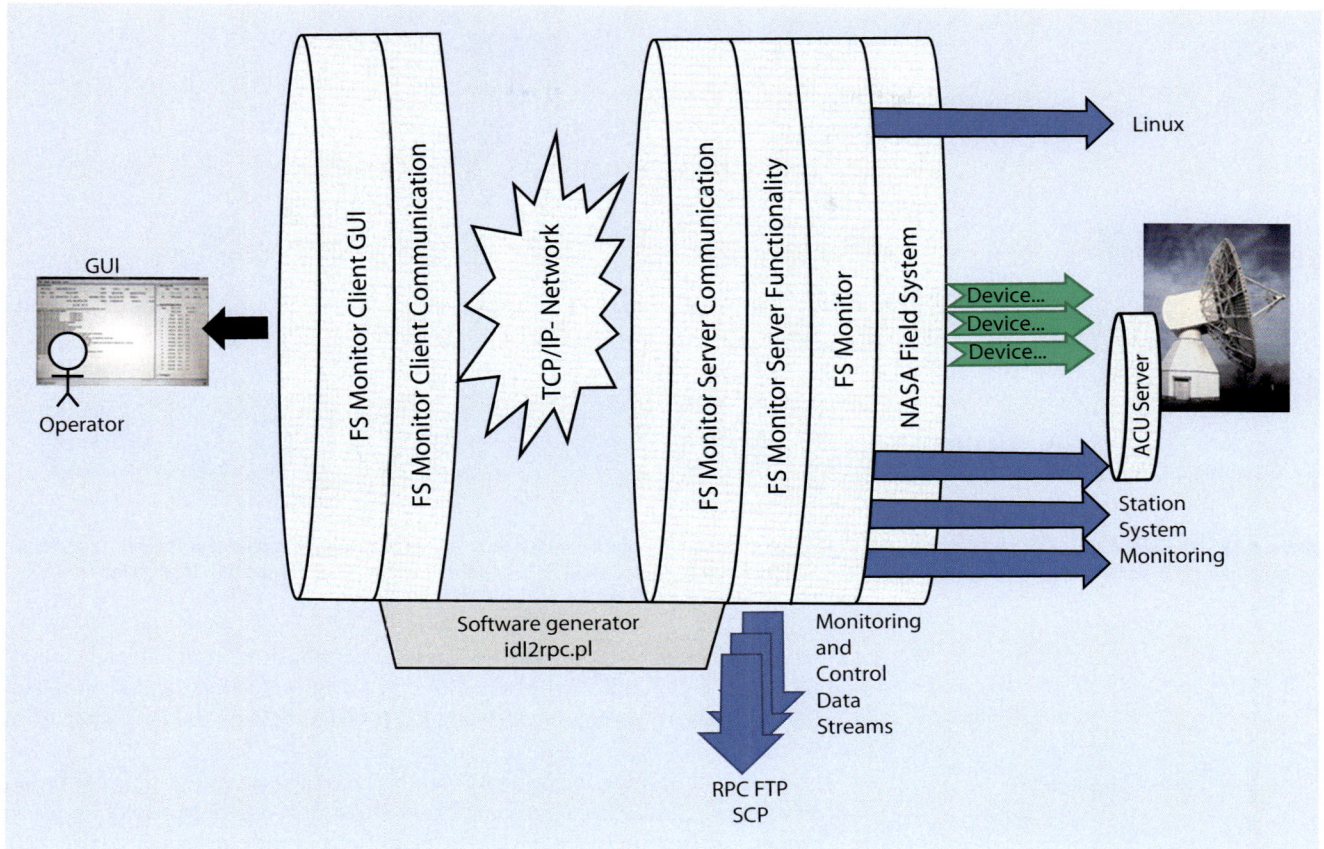

Fig. 5.12 The communication stack of the remote control for the NASA Field System agents

web programming techniques on the basis of HTTP and a «real» web browser.[27] Another disadvantage is that layout and data might conflate again, so that it becomes difficult to draw presentation plots from raw data on the client side, and using HTML increases the data volume. The combination of active buttons is also not solved well if wxWidgets and HTML are mixed. Nevertheless, it offers a very wide variety of possibilities.

With all of the parts described, a communication stack is established for remote control requirements, which is drawn in Fig. 5.12. In this remote control extension, the hardware is mainly controlled by the NASA Field System, except for some additional, station-specific, or safety relevant parts and some operating system interfaces. The data are read from the log file, respectively, from the shared memory, and commands are injected from an external source into the processing queue of the Field System, using the adaptation of the Field System monitor. This adapter functionality is converted into an RPC communication structure for multi-agent systems, where the communication code is generated with «idl2rpc.pl». On the client side, a wxWidgets GUI receives the data and implements the presentation for the operator in classic and graphically sophisticated outputs. This task was originally developed at the Wettzell observatory and then extended for the use within the IVS. Later on, the project NEXPReS for the European VLBI Network (EVN), funded by the European Union in the Seventh Framework Program, was an ideal

Communication Control Stack

27　But with a real web browser and standard HTML, the software implementations, like the automatic safety device, become much more difficult because additional, stateless layers are integrated, so that the server cannot monitor its clients so well.

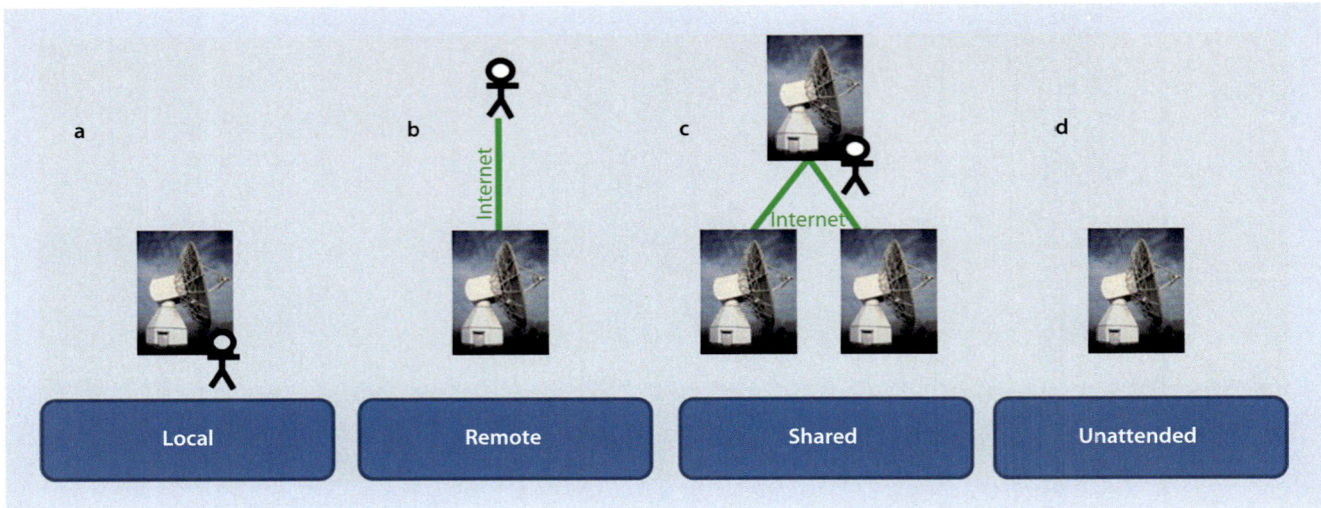

◘ Fig. 5.13   The new operation control strategies for geodetic multi-agent systems: **a** the operator is on location at the telescope (classic operation), **b** the operator controls the telescope remotely, **c** one operator controls several telescopes, and **d** the telescope runs completely unattended

Operation Control Strategies

platform for further progress. The goal was the establishment of safe and secure remote control and monitoring possibilities for astronomical radio telescopes. The result is the software «e-RemoteCtrl», which is distributed by the developer team from Wettzell.

The final goal of these implementations is the establishment of new operation control strategies. These strategies can be split into four groups (see also ◘ Fig. 5.13):

1. *Local operations:* These are the classic operations as they have been performed with the NASA Field System since the beginning of international VLBI. A local operator is available at each site and starts and monitors the observation sessions which follow a schedule offered by an international scheduler. The operation is already automated with the schedule, but the operator is required to interpret errors, to react to them, and to start manual tasks, such as sending status emails.

2. *Remote operations:* With the additional control and coordination layer, which makes decisions autonomously for the system over a specific time period in combination with a sophisticated system monitoring which evaluates system states, it is possible to separate the operator from the location of the telescope. The system monitoring replaces the senses of the operator on site. It is possible to check system states or to command orders from everywhere in an observatory areal and even from anywhere on the Earth via the Internet if the autonomy time can be extended to bypass at least the worst communication round-trip delays or time intervals of network blackouts. The remote operation also allows a form of tele-working, remote diagnosis, assistance of local operators, and the control of very remote telescope locations which are barely accessible to humans. Practical experience has shown that the use of web cameras is essential to check the system health state and critical situations remotely. The remote operation is designed for the telescope teams, so that they have more access possibilities to their system and to optimize their work. This operation strategy is regularly used for weekend shifts at the radio telescope Wettzell.

3. *Shared operations:* With the possibility of remote operations, it is also possible to centralize the control of several telescopes at one place, to share the control between different control sites at the location of the telescopes, or to allow a control participation to VLBI network schedulers or correlators. It might become a reality for schedulers to have direct access to the telescopes to

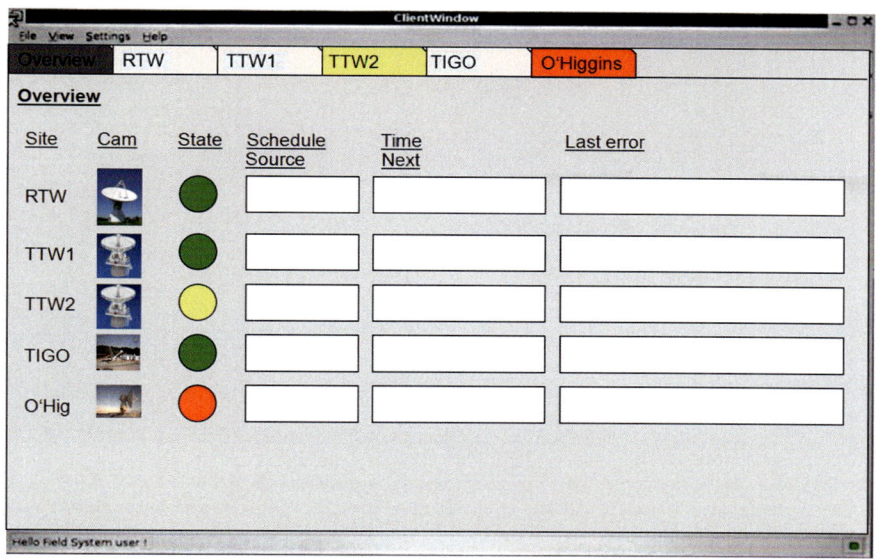

Fig. 5.14 A design sketch for a possible grouping of control information on a one-page status window with web cams, alarm levels, current states, and main outputs

change observation sessions in real-time. This live monitoring and controlling by external people require dedicated access rights and a sophisticated authentication and authorization system in which dynamic user roles can be defined. All activities must be recorded in log files, similar to the flight data recorder in aircraft. The control of several telescopes at the same time also requires special GUI pages, where the main information is grouped. This must represent the hierarchy and must show the alarm states, problems, and maybe web cams (see ☐ Fig. 5.14). Such a one-page window is the only way for an operator to keep an eye on all controlled systems. But then the shared operation strategy also makes it possible to hand over the control to other sites during the day, so that the operator shifts can be optimized for a whole network of stations in different time zones in the world. This operation strategy was demonstrated during the continuous observation session «CONT11», where the radio telescope of the Transportable Integrated Geodetic Observatory (TIGO) in Concepción, Chile, and the 20 meter radio telescope in Wettzell, Germany were controlled by the Wettzell staff during the night shifts at TIGO.

4. *Unattended operations:* Each remote control requires a sophisticated autonomy to bypass time intervals where no operator is available, because of network problems or other error situations. As no local operator is necessarily available, there is no direct way to detect errors or critical situations where the system itself or even human beings are in danger. With adequate safety systems, it is possible to extend the time intervals in which the system runs completely autonomously. Therefore, the next step for operation strategies is to completely automate the system, for example, if the autonomy state can be extended to the complete time interval of one session. A precondition for that is to have excellent maintenance of the hardware and software of the system. Another condition is a verifiable safety system for operations. This strategy is also regularly used for weekend or night shifts at Wettzell.

The hierarchically defined control structures in combination with the previously described operation control strategies allow the extension of the control to more

Hierarchy of Control Centers

Fig. 5.15 The operator room for all three remotely controlled AuScope telescopes at the University of Tasmania, Hobart with the instances of «e-RemoteCtrl» for Hobart, Kathrine, and Yarragadee on the main control screens in the center (April 2013, photo by Jim Lovell; reproduced by permission of Jim Lovell)

than one telescope in parallel. Furthermore, they allow the extension in the control hierarchy from the local system to a control center for one observatory, further to a regional control center for a geographical region, and finally to some redundant worldwide control centers. This opens control strategies and implementations on the basis of cooperating multi-agent systems for future global observing systems, such as the GGOS, or even also for clearly astronomical networks like the Square Kilometer Array (SKA). One example of a control center for one observatory is the newly build control room for the VLBI telescopes of the Geodetic Observatory Wettzell, where SLR systems on site will also be controlled in the future. An example of a regional control center is already implemented for the AuScope VLBI network in Hobart, the capital of Tasmania. From one control room at the University of Tasmania, three VGOS 12 meter dishes and one 26 meter telescope in the Australian region, or more precisely in Kathrine (north of Australia), Yarragadee (west of Australia), and Hobart (Tasmania), are regularly controlled remotely (see ◘ Fig. 5.15). Finally, as an example of the usability and the operativeness of worldwide control centers, the worldwide control experiments at Wettzell can be considered as proof of the concept. During the continuous observation CONT11, the operators in Wettzell took over the control during night shifts of the VLBI system of the TIGO for 15 days (15-SEP-2011 00:00 UT through 29-SEP-2011 24:00 UT; compare also ◘ Fig. 6.4) and were also connected to some AuScope telescopes as an observer (see (Neidhardt et al. 2012b)). Similar tests were also made with the VLBI site Bernardo O'Higgins, Antarctica.

Example: Control of the AuScope Network

According to the additional aspects for remote control, the developers for the AuScope control center created a working remote control example for their VLBI telescopes. The main elements to deal with the additional aspects fulfill the described functional extensions which are required for remote and unattended systems and can be summarized as (see (Lovell and Neidhardt 2013, *l.c. page slide 22 ff.*)):

— A suitable, reliable network connection on fiber optic cables
— A reliable power supply on the basis of a Uninterruptible Power Supply (UPS), and a backup power generator, where switches can be activated with Internet power switches for power cycling
— Reliable computer systems on the basis of Redundant Array of Independent Disks (RAID) systems with at least mirrored contents

- A hardware with «hot» and «cold» spare parts[28]
- Several web and «fish-eye» cams, where special pan, tilt, and zoom cameras are used within the control rooms to watch the hardware racks, antennas, hardware devices, etc., and some IR lights for the antenna illumination at night
- Several network-enabled keyboard, video, and mouse switches to give the operator a low-level console access to the computers in combination with Virtual Network Computing (VNC) connections (see ▶ Sect. 5.5.2 about VNC) to the timing, DBBC, NASA Field System computers, and the spectrum analyzers for system checks
- A centralized software repository to collect the whole source code of all sites
- Additional system monitoring, e.g., with Open-MoniCA (see ▶ Sect. 4.7), which collects receiver and IF data via PIC® microcontroller units with serial and Ethernet ports for control and monitoring to log the data and send short messages to mobile phones or with the system monitoring described (see ▶ Sect. 4.7)
- Reliable remote access for control and monitoring, using «e-RemoteCtrl» with checklists, error handling, alarm tones, and keyboard shortcuts
- Additional one-page summaries per telescope with the main monitoring and control data
- Quality feedback for each individual observation scan, e.g., on the basis of plots from the autocorrelation of a short time interval of a few seconds after every scan for diagnostics and bit statistics

If all operated telescopes of one network are equipped with similar remote control techniques, and if it is possible to standardize access with the multi-agent interfaces described, it is possible to keep the operational costs within manageable limits, while increasing the number of observations required for the new VGOS systems within the GGOS. One operator is then able to take care of more than one telescope somewhere on Earth. It is possible that three, replicating worldwide control and coordination centers in different daylight zones could share the main shifts and offer a reliable control access via web on HTTP or via graphical clients on RPC connections. Through connected sites and also staff with scheduling, correlation, coordination, and analysis tasks can get real-time feedback and can intervene in observation sequences. This improves the central coordination of tasks. Observation sites can either use real-time feedback from the coordination centers to individually control their own equipment or can directly hand over control to the control centers according to the individual safety and security regulations at the sites.

Advantages of Remote Control

A first impression of such a globally available real-time status can be gained with the «IVS Live» web page (see ◘ Fig. 5.16, (Collioud 2011)), to which real-time data from the «e-RemoteCtrl» servers of the different, connected telescopes are forwarded (Collioud and Neidhardt n.d.).[29] This service is called «e-QuickStatus». It is an online stream of status information, which is not really useful for remote control, because there is no guarantee and communication feedback. But it is quite helpful as a general status update for a whole network. The data are coded in the form of status files (see

Globally Available Real-Time Data with e-QuickStatus

---

28 «Hot» spares can be replaced during observations without any losses. «Cold» spares require the processing to stop while the replacement is performed.
29 Unlike the control hierarchy described, where always the next higher level requests data (data pulling) and sends orders, it is also possible to stream data to centralized servers (data pushing). This technique is usually used for the forwarding of monitoring data (see also ◘ Fig. 5.12), but under certain circumstances it can also be used to return feedback to the telescopes, which can be used for control decisions.

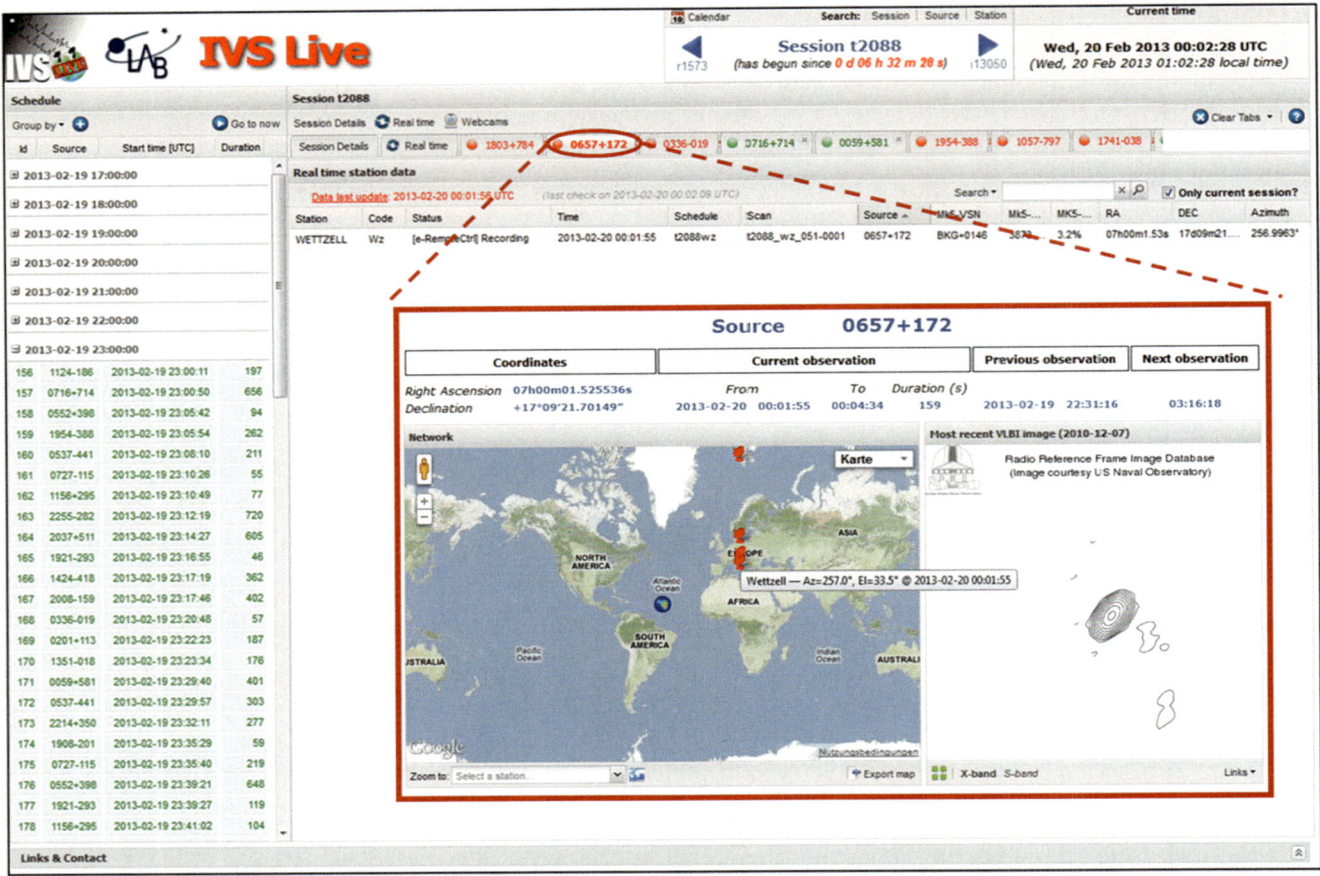

Fig. 5.16 The coordination data as information web page on IVS live (acknowledgement to IVS and A. Collioud of the Laboratoire d'Astrophysique de Bordeaux of the University of Bordeaux/CNRS, France: http://ivslive.obs.u-bordeaux1.fr/)

Example 5.2[30]). After each status change of the connected telescope, this file is created and copied with SCP to the data collecting web server, which is, for example, located at the Technische Universität München (Technical University of Munich), Germany. To enable the secure copy, a key file for the SSH connection is required, which must be requested from the operator of the web server. While all telescopes, using «e-RemoteCtrl», can directly activate this new ability, legacy systems can also join if they implement the file transfer of the status file to the web server themselves. The web server scans the directory for incoming files every second. Each scan updates the local «e-QuickStatus» web page. From there, an additional service of the IVS regularly collects the data and combines them with additional information from central data bases as, for example, images of the observed sources. This service is «IVS Live», which is located at the University of Bordeaux, France. The real-time data are conditioned and presented there on the central web server of «IVS Live». Some of this information is related to the observation (schedule name, scan name, source name, source coordinates) or to the current status of the radio telescopes (station name and code, current state, pointing azimuth and elevation, recording Mark 5 disk number). Others give information about the telescope environment (e.g., temperature, pressure, and humidity). From this central archive, the information can be

---

30 The reason for the use of Secure Copy (SCP) instead of the standard RPC communication is that also legacy systems without «e-RemoteCtrl» can support that service. They just have to send status files with the specified structure to the service, using the requested key file for the authentication. But legacy systems are not allowed to add the «[eRC]» tag to the status text.

downloaded as a one-page summary with the status information of all connected telescopes, operating one dedicated observation session.

■■ **Example 5.2**

The status file structure of the e-QuickStatus service.

```
<eQuickStatusInfo>                                                      1
    Service = IVS                                                       2
    Stationname = Wettzell                                              3
    StationIVSCode = Wz                                                 4
    Schedule = t2088wz                                                  5
    Status = [eRC] Recording # Selection of:                            6
                             # Field system startup                     7
                             # Field system terminated                  8
                             # Staring schedule                         9
                             # Schedule finished                        10
                             # Pointing                                 11
                             # Recording                                12
                             # Recording stopped                        13
                             # e-RemoteCtrl startup                     14
                             # Halt                                     15
                             # Continue                                 16
                             # plus error and warning codes in brackets 17
    Time = 2013.051.14:13:31                                            18
    NextTime = 2013.051.14:14:05                                        19
    Source = 3c418                                                      20
    Scan = t2088_wz_051-1413b                                           21
    Mark5VSN = BKG+0141                                                 22
    Mark5Volume = 2814.575                                              23
    Mark5Used = 70.4                                                    24
    RightAscension = 20h38m37.00s                                       25
    Declination = 51d19m12.3s                                           26
    Azimuth = 299.6721ř                                                 27
    Elevation = 48.5204ř                                                28
    CableDelay = 0.00036                                                29
    SystemTemperatureIFA = 34                                           30
    SystemTemperatureIFB = 68                                           31
    SystemTemperatureIFB = 27                                           32
    SystemTemperatureIFB = 0                                            33
    MeteorologyTemperature = -2.8ř                                      34
    MeteorologyHumidity = 77.6%                                         35
    MeteorologyPressure = 945.2hP                                       36
</eQuickStatusInfo>                                                     37
```

All of these remote techniques use communication channels over the worldwide Internet. Here they open the doors to attackers, who might be able to influence the control, and might destroy hardware in the worst case. For this reason, all remote techniques require the implementation of special security aspects.

5.5 Security for the Controlled Systems

Security in connection with remotely controlled systems has several aspects: it considers the internal security on the machines used, the security to access a machine in use, the security for a distributed system, the security in a tight LAN to which the distributed system is connected, and also the security in global WANs. Security can be defined in the following very rudimentary way:

Aspects

Security

> *DefinitionDEF5.2. of «security»* (see (Langmann 2010, *l.c. page 387*),
> (Tanenbaum 1995, *l.c. page 222*) and (Wagner 2013, *l.c. page 48*))
> Security is the protection from an inexpedient use of a system or its data by an
> unauthorized user. Security protects from willful or unintentional changes to that
> system or its data, and from inexpedient use, which is not specified or desired.
> Security takes care that a system works reliably, with integrity, and is available in
> the way which is expected by the developer and the user.

Security Factors

Security defines what is allowed on a computer and what is not allowed. Usually these conditions are directly related to user rights. Different users have different rights on a system to work with programs and data. To enable suitable regulation, it is necessary that the rights are clearly defined, that the user authenticates its (authentic) identity, and that the rights are checked to authorize users for dedicated activities. This defines the following security factors ((Gerloni et al. 2004, *l.c. page 15*) and (Langmann 2010, *l.c. page 388*)):

— *Availability:* protection of resources from attacks, which block them for authorized users
— *Authentication:* certain, authentic identification of users, computers, or services
— *Authorization and access control:* assignment of dedicated rights to users, computers, or services, which can be used to grant access to resources, like hardware or software functionalities
— *Privacy and integrity:* protection from unauthorized notice or change of data
— *Accountability and non-repudiation:* activities are uniquely accountable to real users, and these users cannot hide their activities

Access Control

The main focus here is suitable access control, as it requires authentication of individuals to offer privacy and integrity. It also reduces the risk of blocking attacks from system outsiders, as they are not granted accesses. Different solutions for access control are in use (Wagner 2013, *l.c. page 40 ff.*):

— *User-based access control:* This is the easiest way to grant system privileges. Administrators (superusers) assign access rights to system users. According to the defined rules, the rights for different accesses (e.g., file access) are checked. These rights can be permitted by the owner of the protected object or by the controlled system.
— *User identity (authorization)-based access control:* This mechanism is similar to the previous one with the only difference that the user is identified with unique patterns which only the real user can have. Such patterns are personal cards, dedicated keys, or even fingerprints. It focuses on a better identification of a real person. The rights of these people are then again permanently assigned.
— *Role-based access control:* While the previous control mechanisms assign fixed rights to a user, the role-based mechanism assigns different roles to each user. Therefore, the access granted is dependent on which current role a user has. This technique allows a more flexible, dynamic rights management, where the different roles have fixed privileges, but a user can change dynamically between these roles.
— *Task-based access control:* This mechanism is more flexible. It grants temporary rights for dedicated tasks or functionalities. If a user has to meet a specified task, he can require the access rights needed. But this access is only valid as long as this task is in process. After finishing, the rights are revoked. It requires an authority who can give and revoke such access rights if needed.

- *Concept-level access control:* Mostly a dedicated task requires a complete context in which it is operated. This is the background for concept-level access. Semantically coherent references, like access to all personal data, get dedicated policies to be accessed.
- *Owner-retained access control:* This mechanism assigns right elements to dedicated data for a digital rights management. All users of the data can assign additional rights but cannot change the already assigned rights of previous users.

These items are already a task within a closed system of a local computer to which different users have access. But this problem is more pressing, if remote access to such a computer is possible. The additional access points besides the local console of a local operator open additional possibilities for unauthorized use. The remote aspect has a further problem which must be considered: it requires communication over a network, in which other participants can spy on and interact with sent messages or have the possibility of manipulating them. This means that interconnectivity and remote control over networks open additional doors for attacks from hostile participants in the network. While it is very easy to control access to a local computer console, for example, with access restrictions to the room in which the computer is located, it is much more difficult to protect the same computer in a network. But as each system is nowadays more or less a set of controlling computers in a distributed system, which requires more or less huge data sets from other distributed systems or external partners, accessibility from outside is already necessary. Therefore, it is very important to offer a specific security strategy on the different layers of such interconnected systems, which begins with the local machine itself.

Security and Interconnectivity

5.5.1 Security for Internal Local Access to a Computer

This section focuses on the real computer system and not on access control to computer rooms or buildings. For local security on a machine, each multi-user operating system offers different mechanisms, which are connected with the security factors discussed. These mechanisms are mainly (Bovet and Cesati 2006, *l.c. page 9*):

Multi-user System Mechanisms

- An authentication mechanism to identify users
- An authorization mechanism to control access
- A protection mechanism against blocking applications
- A protection mechanism against malicious applications which interfere with or spy on other users
- An accounting mechanism to limit resources per user

In such a multi-user system, each user has a private space. This area, for example, files in a home directory tree, is only visible to its owner as long as he does not grant access rights to others. In the Linux system, each user firstly is authenticated with a username (log-in name) and a password. If the user enters a valid pair, the Linux system grants access. Privacy and integrity are only guaranteed if the password is secret and not easy for others to guess. Internally all users are then identified by a unique number, the «user ID». Each user is also a member of one or more user groups to selectively share data with other group members. The user groups are identified by the «group ID» (Bovet and Cesati 2006, *l.c. page 9 f.*). This information about the user, its groups, and the rest of the users on a computer is used to grant access rights to the file system. In Linux, the content of a file is separated from the metadata about the file. The metadata about a file are saved in the «inode» data structure of the file system, where the user and group assignments of the file and the access rights are also stored. Each file belongs to a user and a group. Each user

User ID and Group ID

can then define three types of access rights in the form of read, write, and execute access for himself, for the group, and all others. This means that each user can define nine different binary access flags (e.g., «rwxr-xr--» means read («r»), write («w»), and execute («x») access for the owner (the first three flags), read and write access for the group, and only read access for the rest of the users on the computer[31]) (Bovet and Cesati 2006, *l.c. page 15 f.*). This Linux mechanism implements a simple user-based access control.

Password Encryption

As described, the privacy and integrity for user data are only guaranteed if the password is secret and not easy to guess. But for the authentication of the user during the log-in process, the system must verify the password entered. This means that the password must be stored on the computer for this verification. To avoid direct readable passwords in plain text, the stored passwords are encrypted according to a dedicated ciphering mechanism. Operating systems usually use symmetric block cipher algorithms[32] to encrypt the passwords. In a symmetric cipher, the same secret key (of a size of 64, 128, 192, or 256 bits[33]) is used to encrypt and decrypt a message block of a fixed length (usually 64 or 128 bits) (see ◘ Fig. 5.17, (Gerloni et al. 2004, *l.c. page 84 ff.*)). This mechanism is ideal for passwords, as they do not need to be decrypted again for the verification of a user identity. It is enough to encrypt the password with information derived from the password itself. Then each entered password can be encrypted in that way and can be compared to a saved version of the password to identify a user (Tanenbaum 1995, *l.c. page 232 f.*).

Symmetric Block Cipher Algorithms

Typically used symmetric block cipher algorithms are (Gerloni et al. 2004, *l.c. page 84 ff.*):

– *DES:* The Data Encryption Standard, which has already been mentioned several times, e.g., in combination with RPC, was developed in the 1970s. It uses a 64 bit key (8 bits of the key are used for parity checks, so that the effective key is only 56 bits) to encrypt 64 bit long blocks to a 64 bit long encrypted version. Because of its short key, it is not really safe and usually only useful for password encryptions.

– *Triple-DES:* After attempts to encrypt messages twice with two different keys in sequence, which did not offer a higher security, Triple-DES uses three steps to increase the effective key length to 112 bits. First Triple-DES encrypts the block with the first key to decrypt it again with the second key, which is again a non-plain text. This text is then again encrypted with the first key to get the final encrypted version of the original plain text.

– *AES:* The Advanced Encryption Standard was developed at the end of the 1990s and has been standardized since 2002. AES offers three different key lengths and block sizes of 128, 192, or 256 bits. Because of its huge key sizes, it is currently a safe variant of a ciphering algorithm.

– *Others:* There are several other symmetric ciphers, which are also explicitly available under the SSH, such as «Blowfish», «Twofish», or «CAST-128» (Barrett et al. 2005, *l.c. page 88 f.*). Nevertheless, DES, Triple-DES, and AES are the most well-known block ciphers.

Superuser or Root

To enable the functionality of such a simple access control, a special user is required, which classifies the users and assigns the individual rights. It is also clear that not every user should have access to the file with the encrypted passwords, as it would be easy for a user to check combinations of usernames and passwords against the saved passwords to get access to private area of another user. This means that the operating system requires a «superuser» or «root» user. He is

---

31 The output of the content of a directory with the command «ls -al» prints such flag settings as well as the file mode which defines the file type.

32 A different mechanism is the stream cipher, which encrypts data bit- or byte-wise. An example is the ARCFOUR (RC4) cipher (Barrett et al. 2005, *l.c. page 88*).

33 The size of the secret key defines the security level reached for the encryption.

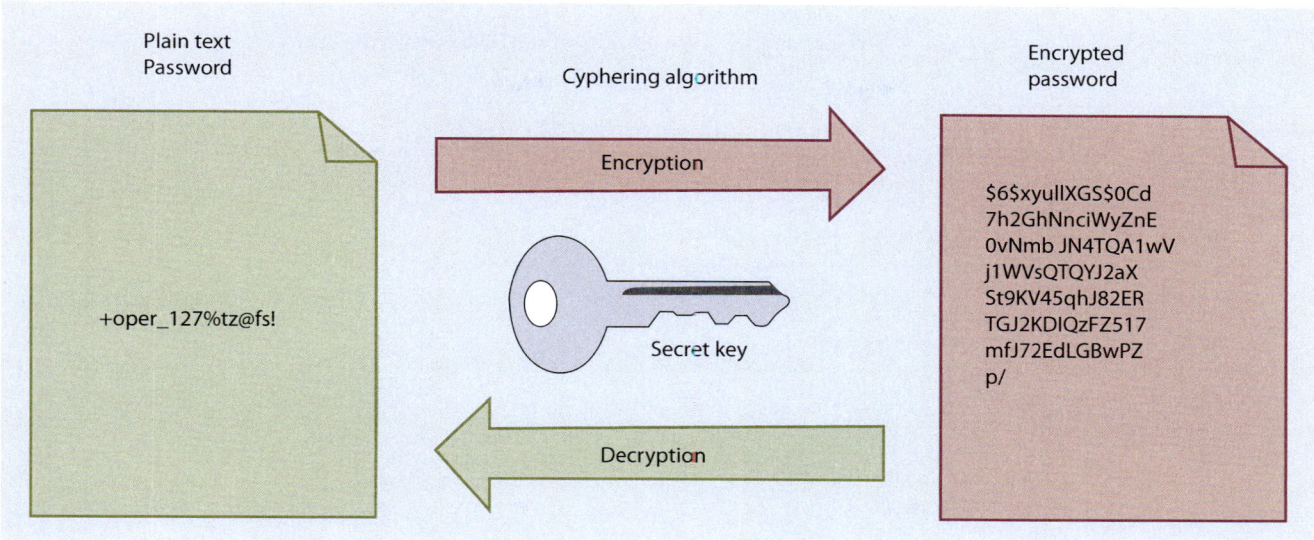

Fig. 5.17 The principle of a symmetric ciphering algorithm (Gerloni et al. 2004, *l.c. page 85*)

intended to be the system administrator, to manage user accounts, and to maintain the system. The superuser has almost all access rights. The root user can access every file and can manipulate every running program (Bovet and Cesati 2006, *l.c. page 10*). He is the only one who has access to the secret password file,[34] so that only a root process can verify user log-ins. Therefore, «root» is the most powerful user on a local system, and it is the most critical situation if an unauthorized user acquires «root» rights.

Nevertheless, in some cases it is necessary that a program which is started by a standard user controls access to different resources especially if the program is used to log in or to perform different user roles to solve a dedicated task. Then it is necessary for the program to run with superuser rights. To limit the risk of complete access by a user, the trick is to reduce the period of time when the program has complete root rights to a minimum. This is possible as each running program is assigned to a real user ID, and also to an effective user ID, which can be changed. Changing that effective ID requires the «`setuid`» or «`setgid`» permission for the executable, which must belong to the root user.[35] Then it is possible that a program which is not started as root can acquire additional rights to authenticate user credentials or to verify access rights for an additional access control (Kofler 2010, *l.c. page 346 f.*). Directly after the start of the program, it changes its effective user ID to the real user ID. Then all operations are processed with the reduced rights of the real user. Only for that short time when the program needs root rights to check access rights does it change its effective user ID to the root ID (which is 0). This is only possible if all file permissions are set correctly. After the secure operation, the program must immediately change the effective user ID back to the real one (Bovet and Cesati 2006, *l.c. page 811 f.*). Example 5.3 demonstrates this mechanism to authenticate standard users with a username and a password.

Access Control in Software

---

34 The password information is split into three files under Linux: «`/etc/password`» with the general user information, «`/etc/group`» with the information about the group assignments, and «`/etc/shadow`» with the encrypted, «shadowed» password (Kofler 2010, *l.c. page 504*).

35 Such a «`setuid`» or «`setgid`» permission flag can be set as superuser with the commands «`chown root <executable>`» and «`chmod 4555 <executable>`», where `<executable>` is the name of the program file. The flags for the file rights are then «`r-sr-xr-x`».

■ ■ **Example 5.3**

The changing of the effectiv euser ID to verify the user authentication.

```
ERoleManagerErrorType simple_role_manager::AuthenticateAtOS(        1
        const std::string &strUserName,                            2
        const CScatteredString &strPasswd) const                   3
{                                                                  4
    /// Get the password information from the password file         5
    /// to request the saved password from the shadow file         6
    // http://www.unix.com/man-page/Linux/3/getspnam_r/            7
    struct passwd SPasswd;                                         8
    struct passwd* tempSPasswdPtr = NULL;                          9
    char pwdbuffer[1024];                                          10
    if(getpwnam_r(strUserName.c_str(),                             11
                &SPasswd,                                          12
                pwdbuffer,                                         13
                sizeof(pwdbuffer),                                14
                &tempSPasswdPtr))                                 15
    {                                                             16
        return ROLE_MAN_ERROR_IN_GET_PWD_NAME; // received an error  17
    }                                                             18
                                                                  19
    /// Acquire superuser rights (User ID = 0),                   20
    /// by changing the effective User ID (EUID)                  21
    seteuid(0);                                                   22
                                                                  23
    /// Get the encrypted shadow password information from the shadow file  24
    struct spwd SShadow;                                          25
    struct spwd* pShadowTemp = NULL;                              26
    char cShadowBuffer[1024];                                     27
    if(getspnam_r(SPasswd.pw_name                                28
            , &SShadow                                           29
            , cShadowBuffer                                       30
            , sizeof(cShadowBuffer)                              31
            , &pShadowTemp))                                     32
    {                                                             33
        // lower rights                                          34
        seteuid(getuid());                                       35
        return ROLE_MAN_ERROR_IN_GET_PWD; // received an error     36
    }                                                             37
    endspent();                                                   38
                                                                  39
    /// Encrypt the entered password                             40
    struct crypt_data SCryptData;                                41
    SCryptData.initialized = 0;                                  42
    std::string strPassword(strPasswd.str().c_str());            43
    char *cpXPassword = crypt_r(strPassword.c_str(), SShadow.sp_pwdp, &SCryptData);  44
    strPassword.clear();                                         45
    if(cpXPassword == NULL && SShadow.sp_pwdp != '\0')           46
    {                                                             47
        // Acquire the usual, lower user rights again to return after error  48
        seteuid(getuid());                                       49
        return ROLE_MAN_ERROR_IN_PWD_CRYPT; // received an error   50
    }                                                             51
                                                                  52
    /// Acquire the usual, lower user rights again,              53
    /// by changing the effective User ID (EUID) back to         54
    /// the real User ID                                         55
    seteuid(getuid());                                           56
                                                                  57
    /// Compare the encrypted password, with the encrypted, saved password  58
    if (strcmp(cpXPassword, SShadow.sp_pwdp) != 0)               59
```

```
    {
        /// Wrong password
        return ROLE_MAN_ERROR_IN_PWD_MISMATCH;
    }

    /// Correct password
    return ROLE_MAN_OK;
}
```

60
61
62
63
64
65
66
67

With this technique, it is possible for the remote control software to use the local standard users, their authentication mechanism, and their rights on the controlled Linux system to enable different access rights for the control, while the program can be started by any user. This means that now user credentials in the operating system get an additional role in the remote control system. The role defines which remote actions are permitted on the general user rights of the operating system user. Roles appear in the system like normal users, but cannot necessarily log in to a shell. They are only used to define access groups or to set file rights explicitly. This role-based mechanism on the basis of a user-based mechanism is a very powerful technique, especially when external users want to log in to control the system (see ▶ Sect. 5.5.2).

Role-Based Security with Standard Users

Before the further requirements for remote control are described, some security risks on the local machine should be mentioned. The most well-known technique is the brute-force attack. It is an attempt of an unauthorized user to guess combinations of usernames and passwords in order to be authenticated as a valid user. This technique can be used on a local log-in shell as well as for remote accesses. But even if it is possible to guess passwords (e.g., with some additional knowledge about the user), most brute-force attacks are automatic (McClure et al. 2006, *l.c. page 293 f.*). Here a robotic process guesses a password and runs a trial. If it fails, the process changes the password slightly and runs another test trial. This method can be continued until a suitable password is found. The bots do not just test passwords with this method, but perform simultaneously a «Denial-of-Service» attack (see ▶ Sect. 5.5.2), because the continuous requests reduce the performance of the system under attack. Strong passwords create a remedy. Additionally, it helps to change passwords regularly. Operating systems such as Linux use «salted» passwords. This means that the real password is concatenated with an n-bit random number, which was generated during the creation of the password. Brute-force attacks must then test 2^n possibilities to find a single password (Tanenbaum 1995, *l.c. page 233 f.*). This stretches the time interval until a password is cracked.

Security Risks: Brute-Force Attacks

Other dangers arise from data-based attacks. They deal with data inputs of unexpected data, which provoke undesirable behavior. Two categories are generally possible: buffer overflows and input validations. Buffer overflows happen if an attacker tries to write more data into a buffer of a fixed length than is allowed in the memory. Usually, this leads to a memory protection violation. The result is a software crash and therefore a reduction of availability. But it is much more dangerous if the attacker is able to write working code («egg») into the overflowed area and if he can redirect the returning program address to that code. If the program uses root privileges, or if it is in the authorization phase with higher privileges, it might grant access for an execution of any dangerous program. Format string attacks are similar. They use programming errors in format string definitions, as used for «printf». The ruse is to hand over tricky text strings, which allow the execution of a dangerous program. Availability attacks are also possible with input validation attacks, where wrong,

Security Risks: Data-Based Attacks

unnecessary, missing, or type-mismatching inputs are not managed correctly. Integer overflow or sign attacks are also very similar. The trick is to bypass checks by using overflowing or wrongly signed numbers. The result might be a reading outside the allowed limits of an array and accessing private data[36] (McClure et al. 2006, *l.c. page 296 f.*). As these attacks are usually based on programming errors, regular static analysis, as offered in the continuous integration method (see ► Sect. 2.8), can improve the situation significantly. But there is no such thing as total security.

As all of the security leaks described are also possible with remote access, it is quite important to build secure local systems and methods to reduce risks from remote sources. Nevertheless, the possibility of security risks increases when users can log in remotely.

5.5.2 Security for External Accesses to a Computer

Security Problem

It might be enough for local users on a multi-user system to restrict their accesses with the described user credential methods. On a computer which controls a geodetic measurement system, it is mostly clear who was and is responsible for the system during the time interval of an operator shift. Even if local operators can also have a destructive influence on the local system, remote access opens additional doors which generate higher risks for the system. The main risk is from attackers. But also poorly trained personnel who do not know the local system and techniques in detail might unintentionally behave in ways which produce security risks for the controlled system. Here it seems likely that external operators have less knowledge about the system than local operators, so that the risk to behave unintentionally wrong is greater. Since this should be avoided, it is necessary to increase the restrictions, so that remote operators are guided only to those tasks they are really allowed to do.

Access Control Lists

The increase in restrictions means the refinement of the right management. Linux provides an option for an Access Control List (ACL) to do that. ACLs make it possible to define any number of additional access rights to file and directories, so that the access can explicitly be permitted or denied for a dedicated user in addition to the standard Linux rights. These ACLs must be activated in the file system and have one big disadvantage: if they are used in the wrong way, they open additional security leaks (Kofler 2010, *l.c. page 349 f.*).

«rbash» or Keep the User in His «Box»

A simpler way is to completely restrict the activities of a remote user, so that he cannot see third-party directories, run programs, or change anything on the machine. As remote control users just want to do their job, which is offered by the remote control software on the controlled computer, their rights can be completely reduced. It is an approach designed to keep the remote user in his «box», defined by the home directory. This can be implemented with «rbash», which is a remote version of the «bash» shell.[37] A shell is the command line user interface between the Linux system and the user. It is processed in each console and log-in window to manage commands from the user (Kofler 2010, *l.c. page 450 f.*). Which standard

---

36 A very well-known example of reading private data with this method is the «Heartbleed» bug in the Open SSL implementation, which was a security leak in several SSH implementations.

37 Originally «bash» comes from «bourne again shell», as it was a follow-on of the original Bourne shell on Unix systems, which was besides Korn and C shell the main commando input terminal.

shell is used by a user is defined in the «/etc/password» file (Kofler 2010, *l.c. page 504*). Changing the entry there changes the shell, which is seen first after a user logs in. Then a user can run programs offered by the Linux system, which are accessible via the directory definition in the «$PATH» variable (Kofler 2010, *l.c. page 453*). Users can change to and read from external directories of other users according to the previously described Linux rights. The «rbash» restricts these possibilities. The following activities are not allowed in the «rbash» (see manual page about «rbash» under Linux[38]):

- *«changing directories with cd»*
- *«setting or unsetting the values of SHELL, PATH, ENV, or BASH_ENV»*
- *«specifying command names containing /»*[39]
- *«specifying a file name containing a / as an argument to the . builtin command»*
- *«specifying a filename containing a slash as an argument to the -p option to the hash builtin command»*
- *«importing function definitions from the shell environment at startup»*
- *«parsing the value of SHELLOPTS from the shell environment at startup»*
- *«redirecting output using the >, >|, <>, >&, &>, and >> redirection operators»*
- *«using the exec builtin command to replace the shell with another command»*
- *«adding or deleting builtin commands with the -f and -d options to the enable builtin command»*
- *«using the enable builtin command to enable disabled shell builtins»*
- *«specifying the -p option to the command builtin command»*
- *«turning off restricted mode with set +r or set +o restricted»*

Using the «$PATH» variable, accessible programs can be defined. The variable definition can be done in the «.bashrc» file in the home directory of the user. If a user is only allowed to get access to a subset of programs offered in the standard program directories, like «/usr/bin» and «/usr/local/bin», it is possible to create other program directories which contain selected copies of or links to the standard programs. If the «$PATH» variable is set to these new, individually selected program directories, a user can only run these executables without a call path. Hosting the authorized copies of programs in the home directory of the user, this restriction can be used in combination with «rbash» to reduce executable programs, as the remote user cannot leave its home directory. To always ensure it is possible to use the latest version of the programs, a symbolic link to the programs instead of a real copy is better, while keeping the same protection. In this setup, the locally hosted or linked programs and the «.bashrc» file are owned by the superuser. All other users have only read rights. So a remote user is not able to apply any changes to the files.[40] The scripts in Example 5.4 and in Example 5.5 demonstrate how to create a restricted «rbash» user and how to delete him. The script offers three security levels for the start of programs: standard Linux behavior, a reduced set of programs only for remote operators, and no permission for the activation of programs at all.

Restrict Allowed Programs

---

38 with «man rbash».
39 «/» is the path delimiter.
40 It is only important to be careful with programs, like the editor «vi», which makes it possible to return to a local shell within the program run, as it might be possible to open a free «bash».

■ ■ Example 5.4

Creating a restricted user, which uses the remote «bash» with three security levels for the start of programs: standard Linux behavior, reduced set of programs only for remote operators and no activation possibility at all.

```bash
#!/bin/bash

# ===============================================
# Create restricted SSH user, using rbash
# ===============================================

# Check command line parameter
if [ $# -ne 1 ]; then
    echo " Usage: addercduser <username>"
    echo "        Add a new e-RemoteCtrl user to system"
    echo " where"
    echo "      <username>: Name of the new user"
    exit 1
fi
#-------------------------------------------------

# Print header
echo "#######################################"
echo "# Add a new e-RemoteCtrl user to system #"
echo "#######################################"
#-------------------------------------------------

# Check if user is in superuser mode
echo "0) Check superuser rights"
SUCHECK=`whoami`
if [ $SUCHECK != "root" ]; then
    echo " => [ERROR] You have no superuser rights!"
    exit 1
else
    echo " => [OK]"
fi
#-------------------------------------------------

SecurityLevel=3
ShellType="bash"
while [ $SecurityLevel -eq 3 ]; do
    echo "1) Enter security level"
    echo "[0] = Standard Linux/Unix user (bash)"
    echo "[1] = Partly restricted remote-user (rbash;"
    echo "      SSH-access with basic programs)"
    echo "[2] = Completely restricted remote-user (rbash;"
    echo "      SSH-access without programs)"
    echo -n "==> "
    read SecurityLevel
    if [ $SecurityLevel -eq 0 ]; then
        ShellType="bash"
    else
        if [ $SecurityLevel -eq 1 ]; then
            ShellType="rbash"
        else
            if [ $SecurityLevel -eq 2 ]; then
                ShellType="rbash"
            else
                SecurityLevel=3
            fi
        fi
    fi
done
#-------------------------------------------------
```

```
# Prepare /bin/rbash as shell                                                 60
BashPath="/bin/"                                                              61
echo "2) Check rbash on system and create link if necessary"                 62
if [ ! -f /bin/bash ]; then                                                  63
    if [ ! -f /usr/bin/bash ]; then                                          64
        echo " => [ERROR] bash shell is not installed!"                      65
        exit 1                                                               66
    else                                                                     67
        BashPath="/usr/bin/"                                                 68
        if [ ! -f /usr/bin/rbash ]; then                                     69
            echo " => [OK] Create link /usr/bin/rbash and register shell"    70
            ln -s /usr/bin/bash /usr/bin/rbash                               71
            echo "/usr/bin/rbash" >> /etc/shells                            72
        else                                                                 73
            echo " => [OK] rbash exists"                                     74
        fi                                                                   75
    fi                                                                       76
else                                                                         77
    if [ ! -f /bin/rbash ]; then                                            78
        echo " => [OK] Create link /bin/rbash and register shell"           79
        ln -s /bin/bash /bin/rbash                                          80
        echo "/bin/rbash" >> /etc/shells                                   81
    else                                                                    82
        echo " => [OK] rbash exists"                                        83
    fi                                                                      84
fi                                                                          85
BashPath=$BashPath$ShellType                                                86
#———————————————————————————————————————                                  87
                                                                           88
                                                                           89
# Check group                                                              90
echo "3) Check ercd user groups and create it, if necessary"              91
GROUPCHECK=`more /etc/group | grep ercd`                                    92
if [ ${#GROUPCHECK} -eq 0 ]; then                                          93
    addgroup ercd                                                          94
    GROUPCHECK=`more /etc/group | grep ercd`                               95
    if [ ${#GROUPCHECK} -eq 0 ]; then                                      96
        echo " => [ERROR] Can't create group ercd!"                       97
        exit 1                                                            98
    else                                                                  99
        echo " => [OK] Group ercd created"                               100
    fi                                                                   101
else                                                                     102
    echo " => [OK] Group ercd exists"                                    103
fi                                                                       104
#———————————————————————————————————————                                105
                                                                         106
# Create new user                                                        107
echo "4) Create new user"                                                108
if [ ! -d /home/$1 ]; then                                               109
    adduser --home /home/$1 --shell $BashPath --ingroup ercd $1          110
else                                                                     111
    echo " => [ERROR] User already exists!"                              112
    exit 1                                                               113
fi                                                                       114
if [ ! -d /home/$1 ]; then                                               115
    echo " => [ERROR] User was not created!"                             116
    exit 1                                                               117
else                                                                     118
    echo " => [OK] User created"                                         119
fi                                                                       120
#———————————————————————————————————————                                121
                                                                         122
# Change access rights for .bashrc                                       123
```

```
echo  "5)  Change  access  rights  for  .bashrc"                                    124
if [ ! −f /home/$1/.bashrc ]; then                                                  125
    echo  " => [ERROR] .bashrc does not exist!"                                      126
    exit 1                                                                           127
fi                                                                                   128
chown bin:bin /home/$1/.bashrc                                                       129
chmod −w /home/$1/.bashrc                                                            130
echo  " => [OK]"                                                                     131
#————————————————————————————————————————————                                       132
                                                                                     133
                                                                                     134
# Define allowed programs for restricted user in .bashrc                            135
mkdir /home/$1/bin                                                                   136
if [ $SecurityLevel −eq 1 ]; then                                                    137
echo  "6) Define allowed programs for restricted user in .bashrc"                   138
cp /bin/cat /home/$1/bin/cat                                                         139
cp /bin/cp /home/$1/bin/cp                                                           140
cp /bin/date /home/$1/bin/date                                                       141
cp /bin/df /home/$1/bin/df                                                           142
cp /bin/echo /home/$1/bin/echo                                                       143
cp /bin/grep /home/$1/bin/grep                                                       144
cp /bin/gunzip /home/$1/bin/gunzip                                                   145
cp /bin/gzip /home/$1/bin/gzip                                                       146
cp /bin/kill /home/$1/bin/kill                                                       147
cp /bin/ls /home/$1/bin/ls                                                           148
cp /bin/mkdir /home/$1/bin/mkdir                                                     149
cp /bin/more /home/$1/bin/more                                                       150
cp /bin/ping /home/$1/bin/ping                                                       151
cp /bin/ps /home/$1/bin/ps                                                           152
cp /bin/pwd /home/$1/bin/pwd                                                         153
cp /bin/rm /home/$1/bin/rm                                                           154
cp /bin/rmdir /home/$1/bin/rmdir                                                     155
cp /bin/tar /home/$1/bin/tar                                                         156
cp /bin/touch /home/$1/bin/touch                                                     157
cp /usr/bin/vi /home/$1/bin/vi                                                       158
chown −R $1:ercd /home/$1/bin                                                        159
chmod −R 110 /home/$1/bin/*                                                          160
echo  " => [OK]"                                                                     161
else                                                                                 162
    if [ $SecurityLevel −eq 3 ]; then                                                163
        echo  "6) Define no programs for restricted user in .bashrc"                164
    else                                                                             165
        echo  "6) Define allowed programs according to standard Linux user rights"  166
    fi                                                                               167
fi                                                                                   168
#————————————————————————————————————————————                                       169
                                                                                     170
# Set program PATH for restricted user in .bashrc                                   171
echo  "7) Set program PATH for restricted user in .bashrc"                          172
cp /home/$1/.bashrc /home/$1/.bashrc_open                                            173
echo  "" >> /home/$1/.bashrc                                                         174
echo  "" >> /home/$1/.bashrc                                                         175
echo  "export PATH=/home/$1/bin" >> /home/$1/.bashrc                                 176
cp /home/$1/.bashrc /home/$1/.bashrc_restricted                                      177
rm /home/$1/.bashrc                                                                  178
if [ $SecurityLevel −eq 0 ]; then                                                    179
    ln −s /home/$1/.bashrc_open /home/$1/.bashrc                                     180
else                                                                                 181
    ln −s /home/$1/.bashrc_restricted /home/$1/.bashrc                              182
fi                                                                                   183
echo  " => [OK]"                                                                     184
#————————————————————————————————————————————                                       185
                                                                                     186
exit 0
```

▪▪ Example 5.5

Deleting a restricted user, which uses the remote «bash».

```bash
#!/bin/bash

# ====================================================
# Create restricted SSH user, using rbash
# ====================================================

# Check command line parameter
if [ $# -ne 1 ]; then
    echo " Usage: delercduser <username>"
    echo "           Delete an e-RemoteCtrl user from system"
    echo " where"
    echo "       <username>: Name of the new user"
    exit 1
fi
#───────────────────────────────────────────────

# Print header
echo "########################################"
echo "# Delete an e-RemoteCtrl user from system #"
echo "########################################"
#───────────────────────────────────────────────

# Check if user is in superuser mode
echo "0) Check superuser rights"
SUCHECK=`whoami`
if [ $SUCHECK != "root" ]; then
    echo " => [ERROR] You have no superuser rights!"
    exit 1
else
    echo " => [OK]"
fi
#───────────────────────────────────────────────

# Delete existing user
echo "3) Delete existing user"
if [ -d /home/$1 ]; then
    userdel -force -remove $1
else
    echo " => [ERROR] User doesn't exist!"
    exit 1
fi
if [ -d /home/$1 ]; then
    echo " => [ERROR] User couldn't be deleted!"
else
    echo " => [OK]"
fi
exit 0
```

The mechanism described tries to keep the effects of user inputs local. Another important security issue is to reduce access points for remote users. These access points are communication ports (compare ▶ Sect. 3.3.1 about communication sockets). Each server which offers a service to the network opens a socket access point on a specific communication port and listens for requests there (see ▶ Sect. 3.3.1). These open ports are potential doors for attacks. Denial-of-Service attacks are very popular. An attacker tries to send such a large amount of requests to an open port that the service is not reachable anymore by regular clients. This means

Reduce Access Points (Ports)

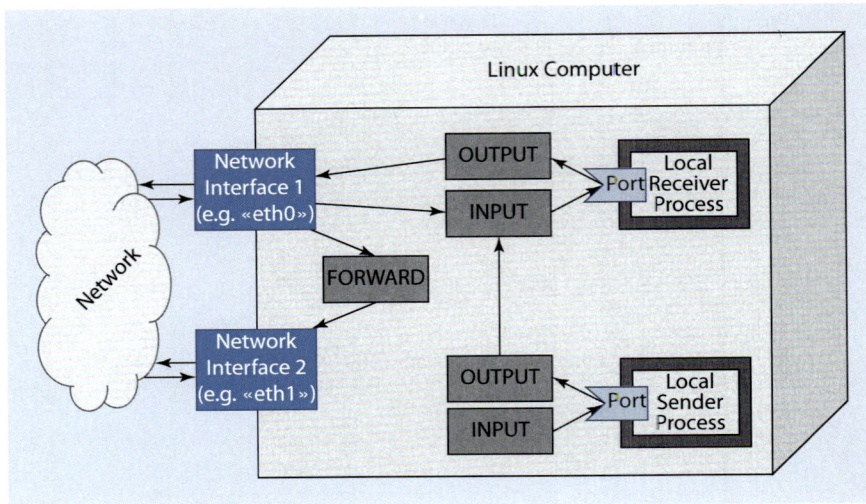

□ Fig. 5.18 The packet flow through the hooks of a filter table

that the flooding stops the availability of a service (McClure et al. 2006, *l.c. page 649*) and also other processes of the computer, because the CPU load is too high. Therefore, it is advisable to block or close unneeded services and their ports.

Filter Tables (Iptables)

This can be implemented with filter tables (under Linux with the «iptables»). Filter tables contain a set of rules describing which action should be done with communication packages from and to dedicated ports on defined machines with specific network interfaces. To allow such rule definitions, filter tables define three hooks (entry points) to set up the processing chains: FORWARD, INPUT, and OUTPUT (see □ Fig. 5.18). The FORWARD chain contains rules for the forwarding of packages from one physical network interface to another. This can be used if the computer works as a router or firewall[41] (see ▶ Sect. 5.5.3). For the protection of a remotely accessible computer, mainly the INPUT and OUTPUT chains are relevant. The INPUT chain uses rules for all packages, entering the computer via one physical network interface and a specific service port. The OUTPUT chain is for all packages sent by a local service to a dedicated physical network interface. The chains themselves are just a list of rules, which are checked sequentially and describe how packages from the network interface to a specific port should be treated. They can be dropped («DROP»), accepted («ACCEPT»), or logged to a log file («LOG»), which is defined in the target parameter of the rule. The setup of the rules is made with the command «iptables». In Example 5.6 is a simple script which deletes previous filter settings, defines general behavior for the local loop-back interface, and defines general policies to drop all packages[42] besides those which are explicitly accepted by the following rules (here for the «ping» command, which is only allowed to be used from the computer to the network and not the other way around). The rules are then checked in the order they were appended.[43] If a matching rule is found, the defined action (named as «target») is processed. The final definition in the example is the rule setting for output loggings, which specifies that all communication attempts to non-permitted ports are logged (see (Kofler 2010, *l.c. page 893 f.*) and (Purdy 2005, *l.c. page 5 ff.*)).[44]

---

41 «iptables» in firewalls are used as a form of ACLs.

42 Two policy approaches are conceivable: everything is allowed what is not prohibited, or everything is prohibited what is not explicitly permitted. The used policy depends on the network owner. But companies usually use the second policy type, so that explicit permissions must be granted for each service. This increases the controlled security (see (Gerloni et al. 2004, *l.c. page 177 f.*).

43 Active rules of filter tables can be printed with the command «iptables –list».

44 The example uses the long variant of command parameters, which intuitively describe its meaning. As the command «iptables» is very complex, only a short overview can be given here.

▪▪ Example 5.6

Filter tables can be manipulated with the command «iptables» in a shell script.

```bash
#!/bin/bash
#————————

LOOPBACK_INTERFACE="lo"
# Computer with only one physical network interface 'eth0'
INTERFACE="eth0"

# ******************************************************************
# ** Flush the iptable                                          **
# ******************************************************************
echo "Flush the iptable"
iptables —flush

# ******************************************************************
# ** Allow everything on the loopback interface                **
# ******************************************************************
echo "Allow everything on the loopback interface"
iptables —table filter —append INPUT    —in-interface  $LOOPBACK_INTERFACE —jump ACCEPT
iptables —table filter —append OUTPUT   —out-interface $LOOPBACK_INTERFACE —jump ACCEPT
iptables —table filter —append FORWARD —out-interface $LOOPBACK_INTERFACE —jump ACCEPT

# ******************************************************************
# ** Define the general policies                               **
# ******************************************************************
echo "Define the general policies"
iptables —policy INPUT DROP
iptables —policy OUTPUT DROP
iptables —policy FORWARD ACCEPT

# ******************************************************************
# ** Define the lists of rules                                 **
# ******************************************************************
echo "Define the single lists of rules"

# —— ICMP (Ping) service ——————————————————————————
iptables —append INPUT \            # Append to INPUT hook
        —in-interface $INTERFACE \  # Physical network interface for input
        —protocol icmp \            # Protocol
        —jump DROP                  # Target (action)
iptables —append OUTPUT \           # Append to OUTPUT hook
        —out-interface $INTERFACE \ # Physical network interface for input
        —protocol icmp \            # Protocol
        —jump ACCEPT                # Target (action)

# ...

# ******************************************************************
# ** Define the logging rules                                  **
# ******************************************************************
echo "Define the logging rules"
iptables —append INPUT  —in-interface $INTERFACE  —jump LOG —log-prefix "INeth0"
iptables —append OUTPUT —out-interface $INTERFACE —jump LOG —log-prefix "OUTeth0"
```

All ports for incoming requests can be closed as long as they are not relevant for basic services (e.g., naming or timing services) and the remote shell. For remote access to the local system, SSH is the appropriate tool. The management of outgoing messages more or less depends on the local security philosophy. Secure systems will also define rules for outgoing ports and will not generally open all communication ports for internal programs on the computer, because then already installed and active viruses can load further dangerous programs.

Secure Shell (SSH) as Remote Shell

Encryption with an SSH

A big problem with the standard remote access methods is that they do not encrypt data transfers. This means that all data can be read as clear text out of the messages sent during the data transfer. As all machines in a network have access to the physical medium and can even read messages not intended for them, this is a general problem. But ▶ Sect. 2.7.2 already described a solution for secure communications with Subversion servers: the use of SSH. SSH offers a way to encrypt the messages of the authentication and even the whole communication. The encryption itself is implemented as public-key encryption (Brookshear 2012, *l.c. page 177 ff.*). This is an asymmetric ciphering method.

Asymmetric Ciphers

For an asymmetric cipher, two keys are involved. The sender of a message holds a public key from the receiver. This key is used to encrypt a message, but cannot be used again for the decryption. The decryption can only be done with a private key kept secret by the receiver. With this key, the encrypted message can only be converted again into a plain one by the receiver (see ◻ Fig. 5.19). This is possible as the public key is calculated with a mathematical one-way function from the secret key (usually of lengths of 1024, 1536, 3072, or 4096 bits). This means that in reverse direction, the private key cannot be calculated from the public key, or it is very time-consuming. The multiplication of prime numbers offers such a one-way functionality, as the factorization of the result is incomparably more complex. As it is not necessary to keep the public key secret, it can also be communicated over insecure communication channels, for example, at the beginning of a communication. This means that everybody with the public key can send encrypted messages to a receiver[45] holding the private key, while nobody else in the communication channel can decrypt them without that private key.[46] If the keys can be exchanged for the cipher process, so that the public key can be used to decrypt encrypted messages which were generated with the private key, the same technique can also be used to sign messages with an identifying fingerprint, as only the one with the private key can be the author of a signed message (Gerloni et al. 2004, *l.c. page 87 f.*).

Asymmetric Cipher Algorithms

Asymmetric ciphers usually use the fact that the calculation of a discrete logarithm is complex and time-consuming. This is the case with integer factorization of big prime numbers (prime factorization). Very big numbers cannot be efficiently decomposed into their products. This is the key technology for almost unbreakable keys. In combination with strong random bit generators to find random key numbers, hash methods (see also ▶ Sect. 2.7.2), and related prime number generators for the creation of suitable prime candidates, the asymmetric cipher algorithms are quite powerful. They can be divided into finite field cryptography and elliptic curve cryptography, where the difference is only the type of math used ((National Institute of Standards and Technology (NIST) n.d.) *l.c. page 46*). In standard environments, the following algorithms are used (Gerloni et al. 2004, *l.c. page 88 ff.*):

- *Diffie-Hellmann:* The Diffie-Hellmann agorithm is not really a ciphering algorithm. It is more often used to negotiate a secret key over a non-secret, untrustworthy channel. It is based on a prime number p and a «primitive root»[47] r of p. These two data can be sent over an untrustworthy network to a receiver. Then both sender and receiver select an individual, secret key (k_i, where $i = s$ is the key of the sender and $i = r$ is the key of the receiver) and calculate locally r^{k_i} modulo p, whose result is sent to the communication partner, respectively. With this result, both can calculate a common key $r^{k_s k_r}$ modulo p, which is then used for the encryption of the following messages.

---

45 Therefore, it is also known as public-key algorithm.
46 The public key of the sender can be automatically transferred to the receiver during the establishment of the communication to encrypt the responding answer messages.
47 A «primitive root», also called a «primitive generator» r for a dedicated prime number p, is given if the result of r^n modulo p goes through all numbers from 1 to $(p-1)$ for the entered numbers $n = 1 \ldots (p-1)$. For example: 3 is a primitive root of the prime number 5 (Morgan 2003).

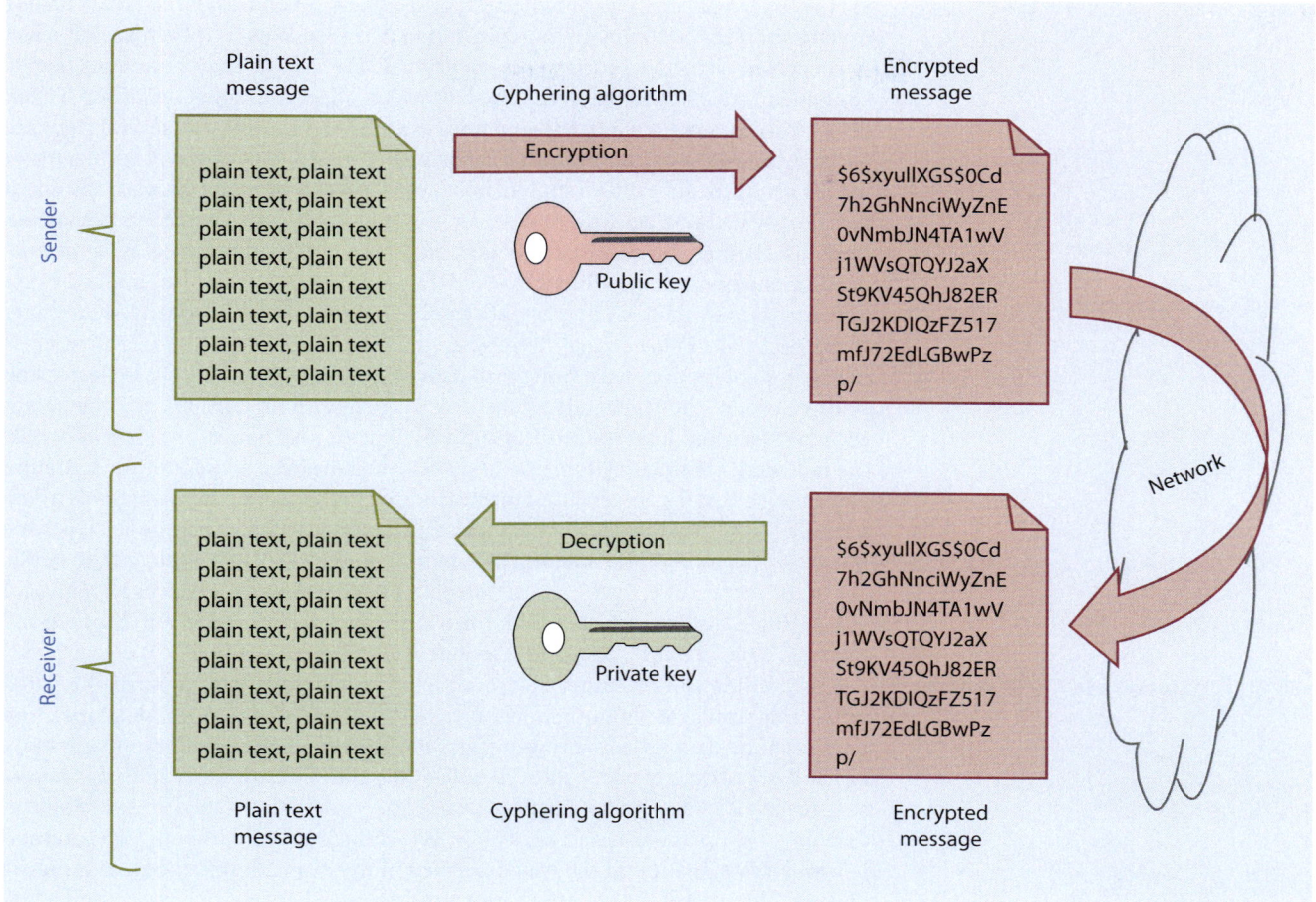

Fig. 5.19 The principle of an asymmetric ciphering algorithm

- *RSA:* RSA was named after its inventors Ron Rivest, Adi Shamir, and Leonard Adleman. It uses two significantly different prime numbers and their product p. In combination with two exponents,[48] it is possible to use the product p and one exponent e_1 as a public key to encrypt a message with $m_{plain}^{e_1}$ modulo p and the other exponent e_2 and the product p to decrypt the encrypted message with $m_{encrypt}^{e_2}$ modulo p .
- *Digital Signature Algorithm (DSA):* DSA uses three domain parameters q, g, and p, which are large integer prime numbers. It is possible to use additional random bits r to calculate the public and private keys. Additionally, the result of the random generator can be tested if the generated bit sequence is a potential candidate. It is then possible to create the private key k_p with $(c \text{ modulo } (q-1)) + 1$ and a public key with g^{k_p} modulo p . The public key is then used for the encryption, which can be reverted with the private key ((National Institute of Standards and Technology (NIST) n.d.) *l.c. page 46 ff.*).
- *ElGamal:* ElGamal also uses the factorization problem for its encryption. It uses a big prime number p and two smaller random numbers r_1 and r_2 to calculate $k = r_1^{r_2}$ modulo p . The public key is then k, r_1, and p and is used together with an additional random number to encrypt messages into two ciphered results, which are both sent to the receiver. The receiver can use r_2 to decrypt the message again. Because of the additional random number, the key is parametrized again and is therefore more secure against spies.

---

48   They must fulfill additional, dedicated rules.

Hybrid Cipher Algorithm in the SSH

SSH uses a hybrid ciphering method. Hybrid ciphers generate a session key, which is used to encrypt the complete communication of one session.[49] Only this session key is encrypted with the primary-key algorithm. The rest of the communication is encrypted with the temporary session key, which offers additional security (Gerloni et al. 2004, *l.c. page 92 f.*). In the concrete case of SSH, it means that several steps are taken during the communication establishment. First the SSH server (or daemon) «sshd» sends its public key («public host key») to the requesting client. If the client has already had contact with that server, the key is already registered in the known hosts list. If not, the client can request additional authentication to avoid attacks from impostors (such as from a man-in-the-middle attack, where an impostor gets between the communication partners and has control over the exchanged messages (McClure et al. 2006, *l.c. page 499*)). Then the server sends supported, favored algorithms for encryption, hash functions, data compression,[50] and so on, in decreasing priority order. The client selects the accepted algorithms and informs the server about its decision. It then starts the Diffie-Hellmann negotiation of a session key for the following communication. The first encrypted message is used for the signature principle, so that the server authenticates himself to the client. The client sends a data package («challenge cookie») which the server encrypts with its private key («private host key»). The encrypted message is sent back to the client, who can use the public key of the server to verify the server's identity by decrypting the challenge cookie and comparing it to the one sent. Then the authentication of the client is processed ((Riedel 2004, *l.c. page 12 ff.*) and (Gerloni et al. 2004, *l.c. page 135 ff.*)).

SSH Authentication and Key Signature

The authentication under SSH is simply implemented with a username and a password (similar to the authentication as a standard user on a local machine). The user account data are already transferred within the encrypted communication, but still have some security risks. To allow this authentication method, it is necessary to set «PasswordAuthentication yes» in the «OpenSSH» configuration file «/etc/ssh/sshd_config». But a much better authentication method is with a key signature in the way described in the explanations about asymmetric ciphers. The user of the client machine must create a key pair of a public and a private key. The public key (usually with the extension «.pub») must then be copied to the server machine into the file «.ssh/authorized_keys» in the home directory of the user, who wants to log in from remote. The location of the public key for the signature of a client is «.ssh/authorized_keys». If a user tries to authenticate with the public-key algorithm, he must show that he is the owner of one of these private keys. Therefore, the client encrypts a message from the server with the private key and sends it back to the server. If the server is able to decrypt it with one public key from the list, the identity of the client is accredited, and the client is allowed to establish a connection to the server with the previously described mechanism. To enable this client key signature at the server, it is necessary to set «PubkeyAuthentication yes» and maybe «RSAAuthentication yes» in the «OpenSSH» configuration file «/etc/ssh/sshd_config». This is a much more secure authentication as long as the private key is secret. It is possible to encrypt the stored private key with a symmetric cipher algorithm (usually DES or Triple-DES), where the symmetric key is the «passphrase». Each time the private key is used, the user has to enter the passphrase to decrypt the key for its use ((Riedel 2004, *l.c. page 101 ff.*) and (Gerloni et al. 2004, *l.c. page 138 ff.*)).[51]

Key Generation

SSH allows DSA and RSA key pairs for the above authentication. DSA pairs can be created using the command «ssh-keygen -t dsa», which creates a key with a

---

49 A session is a communication process which starts when one of the communication partners initiates the communication and ends if one of them closes the communication channels. After a session is closed, a new one must be initiated for a following communication.

50 SSH uses «zlib» for its data compression (Barrett et al. 2005, *l.c. page 91*).

51 There are also other authentication methods available under SSH , for example, host-based or keyboard-interactive, but these are not so relevant here.

size of 1024 bit. An RSA pair can be created with the command «ssh-keygen -t rsa -b 2048», so that the result is a key with the size of 2048 bits. As the key size of RSA can be increased, it is more secure. If the passphrase is left empty, no method is used to encrypt the local private key. Nevertheless, it is possible to combine the key authentication methods with a username and a password. In summary, the following authentication combinations are mainly relevant for the identification of a client:

- Only with a username and a password
- Only with the private key file and without any additional security information
- With the private key file and a passphrase for the private key
- With the private key file, a username, and a password
- With the private key file, the passphrase for the private key, a username, and a password

The SSH techniques described offer a very secure and powerful possibility for remote accesses. Nevertheless, Denial-of-Service attacks cannot be prevented, as an attacker can always fire so many messages to a dedicated port that the service is not reachable anymore. Traffic analysis also makes it possible to collect some background information about user behaviors (Riedel 2004, *l.c. page 6*). Additionally, version 1 of SSH suffers from a buffer overrun error in the CRC calculation, so that an attacker can treat the server machine to run destructive, third-party code. For this reason, version 1 should not be used anymore[52] (Gerloni et al. 2004, *l.c. page 136*).

SSH Limitations

But an advantage of SSH can be utilized: the creation of communication tunnels over an encrypted SSH connection. This means that the complete communication between a client and a server port is forwarded over the secure, encrypted SSH communication. This is only possible if the receiving port is fixed (e.g., not dynamically defined by a portmapper) and not part of the well-known, reserved ports of the operating system. All other ports can be forwarded easily to a remote machine in each user account via the SSH tunnel (see ◘ Fig. 5.20). The generated code from «idl2rpc.pl» (see ▶ Sect. 3.3.3 starting) also offers a calling parameter to use fixed ports instead of dynamic ones, so that the remote procedure streams can be forwarded from a client to a server.

Tunneling with SSH

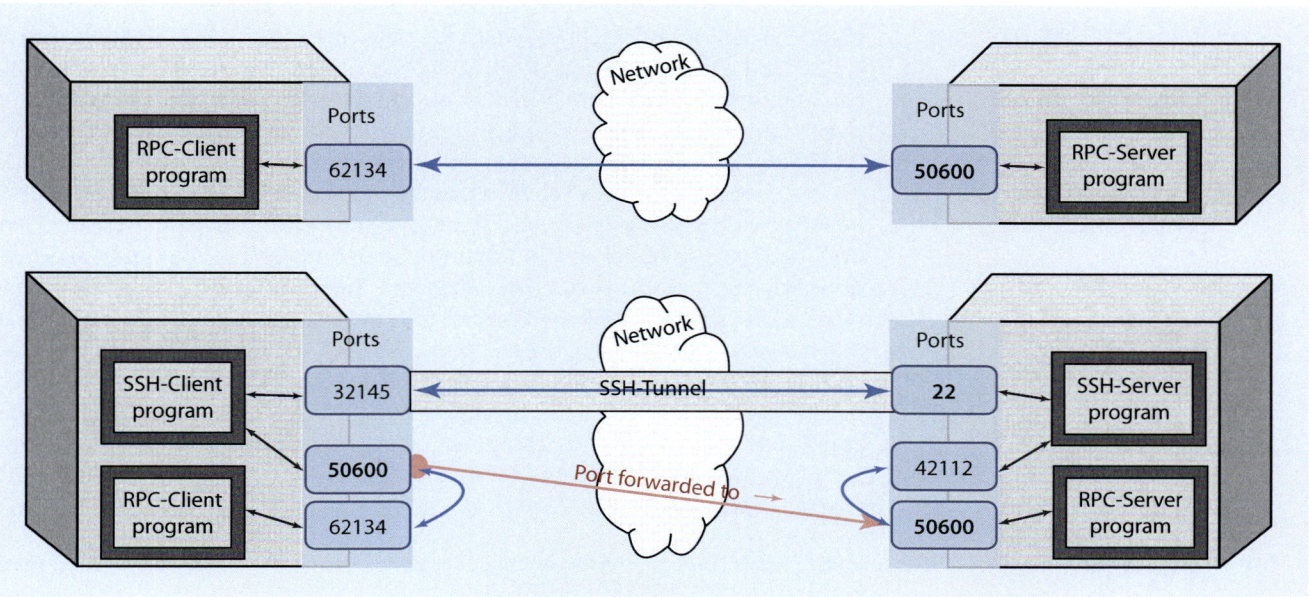

◘ **Fig. 5.20**   The difference between direct connections and connections through an SSH tunnel

---

52 That more modern versions of Open SSL also have such potential risks was shown in the case of the well-known Heartbleed virus (see ▶ Sect. 5.5.1).

Local (Port) Forwarding

With the «-L» argument of the SSH client program «/usr/bin/ssh» (e.g., «-L50600:127.0.0.1:50600»), a local port on the client machine (here «50600») can be connected via a SSH tunnel to a remote port on the server machine (here «127.0.0.1:50600»[53]). A possible, complete setup of the SSH client call is:

```
«ssh -l <username> -p 22
-L50600:127.0.0.1:50600
<IP-address of the server>»[54]
```

The client program then does not communicate directly any longer and uses the local stub port, which is forwarded to the remote server machine. There the SSH daemon operates as client stub and communicates with the server on the same machine, which should receive the communication messages. As the complete communication is encrypted, it is very secure from external spies. But as each message must be encrypted and decrypted again, the communication over a SSH tunnel takes more time than a direct one (Barrett et al. 2005, *l.c. page 349 ff.*). Nevertheless, the technique can be used for the communication with RPC, as defined in ▶ Sect. 3.3.3 to add a standard security layer besides «Secure RPC». This means that almost all external access points (ports) for incoming messages can be blocked for direct access from the external network, using filter tables. Only SSH must be kept open for the SSH tunnel. The additional security by an additional authentication method, the encrypted communication, and the reduction of potential access points, which might be misused by attackers for Denial-of-Service attacks, greatly increases the security for remote control to a system.

Functional Security

The additional technique to keep the remote user within his home box in combination with SSH limits the remote access to the controlled system to one program: the remote server program used, such as «e-RemoteCtrl». The programming of this software is therefore extremely important for the future security of the remotely controlled system. The server must be safe enough but also secure against buffer overruns and overwritings. As each binary stream can be sent to an RPC server, which is then interpreted with the XDR-defined data representation, wrongly used length details for dynamic arrays might lead to server crashes. This can easily happen if, for example, the interface for the «idl2rpc.pl» generated code (see ▶ Sect. 3.3.3) is different between client and server. Because of the watchdog safety used (see ▶ Sect. 3.3.1), the server crashes and restarts again immediately. But this can be quite annoying for other clients, and if such crashes are forced, it leads to a Denial-of-Service attack. This is a point where higher safety also increases the security of a given functionality. Therefore, the «idl2rpc.pl» generated code uses communication checksums (see ▶ Sect. 3.3.3).

Cryptographic Checksums

The reason for the use of sums is that the problem of such crashes can almost[55] be solved with cryptographic checksums (also see ▶ Sect. 3.3.3). They use one-way hash functions which accept an arbitrarily long message block as input and produce a character sequence of a constant length. This is the cryptographic checksum («hash»). It cannot be used to reconstruct the original message. As the checksum is much shorter than the original messages, different messages can create the same hash value. This collision should be avoided. Additionally, each change in the original message also changes the checksum completely. SSH uses such checksums among others for integrity checking and message authentication codes. The following (cryptographic) checksums are used ((Gerloni et al. 2004, *l.c. page 93 ff.*) and (Barrett et al. 2005, *l.c. page 89 ff.*)):

---

53 «127.0.0.1» addresses the local machine of the SSH server, where the tunnel arrives. It is even possible to address another machine in the same network of the server machine. The SSH tunnel would then be used to forward communication streams to this other machine.

54 «-p 22» defines that the SSH daemon on the remote server machine listens on port 22, which is the standard SSH port.

55 It cannot be avoided completely, as the checksum algorithms usually do not map uniquely between an arbitrarily long message block to a smaller, fixed checksum. This can lead to ambiguities.

- *CRC-32:* CRC is not really a cryptographic checksum. But CRC also calculates a mathematical checksum. It uses a generator polynomial, whose result is then divided by an integer division («modulo») to get the final checksum. The collision resistance is defined by the polynomial used. Usually CRC is used for bit streams in a communication message to detect faulty bits and accidental changes (Langmann 2010, *l.c. page 319 f.*).
- *MD5:* MD5 is an advanced hash algorithm. It produces a 128 bit long hash value. Internally it uses nonlinear functions on specially prepared input blocks of the original message. Linux offers the program «md5sum» to calculate MD5 checksums.
- *Secure Hash Algorithm (SHA)-1:* SHA also produces a hash value. It is usually 160 bits long. There are also variants with longer checksums. Linux offers the program «sha1sum» to calculate SHA-1 checksums.
- *RIPE Message Digest (RIPEMD)-160:* This hash algorithm is an advanced version of a previous version of MD5. It was originally used to evaluate ciphering technologies. The first version was not secure, which led to the current RIPEMD-160 with its 160 bit long checksum. There are also variants with longer checksums. Its design criteria and a complete source code are open source. Linux offers the program «gpg» to calculate RIPEMD checksums.[56]

With all this knowledge, it becomes possible to think about a suitable implementation for the scope of duties in the field of secure accesses to controlling systems. The existing e-RemoteCtrl software with its generated communication (see ▶ Sect. 5.4 and ▶ Sect. 3.3.3) must be extended with such secure cipher mechanisms. The combination of all these techniques with a subtle access control in the remote server program makes it possible to integrate most of these solutions for access control according to the individual situation and task. For the start of the client and to primarily authenticate at the server, a username and password are required. Without these data, it is not possible to open the configuration file of the graphical client program and to decrypt the internal keys and authentication data for the communication. These data are stored as AES-encrypted blocks, for example, in the configuration file of the «e-RemoteCtrl» client. A possible entry looks like this

User-Based Access Control

```
«UserName = 0x0D-0x09D-0x8D-0x9C-0x1A-0x81-0x0B-0xBF-
            0xE6-0x48-0x11-0x38-0xE5-0xF1-0x30-0x12»
```

and encrypts the username «oper» with its entered password as key.[57] The decrypted access data, consisting of a username and a password, are also used to authenticate at the server via SSH when a communication is opened. This type of authentication is a representation of a classical «user-based access control».

For remote access to the controlled system, SSH with its provision of communication tunnels is used. It offers the authentication and signature with key files in addition to the user-based access control. As the private key file is kept secret, only the real owner of the access rights can hold it. It uniquely identifies the user for the server and therefore for the remote access. Additionally, the key file can be encrypted with a passphrase to further increase security. The use of such key files forms the «user identity (authorization)-based access control».

User Identity (Authorization)-Based Access Control

When the user is authenticated, he is authorized to do what is allowed by the system. The use of «rbash» in combination with SSH reduces the access rights for users. The remote control server becomes the central instance for all remote

Role-Based Access Control

---

56 «RIPE» stands for «RACE Integrity Primitives Evaluation», where «RACE» are the «Research and development in Advanced Communication tEchnologies», a project of the European Union.

57 The password is also hosted as an encrypted version in the configuration file to validate the user identity.

accesses, which are only controlled by it (central access control). Therefore, it is possible to define specific user roles in the server, which have different rights to request data or to command orders. These roles (see ▢ Fig. 5.21) are administrated in the server program. Each remote user plays a dedicated role which is assigned to him statically during the whole time of access. With these static roles, access control can be arranged like a pyramid. Most users have only a few rights, such as to read the current states or observation data. Then role by role the number of rights increases, while the number of users with these rights decreases. It means that each user on a higher level has the rights of the previous role definition plus additional rights. Finally, there are only a few users (or preferably only one user) who have superuser rights to change the system itself. This defines a «role-based access control», where the VLBI remote control software «e-RemoteCtrl» defines the following roles (see ▢ Fig. 5.21):

1. *Observer role*: An observer is the person with the weakest rights. He is only allowed to request data from the current observation. He cannot send any orders and is not even able to participate in chats with the other users. He is just a passive listener.

2. *Notifier role*: The notifier is an observer with the additional right to participate in the chat or to set some indicator values to notify other users about critical states or about necessary changes.

3. *Scheduler role*: A scheduler is a notifier who can change observation schedules. He has no direct access to the control or to the controlled subsystem, like connected hardware. But he can change parts or the whole sequence of commands in an observation schedule to optimize the observation. These changes have an indirect impact on the controlled devices when the commands are processed. But the commands are mostly procedures with sub-commands, where the required safety of the system is managed. This means that schedulers cannot break the system safety with their changes if everything is in a safe state locally.

4. *Agent role*: An agent is a scheduler with the additional right to change dedicated system parameters which have no direct influence on the hardware safety. An example for such parameters are attenuations for the receiving path, band widths of the receiving channels, or sampling rates in the back end equipment of the VLBI system.

5. *Operator role*: The operator is the first person who can get access to real controlling functionalities of the system. He has all the rights of an agent and additionally the possibility of sending all necessary commands to run the complete system if he gets the permission from a local responsible operator (see «task-based access control» in one of the next sections). Therefore, control rights are just temporarily permitted to him, as given for all external remote operators.

6. *Supervisor role*: The supervisor is an operator who has the rights to control the system permanently. Usually, personnel at the telescope use the supervisor role.

7. *Superuser role*: The superuser has all supervisor rights, and is usually a member of the telescope staff, but is additionally a user with root rights on the controlled system.

8. *Programmer/developer role*: The programmer («prog») or developer has all supervisor rights temporarily. He can change configurations and even programs, but preferably only for a parallel development system and not for the real productive system. Therefore, the role is not a regular user for «e-RemoteCtrl» and exists in parallel for test and development. The programmer usually logs in on a local terminal or console.

Task-Based Access Control

It is recommended to map real user identities to the user roles, so that protocoled interactions of the remote role can directly be assigned to a real person. Additionally, there is one role which requires a separate viewing: the operator role. This is the

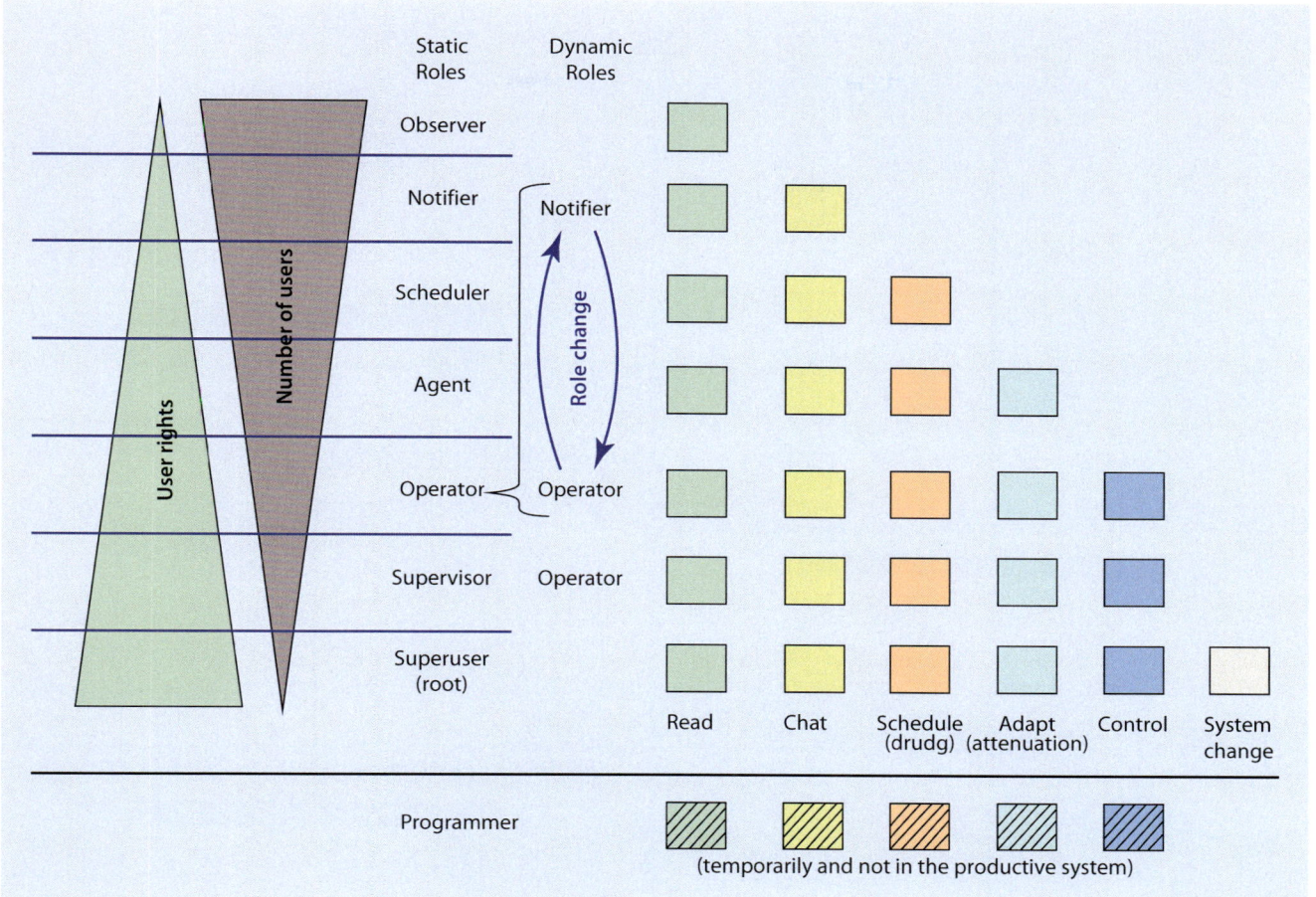

Fig. 5.21 Static and dynamic user roles, such as an authorization pyramid in the «e-RemoteCtrl» server, allow a subtle rights management for remote users

role for all remote operators who do not belong to the staff of the telescope. To avoid accidental interactions to the system by outsiders, which would be a safety risk, the remote operators just get temporary access to the whole functionality. Only when they really need to control the system, for example, if they take over a remote shift from another observatory, does a local operator give them permission to access the control functionality. This permission is only for that task of remote control, includes all necessary authorizations, and can be revoked again by local staff. In the «e-RemoteCtrl» software, there is only one permitted remote operator with the control rights at one time. Without permission, the operator is just a notifier. These dynamic roles implement a «task-based access control» for the remote operator in addition to the «role-based access control».

The handover of the control rights follows a specific workflow, which is comparable to a three-way handshake in the TCP (see ▶ Sect. 3.3.1). It forms the three-way handshake to request remote access rights for control (see ▣ Fig. 5.22). A remote operator, who just has notifier rights because of his dynamic role, requests the control rights by pushing the request button in the GUI. It opens the request dialog, where the operator can send the request or cancel it. The request is registered by the server on the controlled station machine where the NASA Field System runs. Immediately after the registration of the request, all users with higher rights (who are supervisors or superusers, and are therefore members of the local staff, or who are currently permitted to be remote operators) receive a message to inform them about the request. It opens a dialog on each of these user machines, with which they can accept the role change. If one of the users with higher rights

Three-Way Handshake of Remote Access Request

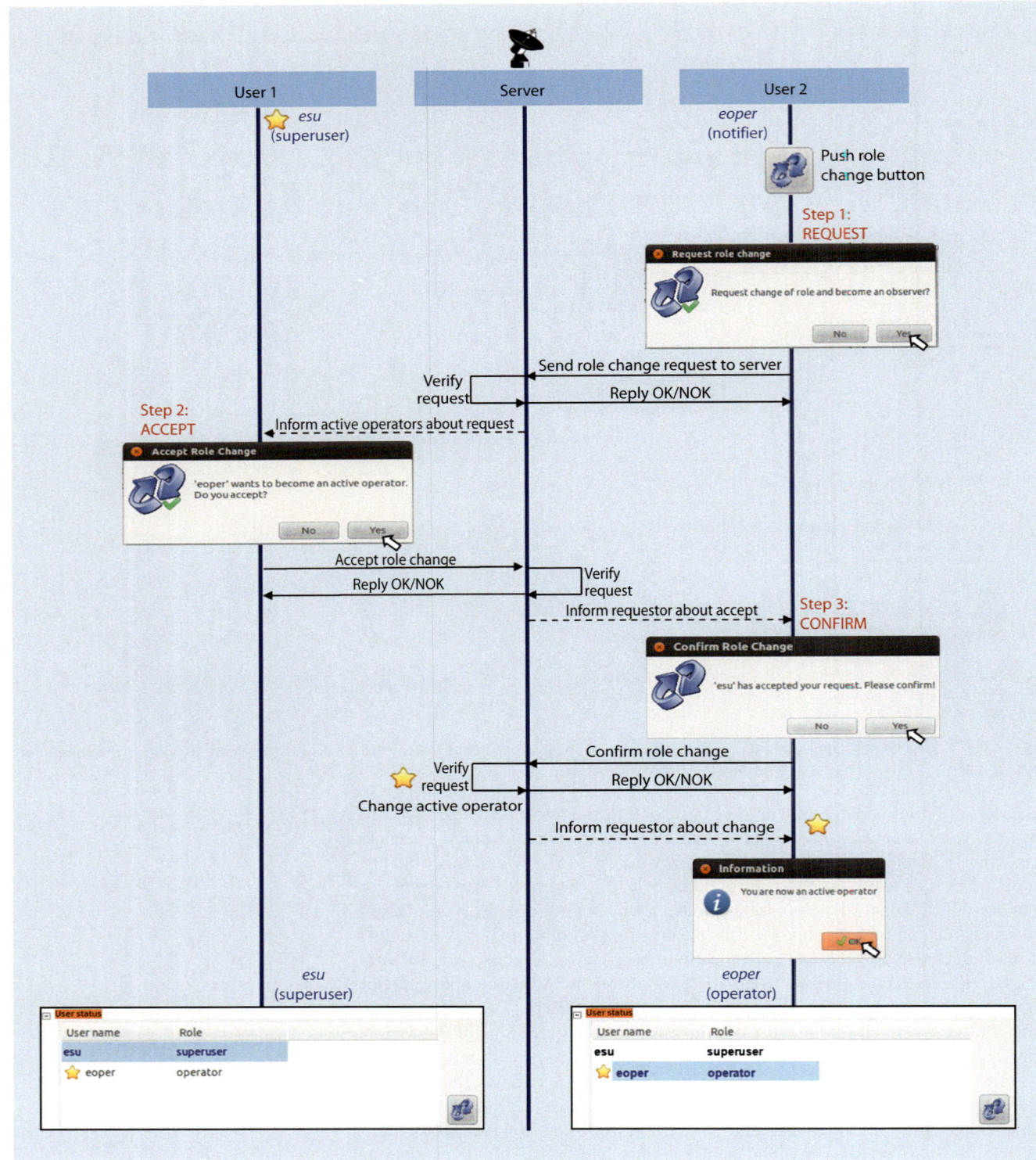

Fig. 5.22 Each task-based role change of an operator requires a three-way handshake between the owners of the rights and the requesting remote operator

accepts, a message is sent to the server. It releases the request dialog on all machines and sends a message to the operator who requested the role change to ask for a confirmation. Only if he confirms this message in another dialog does the server process the role change, and all users are informed about the new owner of the rights. Nevertheless, each supervisor or superuser can revoke the permitted rights from a remote operator with one step, so that local staff can immediately interrupt the remote control, for example, in the case of an emergency.

The guarantee of safety even if there is no responsible remote operator any longer is quite important here. The situation where no remote operator is connected any longer can occur quickly, as the communication medium over the worldwide Internet is not very safe. Each time a responsible, permitted operator does not request information updates within a predefined time interval, the server detects this (e.g., with the automatic safety device of the «idl2rpc.pl» generated code; see ► Sect. 3.3.3) and starts a specific mechanism to inform all other users of the same or higher role category. During the time of the control blackout, the server runs as autonomous controller, where the advantages of an autonomous control cell come to fruition. The different possibilities can be described with a graph of a state machine (see ► Sect. 3.3.1 about finite state machines), where each state shows if the system is controlled by a remote operator and how many different, potential remote operators, supervisors, and superusers[58] are logged in. Each edge in the state graph describes the triggering activity for the state change and the processed role change action. ◘ Fig. 5.23 shows an example for the graph of a state machine, which is used in the «e-RemoteCtrl» server to implement safety during the role changes.[59]

State Machine for Role Change Safety

This role-based access control is only valid within the server program. But usually, a remote operator needs additional programs and access rights. To avoid the complete inclusion of all possibilities into the remote control server, it is better that the role change also activates access to the required, additional programs. This can be implemented with different «.bashrc» files in the home directory of the individual user. These files contain the export of the «$PATH» variable, which points to the directories with the available executables. If these executables are located in the individual home directories, as described for the «rbash» access, and the role change creates symbolic links to the real «.bashrc» file with the individual export for given control situation and roles, the mechanism implements a simple concept-level access control. It is important that the «.bashrc» files and the symbolic links are only writable and can only be changed by the superuser. The management of the links must then be processed after the setting of the effective user identification to the superuser rights in the server program. Then the individual user is unable to change concept levels himself.

Concept-Level Access Control

In addition to the path settings to invoke programs, graphical user interfaces must open their window on the client machine after they have been started remotely. This can be implemented with the «X forwarding» of X Window System[60] sessions over SSH tunnels. If the SSH client is started with the option «-X», the SSH server starts as proxy for the X server and sets the «$DISPLAY» variable for the opened session to point to the proxy. Starting a program with a graphical output will then forward the display to the client machine, where a local X server will receive it for displaying. Another possibility is the use of VNC over SSH. VNC is a free software which provides graphical remote access to Linux and Windows desktops. The whole desktop, as used locally on the machine, appears completely on the client machine. With these powerful VNC techniques over the encrypted SSH, the set of remote possibilities is greatly extended. It is even possible to forward graphical outputs from Keyboard-Video-Mouse (KVM) switches or graphical analyzer outputs from machines which do not offer the possibility to install one's own code (Barrett et al. 2005, *l.c. page 377 ff.*). But the main problems are that the connections are very slow, as they transfer the whole graphics, and that they are not really safe as network blackouts, and so on, are not registered by the servers and the clients (where only the display freezes) to start automatic activities.

X Forwarding

---

58 In the case of remote control safety, supervisors can be treated as superusers.
59 After a longer network blackout, only a supervisor or superuser and the former permitted remote operator can continue the control.
60 The X Window System is the most widely used graphical display system under Linux, which is based on client-server architectures. It uses the X network protocol for the communication.

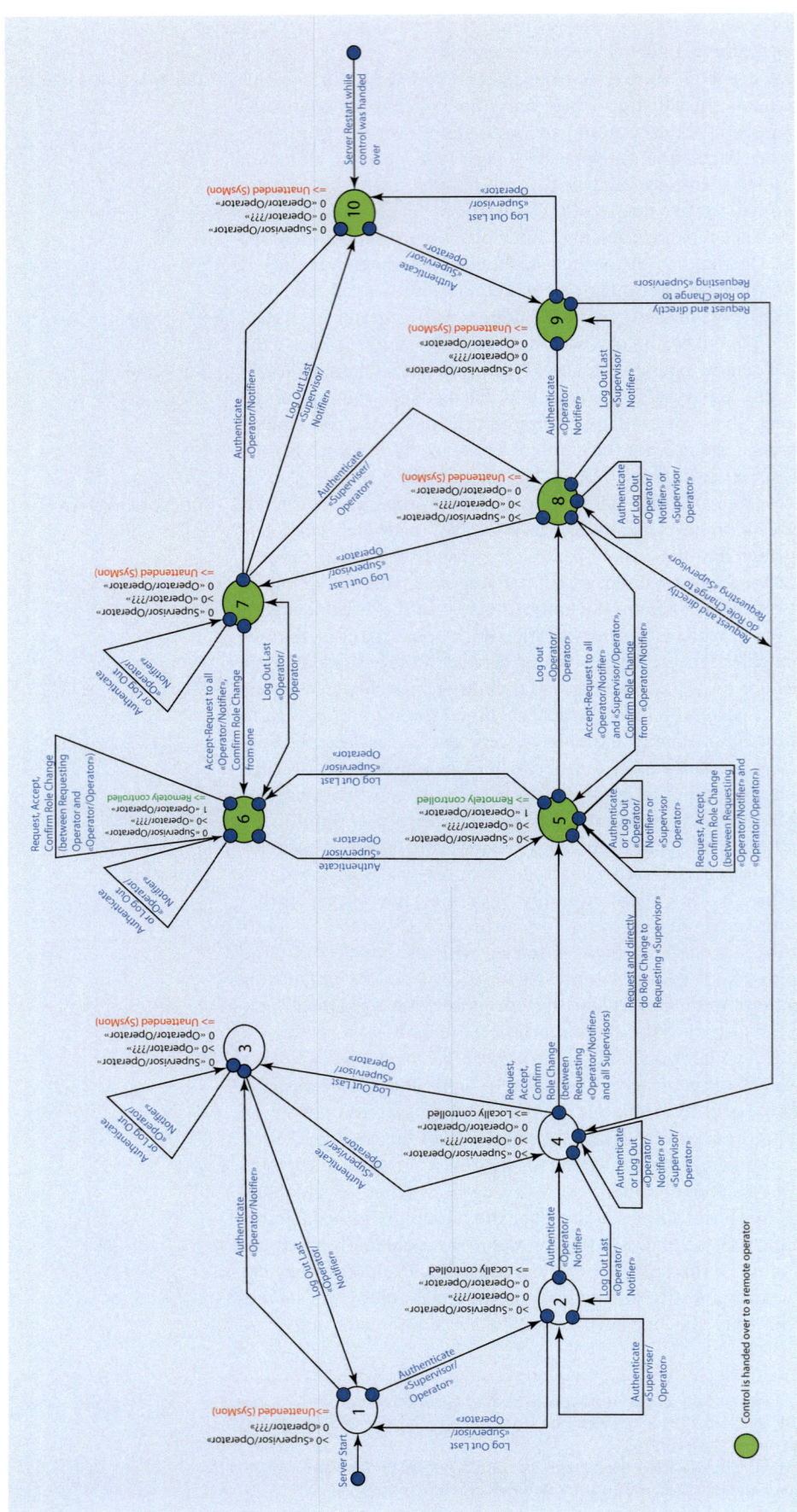

Fig. 5.23 The state machine to guarantee system safety for all situations of a role change

Therefore, it is quite important to use the forwarding of graphical displays very deliberately and to manage the real connections with separate, independent monitoring mechanisms (see ▶ Sect. 5.5.5). Nevertheless, VNC is a good mechanism to follow booting sequences of remote computers if a KVM switch or extender directly supports it.

In combination, all these security aspects together with the security and safety implementations of the communication layer offer all levels of the security factors which are also included in industrial environments, such as for OPC UA (see ((Damm and Winzenried n.d.) *l.c. page 23*) and also ▶ Sect. 4.7):

Comparison to Industrial Standard: OPC UA

— *Confidentiality:* The messages are encrypted within SSH streams to avoid the possibility of the context being monitored by attackers (see ▶ Sect. 5.5.2).
— *Integrity:* The transport layer uses XDR to guarantee the integrity of the data (see ▶ Sect. 3.3.2).
— *Application authentication:* Including MD5 checksums, identifies clients and servers to check if both belong to the same application (see ▶ Sect. 3.3.3).
— *User authentication:* Each user must authenticate himself, and the credentials are then used for each request to identify the account and, ideally, the real person (see ▶ Sect. 5.5.1).
— *User authorization:* Each user has specific access rights defined by its user roles, so that each remote user is only authorized for activities which are permitted to him (see ▶ Sect. 5.5.2).
— *Auditing:* All relevant activities including the security relevant issues are printed to a centralized logbook (see ▶ Sect. 4.4).
— *Availability:* Different mechanisms, such as watchdog processes, guarantee the availability of the data access (see ▶ Sect. 3.3.1).

It is clear that remote access requires many additional mechanisms to enable general security for the system accessed. Nevertheless, programs always contain possible errors, which might also be used for remote attacks. To reduce these risks especially for components and libraries from the operating system, frequent updates of the security patches and bug fixes are suggested. Concerning all of these issues, remote access opens up many new secure ways to run distributed systems remotely, as planned for VLBI or SLR telescopes. Advantageous is that the user community addressed in a network for remote telescope operations is a manageable amount of people.

Security Updates

5.5.3 Security for a Complete Distributed System

All previous sections put the focus on the protection of single computers. But as described in ▶ Sect. 4.2.1, the instruments which should be controlled remotely consist of several different computers and controllers and act together as a distributed system (see also ▶ Sect. 3.3.2 about the principles of distributed systems). This means that each computer would have to be protected in the way described. Even though it is a good advice to increase the protection of single computers, it produces a huge amount of administration work. The problem is that the variety of machines in the network of the distributed system also forms a huge amount of possible access points for potential, deliberately or accidentally executed attacks. To reduce this number of potential security risks for remote access, it is better to physically separate the network of the distributed system from the rest of the networks, which could be considered as potentially «dangerous». Then the rules of each single computer can be greatly simplified, as the accepted messages can be restricted to those from machines of the same network. The rest can be neglected except on the main controlling machine (e.g., the NASA Field System PC), where SSH in particular is accepted for external remote access.

Security Problems for Distributed Systems

Another good reason for a physical separation of the distributed system from the remaining external network is that it reduces the traffic in the control network. Each computer sends several messages to communicate with others or with connected devices. User-friendly operating systems, which offer higher comfort to the

Reduction of Noisy Traffic

user, for example, to search printers and other devices automatically or to regularly check the network for available computers, produce a lot of network traffic. This forms a huge background level of noise especially on company or institute networks with many office applications. But the common bus access protocol Carrier Sense Multiple Access with Collision Detection (CSMA/CD), which manages access to the Ethernet medium, can only deal efficiently with a dedicated amount of traffic. CSMA/CD follows very simple rules. A sender listens on the medium («Carrier Sense»), to ascertain if a transmission is in progress. If the medium is free, it starts transmitting its data, while it continuously listens on the medium. If another sender starts its transmission simultaneously («Multiple Access»), the two data streams will collide. This collision is detected by each sender («Collision Detection»). Then each sender stops its transmission and waits for a random time to reinitiate the transmission again. This method has a disadvantage under a heavy load: competition increases and causes a performance reduction because of many collisions (Singhal and Shivaratri 1994, *l.c. page 84*). Therefore, several required communication attempts from computers of the distributed system interfere with the messages of the background noise. This reduces the availability of the system functions of the controlled distributed system. A filtering, physical separation of the local network for the distributed system and the other systems of an observatory (especially office networks) improves this situation enormously. Experience at Wettzell has shown a significantly higher stability of control communications and a huge reduction of transfer times. The implementation of Virtual Local Area Network (VLAN) would offer a cost-efficient possibility of a quasi-physical separation as they use the same physical network although they define separate networks on them (see (Neidhardt 2005, *l.c. page 179 f.*)). But because of CSMA/CD, this does not help with interference from background noise between the networks on the same physical line.

Creation of Network Enclaves

For an observatory, such as the Geodetic Observatory Wettzell, this means that each distributed system which controls a measuring instrument like SLR or VLBI requires a separate local network. This builds a separated, independent competence area for administration (Neidhardt 2005, *l.c. page 175*). In some cases, it can be extended so that all instruments of one type (e.g., all VLBI telescopes) which need to cooperate are in the same control network. Each network forms a small control entity, which is something like an «enclave» in the network landscape of an observatory (see ◘ Fig. 5.24 as example for an enclave for VLBI in the observatory network of Wettzell[61]).

Network Enclave

> *Definition DEF5.3. of «network enclave»* (Neidhardt 2005, *l.c. page 175*)
> A network or control enclave is a self-contained administration area for the information technology of one distributed system within a computer landscape. It is separate from the networks which are not relevant for the distributed task and which might be potentially dangerous and adversarial because of intended or accidental security intrusions.

Independence of Network Enclaves

The observatory itself is then a composition of such network or control enclaves, to and from which only dedicated data streams are allowed and which are almost independent from external equipment to solve their tasks. The enclaves form a physical implementation of autonomous production cells or logical consolidations of interacting autonomous production cells. They contain all types of cells as described in ▶ Sect. 4.2.2, holding local copies of external data sets which are required for the provision of the production task. This approach can be explained, for example, with meteorological data, which are usually required by all space

---

61   Special thanks to Thomas Klügel for the underlying station map of the observatory Wettzell.

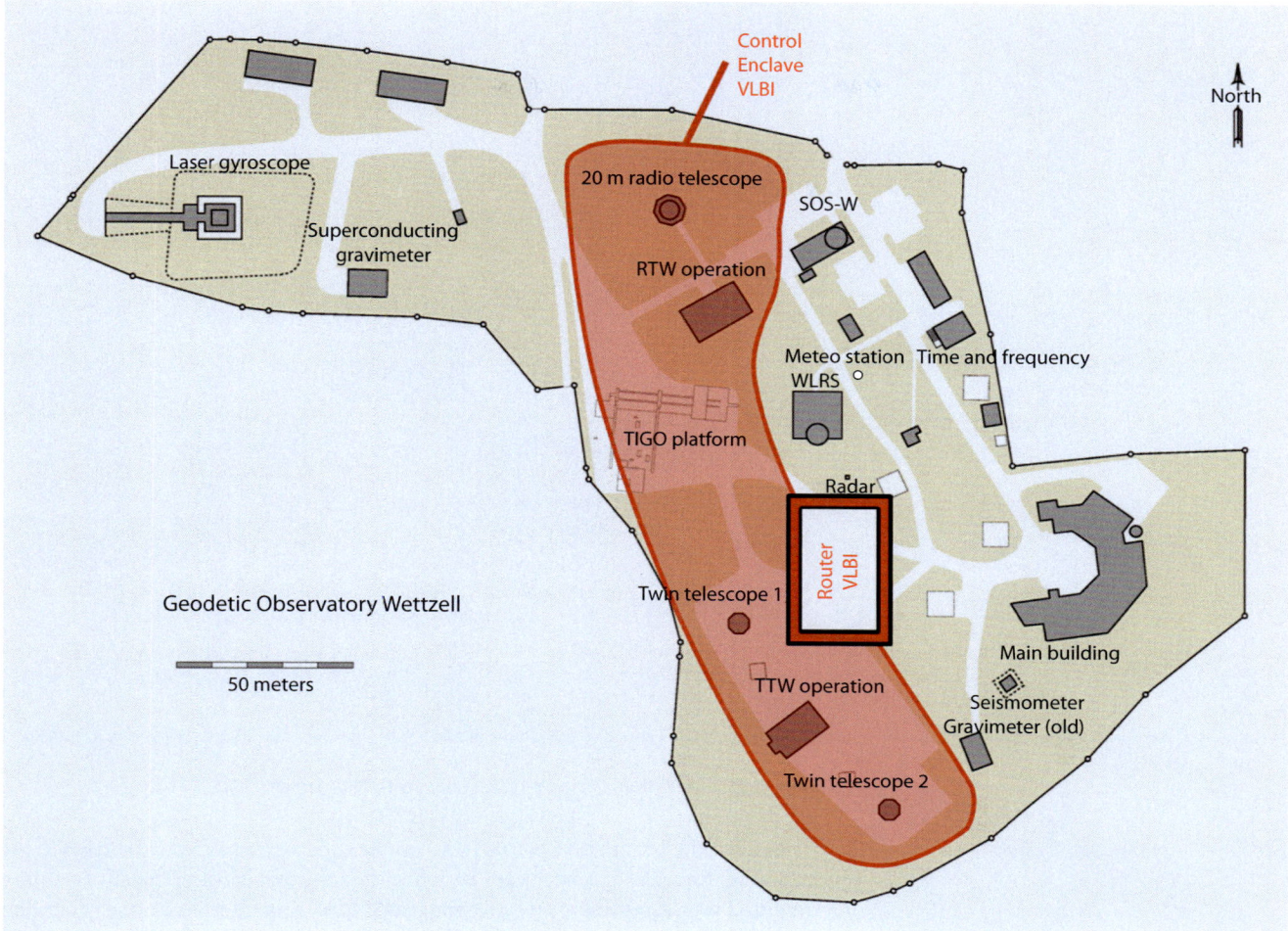

◻ Fig. 5.24 The definition of the control and network enclave for VLBI

geodetic techniques during the observation, but which usually come from the central meteorological station outside the enclaves. If independently local copies of the meteorological data are stored (e.g., by the autonomous data management cell; see ▶ Sect. 4.6), a short blackout of the access to the external meteorological station over a dedicated time period can be absorbed.

Background for implementations of such physical separations are dual-homed hosts (Gerloni et al. 2004, *l.c. page 179*) on the basis of two network interfaces and filter tables («iptables»; see ▶ Sect. 5.5.2). ◻ Fig. 5.25 shows the routing of packages in such a host. It has two physical network interfaces, where one is connected to the internal, trustworthy network, and the other is connected to the external, potentially dangerous network. The forwarding of packages between these two networks is defined with the «FORWARD» hook to the forwarding rule chain of the filter tables. It contains a sequence of rules for forwarding packages from one physical network interface to another (see (Kofler 2010, *l.c. page 893 f.*) and (Purdy 2005, *l.c. page 5 ff.*)). The dual-homed host created in this way is a package filter firewall.

Physical Separation with Dual-Homed Hosts

Definition DEF5.4. of «firewall» (Gerloni et al. 2004, *l.c. page 175*)
A firewall is a computer system which controls and filters the data traffic between two networks. It forms the border between an internal, trustworthy network and an external, potentially dangerous network.

Firewall

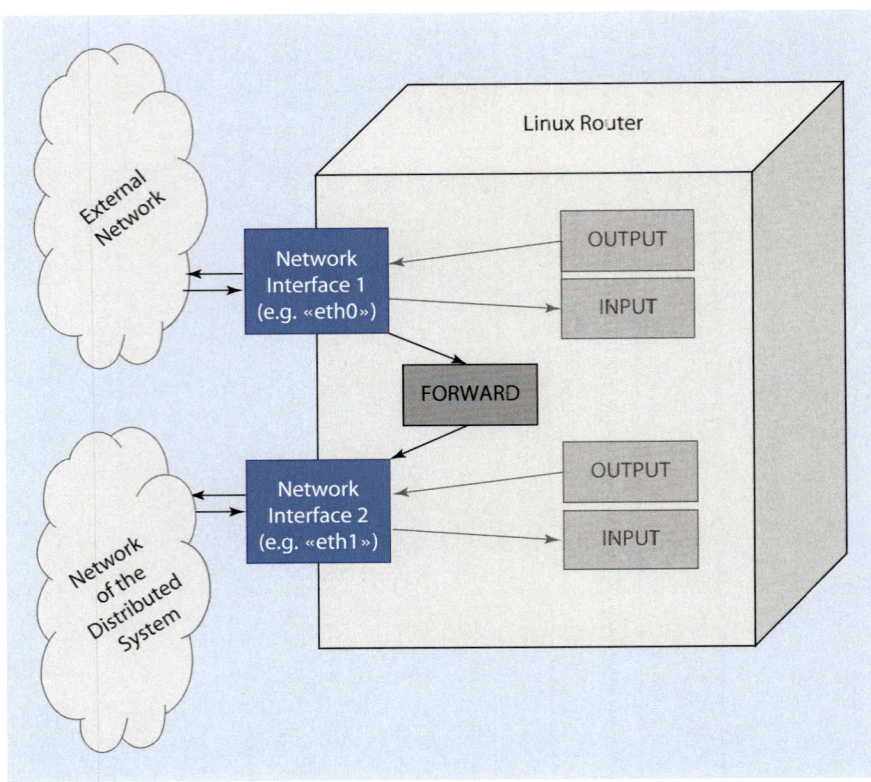

Fig. 5.25   The principle of a dual-homed host with filter tables

Package Filter Firewall Rules

The call of the program «`iptables`» can be used again to create the rules of the package filter firewalls. It is necessary to define the forwarding rules: Usually two rules are required which represent the two directions from one interface to the other and back. Within the rules, it is possible to filter different criteria in the header data of the packages, such as IP addresses, protocol identifier, ports, or flags (similar to the other hooks) (Gerloni et al. 2004, *l.c. page 189*). It is possible to define a stateless, static inspection, where the rules work without memory of previous packages, or a statefull, dynamic inspection. Here there are temporarily additional rules, for example, to allow returning reply messages for the life time of an already accepted communication. Example 5.7 shows a sample rule set of a filter list for the input, output, and forwarding of SSH packages (see ▶ Sect. «Define the Lists of Rules» in the example).

■ ■ **Example 5.7**
Forwarding of SSH packages using network address translation and masquerading.

```
#!/bin/bash                                                              1
#—————                                                                   2
                                                                         3
LOOPBACK_INTERFACE="lo"                                                  4
EXTERNAL="eth0"                                                          5
EXTERNAL_NET="10.10.10.0/24"                                             6
INTERNAL="eth1"                                                          7
INTERNAL_NET="10.10.11.0/24"                                             8
                                                                         9
# ****************************************************************       10
```

```
# ** Flush the iptable                                                    **
# **********************************************************************
echo "Flush the iptable"
# Standard table for IP filtering: checking of rules for packages throught
# the network stack of the kernel to force decisions, how to handle them
iptables —flush
# Table for Network Address Translation (NAT): overwriting of source (SNAT) and
# destination (DNAT) information in the packages
iptables —table nat —flush
# Table for package mangling: mangling is the manipulation of entries in the
# header or user data of packages
iptables —table mangle —flush

# **********************************************************************
# ** Allow everything on the loopback interface                        **
# **********************************************************************
echo "Allow everything on the loopback interface"
iptables —table filter —append INPUT    —in-interface $LOOPBACK_INTERFACE —jump ACCEPT
iptables —table filter —append OUTPUT   —out-interface $LOOPBACK_INTERFACE —jump ACCEPT
iptables —table filter —append FORWARD  —out-interface $LOOPBACK_INTERFACE —jump ACCEPT

# **********************************************************************
# ** Define the general policies                                       **
# **********************************************************************
echo "Define the general policies for all filter queues and hooks"
iptables —policy INPUT DROP
iptables —policy OUTPUT DROP
iptables —policy FORWARD DROP
iptables —table nat —policy PREROUTING ACCEPT
iptables —table nat —policy OUTPUT ACCEPT
iptables —table nat —policy POSTROUTING ACCEPT
iptables —table mangle —policy PREROUTING ACCEPT
iptables —table mangle —policy INPUT ACCEPT
iptables —table mangle —policy FORWARD ACCEPT
iptables —table mangle —policy OUTPUT ACCEPT
iptables —table mangle —policy POSTROUTING ACCEPT

# **********************************************************************
# ** Define the NAT masquerading of internal IPs for the external interface **
# **********************************************************************
echo "Define the NAT masquerading of internal IPs for the external interface"
iptables —table nat              # NAT table
        —append POSTROUTING      # Append to POSTROUTING hook
        —out-interface $EXTERNAL # Physical network interface for output
        —jump MASQUERADE         # Target (action)

# **********************************************************************
# ** Define the lists of rules                                         **
# **********************************************************************
echo "Define the single lists of rules"

# —— SSH service ———————————————————————————————————————
# Accept SSH connections from the internal network to the router
iptables —append INPUT \              # Append to INPUT hook
        —in-interface $INTERNAL \     # Physical network interface for input
        —source $INTERNAL_NET \       # Source address (all internal comp.)
        —destination-port 22 \        # Destination port (SSH of router)
        —jump ACCEPT                  # Target (action)
iptables —append OUTPUT \             # Append to OUTPUT hook
        —out-interface $INTERNAL \    # Physical network interface for output
```

11
12
13
14
15
16
17
18
19
20
21
22
23
24
25
26
27
28
29
30
31
32
33
34
35
36
37
38
39
40
41
42
43
44
45
46
47
48
49
50
51
52
53
54
55
56
57
58
59
60
61
62
63
64
65
66
67
68
69
70

```
              —source $INTERNAL_NET \        # Source address (all internal comp.)      71
              —source—port 22 \              # Source port (SSH)                        72
              —jump ACCEPT                   # Target (action)                          73
# Accept SSH connections from the external network to the router                        74
iptables —append INPUT \                     # Append to INPUT hook                     75
              —in—interface $EXTERNAL \      # Physical network interface for input     76
              —source $EXTERNAL_NET \        # Source address (only computers in        77
                                             # the external network)                    78
              —destination—port 22 \         # Destination port (SSH)                   79
              —jump ACCEPT                   # Target (action)                          80
iptables —append OUTPUT \                    # Append to OUTPUT hook                    81
              —out—interface $EXTERNAL \     # Physical network interface for output    82
              —source $EXTERNAL_NET \        # Source address (only computers in        83
                                             # the external network)                    84
              —source—port 22 \              # Source port (SSH)                        85
              —jump ACCEPT                   # Target (action)                          86
# Routing of SSH connections from the external network to the internal                  87
iptables —append FORWARD \                   # Append to FORWARD hook                   88
              —protocol tcp \                # Protocol                                 89
              —in—interface $EXTERNAL \      # Physical network interface for input     90
              —out—interface $INTERNAL \     # Physical network interface for output    91
              —source—port 1024:65535 \      # Source ports (universal ports)           92
              —destination $INTERNAL_NET \   # Destination address (all internal comp.) 93
              —destination—port 22 \         # Destination port (SSH)                   94
              —jump ACCEPT                   # Target (action)                          95
iptables —append FORWARD \                   # Append to FORWARD hook                   96
              —protocol tcp \                # Protocol                                 97
              —in—interface $INTERNAL \      # Physical network interface for input     98
              —out—interface $EXTERNAL \     # Physical network interface for output    99
              —source $INTERNAL_NET \        # Source address (all internal comp.)     100
              —source—port 22 \              # Source port (SSH)                       101
              —destination—port 1024:65535 \ # Destination ports (universal ports)     102
              —match state \                 # Matching specification                  103
              —state ESTABLISHED,RELATED \   # Matching state                          104
              —jump ACCEPT                   # Target (action)                         105
# Routing of SSH connections from the internal network to the external                 106
iptables —append FORWARD \                   # Append to FORWARD hook                  107
              —protocol tcp \                # Protocol                                108
              —in—interface $INTERNAL \      # Physical network interface for input    109
              —out—interface $EXTERNAL \     # Physical network interface for output   110
              —source $INTERNAL_NET \        # Source address (all internal comp.)     111
              —source—port 1024:65535 \      # Source ports (universal ports)          112
              —destination—port 22 \         # Destination port (SSH)                  113
              —jump ACCEPT                   # Target (action)                         114
iptables —append FORWARD \                   # Append to FORWARD hook                  115
              —protocol tcp \                # Protocol                                116
              —in—interface $EXTERNAL \      # Physical network interface for input    117
              —out—interface $INTERNAL \     # Physical network interface for output   118
              —source—port 22 \              # Source port (SSH)                       119
              —destination $INTERNAL_NET \   # Destination address (all internal comp.)120
              —destination—port 1024:65535 \ # Destination ports (universal ports)     121
              —match state \                 # Matching specification                  122
              —state ESTABLISHED,RELATED \   # Matching state                          123
              —jump ACCEPT                   # Target (action)                         124
# Routing of SSH connections from the internal network to the internal                 125
iptables —append FORWARD \                   # Append to FORWARD hook                  126
              —protocol tcp \                # Protocol                                127
              —in—interface $INTERNAL \      # Physical network interface for input    128
              —out—interface $INTERNAL \     # Physical network interface for output   129
              —source $INTERNAL_NET \        # Source address (all internal comp.)     130
              —source—port 1024:65535 \      # Source ports (universal ports)          131
              —destination $INTERNAL_NET \   # Destination address (all internal comp.)132
              —destination—port 22 \         # Destination port (SSH)                  133
              jump ACCEPT                    # Target (action)                         134
```

```
iptables —append FORWARD \                  # Append to FORWARD hook             135
        —protocol tcp \                      # Protocol                          136
        —in—interface $INTERNAL \            # Physical network interface for input   137
        —out—interface $INTERNAL \           # Physical network interface for output  138
        —source $INTERNAL_NET \              # Source address (all internal comp.)     139
        —source—port 22 \                    # Source port (SSH)                 140
        —destination $INTERNAL_NET \         # Destination address (all internal comp.)  141
        —destination—port 1024:65535 \       # Destination ports (universal ports)     142
        —match state \                       # Matching specification            143
        —state ESTABLISHED,RELATED \         # Matching state                    144
        —jump ACCEPT                         # Target (action)                   145
                                                                                 146
# ...                                                                            147
                                                                                 148
# ***************************************************************************    149
# ** Define destination NAT (DNAT)                                        **     150
# ***************************************************************************    151
echo "Define destination NAT (DNAT) for SSH"                                     152
iptables —table nat \                       # NAT table                         153
        —append PREROUTING \                 # Append to PREROUTING hook         154
        —protocol tcp \                      # Protocol                          155
        —in—interface $EXTERNAL \            # Physical network interface        156
        —source—port 1024:65535 \            # Source ports (universal ports)    157
        —destination 10.10.10.3 \            # Destination address               158
                                             # (external address of router)      159
        —destination—port 2222 \             # Destination port                  160
                                             # (external port of router)         161
        —to—destination 10.10.11.10:22 \ # Internal SSH server                   162
        —jump DNAT                           # Target (action)                   163
iptables —table nat \                       # NAT table                         164
        —append PREROUTING \                 # Append to PREROUTING hook         165
        —protocol tcp \                      # Protocol                          166
        —in—interface $EXTERNAL \            # Physical network interface        167
        —source—port 1024:65535 \            # Source ports (universal ports)    168
        —destination 10.10.11.1 \            # Destination address               169
                                             # (external address of router)      170
        —destination—port 2222 \             # Destination port                  171
                                             # (external port of router)         172
        —to—destination 10.10.11.10:22 \ # Internal SSH server                   173
        —jump DNAT                           # Target (action)                   174
                                                                                 175
                                                                                 176
# ***************************************************************************    177
# ** Define the logging rules                                            **     178
# ***************************************************************************    179
echo "Define the logging rules"                                                  180
iptables —append INPUT  —in—interface $INTERFACE  —jump LOG —log—prefix "INeth0"  181
iptables —append OUTPUT —out—interface $INTERFACE —jump LOG —log—prefix "OUTeth0" 182
```

Usually the forwarding is combined with Network Address Translation (NAT) and IP masquerading. NAT is not really part of the net filters, but it is usually available within the same context. It makes it possible to exchange IP addresses and port numbers, so that the physical send and receiver addresses of the two separated networks can be hidden behind the logical addresses of the router. Static NAT is a static translation of addresses and is separated into Destination Network Address Translation (DNAT), which changes the destination address or port, and Source Network Address Translation (SNAT), which changes the source address or port. Example 5.7 shows how to forward the external port 2222 of the firewall to the SSH daemon of a computer in the internal network with DNAT settings of the «iptables» (see ► Sect. «Define destination NAT (DNAT)» in the example). Masquerading (see ► Sect. «Define the NAT masquerading of internal IPs for the external interface» in the example) is a form of dynamic SNAT, where all addresses of the internal network are hidden behind the official, external address of the firewall, so that

Network Address Translation and Masquerading

Firewall Management

their addresses do not appear outside the protected network[62] (see (Gerloni et al. 2004, *l.c. page 209 ff.*) and (Purdy 2005, *l.c. page 15 f.*)).

The complex commands in Example 5.7 show how difficult it can become to define all necessary rules for all the required protocols, such as for ping, DNS, operating system updates, SSH, HTTP, file transfers, network file systems, mail, NTP, printing, databases, and so on.[63] Therefore, the management of the firewalls is not trivial. A helping hand can be given by supporting programs, such as the Firewall Builder («fwbuilder»; see ▶ http://www.fwbuilder.org/). This program offers a GUI to design package filter firewalls in an object-oriented way. The configurations defined are saved in XML files and can be exported into different formats, such as those required for «iptables», which mostly look very cryptic (Gerloni et al. 2004, *l.c. page 216 f.*). The tool can also be used to administrate the different firewall because it is possible to send and activate the configuration files directly to the target machines. But these are not the only administrative interactions. Besides the implementation of the security policies, all required additional operation data, such as computer aliases, routing information, or DNS entries, must be kept up to date. The logs must be checked and suitable reactions must be performed. Additionally, all cryptographic keys must be managed, and backups must be created or imported if necessary. If these items are managed, the firewall usually runs completely autonomously. But security issues and log information require a receiver, so that alarms can be handled. This is another suitable task for an autonomous system monitoring and safety cell, as described in ▶ Sect. 4.7. Nevertheless, a professional configuration and management of a firewall is not trivial, even with professional systems and their graphical remote interfaces, which use encrypted communications (Gerloni et al. 2004, *l.c. page 185 ff.*), so that designing of control enclaves is challanging.

Passing of Firewalls

The extended security with a firewall also has its price for communications which must pass it, for example, to perform remote control tasks. A suitable concept must be installed in the firewall and in the remote client of the software to simplify the connection to the server despite the firewall. Additionally, the client must be intelligent enough to control and reestablish constructed connections. As all previous considerations focus on SSH, this protocol must be managed. ◘ Fig. 5.26 shows two ways to allow remote access from external clients to internal control servers, such as the NASA Field System PC.

IP (or Port) Forwarding

The standard way has already been described with the forwarding of IP packages in combination with NAT. An external SSH stream connects to one designated port of the firewall or router. This port is internally forwarded to a target machine. Therefore, incoming, external, encrypted SSH packages are directly routed to the target machine, where an SSH daemon manages them. It handles the authentication and authorization, the encryption, and finally the forwarding to the destination service, like the remote control server. This method of IP forwarding takes care (see ◘ Fig. 5.26a) that the external packages arriving at the dedicated external firewall port only appear at the real target machine, where the final security must be guaranteed. All return packages are automatically sent back to the requesting client. Additional security checks at the firewall are not intended. The call of the SSH client is then without a local port forwarding, as the forwarding is automatically handled by the firewall:

```
«ssh -l <username> -p 22000
<IP-address of the router>»64
```

---

62 VLAN setups or suitable Domain Name Service (DNS) entries allow the use of internal addresses also in the external network. But this might be a security risk, as all protected network addresses from the internal network are directly usable and visible in the potentially dangerous, external network.

63 For example, firewalls are also used for the load balancing to different servers which offer the same service.

64 «-p 22000» defines that the dedicated firewall port which is forwarded to the target machine with «iptables» is port number 22000.

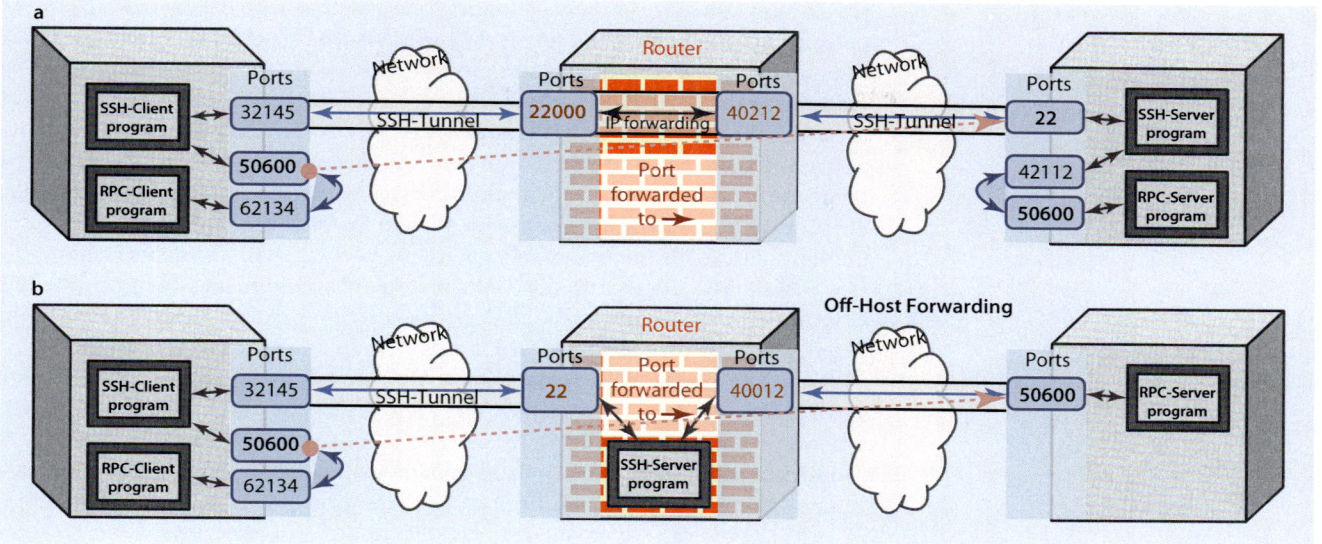

Fig. 5.26 Two ways to pass a firewall with an SSH-encrypted remote control stream: **a** IP (port) forwarding or **b** SSH Off-Host forwarding

Another possibility is Off-Host forwarding with SSH (see Fig. 5.26b). The forwarding of local ports with SSH forwards a local port of a client machine to a service on a target machine which is on the same computer as the SSH server. Thus, the «-L» option of the SSH call on the client machine contains the local host address «local-host» or «127.0.0.1» of the server. But the service requested does not necessarily have to be on the same machine as the SSH server. Off-Host forwarding[65] uses this circumstance. The SSH client connects to an SSH server machine, but forwards a local port of a client to a further computer in the internal network behind the firewall (Barrett et al. 2005, *l.c. page 361 f.*). The advantage is that the complete management of external accesses is on the firewall. The disadvantage is that there are now two user credentials, one for the SSH access to the firewall and one to the final service on the client machine if implemented. Another disadvantage is that the communication between the firewall and the target machine on the internal network is not encrypted anymore, so that other computers in the internal network can read the message contents. This Off-Host forwarding is something like a simple implementation of a higher firewall category, the proxy firewall. It uses a proxy server on the firewall, which represents the real server behind it, and allows additional security mechanisms (for more about firewalls see the following ▶ Sect. 5.5.4).[66] The call of the SSH client in such a scenario is then similar to the previously described call for a local port forwarding, with the exception of the different target address in the «−L» option:

> SSH Off-Host Forwarding

```
«ssh -l <username> -p 22
 -L50600:<IP-address of the target>:50600
 <IP-address of the router>«
```

With the remote control communication with Sun RPC, both techniques to pass a firewall require the target server for remote control to use a fixed port and no portmapper. It is also quite important that no higher protocol layer transfers address information from the internal network (e.g., in the data stream of the remote interface), as those addresses will not be translated correctly without a dedicated proxy on the firewall (see more in (Neidhardt 2005, *l.c. page 123 ff.*)). A problem of the Off-Host forwarding is that neither the RPC client nor the RPC

> Consequences of Forwarding

---

65 «Host» in this context means the SSH server.
66 A real proxy firewall actually separates the two communications from the client to the firewall and from the firewall to the server and is not only a simple forwarding of the same data stream.

server has knowledge of the SSH tunnel. Uncertainties with the tunnel can only be detected by the client or the server over timeouts in the RPC communication. If a timeout expires, a communication partner must assume that something is wrong with the SSH tunnel, even though this might not be true. Usually the client must then close its communication and also the SSH connection to reestablish them again. To do this without interaction from the human user, the SSH client must be controlled automatically by the remote control client as well. Help offer tools, like «expect» (National Institute of Standards and Technology (NIST) n.d.) or a program from the code toolbox at Wettzell (see ► Sect. 3.2) with the name «sshbroker» (see ► Sect. 5.5.5). They run the SSH client in an environment which controls the output and input streams, so that they can react on input requests automatically, like requests for user and password inputs.

Limits of a Firewall

Before proxy firewalls are considered in the following section, a short look should show the limits of a firewall. Firewalls never offer 100% security. User malpractice, communication bypasses, or Denial-of-Service attacks can always be a problem. Usually a firewall is only as good as its design. Therefore, risks which are unknown to the firewall administrator are not prevented by the firewall. Nevertheless, it can log as much as possible to detect security leaks, even though the search for them in the huge amount of data logs is difficult. Additional software on separate computers like «Network Flight Recorders», which act like the flight recorder in an airplane, log everything independently from the firewall as a form of connection tracking and allow search patterns and additional alarms. It can be used as an intrusion detection system or to identify IP spoofing attacks[67] on misconfigured firewalls (Gerloni et al. 2004, *l.c. page 184 f.*).

RPC Security

Firewalls are also not 100% protection for the remote control of geodetic instruments. But they offer a suitable level of security for the RPC communications besides Secure RPC. Originally RPC was not designed for security issues. The problem is that most RPC programs require root rights, so that buffer overflow attacks directly result in access possibilities with superuser rights. The registration at the portmapper additionally simplifies the detection of open RPC ports. Therefore, it is advisable to close all unnecessary RPC ports. Using a firewall and the forwarding of the secure SSH tunneling, no RPC port can be directly accessible from outside the distributed system. The number of ports at the firewall can be reduced to a minimum, and the additional authentication mechanisms offer acceptable protection for the system controlled.

5.5.4 Security for a Complete Observatory

Distributed System of Distributed Systems

As already described, the local network of the distributed system is protected from the remaining Local Area Network (LAN) for office applications. The latter forms the standard network of the company, institute, or observatory and can be connected to other business locations over a Virtual Private Network (VPN).[68] Generally, this network is something like a collection of distributed systems enclaves and office computers, which are combined in a separate administration enclave on an office VLAN. This forms the office enclave of the observatory, even

---

67 During a spoofing attack, an offender sends IP packages with a faked, internal address from the external network or simulates a DNS service to redirect communications (DNS Spoofing) (Gerloni et al. 2004, *l.c. page 203*).

68 While the SSH tunneling method described is an application-specific assurance of the data security, VPN is based on an extension to the IP standard with security mechanisms. This new IPSec standard implements a possibility for authentication with an IP Authentication Header and an encrypted communication with the IP Encapsulating Security Payload functionality on the lower protocol layers. This can be used to create different virtual network topologies over the WAN , like a Gateway-Gateway-VPN between two locations of a company (Gerloni et al. 2004, *l.c. page 284 ff.*), so that computers in both networks seem to be part of only one common network.

though it is not necessarily physically separated. Therefore, an observatory like the Geodetic Observatory Wettzell is nothing more than a distributed system in which the autonomous sub-cells are formed by the distributed systems for the geodetic instruments. This means an observatory is a distributed system of distributed systems (see the explanation for «systems of systems» in ▶ Sect. 6.3). Exactly as defined for the previously described distributed systems with their autonomous sub-cells (see ▶ Sect. 4.2.1), the control enclaves in the observatory network also provide their tasks autonomously for the rest of the other components of the observatory as autonomous production cells. In a similar way to the protection of a single distributed system, the combined distributed system of the single production cells which form the observatory network can be protected with a firewall.

It is possible to split the network situation again into an internal and an external part. The internal observatory LAN must be protected from the potentially dangerous external network. The external network is usually the WAN. Usually the WAN is potentially more risky for the observatory network than the observatory network for an internal enclave. The observatory firewall is something like a «bastion host». It is the only accessible access point and the only gate from the Internet, so that its function is like a castle wall or bulwark of a medieval castle. Usually it is called «gateway» and implements a multi-homed host[69] on the basis of a proxy firewall (often with a multi-level architecture[70]) (Gerloni et al. 2004, *l.c. page 179 ff.*).

Proxy Firewall

As already mentioned in the previous section (see ◘ Fig. 5.26), proxy firewalls contain their own applications as proxies for the services which are offered to the external network. These application level proxies are the most costly but also most secure way to protect internal systems. But each communication protocol which passes the firewall must use its own proxy application. These firewalls combine the package filter rules with a way to understand, analyze, and manipulate the application data on the higher layers of the communication protocol stack. The protection is so great because a client cannot directly communicate with its server. All communications run over the proxy application on the firewall. This applies to communications from the external network as well as to those from the internal network. Unlike with Off-Host forwarding, the proxy firewall really separates the communication from the client to the firewall proxy and from this proxy to the final server in the internal network, so that in fact there are two separate communication connections (Gerloni et al. 2004, *l.c. page 225 ff.*).

Proxy Firewall Principles

Therefore, the proxy is like a server for the real client and like a client for the real server. This means that proxy firewalls offer additional advantages. They can also check and validate application data for faulty information. This can be used to deny dedicated file transfers or access to dangerous web pages. It can also be used to implement content filtering, where dedicated content like commercial content in web pages is not transfered. But this slows the traffic down, and usually wrong matches lead to false positives, so that pages or contents are blocked which should not be. All proxy firewalls implement NAT and IP masquerading. They allow very flexible access control on the basis of application data and tasks. As the whole content must be transcripted by the proxy, it can also cache the data for further requests (Gerloni et al. 2004, *l.c. page 227 ff.*).

Proxy Firewall Advantages

As the creation of such proxy firewalls is quite complex, usually commercial firewalls are used to protect company networks. Nevertheless, there are also very useful open-source firewalls. One of these is «m0n0wall» (pronounced «monowall»). It is a complete, embedded firewall software package for an embedded PC and provides similar features to a commercial firewall box. It uses FreeBSD as an operating system and a web server with a dynamic user interface for the configuration of the firewall, which is stored in an XML file (Kasper n.d.). Such a firewall can be used as a router

Open-Source Firewall: m0n0wall

---

69 A multi-homed host has several physical interfaces with several, different networks behind, which are shielded against each other.

70 For example, with a «demilitarized zone» which is used as an additional network layer between the Internet and the LAN.

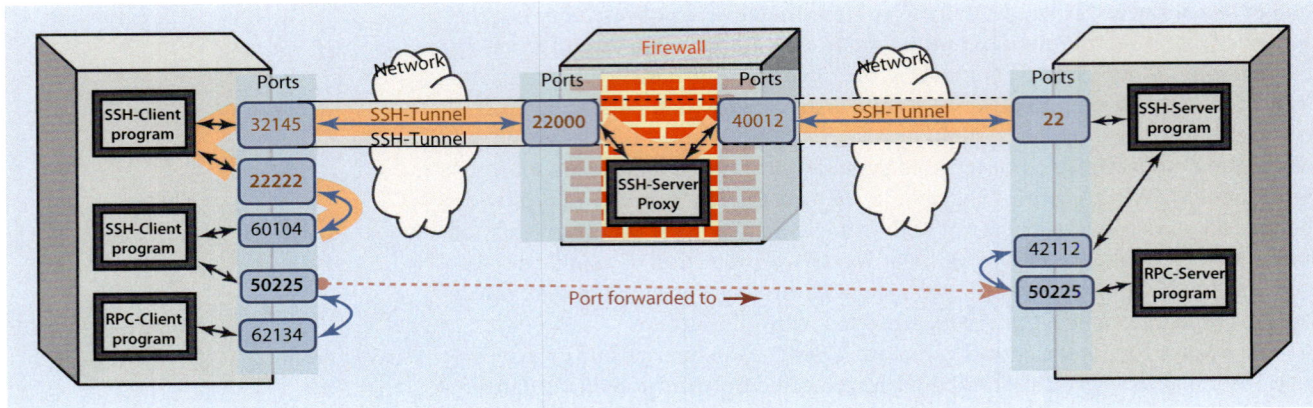

Fig. 5.27 Example for a nested SSH tunnel with one tunnel to the proxy firewall and an internal tunnel with the user communication to the final destination

Nested SSH Tunnels

for a control enclave as well as a simple proxy firewall for small networks. But for the data traffic of a complete observatory like Wettzell, such simple firewalls are too weak.

The additional «bastion host» for an observatory also increases the security for transfers of the remote control, even if the filtering of the encrypted SSH content is limited. But the additional firewall is a potential barrier for attacks and is an additional authentication layer which must be passed. For example, it is possible for a client to first open a secure SSH connection to the proxy firewall, in which another secure SSH transfer to the next firewall level of the enclave or to the real server can be tunneled. Within this second SSH tunnel, the communication between the remote control client and server is transmitted. Firewalls can forward the second SSH tunnel via IP forwarding to any final destination (even over several firewall hierarchies), and clients from the Internet can connect to specific hosts of the internal network, using such nested SSH tunnels. An example of such tunnels is shown in ◘ Fig. 5.27. Such a scenario was used at the Geodetic Observatory Wettzell and can be implemented with two SSH client calls, where one client forwards a local port to the SSH proxy on the firewall, and the second uses this new, local SSH port to forward another local port to the real server:

```
«ssh -l <username> -p 22000
-L22222:<IP-address of the server>:22
<IP-address of the firewall>»
```

and

```
«ssh -l <username> -p 22222
-L50225:<IP-address of the server>:50225
<IP-address of the server>»
```

Nested Enclaves

The goal of these nested tunnels is to manage access to nested control and network enclaves, as defined for a complete observatory. Each enclave is protected against the usual office network and the other enclaves in the observatory. It is additionally protected against the potentially more dangerous Internet by the proxy firewall of the observatory. Nevertheless, it is easily accessible over encrypted and secure SSH connections from everywhere inside the observatory and in the whole WAN when two credentials are checked for one access. ◘ Figure 5.28 shows such a nested access to the VLBI enclave. Each access point requires a separate key file with separate passphrases and user credentials. This offers a very high level of security, even for such a sensitive task as remote control.

But the protection is always only as good as the knowledge of the administrators and the quality of the different firewall layers from simple package filters to higher application level proxies. It is quite important to implement a clean firewall concept and to frequently update all security patches. Another important factor is

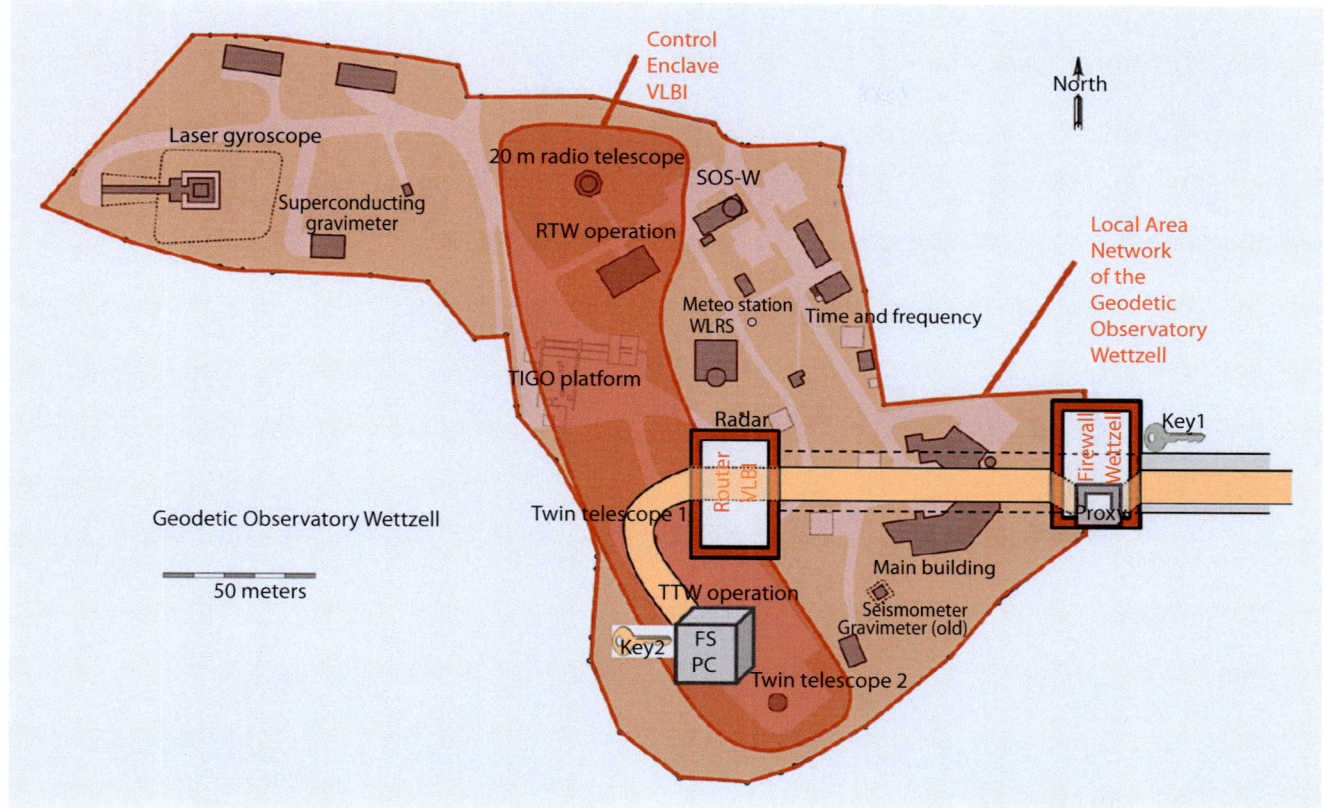

Fig. 5.28 The access to the control and network enclave for VLBI via a proxy firewall

the secure and safe management of access keys, credentials, and communication parameters by the client. Because autonomous reestablishments of remote control communications are required, additional software components are necessary.

5.5.5 On-the-Fly Management of Temporary SSH Tunnels

SSH is quite secure, as it allows almost no external intrusion into the internal workflow. The SSH client is configured so that it reads from and writes to the real terminal[71] in each case, irrespective of whether the standard input or output stream in the application is redirected or not.

    Standard programs use the whole terminal stack. Terminals are full-duplex devices with separated input and output paths. The real terminal hardware (or in the case of terminal emulation of a graphical desktop on the screen) is controlled by the terminal driver (TTY driver). Between the user process and this terminal driver, which also still exists in modern operating systems, is a line discipline module. It controls the displaying of characters («echoing»), concatenation of characters to lines, editing of lines (e.g., pushing back of characters), generation of signals after keyboard buttons are pressed, data flow control, end-of-file management, and character conversions (e.g., a «new line» character into «carriage return» and «line feed» for the output). The access to the line discipline is offered by programming libraries with input and output functions (e.g., the «stdio.h» for C or

SSH and Terminal Security

Terminal Input and Output Streams

---

71 Originally a terminal or console was the physical interface to a command shell. Hardware offered access to enter commands and to receive answers in the same serial text style as earlier with the teletype writers. Therefore, the usual acronym for a terminal is TeleTYpe (TTY). Nowadays, terminals are only software emulations of such former terminals within a graphical desktop on a screen.

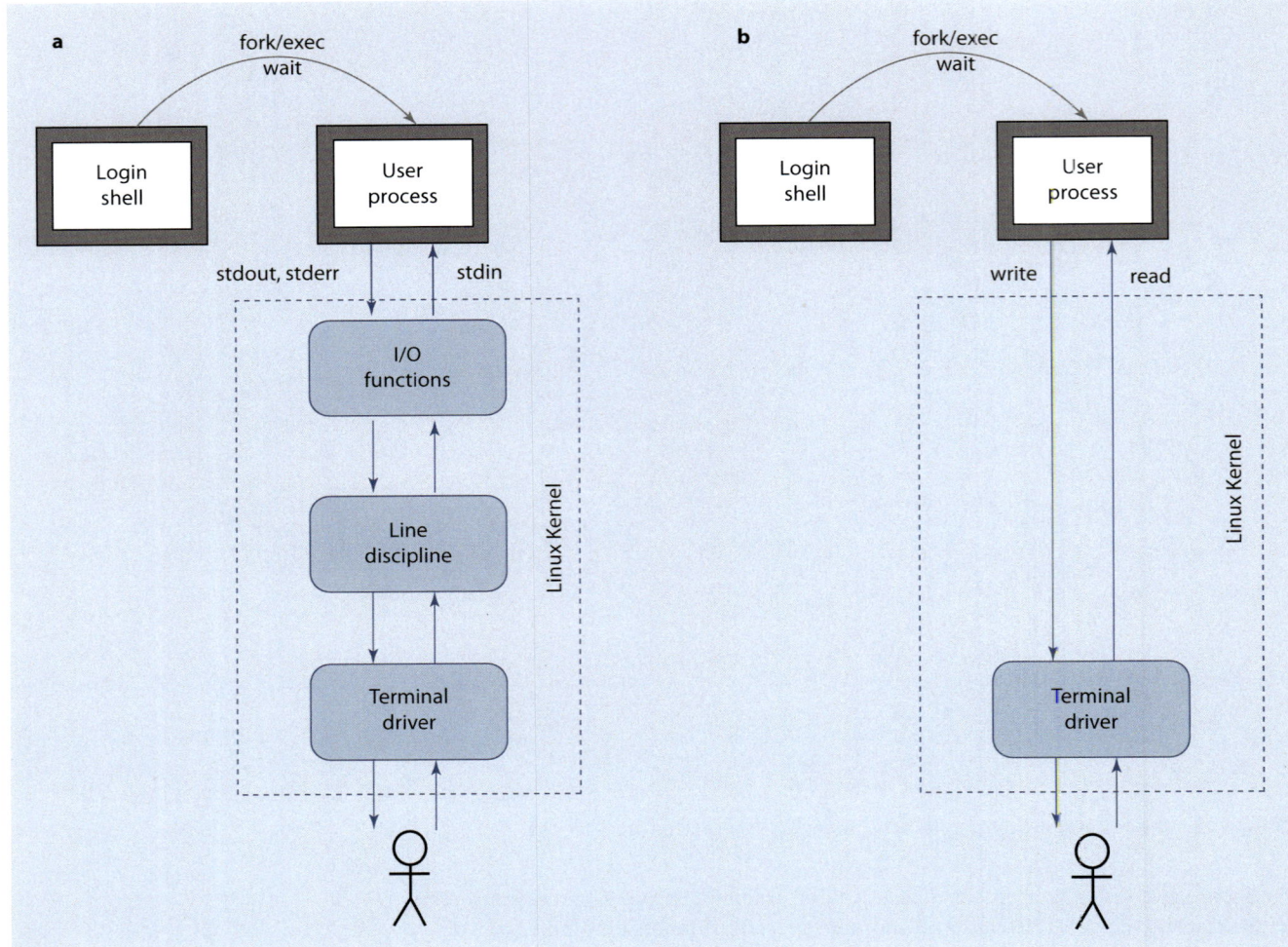

Fig. 5.29 The input and output mechanisms for a terminal: **a** with standard input and output streams or **b** with a direct read and write access to the terminal driver (Adapted and extended from (Stevens 1990, *l.c. page 617*) and (Stevens 1990, *l.c. page 619*))

«iostream» for C++). ◘ Figure 5.29a shows this setup for a standard terminal interaction. The channels for the standard input («stdin» in C, «std::cin» in C++), standard output («stdout» in C, «std::cout» in C++), and standard error output («stderr» in C, «std::cerr» in C++) are inherited by the original login shell (Stevens 1990, *l.c. page 616 ff.*).

Terminal Read and Write Access

Another possibility is a direct reading from and writing to the device. It is necessary to directly open the device driver file «/dev/tty» to get access to the control terminal (Herold 2004, *l.c. page 413*). The control terminal is generated by the process group leader, which is the first process to create a terminal communication (usually the login shell). It is used to communicate with the input and output hardware to the user in front of the screen and keyboard. All processes of the same process group use the same control terminal. Direct access to the device with the «/dev/tty» can be used by a program to guarantee the direct writing to and reading from one's own control terminal independently if the higher input and output streams are redirected to other IPC channels (Stevens 1990, *l.c. page 40*). This is used by the SSH client to read the user login data. Thus, an adversarial intrusion to spy on the password is more difficult. ◘ Figure 5.29b shows this setup with direct access to the control terminal and the terminal driver.

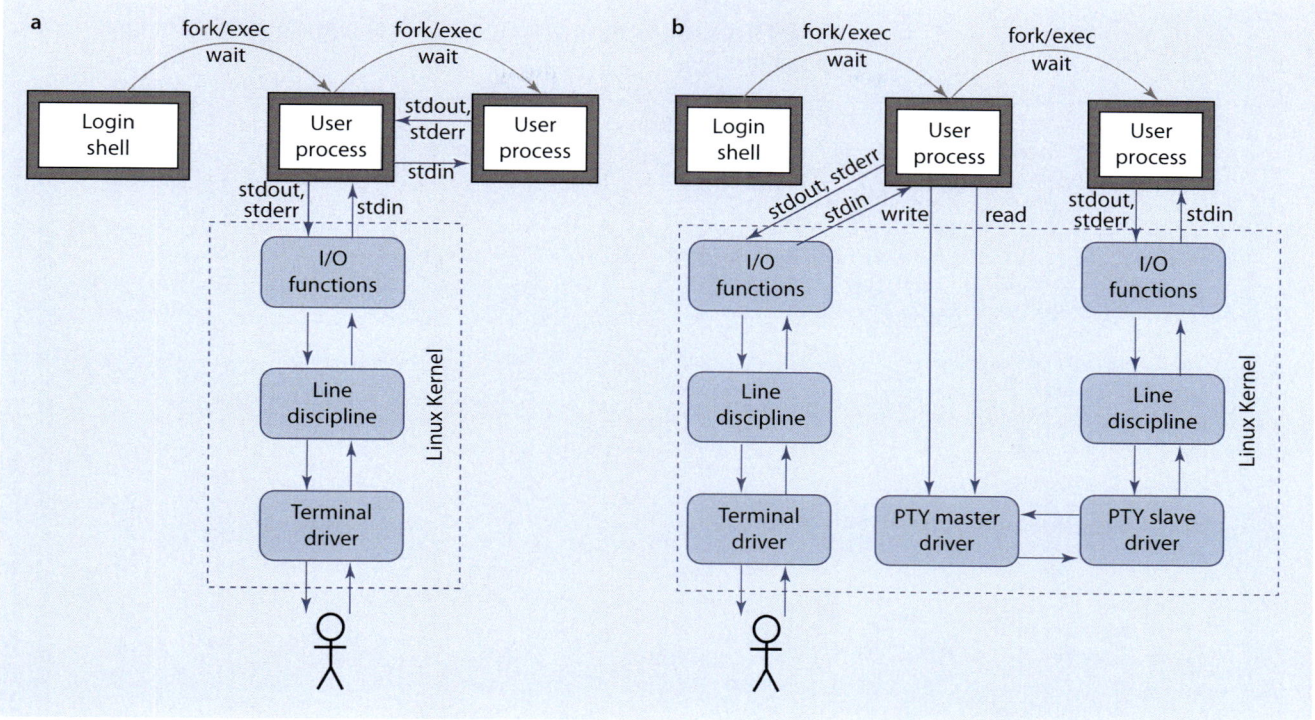

Fig. 5.30 **Fig. 5.30** The redirection of inputs to and outputs from a terminal: **a** for the standard input and output streams or **b** for the direct read and write access to the terminal driver (Adapted and extended from Stevens (1990) *l.c. page 633*)

The big advantage of such intrusion blockage in the SSH client for security issues, on the one hand, is a disadvantage for the automatic management of SSH connections, on the other hand, and therefore for the guarantee of communication safety. If a connection breaks down, it must be possible to reestablish it again without interaction by the user. This means that the SSH connection must be closed or killed. Then a new connection must be started, for which the input of the user login data is required again. But as the SSH client does not use the standard input and output streams but direct TTY control, this automated interaction, such as with the client, is not so simple. This is a point where security works against the safety requirement.

Security Vs. Safety

If the SSH client uses the less secure input and output streams, it would be easy to redirect these streams to IPC channels (see ▪ Fig. 5.30a), like for pipes as an example (or even files), from which the user login data are read and to which the outputs from the SSH client are written. Example 5.8 shows the main parts of such a redirection in a C program, where the standard output «stdout» (file descriptor ID «1»[72]) and the standard error output «stderr» of a created child process are redirected to a previously created pipe using the duplication of the standard file descriptors to the pipe file descriptors with «dup2». Then the child starts a dedicated application program, whose output is forwarded via the pipe streams to a parallel running parent process. This technique can be used to monitor and manipulate interactions of a user with each application program, using the standard input and output streams[73] (see (Stevens 1990, *l.c. page 618 ff.*) in combination with (Herold 2004, *l.c. page 184 ff.*)). But this is not applicable for the automated control of SSH clients.

Redirection of Input and Output Streams

---

72 «stdin» has the file descriptor ID «0».
73 In the case of monitor and control of external applications, this technique is used in combination with «e-RemoteCtrl» in the form of an «e-shell», which offers access to inputs and outputs in the form of Sun RPC servers.

■■ **Example 5.8**

Code snippet with the main parts for the use of pipes to monitor outputs of external programs.

```
int aiPipeDescriptors[2] = { -1, -1}; /// aiPipeDescriptors = pipe descriptors for inter-     1
    process communication between SSH-sender (child) and controller (parent)                  2
                                                                                               3
/// ...                                                                                        4
                                                                                               5
/// Create pipes as interprocess communication between the program, which should be
    monitored and the monitoring process
if (pipe (aiPipeDescriptors) < 0)                                                              6
{                                                                                              7
    priv_bError = true;                                                                        8
    goto Cleanup;                                                                              9
}                                                                                             10
                                                                                              11
/// ...                                                                                       12
                                                                                              13
/// Fork process to run external program                                                     14
if ((priv_ulPIDSSHSenderProcess = fork()) < 0)                                               15
{                                                                                             16
    // Error                                                                                  17
    priv_bError = true;                                                                       18
    goto Cleanup;                                                                             19
}                                                                                             20
else if (priv_ulPIDSSHSenderProcess == 0)                                                    21
{                                                                                             22
    // Child                                                                                  23
    /// Redirect input and output to the pipes                                               24
    close (aiPipeDescriptors[0]);                                                             25
    if (dup2 (aiPipeDescriptors[1], 1) < 0)                                                   26
    {                                                                                         27
        printf ("CHILD: 1 error\r\n");                                                        28
        exit(0);                                                                              29
    }                                                                                         30
    if (dup2 (aiPipeDescriptors[1], 2))                                                       31
    {                                                                                         32
        printf ("CHILD: 2 error\r\n");                                                        33
        exit(0);                                                                              34
    }                                                                                         35
    close (aiPipeDescriptors[1]);                                                             36
    /// Call of the program, which should be monitored                                       37
    (void) execv (strProgramCall.c_str(), apcCallArguments);                                 38
    exit (0);                                                                                 39
}                                                                                             40
                                                                                              41
// Parent                                                                                     42
ulStarttime = time(NULL);                                                                     43
do                                                                                            44
{                                                                                             45
    /// Check if child finished                                                              46
    ulCheckPID = wait(&iChildStateVal);                                                       47
                                                                                              48
    /// Read output from child to stdout                                                     49
    if (read (aiPipeDescriptors[0], acBuffer, sizeof(acBuffer)) != 0)                        50
    {                                                                                         51
                                                                                              52
    /// ...                                                                                   53
                                                                                              54
    }                                                                                         55
}                                                                                             56
```

```
///  ...                                                          57
                                                                 58
                                                                 59
Cleanup:                                                         60
    close  (aiPipeDescriptors[0]);                              61
    close  (aiPipeDescriptors[1]);                              62
                                                                 63
    ///  ...                                                     64
```

To solve that problem, a pseudo-terminal must be used to replace the previous IPC in combination with direct access to the device control terminal, which is then disassociated from the process. A pseudo-terminal is a (virtual) device pair of a «master» and a «slave» device. A process can create such a device pair before it forks to a child and a parent process. Both processes then have access to the pseudo-terminal and can exchange data over it. The slave part of the pair looks like a real terminal, so that a child which uses the slave terminal can interact with it like with a standard terminal device. If the child writes something to the slave, it appears directly at the master and therefore at the parent process which holds the master terminal. This is an IPC in the lower layers of the operating system, directly on the basis of terminal communications. To acquire the originally inherited control terminal, the child additionally must open the «/dev/tty» and therefore the control terminal, before it starts the application program. Then it disassociates itself from that original control terminal and opens the slave part of the pseudo-terminal, which becomes the new terminal for the later started application program. All input and output streams are duplicated and therefore redirected to that slave terminal. On the other side, the parent process reconfigures the master to a raw terminal and then waits for inputs from it to parse the received messages for dedicated information. If defined patterns are found, such as password requests from the SSH client, which can be started as an application program in the child process, the expected answers can be sent (e.g., the password). With this technique, it is possible to redirect the secure terminal inputs for the password requests from the SSH client, so that it can be controlled for a safe communication (compare (Stevens 1990, *l.c. page 628 ff.*) and (Stevens 1990, *l.c. page 642 ff.*). Example 5.9 is an excerpt of the «sshbroker», which uses the mechanism described to monitor one SSH connection for a communication management using a standard SSH client program.

TTY Redirection with a Pseudo TTY

■ ■ Example 5.9

The use of a control terminal to manipulate inputs and outputs to and from the SSH client.

```
/*  ...  */                                                      1
                                                                 2
                                                                 3
#ifdef  __cplusplus                                             4
extern  "C"  {                                                   5
#endif                                                           6
                                                                 7
    /*  ...  */                                                  8
                                                                 9
    static  struct  termio tty_mode; /*! tty_mode = deposit for the original TTY settings */   10
    static  long  g_lPIDChild;       /*! g_lPIDchild = identifier of child process */          11
    static  int  iTTYModeChanged;    /*! iTTYModeChanged = flag to identify, if the settings   12
        were  changed */
                                                                 13
    /*  ...  */                                                  14
                                                                 15
                                                                 16
    /***************************************************************  17
        function   iSetTTYRaw                                    18
        ***************************************************************/  19
```

```
/*!          Sets  raw  mode  for  TTY
 *  \param     int  iFiledescriptor  ->Filedescriptor
 *  \return    int <- Errorcode (0=ok, -1=nok)
 ********************************************************************/
/*  author    Alexander Neidhardt
    date      22.10.2009
    revision  -
    info      Original from Stevens, W.: UNIX Network Programming
 ********************************************************************/
int iSetTTYRaw (int iFiledescriptor)
{
    struct termio temp_mode; /*! temp_mode = temporary mode deposit */

    /*! Read settings */
    if (ioctl(iFiledescriptor, TCGETA, (char*) &temp_mode) < 0)
        return (-1);

    /*! Save for restoring later */
    tty_mode = temp_mode;
    /*! Turn off all input control */
    temp_mode.c_iflag = 0;
    /*! Disable output post-processing */
    temp_mode.c_oflag = (unsigned short)(temp_mode.c_oflag & (~OPOST));
    /*! Disable signal generation, canonical input, echo, upper/lower output */
    temp_mode.c_lflag = (unsigned short)(temp_mode.c_lflag & (~(ISIG | ICANON | ECHO |
        XCASE)));
    /*! Clear char size, disable parity */
    temp_mode.c_cflag = (unsigned short)(temp_mode.c_cflag & (~(CSIZE | PARENB)));
    /*! 8-bit chars */
    temp_mode.c_cflag |= CS8;
    /*! Min #char to satisfy read */
    temp_mode.c_cc[VMIN] = 1;
    /*! 10'ths of seconds between chars */
    temp_mode.c_cc[VTIME] = 1;

    /*! Write settings */
    if (ioctl(iFiledescriptor, TCSETA, (char*) &temp_mode) < 0)
        return (-1);
    iTTYModeChanged = 1;

    return (0);
}

/********************************************************************x
    function  iResetTTY
 ********************************************************************/
/*!          Sets mode back to saved mode for TTY
 *  \param     int iFiledescriptor ->Filedescriptor
 *  \return    int <- Errorcode (0=ok, -1=nok)
 ********************************************************************/
/*  author    Alexander Neidhardt
    date      22.10.2009
    revision  -
    info      Original from Stevens, W.: UNIX Network Programming
 ********************************************************************/
int iResetTTY (int iFiledescriptor)
{
    if (!iTTYModeChanged)
        return (0);

    /*! Write back original settings */
    if (ioctl(iFiledescriptor, TCSETA, (char*) &tty_mode) < 0)
        return (-1);
    return (0);
}

/* ... */
```

```
/* ********************************************************************
    function  main
   *******************************************************************/
#ifdef __USE_SSH_BROKER_AS_LIB__
    int iSSHBroker(int argc, char *argv[])
#else
    int main (int argc, char * argv[])
#endif
    {
        int iMasterTTY = 0;              /*! iMasterTTY = descriptor of master TTY */
        int iSlaveTTY = 0;              /*! iSlaveTTY = descriptor of slave TTY */
        char aczPTTYName[1024];         /*! aczPTTYName = pseudo TTY name */
        char ** pcSSHCommand = NULL;    /*! pcSSHCommand = generated SSH command call */
        int iError = 0;                 /*! iError = internal error number */
        int iTimeoutSec = 0;            /*! iTimeoutSec = defined timeout in seconds */

        /* ... */

        /*! Open pseudo tty */
        if (openpty (&iMasterTTY, &iSlaveTTY, aczPTTYName, NULL, NULL) == -1)
        {
            iError = 121;
            goto MainFinal_Ret;
        }

        /*! Fork: parent = controlling process, child = spying process for execv */
        if ((g_lPIDchild = fork()) < 0)
        {
            /*! Fork error */
            iError = 122;
            goto MainFinal_Ret;
        }
        else if (g_lPIDChild == 0)
        {
            /*! Child: spying process for ssh */
            int iTTYFileDescriptor; /*! iTTYFileDescriptor = real TTY of the program */

            /* ... */
            /* Set signal handlers for the child termination */
            /* Combine the SSH command pcSSHCommand from the program call arguments */
            /* Check if the know hosts file should be deleted */
            /* ... */

            /*! First of all, disassociate from the existing control terminal */
            if (isatty(0))
            {
                if ((iTTYFileDescriptor = open ("/dev/tty", O_RDWR)) >= 0)
                {
                    if (ioctl (iTTYFileDescriptor, TIOCNOTTY, (char *) 0) < 0)
                    {
                        write(iSlaveTTY, "SSHBroker::Error 40\n", 20);
                        iError = 40;
                        goto MainChild_Ret;
                    }
                    close (iTTYFileDescriptor);
                }
                else
                {
                    (void) write(iSlaveTTY, "SSHBroker::Error 40\n", 20);
                    iError = 40;
                    goto MainChild_Ret;
                }
            }

            /*! Close iMasterTTY */
            (void) close (iMasterTTY);
```

```
              /*! Define this process as new session leader */       153
              if ((setsid()) < 0)                                     154
              {                                                       155
                  (void) write(iSlaveTTY, "SSHBroker::Error 41\n", 20);  156
                  iError = 41;                                        157
                  goto MainChild_Ret;                                 158
              }                                                       159
                                                                      160
                                                                      161
              /*! Open iSlaveTTY => this is the new tty */            162
              if ((iSlaveTTY = open (aczPTTYName, O_RDWR)) < 0)       163
              {                                                       164
                  (void) write(iSlaveTTY, "SSHBroker::Error 42\n", 20);  165
                  iError = 42;                                        166
                  goto MainChild_Ret;                                 167
              }                                                       168
                                                                      169
              /*! Redirect stdin, stdout and stderr to the pseudo tty */  170
              if (dup2 (iSlaveTTY, STDIN_FILENO) < 0 ||              171
                  dup2 (iSlaveTTY, STDOUT_FILENO) < 0 ||             172
                  dup2 (iSlaveTTY, STDERR_FILENO) < 0)               173
              {                                                       174
                  (void) write(iSlaveTTY, "SSHBroker::Error 43\n", 20);  175
                  iError = 43;                                        176
                  goto MainChild_Ret;                                 177
              }                                                       178
                                                                      179
              /*! Close it again */                                  180
              (void) close (iSlaveTTY);                              181
                                                                      182
              /*! Run program which should be listened to */         183
              if (execv (pcSSHCommand[0], pcSSHCommand) < 0)         184
              {                                                       185
                  /*! Send error message to the redirected stderr => iSlaveTTY */  186
                  fprintf (stderr, "SSHBroker::Error 44\n");         187
                  iError = 44;                                       188
                  goto MainChild_Ret;                                189
              }                                                       190
MainChild_Ret:                                                        191
              /* ... */                                              192
              /* Clean-up child memory */                            193
              /* ... */                                              194
              return iError;                                         195
          }                                                           196
          else                                                        197
          {                                                           198
              /* Parent: controls the program via the created pseudo tty */  199
              fd_set FDSet;                /*! FDSet = initial set of file descriptor for "  200
                  select" */
              fd_set FDUsedSet;            /*! FDUsedSet = checked set of file descriptor  201
                  for "select" */
              struct timeval Timeout;      /*! Timeout = timeout structure for select */  202
              char acMessageBuffer[512];   /*! acMessageBuffer = buffer for received  203
                  messages from child */
              int iNumberOfReadBufferChar = 0; /*! iNumberOfReadBufferChar = number of received  204
                  bytes in the buffer */
                                                                      205
              /* ... */                                              206
              /* Set signal handlers for the child termination */    207
              /* Print startup messages, if required */              208
              /* Combine the SSH command pcSSHCommand from the program call arguments */  209
              /* Define from where the password should be read: program call arguments or stdin  210
                  */
              /* Set timeouts */                                     211
              /* ... */                                              212
                                                                      213
              /*! Prepare filedescriptor set */                      214
```

```
    FD_ZERO(&FDSet);                                              215
    FD_SET(iMasterTTY, &FDSet);  /* iMasterTTY */                 216
    if (isatty(0))                                                217
    {                                                             218
        FD_SET(0, &FDSet);          /* stdin */                  219
    }                                                             220
                                                                  221
    /*! Check if stdin is a tty */                               222
    if (isatty(0))                                                223
    {                                                             224
        iFlagIsATTY = 1;                                          225
    }                                                             226
    else                                                          227
    {                                                             228
        iFlagIsATTY = 0;                                          229
    }                                                             230
                                                                  231
    /*! Set stdin tty to raw mode */                             232
    if (iFlagIsATTY && iSetTTYRaw (0) < 0)                        233
    {                                                             234
        iError = 1;                                               235
        goto MainParent_Ret;                                      236
    }                                                             237
                                                                  238
    /* ... */                                                     239
                                                                  240
    /*! Read from pseudo tty and define some actions if a specific  241
    tag is found */                                               242
    while (1)                                                     243
    {                                                             244
        /*! Wait on incoming messages with "select" */           245
        if (iTimeoutSec > 0)                                      246
        {                                                         247
            Timeout.tv_sec = iTimeoutSec;                        248
            Timeout.tv_usec = 0;                                  249
            FDUsedSet = FDSet;                                    250
            iSelectRetVal = select (FD_SETSIZE, &FDUsedSet, NULL, NULL, &Timeout);  251
        }                                                         252
        else                                                      253
        {                                                         254
            FDUsedSet = FDSet;                                    255
            iSelectRetVal = select (FD_SETSIZE, &FDUsedSet, NULL, NULL, NULL);  256
        }                                                         257
                                                                  258
        /*! Select error */                                      259
        if (iSelectRetVal < 0)                                    260
        {                                                         261
            iError = 2;                                           262
            goto MainParent_Ret;                                  263
        }                                                         264
                                                                  265
        /*! Select timeout */                                    266
        if (iSelectRetVal == 0)                                   267
        {                                                         268
            /* ... */                                             269
            /* Handle timeouts */                                 270
            /* ... */                                             271
            goto MainParent_Ret;                                  272
        }                                                         273
                                                                  274
        /*! Read and parse incoming messages */                  275
        for (iFD = 0; iFD < FD_SETSIZE; iFD++)                    276
        {                                                         277
            if (FD_ISSET (iFD, &FDUsedSet))                       278
            {                                                     279
                /*! Read from stdin and send to the child process  280
```

```
                                    via the master TTY pseudo device */    281
                           /*! stdin => iMasterTTY */                       282
                           if (iFD == 0 && isatty(0))                       283
                           {                                                284
                               if ((iNumberOfReadBufferChar = read (0,      285
                                           acMessageBuffer, 511)) <= 0)     286
                                                                            287
                               {                                            288
                                   iError = 3;                              289
                                   goto MainParent_Ret;                     290
                               }                                            291
                               acMessageBuffer[iNumberOfReadBufferChar] = '\0';  292
                               if (write(iMasterTTY, acMessageBuffer,       293
                                       iNumberOfReadBufferChar) != iNumberOfReadBufferChar)  294
                               {                                            295
                                   iError = 4;                              296
                                   goto MainParent_Ret;                     297
                               }                                            298
                           }                                                299
                           /*! Read messages from the child from the master TTY  300
                               pseudo device and send it parse it for dedcated   301
                               content as login prompts etc. */             302
                           /*! iMasterTTY => stdout */                      303
                           if (iFD == iMasterTTY)                           304
                           {                                                305
                               if ((iNumberOfReadBufferChar = read (iMasterTTY,  306
                                           acMessageBuffer, 511)) <= 0)     307
                                                                            308
                               {                                            309
                                   iError = 5;                              310
                                   goto MainParent_Ret;                     311
                               }                                            312
                               acMessageBuffer[iNumberOfReadBufferChar] = '\0';  313
                                                                            314
                               /* Write received acMessageBuffer to stdout */  315
                               if (write(1, acMessageBuffer, iNumberOfReadBufferChar)  316
                                       != iNumberOfReadBufferChar)          317
                               {                                            318
                                   iError = 6;                              319
                                   goto MainParent_Ret;                     320
                               }                                            321
                                                                            322
                               /* ... */                                    323
                               /* Parse the received message for  */        324
                               /* - included errors, added by the child process  325
                                               ==> send error message to stderr */  326
                               /* - "Are you sure" ==> register key ==> send "yes" */  327
                               /* - "Enter passphrase" ==> passphrase input */  328
                               /* - "password:" ==> password input */       329
                               /* - SSH errors: "WARNING:", "Command not found",  330
                                               "Connection Refused",        331
                                               "Network is unreachable",    332
                                               "Privileged ports",          333
                                               "unknown error",             334
                                               "Permission denied",         335
                                               "usage: ssh"                 336
                                               ==> send error message to stderr */  337
                               /* - "Last login:" or "permitted by applicable law" or "%"  338
                                               ==> SSH successfully opened */  339
                               /* - "closed" or "100%" ==> SSH closed */     340
                               /* ... */                                    341
                           }                                                342
                       }                                                    343
                   }                                                        344
                                                                            345
               /* ... */                                                   346
           }
```

```
| MainParent_Ret:                                                          347
|            /*  ...  */                                                   348
|            /*  Clean—up  parent  memory  */                             349
|            /*  Print  a  readable  version  of  the  error  codes  from  the  child  */   350
|            /*  ...  */                                                   351
|                                                                          352
|            /*  reset  stdin  tty  */                                     353
|            if  (iFlagIsATTY  &&  iResetTTY  (0)  <  0)                   354
|            {                                                             355
|                iError = 1;                                               356
|                goto  MainParent_Ret;                                     357
|            }                                                             358
|        }                                                                 359
|                                                                          360
| MainFinal_Ret:                                                           361
|            /*  ...  */                                                   362
|            /*  Clean—up  general  memory  */                            363
|            /*  Print  a  readable  version  of  the  general  error  codes  e.g.  help  page  */   364
|            /*  ...  */                                                   365
|            return  iError;                                               366
|        }                                                                 367
| #ifdef  __cplusplus                                                      368
| }                                                                        369
| #endif                                                                   370
```

Such a technique is used as background management of all SSH connections in the «e-RemoteCtrl» client. All the required information for the establishment of SSH connections, the tunneling and handling of encrypted data, can be configured in the GUI (see ▪ Fig. 5.31). The example in ▪ Fig. 5.31 creates a connection to an «e-RemoteCtrl» server on the machine «100.205.23.100»[74] at port 60226. The SSH client forwards the port 50226 to that server port over an encrypted SSH tunnel, which is managed by the client GUI and the underneath sshbroker. The manual call of the SSH client doing the same would be:

sshbroker and Remote Control

```
«ssh -l <username> -p 22
-L50225:127.0.0.1:60225
100.205.23.100»
```

The problem with this redirection of input and output of the SSH client is that it reduces the security on the client side, as it is possible to spy on the password interactions or to inject data from outside. This is a problem, as the «sshbroker» streams can be redirected again using the simple duplication of file descriptors for the streams. Also passwords and passphrases must be saved and accessible by the client management. This discrepancy between security and safety cannot be solved properly. Therefore, it is not advisable to use such a brokerage for a highly secure system. But with some limitations, it is possible to keep a high degree of security, while also allowing the management of communication safety.

Security Problem of a Redirection

It is important that secret information is available in the memory only for as long as required. This means that the passwords, passphrases, and key information are saved encrypted on the machine and that they are only used for the short time needed in the process of an establishment or a reestablishment of a new communication. During the rest of the connection establishment, they are not available anymore. Further on in the communication, everything appears as the standard SSH connection with a standard SSH client.

Reduction of Insecure Times

To enable a primary encryption and decryption, it is necessary to enter something like a master password, which is used for the internal decryption from this

Data Chopping and Steganography

---

74 The address is only an artificial one, and the link to a potentially existing computer with this address might be just accidental.

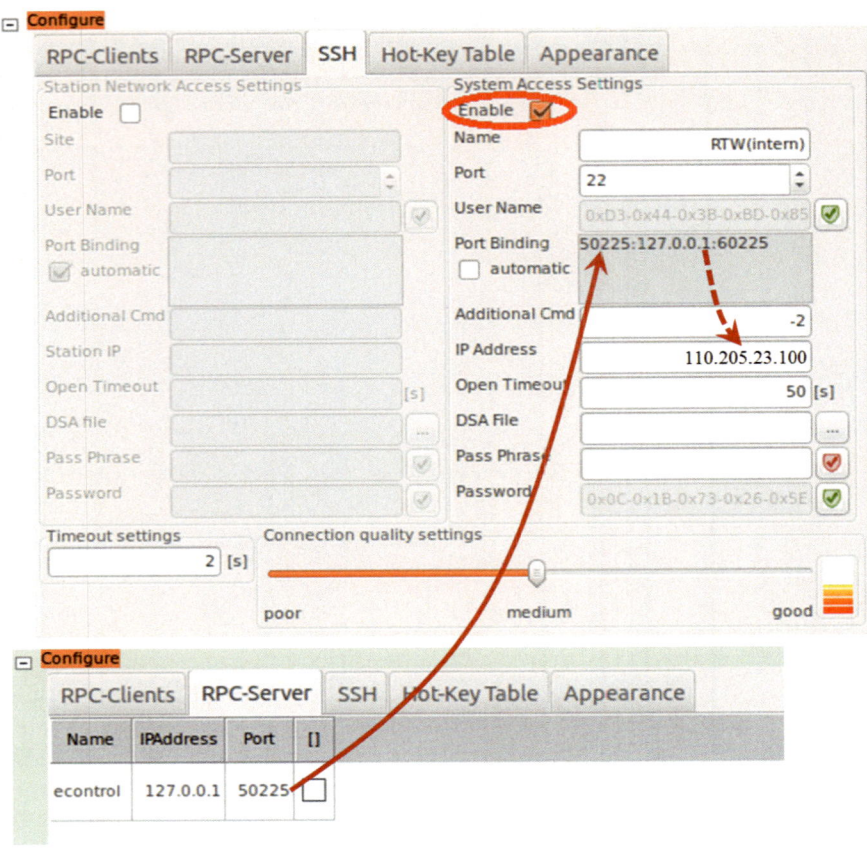

Fig. 5.31 An example for the graphical configuration of the SSH brokerage, which uses the redirection of terminals in the background

time on. The best way would again be to enter such a password each time it is required. But this limits the communication management again. Besides the use of asymmetric ciphering algorithms for that key, another very simple technique of «data chopping» avoids direct readability of the password from a plain memory dump during the short period the password is used. The trick is to avoid standard strings for the password. Instead of this, linked lists are implemented, in which each character of the secret information is connected to each other, but is stored in unpredictable positions. One technique is hashing (see ▶ Sect. 2.7.2). Another is the insertion of random numbers of additional characters to the list in between the characters of the secret information to allocate memory on different, not sequentially ordered addresses. The additional characters can then be deleted again after the complete list is created, so that only the password characters are left which reside on the random addresses. Experience has shown that this is a good technique, which is something like «steganography»[75] in the memory of a process.

All of this shows that safety interacts with security and vice versa. It also shows how quickly one of them is reduced and limited in favor of the other. Security and safety are a wide field. Nevertheless, a basic knowledge is very important for all developers of such remote control interfaces. But a 100% secure and safe system will never be buildable. A system is always only as safe and secure as its design criteria and as the handling by the user. If an operator does not care about security issues, the system accessed will not be secure against attacks. It is always important

---

75 Steganography is a technique to hide secret information cleverly within a harmless carrier data set.

to have responsible handling in combination with a certain degree of distrust, so that unexpected behaviors are detected and investigated. Security mechanisms and regular updates of them on all computers used and continuous personal development are further steps to a reliable use. Then it should be possible to use all the techniques described and similar mechanisms for the development of future observing systems and future observatories. Experience at Wettzell has shown that the risk is calculable, while the opportunities to develop future tasks are immense.

5.6 Summary

5.6.1 What Did We Learn?

Having the autonomous production cells, it is feasible to extend the existing systems with functionalities for remote access. As remote operators are connected via worldwide networks, a permanent access is not guaranteed. Therefore, the systems have to increase their ability to run completely autonomously. They use orders from external partners to optimize their workflows and report results and the current states as feedback to external clients. Such systems use agents which are combined to multi-agent systems.

As VLBI only works with worldwide networks of radio telescopes, the chapter uses the use case of VLBI to explain the techniques required. It explains the principles of Very Long Baseline Interferometry and characterizes the challenges of the future telescope networks.

Most VLBI telescopes of the space geodetic techniques use an external software provided by the NASA: the NASA Field System. For multi-agent systems with remote access and automated control, the Field System is like legacy code from external partners in software projects. Similar techniques which are known for testing scenarios can be used to get access to the functionalities of the existing software. Nevertheless, we saw that the Field System implements similar control structures as discussed in the chapter about the development of control systems.

The chapter also discusses additional requirements if a system is remotely controlled. Web cameras, remotely controlled switches, and a suitable graphical remote interface extend the local architecture to replace the senses of an operator at the system. Using these techniques, additional control strategies can be implemented to run the systems of space geodetic techniques.

The option of remote access to all functionalities requires an expansion of network security to fulfill the security factors. A suitable access control which implements different levels is essential. While usernames and passwords which are encrypted with symmetric ciphering algorithms are suitable for the protection of local computer systems, external access can only be guaranteed if the computer filters arriving requests. The area in which remote users are allowed to be active has to be restricted as much as possible. An individual authentication can be implemented with role-based access control where users only have permission to run specific tasks related to their role. We discussed packet filters using filter tables. The encryption of complete data streams is implemented with asymmetric ciphering algorithms which use a pair of keys to encrypt and decrypt messages. This technique is used by SSH. It provides the option to tunnel one's own communication through encrypted SSH channels.

The final sections showed us how complete networks can be protected. A separation of the networks in an observatory makes it possible to separate between office applications and control systems to improve the stability of the control. The individual networks are separated with firewalls using filter tables. The firewalls can be passed using SSH tunnels. Higher-level proxy firewalls provide acceptable access control to the complete observatory. To simplify the management of system access, redirecting of TTY inputs and outputs was discussed.

5.6.2 How Did We Use the Contents Learned?

The knowledge of testing methods makes it possible to include external software into multi-agent systems. We used stubs and mocks to encapsulate the functionalities of the NASA Field System.

The knowledge of security issues was used to divide observatory networks into suitable control sections. The different techniques of access control are implemented and code snippets show the essential parts in detail. The software used for remote control at the Wettzell observatory implements techniques of TTY redirection and SSH tunneling. Implemented is also a role-based access control. Firewalls and filter tables are one essential technique to reduce the risk of network attacks at the Geodetic Observatory Wettzell.

5.7   Questions

1. Draw a sketch showing the functionalities of a software agent. Explain the single tasks with one sentence.
2. Why is an underlying middleware essential for multi-agent systems?
3. How does VLBI work?
4. Which different types of antenna dishes do you know? What are their characteristics?
5. Which antenna is better, the one with higher SEFD values or the one with lower?
6. What does T_{sys} describe? Which value should it have?
7. Compare the NASA Field System with the control system developed in ► Chap. 4. Which similarities do you find?
8. Which techniques from software testing help to integrate an external software like the NASA Field System into the multi-agent system?
9. Which additional things must be considered when operating large systems remotely?
10. Explain the different control strategies used if large systems are operated remotely.
11. Which security factors do you know?
12. Explain the different solutions for access control.
13. What is the principle of a symmetric ciphering algorithm compared to an asymmetric ciphering algorithm?
14. Check the TCP/IP stack! Up to which layer do «iptables» control the network communication?
15. Why are large prime numbers so important for ciphering algorithms?
16. What are checksums? Why are they also useful for binary communications on serial lines?
17. What is a three-way handshake and why is it essential for the establishment of low-level communications as well as for task-based role changes?
18. What are network enclaves and why do they improve the situation for networks which are used for control tasks?
19. What is the difference between a proxy firewall and a package filtering router?
20. What does «SSH Off-Host forwarding» mean?
21. Why is it not enough to redirect input and output streams of SSH clients to manage the input of passwords to the software?
22. What is a TTY?
23. What does a «line discipline» module do?
24. What is steganography and how can it be used to reduce the chance to read passwords while sending them over a redirected TTY?

Coordination, Communication, and Automation for the GGOS

6.1   The Global Geodetic Observing System (GGOS)
        at a Glance – 484

6.2   Operational Deficits of GGOS – 488

6.3   Coordination, Control, and Operation
        of the Terrestrial GGOS Infrastructure – 491

6.4   Communication on GGOS Networks – 494

6.5   Automation as Key to Deal with 24/7 – 504

6.6   Summary – 507
6.6.1   What Did We Learn? – 507
6.6.2   How Did We Use the Contents Learned? – 507

6.7   Questions – 508

© Springer International Publishing Switzerland 2017
A. Neidhardt, *Applied Computer Science for GGOS Observatories*, Springer Textbooks in Earth Sciences,
Geography and Environment, DOI 10.1007/978-3-319-40139-3_6

6

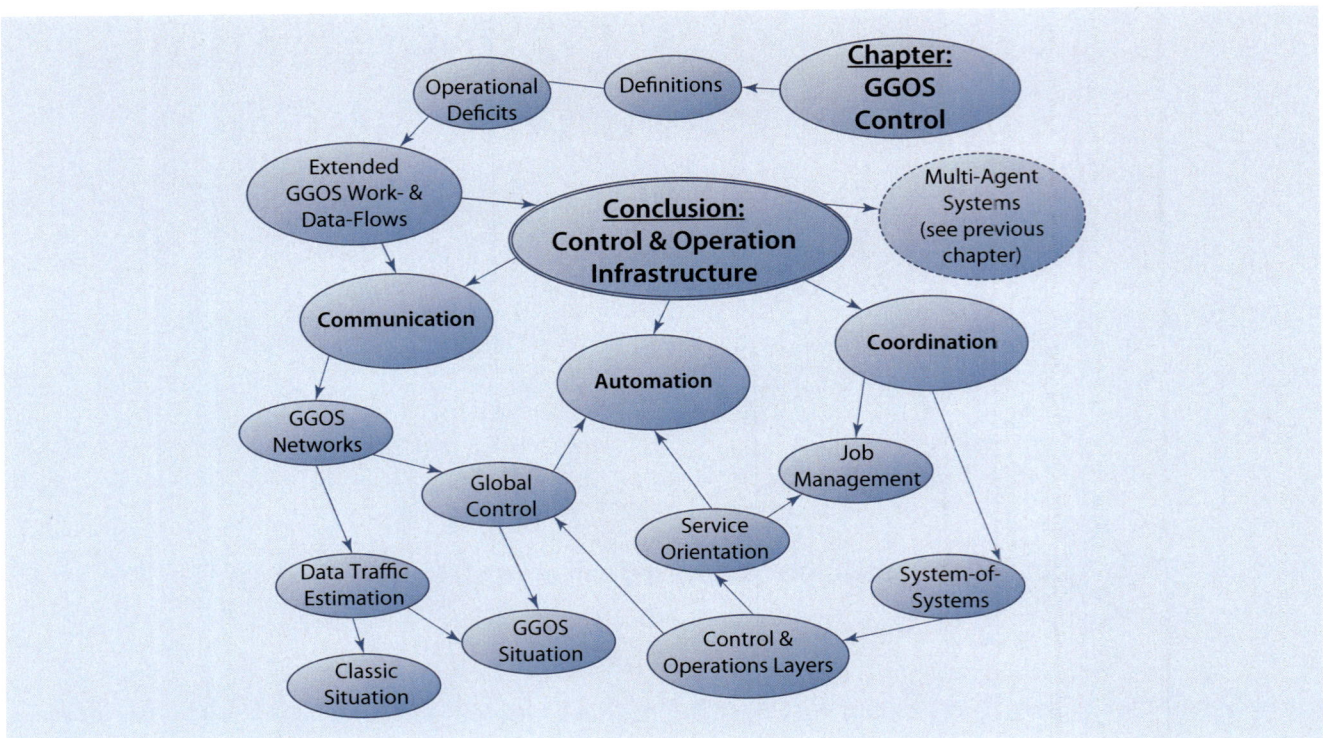

6.1   The Global Geodetic Observing System (GGOS) at a Glance

GGOS History and Purpose

GGOS was established in July 2003 (for the following also see (Rothacher et al. n.d.)). It is the definition of an observing system of the IAG. In the year 2007, it acquired the status of a full component of the IAG after a first design and implementation phase. GGOS is both an organization structure and an observation system (see ◘ Fig. 6.1). One aspect is supported by the Steering Committee, Science Panel, Working Groups, and so on, while the observing is done with an existing, but extended geodetic infrastructure, composed of the systems of the geodetic services. They support the establishment of precise reference systems,[1] which then support the three pillars of geodesy, which are geokinematics, Earth rotation, and the gravity field (see also ◘ Fig. 1.1). Another function of GGOS is to increase public awareness of geodesy through the publicity raised.

GGOS as Unifying Umbrella

In this context, GGOS is a unifying umbrella of all existing and future IAG services as an interface to the «outside world» of geodetic users in the form of an elementary geodetic infrastructure. While networking between the services is highly supported, GGOS is a unique interface of geodesy to science and societal needs. «The implicit vision for GGOS is to empower Earth science to extend our knowledge and understanding of the Earth system processes, to monitor ongoing changes, and to increase our capability to predict the future behavior of the Earth system» (Rothacher et al. n.d.). This includes sea level changes, water storage changes, atmospheric sounding, deformations, surveying, and disaster monitoring, based on

---

1   «[Reference systems] are conventional coordinate systems that include all conventions for the orientation and origin of the axes, the scale, and the physical constants, models, and processes to be used in their realization. Based on observations, these systems can be realized through their corresponding [']reference frames['].» (Rothacher et al. n.d.). The ICRF is an epoch-dependent set of estimated positions of extragalactic reference radio sources. The International Terrestrial Reference Frame (ITRF) is an epoch-dependent set of estimated positions and velocities of globally distributed reference marks on the solid Earth surface (Rothacher et al. n.d.).

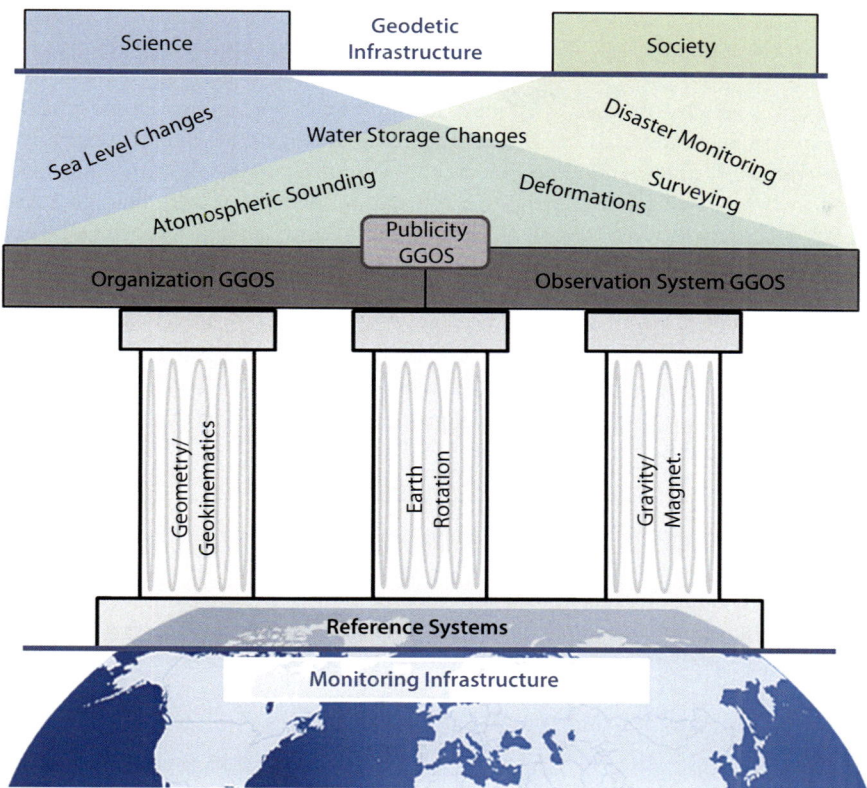

Fig. 6.1 The «big picture» of GGOS: the basis is a continuous monitoring of the Earth, establishing precise reference frames, such as the support of research in and services for geokinematics, Earth rotation, and the gravity field. GGOS itself is the unifying umbrella of these services as an interface of geodesy to science and society

dynamic Earth system processes, whether caused by natural effects or modified by human interactions. GGOS will have a high impact on society, because a growing population is greatly affected by high risks from natural hazards, which affect the economy on national and global levels or cause the displacement of large populations in their aftermath. Precise and continuous monitoring in combination with a better understanding of Earth system processes helps to reduce anthropogenic impacts and to preserve resources for future generations on our restless planet.

GGOS is one of the most advanced and respected visionary descriptions of future geodetic services. A general description of all aspects can be found in the book *Global Geodetic Observing System – Meeting the Requirements of a Global Society on a Changing Planet in 2020* (Plag 2009). It is a collection of ten articles, describing (Plag 2009, *l.c. page xiii ff.*):

GGOS Vision Paper

- «*Living on a dynamic planet – the challenge*»
- «*Geodesy is fundamental in meeting this global challenge*»
- «*Geodesy is in transition*»
- «*International cooperation is essential for geodesy*»
- «*The development of the geodetic observing system needs research*»
- «*The benefits of the global and national geodetic infrastructure are enormous*»
- «*The societal prospects of space geodesy*»
- «*Towards a geodetic Earth system service*»
- «*Geodetic observations and products are crucial for maximizing the benefits of Earth observations*»
- «*Geodesy is essential for exploring the planets, solar system, and beyond*»
- «*User requirements for geodetic observations and products are demanding*»
- «*Towards a modern geodetic reference frame*»
- «*Infrastructure for geodetic Earth system monitoring*»

6

■ «*For a full exploitation of the potential, an operational core component is needed*»
■ «*Implementation of GGOS needs a multi-faceted organizational framework*»

GGOS Main Idea

The book describes the role of geodesy for the future, and its impact on modern society, which is dependent on accurate georeferencing and based on precise reference frames. It allows a continuous monitoring of changes. The three pillars of geodesy, which are geokinematics, Earth rotation, and the gravity field (see also ◘ Fig. 1.1), provide the conceptual and observational basis for those reference frames. Continuous observations, which are geo-tagged in this reference frame, allow a better understanding of the global changes in geodynamics, hazards, water cycles, sea level dynamics, and so on (Plag 2009, *l.c. page 4 ff.*). GGOS is an attempt to identify future requirements in the field of geodesy. On an abstract level, it describes a design, implementation, and administration of a future geodetic Earth observation, which combines all space geodetic techniques, so that they seem to work as one system.

Global Geodetic Observing System (GGOS)

> *Definition DEF6.1. of «Global Geodetic Observing System (GGOS)»:*
> condensed from (Plag 2009)
> GGOS is the abstract description of the design, implementation, and administration of a global geodetic Earth observation, combining all space geodetic techniques. The resulting geodetic products like the geodetic reference frames and the monitored changes seem to be produced by one single, worldwide, distributed observing system.

Operational Specification for GGOS

According to the specific user requirements for societal applications, Earth observations, natural hazards, Earth science, lunar and planetary science, and so on, GGOS creates products with sufficient accuracy over time. In order to do that together with the IAG services, it ensures that global networks of reference stations, gravimeters, and tide gauges are operated within a geodetic observing network. About 30 to 40 globally well-distributed core reference stations co-locate the different space geodetic techniques at one location to allow the integration of the technique-specific results into one terrestrial reference frame and to enable a validation of the quality and accuracy. The determination of the local ties between the co-located techniques is important for that purpose. The observations must then be processed with an accuracy and consistency of 1 part per billion, which are then integrated into a dynamic Earth reference model, including variations of Earth's geometry, gravity field, and rotation. The final products and the data are offered via a database and a GGOS portal, where the procedures, standards, and conventions used are documented. These data sets are continuously kept in a consistent and accurate way (see (Plag 2009, *l.c. page 224*) and (Plag 2009, *l.c. page 241*)).

«Big Picture» of the GGOS Instrumentation

To ensure the operational specification, GGOS defines five levels (Plag 2009, *l.c. page 240 ff.*):

1. *The terrestrial geodetic infrastructure*, which consists of the ground networks of parallel working radio telescopes, laser ranging stations, permanent GNSS receivers, stations of the French Doppler Orbitography and Radiopositioning Integrated by Satellite (DORIS), gravimeters, tide gauges, and geodetic timing stations.

2. *The LEO satellites*, which are the low-orbiting Earth observation satellites or satellite formations for ocean and ice altimetry, gravimetry, or optical imaging, which also carry equipment for the other space techniques like corner cubes for the SLR.

3. *The GNSS and geodetic Lageos-type satellites*, which are the satellites of the different global navigation systems GPS, Globalnaja nawigazionnaja sputnikowaja sistema, (GLONASS), Galileo, or BeiDou (Compass), and so on, in combination with passive (equipped only with retroreflectors) and active

(having additional transponder equipment to amplify the return signal) laser ranging satellites.

4. *The planetary missions and geodetic infrastructures on other planets*, which are orbiters around and landers on the moon and on other planets like Mars and Mercury, to allow the imaging of the surface and the production of new coordinate systems on these celestial bodies.

5. *The extragalactic objects*, which are quasars and other compact radio sources, used to define the celestial reference frame and to monitor changes in it, such as structure variations.

The levels described demonstrate that GGOS combines a complex system of differ- **GGOS Overall Design and Tasks** ent sensors and instruments on Earth, in the air, and in space. This combined system appears to the outside world as one large, comprehensive, global, geodetic instrument for Earth monitoring. Besides this instrumentation, GGOS consists of three further parts: the data infrastructure; the analysis, combination, and modeling centers; and the GGOS portal. The data infrastructure ensures the data transfers and communications between the instruments and the data management and archiving systems. Their different, consistent data processing chains analyze the raw data, combine them, and generate complex numerical models of the Earth system with them. The results are accessible as standardized products on a specific GGOS portal (Plag 2009, *l.c. page 237 ff.*). These crucial components together solve the tasks to define, provide, maintain, and transform the different reference systems, constants, and models. They offer frequent Earth orientation parameters, parameters of the gravity field, and the shape of the land, ice, and ocean. This allows descriptions of mass transports between the atmosphere, oceans, and the land. Additionally, parameters describing the total electron content of the ionosphere and the water vapor content of the troposphere can be derived (Plag 2009, *l.c. page 219*).

In general, GGOS is also a big chance to improve the current situation of geo- **National Requirements** detic observations and to increase their value to society. This is also addressed in the national requirements for the shared resource of a precise geodetic infrastructure in the USA, described in (National Academy of Sciences 2010). The realization of this vision is becoming increasingly important as «[...] the geodetic infrastructure has become increasingly fragile as a consequence of delayed replacement of aging components, lack of redundancy with single-point-of-failure designs, imperfect collaboration among contributors, reductions in the trained geodetic workforce due to retirements, and ongoing tightening of operations and maintenance budgets. These factors combined pose a risk of a sudden, drastic loss of geodetic observing capability [...]» (National Academy of Sciences 2010, *l.c. page 101*). The main recommendation is that the geodetic infrastructure should be maintained and improved also as a matter of national security. This can be done with network upgrades, modernization of the observing systems, and the use of multi-technique capabilities, which require funding opportunities for research, analysis, and education (National Academy of Sciences 2010, *l.c. page 102*).

One main focus of the recommendation report is put on the next generation of **Key Techniques** automated high-repetition-rate laser tracking systems, on the next-generation VGOS techniques, and on a national network of high-precision GNSS stations with real-time, high-rate streaming possibilities. It is also recommended that the global density of the fundamental stations is increased to at least 24. Several nations must cooperate to reach this goal. The support of the international geodetic services plays a key role here that global data can easily be exchanged between national, federal geodetic services (National Academy of Sciences 2010, *l.c. page 104 ff.*). Even though the report is more or less focused on the area of the USA, all of these suggestions can also be partly adapted to the situation in other countries and nations.

To enable the realization of these requirements, one aspect which is not usually paid much attention is also the techniques to control and operate such a global geodetic system. The visions mostly talk about automated systems and do not touch on implementation aspects. The following sections discuss these aspects on the basis of solutions previously discussed. They focus on the first level of GGOS, which is the terrestrial geodetic infrastructure as part of the instrumentation component of GGOS. Aspects of the data infrastructure which is extended with an additional control and operation infrastructure are included. The ideas are based on experience from the operation of the Wettzell geodetic fundamental station and will be extrapolated to the operation of GGOS core sites.

6.2   Operational Deficits of GGOS

Operational Deficits

The problem with most GGOS descriptions is that networks in this implementation-related context only describe geodetic networks at the level of observation combinations and final products. GGOS keeps the separation of the space geodetic techniques at independent observatories, which exist in parallel and separated, and combine only their results into global products. GGOS seems like a global, unified, and common observing system only in the area of analysis and combination and is not a globally cross-linked observing system on the level of instruments and sensors. But as proposed, core stations with co-located instruments offer several advantages to compare and combine data into a unique reference frame. Additionally, they allow the assessment of observation qualities and accuracies (Plag 2009, *l.c. page 241 f.*), especially if they are connected to one timing system and if they can be optimally controlled, so that they support the combination, for example, with parallel observations of the same satellites or with an optimal interleaving between common tasks.

Combined, Globally Cross-Linked Operations and Coordination

With the methods already described in the previous chapters, it is also possible for the GGOS instruments to be operated as one global observing system. The observatories, space geodetic techniques, and control of them have become cross-linked to form a control network, which can be coordinated from centralized operational centers according to the individual requirements for the scientific GGOS goals. With agent systems, it has become possible for techniques on one site, or even in a regional or global operational network, to be combined for one common scientific goal, for example, if VLBI, SLR, and GNSS receivers observe the same satellites at the same time. If, for example, SLR sites do not only work completely independently, but get coordinated for specific tasks, it will also become possible to manage the huge observation load for SLR sites, which will increase, if all GNSS satellites of the American GPS, the Russian GLONASS, the European Galileo, the Chinese BeiDou (formerly Compass), or the locally usable Japanese QZSS are observed with laser ranging techniques.[2] While currently each SLR observatory decides independently only with a limited priority list, which satellite passage is observed, this is not useful anymore if more than 100 GNSS and other satellites are being observed frequently. It makes no sense if two telescopes at a close distance observe the same passage for a routine tracking, while another satellite passage is not observed at all. But if weather conditions on one site prevent an observation, it might be possible at another site which is nearby. Another suitable situation might be the validation of results, so that at exactly the same time, two or more observatories could track the same passage segment. Combined, globally cross-linked operations and coordinations of the instruments support the societal applications formed by GNSS and Earth science tasks (see also (Plag 2009, *l.c. page 210 f.*) about GGOS requirements).

On-Demand Services

Another aspect where the agent systems can support GGOS is the offer of fast, real-time services. The current GGOS papers describe the products and services in such a way that the parallel operated space techniques offer predefined products with predefined accuracies, time intervals, and properties. This seems a little bit

---

2   But not all satellites will have corner cubes or retroreflectors for SLR in the future.

like: «what we do not have, you do not need».[3] But in modern service-oriented societies, users might have the requirements to adapt the products' properties. For example, if planetary or lunar spacecraft are tracked, the operators of such spacecraft require the possibility of requesting an observation when needed by the mission on demand. On the one hand, satellite operators also have some restrictions for operations. Currently, this is more or less weakly solved with a simple Go-Nogo Flag,[4] which defines when a target is not allowed to be observed and whose correct implementation is part of the station's observing software. But remotely accessible agent systems allow the remote activation of restrictions or observation orders directly by external partners, which can check the correct activation of the remotely requested restriction as feedback of the system. On the other hand, satellite Earth observations might necessitate a more precise orbit observation if the satellite is used for measurements during natural hazards (see also (Plag 2009, *l.c. page 211 f.*) about GGOS requirements). In these situations, satellite operators can propose a request for more SLR observations of their satellites directly to the agent systems of the GGOS network, which then optimizes the schedules according to the requests, using the defined metrics.

With the possibility of direct interactions between users and the agents of the GGOS system, it is not only possible for the users to get direct feedback, the systems themselves can also get feedback, as analysis centers and users can manipulate the system behavior according to their quality evaluation. This is a direct feedback to the observatories. For example, if the analysis centers reduce the priority of specific satellite observations or restrict them totally, because the observations only have a limited impact on specific products or requested user demands, the work at the observatories is optimized with this feedback toward the common goal of the complete GGOS system. Nowadays, several Ajisai satellite passages are tracked with SLR as Ajisai is a good target to hit with a laser, while the impact for geodetic research from this satellite is limited. Direct manipulation of system parameters according to a particular evaluation metric, which also appears locally at the observatories, improves quality control and optimizes the use with these additional feedback structures.

Adding these additional data flow requirements to the data flow model in (Plag 2009, *l.c. page 261*), four main categories of dataflows can be identified (see also ◘ Fig. 6.2):

GGOS Feedback Loop, Quality Control, and Optimization of Its Use

Extended GGOS Data flows

1. *Auxiliary and metadata for the control tasks:* These are all data sets (see the dashed lines in ◘ Fig. 6.2, flowing top-down), which are required to plan schedules and observations and which are needed to run an observation (e.g., orbit data, positions of sources in a star catalog, etc.). Such data are mainly prepared by the data centers (GGOS data centers or external data centers, e.g., of the satellite operators). They are uploaded to or fetched by the different space techniques at the observatories if they need them to run the observations. The formats and structures of the auxiliary and metadata are mostly system-specific (e.g., VLBI schedules with the positions of radio sources in a specific format or the CPF for satellite predictions for the SLR), but might be standardized to a general format.

2. *Control data:* These are all data sets, commands, and requests which directly control systems, manipulate the processing during observations, or propose the need for an observation (see the solid lines in ◘ Fig. 6.2, mainly flowing top-down). There are different levels of control influences, which are defined by the different user roles (see ▶ Sect. 5.5.1). On-demand control requests are sent by the GGOS users (if satellite

---

3   This is the title of a German bestseller about living in the former east regions of Germany in the present day, written by Dieter Moor.

4   Such a Go-Nogo Flag is used for specific satellites for which the SLR observation is only allowed within clearly defined time intervals to avoid damages of the on-board instrumentation.

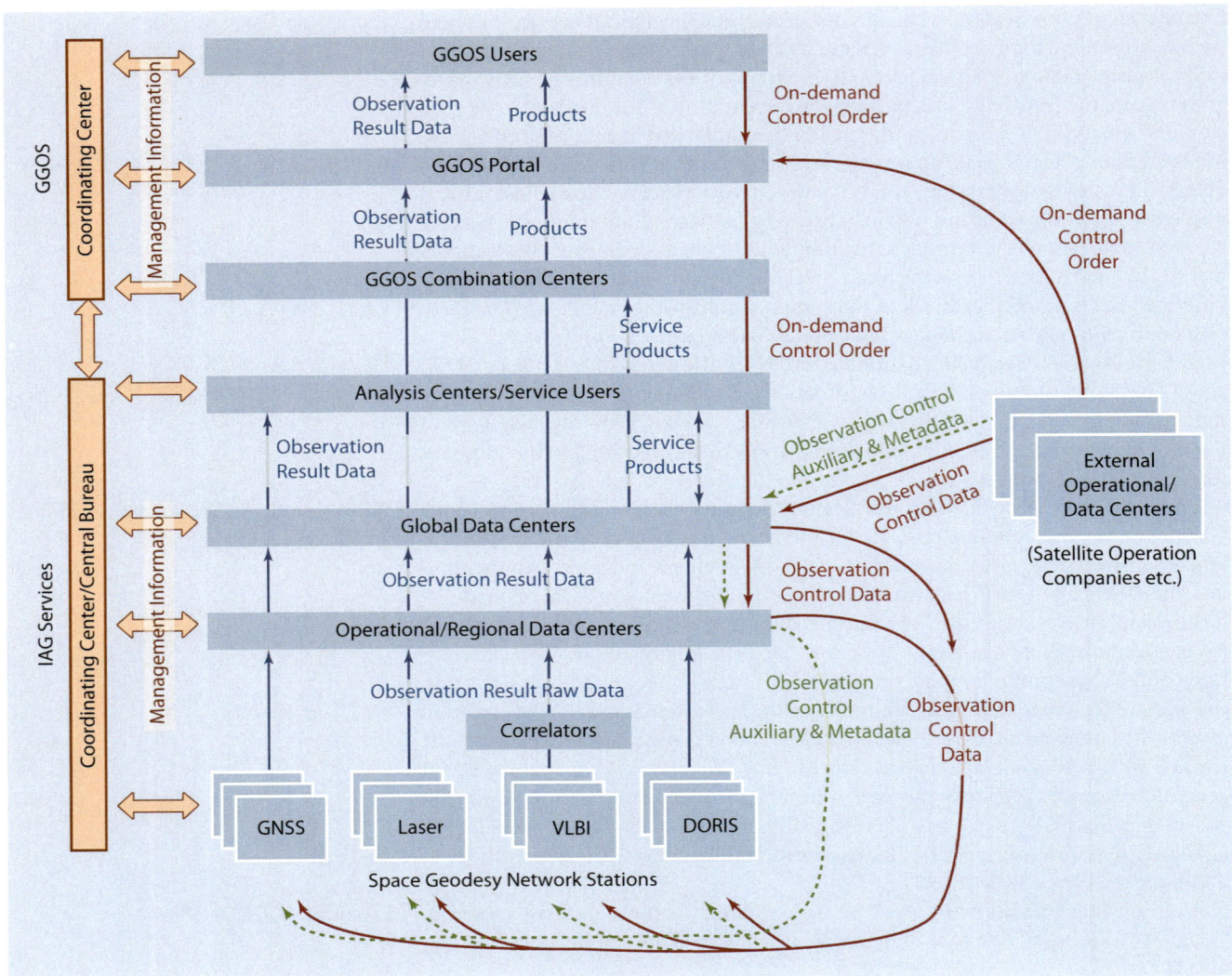

Fig. 6.2 The extended dataflows through the GGOS entities, split into auxiliary and metadata for the control tasks, control data, and observation result data (On the basis of the design in (Plag 2009, *l.c. page 261*))

operators request observations, they act like other GGOS users) via the specific GGOS portal to the dedicated agent systems below, whose first level is the operational data centers. They evaluate the demands according to the metric parameters and create observation control requests to the agents of the space techniques. Additional control data, such as real-time observation restrictions from satellite operators, are included. Then each observatory and each system there decide whether the observation is possible according to the local decision metric and the local situation. External on-demand users do not directly control the observing system. They act as notifiers (which express the desire for an observation to a local staff member), schedulers (which propose complete observation sequences with locally defined control), or agents (which influence specific recording parameters or settings for the observation, but do not directly control the real hardware) according to the user roles defined in
▶ Sect. 5.5.1. The control restrictions are injected as agent inputs, or partly as operator inputs, which can partly cause real-time interrupts.

3. *Observation result data:* These are all data which have already been defined by (Plag 2009, *l.c. page 261*) (see the solid lines in ▶ Fig. 6.2, flowing bottom-up). They are the raw observation data, which are then evaluated and prepared by

the regional and global data centers, so that the analysis centers and combination centers can generate the standard products offered by the GGOS portal.

4. *Management information:* Management information are the horizontal dataflows to the international services, which administrate their subnetworks on a higher level. Management includes general planning and general influence into control streams, so that it triggers specific observations according to a plan or that it manipulates observations. Usually, it is not in real-time. The horizontal management flows focus more on the planning aspects, creating statistics, and performing optimization analysis for the whole network. The results are feedbacks, for example, in the form of metric optimizations or processing and observing guidelines, which must be implemented at the systems of the space geodetic techniques.

6.3 Coordination, Control, and Operation of the Terrestrial GGOS Infrastructure

GGOS is a system consisting of different observing systems. While the single, heterogeneous systems alone offer a well-defined set of possibilities and resulting products, the combination of the systems into a more complex system offers more functionality and options than simply the sum of the results of the single systems. The single subsystems are operationally and managerially independent, geographically distributed, emergent, and evolutionary. They combine command, control, communications, information, and surveillance (Jamshidi 2009, *l.c. page 2 f.*).[5] The complexity of designing this cooperative coordination and interoperability in a system of systems mainly comes from the heterogeneity of the subsystems. For this reason, new design principles are discussed.

System of Systems

A look into the previous chapters shows that already the complete control system for one SLR site is already a system of systems. It is a combination of different, heterogeneous components and sensors, which are combined and controlled together to fulfill a more complex task (see ► Sect. 4.2.1 about the distributed systems). The combination of these components produces more than the sum of the results from each single part. Zooming into the different components and sensors shows that the hardware control is also distributed and builds a system of systems (see ► Sect. 4.2.1). The next levels, which also combine several autonomous production cells into one control enclave and then several control enclaves into one observatory, follow the same structure. Combining these observatories into regional or global control and operation structures is no longer so different. The hierarchical mechanisms of recurring structures with abstracting middleware communications adapt to each level of the different combinations of systems of systems. This means that it might be possible to extend the mechanisms and design criteria described also to a structure, as required for the complete terrestrial geodetic infrastructure of GGOS to form a system of systems of systems, and so on. The result is a GGOS control and operation infrastructure (see ◘ Fig. 6.3).

Recurring Structures

The higher levels of such a system of systems are again similar to the local structures of an observatory. Higher-level control centers combine control of the different observatories, while they have specific access rights to influence the operations there. The GGOS Operation Control Centers can be split into regional control centers, for example, which are responsible for one country or one particular geographical region (see the control of the AuScope network in ► Sect. 5.4), and global control centers, which consolidate the locally operational centers to one

Distributed and Centralized Control Centers

---

5 GGOS is a Global Earth Observation System of Systems, as it combines the space techniques and other sensors to get greater and more accurate output for the Earth models and science (compare (Jamshidi 2009, *l.c. page 21 f.*)).

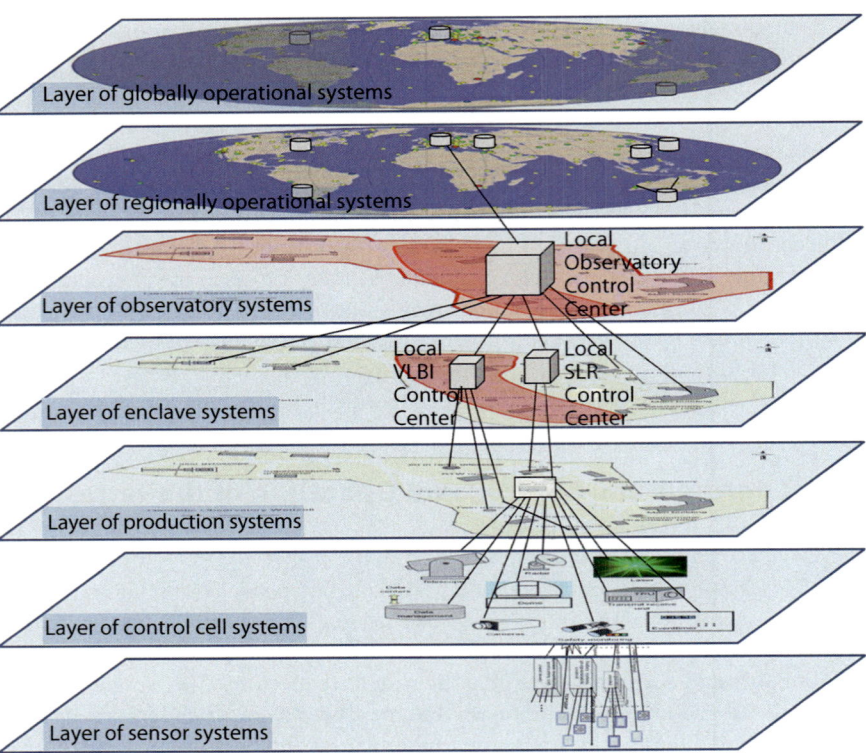

Layer of globally operational systems

Layer of regionally operational systems

Local Observatory Control Center

Layer of observatory systems

Local VLBI Control Center

Local SLR Control Center

Layer of enclave systems

Layer of production systems

Layer of control cell systems

Layer of sensor systems

◻ **Fig. 6.3**   A possible GGOS control and operation infrastructure with different control layers, where each is combined as a system of systems

connection point for the control of the complete GGOS system. Generally, this builds a distributed control system of systems. Nevertheless, it is also possible for a regional control center to act as a global control center. Each layer of the control centers can be seen as a globally designed, horizontal control structure.

Access Rights of the GGOS Control Layers

The access rights and permissions of each layer are strictly determined by its following lower layer on the basis of the defined, hierarchized authorization pyramid (see ▶ Sect. 5.5.2). This guarantees that each system enclave within an observatory can set the permissions for general access for the higher observatory control center. These general static rights can be temporarily changed using the dynamic user roles, which temporarily allow direct control. Once defined at one layer, the user and control roles are forwarded to the higher layers in the GGOS control stack. Each higher center can accept the permissions, and forward them to the next level, or can restrict them. This means that a control center of an observatory can define its own low-order access rights to its systems.

GGOS Job Management System

According to the defined authorizations, the different hierarchical levels with its agent systems can be controlled or influenced. New data streams are formed besides the data streams for the observation results. The new streams contain meta- and auxiliary data for the control tasks with information required for an observation (e.g., passages, EOP, observation restrictions, etc.). Additionally, they contain observation control data with observation requests, task negotiations, direct control orders, or predefined schedules. The data streams flow through the control layers, which form a distributed GGOS job management system. For a user of that system, and especially for a user of the whole GGOS system, it offers the functionalities of a general booking system with control services on demand. The implemented request/reply mechanisms of the agents allow an improvement in the service quality, for example, satellite operators can choose whether an SLR site tracks their satellite or not. Together this implementation allows:

- A fast data propagation of resulting, intermediate analysis, and control data, as the observer role allows a propagation of the data immediately through the whole stack
- A fast troubleshooting and feedback during commissioned observation orders, as the notifier role allows direct notifications on the basis of the control data received if there are errors and an interaction is required
- A common and shared use of the whole network of different space techniques, as the scheduler role enables the specification of a common observation course
- An optimization of the observation results in real-time, as the agent role allows the manipulation of specific observation-relevant control parameters
- A shared real-time control, as the operator role allows complete access to the systems when safety issues are solved

There are numerous use cases for such a «customer-oriented» approach. For example, if several GNSS satellites are observed, different SLR sites can coordinate each other. It is possible to increase the sampling rate of meteorological and other auxiliary data from GNSS sites on demand if there is a temporary requirement for a specific research task, for example, during natural hazards. After natural hazards, such as earthquakes with large ground movements, it becomes possible to focus all techniques to detect the new ties of a site, for example, by releasing specific observation plans, which automatically and immediately influence the different systems of space techniques. Co-location and local ties researches become easier, as a particular scheduler or customer at a university can propose a request for an observation series, for instance, of the same satellite with SLR, VLBI, and GNSS at the same time, while the system of systems decides which observatories are able to support the request. During and after the observation, all data sets can be requested for the analysis in real-time or near real-time. For spacecraft tracking scenarios, it might also be helpful to request observation time on demand. Currently, requests are often sent directly by email to the appropriate people at the observatories. They have to decide if an observation is possible. But this procedure requires sufficient planning time ahead of an observation session. With agent systems in the GGOS job management system, the agents can make their own decisions according to predefined metric parameters and the current situation. This can also help to focus on orbit segments if a more accurate orbit observation is scheduled, for example, for space debris measurements.

Use Cases

Daylight Hopping of Control Responsibilities The agent systems in a layered observing system also support the proposed optimization of observation shifts within a complete network, as suggested in ▶ Sect. 5.4. Shifts in general are very time-consuming for the operational staff at an observatory. Additionally, night shifts are not easy psychologically. The shifts also reduce the time of engineers for other tasks like maintenance, development, and installation of new techniques. A solution for an improvement might be daylight hopping of control responsibility around the world (e.g., (Hase et al. 2010)). As tested for the new observation control strategies in ▶ Sect. 5.4, the implementations make it possible to take over remote shifts for foreign observatories. It is possible for three or more, replicating worldwide control centers in different daylight zones to share the operation shifts. That control center in the daylight zone always controls the complete GGOS network. Then the control rotates with the daylight around the Earth, and each control center only requires observational staff for the day shift, while at the rest of the day, the other centers take over all control and monitoring duties. During these time periods, the inactive centers only have to implement something like an on-call service which is called when critical error situations require a manual, local interaction. These calls can even be made automatically by the control computers without a human interaction, using a modern Internet telephone and a computer-generated voice.

Several remote control tests were performed to show the functionality and abilities of remote control, also using daylight hopping. The longest, continuous test was demonstrated during the continuous VLBI observation CONT11 in September

Proof of Concept

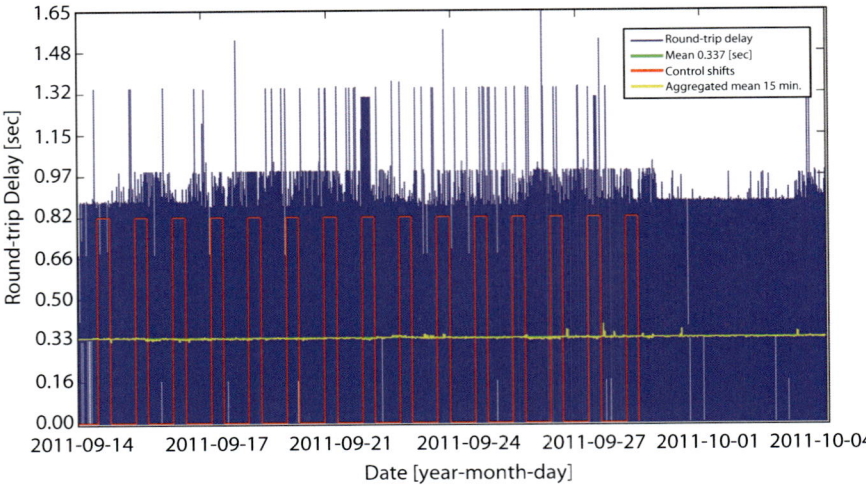

▣ Fig. 6.4 Proof of the concept of daylight hopping of control responsibilities over 15 days during the continuous VLBI observation CONT11 from 15-SEP-2011 00:00 UT through 29-SEP-2011 24:00 UT (The rectangles show night shifts at the observatory TIGO in Chile, remotely operated in the daylight zone at Wettzell)

2011, where Wettzell ran all night shifts of the TIGO VLBI system (see ▣ Fig. 6.4). The long-term test showed that the reestablishment of the connection after network blackouts worked quite well, so that no control gaps appeared. The remote operator always had complete access to the controlled system. The mean round-trip delay was about 1/3 seconds, while the aggregated 15 minute mean value showed no general, longer-lasting, temporary delays to the total, arithmetic mean round-trip delay. Single requests lasted up to 2 seconds, where most of these outliers were shorter than 1.3 seconds. Nevertheless, during this time the antenna was completely unattended and had to make its own decisions.

To implement such ideas, it is important to follow the suggestions of the previous chapters. To optimally support the idea of the GGOS control structures, the data transfer and network design for an observatory can be refined again.

6.4   Communication on GGOS Networks

Data and Control Traffic

The basis for all transfer and control requests is the defined multi-agent systems and their specific agents for each observation system (see ► Sect. 5.1). They use middleware structures (see ► Sect. 3.4) to create hierarchical control centers at the observatories and over the whole GGOS network (see ► Sect. 5.4). Suitable network enclaves can ensure enough security to protect the observation process and the controlled systems themselves (see ► Sect. 5.5.3). The access point to each enclave is a package filter firewall or router, using access rules with «iptables» (see ► Sect. 5.5.2). The whole observatory is protected by a proxy firewall (see ► Sect. 5.5.4). The problem is that firewalls generally reduce or limit the throughput, as they require time to check the rules or to process the communication over the proxy server. Thus, they limit the data and control traffic. Usually, this is not too relevant for the control traffic, as the repetition rate of control requests can be reduced due to the autonomy of an autonomous production cells. But for the resulting observation data transfers, it might become a problem depending on the volume of data transmitted within a specific time interval.[6]

---

6   Most of the following parts were collected during the work for the internal, unofficial GGOS working group for coordination, communication, and automation as part of the GGOS Bureau of Networks and Communication (now: GGOS Bureau of Networks and Observations).

Usually the data transfer can be split into real-time and not real-time data. The classic way of transferring GNSS data is the transmission of hourly and daily files of the 30 seconds sampling data from the GNSS receiver. This means that every hour a compressed file with a size between 30 and 80 kilobytes (in the medium currently about 35 kilobytes), which depends on the received navigation satellite systems (e.g., GPS, GPS/GLONASS, GPS/GLONASS/Galileo, etc.), is sent from a permanently installed receiver location to a centralized data center. There, for example, at the location of an observatory, the data files are validated and verified. After these checks, they are sent to the data centers, from where they are fetched by the analysis centers. Once per day a concatenation of all hourly files is sent as a daily file from the receiver site to the data center in the same way as the hourly files. As the daily file is a sequence of the hourly files, its file size is 24 times the hourly sizes and therefore between 720 kilobytes and 2 megabytes (in medium usually lower than 1 megabyte). As these files should arrive at the data center within 10 minutes after the sending, a theoretical maximum data rate of about 1 kilobit per second is required for the hourly data, while about 30 kilobits per second is necessary for the daily data files. These transfer results in a data volume of 80 megabytes per day if the daily and also the hourly data of about 30 permanently installed GNSS sites are collected (as performed by the Geodetic Observatory Wettzell). This data amount corresponds to about seven JPG images from a current middle-class single-lens reflex camera with a resolution of 18 megapixels.

Asynchronous Data Traffic for GNSS

In addition to the classic GNSS data transfer, the real-time data stream with Networked Transport of RTCM via Internet Protocol (NTRIP) is available. NTRIP is a standard of the Radio Technical Commission for Maritime Services (RTCM) and is something like an «Internet radio stream» with GNSS data in the RTCM-102 format or as raw data (e.g., in the National Marine Electronics Association (NMEA) format). The purpose is to offer real-time corrections for the Differential Global Positioning System (DGPS)/ Differential Global Navigation Satellite Systems (DGNSS) of stationary or mobile users. «A DGNSS reference station in its simplest configuration consists of a GNSS receiver, located at a well-surveyed position. Because this stationary-operated GNSS receiver knows where the satellites are located in space at any point in time, as well as its own exact position, the receiver can compute theoretical distance and signal travel times between itself and each satellite. When these theoretical values are compared to real observations, differences represent errors in the received signals. RTCM corrections are derived from these differences. Making these corrections available in real-time for mobile users is the major purpose of the Ntrip system elements [...]» (Bundesamt für Kartographie und Geodäsie (BKG) 2004, *l.c. page 3–1*).[7] NTRIP is an application-level, generic, stateless protocol on the basis of HTTP and usable on all IP networks (GNSS Data Center n.d.). NTRIP uses three layers (see ▫ Fig. 6.5, (Bundesamt für Kartographie und Geodäsie (BKG) 2004, *l.c. page 2–1 ff.*) and compare also the multilayered data propagation in the autonomous system monitoring and safety cell in ▶ Sect. 4.7):

NTRIP

1. *The NtripServer with a NtripSource:* This is a permanently installed GNSS receiver at a continuously operating reference station, which provides GNSS data as data stream for a specific location. The source is described with specific parameters, like the format used, the received navigation system, the coordinates of the location, and so on. All of these GNSS data from the NtripSource are read by the NtripServer, using a serial or an Ethernet connection. The received data are then transferred by the NtripServer to a specific NtripCaster at a GNSS Data Center, where each source must be registered. For the NtripCaster, each NtripServer is just a client program sending data.

---

7   Reproduced by permission of Bundesamt für Kartographie und Geodäsie.

6

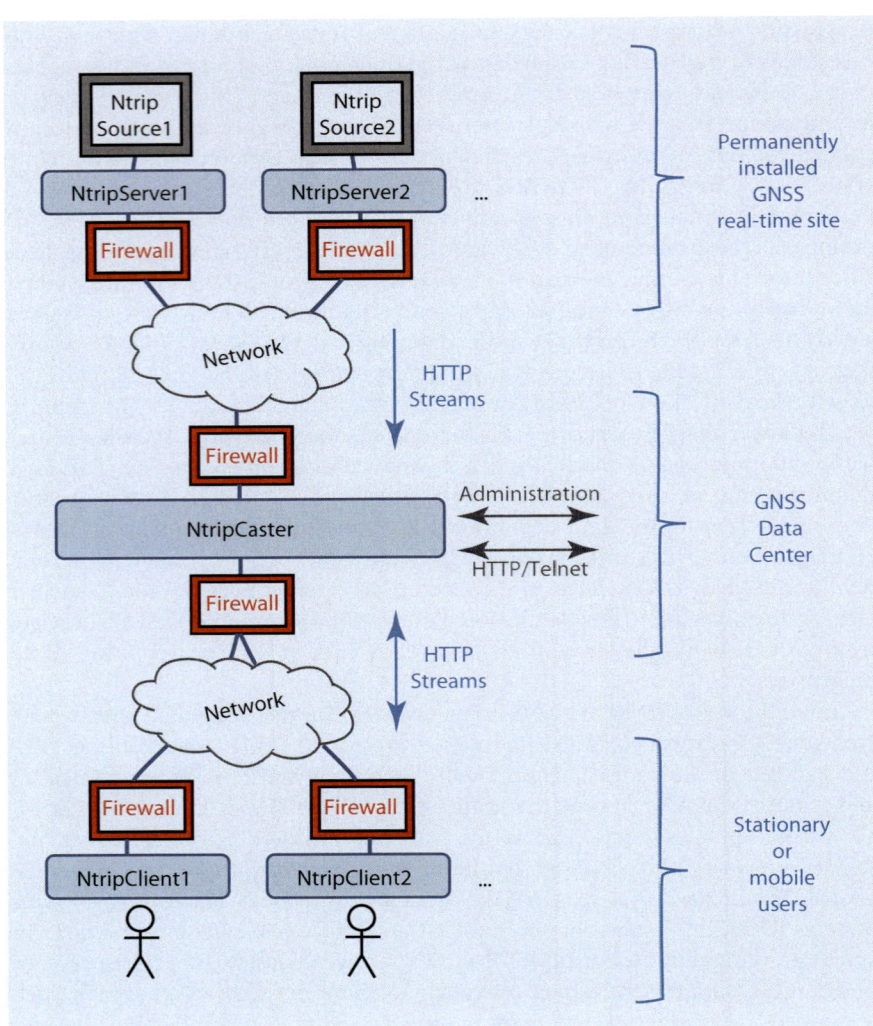

■ **Fig. 6.5** The layers of a regular NTRIP communication from the GNSS site to the potential users of the correction data (Adapted from (Bundesamt für Kartographie und Geodäsie (BKG) 2004, l.c. page 2–1 ff.))

2. *The NtripCaster:* Each single NtripSource is a unique, possibly password-protected mount point on an NtripCaster. It is an HTTP server for low-bandwidth streams with 50 to 500 Bytes per second and per stream. The NtripCaster accepts requests from either NtripClients or NtripServers, while both are only clients. According to their functionality, the caster decides if it is an incoming or outgoing data stream. The caster is administrated by the GNSS Data Center, where also each new source must be registered.

3. *The NtripClient:* The NtripClient is a client program, which requests data streams for local GNSS corrections. The NtripCaster supports these requests and forwards the data over a fully HTTP 1.1 compliant, but not persistent connection.

Real-Time Data Traffic for GNSS

The sent NTRIP data are usually GNSS data sampled with one second. They are used for different purposes, which cause different data rates. Real-time kinematics require a data rate of about 5 kilobits per second, while the DGPS just needs 0.5 kilobit per second with a latency of maximum 2 seconds (Weston 2005). The permanent sites of the observatory use a rate of 5 to 8 kilobits per second.[8] Some permanent sites also offer meteorological data with similar sampling rates, so that the required data rate is slightly higher. As the data are sampled in 1 second, the data volume per day is at least 30 times higher than for the 30 seconds sampled hourly files, which means about 1.2 gigabytes per site. This corresponds to about 200 JPG images from a current middle-class single-lens reflex camera with a resolution of 18 megapixel.

The future of GNSS includes more navigation satellite systems, such as a fully operative Galileo. The data rates will increase, but will not become much higher than a few times of 10 kilobits per second. Increasing the sampling rate for particular topics might have a greater effect. Usually, it is also not necessary to use additional control streams as the permanent sites run unattended. From time to time, maintenance accesses via SSH are required. But already current networks can easily deal with all such data rates and data volumes whether real-time or classic.

Future Data Traffic for GNSS

Unlike GNSS, SLR systems need additional data prior to the observation (e.g., orbit predictions, Earth orientation parameter, etc.; see ◘ Fig. 4.1), which must be requested from or sent by an external data provider. Additionally, the data transfers of an SLR site are usually always asynchronous and never in real-time, because the NP algorithm needs a completely observed orbit segment for the statistics. The only time constraints are that the data required for the observation must be available and converted into the internal data structures before the observation starts and that the resulting observation data must be sent in an acceptable time interval (within hours or on the same day) to the according data centers after the observation.

Data Traffic for SLR

The predictions for currently about 66 satellites (including the lunar reflectors) come every day as CPF files and result in less than 500 megabyte per month of data volume. Each file is between a few times of 10 to a few times of 100 kilobytes, depending on the contained prediction interval (usually about 1 week). This means that a transfer rate of 70 to 100 kilobits per second is enough to transfer all prediction files for 1 day within 30 minutes from the data center EDC or CDDIS to a specific site. All other required data (e.g., EOP: 13 kilobyte, clock corrections < 1 kilobyte, meteorology << 1 kilobyte, etc.) are far below those data rates.

Data Traffic to an SLR Site

The resulting data sent by an SLR site are mainly the NP files and the full-rate files in the CRD format for each successful observation. The data volume and rates can theoretically be calculated on the basis of the file format definition in comparison with real observations. For NP files each entry of an NP requires 106 bytes, and each file contains additional bytes in a header and trailer of about 681 bytes. If an NP is build every 15 seconds, files with sizes of up to a few times of 10 kilobytes are produced independently from the repetition rate of the laser (e.g., a 20 minute passage will result in about 9 kilobytes of data). As the files are transferred asynchronously to the receiving data centers, there are no hard time constraints. But if a maximum delay of 10 minutes after the observation might theoretically be defined (e.g., as given for the GNSS data), a theoretical data rate of very far below 1 kilobit per second would be enough. Therefore, the data traffic from NP data is not relevant in current networks.

Data Traffic from an SLR Site: Normal Points

Unlike the NP data, which are independent from the laser repetition rate, and only defined by the interval for one normal point, full-rate data increase with the higher repetition rates. A 10 hertz laser will produce 4800 bits per second in the CRD format, while a kilohertz laser will write about 480 kilobytes per second. Including the CRD header again, this would mean a theoretical «real-time» transfer rate of about 5 kilobits per second for a 10 hertz laser and of about 500 kilobits per second for a 1 kilohertz laser. But currently all data are sent as complete files after the observation. Therefore a 20 minute observation will produce a file with

Data Traffic from an SLR Site: Full Rate

---

8   Special thanks should be given here to Uwe Hessels for the provision of the statistical load data of the GNSS streams!

700 kilobytes of data at a 10 hertz system and about 70 megabytes at a kilohertz system (this maximum data amount is again corresponding to about 6–7 JPG images from a current middle-class single-lens reflex camera with a resolution of 18 megapixels). With the same 10 minute constraint for the transfer time as for GNSS data, 10 kilobits per second is enough for a 10 hertz laser system to send the file right on time. The kilohertz system would need about 1 megabit per second to do the same.

Future Data Traffic for SLR

Future data traffic will include an increasing number of satellite predictions, which must be fetched, and also higher return rates, because of higher repetition rates of new lasers (e.g., kilohertz laser or multicolor systems). As this increase mainly concerns the size of full-rate data, they mostly define the requirements. In other words, a 10 kilohertz laser with two colors will produce full-rate data with about 10 megabits per second. This is also enough to stream the file within 10 minutes to the data center after the observation. Current networks can already deal with the future load of data to and from SLR sites.

Data Traffic for VLBI

While GNSS and SLR can already be easily supported with current network scenarios, VLBI is increasingly becoming a challenge. The sizes of the files required for a VLBI observation (schedule files, meteorological data, etc.) are here comparable with the sizes received from an SLR site (e.g., usually lower than 200 kilobytes for SKD files and lower than 500 kilobytes for VEX files; see ▶ Sect. 5.3). But as the recorded raw data of VLBI consist of noise, which cannot be compressed and which is currently sampled with up to 256 megabits per second, huge data volumes are created (for each observation, a log file is also written, containing all commands and outputs, which is usually of a size of a few 100 kilobytes to a few megabytes, depending on the local output). Because incoming data to a VLBI site are not really significant, the main focus must be laid on traffic from the VLBI site to a correlation center.

Data Traffic from a VLBI Site

As the resulting data volume is so high (several terabytes; besides the small log files with a size of about a few 100 kilobytes), a lot of VLBI sites send their data on the Mark5 modules, using a carrier service, which ships the eight-pack hardware to the correlator. But at several sites, e-VLBI techniques (see ▶ Sect. 5.2) are becoming increasingly common. Nowadays, it means that geodetic VLBI experiments are usually completely recorded and sent afterward via the Internet to the correlator (e-Transfer). Standard data rates for a single peer-to-peer WAN connection are here about 800 megabits per second (maximum over 900 megabits per second) on a 1 gigabit per second local connection (see also (Wagner n.d.)). The limiting factor is the connection bandwidth at the correlator, where several data streams must be received in parallel. Experience at the Wettzell observatory has shown that usually streaming, for example, to the correlator at Bonn, can be done with 400 megabits per second, which means that a standard 24 hours observation with about 1.5 terabytes of data (this corresponds to about 60 video films on Blu-ray disks with all possible features) can be transferred within 9 hours to the correlation center.[9]

Tsunami UDP Protocol

To run such transfers, the Tsunami UDP Protocol was established. The Tsunami protocol is «[a] fast user-space file transfer protocol that uses TCP control and UDP data for transfer over very high speed long distance networks» (Wagner n.d.). The origins of the current version can be found in a project of the Indiana University 2002 Tsunami protocol. The current protocol transfers large files by splitting them into (usually) 32 kilobytes blocks, which are numbered. All of the communication control is done by the client, such as setting the communication parameters, defining the number of blocks, and requesting retransmissions. The server only adjusts the transfer rate according to the commit error rate information from the client or throttles down the defined limiting maximum rate by pausing the sending for specific time intervals. While transferring data blocks, the client notes all missing

---

9   The Wettzell observatory currently has a connection with a maximum rate of one gigabit per second.

blocks in a list to request them again from time to time from the server for a retransmission. If too many retransmissions are required, the communication is completely reinitiated and starts with the transmission of all blocks again from the first one missing. The protocol is mainly used for e-Transfers (after the observation) but can also be used for real-time transfers if a special hardware is installed (VSI-B card and driver, developed at the Metsahovi observatory) (Wagner n.d.).

The near future of the upcoming VGOS already means increasing the sampling rates on four broadband channels. This means for the transfer that four streams must be transferred with more than 1 gigabit per second. Including retransmissions, losses because of the CSMA/CD, and so on, the next possible connection with 10 gigabits per second will only fulfill the requirements. But this rate is not only necessary for local networks but also for the WAN transfers. An upgrade of the correlator connectivity is here much more essential. The storage technology must also be improved. VGOS suggests 30 seconds of observation time per source with 1 gigabit per second data rate per channel. For every 30 seconds of slewing time, a new source should be observed again. In total, this means for a 24 hour session with 1,440 scans, a single antenna will produce about 21 terabytes of data in single-bit mode and in only one signal polarization (14 times more than the 1.5 terabytes of today). This data volume currently corresponds to 860 video films on Blu-ray disks with all possible features. The suggested TWIN radio telescopes will produce double this volume (42 terabytes). The two-bit mode would double this amount, and if also the currently executed tests with 16 gigabit per second (e.g., for four bands with two signal polarizations) are also comprised, completely new dimensions of transfer rates and data volumes will break new ground (e.g., it can be estimated that one single antenna might produce about 80 terabytes per day in such modes).

To deal with such new challenges, it is necessary to use new network technologies. Besides the use of dark fibers,[10] projects dealing with high bandwidth on demand are becoming increasingly important. The background of such high-speed connections is a suitable high-speed infrastructure for the WAN. To enable such networks, several initiatives in Europe are currently funded by the European Union Framework program, such as for the GÉANT project, which focuses on the creation of a pan-European research and education network that interconnects a set of National Research and Education Network (NREN) of the participating European countries with a connectivity speed up to 500 gigabits per second (GÉANT n.d.). On the basis of that, Ethernet network infrastructure provisioning tools of the NRENs can find a suitable, ideal path through the network nodes to create an end-to-end, point-to-point bidirectional connection. Then end users can request a direct connection between two end points with guaranteed transport capacities from 1 megabit per second up to 10 gigabits per second over the whole WAN path (DANTE n.d.). The first tests for VLBI were performed for the NEXPReS project (see also ▶ Sect. 5.4). Here the network services infrastructure of GÉANT was used to dynamically request a bandwidth up to 10 gigabits per second between telescopes (e.g., Jodrell Bank of the University of Manchester in Lower Withington, England) and the correlator of the Joint Institute for VLBI in Europe (JIVE)[11] in Dwingeloo, Netherlands (Langevelde 2013, *l.c. page A.7 f.*). This has already enabled real-time correlations, which means that the data are directly transferred to the correlator during the data acquisition at the telescopes, so that the processing can be done live.

The previous qualitative discussion shows that VLBI currently already defines and will define even more the required connection type and transfer rate of future GGOS sites. The near future vision of VLBI is only possible if all telescopes are connected to the WAN at least with 1 gigabit per second (up to 4

Future Data Traffic for VLBI

Bandwidth on Demand

External Connectivity of GGOS Observatories

---

10 Dark fibers are fiber optics or glass fiber cables, which are already installed in the ground but not yet used, so that one can hire them (e.g., also temporarily) for network setups for the peer-to-peer connection in the WAN.

11 Now JIVE European Research infrastructure Consortium (JIVE ERIC).

gigabits per second for real-time transfers using bandwidth on demand) and if all correlators have a connectivity of 10 gigabits per second and enough data storage capacities for the huge data amounts. The multi-agent systems (see ▶ Sect. 5.1) then must not only negotiate the observation times but also the bandwidth on demand for the connectivity between the observatory and the correlator. In the more distant future, these rates must be topped, so that 10 gigabits per second can be guaranteed from each observatory to a correlation center, where the connected load must be several times of 10 gigabits per second. These requirements will also have a huge impact on the operational costs of an observatory.

General Situation

Because of the high data rates with VLBI and the future requirements to stream or send VLBI data via the Internet or bandwidth-on-demand connections directly to the correlator, it is not suitable to use recording systems behind extended firewall systems (see ▶ Sect. 5.5 starting). Currently, each antenna has its own control system (NASA Field System PC or FS PC) in combination with DBBC or Mark4 systems, which are connected to a local data recorder like the Mark5 using VSI (see ▶ Sect. 5.2). All of these components are behind firewalls in one network, usually the VLBI control enclave (see ▶ Sect. 5.5.3). Because firewalls and especially gateways check package contents of the higher network stack layers, which is time-consuming, the data rates are limited. One possibility is to take the exchangeable data modules out of the recording system and put them into a separate Mark5 machine after a session, which is directly connected to a network behind the (gigabit-) switch. It offers high data rates, but only IP filtering as protection. From here the data can be streamed quite fast to the correlators. This method uses «e-Transfer». A big disadvantage of this practice is that it is combined with several manual interactions and that streaming is only possible after a session, which prevents the ideas of e-VLBI, where the data should be streamed in real-time or scan by scan.[12]

Physically Separated Transfer Network

Because of the limiting factor of standard firewalls, direct streaming is difficult and the required hardware is quite expensive. A simple and cheap solution is the establishment of a high-speed transfer network between the digitization and formatting equipment and the recorders (see ◘ Fig. 6.6). A physical separation between the internal control network and the transfer network with connection to the insecure Internet is essential to prevent security attacks. But this physical separation is offered by the VSI as connection, which is a separate cable between a DBBC and a Mark5B+ in classic VLBI system. The newer 10 gigabits per second refer to connections (or multiple 1 gigabit per second connections, which are optimally used with a form of round-robin allocation; see ▶ Sect. 2.6.4) between a DBBC with a FILA10G extension[13] and the high-speed recorders like Mark6 or Flexbuff also support this physical separation. The electrical design does not enable a forwarding in the FILA10G. Thus, there is physically no backdoor between the two networks in the hardware, which prevents the control network from hacker attacks. The transfer network can also be optimized for the transfer of large data junks, for example, by increasing the Maximum Transmission Unit (MTU) size, which is usually 1500 bytes in regular networks. The use of «jumbo frames» with several kilobytes or even several megabytes reduces the wire time proportional to the payload enormously, but requires the transport protocol (see ▶ Sect. 3.3.1) to handle the huge packages which are on their way to their destination.

Currently three possibilities can be considered to ship the data from the formatter (over the transfer network) to the recorder (see ◘ Fig. 6.6):

- The first method does not use a transfer network, because it directly connects the Mark5B+ to the e-VLBI network. The Mark5 recorder must then be at a close distance to the formatter, like the DBBC to which it is connected via VSI.

---

12 One scan is one observation interval of one single source.
13 FILA10G is a double port ten gigabit per second VLBI-Ethernet interface, formatting raw VLBI for the recording on high-speed recording systems.

— Another technique uses a transfer network between two Mark5B+, where one is in close connection to the formatter using VSI and the other implements the connection to the e-VLBI system. This has the additional benefit that the data are redundantly stored in case of recording problems.

— Real benefits are given if high-speed transfer networks are used to connect fast recorders, like Mark6 systems, with modern digitizers/formatters, like the DBBC 2, in combination with an external FILA10G device or the DBBC 3 with an included FILA10G board (HATLab n.d.). The recorders are usually connected to the FILA10G device with two 10 gigabits per second interfaces and enable transfer rates of several gigabits per second from the DBBC to the storing hard drives.

All of the network solutions make it possible for the recorder to be located remotely, for example, in a data center to where all data are sent. They also allow the complete high-speed streams from modern broadband systems. Nevertheless, the data recorders themselves are inviting targets for hackers, because they are connected to a high-speed Internet and might be used as origin platform for Denial-of-Service attacks on other computers worldwide. But this risk can be reduced using the «iptable» described (see ▶ Sect. 5.5.2). They can be used to reduce the open ports in the e-VLBI network and for the Internet to those which are really necessary for the

Security in the e-VLBI Network

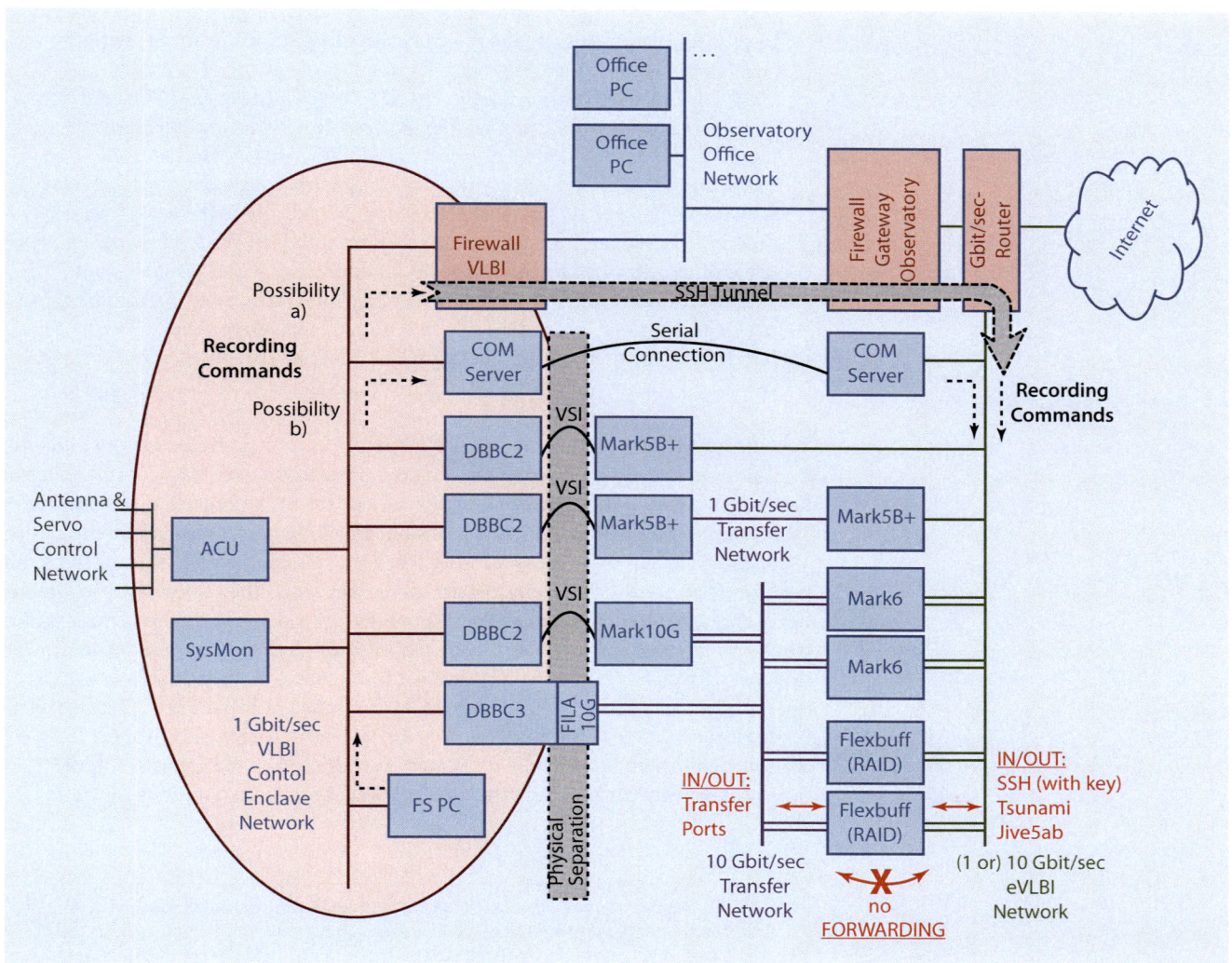

◻ Fig. 6.6 The possibilities for high-speed VLBI data acquisition, using a physically separated transfer network and a secure command forwarding via **a** tunneled communication through the firewalls or **b** with an additional, traditional serial communication line

transfer protocols of «Jive5ab»[14] or «Tsunami». The only user access allowed uses SSH with a suitable key file (protection only with passwords is too weak) from the observatory network and not from the Internet. Forwarding to other network interfaces must completely be blocked. The network connectivity in the transfer network can also be reduced to the transfer protocols to reduce propagating attacks to other machines if a single machine has already been hijacked. It is important that no other (virtual) networks run over the same physical connection and that all machines use fixed IP addresses and not Dynamic Host Configuration Protocol (DHCP).[15] An active monitoring and Intrusion Detection System (IDS) are another important installation which should be considered. It monitors network activities and checks violations against predefined security policies.

Networks for an Observatory

Taking all this into account, the following networks can be identified for a GGOS observatory:

— *Several control networks (control enclaves):* These are the standard, already described control areas for each system and for the observatory itself. They use a standard Ethernet technology and are separated from each other with a package filtering router, a dual-homed host or a proxy firewall (see ► Sect. 5.5 starting).

— *One or more high-speed data transfer networks (transfer enclaves):* The high-speed transfer networks are direct peer-to-peer links or connections via 10 gigabits per second switches with high bandwidth and an optimized communication, e.g., by increasing the MTU size. Several tests are currently in progress to run such real-time transfers in LANs, for example, with the additional software «Jive5ab» with its use of the high-performance UDP-based Data Transfer Protocol (UDT)[16] (JIVE (Eldering, Bob) n.d.).

— *External network connection with high-speed bandwidth-on-demand capabilities:* This is the e-VLBI connection to the world. The external connection to a GGOS observatory must support communications at least with a transfer rate of 1 gigabit per second, but ideally with 10 gigabits per second. As the data transfer is located in the standard WAN networks, it is difficult to optimize the MTU size, as all intermediate distributors must support the settings. UDT is again very useful. An alternative might be the use of a dark fiber. A solution which is already available is the bandwidth-on-demand service of several NRENs. Usually the data are nowadays streamed with optimized protocols, like the Tsunami protocol.

Command Tunneling

When dealing with equipment in separate networks, control becomes more difficult if additional backdoors to internal enclaves need to be avoided. Commanding of hardware is quite important, because the central coordination cell, or in the area of VLBI, the NASA Field System, is responsible for all connected devices. In the case of VLBI, the Field System tells the recorder when the recording should start or stop. This is no problem in classic systems, where all devices are in the same network domain. But if they are in different Local Area Networks (LANs), the commands must tunnel from one to the other network, while the answers and responses must return the same way. Two possibilities are planned for the Geodetic Observatory Wettzell:

— A classic SSH tunnel (see ► Sect. 5.5.2) from the control enclave through all firewalls to the e-VLBI network (see ◘ Fig. 6.6a). Commands tunnel through this channel to the dedicated recorder, while the answers follow the same way back. A big disadvantage is that many components must be

---

14 «Jive5ab» resides on the data recorder Mark5 and replaces the classic programs, having the same functionalities and interfaces (JIVE (Eldering, Bob) n.d.).
15 DHCP is a network protocol to distribute and dynamically assign network configurations like IP addresses.
16 UDT is a high-performance data transfer protocol designed for transferring huge data volumes over the long-haul connections in the WAN.

passed which have to be kept fail-safe. It also requires active management of the tunnel itself (see ► Sect. 5.5.5).
- An easier solution uses a classic serial communication between two communication servers (COM Servers (see ◼ Fig. 6.6b)). All commands are converted into messages which pass a two-wire connection between the servers, which can be one's own developments or industrial products. The advantage is that there are no IP message stack and also no IP forwarding on this connection. The endpoints are also clearly defined, where the messages are interpreted. Such a solution is usually installed for the configuration of hardware switches, which implement the connection to the Internet. Individually programmed serial connections also provide an active switching between different receivers, so that the recording can be switched between different data recorders.

All of these scenarios enable another important possibility: spreading the data over the recorders with round-robin access (see ► Sect. 2.6.4). The high data volumes of future VLBI systems make it impossible to run a manual logistics. It must be organized to automatically record data on the recorders and transfer data to the correlation centers. Intelligent switching between hardware recorders reduces the requirements within the recorders themselves for a parallel recording and streaming access. A coordination communications server can organize the data logistics, while selecting the suitable storage system according to a predefined metrics. It is important that enough space is available and distributed in an intelligent way to keep the data stored until the correlation is finished and to organize a suitable round-robin mechanism between the different data pools. First estimations tend to a mixture of recorders with removable media (Mark6) and static data raids (hard drive stacks with fixed media sizes; Flexbuff). If the correlation capacities and also the network capabilities to the correlators are given, so that maximum delays of 1 week appear, three storage systems per antenna with a capacity of 1 week of data produced plus about 10% overhead in combination with several removable medias should be enough for a trouble-free handling.

Round-Robin Recording

Last but not least, it is also necessary to take a look at the data rates required for the remote control of a system, using, for example, «e-RemoteCtrl», for VLBI. The following data were taken during a few intensives on weekends between 8:30 a.m. and 9:30 a.m. Central European Time, where the 20 meter antenna of the Geodetic Observatory Wettzell was controlled from the home of the operators at a distance of about 24 kilometers from the observatory and over telecommunication networks and the German NREN. The update rate was not managed, so that after each returned request, a new one was initiated, which leads to the worst-case scenario and best-time update rates for information. In this case, the client receives about 150 megabytes and sends about 15 megabytes of data in 1 hour. This corresponds to about 19 megabits per second receiving and about 2 megabits per second for sending.[17] But as the updates are only repeated after a return of a request, longer-lasting requests just reduce the nominal data rate but do not influence the functionality of the software. Successful tests were also performed to Antarctica with its satellite links and a net speed of 256 kilobits per second. A problem with these unmanaged update rates is that they always produce maximum performance, which is not desirable if a volume tariff must be paid for the Internet connection. Future releases of the «e-RemoteCtrl» will also take care of this aspect.

Data Traffic for e-RemoteCtrl

VLBI is, as shown, a special case in the transfer scenario of an observatory. After the correlation, the data volume reduces to comparable data sizes, as given for the other space techniques. The high-speed problem is mainly located between the observatories and the correlation centers. Later on, the standard network capabilities of the WAN are sufficient to support the GGOS services.

Further Processing

---

17 As with all data rates, this is just a theoretically estimated rate.

6.5   Automation as Key to Deal with 24/7

Potential of Automation

Already in the early books about fundamental stations, widely automated measuring processes were proposed as the key to running the observatories with only a few operators (see (Schneider 1990, *l.c. page 38*); proposed for the mobile, integrated, geodetic observatories). The technique of multi-agent systems with unmanned, autonomous production cells which offer stable, safe, and notifying operations explicitly supports all mechanisms of automation, autonomy, remote access, and remote control, as mentioned in ► Sect. 5.2. Real-life examples impressively demonstrate the possibilities (see ► Sect. 4.2.2 about the experience with the Liverpool telescope). Having an additional look at the requirements for the VGOS, it is also obvious that «continuous measurements for time series of station positions and Earth orientation parameters» (Niell et al. 2005, *l.c. page i*) are planned. This calls for reliable 24/7 operations. But this contrasts with the operational costs and required manpower at the observatories. The prognosis is that a higher degree of automation offers a lot of new possibilities and is the only chance to manage the workload generated by operations which run for 24 hours, 7 days a week (24/7).

Workload Estimation

A local estimation of operational manpower for VLBI at the Geodetic Observatory Wettzell shows that the workload of VGOS is not manageable without an increase of automated processes. The internal estimation defines as preconditions that the staff will not be reduced, that there is at least one position funded by third-party funds per year, that student and pure operator contracts will increase, and that the operational staff can operate from one control room or can be shared between the different space techniques. The main focus was laid on continuous automation of the systems. Nevertheless, the assumptions made are very optimistic.

Workload Splitting

For the estimation, the workload of 100% available personnel was split into the following parts:

- *Maintenance work:* Covers controlling and maintenance of equipment, for example, greasing the bearings or maintaining the cryo-system. Also updates and upgrades of hardware and software are part of this work.
- *Repair work:* Covers mechanical repairs after defects. Changing hardware or electronic boards, or replacing defective control units and software, is another aspect of this work.
- *Development and research work:* Includes software and hardware developments, testing new systems and techniques, and supporting new research goals including the writing of publications and papers (mainly funded by third-party funds).
- *Observation and operations work:* Includes all operational tasks, such as preparing observations, running observations, quality control of results, and the preparation of the shipment of the data. The observations include interventions if there are critical situations or problems during the observation itself. Currently three 8 hour time intervals («shifts») are required for one 24 hour observation day.
- *Administrative work:* Preparing the ordering of equipment, time accounting, commissioning hardware and software, project planning, and meetings are part of the administrative work.
- *Support of other instruments:* Required when the VLBI system personnel work for other space techniques or run observation campaigns at other observatories, for example, at the German Antarctic Receiving Station O'Higgins on the Antarctic Peninsula.
- *Other:* Lectures at universities, public tours, youth development in school labs, attending workshops, and writing reports are allocated to this item.

Current Workload Splitting

The starting point of the estimation is the current splitting of manpower for the different working tasks, which gives the initial situation of 100%. Currently (in the year 2013/2014) it means (see also the graphical presentation in ▨ Fig. 6.7) that

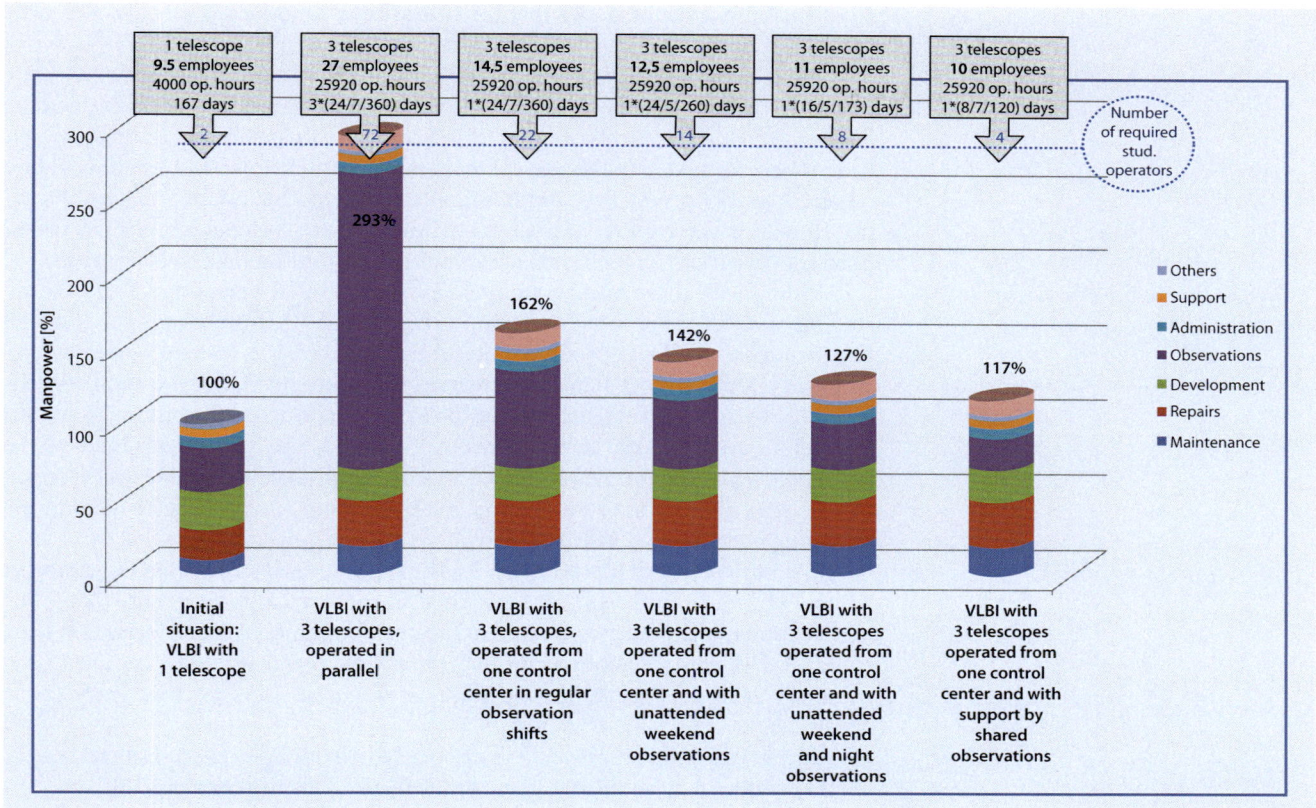

● **Fig. 6.7**   The workload splitting for VLBI as it is currently and as it would be for 24/7 operations with and without automation of observations

10% of the total available manpower is employed on the maintenance of the current 20 meter VLBI system and its components. Repair work claims 20%. About 25% of the manpower is employed in development. On average 7% of the total manpower is employed in the administration. About 8% of the total manpower is employed on other tasks, including the cooperative support of other measuring systems at the observatory. But already, about 30% of the staff's time is taken up with observations and operations. Currently, the system is in use on about 4,000 hours per year.

While only one VLBI system is currently in operation, three telescopes will be used at the Wettzell observatory in the future. Estimating the numbers for this new situation, different attempts were considered: approaches with and without automation and operator sharing (see ● Fig. 6.7). It is quite obvious that the basic load of available manpower will still be employed on maintenance, repairs, development, research, and administration respectively support of other instruments. Maintaining the additional telescopes doubles the required time. Even if three systems need to be handled, the required time only doubles, as new techniques, such as an automatic lubrication system, reduce the working time per telescope. Standardization of maintenance with maintenance guides and manuals improves the situation and reduces unnecessary repairs, as error-prone techniques can be replaced on time. The costs for repair work are estimated as 30% of the total manpower. Another cause of this moderate increase is that new memory media might reduce the hardware work for data storage (e.g., for changing VLBI storage packs). Optimistically, the time spent on administration, support, and other issues will remain as it is now. Another assumption is that there will always be one scientist on a third-party funded position for the research and development tasks. Taking

Future Basic Workload Splitting for 24/7

6

Future Observation Workload for 24/7

all of this into account shows that the future basic workload is already 30% over the current situation.

But calculating the required time for observation in the 24/7 mode (24 hour operations on all days of the week) without any automation and with an operator in an independent control room for each telescope would show an increase, which cannot be managed with the current team (almost three times more staff would be required; see ◻ Fig. 6.7). Even if this is more or less hypothetical for a single observatory, it reflects the increase in required manpower from a global perspective, where different countries run their own observatories, supporting the GGOS services.

Future Observation Workload for 24/7 with Automation

Dealing with these numbers immediately shows that operating GGOS is only suitable if an increase in automation is encouraged. This means for the Wettzell observatory that all telescopes can be controlled from one operator room by one single operator and that some observations run unattended. But this requires reliable remote control and autonomous processes. Increasing the automation with the techniques described in the previous chapters is the only way to manage the new challenges. The estimation for VLBI at the geodetic observatory was changed step by step, adding more automation and operator sharing, which resulted in the following numbers (see ◻ Fig. 6.7):

— *Centrally controlled observations:* Running all observations with an operator in one central control room on location at the observatory will require 62% additional manpower compared to the current situation. The observation load takes 66% of the manpower for the entire work. If this would be compensated for by using cheaper student operators, it would be necessary to employ 22 students.

— *Automation of the weekend observations:* Unattended weekend shifts reduce the workload by about 20% for a 24/7 situation (for the current VLBI observation plan, it would just be reduced by 1%, because currently there are only short sessions of 1 hour on Saturday and on Sunday). The observation load for a 24/7 situation requires 46% of the manpower for the whole work. If this would be compensated for by using student workers, it would mean employing 14 students.

— *Automation of the weekend and night observations:* Besides the weekends, observations during the night hours could be automated. This also includes night observations controlled remotely by another observatory. The estimated number of staff required decreases by 15% for a future 24/7 scenario. It would also have a huge impact on the current VLBI observation plan, where it would reduce the required manpower for observations to 34% of the manpower required for the entire work. The observation load for a 24/7 situation requires 31% of the manpower for the entire work. If this would be compensated for by using student operators, it would mean employing eight students.

— *Shared observations between different space techniques:* The most improvement is given if the human resources of a complete worldwide network can be shared. With the new 24/7 scenarios, SLR and VLBI must deal with similar burdens of observations. Optimizing the daylight shifts and controlling different systems of the different space techniques in a more centralized way can reduce the costs for observations to 21% in a 24/7 scenario. This can also be reached if two-thirds of the day shift can be operated autonomously or by external partners in the form of shared control. Nevertheless, altogether 17% more manpower will be needed compared to now. If this would be compensated for by employing cheaper student operators, it would require employing four students. But this is just two more than at present.

Disadvantages

The estimation makes it clear that automation has no impact on the required maintenance, repairs, developments, or scientific workloads. Usually it will not result in the loss of scientific jobs or technical jobs. It will change the daily duties of station staff from routine things to more scientific and maintenance activities. An example is the new role of data logistics. Experience in the Australian AuScope network with four telescopes shows that one person is completely occupied with data module shipping or data transfers. These additional, new tasks have not yet been taken into account in the estimation above. This also might change the job landscape at the observatories, which might be split into technical and operational jobs. While the technical staff have a more or less full overview of the system and technique, operational staff are usually only trained to run the operations. This disadvantage might lead to downtimes, as the operational staff are not able to repair things themselves. New labor legislation is necessary to enable stand-by for emergency and paid remote access in the form of tele-working. Also agreements about the responsibilities of the different jobs and accessing remote partners are necessary, which will require additional work to prepare the new techniques.

In general, it should be clear that remote control and automation do not really entail a reduction of personnel. It should only result in a rearrangement of the operational tasks. Regular maintenance with clearly defined checklists is still essential and requires well-qualified and responsible people at the observatories. Nevertheless, all the estimations and the developments in recent years point in one direction: without automation, a suitable communication structure, and the use of modern service-oriented production systems, it is difficult to coordinate, operate, and implement the GGOS infrastructure without an increase in personnel costs.

6.6 Summary

6.6.1 What Did We Learn?

This chapter completed the discussions about suitable techniques for operations of space geodetic techniques in the context of GGOS operations. Starting with a general definition of what GGOS is, operational deficits were identified. These are missing on-demand services, missing feedback loops to customers, and missing control structures. They can be reduced if methods and techniques of the previous chapters are implemented. The result is a possible implementation of a structure of systems of systems. The background is a consequent implementation of multi-agent systems where each agent represents an autonomous production cell. Each cell consists of further production cells or finally contains the individual implementation of a control system for a space geodetic system. The global architecture forms a hierarchical control system with several horizontal coordination layers.

To deal with future challenges, network capacities are essential. The chapter discussed the current and expected data volumes and rates to and from the different systems of the space geodetic techniques. Real-time and non-real-time streams have to be provided to create the GGOS products. VLBI is here a real challenge because of its huge data volumes which need to be transferred to the correlation centers. Bandwidth on demand over dark fibers might offer a solution to deal with these volumes. Specially protected high-speed network enclaves need to be installed at the observatories to implement secure and fast transfers.

Another challenge is the operation of the future core sites. The implementations of autonomous production cells proved the necessary level of automation to deal with future workload. We saw a real-life calculation using numbers from Wettzell as example of manpower required to run the operation shifts of 24/7. A higher degree of automation is the key technology.

6.6.2 How Did We Use the Contents Learned?

The chapter combined the techniques of all previous chapters. The contents learned made a test observation possible in which shifts were operated remotely over the worldwide networks.

Methods from the system monitoring and safety cell are used to design the data propagation of auxiliary seamless data to the data centers.

The knowledge of security is the basis of a design for a transfer network which connects equipment for transferring VLBI data.

The technique of autonomous production cells and the higher degree of automation using agents are the key features to design systems which can deal with future workload.

6.7   Questions

1. What is the background of GGOS? Why are the suggested goals so essential for Earth science?
2. What are the three pillars of geodesy?
3. What is a GGOS core station or fundamental station? Focus on the specific characteristics of such a core station compared to a single space geodetic technique.
4. What are the five levels for GGOS operations?
5. What are auxiliary data related to geodetic products? What are metadata?
6. What are the big challenges for future data communications in a GGOS network?
7. Compare the discussions about security in ▶ Chap. 5. Why is it essential to physically separate the high-speed transfer network from the rest of the networks of an observatory?
8. What are «jumbo frames»?
9. How can the operating system technique of «round robin» improve the recording of high-sampled data?
10. Why are the different control strategies from ▶ Chap. 5 quite helpful to run the future GGOS observatories?
11. What would you expect for the situation of employment at geodetic observatories in the future if several space techniques must be operated for GGOS?

Outlook

7.1   GGOS Operations as a Revolution – 510

© Springer International Publishing Switzerland 2017
A. Neidhardt, *Applied Computer Science for GGOS Observatories*, Springer Textbooks in Earth Sciences,
Geography and Environment, DOI 10.1007/978-3-319-40139-3_7

7

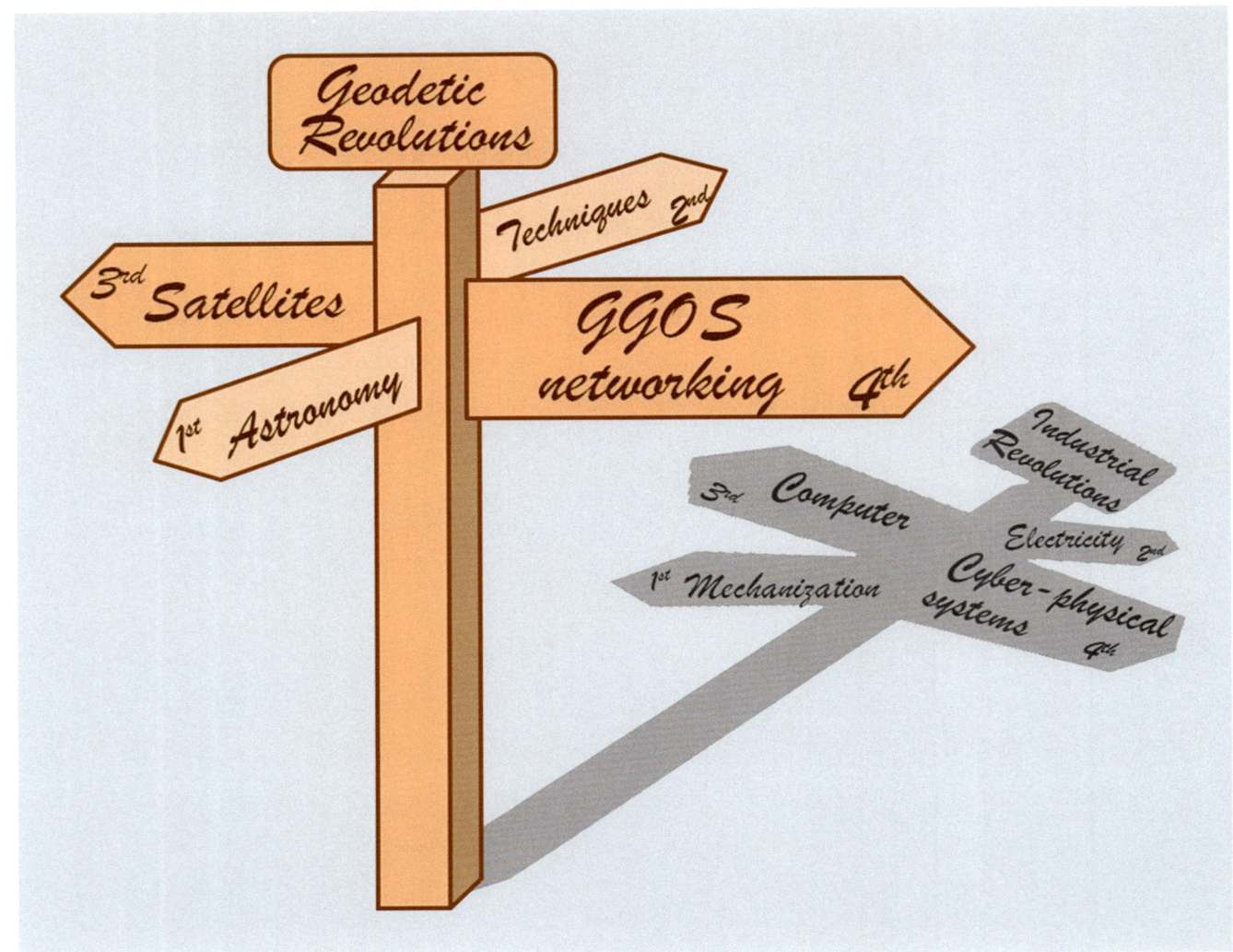

7.1   GGOS Operations as a Revolution

GGOS as a Revolutionary Step

An increasing number of space geodetic systems, more telescopes which are globally distributed, diversification of hardware and equipment, an increasing workload, new scientific questions, a direct impact on disaster monitoring, social consequences, and so on, are the current and future challenges for global geodesy and all interdisciplinary fields supporting global geodetic solutions. A complete implementation of all aspects requires a paradigm change and new approaches. It is a further revolutionary step to monitor the Earth. Looking at the history of geodetic techniques and the monitoring of Earth, several revolutionary steps can be identified. They cannot be dated exactly and describe whole epochs of evolution.

First Revolution: Astronomical Findings and Geodetic Surveys

Discoveries in the field of astronomy made it possible to get an extended view of the world. Eratosthenes already used observations of the sun in Alexandria and Syene to determine the circumference of the Earth about 200 years before the birth of Christ. The heliocentric world view of Nicholas Copernicus, its evidence by Galileo Galilei by finding moons on orbits around Jupiter, and improvements in the calculation of star positions and planetary orbits are also the foundation for extended geodetic surveys using extraterrestrial sources. Surveys, for example, to measure and define the meridian of origin, resulted in increasingly detailed maps and affected social progresses and military conflicts.

Second Revolution: Improvements of Observing Technique

The increasing knowledge about astronomy was supported by an improvement of technique used to observe and measure the natural environment. The use of telescopes, such as those of Sir Isaac Newton, and stable clocks which were invented by John Harrison made global navigation possible. These techniques enhanced the drawing of maps and were the basis for maritime transport. Its social impact was defined by global trade and the age of the great discoveries of the Earth, which «colored» the existing «white spots» on the maps.

Navigation also plays a major role for our age. The key technologies are astronautics, space, and satellite techniques. Mobile navigation with mobile phones is an indispensable part of everyday life. Hundreds of satellites are on orbits around the Earth and offer TV, telephone, and navigation. Everything started with a competition between the two superpowers, the USA and the former USSR. Sputnik and the moon landing paved the way for global navigation satellite systems. Permanent receiving stations for positioning data from the satellites can be used to monitor tectonic plates, earthquakes, and other global effects. Satellites return detailed images of the Earth, and extended data sets to determine sea level rise or melting glaciers.

Third Revolution: Space Techniques

While all the data are increasingly composed to a common view, geodesy and its impact on the society will benefit from increasing networking of all components. Improvements in global communication, faster networks, and autonomous computer techniques provide the components which are required to create a global observing system, as proposed for GGOS, which is able to quickly adapt to future requirements of the society. A changing planet, suffering from global warming and increasing populations, requires a better adaptable global monitoring of Earth effects to protect people from disasters and hazards.

Fourth Revolution: Global Networking

The chapters of this book explain technical and theoretical background of computer science supporting technical solutions to deal with challenges of the current age of a geodetic revolution on the way to GGOS (see ▶ Chap. 1). A solid knowledge of software development methods with all aspects of writing programs, which fulfill several quality factors for high-quality code, is the foundation for applicable software components. This was shown in ▶ Chap. 2. Prefabricated, well-tested components and generated programs which can ideally be used to parametrize individual applications while reducing the development workload provide a technique to decompose complex systems into «off-the-shelf» building blocks. They are used to create a standardized toolbox and communication middleware, which was shown in ▶ Chap. 3. Autonomous coordination and well-structured controlling and monitoring techniques for space geodetic systems improve the operation of geodetic measurement systems. Autonomous production cells use hierarchical designs to decompose the communication for control and feedback mechanisms to manageable feedback loops. The result is an increase of automation in combination with a sophisticated interaction to the operator and the user of the system. ▶ Chapter 4 presented these aspects of implementing modern control systems. Finally, global networking requires an increase of remote access. Multi-agent systems interact on the basis of orders and reports. Network and system security plays a central role to prevent attacks from the worldwide network. ▶ Chapter 5 examines these facets of global networking. The abstraction and generalization of the found solutions result in extended control and data flows which combine the different techniques to a system of systems, as proposed for GGOS. This was shown in ▶ Chap. 6.

Supporting Solutions for the Challenges

But not only the geodetic science has do deal with new challenges. It finds its analogon in industrial development. A keyword here is «Industry 4.0», or the fourth industrial revolution (see (Bundesministerium für Bildung und Forschung n.d.) for the following sections). «Industry 4.0» is a vision for the industry in the year 2025. It is based on a large network of communicating and interacting sensor and actor systems which are embedded into materials, machines, computers, products, and so on. They are composed to a Cyber-physical system (CPS). The small chips and devices use the Internet as a communication platform to optimize and control logistic and production processes which are globally distributed. These processes have to be standardized and modularized and can adapt autonomously to different workflows. CPS form vertical coordination layers and virtual collectives between the interacting devices and control levels to support the whole product workflow of order inflow, resource planning, production, delivery, and life cycle of goods. A complete implementation of such autonomous production networks also means a change of the working environment for the employees. Frequent, uniform,

Fourth Industrial Revolution

7

Revolutionary Steps in Industry

Individualization

Adapting Skills of Employees

parameterizable tasks will be overtaken and organized by machines. Flexible and creative jobs will further be in the hands of people. This requires a suitable interfacing between machines and humans with intelligent assistant systems.

Such an implementation can be seen as another revolution of industrial processes. The first revolution was the use of steam and water power, which provided a higher degree of mechanization of manufacturing. The second revolutionary step was given by the establishment of mass production using assembly lines. Electricity is the basis for this step. With the invention and technical implementation of computers, electronic data processing, and industrial robots, the assembly lines were increasingly automated. Similar to the social changes for the weavers and textile industry after the mechanization of manufactories in the nineteenth century, the automation of production lines was one of the huge paradigm changes with many consequences for workers and employees in the twentieth century. It was the third industrial revolution. Nowadays, increasing high-speed networks, global connectivity, and mobile communication make it possible that devices continuously exchange status data. For example, mobile phones automatically update positions in social networks and printers order spare parts completely autonomously. The availability of these techniques can be used to perform the fourth industrial revolution, where CPS plays a decisive role.

Industry 4.0 is a possible vision to deal with future challenges of industrial processes. Individual products will have to adapt to customer requirements. Nowadays, customers can select between standardized products and have the choice to select specific extras, for example, they can select between different packages as add-ons to a basic configuration of a car. In the future, customers can create their individual designs and ideas. Global facilities with 3D printers and other CPS assembly lines are flexibly selected to produce the individual product related to their individual workflow. Production data and status information are available all the time and can be used to optimize the production in real-time or to continue the assembly at another facility after a failure or blackout in the factory which was originally commissioned. This creates virtual ad-hoc organizations in which companies coordinate and organize their resources. Even the products themselves carry small chips and influence the production by own control commands to implement a situation-dependent fine-tuning. This means that not the products have to be standardized anymore. Production processes, interfaces, and workflows must be standardized and well designed with virtual process models. It is a trend away from off-the-shelf building blocks for products to off-the-shelf coordination blocks for production lines, where free capacities can be coordinated by central capacity brokers.

This trend of changes will influence the daily work of employees. The intelligence of machines is limited to predefined combinations of existing solutions. Heavy work can increasingly be reduced because specialized machines are predestined for these tasks. Routinely jobs will be done by machines. All jobs which can only be solved with creativity and aesthetic feeling require well-trained specialists. Future employees will take over control, maintenance, and creative developments. Assisting interactions between machines and humans support older employees in their daily work. Teams in factories will have to use social networking between different scopes of competence. Flexible working hours and tele-working capabilities provide a basis for families where both parents have to work. Therefore, CPS can be a great chance supporting different employees with their different individual abilities.

Whether industry or geodesy, both have to deal with new demands and challenges. They will influence social behavior but will also have great benefit for society. But only the future will show if the visions will become true and if the advised solutions and implementations can deal with all future aspects.

Looking at the prognoses described, a solid education in computer science and knowledge of information technology are key skills. Their interdisciplinary use in other fields, such as electronics, mechanics, and geodesy, is essential and the foundation to support solutions for current and future challenges. This book provides the necessary theoretical basis combined with several real-life examples. All aspects are applied to the interesting field of geodesy which is currently at the edge of a new revolution. May the ideas shown guide future engineers and scientists to further solutions on their own scientific ways.

Service Part

Appendix : A Style Guide for Geodetic Software in C and C++ – 514

References – 524

Index – 531

© Springer International Publishing Switzerland 2017
A. Neidhardt, *Applied Computer Science for GGOS Observatories*, Springer Textbooks in Earth Sciences,
Geography and Environment, DOI 10.1007/978-3-319-40139-3

Appendix: A Style Guide for Geodetic Software in C and C++

Main Topics in This Chapter
The following chapter contains a reduced selection of important and helpful rules, mostly taken from the style guide and programming policies which are used at the Geodetic Observatory Wettzell. The following suggestions do not claim to be complete, but are a collection of hints, tricks, and definitions, which simplify the common development of software in a team. The chapter is split into sections about:
- General structures
- Preprocessor directives
- Plain coding
- Functions and methods
- Libraries

A.1   Introduction

The following sections describe the most useful or critical rules based on experience at the Wettzell observatory, and which are collected in the local design rules definitions (see (Dassing et al. 2008)). The following rules make no claim to be complete, but are a minimal set of guidelines which help to write better code.

A.2   General Structures

■ **Rule 1: Project Structure**

Languages	Relevance
C/C++	+

Software is developed within a project to do a specific set of tasks. For this reason, it helps to create a suitable directory tree for each project, as this can also be used for a repository in a version control system (see ▶ Sect. 2.7.). It makes sense to organize the directories hierarchically. The first level of this hierarchy represents the repository view (Wassermann 2006, *l.c. page 47*). The next level of the tree contains a general view of the project and can be separated into directories and subdirectories, as shown in

▶ Sect. 2.7.1. Not all the directories suggested are mandatory. If they contain no files, they can be left out and created when needed.

■ **Rule 2: File Names and Internal File Structure**

Languages	Relevance
C/C++	+

Code files are collections of function sets with the same purpose or characteristics. File names should be descriptive. For portability, the names are lower case, but can contain «_» as separation between the logical word parts. Files with pure C code have the extension «.c» for the source file and «.h» for the header file. In the same way C++ code uses the extensions «.cpp» and «.hpp». A source/header-tuple defines a module or component. There is a strict separation between the declarations (function/method head) as interface to the module in the header file and the definitions (body with functional code) in the source file. A source file with its related header file is a software module. There is only one exception to this strict separation: templates. The function body of template functions must be located in the header file because of the language syntax. Relations between modules are organized using the related header files in precompiler include directives. External variables and functions (defined with the tag «extern») are not allowed. The internal structure is a sequence of (see also ◘ Fig. A.1):
- Comment header (version control tags, general info, interface info, authors, etc.) (see ▶ Sect. 2.5.1 and ▶ Sect. 2.7.2)
- Define or include guard «begin» directive (only in the header files, see ▶ Sect. 2.3.2)
- Preprocessor directives in a particular order (should be used only in reduced form in header files to keep includes local):
  - All «define» statements
  - All «include» statements for standard libraries in the standard library path (such as <stdio.h> (with extension «.h») for C code libraries (Kernighan and Ritchie 1990, *l.c. page 86*) or <string> (without extension) for C++ code libraries)
  - All «include» statements for modules in the local work path (identified with quotes) (Kernighan and Ritchie 1990, *l.c. page 86*)

Fig. A.1 The structure of a software module with a header and a source code file

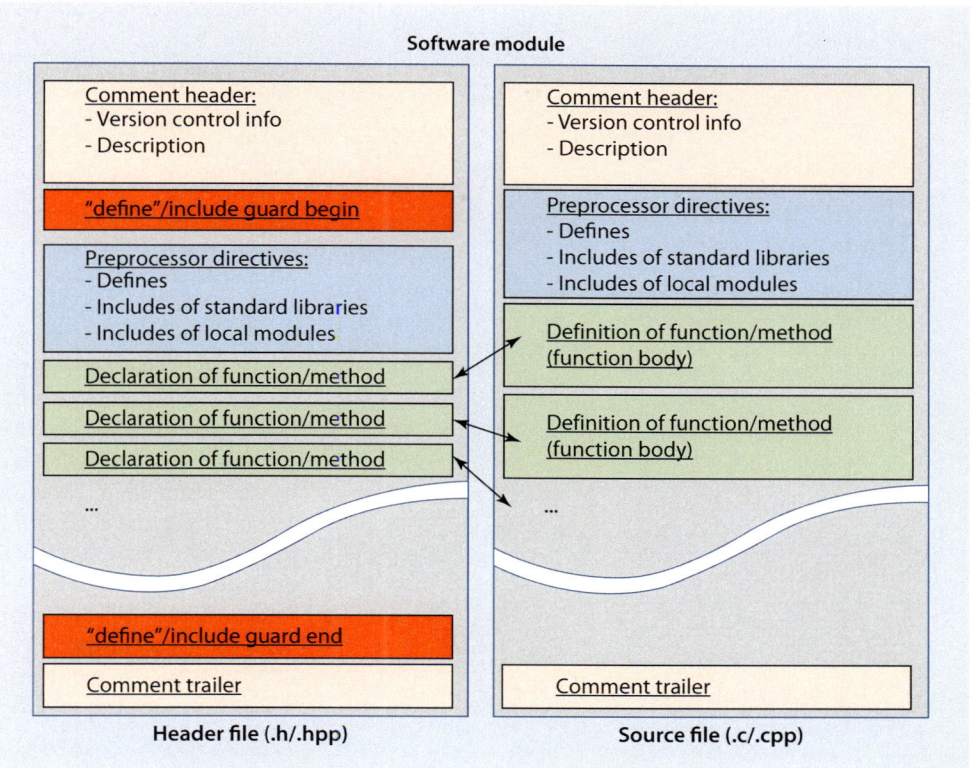

Software module

Header file (.h/.hpp)

Comment header:
- Version control info
- Description

"define"/include guard begin

Preprocessor directives:
- Defines
- Includes of standard libraries
- Includes of local modules

Declaration of function/method

Declaration of function/method

Declaration of function/method

...

"define"/include guard end

Comment trailer

Source file (.c/.cpp)

Comment header:
- Version control info
- Description

Preprocessor directives:
- Defines
- Includes of standard libraries
- Includes of local modules

Definition of function/method (function body)

Definition of function/method (function body)

...

Comment trailer

— Sections with logically combined declarations (in header files) or definitions (in source files) of functionalities[1] of the same characteristics
— Define or include guard «end» directive (only in the header files)
— Comment trailer (see ► Sect. 2.7.2)

A very important instrument in header files to avoid multiple includes, and therefore multiple declarations, is the include guard (see ► Sect. 2.3.2).

Additionally, line length is normally standardized by style guides, which help to print the code properly onto standard paper sizes. Normally, large code projects are no longer printed. Therefore, this rule is not as important as it was. Nevertheless, it is quite helpful to standardize on a normal print-width (about 80 characters), as it makes the code more readable.

■ **Rule 3: Class Structure and Obligatory Members**

Languages	Relevance
C++	++

In general, a class can be related to a specific module. The class itself contains member elements like attributes (variables) and methods (functions). The class structure is ordered from the highest restricted parts (which are only accessible within the class object and defined with the keyword «`private`»), followed by the inheritable parts (which are accessible within the class object, from all derived class objects, and defined with the keyword «`protected`») and the public interface parts (which are accessible from outside as interface methods to the class functionality, and are defined with the keyword «`public`») as shown in ◘ Fig. A.2. The use of «friend» methods or classes should be avoided to keep clear access structures. Each restriction block should be ordered as well. The first section defines constants and enumerations, followed by internal new data types defined by class descriptions such as for exceptions. The next section defines member variables followed by class methods. Each class must have the standard constructor (which is called each time an object of this class is created without specific definitions), a copy constructor (used to create a new object containing the values from an existing one), the destructor (called each time an object is destroyed, for example, at the end of a local scope), and the copy assignment operator (operator =, which is used to assign the content of an existing object to another existing

---

1 Functionalities are called «functions» in C and «methods» in C++, as they describe the behavior of a class.

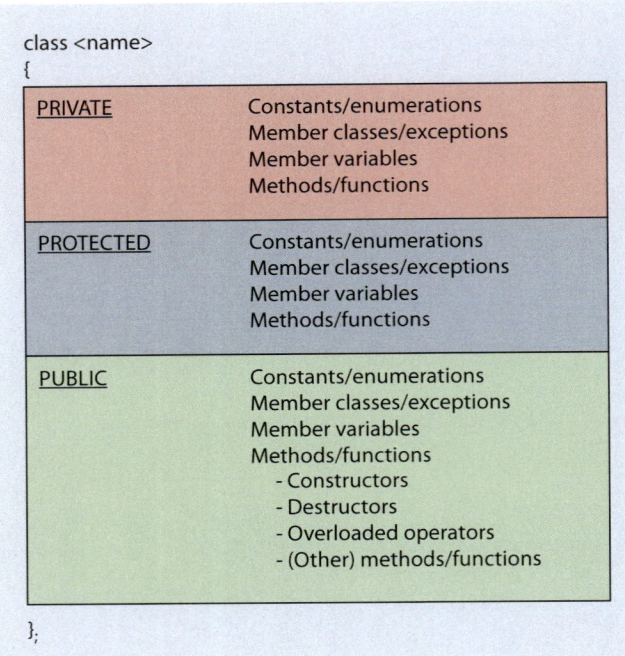

```
class <name>
{
    PRIVATE          Constants/enumerations
                     Member classes/exceptions
                     Member variables
                     Methods/functions

    PROTECTED        Constants/enumerations
                     Member classes/exceptions
                     Member variables
                     Methods/functions

    PUBLIC           Constants/enumerations
                     Member classes/exceptions
                     Member variables
                     Methods/functions
                       - Constructors
                       - Destructors
                       - Overloaded operators
                       - (Other) methods/functions

};
```

■ **Fig. A.2**   The structure of a class in a software module

object). As these functions are called implicitly, for example, if the new class objects are used within lists or maps, they are essential. If one of the functions or operations is not allowed in the context of a specific class, the individual function must be blocked by defining it as private.

■■ **Example A.1**
Obligatory member functions

```
class callisto_dewar
{
    private:
        // ...
    public:
        // General class functions
        // Constructor
        callisto_dewar ();
        // Copy-constructor
        callisto_dewar    (const    callisto_
        dewar & CIn);
        // Destructor
        virtual ~callisto_dewar ();
        // copy assignment operator
        callisto_dewar  &  operator=  (const
        callisto_dewar & CIn);
        // …
};
```

■ **Rule 4: Naming Conventions**

Languages	Relevance
C/C++	+++

Even though naming is quite arbitrary, it is useful to define some conventions. Names should directly provide information about the characteristics of a named element. It should directly be visible whether it is a new defined type, a function, or a variable, and the name should make clear which type of variable it is, or even what the return value of a function is. Combinations of variables such as arrays or higher sophisticated class types should also be derivable from the name. An early definition to support such ideas was given by the Hungarian Notation of Charles Simonyi (Maguire 1993, *l.c. page xxv–xxviii*). As this notation helps to visually identify, for example, type mismatches in the code, a similar definition which also includes C++ elements is effective. The following notation is a step-by-step extended heading character set in front of a descriptive identifier. It offers more details (as type or structure) and classifies the named identifier in the following way for variables and functions:

> **4. Notation Prefix: memory class**
> **+ 3. Notation Prefix: access**
> **+ 2. Notation Prefix: combination**
> **+ 1. Notation Prefix: type**
> **+ Descriptive identifier name**
>
> and in the following way for new type definitions:
> **Descriptive type name**
> **+ Notation Postfix: constitution**

The identifier and type names themselves should be descriptive and should avoid acronyms. Variable names and type names are a combination of nouns (e.g., «MinIndex» or «IndexCounter»). Function names should be an imperative combination of nouns and verbs which describe the activity done within the function body (e.g., «GetFamilyName» or «SumIntegers»). The logical patterns in the names are separated by capital letters (e.g., «CalculateMeanElevation»). Together with the type codings from the notation, the name follows the «camel case» style (e.g., «uiArrayIndex»). Only constants are named in a different way to the pattern highlighting described. As is often the case with classical program styles, it is advisable that they are completely written in capital letters (as shown in (Kernighan and Ritchie 1990, *l.c. page 37 f.*)), which is the «upper case» style. Constants may optionally also carry the extension for a type, and their logical patterns may be separated, using an underscore «_».

New individual variable types should be extended with trailing extensions as postfix. Possibilities are:

Postfix	Constitution	Example
Type	New type definition	`typedef int * IntPtrType;`

Struct	Structure definition	`struct SensorStruct {...}; or`
		`typedef struct SensorStruct {...}`
		`    SensorStructType;`

Union	Union definition	`union ConverterUnion {...}; or`
		`typedef struct ConverterUnion {...}`
		`    ConverterUnionType;`

Enum	Named enumeration definition	`enum SwitchStatesEnum` `    {uiOFF=0, uiON=1};`
Class	Class definition	`class MeteoClass {...};`
TClass	Template class definition	`template <class StackTType>` `class StackTClass {...};`
TType	Template type	`template <class StackTType>` `    class StackClass {...};`
Namespace	Namespace definition	`namespace SerialDeviceNamespace {...};`

In front of such a named identifier, the different notation extensions are added as prefixes:

1. *Notation Prefix: type* A coded description for the different types is added directly in front of the descriptive identifier name. Suggested classification character sets for C types, C++ types, and STL types are:

Prefix	Type	Example
b	Boolean	`boolean bExit;`
c	Character	`char cLetter;`
uc	Unsigned character (Hungarian: by)	`unsigned char ucSign;`
f	Float	`float fCash;`
d	Double	`double dPrecision;`
s	Short integer (Hungarian: n)	`short sIndex;`
us	Unsigned short integer (Hungarian: w)	`unsigned short usCount;`
i	Integer	`int iSum;`
ui	Unsigned integer	`unsigned int uiSecondsOfDay;`
l	Long integer	`long lSeconds;`
ul	Unsigned long integer (Hungarian: dw)	`unsigned long ulMilliSec;`
v	Void	`void vReadLine (...);`
F	FILE	`FILE * pFInputFile;`
E	Named enumeration	`enum SwitchStatesEnum eSwitchStates;`
S	Structure	`struct SensorStruct SSensor;`
U	Union	`union ConverterUnion UConvert;`
C	Class	`MeteoClass CMeteorology;`

Prefix	Type	Example
T	Template variable	`StackTClass<int> TCStack;`
TC	Template class	`template <class StackTType> class {StackTType TStack; ...};`
IO	I/O Stream	`stream IOLogFile;`
O	Output stream	`ostream OResultFile;`
I	Input stream	`istream IResultFile;`
FIO	Filestream	`fstream FIOLogFile;`
FO	Output filestream	`ofstream FOResultFile;`
FI	Input filestream	`ifstream FIResultFile;`
str	STL string	`std::string strName;`
it	STL iterator	`map<char,int>::iterator itNamesMap;`
bst	STL bitset (set of bits)	`std::bitset<8> bstByte;`

2. *Notation Prefix: combination* (added in front of 1. Notation Prefix) This prefix codes if the named element is a multielement combination like an array or a list.

Prefix	Type	Example
a	Array in the memory stack	`long alKeys[128];`
ac	Character array in the memory stack (Hungarian: s)	`char acPhrase[128];`
az	Zero-terminated array in the memory stack (Hungarian: sz)	`char azcName[128];`
ma	Multidimensional array in the memory stack	`char macPhrase[128][256];`
pp	C++ pointer (dynamic heap, which is created with new/delete)	`long * pplAddress;`
p	C pointer (dynamic heap with malloc/free)	`long * plAddress;`
lst	STL list	`std::list<int> lstiSizes;`
map	STL map (hash array with unique keys)	`std::map<int,std::string> mapistrHash;`
mmp	STL multimap (hash array without unique keys))	`std::multimap<int,std::string> mmpistrHash;`
vec	STL vector	`std::vector<int> veciX;`
set	STL set with unique elements	`std::set<int> setiNumbers;`
mst	STL multiset without unique elements	`std::multiset<int> mstiNumbers;`
stk	STL stack	`std::stack<int> stkiStates;`
que	STL queue	`std::queue<int> queiJobNumbers;`
pqu	STL prioritized queue	`std::priority_queue<int> pquiOrders;`
dqu	STL double-ended queue	`std::deque<int> dquiLogIDs;`

3. *Notation Prefix: access* (added in front of 2. Notation Prefix) The final notation describes which access possibilities, such as global or private availability in a class, are given.

Prefix	Type	Example
g_	Globally accessible	`int g_iPort; int main {...}`

pri_	Private class member	`class SensorClass`
		`{`
		    `private:`
		       `int pri_iSensorNumber;`
		`};`

pro_	Protected class member	`class SensorClass`
		`{`
		    `protected:`
		       `int pro_iSensorNumber;`
		`};`

pub_	Public class member (optional, as all not labeled members should be public)	`class SensorClass`
		`{`
		    `public:`
		       `int pub_iSensorNumber;`
		`};`

_	If 4. Notation Prefix is needed and no 3. Notation Prefix is used.	

4. *Notation Prefix: memory class (optional)* (added in front of 3. Notation Prefix). Additionally, it is recommended that «sta» is added in front of the identifier names for static memory elements (e.g., «`stapriv_pFInputFile`») which are uniquely located at one specific memory element and generated only once (Kernighan and Ritchie 1990, *l.c. page 81*). Equivalent «reg» can be added for memory elements which should be optimized to be kept in the fast registers (e.g., «`reg_iCounter`») (Kernighan and Ritchie 1990, *l.c. page 81 f.*).

Prefix	Type	Example
sta	Static memory element	`static int * sta_piPort;`
reg	Register element	`register int reg_iCounter;`

A.3 Preprocessing Directives

■ **Rule 5: Define or Include Guard**

Languages	Relevance
C/C++	+++

To avoid multiple includes, and therefore multiple declarations, it is necessary to use include guards. They are preprocessing directives which prevent the inclusion of code which has already been included.

■ ■ **Example A.2**

Define/include guard

```
#ifndef __FILEIO__
#define __FILEIO__
... // Header code
#endif //__FILEIO__
```

■ **Rule 6: Include Paths and Include Order**

Languages	Relevance
C/C++	+++

Relations between the modules are organized with header file includes. The declarations in the header files are used for type checking. Includes in header files should be avoided to keep the included parts as local as possible. In any case, it is advisable to follow a specific include-order (if there are no other constraints from external legacy code). First all «include» statements for standard libraries in the standard library path (such as <stdio.h> (with extension «.h») for C code libraries (Kernighan and Ritchie 1990, *l.c. page 86*) or <string> (without extension) for C++ code libraries) should be written. These statements are followed by all «include» statements for modules in the local work path (identified with quotes) (Kernighan and Ritchie 1990, *l.c. page 86*). Relative paths are not allowed. The search path to find the header files should be set with compiler parameters (such as the «-I» parameter of the GNU gcc and g++ compiler). There is only one exception to this. If the header files are collected in one specific subdirectory tree, it is possible to use one hierarchical level of the following form: «`#include <sys/types.h>`».

■ Rule 7: Avoid Preprocessing Macros and Conditions

Languages	Relevance
C/C++	++

As preprocessing macros are simple code replacements without any type checks, it is advisable to avoid these macros completely. Usually macros can easily be replaced by simple functions or, in the case of constants, with equivalent language elements, like «enum» or «const» declarations. To make the code more understandable, another helpful hint is to reduce the use of preprocessing conditions (such as «#ifdef», «#else» and «#endif»). Normally, they must be used to separate the different system function calls of the different operating systems in a file. But its area of influence should be as clear as possible, and should also be kept as small as possible. Unused code («dead instructions») should be deleted, instead of being excluded using preprocessing conditions.

■ Rule 8: Constants with «const» or «enum»

Languages	Relevance
C/C++	+++

To provide a complete type checking of the values assigned to constants, it is better to use the «const» attribute in combination with variables (e.g., «const double dPI = 3.141592») instead of preprocessing defines. For the definition of error codes, return values, or states of finite state machines, it is also suitable to define specific enumerations (e.g., «enum SwitchStatesEnum uiOFF=0, uiON=1;»). Return codes of functions are of the type «unsigned integer» to code the different error situations and causes. Here «0» represents a successful run, while all numbers greater than «0» are error codes.

A.4   Plain Coding

■ Rule 9: Indents with White Spaces

Languages	Relevance
C/C++	+++

Different editors replace tabulator stops with a different number of white spaces. The code becomes more readable if indents are created with blanks in front of the instructions. Experience shows that four blank signs per indent are ideal to have a good visual separation, but also avoid excessively long line lengths.

■ Rule 10: Parentheses

Languages	Relevance
C/C++	++

In classic programs, the starting curly bracket of a code block was directly behind the instruction, which belonged to this code segment (e.g., the curly bracket for the function body was directly behind the function head). But this makes it difficult to find the related start and end points of longer code passages with several code blocks. Parentheses should be used to enhance readability and reduce mistakes. Therefore, it is better if the start and the end curly brackets are in the same column as the first instruction (ANSI style, see example). In combination with generative programming, it is desirable to use parenthesis all the time, as search patterns for replacements can be defined more easily. It is advisable to use parenthesis in the conditional states of a «switch/case» condition to enhance the clearness of instructions which belong together. In this context, each «case» section which processes its code should have its own «break» as stop condition. The instruction flow should not automatically continue with the code of the next «case» statement, which would happen if no «break» is used.

■ ■ Example A.3
Parenthesis

```
// Parentheses in combination with loops
for  (iIndex  =  0;  iIndex  <  iMaxIndex;
iIndex++)
{
    ... // loop code
}
// Parentheses in combination with if
if (uiIndex <= 100)
{
    ... // if branch
}
else
{
    ... // else branch
}
// Parentheses in combination with
// Switch/case
switch (uiState)
{
    case 1:
    {
        ... // first case
        break;
    }
    case 2:
    {
        ... // second case
        break;
    }
```

```
    ...
    default:
    {
            ... // default case
            break;
    }
}
```

Rule 11: Avoid Global and Static Definitions

Languages	Relevance
C/C++	+++

Global variables are difficult to handle, because it is not always clear where they are defined in large programs. In this combination, attention should be paid to local scoping. Variables, should be kept as local as possible and should be encapsulated. C++ namespaces are sometimes useful to protect complete, one's own module spaces. But it can also make programming long-winded. The best way is to avoid global definitions, and also to avoid the combination of statements which do not belong together (e.g., do not write additional instructions in «for» loop heads besides the start counter definition, the stop criterion, and the iteration operator).

Also static definitions can cause some problems in systems with parallel threads. As static memory exists only once at a specific address, and all functions with this memory have the same access rights, it is a critical section. Therefore, race conditions where results are dependent on the access order and time can be the consequence if this critical section is not protected, for example, by semaphores which handle the read and write access of the competing threads.

Rule 12: Jump Instructions Are Prohibited

Languages	Relevance
C/C++	+++

Jump instructions make it quite difficult to understand large programs, especially when many of the jump locations are far apart. As jumps are a construct in unstructured programming languages to repeat or process conditional code, but can be replaced by equivalent instructions such as loops and conditions in structured programming languages, «goto» instructions are generally prohibited. Therefore, exceptions should also be avoided, because these OOP constructs are higher level jumps, which can hardly be understood in the workflow during runtime.

The exception to this rule is for jumps which make it possible to reduce several, individual function exits. In this case, well-defined jump labels are used at the end of a function to which the workflow can jump if an error occurs. Using this method, there is only one function return, located at the end of the function. In general, these jumps are only forward jumps to the following code lines and not backwards. In addition, they are kept quite local within one function. After the jump label of the single return, memory cleanups can be processed. This technique allows exactly one entry and one exit point in a function.

Rule 13: Variable Initialization

Languages	Relevance
C/C++	+++

All variables must be initialized with a start value. This reduces errors, such as the use of a pointer pointing to wrong memory addresses or index variables with the wrong start value. «NULL» is the correct start value for pointers. Call-by-reference parameters of functions which return result values to the calling instance also have to be initialized. A disadvantage is that the warning system of some compilers does not check the initialized variables anymore if they are used further on or not. Another disadvantage is that overwriting the value, which can occur shortly after initialization, results in additional runtime.

Rule 14: Dynamic Memory

Languages	Relevance
C/C++	+++

Even if variables in the memory stack of the program are much less error-prone (if index overflows are checked and avoided) and do not consume so much runtime, the reduction to non-dynamic memory would reduce the possibilities of dynamic adaption of the data to requirements during the runtime of the program. To reduce the memory consumption in an optimal way to the requirements, the use of heap memory is effective. Therefore, it is necessary to administrate dynamic memory. This can be done by using «malloc/calloc/realloc/free» in C and «new/delete» in C++. These methods should not be mixed. In C++, it is better to use the operators «new» and «delete». It is advisable to check memory consumption during runtime with appropriate tools to find memory leaks. To get a proper workflow in low-memory areas as well, it is necessary to check the success of a memory allocation by evaluating if the returned memory address is not NULL, and/or if the exception «bad_alloc» is thrown under C++. This is often neglected in open-source OOP projects, as it takes more programming time to check all allocations. But this can lead to unpredictable situations if several memory-hungry programs share an increasingly shrinking memory.

Multiple deletes of memory can also lead to unpleasant crashes. Therefore, it is advisable to set the value of a pointer to NULL after a delete. Multiple deletes on a NULL pointer have no influence anymore.

The OOP world also offers programming patterns which help to deal with such requirements automatically. In the case of controlling heap memory, the pattern is described as a «smart pointer». Smart pointers are objects which act as standard pointers but manage the underlying memory automatically, including things such as garbage collecting (cleaning up unused code) on the basis of ownership logging. Used in the right way (checking on each touch if exceptions are thrown), smart pointers can make life easier. They look like normal pointers and the handling does not differ from standard behavior. But this can lead to a kind of «laissez-faire» attitude, for example, not taking care of exceptions anymore, which lead to unpredictable behavior. Sometimes smart pointers can also be non-obvious and confusing (see also ▶ Sect. 3.1).

■ Rule 15: Types and Type Casting

Languages	Relevance
C/C++	+++

Casts (conversions) between types should be avoided. Where a type conversion is necessary, only an explicit cast should be done to avoid non-obvious, implicit type changes. Compiler switches or preprocessor switches which change between variable types are not allowed. In C++, the specific cast operators, such as static, dynamic, const, and reinterpret cast should be used in addition to the traditional-style cast from C. The most important cast operators are dynamic and static. Hereby a dynamic cast («pCMeteoSensor = dymanic_cast<MeteoSensorClass *> pCSensor;») performs a verification and validation of the cast during runtime, while a static cast («lId = static_cast<long> iId;») is very similar to the traditional one. It is a non-polymorphic standard conversion without runtime checks. The const cast allows a conversion of constant variables into nonconstant access (therefore, it should be avoided), and the reinterpret cast can also convert between completely incompatible types (which should also be handled carefully) (Schildt 1998, *l.c. page 580 ff.*).

In the context of mathematical calculations, it is useful to use double precision types instead of single. For those cases where the valid decimal places are not enough, there are appropriate multiprecision libraries.

For text strings, it is better to use UTF types and libraries instead of the standard ASCII types, because then all language-specific characters can be represented without additional work (see ▶ Sect. 3.3.1).

■ Rule 16: Comments

Languages	Relevance
C/C++	+

Comment lines are used to make code more comprehensible. Comments are additional information from the programmer or developer of the source. It is essential to add as many comments as possible, but to avoid statements which are directly visible from the code (e.g., it is not necessary to explain that a loop follows without saying what the loop does). Comments are placed in front of the instructions which they explain. Only for variable explanations, they can be added at the end of the same line where the variable is located.

In C++, it is useful to use C++ single-line comments («//…») instead of the start and end comment tags («/* ... */») in C to mark the comments. Start and end comment tags help to hide sequences of command instructions for testing (but such code should not be committed to the code repository).

Comments are also used to generate the developer documentation automatically. One tool for this is «Doxygen», which can produce documentation in different styles, such as HTML or LATEX. For more information see also ▶ Sect. 2.5.1.

■ Rule 17: 64-bit/32-bit Friendliness

Languages	Relevance
C/C++	+++

The code should be general enough to run on 64-bit and 32-bit machines. As long as only the standard types are used, all types should work properly (special types such as «size_t» from the operating system should be handled carefully; therefore, also avoid «memcpy» and «memmove» if possible). 64-bit variables should also be used in the preprocessor conditional definition «#ifdef _LP64». The definitions in «inttypes.h», and other type header files can be used for the new types and to output the values correctly in C. The C++ streams should automatically be able to print the values correctly. To avoid problems with the sizes of the type, it is better to calculate the size of the variable instance (e.g., «sizeof(iIndex)») instead of the type. Special care must be taken on bit shifts, arithmetic operations, and within unions in combination with pointers (see also (Karpov and Ryzhkov n.d.) and ▶ Sect. 2.3.2).

A.5   Functions/Methods

■ Rule 18: Function Parameters and Return Values

Languages	Relevance
C/C++	+++

Parameters to C/C++ functions are either input to the function, output from the function, or both. Therefore, it makes the function body more readable if the order of the parameters is from input, input/output, to output. New parameters should be placed in the corresponding position. In general, it is useful in C++ to write all the output arguments as call-by-reference, using the «&» sign (e.g., «`int & iResult`»). It ensures that the functions process on the original data without creating a copy of an external variable. In C++ the call-by-reference variables can be used as normal local variables (there is no need for a dereferencing with an asterisk, as needed in C) with the difference that all changes are visible outside the function. Even input parameters can be defined by using the call-by-reference method to make the processing much faster, as no copying is needed. But to avoid changes in input parameters, it is necessary to define them as constant references (e.g., «`const int & iSummand`»).

Returned error codes should not only be Boolean values, which only provide information about whether there has been a failure or not. They should also be enum or at least unsigned integer values, which code the type of error, for example, with different «enum» numbers. By using this error coding, it is possible to propagate error states to higher levels, keeping the information about which error caused the current failure return. It is important in this context to create defined states with correct initialized return variables also in case of an error. Therefore, all return parameters should be initialized first before any other instruction is processed. Memory which is allocated during the processing should be freed again and reset to NULL after an error to avoid adjacent activities outside the function which could lead to memory leaks.

Rule 19: Reduced Use of Inline Code

Languages	Relevance
C/C++	+

In C++, it is possible to define functions with inline code by writing the function body directly into the class declaration in the header file. The compiler expands this code directly as inline sequence instead of calling the function conventionally. This increases performance as long as the inline code is short enough. Therefore, inline code is only allowed for very short functions (a few lines of code). For standard applications in modern architectures, it is also no problem to abstain from inline code, which makes the sources more readable, as all function bodies can only be found in the source files and not in the header files.

Rule 20: Const Correctness

Languages	Relevance
C/C++	++

The compiler checks if changes are made where they should not be made. Therefore, it is advisable to protect from changes with «const». In particular, functions which do not make any changes to member attributes should have the «const» qualifier (e.g., «`int iReturnState () const;`»). It helps to detect misuses during the compilation.

Rule 21: Thread Safety

Languages	Relevance
C/C++	++

Threads simplify the programming of deadline-dependent, parallel tasks, while keeping the possibility of directly exchanging data in the memory. Therefore, all operating systems also support threads (e.g., POSIX threads under Linux) besides the standard processes, which require a special IPC for the data exchange. But threads have their pitfalls. As two tasks can manipulate memory elements in parallel, the wrong usage can lead to critical sections. To avoid them, semaphore or mutex constructions are required.

Additionally, some libraries are not thread safe. The STL has this problem, which means that workarounds are necessary. Copying strings should always follow the «deep copy» methods (e.g., in the form of «`strText1 = strText2.c_str();`»), which really forces a copying and not just a referencing with an internal reference counting. The same style can be used for deleting, where an empty string can be copied into a string which should be deleted using the «deep copy» method. Special care must be taken if strings are used in further container classes, such as lists, vectors, and so on, because they do not process such deep copies. In this case, it is better to overload the copy functions and redesign them with «deep copy» techniques. For more information see ▶ Sect. 2.6.4.

A.6 Libraries

Rule 22: Usage of Libraries and Toolboxes

Languages	Relevance
C/C++	++

In any case it is advisable to reduce the manually written «glue code». This can only be done if existing code libraries are used. These libraries can be external repositories. But it is important here that updates and upgrades are possible without problems and that the code should be downward compatible to older versions. A suitable library is the STL if the aspect of thread-safety is considered. An institute's internal toolboxes, which should follow the aspects described in ▶ Sect. 3.1 are much better than external libraries. With these toolboxes, decisions about the code are always in the hands of the local development team. Experience shows that even a minimal set of modules helps to improve code development. But it is important to ensure that the tools in this toolbox are regularly inspected using CI methods (see ▶ Sect. 2.8).

References

Aho, Alfred V; Sethi, Ravi; Ullman, Jeffrey D: *Compilerbau*. Teil 1, Addison-Wesley (Deutschland) GmbH, 1997. (Original title: COMPILERS. Principles, Techniques and Tools. Bell Telephone Laboratories, Inc., 1986).

Alexandrescu, Andrei: *Modernes C++ Design. Generische Programmierung und Entwurfsmuster angewendet*. mitp-Verlag Bonn 2003. (Original title: Modern C++ Design: Generic Programming and Design Patterns Applied. Pearson Education, Inc., Addison Wesley Professional 2001).

Allan, Peter M; Chipperfield, Alan J; Warren-Smith, R. F; Draper, Peter W: *CNF and F77. Mixed Language Programming – FORTRAN and C*. Version 4.3, Starlink User Note 209.10, CCLRC/Rutherford Appleton Laboratory 2008. http://star-www.rl.ac.uk/star/docs/sun209.ps, Download 2011-07-29.

Arduino team: *Arduino*. http://www.arduino.cc/, Download 2012-09-18.

ASCII-Table.com: *ANSI Escape sequences (ANSI Escape codes)*. http://ascii-table.com/ansi-escape-sequences.php, Download 2012-09-26.

Atmel Corporation: *Atmel 8-bit Microcontroller with 4/8/16/32KBytes In-System Programmable Flash. ATmega48A; ATmega48PA; ATmega88A; ATmega88PA; ATmega168A; ATmega168PA; ATmega328; ATmega328P*. Atmel Corporation 2012. http://www.atmel.com/Images/8271S.pdf, Download 2012-09-18.

Barker, Thomas T: *Writing Software Documentation. A Task-Oriented Approach*. Second Edition, Pearson Education, Inc. 2003.

Barrett, Daniel; Silverman, Richard; Byrnes, Robert: *SSH, The Secure Shell: The Definitive Guide*. O'Reilly Media, Inc. 2005.

Bauer, Sebastian: *Eclipse für C/C++-Programmierer. Handbuch zu den Eclipse C/C++ Development Tools (CDT)*. 2. Auflage. dpunkt.verlag GmbH 2011.

Beaudoin, Christopher; Whittier, Bruce: *Sensitivity evaluation of two VLBI2010 candidate feeds*. https://www.mpifr-bonn.mpg.de/1263470/Beaudoin_VLBI2010FeedDev.pdf, Download 2016-03-31.

Beck, Kent; Beedle, Mike; van Bennekum, Arie; Cockburn, Alistair; Cunningham, Ward; Fowler, Martin; Grenning, James; Highsmith, Jim; Hunt, Andrew; Jeffries, Ron; Kern, Jon; Marick, Brian; Martin, Robert C; Mellor, Steve; Schwaber, Ken; Sutherland, Jeff; Thomas, Dave: *Manifesto for Agile Software Development*. http://www.agilemanifesto.org/ 2001, Download 2011-12-08.

Beebe, Nelson H. F: *Using C and C++ with Fortran*. Department of Mathematics, University of Utah 2001. http://www.math.utah.edu/software/c-with-fortran.html, Download 2011-07-30.

Bernese: *Bernese GPS Software*. http://www.bernese.unibe.ch/, Download 2011-06-18.

Bundesamt für Kartographie und Geodäsie (BKG): *Networked Transport of RTCM via Internet Protocol (Ntrip). Version 1.0*. Bundesamt für Kartographie und Geodäsie Frankfurt am Main 2004. http://igs.bkg.bund.de/root_ftp/NTRIP/documentation/NtripDocumentation.pdf, Download 2013-11-25.

Bloomer, John: *Power Programming with RPC*. O'Reilly & Associates, Inc. Sebastopol 1991.

Bundesministerium für Bildung und Forschung. (editor): *Zukunftsbild «Industrie 4.0». HIGHTECH-STRATEGIE*. Bundesministerium für Bildung und Forschung. https://www.bmbf.de/pub/Zukunftsbild_Industrie_40.pdf, Download 2016-03-28.

Bovet, Daniel P; Cesati, Marco: *Understanding the LINUX KERNEL*. O'Reilly Media, Inc. 2006.

Brookshear, J. Glenn: *Computer Science. An Overview*. Pearson Education, Inc. 2012.

Brown, Martin C: *Perl: The Complete Reference*. Second Edition. Osborne/McGraw-Hill Companies Berkeley 2001.

Clang team: *clang: a C language family frontend for LLVM*. http://clang.llvm.org/, Download 2013-01-09.

Clark, Mike: *Projekt-Automatisierung. Pragmatisch Programmieren*. Carl Hanser Verlag München Wien 2006. (Original title: Pragmatic Project Automation. How to Build, Deploy, and Monitor Java Applications. The Pragmatic Programmers, LLC. 2004).

A. Collioud: *IVS Live: All IVS on your desktop*. In: Alef, W; Bernhart, S; Nothnagel, A. (eds.): *Proceedings of the 20th Meeting of the European VLBI Group for Geodesy and Astronomy (EVGA)*. Schriftenreihe des Instituts für Geodäsie und Geoinformation. Heft 22. pp 14–18, Universität Bonn 2011.

Collioud, A; Neidhardt, A: *When «IVS Live» meets «e-RemoteCtrl» real-time data ....* In: Zubko, N; Poutanen, M. (eds.): *Proceedings of the 21st meeting of the European VLBI Group for Geodesy and Astrometry*. pp 17–20. Finnish Geodetic Institute 2013.

Colucci, G: *Matera VLBI station report on the operational and performance evaluation activities from January to December 1998*. telespazio spa 1999. ftp://geodaf.mt.asi.it/GEOD/INFO/matvlbi98.pdf, Download 2013-01-12.

Cousot, Patrick: *Abstract Interpretation*. In: ACM Computing Surveys, Vol. 28, No. 2, June 1996., http://cs.nyu.edu/pcousot/publications.www/Cousot-ACM-Computing-Surveys-v28-n2-p324-328-1996.pdf, Download 2012-06-07.

Cppcheck: *Cppcheck. A tool for static C/C++ code analysis*. http://cppcheck.sourceforge.net/, Download 2012-06-07.

Cppcheck: *Cppcheck 1.55. Manual*. http://cppcheck.sourceforge.net/manual.pdf, Download 2012-06-07.

CppUnit: *CppUnit. Main Page. Welcome! to CppUnit Wiki*. http://sourceforge.net/apps/mediawiki/cppunit/index.php?title=Main_Page, Download 2012-06-06.

Crowther, Mark: *Software Test Metrics. Key metrics and measures for use within the test function. Discussion document*. http://www.cyreath.co.uk/papers/Cyreath_Software_Test_Metrics.pdf, Download 2011-07-30.

CSIRO: *open-monica. Engineering tool for real-time data collection, control, archiving and display*. https://code.google.com/p/open-monica/, Download 2013-05-13.

Czarnecki, Krzysztof; Eisenecker, Ulrich W: *Generative Programming. Methods, Tools, and Applications*. Addison-Wesley 2000.

Damm, Matthias; Winzenried, Oliver: *Security in OPC UA*. In: WEKA FACHMEDIEN GmbH: *Computer & AUTOMATTION. Fachmedium der Automatisierungstechnik*. Issue 12–2013.

DANTE: *GE'ANT Bandwidth on Demand*. http://www.dante.net/DANTE_Network_Projects/GEANT/Pages/Home.aspx, Download 2013-12-01.

Dassing, Reiner; Lauber, Pierre; Neidhardt, Alexander: *Design-Rules für die strukturierte Programmierung unter C und die objektorientierte Programmierung unter C++*. Revision 28.02.2008, Geodtätisches Observatorium Wettzell, 2008.

Davidson, Tal; Pattee, Jim: *Artistic Style 2.02. A Free, Fast and Small Automatic Formatter for C, C++, C#, and Java Source Code*. http://astyle.sourceforge.net/ and http://astyle.sourceforge.net/astyle.html, Download 2012-07-16.

Dawes, Beman; Abrahams, David: *Boost C++ Libraries*. http://www.boost.org/, Download 2011-06-03.

Deutsche Forschungsgemeinschaft Geschäftsstelle (DFG): *Proposals for Safeguarding Good Scientific Practice. Recommendations of the Commission on Professional Self Regulation in Science*. Wiley-VCH Verlag GmbH Weinheim 1998., http://www.dfg.de/download/pdf/dfg_im_profil/reden_stellungnahmen/download/empfehlung_wiss_praxis_0198.pdf, Download 2011-12-08.

DESCA: *DESCA 2020 Model Consortium Agreement. NEW: DESCA 2020 version 1.2, February 2016*. http://www.desca-2020.eu/, February 2016, Download 2016-03-11.

Diedrich, Olive: *Sun RPC-Code steht jetzt unter BSD-Lizenz*. heise online 2009., http://www.heise.de/newsticker/meldung/Sun-RPC-Code-steht-jetzt-unter-BSD-Lizenz-197329.html, Download 2012-08-19.

Dominguez, Jorge: *The Curious Case of the CHAOS Report 2009*. http://www.projectsmart.co.uk/the-curious-case-of-the-chaos-report-2009.html, Download 2011-06-09.

Duvall, Paul M; Matyas, Steve; Glover, Andrew: *Continuous Integration. Improving software quality and reducing risk*. Sixth printing. Rearson Education, Inc. 2011.

Eclipse Foundation, The: *Eclipse Open Source Developer Report*. http://www.slideshare.net/IanSkerrett/eclipse-survey-2012-report-final, Download 2012-07-17.

Enste, Udo; Müller, Jochen: *Datenkommunikation in der Prozessindustrie. Darstellung und anwendungsorientierte Analyse*. Oldenbourg Industrieverlag GmbH 2007.

ESA: *ARIANE 5. Flight 501 Failure. Report by the Inquiry Board*. ESA, http://esamultimedia.esa.int/docs/esa-x-1819eng.pdf, Download 2012-01-02.

Ettl, M: *Entwicklung einer Methodenbibliothek zur genauen Positionsbestimmung beliebiger Objekte mit echtzeitnaher Bildverarbeitung*. Diploma thesis. FH Regensburg 2006.

Ettl, Martin: *Hochgenaue numerische Lösung von Bewegungsproblemen mit frei wählbarer Stellengenauigkeit*. Dissertation. Fakultät für Bauingenieur- und Vermessungswesen. Technische Universität München 2012., http://d-nb.info/1030100101/34, Download 2013-05-13.

Favre-Bulle, Bernard: *Automatisierung komplexer Industrieprozesse. Systeme, Verfahren und Informationsmanagement*. Springer-Verlag/Wien 2004.

Feathers, Michael C: *Effektives Arbeiten mit Legacy Code. Refactoring und Testen bestehender Software*. mitp, Verlagsgruppe Hüthig Jehle Rehm GmbH 2011. (Original title: Working Effectively with Legacy Code. Feathers, M., Pearson Education, Inc., Prentice Hall PTR 2005).

Feister, Uwe: *Nubiscope – ein neues Messgerät für die Wolkendokumentation*. MOL-RAO Aktuell. Deutscher Wetterdienst, Meteorologisches Observatorium Lindenberg – Richard-Aßmann-Observatorium 2/2013. http://www.dwd.de/bvbw/generator/DWDWWW/Content/Forschung/FELG/Download/aktuell__022013, templateId=raw,property=publicationFile.pdf/aktuell_022013.pdf, Download 2015-09-17.

Feldman, S. I; Gay, David M; Maimone, Mark W; Schryer, N. L: *A Fortran to C Converter*. Computing Science Technical Report No. 149, AT&T Bell Laboratories 1995., http://www.netlib.org/f2c/f2c.pdf, Download 2011-07-29.

Flightradar24 AB: *flightradar24. Live Air Traffic*. http://www.flightradar24.com/, Download 2012-09-17.

Forbrig Peter; Kerner, Immo O: *Softwareentwicklung. Lehr- und Übungsbuch*. Fachbuchverlag Leipzig 2004.

Fowler, Martin; Scott, Kendall: *UML konzentriert. Eine strukturierte Einführung in die Standard-Objektmodellierungssprache*. 2. Auflage. Addison-Wesley Verlag, Pearson Education Deutschland GmbH 2000. (Original title: UML Distilled. Second Edition. A brief guide to the standard object modelling language. Addison-Wesley).

Field System Team: *Mark IV Field System*. Volume 1. Space Geodesy Program. NASA/Goddard Space Flight Center 1993/1997a.

Field System Team: *Mark IV Field System*. Volume 2. Space Geodesy Program. NASA/Goddard Space Flight Center 1993/1997b.

Springer Gabler Verlag (Herausgeber): *Gabler Wirtschaftslexikon. Keyword: Projekt*. Springer Gabler. Springer Fachmedien Wiesbaden GmbH, http://wirtschaftslexikon.gabler.de/Archiv/13507/projekt-v7.html, Download 2015-09-11.

GCC team: *GCC, the GNU Compiler Collection*. Free Software Foundation, Inc.. http://gcc.gnu.org/, Download 2012-06-05.

GDB team: *GDB: The GNU Project Debugger*. Free Software Foundation, Inc.. http://sources.redhat.com/gdb/, Download 2012-06-11.

GÉANT: *GÉANT. The pan-European research and education network that interconnects Europe's National Research and Education Networks (NRENs)*. http://www.geant.net/Services/ConnectivityServices/Pages/Bandwidth_on_Demand.aspx, Download 2013-12-01.

Cerloni, Helmar; Oberhaitzinger, Barbara; Reiser, Helmut; Plate, Jürgen: *Praxisbuch Sicherheit für Linux-Server und -Netze*. Carl Hanser Verlag München 2004.

Golem.de: *golem.de. IT-News für Profs. Die ferngesteuerte Fabrik*. http://www.golem.de/news/steuerungstechnik-die-ferngesteuerte-fabrik-1207-93515.html, Download 2013-01-10.

Geodätisches Observatorium Wettzell: *Lokale Meßverfahren. Wasserdampfradiometer Wettzell*. Hompage of the Geodätisches Observatorium Wettzell. http://www.wettzell.ifag.de/WATER/water_start.html, Download 2012-12-31.

Gierhardt, Horst: *Informatik in der Oberstufe. Informatik in der 12 mit Java. Acht-Damen-Problem*. http://www.gierhardt.de/informatik/info12/Damen.pdf, Download 2011-06-28.

Gigantic Software: *logog – logger optimized for games*. http://johnwbyrd.github.io/logog/, Download 2014-04-01.

GNU: *The GNU Fortran Compiler*. http://gcc.gnu.org/onlinedocs/gfortran/index.html#Top, Download 2011-06-18.

GNU: *GNU Operating System. GNU Make*. Free Software Foundation, Inc. http://www.gnu.org/software/make/, Download 2012-06-05.

GNSS Data Center: *about NTRIP. Networked Transport of RTCM via Internet Protocol*. Bundesamt für Kartographie und Geodäsie Frankfurt am Main. http://igs.bkg.bund.de/ntrip/about, Download 2013-11-25.

Güting, Ralf Hartmut; Erwig, Martin: *Übersetzerbau. Techniken, Werkzeuge, Anwendungen*. Springer-Verlag Heidelberg 1999.

Gutiérrez-Naranjo, Miguel A; Martínez-del-Amor, Miguel A; Pérez-Hurtado, Ignacio; Pérez-Jiménez, Mario J: *Solving the N-Queens Puzzle with P Systems*. http://www.gcn.us.es/7BWMC/volume/21_queens.pdf, University of Sevilla, Download 2011-06-28.

Graphviz: *Graphviz – Graph Visualization Software*. http://www.graphviz.org/, Download 2012-01-16.

Halsall, Fred: *Data Communications, Computer Networks and Open Systems*. Fourth Edition. Addison-Wesley Company Inc. 1996.

Hase, Hayo; Behrend, Dirk; Ma, Chopo; Petrachenko, Bill; Schuh, Harald; Whitney, Alan: *VLBI2010 – An International VLBI Service Project in Support of the Global Geodetic Observing System*. REUNION SIRGAS2010, celebrada en el marco de la 42 Reunión del Consejo Directivo del Instituto Panamericano de Geografía e Historia (IPGH). Instituto Geográfico Nacional, Lima, Perú 2010. http://www.sirgas.org/fileadmin/docs/Boletines/Bol15/17_Hase_VLBI2010.pdf, Download 2013-11-21.

HATLab: *FiLa 10G Board. Connection and Service*. HAT-Lab s.r.l.. http://www.hat-lab.com/hatlab/component/content/article/31-generale/50-fila-10g-board, Download 2013-12-01.

van Heesch, Dimitri: *Doxygen Documentation. Generate documentation from source code*. http://www.stack.nl/dimitri/doxygen/, Download 2011-12-08.

Helmke, Hartmut; Höppner, Frank; Isernhagen, Rolf: *Einführung in die Software-Entwicklung. Vom Programmieren zur erfolgreichen Software-Projektarbeit. Am Beispiel von Java und C++*. Carl Hanser Verlag München 2007.

Hermes, Thorsten: *Digitale Bildverarbeitung. Eine praktische Einführung*. Carl Hanser Verlag München Wien 2005.

Herold, Helmut: *Linux/Unix Systemprogrammierung*. Addison-Wesley Verlag München 2004.

Hiener, Michael: *Entwicklung eines Programmmoduls zur automatischen Extraktion gültiger Messwerte einer Satelliten-Entfernungsmessanalge in Anwesenheit von starkem Rauschen*. Diplomarbeit der Forschungseinrichtung Satellitengeodäsie und des Lehrstuhls für Raumfahrttechnik. Technische Universität München 2006.

Hofmann, Fridolin: *Betriebssysteme: Grundkonzepte und Modellvorstellungen*. B. G. Teubner Stuttgart 1991.

ILRS: *Data and Product Formats*. http://ilrs.gsfc.nasa.gov/data_and_products/formats/index.html, Download 2012-10-04.

Ippolito, Greg: *Using C/C++ and Fortran together:*. YoLinux.com Tutorial, YoLinux.com: Linux Information Portal 2008 http://www.yolinux.com/TUTORIALS/LinuxTutorialMixingFortranAndC.html, Download 2011-07-30.

ITWissen: *Information*. ITWissen. Das große Online-Lexikon für Informationstechnologie. http://www.itwissen.info/definition/lexikon/Information-information.html, Download 2011-11-25.

Jamshidi, Mo: *Systems of Systems Engineering. Principles and Applications*. CRC Press. Taylor & Francis Group 2009.

JIVE (Eldering, Bob): *Jive5ab*. Joint Institute for VLBI in Europe. http://www.jive.nl/jive_cc/sin/sin23/node3.html, Download 2013-12-01.

JMU, Liverpool: *The Liverpool Telescope*. Liverpool John Moores University, http://telescope.livjm.ac.uk/, Download 2013-01-10.

Jones, Allen; Freeman, Adam: *Visual C# 2010 Recipies. A Problem-Solution Approach*. apres Springer-Verlag New York, Inc. 2010.

Jet Propulsion Laboratory: *JPL Institutional Coding Standard for the C Programming Language*. Jet Propulsion Laboratory. California Institute of Technology 2009. http://lars-lab.jpl.nasa.gov/JPL_Coding_Standard_C.pdf, Download 2012-11-25.

Kan, Stephen H: *Metrics and models in software quality engineering*. Second Edition. Pearson Education, Inc. 2003.

Karpov, Andrey; Ryzhkov, Evgeniy: *20 issues of porting C++ code on the 64-bit platform*. http://archive.gamedev.net/reference/programming/features/20issues64bit/, Download 2011-07-07.

Kasper, Manuel: *m0n0wall*. http://m0n0.ch/wall/, Download 2013-05-30.

Kazakos, Wassilios; Schmidt, Andreas; Tomczyk, Peter: *Datenbanken und XML*. Springer Verlag Berlin Heidelberg 2002.

Linux Kernel Organization, Inc: *The Linux Kernel Archives*. http://kernel.org/, Download 2011-06-18.

Kernighan, Brian W; Ritchie, Dennis M: *Programmieren in C. Mit dem C-Reference Manual in deutscher Sprache*. 2. Ausg., ANSI-C, Carl Hanser Verlag München Wien 1990. (Original title: The C Programming Language. Second Edition. ANSI-C. Bell Telephone Laboratories, Inc., Prentice-Hall International Inc. London 1990).

Kilger, Richard: *Das 10-m-Radioteleskop*. In: Schneider, Manfred (ed.): *Satellitengeodäsie. Ergebnisse aus dem gleichnamigen Sonderforschungsbereich der Technischen Universität München*. Sonderforschungsbereich. Deutsche Forschungsgemeinschaft. VCH Verlagsgesellschaft 1990.

Klar, Rainer: *Messung und Modellierung paralleler und verteilter Rechensysteme*. B. G. Teubner Stuttgart 1995.

Klar, Michael; Klar, Susanne: *Einfach generieren. Generative Programmierung verständlich und praxisnah*. Carl Hanser Verlag München 2006.

Kofler, Michael: *Linux 2010. Debian, Fedora, openSUSE, Ubuntu*. 9. Auflage. Addison-Wesley Verlag 2010.

Kopp, Herbert: *Bildverarbeitung interaktiv. Eine Einführung mit multimedialem Lernsystem auf CD-ROM*. B. G. Teubner Stuttgart 1997.

Kretschmar, O; Dreyer, R: *Mediendatenbank- und Medien-Logistik-Systeme. Anforderungen und praktischer Einsatz*. Oldenbourg Wissenschaftsverlag GmbH 2004.

van Langevelde, Huib Jan [coordinator]: *NEXPReS Final Report*. Joint Institute of VLBI in Europe Dwingeloo 2013. http://www.jive.nl/nexpres/lib/exe/fetch.php?media=nexpres:nexpres-final-report-rev_2013-09-04.pdf, Download 2013-12-01.

Langmann, Reinhard: *Taschenbuch der Automatisierung*. Carl Hanser Verlag Leipzig 2010.

LangPop.com: *Programming Language Popularity*. http://langpop.com/, Download 2011-06-18.

Leffingwell, Dean; Widrig, Don: *Managing Software Requirements: A Use Case Approach*. Pearson Education, Inc. 2003.

Lévénez, Éric: *Computer Languages History*. http://www.levenez.com/lang/, Download 2011-06-19.

Li, Qingyong; Lu, Weitao; Yang, Jun: *A Hybrid Thresholding Algorithm for Cloud Detection on Ground-Based Color Images*. In: American Meteorological Society: *Journal of Atmospheric and Oceanic Technology*. Volume 28. Issue 10. Boston 2011. http://journals.ametsoc.org/doi/pdf/10.1175/JTECH-D-11-00009.1, Download 2015-09-13.

Linux Foundation, The: *API Sanity Checker*. LSB Infrastructure Program. The Linux Foundation. http://ispras.linuxbase.org/index.php/API_Sanity_Checker, Download 2012-11-25.

Lockheed Martin Corporation: *JOINT STRIKE FIGHTER AIR VEHICLE C++ CODING STANDARDS FOR THE SYSTEM DEVELOPMENT AND DEMONSTRATION PROGRAM*. Document Number 2RDU00001 Rev C. Lockheed Martin Corporation, Dec. 2005., http://www2.research.att.com/bs/JSF-AV-rules.pdf, Download 2011-06-28.

Loeliger, Jon: *Versionskontrolle mit Git*. O'Reilly Verlag Köln 2010. (Original title: Version Control with Git. O'Reilly Media, Inc. 2009).

Lovell, Jim; et al: *AuScope VLBI Operations Wiki*. http://auscope.phys.utas.edu.au/opswiki/doku.php, Download 2012-01-20.

Lovell, Jim; Neidhardt, A: *Field System Remote Operations. Operation of the AuScope VLBI Array*. Course script of the Seventh IVS Technical Operations Workshop. Haystack 2013.

Luiz, Sylvio; Neto, Mantelli; von Wagenheim, Aldo; Pereira, Enio Bueno; Comunello, Eros: *The Use of Euclidean Geometric Distance on RGB Color Space for the Classification of Sky and Cloud Patterns*. In: American Meteorological Society: *Journal of Atmospheric and Oceanic Technology*. Volume 27. Issue 9. Boston 2010. http://sonda.ccst.inpe.br/publicacoes/periodicos/Euclidean_Geometric_Distance_on_RGB_for_Sky_and_Cloud_Patterns.pdf, Download 2015-09-14.

Maguire, Steve: *Writing solid code. Microsoft's Techniques for Developing Bug-Free C Programs*. Microsoft Press Redmond, Washington (USA) 1993.

Mahnke, Wolfgang; Leitner, Stefan-Helmut; Damm, Matthias: *OPC Unified Architecture*. Springer-Verlag Berlin Heidelberg 2009.

Marjamäki: *Writing Cppcheck rules. Part 1 – Getting started*. http://heanet.dl.sourceforge.net/project/cppcheck/Articles/writing-rules-1.pdf, Download 2012-06-07.

The MathWorks, Inc: *PolySpace for Code Verification*. Training Course Notebook. The MathWorks, Inc. 2010.

McClure, Stuart; Scambray, Joel; Kurtz, George: *Das Anti-Hacker-Buch*. 5. Auflage. bhv, Redline GmbH Heidelberg 2006. (Original title: Hacking Exposed™. Fifth Edition. Network Security Secrets & Solutions 2005).

Meek, Brian L; Heath, Patricia L. (eds.): *Guide to Good Programming Practice*. Wiley & Sons Inc 1983. In: Chapman, Alan: *tree swing pictures. the tree swing or tire swing funny diagrams – for training, presentations, etc.* 2014. http://www.businessballs.com/treeswing.htm, Download 2014-03-24.

MISRA (The Motor Industry Software Reliability Association): *MISRA-C:2004. Guidlines for the use of the C language in critical systems*. MIRA Limited Warwickshire (UK) 2004.

MISRA (The Motor Industry Software Reliability Association): *MISRA C++:2008. Guidlines for the use of the C++ language in critical systems*. MIRA Limited Warwickshire (UK) 2008.

Morgan, David: *VPNs: Virtual Private Networks. Primitive roots and Diffie-Hellman key exchange*. Computer Science Department. Santa Monica College 2003. http://homepage.smc.edu/morgan_david/vpn/assignments/assgt-primitive-roots.htm, Download 2015-09-21.

Müller, Andrea: *Erweiterte Systemüberwachung mit rsyslog». rsyslog überwacht unter Linux Anwendungen und Betriebssystem*. heise Netze 1999. http://www.heise.de/netze/artikel/Erweiterte-Systemueberwachung-mit-rsyslog-846750.html, Download 2014-04-01.

MSDN Microsoft: *Design Guidelines for Class Library Developers. .NET Framework 1.1*. http://msdn.microsoft.com/en-us/library/czefa0ke%28vs.71%29.aspx, Download 2011-11-18.

National Academy of Sciences: *Precise Geodetic Infrastructure. National Requirements for a Shared Resource*. National Academies Press 2010. http://www.nap.edu/catalog.php?record_id=12954, Download 2013-12-20.

National Aeronautics and Space Administration (NASA), Goddard Space Flight Center: *GMSEC. Goddard Mission Service Evolution Center*. http://gmsec.gsfc.nasa.gov/index.php, Download 2014-03-30.

Ncurses team: *Announcing ncurses 5.9* Free Software Foundation, Inc. http://www.gnu.org/software/ncurses/, Download 2012-09-26.

Neidhardt, Alexander: *Verbesserung des Datenmanagements in inhomogenen Rechnernetzen geodätischer Messeinrichtungen auf der Basis von Middleware und Dateisystemen am Beispiel der Fundamentalstation Wettzell.* Mitteilungen des Bundesamtes für Kartographie und Geodäsie. Band 37. Verlag des Bundesamtes für Kartographie und Geodäsie Frankfurt am Main 2005.

Neidhardt, A; Ettl, M; Lauber, P; Leidig, A; Eckl, J; Riederer, M; Dassing, R; Schönberger, M; Plötz, Ch; Schreiber, U; Steele, I: *Automation and remote control as new challenges on the way to GGOS.* In: Proceedings of the 17th International Workshop on Laser Ranging, Nr. 48, pp 273–278, Verlag des Bundesamtes für Kartographie und Geodäsie 2012a.

Neidhardt, A; Ettl, M; Mühlbauer, M; Plötz, C; Hase, H; Sobarzo, S; Herrera, C; Onate, E; Zaror, P; Pedreros, F; Zapato, O; Lovell, J: *Two weeks of continuous remote attendance during CONT11.* Poster at the 7th IVS General Meeting. Madrid 2012b.

Neumann, Thomas: *bcov.* http://bcov.sourceforge.net/, Download 2012-06-06.

Niell, Arthur; Whitney, Alan; Petrachenko, Bill; Schlüter, Wolfgang; Vandenberg, Nancy; Hase, Hayo; Koyama, Yasuhiro; Ma, Chopo; Schuh, Harald; Tuccari, Gino: *VLBI2010: Current and Future Requirements for Geodetic VLBI Systems.* Report of Working Group 3 to the IVS Directing Board. GSFC/NASA 2005. http://ivscc.gsfc.nasa.gov/about/wg/wg3/IVS_WG3_report_050916.pdf, Download 2013-01-19.

Niemann, Heinrich: *Pattern Analysis and Understanding.* Second Edition. Springer-Verlag Berlin Heidelberg New York 1990.

National Institute of Standards and Technology (NIST): *Digital Signature Standard (DSS).* FEDERAL INFORMATION PROCESSING STANDARDS PUBLICATION. FIPS PUB 186–3. http://csrc.nist.gov/publications/fips/fips186-3/fips_186-3.pdf, Download 2013-05-19.

National Institute of Standards and Technology (NIST): *Expect.* http://www.nist.gov/el/msid/expect.cfm, Download 2012-09-06.

Oestereich, Bernd: *Analyse und Design mit UML 2. Objektorientierte Softwareentwicklung.* 7. Ausgabe, Oldenbourg Wissenschaftsverlag GmbH München 2005.

Oracle Sun Developer Network (SDN): *Standard Library, STL and Thread Safety.* SDN April 2000., http://developers.sun.com/solaris/articles/stl-new.html, Download 2011-11-24.

Oracle: *System Administration Guide: Security Services.* Oracle and/or its affiliates 2011, http://docs.oracle.com/cd/E23823_01/html/816-4557/auth-2.html and http://docs.oracle.com/cd/E23823_01/pdf/816-4557.pdf, Download 2013-05-14.

Osherove, Roy: *The Art of Unit Testing. Deutsche Ausgabe.* mitp, Verlagsgruppe Hüthig Jehle Rehm GmbH München 2010. (Original title: The Art of Unit Testing with Examples in .NET. Manning Publications Co., Greenwich, Connecticut 2009).

Osovlanski, Doron; Nissenbaum, Baruch: *baruch.c. baruch.hint.* In: Noll, Landon Curt; Cooper, Simon; Seebach, Peter; Broukhis, Leonid A: *The International Obfuscated C Code Contest. Winning entries. 1990. 7th International Obfuscated C Code Contest.* 2003., http://www.de.ioccc.org/years.html#1990, Download 2011-08-04.

Peitgen, Heinz-Otto; Jürgens, Hartmut; Saupe, Dietmar: *Bausteine des Chaos. Fraktale.* Rowohlt Taschenbuch Verlag GmbH Hamburg 1998. (Original title: Fractals for the Classroom. Part 1. Springer Verlag New York 1992).

Pfeifer, Tilo; Schmitt, Robert (ed.): *Autonome Produktionszellen. Komplexe Produktionsprozesse flexibel automatisieren.* Springer-Verlag Berlin Heidelberg 2006.

Pilato, C. Michael; Collins-Sussman, Ben; Fitzpatrick, Brian W: *Versionskontrolle mit Subversion. Software-Projekte intelligent koordinieren.* 3. Auflage. O'Reilly Verlag GmbH & Co. KG Köln 2009. (Original title: Version Control with Subversion. The Standard in Open Source Version Control. Second Edition. O'Reilly Media, Inc. Sebastopol 2008).

Plag, Hans-Peter; Pearlman, Michael: *Global Geodetic Observing System. Meeting the Requirements of a Global Society on a Changing Planet in 2020.* Springer-Verlag Berlin Heidelberg 2009.

PostgreSQL Global Development Group: *PostgreSQL. Das offizielle Handbuch.* mitp-Verlag Bonn 2003.

Precht, Manfred; Meier, Nikolaus; Kleinlein, Joachim: *EDV-Gundwissen. Eine Einführung in Theorie und Praxis der modernen EDV.* 2. Auflage. Addison-Wesley (Deutschland) GmbH 1994.

Press, William H; Teukolsky, Saul A; Vetterling, William T; Flannery, Brian P: *Numerical Recipes in C++. The Art of Scientific Computing.* Second Edition, Cambridge University Press New York 2005.

Puder, Arno; Römer, Kay: *Middleware für verteilte Systeme.* 1. Auflage. dpunkt.verlag GmbH Heidelberg 2001.

Purdy, Gregor N: *LINUX iptables. KURZ & GUT.* 1. Auflage. O'Reilly Verlag GmbH & Co. KG 2005. (Original title: lnux iptables Pocket Reference. O'Reilly Media, Inc. 2004).

Python Software Foundation: *Python v2.7.3 documentation. The Python Standard Library. Built-in Functions.* http://docs.python.org/library/functions.html, Download 2012-09-08.

Rawashdeh, Adnan; Matalkah, Bassem: *A New Software Quality Model for Evaluating COTS Components.* In: Journal of Computer Science 2 (4), pp 373–381, ISSN 1549–3636, 2006.

Rembold, Ulrich (ed.); Blume, Christian; Epple, Wolfgang; Hagemann, Manfred; Levi, Paul: *Einführung in die Informatik für Naturwissenschaftler und Ingenieure.* Carl Hanser Verlag München Wien 1991.

Ricklefs, R. L: *Consolidated Laser Ranging Prediction Format. Version 1.01.* Feb. 2006. http://ilrs.gsfc.nasa.gov/docs/cpf_1.01.pdf, Download 2012-09-16.

Ricklefs, R. L; Moore, C. J: *Consolidated Laser Ranging Data Format (CRD). Version 1.01.* Oct. 2009. http://ilrs.gsfc.nasa.gov/docs/crd_v1.01.pdf, Download 2012-09-16.

Riedel, Sven: *SSH. kurz & gut.* 1. Auflage. O'Reilly Verlag GmbH & Co. KG Köln 2004.

Ristanovic CASE: *Software Metrics. Software Metrics Parameters.* http://www.ristancase.com/html/dac/manual/2.12.01%20Software%20Metrics%20Parameters.html, Download 2012-07-02.

Rothacher, Markus: *Towards a Rigorous Combination of Space Geodetic Techniques.* In: Richter, Bernd; Schwegmann, Wolfgang; Dick, Wolfgang R: *Proceedings of the IERS Workshop on Combination Research and Global Geophysical Fluids.* International Earth Rotation and Reference Systems Service (IERS). IERS Technical Note No. 30. Bavarian Academy of Sciences Munich 2002. http://www.iers.org/SharedDocs/Publikationen/EN/IERS/Publications/tn/TechnNote30/tn30.pdf;jsessionid=9E49A13FF5F13BCC32A093BFBB613DEB.live1?__blob=publicationFile&v=1, Download 2015-09-06.

Rothacher, Markus; Plag, Hans-Peter; Neilan, Ruth: *The Global Geodetic Observing System. Geodesy's contribution to Earth Observation.* GGOS. International Association of Geodesy. http://www.ggos-portal.org/lang_en/nn_261454/SharedDocs/Publikationen/EN/GGOS/Introduction__to__GGOS,templateId=raw,property=publicationFile.pdf/Introduction_to_GGOS.pdf, Download 2015-09-27.

RTAI Team: *RTAI – the Real-Time Application Interface for Linux from DIAPM. RTAI – Real-Time Application Interface Official Website.* Politecnico di Milano – Dipartimento di Ingegneria Aerospaziale 2010. https://www.rtai.org/, Download 2012-11-25.

Rüping, Andreas: *Agile Documentation. A Pattern Guide to Producing Lightweight Documents for Software Projects.* John Wiley & Sons Ltd. The Atrium Chichester UK 2003.

Saake, Gunter; Sattler, Kai-Uwe; Heuer, Andreas: *Datenbanken. Konzepte und Sprachen.* 3. Auflage. mitp/Redline GmbH 2008.

Savage, John E: *Models of computation: exploring the power of computing.* Addison Wesley Longman, Inc. 1998.

Schicker, Edwin: *Datenbanken und SQL.* B. G. Teubner Stuttgart 1996.

Schildt, Herbert: *C++: The Complete Reference.* Third Edition. Osborne/McGraw-Hill Companies Berkeley 1998.

Schlüter, Wolfgang; Brandl, Nikolaus; Dassing, Reiner; Hase, Hayo; Klügel, Thomas; Kilger, Richard; Lauber, Piere; Neidhardt, Alexander; Plötz, Christian; Riepl, Stefan; Schreiber, Ulrich: *Fundamentalstation Wettzell – ein geodätisches Observatorium.* zfv – Zeitschrift für Geodäsie, Geoinformation und Landmanagement. Volume 132.

Number 3. pp 158–167. Wißner-Verlag 2007. http://www.iapg.bgu.tum.de/mediadb/143141/143142/Schlueter_et-al_zfv_3_2007.pdf, Download 2014-03-25.

Schmid, Dietmar; Kaufmann, Hans; Pflug, Alexander; Strobel, Peter; Baur, Jürgen: *Automatisierungstechnik. Mit Informatik und Telekommunikation. Grundlagen, Komponenten, Systeme.* Verlag Europa-Lehrmittel, Nourney, Vollmer GmbH & Co. KG Haan-Gruiten 2011.

Schneider, Manfred: *Fundamentalstation Wettzell. Konzept und Rolle von Fundamentalstationen.* In: Schneider, Manfred (ed.): *Satellitengeodäsie. Ergebnisse aus dem gleichnamigen Sonderforschungsbereich der Technischen Universität München.* Sonderforschungsbereich. Deutsche Forschungsgemeinschaft. VCH Verlagsgesellschaft 1990.

Schöning, Uwe: *Theoretische INformatik – kurzgefaßt.* 3. Auflage. Spektrum, Akad. Verl. Berlin 1997.

Schüler, Torben; Kronschnabl, Gerhard; Plötz, Christian; Neidhardt, Alexander; Bertarini, Alessandra; Bernhart, Simone; la Porta, Laura; Halsig, Sebastian; Nothnagel, Axel: *Initial Results Obtained with the First TWIN VLBI Radio Telescope at the Geodetic Observatory Wettzell.* Sensors — Open Access Journal. MDPI AG. ISSN (Online) 1424–8220, 2015.

Seeber, Günter: *Satellitengeodäsie.* Walter de Gruyter & Co. 1988.

Seward, Julian; Valgrind developers: *Valgrind Documentation. Release 3.7.0 2 November 2011.* 2011. http://valgrind.org/docs/manual/valgrind_manual.pdf, Download 2012-07-11.

Shaffer, D. B; Vandenberg, Nancy R: *Antenna Performance. Operations Manual.* NASA/Goddard Space Flight Center. Space Geodesy Project 1993. ftp://gemini.gsfc.nasa.gov/pub/fsdocs/antenna.pdf, Download 2013-01-19.

Singhal, Mukesh; Shivaratri, Niranjan G: *Advanced concepts in operating systems. Distributed, database, and multiprocessor operating systems.* McGraw-Hill, Inc. 1994.

Smart, Julian; Hock, Kevin; Csomor, Stefan: *Cross-Platform GUI Programming with wxWidgets.* Pearson Education, Inc. 2006.

Smith, Greg; Sidky, Ahmed: *Becoming agile in an imperfect world.* Manning Publications Co. 2009.

Sneed, Harry M; Seidl, Richard; Baumgartner, Manfred: *Software in Zahlen. Die Vermessung von Applikationen.* Carl Hanser Verlag München 2010.

Sommerville, Ian: *Software Engineering 8.* Addison-Wesley Publishers Limited and Pearson Education Limited 2007.

Souza-Echer, M. P; Pereira, E. B; Bins, L. S; Andrade, M. A. R: *A Simple Method for the Assessment of the Cloud Cover State in High-Latitude Regions by a Ground-Based Digital Camera.* In: American Meteorological Society: *Journal of Atmospheric and Oceanic Technology.* Volume 23. Issue 3. Boston 2006. http://journals.ametsoc.org/doi/pdf/10.1175/JTECH1833.1, Download 2015-09-13.

Spillner, Andreas; Roßner, Thomas; Winter, Mario; Linz, Tilo: *Praxiswissen Softwaretest. Testmanagement. Aus- und Weiterbildung zum Certified Tester – Advanced Level nach ISTQB-Standard.* 3. Auflage, dpunkt.verlag GmbH 2011.

Stallman, Richard M; McGrath, Roland; Smith, Paul D: *GNU Make. A Program for Directing Recompilation.* GNU make Version 3.82. Free Software Foundation, Inc. 2010., http://www.gnu.org/software/make/manual/make.pdf, Download 2012-06-05.

Stallman, Richard M; GCC developer team: *Using the GNU Compiler Collection. For gcc version 4.7.1.* Free Software Foundation, Inc., http://gcc.gnu.org/onlinedocs/gcc-4.7.1/gcc.pdf, Download 2012-06-05.

The Standish Group: http://www1.standishgroup.com/newsroom/chaos_2009.php, Download 2011-06-09.

The Standish Group: *CHAOS MANIFESTO 2013. Think Big, Act Small.* The Standish Group. http://www.versionone.com/assets/img/files/ChaosManifesto2013.pdf, Download 2014-05-04.

Starlink Joint Astronomy Centre: *Starlink.* Webpage of The Starlink Project, http://starlink.jach.hawaii.edu/starlink, Download 2011-07-29.

Steffens, Guillaume: *SMART CRITERIA.* 50MINUTES.com 2015.

Stevens, W. Richard : *Programmieren von UNIX-Netzen. Grundlagen, Programmierung, Anwendung.* Carl Hanser Verlag München and Prentice-Hall International, Inc. London 1990. (Original title: UNIX Network Programming. Prentice-Hall, Inc. 1990).

Stroustrup, Bjarne: *Die C++-Programmiersprache.* 2. Auflage. Addison-Wesley (Deutschland) GmbH München 1995. (Original title: The C++ programming language. Second Edition. At&T Bell Telephone Laboratories, Inc. 1991).

Sun Microsystems, Inc: *RFC: Remote Procedure Call. Protocol Specification.* Version 2. Network Working Group RFC 1057 June 1988., http://www.ietf.org/rfc/rfc1057.txt, Download 2012-08-19.

Takahashi, Fujinobi; Kondo, Tetsuro; Takahashi, Yukio; Koyama, Yasuhiro: *Very Long Baseline Interferometer.* IOS Press 2000.

Tanenbaum, Andrew S: *Moderne Betriebssysteme.* 2. Auflage. Carl Hanser Verlag München 1995. (Original title: Modern operating systems. Prentice-Hall, Inc. 1992).

TIOBE Software: *TIOBE Programming Community Index for April 2014. April Headline: Perl hits all-time low (position 13).* http://www.tiobe.com/index.php/content/paperinfo/tpci/index.html, Download 2014-05-04.

Toal, Ray: *Compiler Architecture.* Loyola Marymount University, Los Angeles. http://cs.lmu.edu/ray/notes/compilerarchitecture/, Download 2016-02-29.

Unidata: *The NetCDF Users' Guide. The Extended XDR Layer.* http://www.unidata.ucar.edu/software/netcdf/docs/netcdf/XDR-Layer.html, Download 2014-06-22.

Valgrind developers: *Valgrind.* http://valgrind.org/, Download 2012-07-11.

Vandenberg, Nancy R: *Calibration Data. Operations Manual.* NASA/Goddard Space Flight Center. Space Geodesy Project 1993. ftp://gemini.gsfc.nasa.gov/pub/fsdocs/calibrat.pdf, Download 2013-01-19.

Vandenberg, Nancy R: *drudg: Experiment Preparation Drudge Work.* NASA/Goddard Space Flight Center. Space Geodesy Project 2000. ftp://gemini.gsfc.nasa.gov/pub/fsdocs/drudg.pdf, Download 2013-02-12.

Vandenberg, Nancy R; Rogers, E. E; Himwich, W. E: *Phase Cal System. Operations Manual.* NASA/Goddard Space Flight Center. Space Geodesy Project 1993. ftp://gemini.gsfc.nasa.gov/pub/fsdocs/phase.pdf, Download 2013-01-19.

Vandenberg, Nancy R: *Standard Schedule File Format.* NASA/Goddard Space Flight Center. Space Geodesy Project 1997. ftp://gemini.gsfc.nasa.gov/pub/fsdocs/skfile.pdf, Download 2014-06-27.

Vandevoorde, David; Josuttis, Nicolai M: *C++ Templates. The Complete Guide.* Pearson Education, Inc. 2003.

Vigenschow, Uwe: *Objektorientiertes Testen und Testautomatisierung in der Praxis. Konzepte, Techniken und Verfahren.* 1. Auflage, dpunkt.verlag GmbH 2005.

VMware: *vmware.* VMware Inc. http://www.vmware.com/de/, Download 2012-07-25.

Vogel, Oliver; Arnold, Ingo; Chughtai, Arif; Kehrer, Timo: *Software Architecture. A Comprehensive Framework and Guide for Practitioners.* Springer 2011.

Wagner, Jan: *Tsunami UDP Protocol.* http://tsunami-udp.sourceforge.net/, Download 2012-08-14.

Wagner, Kai: *Selbstdatenschutz durch präventive Verarbeitungskontrolle.* Dissertation. Universität Hamburg 2013.

Warner, Daniel: *Advanced SQL. Studienausgabe. SQL für Praxis und Studium.* Franzis Verlag GmbH Poing 2007.

Wassermann, Tobias: *Versionsmanagement mit Subversion. Installation, Konfiguration, Administration.* mitp, REDLINE GmbH, Heidelberg 2006.

Weinberger, Benjy; Silverstein, Craig; Eitzmann, Gregory; Mentovai, Mark; Landray, Tashana: *Google C++ Style Guide.* http://google-styleguide.googlecode.com/svn/trunk/cppguide.xml, Revision 3.188, Download 2011-06-28.

Wells, Donovan: *Extreme Programming: A gentle introduction.* September 28, 2009. http://www.extremeprogramming.org/, Download 2012-07-31.

Werner, Matthias: *Algorithmen und Programmierung. Skript zur Vorlesung.* http://osg.informatik.tu-chemnitz.de/lehre/old/ws1011/aup/AuP-script.pdf, Download 2011-08-05.

Weston, Neil: *NGS Real-Time Network. Status & Future Prospects.* National Oceanic and Atmospheric Administration. National Geodetic Survey. Presentation at the NGS Convocation Silver Spring 2005. http://www.ngs.noaa.gov/Convocation2005/Presentations/NGSRealTime_Weston.ppt, Download 2013-11-25.

Whitney, Alan; Lonsdale, Colin; Himwich, Ed; Vandenberg, Nacy; van Langevelde, Huib; Mujunen, Ari; Walker, Craig: *VEX File Definition/Example.* http://www.vlbi.org/vex/docs/vex%20definition%2015b1.pdf, Download 2014-06-27.

Wiest, Simon: *Continuous Integration mit Hudson. Grundlagen und Praxiswissen für Einsteiger und Umsteiger.* 1. Auflage. dpunkt.verlag GmbH 2011.

Wolf, Henning; Bleek, Wolf-Gideon: *Agile Softwareentwicklung. Werte, Konzepte und Methoden.* 2. Auflage. dpunkt.verlag GmbH 2011.

Wooldridge, Michael: *An Introduction to MultiAgent Systems.* Second Edition. John Wiley & Sons Ltd. 2009.

Wunderlich, Lars: *AOP. Aspektorientierte Programmierung in der Praxis.* entwickler.press Software & Support Verlag GmbH 2005.

Xen: *Xen . What is Xen ?.* Citrix Systems, Inc. http://xen.org and http://xen.org/files/Marketing/WhatisXen.pdf, Download 2012-07-25.

Yadav, Abhishek: *Analog Communication System.* University Science Press New Delhi 2008.

Zabbix LLC: *ZABBIX. The Enterprise-class Monitoring Solution for Everyone.* http://www.zabbix.com/, Download 2015-09-19.

Zavodnik, Raymond; Kopp, Herbert: *Graphische Datenverarbeitung. Grundzüge und Anwendungen.* Carl Hanser Verlag München Wien 1995.

(Notice: It was necessary to cite web content and web pages to write this book. Even if the links are correctly referenced, they often only represent a snapshot of the web content taken at a specific date. If the pages are not available anymore, it is not in the hands of the author of this work. In these cases, it might be useful to search the web archives, which are web pages offering archived web content taken from time to time as a backup. One of these archives is http://archive.org/web/web.php.)

Index

A

Abstraction 137
Acceptance tests 83
Access control 432
– in software 435
Access Control List (ACL) 438, 444
Access rights 355
– of the GGOS control layers 492
Accuracy 17
Acknowledgment 12
Actuators 254–256, 279, 390, 397
ACU. See Antenna Control Unit (ACU)
Adapting skills of employees 512
Adaption from the industry 277
Additional aspects for remote control 422
Additional CI jobs 112
Additional sensor net 388
ADIZ. See Air Defense Identification Zone (ADIZ)
ADS-B. See Automatic Dependent Surveillance Broadcast (ADS-B)
Advanced Encryption Standard (AES) 146, 434, 451
Advantages of remote control 429
Advantages of the iterative process 119
Agent 397
– autonomy 397
– networks 399
– properties 399
Agile 8, 47–49, 57, 59, 63–62, 66, 67, 69, 81, 83, 107, 116, 120–126, 128–130, 132, 144, 321
– documentation 47, 48, 56, 121, 128
– iterative process 121
– meetings 124
– philosophy 150
– principles 120
Agile software development
– advantages of the iterative process 119
– agile iterative process 121
– agile meetings 124
– agile philosophy 120
– agile principles 120
– agile team 123
– agile team leader and core team 122
– agile values 120
– classic models, sequential, V, W or spiral 117–118
– development process disciplines/workfows 118
– development process phases 119
– elevator statement 121
– example
  – students project and agile training 126
  – svn diff 125
– iteration plan 123
– iterative analysis and design, agile meetings 124
– iterative coding and private testing 125
– iterative committing, continuous integration and deployment 126
– iterative software development processes 118
– need for agile practices 121
– planning of the iterative construction 123
– release plan 123
– requirements of scientific software 117
– SMART criteria for iteration goals 119
– use cases 119
– user story cards 119
– virtual activities of one construction iteration 124
– virtual phases 121
Agile team leader and core team 122, 123
Agile values 120
Air Defense Identification Zone (ADIZ) 4
Alarm levels 377
Alphabet, syntax, semantics 182
Alternative GUI, HTML 293
American National Standards Institute (ANSI) 162, 165, 284, 292
American Standard Code for Information Interchange (ASCII) 162, 165, 173, 243, 292, 306, 308, 323, 338, 377, 420, 522
Analog-to-digital converters 254
Analyzer checks and issues 85
Analyzer tools 84
ANSI. See American National Standards Institute (ANSI)
Antenna Control Unit (ACU) 265, 420
AOP. See Aspect Oriented Programming (AOP)
AOP solutions for tracing and its interpretation 93
Apache 102, 114, 379
Apache Ant™ 114
Apache Maven 114
APD. See Avalanche Photo Diode (APD)
API Sanity Checker 82
Appendices 10
Application layer 150
Application Programming Interface (API) 284, 372, 376, 380
Applied Computer Science as essential part 5
ARIANE 5 type conversion error 28
Artistic Style 2.02 (astyle) 66, 115
ASCII. See American Standard Code for Information Interchange (ASCII)
Aspect Oriented Programming (AOP) 92, 93
Aspects 431
Asterisk 378
Asymmetric cipher algorithms 446
Asymmetric ciphers 446
Asynchronous data traffic for GNSS 495
Atomic 89, 103, 251, 323, 355, 356, 387
Audit 29
AuScope 57, 428, 491, 506
Authorization pyramid 451
Automated observation of mount model corrections 317
Automated unit tests 78
Automatic Dependent Surveillance Broadcast (ADS-B) 251
Automatic sequence control 260
Automation 6, 62, 253, 262, 277, 279, 335, 370, 378, 397, 407–409, 422, 504–507, 511, 512
Autonomous control cells 280
Autonomous controlling 366
Autonomous failure management 368
Autonomous Field System hardware control cells 420
Autonomous functionalities
– control abstraction 335
– control data propagation 334
– controlling 323, 333, 378
– controlling tasks 324
– controlling tasks for the mount model 327
– detailed planning of actions 308
– failure management 330, 337, 389
– hardware driving 328, 337, 386
– information abstraction 335
– intelligent planning systems 322
– planning 333, 378
  – adjustment 322
  – analogies to operating systems 318
  – assignment strategies 318
  – evaluation metrics factors 310
  – priority evaluation 321
  – strategies 319
– roughcut planning of main tasks 308
– simulation of failures and functionalities 338
– update constraints for controlling data 334
– user interfacing 336, 378
  – and main roles 327
  – to the outer world 327
Autonomous hardware driving 367
Autonomous planning 364
Autonomous production cell 278
Autonomous user interfacing 367
Autonomy and automation 279
Availability of the server 166
Avalanche Photo Diode (APD) 250, 324, 335
Avoidance of documentation redundancies 54

B

Background 10, 11
Backus-Naur-Form (BNF) 182
Bandwidth on demand 499
Basic Input Output System (BIOS) 260
bcov 80, 81, 115
Benchmark 66
Berkeley Software Distribution (BSD) 170
Better understanding 7
Big picture 18, 49, 50, 54, 60, 119, 121, 124, 127, 279, 282, 327
– of the GGOS instrumentation 486
Binary Large Object (BLOB) 108, 363
Binary messages 162
Binary protection mask and observation planning 316
Binary search 352, 392
BIOS. See Basic Input Output System (BIOS)
bison 190
Bit order 162
Black box tests 68
BLOB. See Binary Large Object (BLOB)
BNF. See Backus-Naur-Form (BNF)
Book about coordination, communication and automation 5
Boolean algebra 256

Boost C++ Libraries 23, 141
Bottom-up 197, 490
Branches 103
Brief information at the margin 12
BSD. *See* Berkeley Software Distribution (BSD)
Buddy testing and pair programming 83
Build and release management software 114
Build output as test feedback 76
Bypass 420

C

C 6, 20, 22, 24, 30, 31, 41–44, 46, 50, 116, 138,
    258, 412, 420, 469, 471
C# 23, 26, 50
C++ 6, 20, 22, 23, 41–43, 46, 50, 116, 138,
    142, 258, 420, 470
C++ Coding Style Guide 514–523
Cable delay 406
Call-by-reference 21, 33, 44, 46, 85, 173
Call-by-value 21, 33, 46, 85
Call correctness 174
Call-graphs 50
Call semantics 173
Camel case 29
Camera selection 313
Carrier Sense Multiple Access with Collision
    Detection (CSMA/CD) 458, 499
Categories 47
Categorization of toolbox elements 134
C Coding Style Guide 514–523
CDDIS. *See* Crustal Dynamics Data Information
    System (CDDIS)
Celestial reference frame 148, 487
Cell structure 369
Central coordination
– automated observation of mount model
    corrections 217
– autonomous functionalities
    – controlling 323
    – controlling tasks 324
    – controlling tasks for the mount model 327
    – detailed planning of actions 308
    – failure management 320
    – hardware driving 328
    – intelligent planning systems 322
    – planning adjustment 322
    – planning analogies to operating systems 318
    – planning assignment strategies 318
    – planning evaluation metrics factors 310
    – planning priority evaluation 321
    – planning strategies 319
    – roughcut planning of main tasks 308
    – user interfacing and main roles 327
    – user interfacing to the outer world 327
– binary protection mask and observation
    planning 316
– camera selection 313
– client selection for the data propagation 306
– coordination functionalities 300
– developer role 327
– digital image formats, characteristics
    and sources 311
– digital images 311
– example
    – iteration over all interfaces 305
    – VLBI log file 331
    – VLBI snap file 322

– first-come first-served 318
– fuzzy logic 315, 321
– genetic algorithms 322, 326
– image processing for metric parameters 310
– image processing tools and libraries 318
– image textures 312
– laser-ceilometer 311
– local thresholding 315
– logbook as a proprietary solution 330
– lucky imaging 317
– naming service 328
– neuronal nets 322, 326
– nubiscope 311
– planning evaluation metrics 310
– processing chain 314
– segmentation 311
– separation of the workflows 308
– shortest job first 319
– technical realization of the functionalities 302
– threshold segmentation 312
– topology 300
Centralized code assets 144
Centralized continuous integration build 109
Centralized module base 106
Centralized software assets 109
Central Processing Unit (CPU) 95, 258, 260,
    318, 444
CGI. *See* Common Gateway Interface (CGI)
Changeability 18
Chaos report 16
Characteristic factors for scientific
    developments 65
Characteristic sensor profiles 386
Chomsky hierarchy 162, 184
CI. *See* Continuous integration (CI)
Classes 15
Classic models, sequential, V, W or spiral 117
Classification 312
– of user documentation 57
Client 100–103, 105, 106, 113, 127, 148, 154, 155,
    163, 166, 169–179, 188, 202, 206–211,
    216–217, 232–238, 243–244, 269, 282, 300,
    302–304, 306, 353, 362, 378, 391, 397, 415,
    420, 423–425, 429, 443, 448–451, 455, 457,
    464, 465, 467–473, 479, 481, 495, 498, 503
Client application
– administration methods 210
– closing and copying 210
– opening 210
– sample 211
Client safety 216
Client selection for the data propagation 306
Client-server model 151
CMS. *See* Content Management System (CMS)
Codan 116
Code 6, 8, 10–12, 18–31, 33–35, 39–46, 48–52, 56,
    59, 60, 62–72, 77, 78, 80, 83–86, 92, 93, 95, 98,
    100, 101, 105, 109, 110, 112–119, 125–130,
    132–145, 147, 148, 153, 155, 162, 165, 167, 168,
    171, 175–179, 181, 184, 188–190, 197–208,
    211–213, 216–217, 221–223, 234, 236, 238,
    242, 244, 246, 257, 263, 269, 277, 302, 305, 310,
    322, 323, 333, 338, 340, 349, 362, 377, 388, 391,
    409, 412, 414–415, 420, 424, 425, 429, 437,
    449–450, 455, 466, 481, 482, 511, 514
– beautifier 66
– generation 204
– as integral part of research 6

– layout 26
– snippets and illustrations 10
codespell 115
Code toolbox 134
– categorization of toolbox elements 134
– cohesion 133
– coupling 133
– elements of a code toolbox 144
– example
    – component interface 137
    – module interface 135
    – template 139
    – template metaprogramming 142
– generative programming 143
– generative template components 138
– high parametrization 134
– information hiding 134
– object-oriented components 137
– object oriented programming 137
– parametrization 134
– sharing prefabricated off-the-shelf building
    blocks 132
– STL and design patterns 141
– structural modules 135
– template metaprogramming 141
– turing completeness 142
Code version management tools 100
Coding layout 26–30, 35, 98, 129
– camel case 29
– disadvantages of coding style guides 30
– Hungarian notation 28
– Pascal case 29
– rules for code instructions 27
– rules for functions and methods 28
– rules for names 28
– rules for project and source file 27
– rules for the indenting 28
– rules for whitespaces use 27
– self-documenting code 26
– upper case 29
Coding policies 26, 30, 31, 33–35, 66, 129
– compiling workflow 31
– conditional statements 33
– const-correctness 33
– deep copy 34
– early code quality 35
– evaluation order of operators 34
– example
    – code, using styles and policies 36
    – define/include guard 31
    – self-documenting code 36
– exploitation of compiler abilities 35
– function arguments 33
– include guard 33
– jumps 33
– policies with STL 34
– policy dependencies 34
– preprocessor defines *vs.* compiler checkable
    defines 31
– program workflow 33
– quality profiling 35
– reference counting 34
– scoping of variables 31
– shallow copy 35
– try/catch exceptions 33
– type casting 32
– type compatibility of 32-bit and 64-bit 32
Coding style and programming language 11

Coding style guide 26
 – ARIANE 5 type conversion error 28
 – avoid global and static definitions 521
 – avoid preprocessing macros and
   conditions 519
 – class structure and obligatory members 515
 – coding layout 26, 27
 – coding policies 26, 30
 – comments 522
 – constants with const or enum 520
 – const correctness 523
 – define or include guard 519
 – dynamic memory 521
 – existing style guidelines 26
 – file names and internal file structure 514
 – function parameters and return values 522
 – fundamental layout styles and programming
   policies 27
 – for Geodetic Software 514
 – Google C++ style guide 27
 – include paths and include order 519
 – indents with white spaces 520
 – JOINT STRIKE FIGHTER AIR VEHICLE C++
   CODING STANDARDS 26
 – JPL Institutional Coding Standard 27
 – jump instructions are prohibited 521
 – MISRA-C:2004 27
 – MISRA C++:2008 27
 – naming conventions 516
 – N-queens puzzle 25
 – parentheses 520
 – project structure 514
 – psychology behind style guide rules 26
 – queens puzzle 25
 – reduced coding style guide 27
 – reduced use of inline code 523
 – 64-bit/32-bit friendliness 522
 – thread safety 523
 – types and type casting 522
 – unreadable code 25
 – usage of libraries and toolboxes 523
 – variable initialization 521
 – Wettzell's design-rules 27
Cohesion 133
Co-location of space geodetic techniques 4
Combined, globally cross-linked operations
   and coordination 488
Command line abstraction 243
Command line call 207
Command line help 242
Command line language 237
Command line parser 242
Command line scanner 242
Commands 103
Command tunneling 502
Commits 103
Common Gateway Interface (CGI) 293
Common Object Request Broker Architecture
   (CORBA) 237
Commonwealth Scientific and Industrial
   Research Organisation (CSIRO) 372
Communication 5, 9, 10, 19, 33, 46, 48, 61, 64, 88,
   93, 101, 102, 107, 108, 116, 120, 127–128, 135,
   145, 146, 148–151, 153–156, 161–162,
   164–171, 173, 174, 177, 180, 184, 188, 201, 202,
   205, 206, 209, 216–217, 222, 223, 234–237,
   243–244, 246, 253, 262–263, 269, 275, 279,
   282–283, 332, 337, 338, 364, 378, 386, 388, 397,
   399, 400, 409, 411, 414, 420, 425, 426, 429, 431,
   433, 443, 445–451, 455, 457, 458, 460, 464–470,
   473, 479–482, 487, 491, 494–503, 506–508,
   511–512
 – checksums 217
 – control stack 425
 – interface layers 262
 – middleware 237
 – protocol 101, 127, 145, 149, 161, 168, 170,
   175, 243, 269, 467
 – safety 217
 – socket 149
 – user credentials 101
Comparison to industrial standard, OPC UA
   378, 457
Compilation 12, 20, 21, 23, 31, 41, 44, 74–77, 109,
   129, 139, 141, 143, 190, 197, 201, 203, 245, 305
Compiler 12, 20, 21, 24, 26, 31–33, 36, 41–46, 52,
   54, 62, 64, 76, 85, 94, 95, 109, 110, 128, 129,
   139–143, 147, 175, 177, 181, 190–192, 201,
   204, 242, 244, 420
 – vs. interpreter 20
 – outputs 84
 – phases 190
Compiler checkable 31
Compiling workflow 31
Complete system tests 83
Complexity 18
Complex system behavior 15
Component 133
Compositional and transformational
   refinements 202
Comprehensability 18
Computer 258
Computer program 14
Concept-level access control 455
Concurrent Version System (CVS), 100
Conditional statements 33
Configurability 18
Configuration 373
Consequences of forwarding 465
Consolidated Laser Ranging Data Format
   (CRD) 250, 340, 366, 497
Consolidated Laser Ranging Prediction Format
   (CPF) 147, 251, 340, 347, 354, 359, 364,
   489, 497
Const-correctness 33
Content Management System (CMS) 61
Context-free languages 184
Context-sensitive languages 185
Continuous deployment 112
Continuous documentation 112
Continuous feedback 111
Continuous integration (CI) 8, 30, 35, 77,
   109–117, 123, 124, 126, 129–130, 134, 144,
   146, 148, 203, 438, 523
 – additional CI jobs 112
 – build and release management software 114
 – centralized continuous integration build 109
 – centralized software assets 109
 – continuous deployment 112
 – continuous documentation 112
 – continuous feedback 111
 – continuous self-testing and self-inspecting 110
 – developer practices 115
 – Eclipse 116
 – mechanisms for continuous integration
   builds 111
 – public project status displays 111
 – useful tools 114
 – version control system 109
 – virtualization 110
Continuous self-testing and self-inspecting 110
Control layers 409
Controlling layers from signal circuits to genetic
   algorithms 275
Control systems (laser ranging) 9
Control terminal 470
 – SSH tunnel management 473
Conventions
 – acknowledgment 12
 – brief information at the margin 12
 – coding style and programming language 11
 – explanation of theory, technical terms and
   acronyms 11
 – language and citation 11
 – use of source code 11
Converter classes for external code 40
Coordination 5, 9, 278, 280–282, 300–302,
   304–306, 311, 317–319, 323–325, 327, 328,
   332–336, 366, 368, 378, 388, 390, 391, 397, 399,
   400, 405, 409, 412, 424, 426, 429, 488, 491–494,
   502, 503, 507, 511, 512
Copy-modify-merge solution 101
CORBA. See Common Object Request Broker
   Architecture (CORBA)
Correctness 17
Costs and benefits in the risk management 101
Counts, metric parameters, measures 65
Coupling 133
Cover and modify 40
CPF. See Consolidated Laser Ranging Prediction
   Format (CPF)
Cppcheck 78, 84, 115
CPS. See Cyber-physical system (CPS)
CPU. See Central Processing Unit (CPU)
CRC. See Cyclic redundancy check (CRC)
CRD. See Consolidated Laser Ranging Data
   Format (CRD)
Create rbash user, remote security 440
Creation of network enclaves 458
Criteria to select a programming language 19
Critical section 88–90, 128, 166, 204, 216, 222,
   223, 269, 292
Cron job 111, 114, 378
Crustal Dynamics Data Information System
   (CDDIS) 364, 497
Cryostat 401
Cryptographic checksums 450
CSIRO. See Commonwealth Scientific and
   Industrial Research Organisation (CSIRO)
CSMA/CD. See Carrier Sense Multiple Access with
   Collision Detection (CSMA/CD)
Culture for the constructive handling of errors 97
Current workload splitting 504
CVS. See Concurrent Version System (CVS)
Cyber-physical system (CPS) 511, 512
Cyclic redundancy check (CRC) 146, 162,
   449, 451

D

Daemon 166, 171, 330, 448, 450, 463, 464
Dangling else 183
Data and control traffic 494

Database 353
– design
  – ERM 358
  – normalization 356
  – proprietary documentation style 359
  – recipe to avoid redundancies 357
– metadata tables 376
– principles 354
– SQL interface 360, 362
– tables 376, 377
Database Management System (DBMS) 353,
  363, 371
– PostgreSQL 362
Data chopping and steganography 479
Data Encryption Standard (DES) 175, 235,
  434, 448
Data for laser ranging 250
Data management
– access rights 355
– atomic operations 355
– autonomous controlling 366
– autonomous failure management 368
– autonomous hardware driving 367
– autonomous planning 364
– autonomous user interfacing 367
– binary search 352
– database 353
– database design
  – ERM 358
  – normalization 356
  – proprietary documentation style 359
  – recipe to avoid redundancies 357
– Database Management System,
  PostgreSQL 362
– database principles 354
– direct addressing 339
– example
  – configuration file format 349
  – CPF parsing correctness 341
  – CPF prediction file 340
  – CPF regular expressions 341
  – SQL driver module 362
  – SQL statements 361
  – XML passage 347
  – XML passage info 348
– fail-safe 356
– file format 339
  – ILRS CPF 340
  – proprietary configuration file format 348
  – XML 346
– file manipulations 338
– file systems 338
– hash function 339
– hybrid data management 363
– integrity 355
– limits of database 363
– limits of directories and file formats 352
– memory space vs. access speed 339
– redundancy 355
– relational database concept 353
– single-level and multi-level data
  structures 351
– SQL database interface 360, 362
– toolbox module for configurations 349
– transactions 356
– use of files for specific laser ranging data 352
– user roles 356
Data representation, XDR 171

Data traffic
– for e-RemoteCtrl 503
– for SLR 497
– for VLBI 498
Daylight hopping of control responsibilities 493
DBBC. See Digital Baseband Converter (DBBC)
DBE. See Digital Back End (DBE)
DBMS. See Database Management System
  (DBMS)
DCE. See Distributed Computing Environment
  (DCE)
Deadlock 90, 94, 95, 130, 166, 222
Dealing with software errors 62
Debugger 93
Debugging 31, 69, 92–95, 116, 125, 141, 145, 330
Deep copy 34, 91
Definition 169
– agent 397
– agile documentation 48
– autonomous control cells 280
– autonomous production cell 278
– classification 312
– client-server model 151
– code toolbox 134
– communication middleware 237
– communication protocol 149, 175
– communication socket 149
– component 133
– continuous integration 109
– critical section 88
– database 353
– distributed system 180
– documentation landscape 61
– dynamic code analysis 92
– elementary segmentation 312
– firewall 459
– fundamental station 4
– Geodesy (the classic definition) 2
– Global Geodetic Observing System
  (GGOS) 486
– grammar 184
– integration tests 83
– legacy code 40
– in a makefile 76
– modularity 133
– module 133
– network enclave 458
– Nyquist-Shannon sampling theorem 334
– processes 87
– program 14
– real-time systems 276
– refactoring 41
– Remote Procedure Call (RPC) 169
– safety 216
– security 432
– semaphore 89
– software design rules 19
– software project 15
– software system 19
– software tests 68
– static code inspections 84
– testing landscape 96
– test metrics 64
– threads 87
– unit tests 69
– version control system 97
Delete rbash user, remote security 443
Deliverables 60

Denial-of-Service-attacks 443, 449, 450
DES. See Data Encryption Standard (DES)
DESCA. See Development of a Simplified
  Consortium Agreement (DESCA)
Design issues 171
Design patterns 141
Design rules 19
Destination Network Address Translation
  (DNAT) 463
Developer practices 115
Developer role 327
Development of a Simplified Consortium
  Agreement (DESCA) 60
Development process disciplines/workflows 119
Development process phases 119
Development steps 206
Device control loop 254
Device Independent file format (DVI) 54
Dewar 401
DHCP. See Dynamic Host Configuration Protocol
  (DHCP)
Differential Global Navigation Satellite Systems
  (DGNSS) 495
Differential Global Positioning System
  (DGPS) 495, 496
Different mechanisms 149
Different physical communication lines 262
Digital Back End (DBE) 404
Digital Baseband Converter (DBBC) 404, 408,
  429, 500–501
Digital image formats, characteristics
  and sources 311
Digital images 311
Digital Signature Algorithm (DSA) 447, 448
Digital-to-analog converters 254
Direct addressing 339
Directed state graphs 164
Disadvantages 506
– of coding style guides 30
– of the usual testing 62
Distributed and centralized control centers 491
Distributed Computing Environment (DCE) 170
Distributed hardware control 274
Distributed system 180, 466
Distribution platform 237
DNAT. See Destination Network Address
  Translation (DNAT)
DNS. See Domain Name Service (DNS)
Documentation 8, 26, 47–52, 54–57, 60–63, 65,
  95, 98, 107, 109, 112, 113, 116, 117, 119,
  127–130, 367, 409
– avoidance of documentation redundancies 54
– big picture 49
– call-graph 50
– categories 47
– classification of user documentation 57
– deliverables 60
– detailed, code-related, fast-changing 50
– documenting mathematic formulas 54
– documents for administration and other
  scientist 60
– documents for users 56
– Docu-Wiki 61
– Doxygen
  – in the development process 56
  – on Doxygen-commented code 50
  – function header 52
  – inline comments 52

– on plain code 50
– source file header 51
– example
  – file header with Doxygen comments 51
  – function with Doxygen comments 53
  – function with non-redundant Doxygen comments 56
– focus 47
– guidance 57
– guides 57
– include documenting images and graphics 52
– individual documents 49
– as information carrier 47
– landscape 60
– long-term
  – design describing overview 48
  – vs. fast-changing documents 48
– notification 62
– offering documentation 60
– organizing documentation 61
– organizing document versions 61
– readership 47
– references 57
– requirement of documentation 47
– The Right Language 59
– scientific project papers 60
– scientific result papers 60
– sequence charts 50
– structure and formats 49
– text-highlighting 51
– tutorials 57
– user documentation in scientific projects 56
– user's manuals 59
– written documentation 48
Documentation systems, Doxygen 50
Documenting mathematic formulas 54
Document Object Model (DOM) 346
Documents for administration and other scientist 60
Documents for users 56
Document Type Definition (DTD) 346
Docu-Wiki 61
DOM. See Document Object Model (DOM)
Domain Name Service (DNS) 464, 466
Domain Specificatio Language (DSL)
– alphabet, syntax, semantics 181
– Chomsky hierarchy 184
– context-free languages 184
– context-sensitive languages 185
– dangling else 183
– domain vs. application engineering 181
– DSL for a generator system 181
– example, interface definition file 189
– free-format language 187
– generator development 181
– grammar 183
– IDL-constants definitions 188
– IDL-file 188
– IDL-grammar 187
– IDL-interface definition 188
– IDL-type definitions 188
– Interface Definition Language (IDL) 186
– recursively enumerable languages 186
– regular languages 184
– syntax tree 182
– textual syntax notation 182
Domain Specification Language (DSL) 143, 181, 183, 186, 188, 190, 340

Domain vs. application engineering 181
Doppler Orbitography and Radiopositioning Integrated by Satellite (DORIS) 486
Doxygen 50, 115, 117, 148, 522
– in the development process 56
– on Doxygen-commented code 50
– function header 52
– inline comments 52
– on plain code 50
– source file header 51
DSA. See Digital Signature Algorithm (DSA)
DSL. See Domain Specification Language (DSL)
DTD. See Document Type Definition (DTD)
DVI. See Device Independent file format (DVI)
Dynamic code analysis 92
– AOP solutions for tracing and its interpretation 93
– Aspect oriented programming (AOP) 92
– debugger 93
– debugging tools 93
– deep copy 91
– dynamic heap elements in parallel systems 90
– false positives 95
– multitasking 86
– mutex 90
– mutual exclusion 89
– non-intrusive checks 95
– operating systems 86
– pitfalls with STL-strings 90
– preemptive scheduling 87
– processes 87
– profiler 94
– profiling tools 94
– race condition 94
– reference counting 90
– round-robin-scheduling 87
– scheduling 87
– semaphore 89
– shallow copy 90
– shared memory 87
– threads 87
– valgrind 94
Dynamic code optimization 203
Dynamic heap elements in parallel systems 90
Dynamic Host Configuration Protocol (DHCP) 502

E

Early code quality 35
Earth Orientation Parameters (EOP) 251, 364, 405, 407, 409, 492, 497
Eclipse 116, 190
EDC. See European Data Center (EDC)
EEPROM. See Electrically Erasable Programmable Read-Only Memory (EEPROM)
Effective user ID, local security 433
Efficiency 18
Electrically Erasable Programmable Read-Only Memory (EEPROM) 258, 260
Elementary segmentation 312
Elements of a code toolbox 144
Elevator statement 121
Encapsulation 24, 41, 43, 44, 46
– of C in C++ 41
– of C++ in C 41
– of compiler and interpreter code 46

– of other languages 43
Enclave 10, 458, 459, 464, 466–468, 482, 491, 492, 494, 500, 502, 507
Encryption with a SSH 446
Endianess 162
Entity Relationship Model (ERM) 358, 359
EOP. See Earth Orientation Parameters (EOP)
Equivalence to software test systems 415
e-RemoteCtrl 58, 426, 428–430, 450–453, 455, 479, 503
ERM. See Entity Relationship Model (ERM)
Error 5, 6, 16, 17, 20, 23, 26–30, 33–35, 40, 41, 57, 62–70, 72, 77, 83–86, 88, 90, 92–97, 110, 111, 115, 116, 121, 125, 126, 128, 130, 141, 147, 149, 150, 153, 156, 161, 162, 164–168, 171, 174, 175, 178, 179, 190, 191, 201, 204, 210, 217, 220, 223, 232, 233, 243, 244, 251, 269, 277, 279–281, 283, 284, 300, 302, 304, 305, 310, 317, 323, 325, 330, 337, 338, 369, 373, 379, 387–389, 397, 407, 409–412, 414, 422, 424, 426, 427, 429, 437, 438, 449, 457, 470, 471, 493, 495, 498, 505, 520, 521, 523
Error handling 173, 204
ESA. See European Space Agency (ESA)
Escape sequences and ncurses 284
e-shell 471
Ethernet 149
European Data Center (EDC) 364, 497
European Space Agency (ESA) 28
European VLBI Network (EVN) 425
Evaluation order of operators 34
Example
– client stub function call 178, 179
– code, using styles and policies 36
– command line interpreter 238
– component interface 137
– configuration file format 348
  – for data servers 375
  – for sensor control points 373
  – for sub-cells 374
– control of the AuScope network 428
– CPF parsing correctness 341
– CPF prediction file 340
– CPF regular expressions 341
– deep copy 91
– define/include guard 31
– dome IDL file 271
– extend existing systems 418, 431, 472
– file header with doxygen comments 51
– function stabilization 81
– function with goxygen comments 53
– function with non-redundant doxygen comments 56
– hardware driver module 265
– HTML template 294
– IDL file 238
– idl2rpc.pl application engineering 218, 223, 229
– interface definition file 189
– iteration over all interfaces 305
– local security 436
– makefile 74
– makefile dependencies 76
– MCI functionalities 381
– module interface 135
– nested FOR-loop generation 200
– periodic control loop 269
– process watchdog 167

Example (*cont.*)
- remote security 440, 443, 445, 460
- scanner 192
- self-documenting code 37–39
- semantic decisions 202
- server header skeleton 213
- server source skeleton 214
- socket connection timeout 156
- socket select 159
- SO_LINGER 154
- SQL driver module 362
- SQL statements 361
- SSH tunnel management 473
- students project and agile training 126
- sun RPC specification file 176
- SVN commands 104
- svn diff 125
- SVN file header/trailer 107
- SVN properties 106
- template 139
- template metaprogramming 142
- test case 72
- test class 79
- textual RPC instructions 241
- unreadable code 25
- VLBI log file 331
- VLBI snap file 324
- wxWidgets windows 287
- XML passage 347
- XML passage info 348
Example environments, laser ranging
    and radio interferometry 248
Existing style guidelines 26
expect 466
Expert system 276
Explanation of theory, technical terms
    and acronyms 11
Exploitation of compiler abilities 35
Extended GGOS data flows 489
Extend existing systems
- additional aspects for remote control 422
- advantages of remote control 429
- autonomous field system hardware control
    cells 420
- bypass 420
- communication control stack 420
- equivalence to software test systems 415
- example control of the AuScope network 428
- field system monitor 428
    - restrictions 420
- functional extensions 424
- globally available real-time data with
    e-QuickStatus 429
- graphical user interface 431
- GUI variants 417
- hierarchy of control centers 427
- importance of parallel remote equipment 422
- importance of system monitoring 422
- look-and-feel with new graphic
    extensions 423
- operation control strategies 426
- optimization for C++ 420
- system health and human safety 420
Extensible Markup Language (XML) 347, 348,
    464, 467
External code libraries 40
External connectivity of GGOS observatories 499

External Data Representation (XDR) 171, 172,
    217, 308, 351, 450, 457
Extreme Programming (XP) 120

F

Factors of success 18
Fail-safe 356
Failure Mode and Effects Analysis (FMEA) 62, 63
False positives 86, 95
Fault tolerance 17
f2c 43
Feedback control loop 255
Field Programmable Array (FPGA) 256, 404
Field System Monitor 415
Field System Monitor Code, Extend existing
    systems 418
Field System Monitor restrictions 420
File format, example 340
- ILRS CPF 340
- proprietary configuration file format 348
- XML 346
File manipulations 338
File systems 338
File Transfer Protocol (FTP) 330, 364, 366, 371
Filter tables (iptables) 444
Finite state machine and regular expressions 164
Firewall 10, 102, 107, 209, 459, 463–468, 481,
    482, 494, 500, 502
First-come first-served 318
First revolution, astronomical findings
    and geodetic surveys 510
Flawfinder 115
Flexbuff 408, 500, 503
Flexibility 18
FMEA. *See* Failure Mode and Effects Analysis (FMEA)
FORTRAN 20–22, 43–46, 50, 409
- encapsulation
    - of arrays 46
    - with the converter module f77.h 44
    - with f2c 43
    - of function arguments 46
    - restrictions 46
    - of standard types 44
- former de-facto standard 21
Fourth industrial revolution 511
Fourth revolution, global networking 511
FPGA. *See* Field Programmable Array (FPGA)
Free-format language 187
FTP. *See* File Transfer Protocol (FTP)
Functional decomposition 15
Functional extensions 424
Functionality 17
Functional security 450
Function arguments 33
Functions 15
Fundamental layout styles and programming
    policies 27
Fundamental station 4
Further parameterizations 207
Further processing 503
Future basic workload splitting for 24/7 505
Future data traffic
- for GNSS 497
- for SLR 498
- for VLBI 499
Future observation workload for 24/7 506

Future techniques VGOS 407
Fuzzy logic 276, 315, 321
fwbuilder 464

G

Galvanic isolators 263, 388
GCC. *See* GNU Compiler Collection (GCC)
GDB. *See* GNU Project Debugger (GDB)
General Purpose Interface Bus (GPIB) 411
General situation 500
General structure 332
Generated modules and components 148, 205
Generating the protocol modules 177
Generative domain model 190
Generative programming 92, 134, 143, 148, 181,
    236, 243, 244, 302, 305
Generative template components 138
Generator 9, 26, 82, 92, 107, 128, 134, 135, 143,
    148, 181, 183, 186, 188, 190, 202–207, 209,
    210, 217, 220, 234, 238, 242, 263, 271, 305,
    330, 428, 446, 447, 451
Generator development 181
- Chomsky hierarchy 162
- regular languages 162
Genetic algorithms 276, 322, 326
Geodesy (the classic definition) 2
Geodetic infrastructure 3
Geodetic Observatory Wettzell 4–6, 11, 12, 19,
    27, 80, 85, 90, 99, 100, 103, 110, 114, 121, 125,
    145, 148, 156, 167, 253, 277, 318, 347, 390,
    391, 428, 458, 466, 468, 482, 495, 502–504
Geodetic VLBI
- cable delay 406
- future techniques, VGOS 407
- meteorology and calibration 407
- phase calibration 406
- principles of VLBI 400
- system efficiency and calibration 405
- system time 406
- technical VLBI aspects
    - down-conversion and sampling 403
    - reception 401
    - shipment and correlation 404
- VLBI observation results 405
- VLBI observation workflow 404
GIF. *See* Graphics Interchange Format (GIF)
Git 100, 108
Given situation and risks 39
Global Geodetic Observing System (GGOS) 5, 8,
    10, 11, 400, 428, 429, 484–495, 499, 502, 503,
    506–507, 510, 511
- automation
    - current workload splitting 504
    - disadvantages 506
    - future basic workload splitting for 24/7 505
    - future observation workload for 24/7 506
    - potential of automation 504
    - workload estimation 504
    - workload splitting 504
- big-picture of the GGOS instrumentation 486
- Control Infrastructure 10
- coordination, control and operation
    - access rights of the GGOS control
        layers 492
    - daylight hopping of control
        responsibilities 493

– distributed and centralized control centers 491
– job management system 492
– proof of concept 493
– recurring structures 491
– system of systems 491
– use cases 493
– feedback loop, quality control and optimization of its use 489
– history and purpose 484
– key techniques 487
– main idea 486
– national requirements 487
– networks
  – asynchronous data traffic for GNSS 495
  – bandwidth on demand 499
  – command tunneling 502
  – data and control traffic 494
  – data traffic for e-RemoteCtrl 503
  – data traffic for SLR 497
  – data traffic for VLBI 498
  – data traffic from a SLR site 497
  – data traffic from a VLBI site 498
  – data traffic to a SLR site 497
  – data traffic to a VLBI site 498
  – external connectivity of GGOS observatories 499
  – further processing 503
  – future data traffic for GNSS 497
  – future data traffic for SLR 498
  – future data traffic for VLBI 499
  – general situation 500
  – networks for an observatory 502
  – NTRIP 495
  – physically separated transfer network 500
  – real-time data traffic for GNSS 496
  – round-robin recording 503
  – security in the e-VLBI network 501
  – tsunami UDP Protocol 498
– operational specification for GGOS 486
– overall design and tasks 487
– as a revolutionary step 510
– as unifying umbrella 484
– vision paper 485
Globally available real-time data with e-QuickStatus 429
Globalnaja nawigazionnaja sputnikowaja sistema (GLONASS) 486, 488, 495
Global Navigation Satellite Systems (GNSS) 4, 310, 486–488, 493, 495–498
Global Positioning System (GPS) 147, 400, 406, 486, 488, 495
GMSEC. See Goddard Mission Services Evolution Center (GMSEC)
GNU. See GNU's Not Unix (GNU)
GNU Build System 73
GNU Compiler Collection (GCC) 12, 20, 21, 94, 95, 110
GNU Lesser General Public License (LGPL) 12, 284
GNU make 74, 114, 115
GNU Project Debugger (GDB) 94
GNU's Not Unix (GNU) 12, 44, 74, 95, 114, 115, 519
Goal of observations 253
Goddard Mission Services Evolution Center (GMSEC) 372
Google C++ Style Guide 27
gpg 451
GPIB. See General Purpose Interface Bus (GPIB)

GPS. See Global Positioning System (GPS)
Grammar 184
Graphical user interface (GUI) 50, 57, 83, 90, 110, 116, 127, 217, 280, 283–284, 292–294, 300, 325, 327, 334–337, 363, 367, 378, 380, 397, 398, 423–427, 453, 455, 464, 479
Graphics Interchange Format (GIF) 49, 311
Graphviz 50
Gray-box-tests 68
Group ID 433
GTK+ 284
GUI. See Graphical user interface (GUI)
Guidance 57
Guides 57
GUI variants 424

H

HALCON 318
Hardware control
– autonomous functionalities
  – control abstraction 335
  – control data propagation 334
  – controlling 333
  – failure management 337
  – hardware driving 337
  – information abstraction 335
  – planning 333
  – simulation of failures and functionalities 338
  – update constraints for controlling data 334
  – user interfacing 336
– cells 333
– general structure 332
– hardware control cells 333
– Nyquist-Shannon sampling theorem 334
– real-time 333
– timing 333
Hardware driver module/component 263
Hardware layers 369
Hardware Virtual Machine (HVM) 111
Hash function 113, 339, 450
Hierarchical document tree 103
Hierarchy of control centers 427
High Earth Orbit (HEO) 248, 324, 335
High parametrization 134
Histogram 312, 335, 336
Hot pluggable 422
Hot standby 422
HTML. See Hypertext Markup Language (HTML)
HTTP. See Hypertext Transfer Protocol (HTTP)
HTTPS. See Hypertext Transfer Protocol Secure (HTTPS)
Human skills 108
Hungarian notation 28
HVM. See Hardware Virtual Machine (HVM)
Hybrid cipher algorithm in the SSH 448
Hybrid data management 363
Hypertext Markup Language (HTML) 49–51, 54, 57, 59, 61, 77, 78, 80, 129, 293–295, 300, 319, 346, 366, 367, 370, 424, 522
– limitations 300
– templates 294
Hypertext Transfer Protocol (HTTP) 102, 151, 293, 370, 425, 429, 464, 495, 496
Hypertext Transfer Protocol Secure (HTTPS) 102

I

IAG. See International Association of Geodesy (IAG)
ICRF. See International Celestial Reference Frame (ICRF)
Identification in a network 375
IDL. See Interface Definition Language (IDL)
IDL-constants definitions 188
IDL-file 188
IDL-grammar 187
IDL-interface definition 188
idl2rpc.pl application engineering
– client application
  – administration methods 210
  – closing and copying 210
  – opening 210
  – sample 211
– client safety 216
– command line call 207
– communication checksums 217
– communication safety 217
– development steps 206
– example
  – client main 211
  – server header skeleton 213
  – server source skeleton 214
– further parameterizations 207
– generating the modules 205
– possible error sources 220
– server application sample 213
– server application skeleton 212
– server safety 220
  – automatic safety device 213
  – demonstration 223
  – multi-threading 222
  – parallel safety device thread 233
  – persistence and idempotence 232
  – semaphores 222
  – startup control 222
  – time limitations 222
  – watchdog 232
– server security
  – authentication 234
  – authorization 235
  – in the LAN and WAN 234
  – secure RPC 235
Idl2rpc.pl phases 204
IDL type definitions 188
IDS. See Intrusion Detection System (IDS)
IEC. See International Electrotechnical Commission (IEC)
IEEE. See Institute of Electrical and Electronics Engineers (IEEE)
IEEE 754 172
IF. See Intermediate Frequency (IF)
IGS. See International GNSS Service (IGS)
ILRS. See International Laser Ranging Service (ILRS)
Image processing for metric parameters 310
Image processing tools and libraries 318
Image textures 312
Importance of parallel remote equipment 422
Importance of software 15
Importance of system monitoring 422
Include documenting images and graphics 52
Include guard 31

Independence of network enclaves 458
Index Sequential Access Method (ISAM) 355
Individual documents 49
Individualization 512
Industrial remote control 397
Industrial PC 260
Informal reviews 84
Information hiding 134
Infrared (IR) 311–314, 390, 429
Inheritance 137
Initialization of variables, type casting 32
Initial situation
– applied computer science as essential part 5
– book about coordination, communication
  and automation 5
– co-location of space geodetic techniques 4
– geodetic infrastructure 3
– geodetic observatory Wettzell 4
– new systems for SLR and VLBI 5
– satellite geodesy 2
– space geodesy 3
– terrestrial geodesy 2
– terrestrial infrastructure 4
– on the way to a GGOS core site 4
InSAR. See Interferometric Synthetic Aperture
  Radar (InSAR)
In-sky and on-ground laser ranging safety with
  additional data 251
In-sky laser ranging safety with radars 251
In-sky safety 389
Institute of Electrical and Electronics Engineers
  (IEEE) 173, 377, 411
Institute's construction kit 145
Integration testing 83
– buddy testing and pair programming 83
– complete system tests 83
– integration test 83
– pair programming 83
– reviews 83
– static code inspections 84
– system tests 83
Integrity 18, 355
Intelligent control units 254
Intended readers
– background 11
– observatory staff 11
– students 11
Interaction of the hardware 249
Interface Definition Language (IDL) 186–191,
  197, 201, 204–206, 209, 210, 213, 217, 233,
  238, 242, 244, 271, 302, 305, 378
Interferometric Synthetic Aperture Radar
  (InSAR) 3
Intermediate code generation 201
Intermediate Frequency (IF) 403, 404, 406, 408, 429
International Association of Geodesy (IAG) 5,
  484, 486
International Celestial Reference Frame
  (ICRF) 405, 484
International Electrotechnical Commission
  (IEC) 378
International GNSS Service (IGS) 5
International Laser Ranging Service (ILRS) 5, 148,
  253, 340, 366
International Standards Organization (ISO) 149,
  150, 360
International Terrestrial Reference Frame
  (ITRF) 484

International VLBI service for geodesy and
  astrometry (IVS) 4, 369, 407, 413, 414, 425,
  429, 430
Internet Protocol (IP) 149–151, 153, 155, 162,
  171, 178, 179, 210, 237, 262, 330, 364, 375,
  378, 460, 463, 464, 466–468, 495, 500, 502, 503
– addresses and port numbers 150
– forwarding 464
– spoofing 466
Internetwork layer 150
Interoperability 18
Interpiler 20, 21
Interpretation of test results 67
Interpreter 20–21, 24, 46, 128, 129, 165, 182,
  238, 241–244, 306, 308, 375
Interprocess Communication (IPC) 87, 148, 151,
  156, 161, 162, 166–168, 174, 180, 181, 183,
  190, 236, 409, 411, 415, 470, 473, 523
– application layer 150
– availability of the server 166
– binary messages 162
– bit order 162
– client-server model 151
– communication protocol 149
– communication sockets 149
– different mechanisms 149
– directed state graphs 164
– endianness 162
– ethernet 149
– example
  – process watchdog 167
  – socket connection timeout 156
  – socket select 159
  – SO_LINGER 154
– finite state machine and regular
  expressions 164
– internetwork layer 150
– IP addresses and port numbers 150
– memory and processing time consumption
  for text messages 165
– message and dataflow correctness 162
– module for socket-based IPC 156
– network interface layer 150
– open system standards 149
– performance improvements with parallel
  threads 166
– proprietary communication language 161
– proprietary text messages 163
– text messages 162
– three-way-handshake of TCP 153
– transport layer 150
– use of connectionless sockets 155
– use of connection-oriented sockets
  – close communication 154
  – create socket 151
  – data exchange 153
  – initiate communication 153
– watchdog process 166
Introducing mind maps 10
Intrusion detection 466
Intrusion Detection System (IDS) 502
IP. See Internet Protocol (IP)
Iptables 444, 459, 463, 482, 494
IR. See Infrared (IR)
ISAM. See Index Sequential Access Method
  (ISAM)
ISO. See International Standards Organization
  (ISO)

Iteration plan 123
Iterative analysis and design, agile meetings 124
Iterative coding and private testing 125
Iterative committing, continuous integration
  and deployment 126
Iterative software development processes 118
ITRF. See International Terrestrial Reference
  Frame (ITRF)
IVS. See International VLBI service for geodesy
  and astrometry (IVS)

J

Java 20, 22, 50, 116, 372
JavaScript 293
Jenkins 114
Jive5ab 502
Joint Institute for VLBI in Europe (JIVE) 499
Joint Photographic Experts Group (JPG) 49, 52,
  311, 495, 497
JOINT STRIKE FIGHTER AIR VEHICLE C++ CODING
  STANDARDS 26
JPL Institutional Coding Standard 27
Jumps 33

K

Karnaugh-Veitch-diagram 256
Keyboard-Video-Mouse (KVM) 455
Key generation 448
Key techniques 487
Key technology for a command
  interpretation 243
KVM. See Keyboard-Video-Mouse (KVM)

L

LAN. See Local Area Network (LAN)
Land mines in test results 67
Language and citation 11
LASER. See Light Amplification by Stimulated
  Emission of Radiation (LASER)
laser-ceilometer 311
Laser ranging
– data for laser ranging 250
– goal of observations 253
– in-sky and on-ground laser ranging safety
  with additional data 251
– in-sky laser ranging safety with radars 251
– interaction of the hardware 249
– perturbations, calibration and quality
  control 253
– technical principles 248
– telescope and reception 249
– workflow of observations
  – after the observation 252
  – before the observation 252
  – during the observation 252
Laser ranging control
– actuators 254
– adaption from the industry 277
– alternative GUI, HTML 293
– analog-to-digital converters 254
– automatic sequence control 260
– autonomous functionalities 278
– autonomous production 278
– autonomy and automation 279

- Boolean algebra 256
- communication interface layers 262
- computer 258
- controlling layers from signal circuits to genetic algorithms 275
- device control loop 254
- different physical communication lines 262
- digital-to-analog converters 254
- distributed hardware control 274
- escape sequences and ncurses 284
- example
  - dome IDL file 271
  - hardware driver module 265–268
  - HTML template 293
  - periodic control loop 269
  - wxWidgets windows 287–291
- expert system 276
- feedback control loop 255
- fuzzy logic 276
- galvanic isolators 263
- genetic algorithm 276
- hardware driver module/component 263
- HTML limitations 300
- HTML templates 293
- industrial PC 260
- intelligent control units 254
- Karnaugh-Veitch-diagram 256
- line driver 263
- Linux-GUI system X11 284
- logic gates 259
- manufacturing cells production 278
- microcontroller 256
- multi-layered autonomous production cell of a laser ranging system 282
- multi-layer model of an autonomous production cell 281
- neuronal network 276
- periodic control loop for devices 269
- periodic control loop thread 269
- programmable logic controller 260
- proportional controller 255
- proportional-derivative controller 255
- proportional-integral controller 255
- RPC device interface 281
- RPC server as device driver 263
- self-similar decomposition 280
- sensors 254
- shell interface 283
- signal processing circuit with logic gates 256
- situation of employees 279
- supporting production cell structures 281
- thin clients and fat servers 292
- timing 276
- timing layers from real-time to normal time 276
- two-layer model of an autonomous production cell 281
- types of autonomous control cells 280
- types of supporting production cells 282
- von Neumann architecture 256, 260
- White Rabbit Solution 276
- window designing 285
- wxWidgets
  - advanced 291
  - architecture 284
  - event-driven programming 286
  - framework 285
  - program elements and flow 286
  - rapid application development 291

Legacy 8, 22, 39, 41, 43, 78, 110, 128, 129, 415, 420, 430, 481
Legacy code 40
- converter classes for external code 40
- cover and modify 40
- encapsulation of C in C++ 41
- encapsulation of C++ in C 41
- encapsulation of compiler and interpreter code 46
- encapsulation of other languages 43
- external code libraries 40
- FORTRAN encapsulation
  - of arrays 46
  - with the converter module f77.h 44
  - with f2c 43
  - of function arguments 46
  - restrictions 46
  - rudimentary rules for 44
  - of standard types 44
- given situation and risks 39
- thread-safe encapsulation 42
- three steps of use 41
- types 40
- unit tests and refactoring 41
LEO. See Low Earth Orbit (LEO)
Lexical analysis 190
Lexical error handling 191
LGPL. See GNU Lesser General Public License (LGPL)
LIDAR. See Light detection and ranging (LIDAR)
Light Amplification by Stimulated Emission of Radiation (LASER) 249
Light detection and ranging (LIDAR) 311, 390
Limits and extreme situations 72
Limits of a firewall 466
Limits of database 363
Limits of directories and file formats 352
Limits of simple solutions 73
Line driver 263
Linux 12
Linux-GUI system X11 284
LLR. See Lunar Laser Ranging (LLR)
Local Area Network (LAN) 149, 234, 370, 431, 467, 502
Local (port) forwarding 450
Local function calls 168
Local security
- access control in software 435
- multi-user system mechanisms 433
- password encryption 434
- role based security with standard users 437
- security risks
  - brute-force-attacks 437
  - data based attacks 437
- superuser or root 434
- symmetric block cipher algorithms 434
- user ID and group ID 433
Local thresholding 315
Lock-modify-unlock solution 100
Logbook as a proprietary solution 330
Logical paths 69
Logic gates 256
logog 330
Long-term vs. fast-changing documents 48
Look-and-feel with new graphic extensions 423
Low Earth Orbit (LEO) 248, 250, 276, 324, 335, 486
Lucky imaging 317
Lunar Laser Ranging (LLR) 9, 248

M

Macros 20
Main deficits in scientific software 15
Maintainability 18
Makefile 50, 73–77, 99, 110
Man-in-the-middle attack 448
Manufacturing cells 278
Masquerading 463
MAT. See Microprocessor ASCII Transceiver (MAT)
MathWorks®Matlab® 6
Maturity 17
Maximum Transmission Unit (MTU) 500, 502
MCB. See Monitor and Control Bus (MCB)
MCI standardization 369
- md5sum 451
MCP. See Micro-Channel Plate (MCP)
MD5. See Message Digest Nr. 5 (MD5)
Measuring process and its classification 66
Mechanisms for continuous integration builds 111
Medium Earth Orbit (MEO) 335
Memory and processing time consumption for text messages 165
Memory space vs. access speed 339
Merge conflicts 101
Message and data flow correctness 162
Message Digest Nr. 5 (MD5) 146, 217, 451, 457
Meteorology and calibration 407
Metrics
- categories for scientific developments 64
- parameter severities 67
- planning evaluation metrics 310
- software metrics 18, 64
- test metrics 64
Micro-channel Plate (MCP) 250, 324, 335
Microcontroller 256
Microprocessor ASCII Transceiver (MAT) 411
Middleware
- command line abstraction 243
- command line help 242
- command line language 237
- command line parser 242
- command line scanner 242
- communication middleware 237
- distribution platform 237
- example
  - command line interpreter 238
  - IDL file 237
  - textual RPC instructions 241
- key technology for a command interpretation 243
- principles 236
Minimal invasive concept 386
MISRA-C:2004 27
MISRA C++:2008 27
m0n0wall 467
Mock 78, 130, 415, 482
Modifiability 18
Modularity 18, 133
Module 15, 18, 20, 21, 23, 24, 28, 31, 39–42, 44–46, 52, 56, 57, 64, 65, 67, 68, 70, 72, 73, 78, 83, 86, 98, 105, 106, 116, 127–128, 133–138, 143–147, 156, 161, 168, 178, 181, 188, 202–205, 207–209, 212, 213, 216, 222, 236, 237, 243, 244, 263, 269, 276, 302, 337, 338, 349, 362, 372, 375, 376, 386, 404, 409, 415, 420–422, 469, 482, 498, 500, 507
Module for socket-based IPC 156

Monitor and Control Bus (MCB) 411
Motivation
– better understanding 7
– code as integral part of research 6
– purpose of this work 6
– reviewing techniques for code 6
– solution examples in programmed form 7
– tree swing comic is up-to-date 6
MTU. *See* Maximum Transmission Unit (MTU)
Multi-agent systems
– agent autonomy 397
– agent networks 399
– agent properties 399
– industrial remote control 397
– remote control idea 396
Multi-layered autonomous production cell
    of a laser ranging system 282
Multilayer model of an autonomous production
    cell 281
Multitasking 86
Multi-user system mechanisms 433
Mutex 90
Mutual exclusion 90

N

Nagios 370
Naming service 328
NASA Field System 323, 405, 409
– programs 411
– tasks 410
NAT. *See* Network Address Translation (NAT)
National Aeronautics and Space Administration
    (NASA) 147, 302, 323, 330, 372, 405, 409–415,
    420, 423, 425, 429, 453, 457, 464
National Marine Electronics Association
    (NMEA) 495
National requirements 487
National Research and Education Network
    (NREN) 499, 502, 503
Need for agile practices 21
Nested enclaves 468
Nested SSH tunnels 468
Network address translation (NAT) 463, 464, 467
Network Common Data Format (NetCDF) 351
Networked Transport of RTCM via Internet
    Protocol (NTRIP) 495, 496
Network enclave 458
Network File System (NFS) 371
Network Flight Recorder 466
Network identification example 375
Network Information System (NIS) 235
Network interface layer 150
Networks for an observatory 502
Network Time Protocol (NTP) 277, 411, 464
Neuronal network 276, 322, 326
New systems for SLR and VLBI 5
NEXPReS. *See* Novel EXploration Pushing
    Robust e-VLBI Services (NEXPReS)
NFS. *See* Network File System (NFS)
NIS. *See* Network Information System (NIS)
NMEA. *See* National Marine Electronics
    Association (NMEA)
Nonintrusive checks 95
Non-recursive, predicative parser 197
Normal form 256, 356, 357, 392
Normal Points (NP) 250, 252, 326, 366, 497
Novel EXploration Pushing Robust e-VLBI
    Services (NEXPReS) 128, 426, 499

N-queens puzzle 25
NREN. *See* National Research and Education
    Network (NREN)
nsiqcppstyle 115
NTP. *See* Network Time Protocol (NTP)
NTRIP. *See* Networked Transport of RTCM via
    Internet Protocol (NTRIP)
Nubiscope 311
Nyquist-Shannon sampling theorem 334

O

Objective-C 23, 26, 50
Object-oriented components 137
Object Oriented Programming (OOP) 15, 22, 23,
    33, 41, 69, 73, 90, 133, 134, 137, 138, 141, 145,
    180, 181, 204, 216, 258, 284, 285, 303, 354, 521
– abstraction 137
– inheritance 137
– overloading 137
– polymorphism 137
– virtualization 137
Objects 15
Oblique transformation 202
Observatory staff 11
Observatory Web page 12
Offering documentation 60
ONC. *See* Open Network Computing (ONC)
On-demand services 488
On-ground safety 390
On the way to a GGOS core site 20
OPC United Architecture (OPC UA) 378, 457
Open-MoniCA 372
Open Network Computing (ONC) 170, 173
Open Software Foundation (OSF) 171
Open source firewall, m0n0wall 467
Open System Interconnection (OSI) 149, 150
Open system standards 149
Operability 17
Operating system 21, 23, 24, 32, 62, 64, 86, 90,
    97, 101, 109, 110, 130, 146, 147, 154, 162, 166,
    167, 170, 175, 178, 179, 222, 237, 260, 276,
    283, 318, 319, 327, 332, 333, 338, 339, 367,
    391, 413, 425, 433, 434, 437, 449, 457, 464,
    467, 469, 473, 508
Operational deficits of GGOS
– combined, globally cross-linked operations
    and coordination 488
– extended GGOS data flows 489
– feedback loop, quality control and
    optimization of its use 489
– on-demand services 488
– operational deficits 488
Operational specification for GGOS 486
Operation control strategies 426
Optimization for C++ 420
Oracle's Hudson 114
Organizing documentation 61
Organizing document versions 61
OSF. *See* Open Software Foundation (OSF)
OSI. *See* Open System Interconnection (OSI)
Other fields for a safety system 390
Outlook
– adapting skills of employees 512
– first revolution, astronomical findings
    and geodetic surveys 510
– fourth industrial revolution 511
– fourth revolution, global networking 506, 511
– GGOS as a revolutionary step 510

– individualization 512
– revolutionary steps in industry 511
– second revolution, improvements of
    observing technique 510
– supporting solutions for the challenges 511
– third revolution, space techniques 510
Overloading 137

P

Package filter firewall rules 459
Pair programming 83
Pandora FMS 370
Parameter and result passing 173
Parameter interpretation 67
Parametrization 134
Parser 146, 197, 199, 201, 242, 244, 340, 346,
    349, 373
Pascal case 29
Passage 12, 59, 96, 252, 302, 310, 317–326, 347,
    352, 363–366, 489, 492, 497
Passing of firewalls 464
Password encryption 434
PC. *See* Personal Computer (PC)
PDF. *See* Portable Document Format (PDF)
PDU. *See* Protocol Data Unit (PDU)
Performance 18, 174
Performance improvements with parallel
    threads 166
Periodic control loop for devices 269
Periodic control loop thread 269
Periodic function with semaphore, idl2rpc.pl
    application engineering 223
Perl 21, 24, 50, 165, 190, 191
Perl-based IPC generator
– code generation 204
– compiler phases 190
– compositional and transformational
    refinements 202
– dynamic code optimization 203
– error handling 204
– example
    – Nested FOR-loop generation 200
    – Scanner 192
    – Semantic decisions 202
– generative domain model 190
– idl2rpc.pl phases 204
– intermediate code generation 201
– lexical analysis 190
– lexical error handling 191
– non-recursive, predicative parser 197
– oblique transformation 202
– parser 197
– scanner 190, 191
– semantic analysis 201
– static code optimization 203
– symbol table 197
– syntax analysis 197
– syntax graph 197
Perl details 24
Personal Computer (PC) 260–263, 276, 332, 370,
    409, 412, 424, 457, 464, 467
Perturbations, calibration and quality
    control 253
PFB. *See* Polyphase Filter Bank (PFB)
Phase calibration 406
PHP 24, 50
Physically separated transfer network 500
Physical separation with dualhomed hosts 459

Pipes to monitor program outputs, extend existing systems 471
Pitfalls with STL-strings 90
Planning evaluation metrics 310
Planning of the iterative construction 123
PLC. See Programmable Logic Controller (PLC)
pmap_dump 179
pmap_set 179
PNG. See Portable Network Graphic (PNG)
Pointer 23, 26, 30, 31, 34, 42, 44, 57, 85, 90, 129, 138, 141, 173, 178, 179, 187, 188, 222, 304, 338, 358, 420
Policy(ies)
– for codes 30
– dependencies 34
– with STL 34
Polymorphism 137
Polyphase Filter Bank (PFB) 408
PolySpace® 84
POP3. See Post Office Protocol Version 3 (POP3)
Popularity of compiler languages 22
Popularity of interpreter languages 24
Portability 18
Portable Document Format (PDF) 50, 61
Portable Network Graphic (PNG) 49
Portmap 171
Portmapper 171, 179, 180, 209, 222, 232, 236, 328, 449, 465, 466
Possible error sources 220
PostgreSQL 146, 330, 362, 371
Post Office Protocol Version 3 (POP3) 150
Potential of automation 504
PPS. See Pulse Per Second (PPS)
Praxis interwoven with theory 10
Precision Time Protocol (PTP) 276
Preemptive scheduling 87
Preprocessor 20, 23, 29, 31, 420
Principles 236
Principles of VLBI 400
Prioritization of tests 96
Process 6, 8, 9, 15, 18, 20, 23, 24, 26, 30, 34, 44, 46, 48–50, 56, 60, 62, 66, 71–78, 83–92, 95, 99, 103, 106, 110, 111, 114, 115, 117–128, 130, 132, 141–143, 149, 151, 154, 156, 164, 166–171, 173, 177–179, 181, 184–185, 190, 197–207, 210, 212, 217, 220–223, 232–234, 236, 246, 250, 252, 254–263, 269, 275, 277–279, 281, 300, 302, 305, 308, 311, 314, 317–319, 322–328, 332–334, 336, 352, 356–358, 360–366, 369, 370, 372, 373, 378, 388, 390, 393, 400, 404, 407–409, 412–415, 423, 425, 432–438, 444, 446, 448, 452, 454, 455, 457, 469–473, 479, 480, 485–487, 489, 491, 494, 499, 504, 506, 511, 512
Processing chain 314
Processor 16, 76, 87, 94, 258, 276, 318, 319
Profiler 94
Profiling tools 94
Program 14
Program directories 98
Programmable logic controller (PLC) 260, 262, 263, 276, 388, 412, 422
Programming Community Index 22
Programming languages
– boost C++ Libraries 23, 141
– C 6, 20, 22, 24, 30, 31, 41–44, 46, 50, 116, 138, 258, 412, 420, 469, 471
– C# 23, 50

– C++ 6, 20, 22, 23, 30, 41–43, 46, 50, 116, 138, 142, 258, 420, 470
– C++ Coding Style Guide 514
– call-by-reference 21
– call-by-value 21
– C Coding Style Guide 514
– coding style guide for Geodetic Software 514
– compiler 20
– criteria to select a programming language 19
– former de-facto standard 21
– FORTRAN 20–22, 43–46, 50, 409
– interpiler 20, 21
– interpreter 20
– Java 20, 22, 50, 116, 372
– JavaScript 239
– Macros 20
– MathWorks® Matlab® 6
– Objective-C 23, 50
– perl 21, 24, 50, 165, 190, 191
– PHP 24, 50
– popularity of compiler languages 22
– popularity of interpreter languages 24
– programming paradigm 22, 40, 41, 134, 135, 303
– python 21, 24, 50, 243
– regular expressions 21
– structured programming 22
– VHDL 50
Programming paradigm 22, 40, 41, 134, 135, 303
Program workflow 33
Project 4, 5, 7, 8, 11, 15–16, 26, 27, 31, 40, 43, 44, 47–49, 51, 54, 56, 60–70, 74, 77, 78, 80, 83–85, 90, 92, 96–98, 100, 101, 103–111, 115–129, 132, 143–148, 190, 317, 318, 425, 481, 498, 499, 504
Proof of concept 493
Properties to define externals 105
Properties to replace keywords 106
Proportional controller 255
Proportional-derivative controller 255
Proportional-integral controller 255
Proprietary communication language 161
Proprietary text messages 163
Pros and cons of SVN, 107
Protocol 63, 83, 93, 101, 102, 111, 149–151, 153, 155, 162, 164, 168, 170, 171, 173, 175, 177, 179–181, 209, 210, 217, 243, 262, 263, 386, 452, 458, 464–465, 499, 500, 502
Protocol Data Unit (PDU) 150, 151, 155, 165, 170, 173
Proxy firewall
– advantages 467
– principles 467
PScan 115
Psychology behind style guide rules 26
PTP. See Precision Time Protocol (PTP)
Public project status displays 111
Pulse Per Second (PPS), 276, 401
Purpose of metrics 64
Purpose of this work 6
Python 21, 24, 50, 243

Q

Qt 284
Quadridge Feed Horn (QRFH) 407
Quality 17
– factors 17

– metrics 18
– profiling 35
Quasar 400–401, 405, 408, 487
Quasi-Zenith Satellite System (QZSS) 488
Queens puzzle 25

R

Race conditions 88, 94
RAD. See Rapid Application Development (RAD)
Radio Detection and Ranging (RADAR) 252, 311, 333, 335, 370, 388, 389
Radio Frequency Interference (RFI) 260, 262, 401, 407
Radio Technical Commission for Maritime Services (RTCM) 495
RAID. See Redundant Array of Independent Disks (RAID)
Rapid Application Development (RAD) 291
Readership and focus 47
Real-time 333
Real-Time Application Interface (RTAI) 89
Real-time data trafic for GNSS 496
Real-time systems 276
Recurring structures 491
Recursively enumerable languages 193
Red, Green and Blue (RGB) 311–312
Redirection of input and output streams 471
Reduce access points (ports) 443
Reduced coding style guide 27
Reduction of insecure times 479
Reduction of noisy traffic 457
Redundancy 355
Redundant Array of Independent Disks (RAID) 428
Refactoring 41
References 57
– counting 34, 35, 90, 95, 141
– frame 3, 147, 400, 485, 486, 488
– point 4, 401
Reflecting the scientific needs 17
Regular expression 21, 24, 84, 92, 164, 165, 182, 184, 187, 190, 244, 340
Regular languages 162, 184
Relational database concept 353
Release plan 123
Reliability 17
Remote control 479
Remote Control (VLBI) 10
Remote control idea 396
Remote function calls 168
Remote Operations Center (ROC) 397
Remote Procedure Call Language (RPCL) 175, 203
Remote Procedure Calls (RPCs) 167–181, 186–188, 190, 200, 202, 204–210, 216–217, 222–223, 232, 234–238, 242, 244, 263, 269, 271, 274–277, 280–282, 293, 300, 302, 306, 308, 323–325, 327, 332, 334, 337, 338, 364, 367–371, 375–376, 378, 396, 415, 420, 423, 425, 429, 430, 434, 450, 466, 471
– application parametrization 175
– call correctness 174
– call semantics 173
– communication protocol 175
– daemon 171
– data representation, with XDR 171
– definition 169

Remote Procedure Calls (RPCs) (cont.)
- design issues 171
- device interface 271
- distributed system 180
- error handling and safety 173
- example
  - client stub function call 178, 179
  - sun RPC specification file 176
- function with semaphore, idl2rpc.pl application engineering 233–235
- generating the protocol modules 177
- IEEE 754 172
- local function calls 168
- parameter and result passing 173
- performance 174
- realizations 170
- remote function calls 168
- round-trip-delay 178
- running server and client 186
- security 174, 466
- server as device driver 263
- simplify the development of distribute systems 180
- stubs and their use 170
- sun RPC 170
- transport protocols 173
- writing and compiling the client 178
- writing and compiling the server 178
Remote security
- access control lists 438
- asymmetric cipher algorithms 446
- authorization pyramid 451
- comparison to industrial standard, OPC UA 457
- concept-level access control 455
- cryptographic checksums 450
- denial-of-Service-attacks 443, 449–450
- encryption with a SSH 446
- filter tables (iptables) 444
- functional security 450
- hash function 450
- hybrid cipher algorithm in the SSH 448
- key generation 448
- local (port) forwarding 450
- man-in-the-middle attack 448
- reduce access points (ports) 443
- restrict allowed programs 439
- role based access control 451
- Secure Shell (SSH)
  - authentication and key signature 448
  - limitations 449
  - as remote shell 445
- security problem 438
- security updates 457
- Setup script for iptables! 445
- Setup script for iptables! 445
- Setup script for NAT iptables! 461
- state machine for role change safety 455
- task based access control 452
- three-way-handshake of remote access request 453
- tunneling with SSH 449
- user based access control 451
- user identity (authorization) based access control 451
- X forwarding 448
- X window system 455
Repository 100

Repository project directories 98
Request for Comments (RFC) 171
Requirement of documentation 47
Requirements of scientific software 117
Residual 250, 252, 325, 335, 366, 414
Restrict allowed programs 439
Reusability 18
Reviewing techniques for code 6
Reviews 83, 84
Revisions 103
Revolutionary steps in industry 511
RFC. See Request for Comments (RFC)
RFI. See Radio Frequency Interference (RFI)
RGB. See Red, Green and Blue (RGB)
RGO. See Royal Greenwich Observatory (RGO)
The Right Language 59
Right method of testing 63
RIPE Message Digest (RIPEMD) 451
Risk analysis 63
Robustness 17
ROC. See Remote Operations Center (ROC)
Role based access control 451
Role based security with standard users 437
Round-robin recording 503
Round-robin-scheduling 87
Round-trip-delay 178
Royal Greenwich Observatory (RGO) 366
rpcbind 171
rpcgen 175
RPCL. See Remote Procedure Call Language (RPCL)
RPCs. See Remote Procedure Call (RPCs)
rsyslog 330
RTAI. See Real-Time Application Interface (RTAI)
RTCM. See Radio Technical Commission for Maritime Services (RTCM)
Rudimentary rules for a FORTRAN encapsulation 44
Rules for code instructions 27
Rules for functions and methods 28
Rules for names 28
Rules for project and source file 27
Rules for the indenting 28
Rules for whitespaces use 27
Running server and client 179

S

Safe systems with separate channels 369
Safety 23 173
Safety critical approach with FMEA 62
Safety tests 72
SAP. See Service Access Point (SAP)
SASL. See Simple Authentication and Security Layer (SASL)
Satellite 2–4, 11, 12, 96, 147, 148, 248–253, 276–278, 282, 308, 310, 315–322, 324–327, 335, 347, 352, 354, 358, 364, 366, 367, 400, 408, 486, 488–490, 493, 495, 497, 498, 510, 511
Satellite geodesy 2–4, 11, 12
Satellite Laser Ranging (SLR) 4, 9, 63, 248, 410, 428, 457, 458, 486, 488–489, 491–493, 497, 498, 506
SAX. See Simple API for XML (SAX)
Scalability 18
Scanner 190–192, 197, 204, 242, 244, 390
Scheduling 87

Scientific construction kit 145
Scientific project papers 60
Scientific result papers 50
Scientific software 8, 14
- accuracy 17
- changeability 18
- CHAOS report 16–17
- classes 18
- complexity 18
- complex system behavior 15
- comprehensibility 18
- computer program 14
- configurability 18
- correctness 17
- design rules 19
- effciency 18
- factors of success 18
- fault tolerance 17
- flexibility 18
- functional decomposition 15
- functionality 17
- functions 15
- importance of software 15
- integrity 18
- interoperability 18
- main deficits in scientific software 15–16
- maintainability 18
- maturity 17
- modifiability 18
- modularity 18
- modules 15
- objects 15
- operability 17
- performance 17
- portability 18
- program 14
- quality 17
  - factors 17
  - metrics 18–19
- reflecting the scientific needs 17
- reliability 17
- reusability 18
- robustness 17
- scalability 18
- security 18
- software design rules 19
- software metrics 64
- software quality 17
- software quality factors 17–18
- stability 17
- Standish Group, The 16
- status 15
- supportability 18
- testability 18
- understandability 18
- usability 17
- valuation of scientific software 14
Scoping of variables 31
Script language tools and vendor tools 148
Second revolution, improvements of observing technique 510
Secure Copy (SCP) 430
Secure File Transfer Protocol (SFTP) 371
Secure Hash Algorithm (SHA) 451
Secure RPC 235
Secure Shell (SSH) 102, 148, 180, 209, 236, 330, 415, 430, 434, 438, 445, 448–451, 455, 457, 460, 463–473, 479, 481, 497, 502

– authentication and key signature 448
– limitations 449
– Off-Host forwarding 457
– and terminal security 469
– tunnel management
    – control terminal 470
    – data chopping and steganography
        479–181
    – redirection of input and output
        streams 471
    – reduction of insecure times 479
    – security problem of a redirection 479
    – security vs. safety 471
    – SSH and terminal security 469
    – sshbroker and remote control 479
    – steganography 480
    – terminal input and output streams 469
    – terminal read and write access 470
    – TTY redirection with a pseudo TTY 473
Secure Socket Layer (SSL) 102, 438, 449
Security 18, 174–175, 432
– access control 432–433
– aspects 431
– definition 432
– for distributed systems
    – consequences of forwarding 465–466
    – creation of network enclaves 458
    – firewall management 464
    – independence of network enclaves
        458–459
    – intrusion detection 466
    – IP (or port) forwarding 464
    – IP Spoofing 466
    – security problems 457
    – SSH Off-Host forwarding 465
– in the e-VLBI network 501–502
– factors 432
– interconnectivity 433
– for an observatory
    – distributed systems 466–467
    – limits of a firewall 466
    – nested enclaves 468–469
    – nested SSH tunnels 468
    – network address translation
        and masquerading 463
    – Network Flight Recorder 466
    – open source firewall m0n0wall 467–468
    – package filter firewall rules 460
    – passing of firewalls 464
    – physical separation with dualhomed
        hosts 459
    – proxy firewall advantages 467
    – proxy firewall principles 467
    – reduction of noisy traffic 457–458
    – RPC security 466
    – security problems for distributed
        systems 457
– vs. safety 471
– updates 457
Security problems 438
– for distributed systems 457–458
– of a redirection 479
Security risks
– brute-force-attacks 437
– data based attacks 437–438
SEFD. See System Equivalent Flux Density (SEFD)
Segmentation 311–312
Self-similar decomposition 280

Semantic analysis 201
Semaphore 89
Sensor(s) 98, 127, 254–256, 279, 280, 311,
    369–371, 378–380, 386–391, 397, 399, 422,
    487, 488, 491, 492, 511
– control points and data injection 372–373
– data categories 388–389
– measures 388
Separation of the workflows 308
Sequence charts 50
Server 77, 97, 100–103, 109, 110, 112, 114, 116,
    127, 128, 148, 151, 153–156, 159, 163, 166,
    167, 169–171, 173–180, 188, 202, 204,
    206–210, 212–214, 216, 217, 220–223,
    232–236, 238, 243, 244, 263, 269, 271, 274,
    277, 282, 286, 292, 293, 300, 302, 308, 327,
    330, 332, 333, 362, 366, 368, 372, 373, 375,
    376, 400, 414, 423, 424, 430, 448–456, 464,
    466–468, 479, 496, 498
Server application sample 213
Server application skeleton 212–213
Server safety 220
– automatic safety device 233
– demonstration 223
– multi-threading 222
– parallel safety device thread 233–234
– persistence and idempotence 232–233
– semaphores 222
– startup control 222
– time limitations 222–223
– watchdog 232
Server security
– authentication 234–235
– authorization 235
– in the LAN and WAN 234
– Secure RPC 235–236
Service Access Point (SAP) 175
Session Initiation Protocol (SIP) 377
SFD. See Source Flux Density (SFD)
SFTP. See Secure File Transfer Protocol (SFTP)
SHA. See Secure Hash Algorithm (SHA)
Shallow copy 35, 90
Shared memory 87–88
Sharing prefabricated off-the-shelf building
    blocks 132–133
sha1sum 451
Shell interface 283–284
Shortest job first 319
Signal processing circuit with logic gates 256
Signal to Noise Ratio (SNR) 406
Simian 115
Simple API for XML (SAX) 346
Simple Authentication and Security Layer
    (SASL) 101
Simple Mail Transfer Protocol (SMTP) 150
Simple Network Management Protocol
    (SNMP) 375
Simplify the development of distribute
    systems 180–181
Single-level and multi-level data structures
    351–352
SIP. See Session Initiation Protocol (SIP)
Situation of employees 279
Six incremental pillars 8
SKA. See Square Kilometer Array (SKA)
SLR. See Satellite Laser Ranging (SLR)
SMART criteria for iteration goals
    119–120

SMTP. See Simple Mail Transfer Protocol (SMTP)
SNAP. See Standard Notation for Astronomical
    Procedures (SNAP)
SNMP. See Simple Network Management Protocol
    (SNMP)
SNR. See Signal to Noise Ratio (SNR)
Software creation workflow 77
Software design rules 19
Software development 6–8, 12, 15, 16, 19, 30, 35,
    37, 49, 56, 62, 77, 81, 84, 97, 103, 109,
    117–128, 132, 143, 284, 336, 511
Software Failure Mode and Effects Analysis
    (SFMEA) 62
Software metrics 18, 64
– benchmark 66
– characteristic factors for scientific
    developments 65–66
– code beautifier 66–67
– counts, metric parameters, measures 65
– interpretation of test results 67
– land mines in test results 67–68
– measuring process and its classification 66
– metric categories for scientific
    developments 64
– metric parameter severities 67
– purpose of metrics 64–65
– test metrics 64
– three main code metric dimensions 64
Software project 15
Software quality factors 17
– accuracy 17
– changeability 18
– complexity 18
– comprehensability 18
– configurability 18
– correctness 17
– efficiency 18
– fault tolerance 17
– flexibility 18
– functionality 17
– integrity 18
– maintainability 18
– maturity 17
– modifiability 18
– modularity 18
– operability 17
– performance 18
– portability 18
– reliability 17
– reusability 18
– robustness 17
– scalability 18
– security 18
– software metrics 18
– stability 17
– supportability 18
– testability 18
– understandability 18
– usability 17
Software requirements
– CCNU GCC 12
– Linux 12
– Observatory Web page 12
– TUM Web page 13
Software structure 370
Software system 19
Software tests 68
Software toolbox 8

Software tools
- Apache Ant™ 114
- Apache Maven 114
- API Sanity Checker 82
- Artistic Style 2.02 (astyle) 66, 115
- asterisk 378
- bcov 80, 115
- bison 190
- Codan 116
- codespell 115
- Cppcheck 78, 84, 115
- Doxygen 52, 117, 129, 148, 202, 522
- Eclipse 116, 190
- e-RemoteCtrl 58, 426, 428–430, 450–453,
  455, 471, 479
- e-shell 471
- expect 466
- f2c 43
- flawfinder 115
- fwbuilder 464
- GCC 94, 95
- GDB 94
- Git 100, 108
- GNU make 74, 115
- gpg 451
- Graphviz 50
- GTK+ 284
- HALCON 318
- Jenkins 114
- Jive5ab 495, 496
- logog 329
- md5sum 451
- m0n0wall 467
- Nagios 370
- NASA Field System 323, 405, 409
- nsiqcppstyle 115
- Open-MoniCA 372
- Oracle's Hudson 114
- Pandora FMS 370
- pmap_dump 179
- pmap_set 186
- PolySpacer 84
- portmap 171
- PostgreSQL 146, 330, 358, 360, 362, 371
- PScan 115
- Qt 284
- rpcbind 171
- rpcgen 175
- rsyslog 330
- secure RPC 235
- sha1sum 451
- Simian 115
- StatSVN 114
- Subclipse 116
- Subversionr 100, 116
- syslog 330
- Valgrind 94, 115, 116
- VMware 110
- Wiki 57
- wxFormBuilder 291
- wxWidgets 291, 423
- Xen 110
- yacc 190
- Zabbix 370, 379
Solution examples in programmed
  form 7–8
Source code 6, 9, 11, 15, 17, 20, 21, 27, 31, 44, 50,
  51, 56, 61, 62, 66, 68, 70, 76, 92, 95, 98, 99, 103,
  105, 107, 126, 128, 129, 132, 134, 135, 137, 139,
  141, 148, 168, 178, 181, 190, 191, 197, 244, 362,
  376, 411, 414, 429
Source Flux Density (SFD) 406
Source Network Address Translation (SNAT) 463
Space debris 308, 493
Space geodesy 3
Space geodetic techniques 3–6, 9, 10, 19, 277,
  369, 399, 481, 486, 488, 491, 507
Space techniques 4, 248, 327, 351, 486, 488–490,
  493, 504, 506, 510
SQL. See Structured Query Language (SQL)
Square Kilometer Array (SKA) 428
SSFMEA. See Software Failure Mode and Effects
  Analysis (SFMEA)
SSH. See Secure Shell (SSH)
sshbroker 479
SSL. See Secure Socket Layer (SSL)
SSRAM. See Static Random Access Memory
  (SRAM)
Stability 17
Standard Notation for Astronomical Procedures
  (SNAP) 323, 331, 405, 409, 412–415, 420
Standards 20, 21, 23, 26, 27, 30, 33, 41, 44, 65, 84,
  107, 127, 148–150, 162, 171, 279, 378, 387,
  446, 447, 466, 486
Standard Template Library (STL) 23, 34, 35, 91,
  95, 141, 146, 303–305, 517, 523
Standish Group, The 16
Starvation 87, 90, 166, 208, 318, 319
State machine for role change safety 455
Static code analysis 35, 116
Static code inspections
- analyzer checks and issues 85
- analyzer tools 86
- compiler outputs 84
- false positives 86
- informal reviews 83
- reviews 83
- statistics as metrics parameter 85
- technical reviews 84
Static code optimization 203
Static Random Access Memory (SRAM), 258
Station specific parts 409
Statistics as metrics parameter 85
StatSVN 114
Status 15
Status file, extend existing systems 431
Steganography 480
STL. See Standard Template Library (STL)
Structural modules 135
Structured programming 24
Structured Query Language (SQL) 99, 146,
  360–362, 367
Stub 77, 130, 170, 178–179, 234, 390, 409, 415,
  420, 450, 482
Students 11
Style guide. See Coding style guide
Subclipse 116
Subversion® 100, 116
Sun RPC 170
Superuser or root 434
Supportability 18
Support for automated, complete builds 76
Supporting environment, Eclipse 116
Supporting production cell structures 281
Supporting solutions for the challenges 511
SVN file header with keywords 107
Symbol table 197
Symmetric block cipher algorithms 434

Syntax analysis 197
Syntax graph 197
Syntax tree 183
Syslog 330
System efficiency and calibration 405
System Equivalent Flux Density (SEFD) 406, 412
System health and human safety 420
System health states 388
System monitoring and safety
- additional sensor net 388
- alarm levels 377
- autonomous functionalities
  - controlling 378
  - failure management 389
  - hardware driving 386
  - planning 378
  - user interfacing 378
- cell structure 369
- characteristic sensor profiles 386
- comparison to industrial standard, OPC
  UA 378
- configuration 373
- database tables 376, 377
- example
  - configuration file format for data
    servers 375
  - configuration file format for sensor control
    points 373
  - configuration file format for sub-cells 375
  - MCI functionalities 381
- galvanic isolators 387
- hardware layers 369
- identification in a network 375
- in-sky safety 389
- MCI standardization 369
- minimal invasive concept 386
- network identification example 375
- on-ground safety 390
- other fields for a safety system 390
- safe systems with separate channels 369
- sensors 386
  - control points and data injection 372
  - data categories 388
  - measures 388
- software structure 370
- system health states 388
- Zabbix 379
System of systems 491
System tests 83

T

Tags 103
Targets in a Makefile 76
Task based access control 452
TCP. See Transmission Control Protocol (TCP)
Technical principles 248
Technical realization of the functionalities 302
Technical reviews 84
Technical VLBI aspects
- down-conversion and sampling 403
- reception 401
- shipment and correlation 404
Telescope and reception 249
TeleTYpe (TTY) 469, 471, 481
Template metaprogramming 141
Terminal input and output streams 469
Terminal read and write access 470
Terrestrial geodesy 2

Terrestrial infrastructure 4
Terrestrial reference frame 3, 4, 407, 486
Test(ing) 4, 6, 8, 12, 16, 19, 26, 30, 35, 39, 41, 44, 46, 62–72, 76–84, 86, 92, 95–97, 103, 109–112, 115–120, 125–130, 133, 170, 254, 310, 327, 338, 340, 390, 400, 406, 415, 428, 437, 447, 452, 481, 482, 493, 499, 502–504, 507
– acceptance tests 83
– black-box-tests 68
– case generators 82
– case structure 72
– coverage as metrics parameter 80
– data 68
– dealing with software errors 62
– debugging 69
– disadvantages of the usual testing 62
– driven development 70
– feedback history 80
– friendliness 70
– gray-box-tests 68
– integration tests 83
– metrics 63
– protocol 63
– reviews 83
– right method of testing 63
– risk analysis 63
– safety critical approach with FMEA 62
– static code inspections 84
– system tests 83
– test protocol 63
– verification and validation 68
– white-box-tests 68
Testability 18
Testing landscape 96
– costs and benefits in the risk management 96
– culture for the constructive handling of errors 97
– prioritization of tests 96
Text-highlighting 51
Text messages 162
Textual syntax notation 182
Thin clients and fat servers 292
Third revolution, space techniques 510
Thread 16, 22, 23, 31, 34, 35, 87–90, 93, 94, 128, 166, 206, 208, 216, 220–223, 233, 269, 306, 308, 332, 334, 373
Thread-safe encapsulation 42–43
Three main code metric dimensions 64
Three steps of use 41
Three-way-handshake of remote access request 453–455
Three-way-handshake of TCP 153
Threshold 110, 312–316, 322, 330, 336, 377, 393
TIGO. See Transportable Integrated Geodetic Observatory (TIGO)
Timing 276, 333
Timing layers from real-time to normal time 276–277
Toolbox 8, 106, 128, 134–135, 137, 141, 144–145, 148, 156, 235, 236, 263, 302, 349, 466, 511
Toolbox module for configurations 349
Top-down 130, 197, 400, 489
Topology 300
Transaction 94, 174, 233, 235, 236, 356, 360, 364, 392
Transmission Control Protocol (TCP) 146, 149–153, 156, 170, 173, 174, 178–180, 208–210, 217, 237, 262, 330, 364, 378, 453, 498

Transportable Integrated Geodetic Observatory (TIGO) 427, 428, 494
Transport layer 150
Transport protocols 173
Tree swing comic is up-to-date 6
Try/catch exceptions 33
Tsunami UDP Protocol 498–499
TTY. See TeleTYpe (TTY)
TTY redirection with a pseudo TTY 473
TUM Web page 12
Tunneling with SSH 449
Turing 142, 186, 187, 201
Tutorials 57–59
Two-layer model of an autonomous production cell 281
Types 40
– of autonomous control cells 280
– cast(ing) 31–33
– compatibility of 32-bit and 64-bit 32–33
– of supporting production cells 282

U

UDP-based Data Transfer Protocol (UDT) 502
Understandability 18
Unicode Transformation Format (UTF) 163, 292, 522
Unified Modelling Language (UML) 49, 50, 93
Uniform Resource Locator (URL) 54, 102, 103, 105, 107, 293, 294
Uninterruptible Power Supply (UPS) 428
Unit test(ing) 35, 41, 44, 46, 66, 69, 70, 77, 78, 80–83, 86, 110, 115, 116, 125, 128–130, 134, 144, 148
– automated unit tests 78
– build output as test feedback 76
– definitions in a makefile 76
– function stabilization 80
– GNU Build System 74
– limits and extreme situations 72
– limits of simple solutions 73
– logical paths 69
– makefile 73, 74, 76
– mocks for system imitations 78
– safety tests 72
– software creation workflow 77
– stubs for external dependencies 77
– support for automated, complete builds 76
– targets in a makefile 76
– test case generators 82
– test case structure 72
– test class 79
– test coverage as metrics parameter 80
– test driven development 70
– test feedback history 80
– test friendliness 70
– unit tests 60
– usual test culture 69
– wettzell unit test suite 78
Universal Serial Bus (USB) 370
Universal Time Coordinated (UTC) 107, 147, 340, 401, 407, 411
Upper case 29
UPS. See Uninterruptible Power Supply (UPS)
URL. See Uniform Resource Locator (URL)
Usability 17
USB. See Universal Serial Bus (USB)

Use cases 3, 10, 18, 49, 56, 59, 119, 135, 181, 276, 312, 357, 481, 493
Used time 411
Use example of the wettzell simple unit test suite 78
Useful tools 114
Use of connectionless sockets 155
Use of connection-oriented sockets
– close communication 154
– create socket 151
– data exchange 153
– initiate communication 153
Use of files for specific laser ranging data 352
Use of source code 11
User based access control 451
User Datagram Protocol (UDP) 146, 150, 155, 170, 173, 174, 178–180, 208–210, 217, 498
User documentation in scientific projects 56
User ID 433
User identity (authorization) based access control 451
User roles 356
User's manuals 59
User story cards 119
Usual test culture 69
UTC. See Universal Time Coordinated (UTC)
UTF. See Unicode Transformation Format (UTF)

V

Valgrind 94, 115, 116
Valuation of scientific software 14
Verification and validation 68
Version checksums, idl2rpc.pl application engineering 217
Version control system 97, 109
– branches 103
– centralized module base 106
– code version management tools 100
– commands 103
– commits 103
– communication, protocols and user credentials 101
– copy-modify-merge solution 101
– example
  – SVN commands 104
  – SVN file header/trailer 107
  – SVN properties 106
– Git and its differences 108
– hash function 108
– hierarchical document tree 98
– human skills 108
– lock-modify-unlock solution 101
– merge conflicts 101
– program directories 98
– properties to define externals 105
– properties to replace keywords 106
– pros and cons of SVN 107
– repository 100
– repository project directories 98
– revisions 103
– SVN file header with keywords 107
– tags 103
– URLs 102
– version control 97
– working copies and revisions checkouts 102, 103
Very Long Baseline Array (VLBA) 411, 412

Very Long Baseline Interferometry (VLBI), 4–6, 9,
    21, 57, 223, 323–326, 330, 372, 389, 400–409,
    415, 422, 424, 426–428, 452, 468, 481, 488,
    489, 493, 498, 500, 502–506, 508
– control system
    – control layers 409
    – NASA Field System 409
    – NASA Field System programs 411
    – NASA Field System tasks 410
    – station specific parts 409
    – used time 411
– observation results 405
– observation workflow 404
VEX. *See* VLBI EXperiment format (VEX)
VGOS. *See* VLBI Global Observing System (VGOS)
VGOS Monitor and Control Infrastructure
    (MCI) 369–372, 375, 378, 380, 388
VHDL 50
Virtual activities of one construction
    iteration 124
Virtualization 110, 137
Virtual Local Area Network (VLAN) 458, 464, 466
Virtual Network Computing (VNC) 429, 455
Virtual phases 121
Virtual Private Network (VPN) 466
VLAN. *See* Virtual Local Area Network (VLAN)
VLBA. *See* Very Long Baseline Array (VLBA)
VLBI. *See* Very Long Baseline Interferometry
    (VLBI)
VLBI EXperiment format (VEX) 413, 498
VLBI Global Observing System (VGOS), 5, 10, 372,
    405, 407, 428, 429, 487, 499, 504
VLBI Standard Interface (VSI), 404, 499–500
VMware 110
VNC. *See* Virtual Network Computing (VNC)
Voice over IP (VoIP), 377

von Neumann architecture 256, 260
VPN. *See* Virtual Private Network (VPN)

W

WAN. *See* Wide Area Network (WAN)
Watchdog 166, 180, 204, 209, 210, 217, 220, 221,
    232, 244, 246, 277, 282, 305, 332, 368, 372,
    373, 450, 457
Watchdog process 166
Water Vapor Radiometer (WVR), 311, 388
Web-based Distributed Authoring and
    Versioning (WebDAV), 102
Well-tested modules and components
– centralized code assets and continuous
    integration 144
– generated modules and components 148
– institute's construction kit 145
– scientific construction kit 145
– script language tools and vendor tools 148
– wettzell's simple code toolbox 145
Wettzell's design-rules 27
Wettzell's simple code toolbox 145
Wettzell unit test suite 78
White-box-tests 68
White Rabbit Solution 276
White space 24, 27, 30, 67, 146, 187, 190,
    192, 340
Wide Area Network (WAN) 149, 234, 431, 467,
    468, 498–499, 502, 503
Wiki 22, 57, 61, 128, 367
Window designing 285
Workflow of observations
– after the observation 252
– before the observation 252

– during the observation 252
Working copies and revisions, checkouts
    102, 103
Workload estimation 504
Workload splitting 504
Writing and compiling the client 178
Writing and compiling the server 178
Written documentation 48
WVR. *See* Water Vapor Radiometer (WVR)
wxFormBuilder 291
wxWidgets 284, 423–425
– advanced 291
– architecture 284
– event-driven programming 286
– framework 284
– program elements and flow 286
– rapid application development 291

X

XDR. *See* External Data Representation (XDR)
Xen 110
X Forwarding 455
XML. *See* Extensible Markup Language (XML)
XP. *See* Extreme Programming (XP)
X Window System 455

Y

yacc 190

Z

Zabbix 370, 378

Printed in the United States
By Bookmasters